Handbook of Experimental Pharmacology

Continuation of Handbuch der experimentellen Pharmakologie

Vol. 79

New Neuromuscular Blocking Agents

Basic and Applied Aspects

Contributors

S. Agoston · K. Birò · W. C. Bowman · A. A. Bunatian
V. V. Churyukanov · D. Colquhoun · A. F. Danilov · A. Deery
D. Duncalf · V. P. Fisenko · F. F. Foldes · A. J. Gibb · A. L. Harvey
R. Hughes · E. Kárpáti · D. A. Kharkevich · N. V. Khromov-Borisov
A. A. Ķimenis · I. G. Marshall · M. D. Mashkovsky · R. D. Miller
H. Nagashima · D. S. Paskov · J. P. Payne · M. Riesz · V. A. Shorr
A. P. Skoldinov · J. B. Stenlake · G. A. Sutherland · G. Szabó
L. Szporny · E. Tassonyi · M. B. Tyers · L. Vimláti · L. P. Wennogle
L. N. Yakhontov

Editor

D. A. Kharkevich

Springer-Verlag Berlin Heidelberg New York Tokyo

DIMITRY A. KHARKEVICH, M.D., D. Sc.

Professor and Chairman of the
Department of Pharmacology
Academician of the USSR
Academy of Medical Sciences
First Medical Institute
Department of Pharmacology
2/6 Pirogovskaya Street,
Moscow II9435, USSR

With 128 Figures

ISBN-13: 978-3-642-70684-4 e-ISBN-13: 978-3-642-70682-0
DOI: 10.1007/978-3-642-70682-0

Library of Congress Cataloging-in-Publication Data. Main entry under title: New neuromuscular blocking agents. (Handbook of experimental pharmacology; vol. 79) Includes bibliographies and index. 1. Neuromuscular blocking agents – Physiological effect. 2. Myoneural junction – Effect of drugs on. I. Agoston, Aleksandar. II. Kharkevich, D.A. (Dmitriĭ Aleksandrovich) III. Series: Handbook of experimental pharmacology; v. 79. [DNLM: 1. Neuromuscular Blocking Agents – pharmacodynamics. 2. Neuromuscular Junction – drug effects. 3. Neuromuscular Junction – physiology. W1 HA51L v. 79/QV 140 N532] QP905.H3 vol. 79 [RM312] 615'.1 s [615'.77] 85-20809

Softcover reprint of the hardcover 1st edition 1986

2122/3130-543210

List of Contributors

S. Agoston, Research Group of the Institutes of Anesthesiology and Clinical Pharmacology, University of Groningen, University Hospital, Oostersingel 59, 9713 EZ Groningen, The Netherlands

K. Birò, Pharmacological Research Center, Chemical Works of Gedeon Richter Ltd., P.O. Box 27, 1475 Budapest 10, Hungary

W. C. Bowman, Department of Physiology and Pharmacology, University of Strathclyde, Royal College, 204 George Street, Glasgow G1 1XW, Great Britain

A. A. Bunatian, Department of Anesthesiology, National Research Centre of Surgery, USSR Academy of Medical Sciences, 2 Abrikosovsky per., Moscow 119874, USSR

V. V. Churyukanov, Department of Pharmacology, First Medical Institute, 2/6 Bolshaya Pirogovskaya Street, Moscow 119435, USSR

D. Colquhoun, Department of Pharmacology, University College London, Gower Street, London WC1E 6BT, Great Britain

A. F. Danilov, Department of Pharmacology, Sechenov Institute of Physiology and Biochemistry, USSR Academy of Sciences, 44 M. Thorez Prospect, Leningrad 194223, USSR

A. Deery, Department of Anesthesiology, Montefiore Medical Center and Albert Einstein College of Medicine, 111 East 210th Street, Bronx, NY 10467, USA

D. Duncalf, Department of Anesthesiology, Montefiore Medical Center and Albert Einstein College of Medicine, 111 East 210th Street, Bronx, NY 10467, USA

V. P. Fisenko, Department of Pharmacology, First Medical Institute, 2/6 Bolshaya Pirogovskaya Street, Moscow, 119435 USSR

F. F. Foldes, Department of Anesthesiology, Montefiore Medical Center and Albert Einstein College of Medicine, 111 East 210th Street, Bronx, NY 10467, USA

A. J. Gibb, Department of Physiology and Pharmacology, University of Strathclyde, Royal College, 204 George Street, Glasgow G1 1WX, Great Britain

A. L. Harvey, Department of Physiology and Pharmacology, University of Strathclyde, Royal College, 204 George Street, Glasgow G1 1XW, Great Britain

R. Hughes, Clinical Investigation Department, The Wellcome Research Laboratories, Langley Court, Beckenham, Kent BR3 3BS, Great Britain

E. Kárpáti, Pharmacological Research Center, Chemical Works of Gedeon Richter Ltd., P.O. Box 27, 1475 Budapest 10, Hungary

D. A. Kharkevich, Department of Pharmacology, First Medical Institute, 2/6 Bolshaya Pirogovskaya Street, Moscow 119435, USSR

N. V. Khromov-Borisov, Pr. Korablestroitelei 23-1, 214 Leningrad 199226, USSR

A. A. Ķimenis, Biological and Medical Research Division, Institute of Organic Synthesis of the Latvian SSR Academy of Sciences, 21 Aizkraukles Street, Riga 226006, USSR

I. G. Marshall, Department of Physiology and Pharmacology, University of Strathclyde, Royal College, 204 George Street, Glasgow G1 1XW, Great Britain

M. D. Mashkovsky, Department of Pharmacology, All-Union Chemical Pharmaceutical Research Institute, 7 Zubovskaya Street, Moscow 119815, USSR

R. D. Miller, Department of Anesthesia, University of California, School of Medicine, San Francisco, CA 94143, USA

H. Nagashima, Department of Anesthesiology, Montefiore Medical Center and Albert Einstein College of Medicine, 111 East 210th Street, Bronx, NY 10467, USA

D. S. Paskov, 5 Razlatiza Street, 1463 Sofia, Bulgaria

J. P. Payne, Research Department of Anesthesia, Royal College of Surgeons of England, Lincolns Inn Fields, London WC2, Great Britain

M. Riesz, Pharmacological Research Center, Chemical Works of Gedeon Richter Ltd., P.O. Box 27, 1475 Budapest 10, Hungary

V. A. Shorr, Department of Pharmacology, First Medical Institute, 2/6 Bolshaya Pirogovskaya Street, Moscow 119435, USSR

A. P. Skoldinov, Department of Organic Chemistry, Institute of Pharmacology of the USSR Academy of Medical Sciences, 8 Baltijskaya Street, Moscow 125315, USSR

J. B. Stenlake, Department of Pharmacy, University of Strathclyde, Royal College Building, 204 George Street, Glasgow G1 1XW, Great Britain

G. A. Sutherland, Division of Anesthesia, Royal Infirmary, Castle Street, Glasgow G4, Great Britain

G. Szabó, Postgraduate Medical School, Central Research Division, P.O. Box 112,1389 Budapest 62, Hungary

L. Szporny, Pharmacological Research Center, Chemical Works of Gedeon Richter Ltd., P.O. Box 27, 1475 Budapest 10, Hungary

E. Tassonyi, Postgraduate Medical School, 2nd Department of Surgery, P.O. Box 112, 1389 Budapest 62, Hungary

M. B. Tyers, Neuropharmacology Department, Glaxo Group Research Ltd., Ware Hertfordshire, SG12 ODJ, Great Britain

L. Vimláti, Postgraduate Medical School, 2nd Department of Surgery, P.O. Box 112, 1389 Budapest 62, Hungary

L. P. Wennogle, Neurosciences Cardiovascular Research, Ciba-Geigy Corp., Summit, NJ 07901, USA

L. N. Yakhontov, Department of Organic Chemistry, All-Union Chemical Pharmaceutical Research Institute, 7 Zubovskaya Street, Moscow 119815, USSR

Preface

The problems associated with the pharmacologic and physiologic regulation of neuromuscular transmission and of the morphofunctional organization of neuromuscular junctions have attracted a wide range of investigators. Numerous handbooks, monographs, and reviews are devoted to this subject. At the same time, many fundamental and applied aspects of this trend continue to progress succesfully. In recent years, new experimental and clinical data on the structure and function of neuromuscular junctions have been gained, and new, more perfect neuromuscular blocking agents have been designed. It is these data that the present handbook mainly deals with.

A considerable number of chapters have been written by authors from eastern Europe. This was done intentionally since much of their work has previously been published only in their own languages, and is thus inaccessible to most Western readers. This is why some of the data included in the volume are not quite the latest, but they contain fruitful ideas or important results and are of value for further progress in the pharmacology of neuromuscular transmission. Naturally, the methodological level of the investigations differs, depending on when they were carried out.

The handbook contains a number of selected chapters on the pharmacology of neuromuscular junctions; they comprise data otherwise insufficiently reviewed or not dealt with at all. They furthermore reflect the up-to-date state of the problem and probable directions of further developments in this field.

D. A. KHARKEVICH

Preface

Contents

Pharmacodynamics and Pharmacokinetics of Neuromuscular Blocking Agents

CHAPTER 3

On the Principles of Postsynaptic Action of Neuromuscular Blocking Agents
D. COLQUHOUN. With 4 Figures

CHAPTER 4

On the Hydrophobic Interaction of Neuromuscular Blocking Agents with Acetylcholine Receptors of Skeletal Muscles
D. A. KHARKEVICH and A. P. SKOLDINOV. With 11 Figures

CHAPTER 5

Prejunctional Actions of Cholinoceptor Agonists and Antagonists, and of Anticholinesterase Drugs. W. C. BOWMAN, A. J. GIBB, A. L. HARVEY, and I. G. MARSHALL. With 8 Figures

CHAPTER 6

On the Comparative Sensitivity of Acetylcholine Receptors of Various Groups of Skeletal Muscles to Neuromuscular Blocking Agents
D. A. KHARKEVICH and V. P. FISENKO. With 7 Figures

CHAPTER 9

On the Effect of Neuromuscular Blocking Agents on the Central Nervous System. D. A. KHARKEVICH. With 5 Figures

CHAPTER 10

Biodegradation and Elimination of Neuromuscular Blocking Agents
J. B. STENLAKE. With 5 Figures

On the Relationship Between the Chemical Structure and the Neuromuscular Blocking Activity

CHAPTER 11

Methods for the Experimental Evaluation of Neuromuscular Blocking Agents
D. A. KHARKEVICH. With 2 Figures

CHAPTER 12

Steroid Derivatives. M. RIESZ, E. KÁRPÁTI, and L. SZPORNY
With 2 Figures

CHAPTER 13

The Derivatives of Carboxylic Acids
D. A. KHARKEVICH and A. P. SKOLDINOV

CHAPTER 14

Quinuclidinium Compounds
M. D. MASHKOVSKY, L. N. YAKHONTOV, and V. V. CHURYUKANOV

CHAPTER 18

Vecuronium (ORG-NC-45). W. C. BOWMAN and G. A. SUTHERLAND
With 8 Figures

CHAPTER 19

The Derivatives of α-Truxillic Acid. D. A. KHARKEVICH. With 4 Figures

CHAPTER 26

Neuromuscular Blocking Agents of Different Chemical Structure
A. A. BUNATIAN. With 11 Figures

CHAPTER 27

Pipecuronium Bromide (Arduan). E. TASSONYI, G. SZABÓ, and L. VIMLÁTI
With 5 Figures

CHAPTER 28

Vecuronium (ORG-NC-45). R. D. MILLER. With 1 Figure

CHAPTER 29

Fazadinium Dibromide. M. B. TYERS. With 1 Figure

CHAPTER 30

Atracurium. R. HUGHES and J. P. PAYNE. With 5 Figures

Antagonists of Neuromuscular Blocking Agents (Pharmacology and Clinical Use)

CHAPTER 31

Galanthamine. D. S. Paskov. With 5 Figures

CHAPTER 32

Chinothylinum. A. A. Ķimenis

CHAPTER 33

4-Aminopyridine Hydrochloride (Pymadin)
D. S. Paskov, S. Agoston, and W. C. Bowman. With 5 Figures

CHAPTER 1

Neuromuscular Blocking Agents: General Considerations

D. A. KHARKEVICH

Introduction

Neuromuscular blocking agents are of great importance in anesthetic practice. They are represented by a series of compounds which differ by their chemical structure, mechanism of neuromuscular block, potency, duration of action, safety margin, order of muscle relaxation, effectiveness of various administration routes, side effects, available antagonists, etc. (FOLDES 1957; KHARKEVICH 1969, 1970a, 1983; CHEYMOL 1972; FELDMAN 1973; KATZ 1975; ZAIMIS 1976; AHNEFELD et al. 1980; BOWMAN 1980; BUZELLO 1981).

Chemical Structure

Defined in terms of chemical structure, neuromuscular blocking agents are presented by three groups (Table 1). The majority of drugs contain two cationic centers. Of these, only mellictinum and condelphinum have one tertiary nitrogen atom, and gallamine is a trisquaternary ammonium compound.

1. Bis- and trisquaternary ammonium compounds

Succinylcholine	Metocurine
Diadonium	Alcuronium
Anatruxonium	Pancuronium
Cyclobutonium	Pipecuronium
Tercuronium	Fazadinium
Dioxonium	Atracurium
Diplacinum	Gallamine
Qualidilum	

2. Compounds with one quaternary and one tertiary nitrogen atom
   Tubocurarine chloride
   Vecuronium
3. Salts of tertiary amines
   Pyrocurinum
   Mellictinum
   Condelphinum

The majority of the neuromuscular blocking agents are synthetic alkaloids. Tubocurarine, methyllicaconitine (methyllicaconitine hydroiodide is mellictinum), and condelphinum are extracted from plants; metocurine, alcuronium, and diplacinum are semisynthetic agents.

Table 1. Chemical structure of some neuromuscular blocking agents

Name (synonyms)	Chemical structure
Bis and trisquaternary ammonium compounds	
Succinylcholine chloride (Ditilinum, Suxamethonii chloridum, Lysthenon, Myo-Relaxin, etc.)[a]	$(CH_3)_3\overset{+}{N}{-}CH_2{-}CH_2{-}OOC{-}CH_2{-}CH_2{-}COO{-}CH_2{-}CH_2{-}\overset{+}{N}(CH_3)_3 \cdot 2Cl^-$
Diadonium	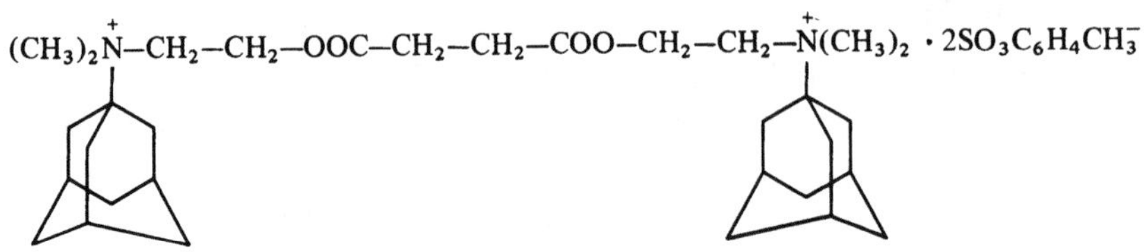
Atracurium (BW 33A)	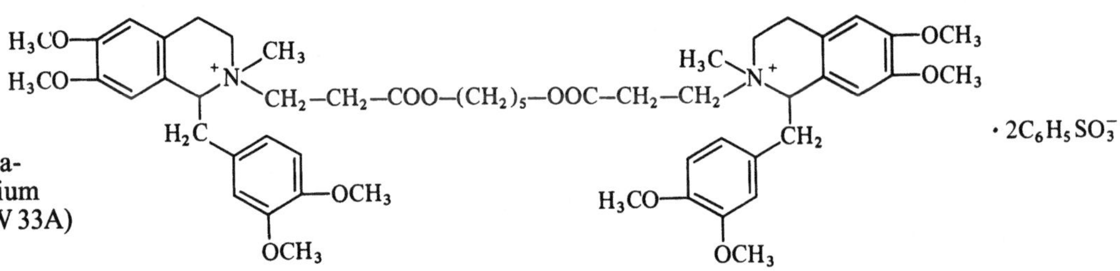
Anatruxonium (Truxipicurium iodide)	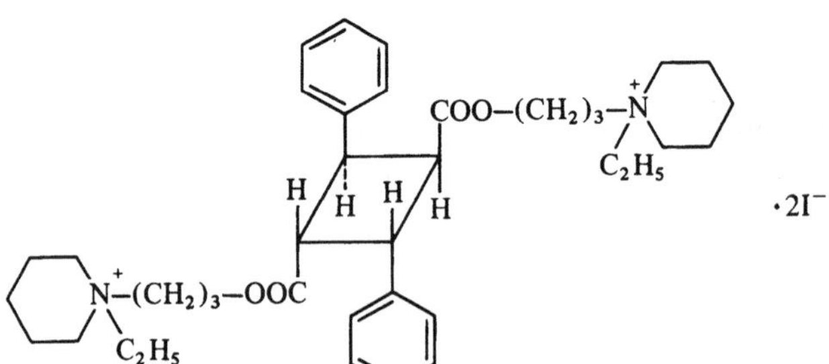
Cyclobutonium (Truxicurium iodide)	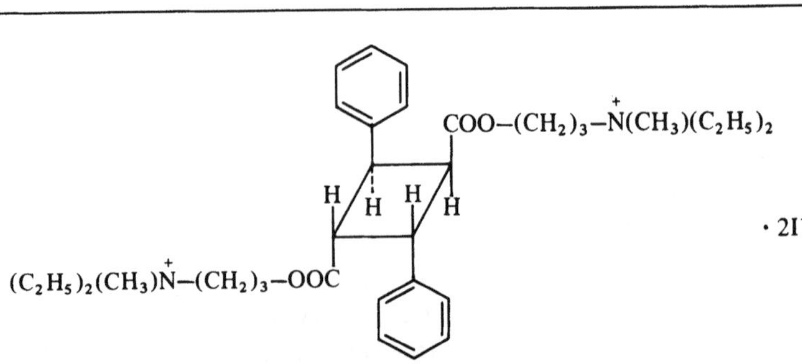

[a] Synonymes indicate drugs with the same cations and identical action, but different anions

Table 1 (continued)

Name (synonyms)	Chemical structure	
Qualidilum	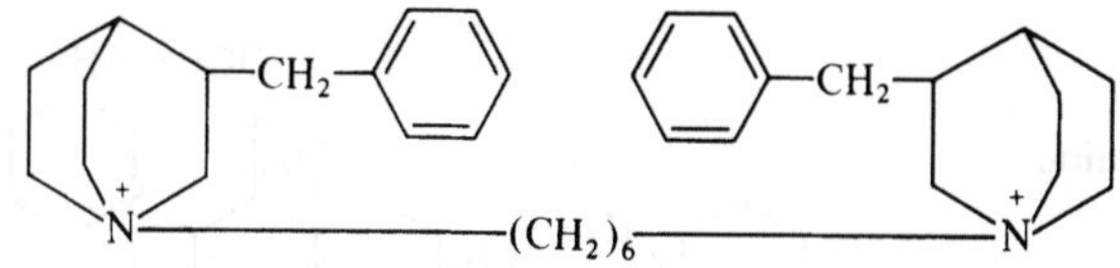	$\cdot 2Cl^- \cdot 4H_2O$
Diplacinum (Diplacini dichloridum)	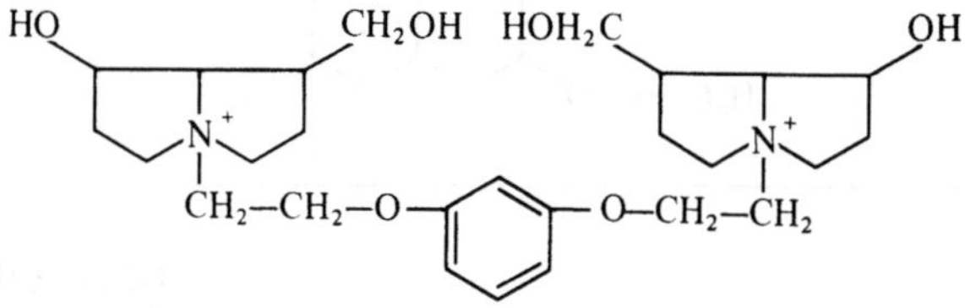	$\cdot 2Cl^-$
Tercuronium	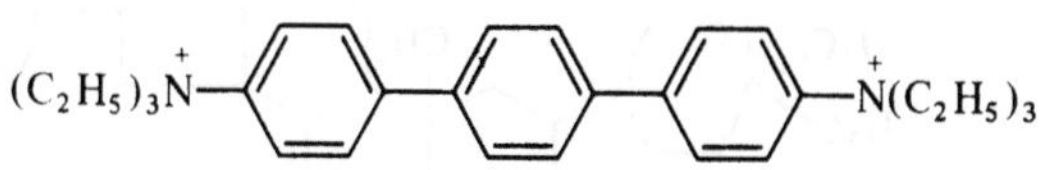	$\cdot 2C_6H_5SO_3^-$
Dioxonium	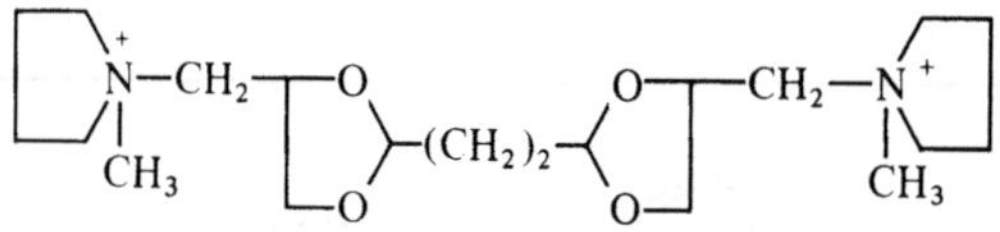	$\cdot 2I^-$
Metocurine (Dimethyl-tubocurarine iodide, Diamethine bromide, Metubine iodide, Mecostrin chloride)[a]	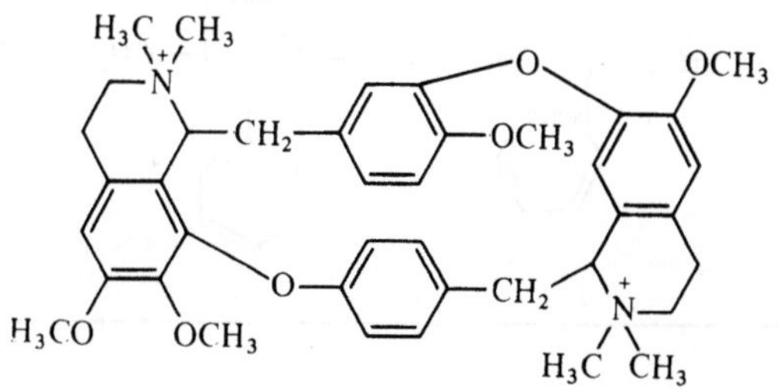	$\cdot 2I^-$
Alcuronium chloride (Ro-4-3816, Alloferin, Toxiferine)	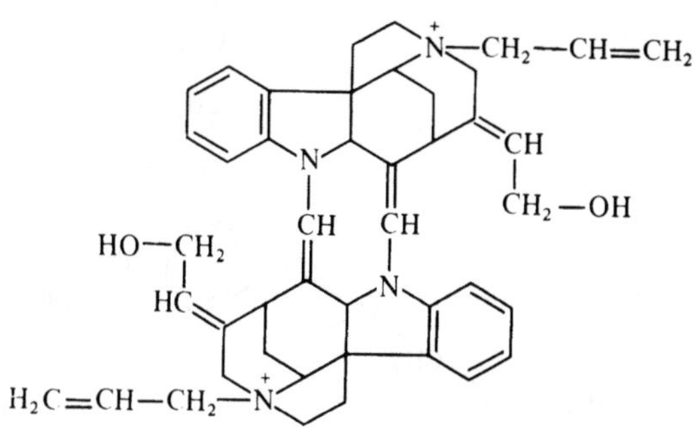	$\cdot 2Cl^- \cdot 5H_2O$

Table 1 (continued)

Name (synonyms)	Chemical structure
Pancuronium bromide (NA 97, Pavulon, Myoblock)	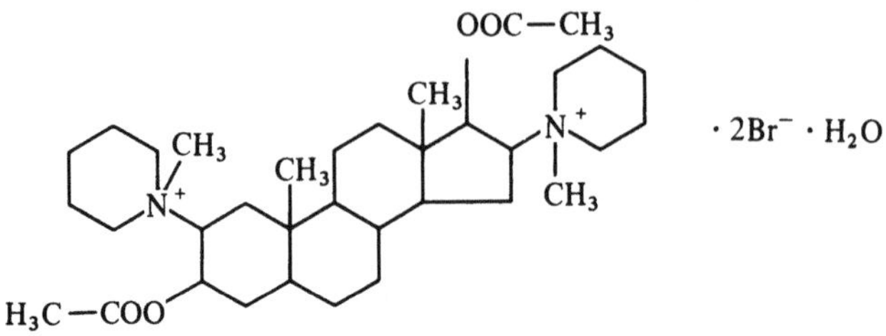
Pipecuronium (Arduan, Pipecurium)	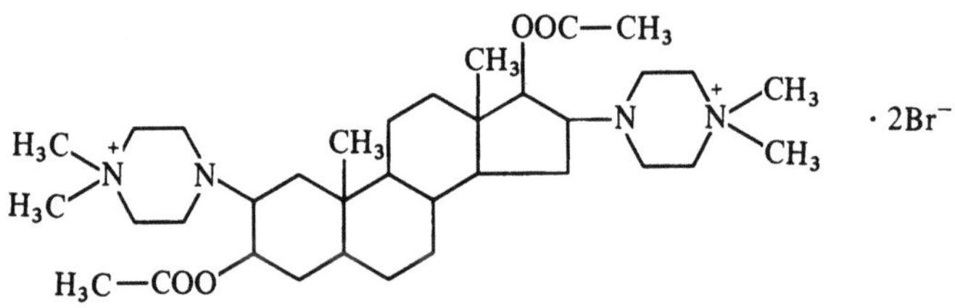
Fazadinium (AH 8165)	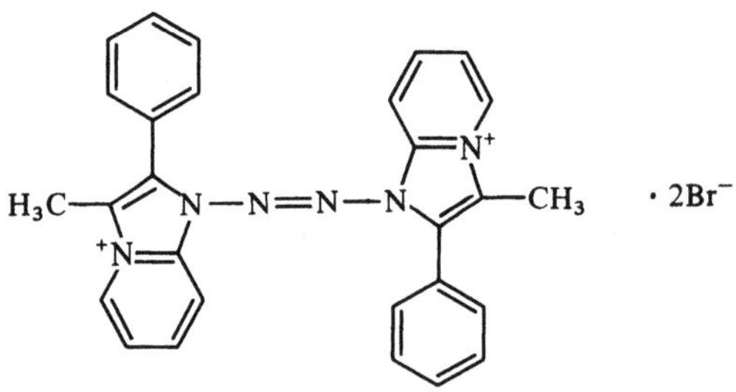
Gallamine triethiodide (Flaxedil, Relaxan, Pirolaxonum, Sincurarin, Syntubin, Tricuran)	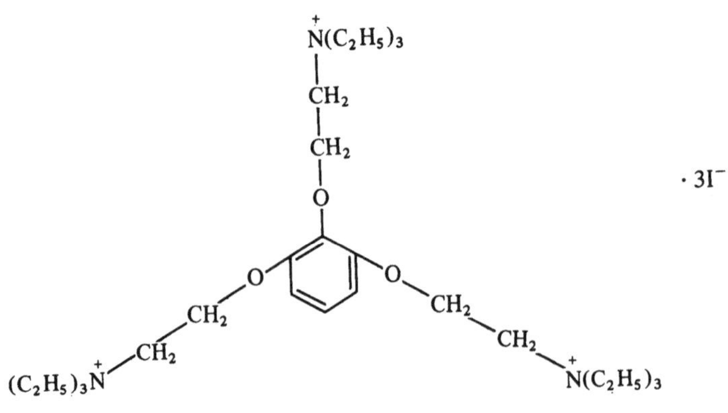

Table 1 (continued)

Name (synonyms)	Chemical structure	
Monoquaternary – monotertiary amines (+)-Tubocurarine chloride (Tubarine, Tubadil, Intocostrin-T)	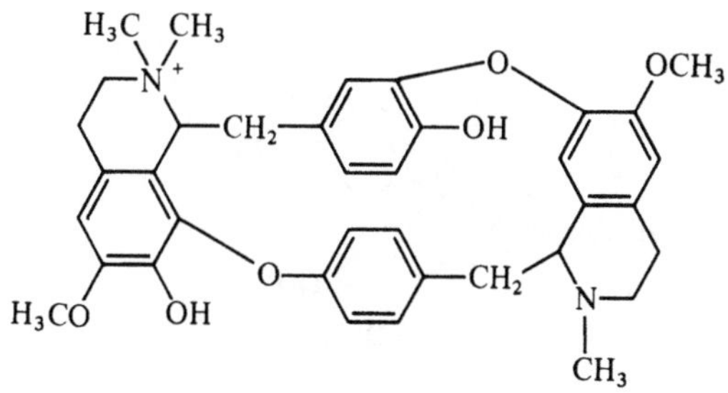	$\cdot HCl \cdot Cl^- \cdot 5H_2O$
Vecuronium (Norcuron, Org-NC-45)	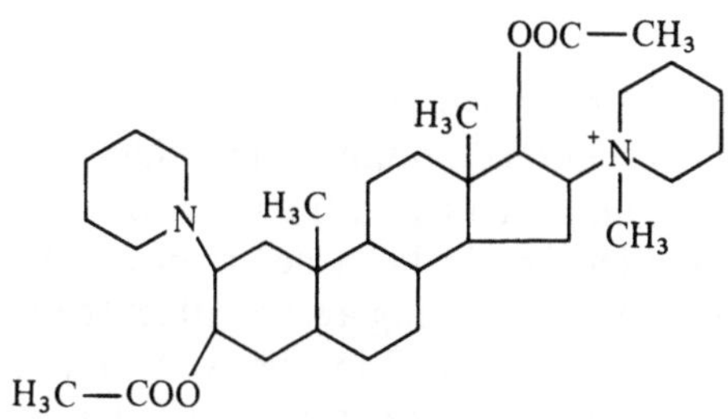	$\cdot Br^-$
Tertiary amines Pyrocurinum	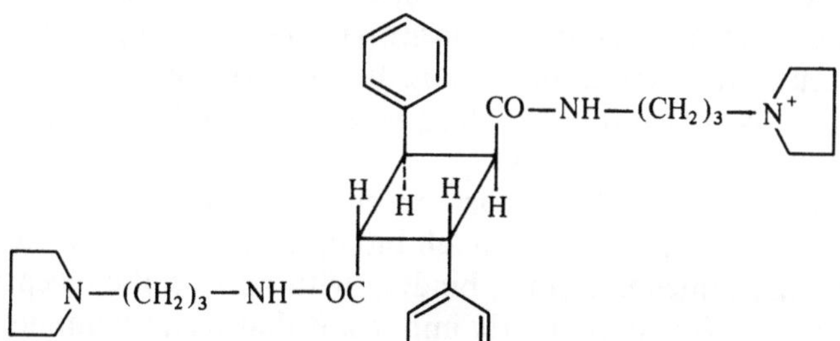	$\cdot 2HCl \cdot 2H_2O$
Mellictinum	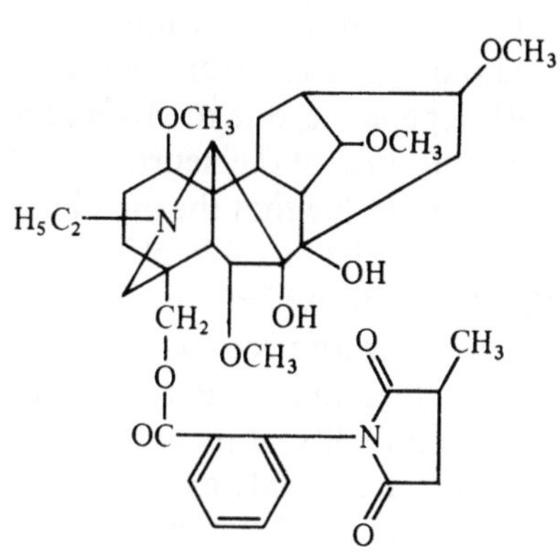	$\cdot HI$

Table 1 (continued)

Name (synonyms)	Chemical structure
Condelphinum	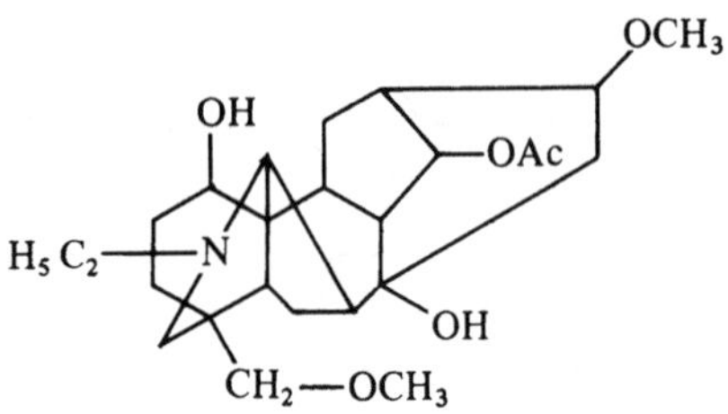

The dissimilarity in the chemical structure of the neuromuscular blocking agents lies in the size of the charged onium atoms, charge distribution, nature of the substituents at the onium centers, and the interonium structure of the molecule. These and other physicochemical characteristics as well as the spatial configuration of the neuromuscular blocking agents affect their interaction with acetylcholine receptors which is manifested by the contribution of various types of molecular bonds in this process (STENLAKE 1979, 1980; CAVALLITO 1980; KHARKEVICH and SKOLDINOV 1980). In its turn, this determines the mechanism of the block, the rate of drug action (and in large part its duration), and the rate of restoration of neuromuscular transmission. In this case, we refer to the rate and character of the association of the drug-receptor complex, its stability, and the rate of dissociation of the complex.

The chemical structure of the neuromuscular blocking agents also affects their pharmacokinetics: permeability through biologic membranes, including absorption from the gastrointestinal tract, binding with nonspecific receptors, metabolism, and excretion. It is particularly important that tertiary amines easily penetrate biologic membranes and in some cases can be prescribed for oral administration (for example, mellictinum and condelphinum). Chemical stability of the agents is a significant factor, too. The transformation of the agents under the influence of the enzymes (e.g., cholinesterases, liver microsomal enzymes) or their nonenzymatic degradation (pH dependent, due to Hofmann elimination) are decisive factors responsible for the duration of the neuromuscular block (see following discussion and Chap. 10). It is desirable that the metabolites formed be devoid of neuromuscular blocking activity, with no side effects, and minimal toxicity. One should not disregard the degree to which chemically different agents may bind with nonspecific receptors (e.g., with proteins of the blood serum, acid mucopolysaccharides of the tissues) which might be important not only for the duration of action, but also for the possibility of recurarization.

Neuromuscular blocking agents are mainly excreted by the kidneys and partially with bile. Their chemical structure and stability are responsible for whether they are excreted predominantly unchanged (tubocurarine, gallamine), in the form of metabolites (succinylcholine), or as a combination of unchanged drugs and their metabolites (pancuronium, vecuronium). Naturally, the excretion kinetics significantly affects the duration of action.

Mechanism of Action

According to the principles of interaction with end-plate acetylcholine receptors, neuromuscular blocking agents are usually divided into three groups:

1. Nondepolarizing (antidepolarizing) agents
   (a) Competitive action
   (b) Noncompetitive action
2. Depolarizing agents
3. Agents of mixed action

However, this classification is likely to require a certain revision in the near future, since there are data indicating that tubocurarine and gallamine (and apparently, other neuromuscular blocking agents) interact not only with acetylcholine receptor sites, but also with ionic channel sites (KATZ and MILEDI 1978; COLQUHOUN et al. 1979; DREYER 1982; PEPER et al. 1982; SHAKER et al. 1982; TRAUTMANN 1983; see also Chap. 3). For the present, no new, consistent classification has been suggested. However, it should be kept in mind that channel blockade may be an important component in the mechanism of action of neuromuscular blocking agents.

According to the classification, the majority of known drugs are nondepolarizing competitive agents (tubocurarine, anatruxonium, cyclobutonium, pancuronium, pipecuronium, tercuronium, diadonium, diplacinum, etc.). Prestonal exemplifies noncompetitive nondepolarizing agents. Succinylcholine represents depolarizing agents. Dioxonium is a neuromuscular blocking agent of mixed action (KIMENIS et al. 1976).

Nondepolarizing competitive agents are of exceptional interest for use in patients, since they have effective pharmacologic antagonists (anticholinesterase agents, Pymadin). Depolarizing agents are devoid of pharmacologic antagonists that might be used in anesthesiology. Furthermore, the depolarizing agents induce a release of potassium ions from the skeletal muscles into blood plasma (resulting from persistent depolarization of the subsynaptic membrane). An increased concentration of potassiums ions in blood may induce several side effects (see following discussion). Today, the short-acting succinylcholine is the only depolarizing agent used in patients. Intubation is the main indication for its application. The use of succinylcholine throughout the whole operation is inexpedient. Nondepolarizing (competitive) agents of high potency which have practically no side effects and possess effective antagonists, are preferable for such purposes.

Onset of Action

Neuromuscular blocking agents causing rapid complete paralysis with apnea are required for practical use. They provide a rapid intubation of the patient. The optimal onset of action is 30–45 s. In this respect, succinylcholine is the best drug.

Fazadinium was observed to have a similarly rapid onset of neuromuscular blocking action (see Chaps. 23 and 29). A short-acting nondepolarizing analog of succinylcholine, diadonium, creates the necessary conditions for intubation only after 90–120 s. If a nondepolarizing agent is devoid of marked side effects, the onset of neuromuscular block can be made more rapid by increasing the dose. However, the latter inevitably results in the prolongation of the neuromuscular block. It is quite obvious that rapid onset of action is one of the desired properties of new nondepolarizing neuromuscular blocking agents, especially short-acting ones.

Duration of Action

Duration of action of the neuromuscular blocking agents is one of the principal practical criteria. Anesthesiology needs a set of agents with a range of duration of action. They can be classified as: (a) short-acting (5–15 min); (b) medium-acting (20–50 min); and (c) long-acting (60 min and more).

Short-acting neuromuscular blocking agents are of particular interest for anesthesiologists. Theoretically, were they highly potent, sufficiently short-acting, and causing no cumulation or side effects, they could be used for controlled neuromuscular block via intravenous infusion. However, there are no such agents. The two available short-acting agents: succinylcholine (depolarizing) and diadonium (nondepolarizing), are not fit for prolonged intravenous infusion because of their side effects. In addition, succinylcholine has no antagonists and diadonium is insufficiently active and might cumulate, though to a low extent, aggravating the side effects. Thus, the design of a more perfect antidepolarizing short-acting neuromuscular blocking agent is still necessary.

The quest for such agents is conducted among compounds with the following properties: (a) rapid enzymatic or nonenzymatic biodegradation; and (b) unstable binding with acetylcholine receptors of the end-plate. The first group includes the bisonium esters which are rapidly hydrolyzed by plasma cholinesterase: diadonium (Kharkevich 1970 b, 1973), compounds of structure I (Ginsburg et al. 1971; Savarese et al. 1979), and among these compound BW 785U, which has been tested clinically (Savarese and Wastila 1979; Wastila and Savarese 1979; Savarese et al. 1980).

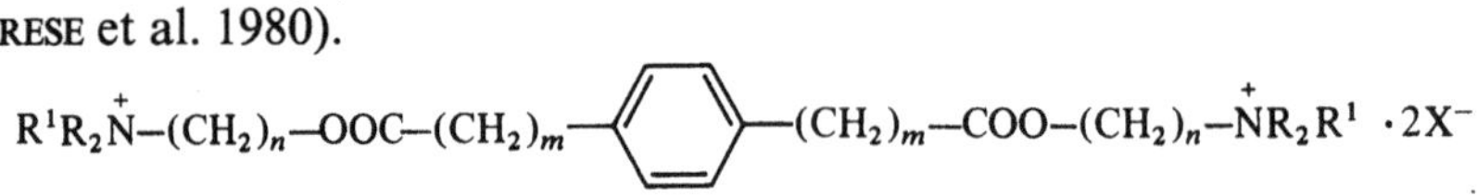
(I)

The azobisbenzimidazolinium salt AH 10407 (structure II), also short-acting, is very rapidly degraded in plasma by bicarbonate ion (Glover and Yorke 1971; Blogg et al. 1975).

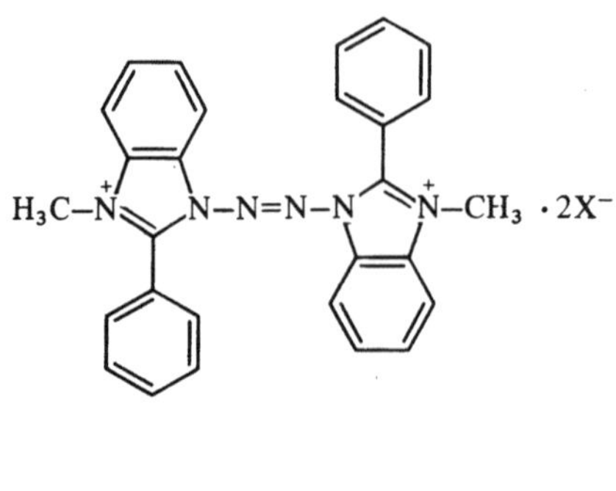
(II)

On the basis of experimental data, the steroid drug Duador (RGH-4201) (structure III; BIRÓ and KÁRPÁTI 1981) has been suggested as a likely short-acting drug.

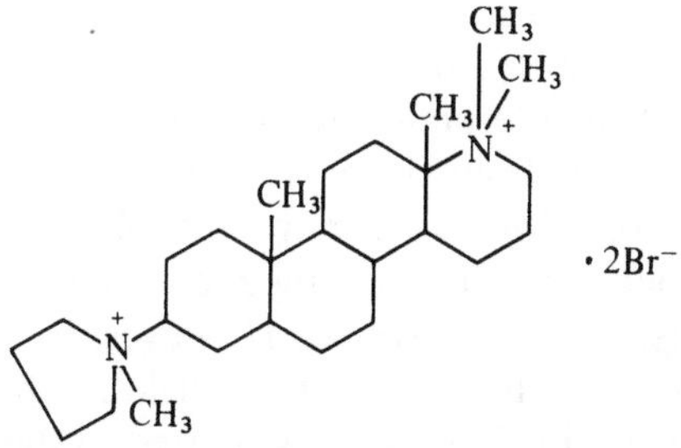

(III)

The routes of synthesis of compounds that bind unstably with acetylcholine receptors of the end-plate are not so clear. It is not excluded that these agents, like tubocurarine, might possess one quaternary and one tertiary atom. The search among bistertiary amines is quite practicable and investigations in the series of mono-quaternary ammonium compounds are promising, too. The latter have less stable bonds with acetylcholine receptors of skeletal muscles than bisquaternary ammonium compounds. Furthermore, in the series of monoquaternary ammonium salts, it is possible to synthesize rapidly hydrolyzed esters. Those investigations resulted in the synthesis of highly active compounds, but with a depolarizing action (KHARKEVICH et al. 1967; KITZ et al. 1969; GINSBURG et al. 1971; KHARKEVICH 1973). Monoquaternary ammonium salts with nondepolarizing action turned out to be insufficiently active in blocking the neuromuscular transmission.

Neuromuscular blocking agents of medium duration of action are numerous. They include tubocurarine, diplacinum, anatruxonium, cyclobutonium, tercuronium, dioxonium, gallamine, fazadinium, pancuronium, pipecuronium, vecuronium, and some others. The duration of their action after single administration is about 20–50 min. Many of them are very potent and have a selective blocking action on the end-plate acetylcholine receptors, side effects being only slight. Basically, the available neuromuscular blocking agents are quite sufficient to meet anesthesiologic needs. Further research should aim at the improvement of some properties of these drugs. In this framework, the synthesis of the steroid compound vecuronium is an advance (DURANT et al. 1979; AGOSTON et al. 1980; SAVAGE et al. 1980(. It has already been mentioned that vecuronium is an analog of pancuronium, but it has one quaternary nitrogen atom; the second cationic center is a tertiary nitrogen. Its specific feature, in addition to its selectivity, medium duration of action, and potency, is rapid restoration of neuromuscular transmission after the block (see following discussion). Vecuronium is deacetylated at the expense of acetylcholine-like groups, more rapidly than pancuronium. Vecuronium degradation yields a poorly active metabolite (diol-Org 7402; SAVAGE et al. 1980). This is mainly responsible for the shorter action of vecuronium as compared with pancuronium.

Carbolonium (Imbretil) can be grouped with the long-acting neuromuscular blocking drugs. In addition, a long-lasting neuromuscular block can be caused by increasing the dose of medium-acting agents (if side effects do not interfere).

Rate of Restoration of Neuromuscular Transmission

This value is characterized by the time between onset and essentially complete restoration of neuromuscular transmission after its complete block. The rate of restoration depends on the chemical stability of the neuromuscular blocking agents and on the stability of their interaction with acetylcholine receptors of the endplate. For practical anesthesiology, drugs with rapid restoration of neuromuscular transmission (within 10–15 min) are more suitable. This facilitates the control over neuromuscular block. For short-acting agents, like diadonium, rapid restoration of the initial muscle tone is normal. However, for agents of medium and long duration of action, the rate of restoration varies. After administration of vecuronium, restoration is rapid. This is likely related to the biodegradation of the drug and also to the presence of a tertiary nitrogen atom in the second cationic center which seems to decrease the stability of the drug-receptor complex, thus facilitating a more rapid restoration of the neuromuscular transmission after its block. This is a great advantage of vecuronium, which facilitates the postanesthetic treatment of the patient. Rapid restoration of the muscle tone after the block is also typical of pyrocurinum which has two tertiary nitrogen atoms as cationic centers. After administration of bisquaternary ammonium derivatives of α-truxillic acid (anatruxonium, cyclobutonium), neuromuscular transmission is restored relatively slowly. Tubocurarine occupies an intermediate position. The duration of the block and the rate of restoration are interrelated. However, equal durations of complete block can be associated with different rates of restoration.

Safety Margin

It is generally assumed that this parameter is determined by the range between the doses in which the drugs evoke relaxation of body muscles and those in which they paralyze respiratory muscles and induce apnea. The safety margin is dissimilar in drugs with different chemical structure. This seems to be related to unequal sensitivity of the acetylcholine receptors of different groups of muscles to neuromuscular blocking agents. Some agents have small safety margins (e.g., tubocurarine, pancuronium, pipecuronium) and some have comparatively large safety margins (diplacinum, cyclobutonium).

Anesthesiologists are in need of both types of agents. The neuromuscular blocking agents of moderate and long duration of action used in operations on the organs of the chest and the upper part of the stomach, should abolish respiration rapidly, i.e., they must have a minimal safety margin. At the same time, in operations on abdominal organs, small pelvis, and limbs, the neuromuscular blocking agents which do not affect spontaneous respiration can be of great use. This is also true for neuromuscular blocking agents prescribed for parenteral use in neurology and drugs applied for the treatment of tetanus. There is a significant number of neuromuscular blocking agents with a narrow safety margin. Therefore, the design of neuromuscular blocking agents which effectively relax the body muscles and have minimal effect on respiration is of great interest. The number of such agents is quite limited.

Order of Relaxation of Different Groups of Muscles

For many years it was generally assumed that the order of relaxation of various groups of muscles established for tubocurarine was the same for all other neuromuscular blocking agents. However, it was shown that the order of muscle relaxation induced by agents with different chemical structure can be dissimilar. There are reasons to believe that this is mainly related to peculiarities of molecular structure. At the same time, the mechanism of action of neuromuscular blocking agents has no influence on the order of muscle relaxation. Indeed, the depolarizing agent, succinylcholine and its nondepolarizing bis-*N*-adamantyl analog, diadonium induce the same order of neuromuscular block in various groups of skeletal muscles (LEPAKHIN and FISENKO 1970; KHARKEVICH and FISENKO 1981). The data obtained (Chap. 6) indicate certain differences in morphofunctional organization of acetylcholine receptors in the skeletal muscles at various sites. Furthermore, this confirms the possibility of creating neuromuscular blocking agents with predominant action upon certain groups of skeletal muscles.

It should be taken into account that predominant action on certain muscles may be manifested not only by the order of their relaxation, but also by a longer duration of the neuromuscular block of a particular group of muscles. Thus, for instance, in anesthesiology, it is known that cyclobutonium causes an especially long-lasting block of the abdominal wall muscles (BARSUKOV 1970), the tone of other muscles, respiratory muscles included, being to a certain extent restored.

Cumulation

Nondepolarizing neuromuscular blocking agents used in anesthesiology are more or less cumulated, as evident during their repeated administration. Naturally, the lower the cumulation, the easier it is to control the degree and the duration of the neuromuscular block. Of the known neuromuscular blocking agents, vecuronium and pyrocurinum are cumulated to a relatively small extent. Little or no cumulation is one of the important requirements to be met by new neuromuscular blocking agents.

Effectiveness of Various Administration Routes

The agents considered so far belong to the neuromuscular blocking drugs used intravenously in anesthesiology. However, orally effective drugs are available too. Among these are tertiary amines, e. g., delphinium alkaloids such as condelphinum and mellictinum. They can be used in some disturbances of the central nervous system to decrease the elevated tone of the skeletal muscles. Such neuromuscular blocking agents should be well absorbed from the gastrointestinal tract and have effective and promptly acting antagonists. It is particularly necessary that such agents have a maximum safety margin. The delphinium alkaloids do not sufficiently meet the latter requirement, which significantly restricts their application in neurology.

Side Effects

Side effects are quite important for evaluation of neuromuscular blocking agents. The ideal neuromuscular blocking agent should possess a high selectivity of action on the acetylcholine receptors of the skeletal muscles, and have no side effects. There are a few nodepolarizing agents which partially meet this requirement: pyrocurinum (KHARKEVICH 1983); tercuronium (DANILOV et al. 1979), pipecuronium (KÁRPÁTI and BIRÓ 1980), and vecuronium (I. G. MARSHALL et al. 1980; R. J. MARSHALL et al. 1980). The majority of other neuromuscular blocking agents are known to possess various side effects (MARSHALL 1980; BOWMAN 1982).

As mentioned previously, in depolarizing neuromuscular blocking agents, e. g., succinylcholine, many side effects are determined by their mechanism of action. It is known, that the depolarization of the subsynaptic membrane and probably microinjuries of muscle fibers in the course of their fasciculations are followed by potassium ion loss from the skeletal muscles. This results in an increased concentration of potassium ions in the blood plasma. Hyperkalemia is most pronounced during muscle denervation. In patients, this is observed after muscle injuries, burns, or neurologic injuries (SMITH 1976). Such dissimilarities in plasma potassium ion elevation are due to the fact that, under normal conditions, succinylcholine causes depolarization of only the subsynaptic membrane of the endplate, while during denervation it leads to depolarization of a significant part of muscle fiber membrane. An increased concentration of potassium ions in the blood plasma may induce cardiac arrthythmias. During the postanesthetic period, patients may sometimes suffer from muscular pain. It was suggested that, like one of the hyperkalemia mechanisms, muscular pains were caused by noncoordinated fasciculations and the resulting microinjuries of muscle fibers (COLLIER 1975; BALI et al. 1975). Furthermore, succinylcholine elevated intraocular pressure, most likely by increasing the tone of the external muscles of the eye, and probably the ganglion-stimulating action. It also causes an increase in the intragastric pressure. In the case of genetically or otherwise conditioned insufficiency of plasma cholinesterase (WHITTAKER 1980), a prolonged apnea may be observed (for details of side effects of succinylcholine see review article by DURANT and KATZ 1982).

The non-depolarizing neuromuscular blocking agents may cause various side effects: tubocurarine – hypotension and bronchospasm; anatruxonium, cyclobutonium, and diadonium – tachycardia and slight fall of blood pressure; pancuronium and gallamine – tachycardia and hypertension, etc. These side effects are due to ganglion-blocking action (hypotension), histamine release (hypotension, bronchospasm), inhibition of cardiac output (hypotension), blockade of muscarinic acetylcholine receptors of the heart (tachycardia), block of neuronal uptake of norepinephrine (tachycardia, hypertension), or inhibition of acetylcholinesterase (increased secretory activity of glands). Probable inhibition of plasma cholineterase can result in the disturbed hydrolysis of esters (e. g., succinylcholine).

Antagonists

Safe application of the neuromuscular blocking agents in anesthesiology requires effective antagonists. At present, only the antagonists of nondepolarizing agents are well known and widely used (see review by CRONNELLY and MORRIS 1982). They are particularly effective with respect to the neuromuscular blocking agents with a competitive type of action (devoid of antiacetylcholinesterase activity). The depolarizing agents have no pharmacologic antagonists which can be used in patients. The most widely used antagonists of nondepolarizing neuromuscular blocking agents can be represented by the following groups:

1. Agents which inhibit the hydrolysis of endogenous acetylcholine: anticholinesterase agents (neostigmine, galanthamine, pyridostigmine, edrophonium)
2. Agents which activate acetylcholine release from motor nerve terminals (Pymadin)

Consequently, the principle of action of both groups of antagonists lies in the increase of concentration of acetylcholine in the junctional gap which competes for acetylcholine receptor sites with the neuromuscular blocking agents, thus promoting the restoration of neuromuscular transmission.

Another principle for the design of specific antagonists has been proposed. This is the chemical inactivation of neuromuscular blocking agents of a certain structure. For instance, KHROMOV-BORISOV et al. (1969) used sodium sulfite, unithiol, or cysteine as effective antagonists of a bisquaternary ammonium derivative of diphenylsulfide. The authors believe that the inactivation results from nucleophilic substitution at the sulfur atom in the disulfide group and the formation of a monoquaternary salt. None of the antagonists tested affected the blocking action of tubocurarine, succinylcholine, decamethonium, or analogs of compound IV without a disulfide bond.

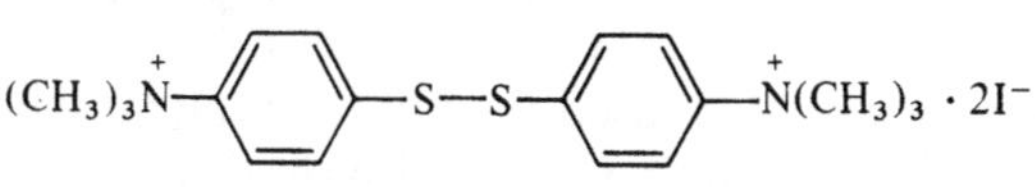

(IV)

However, this principle is significantly less universal than the application of anticholinesterase agents or Pymadin. However, chemical antagonists may act more selectively, causing no side effects.

Main Trends in the Search for New Agents

All these observations confirm that there is widespread interest in selectively acting, nondepolarizing, competitive neuromuscular blocking agents which have antagonists and cause no side effects. In spite of the considerable number of neuromuscular blocking agents adopted into practical medicine, there is still a need for new agents of a certain type. One of the most important trends in the design of new neuromuscular blocking agents is the quest for nondepolarizing short-acting drugs. Such a drug should have the following properties:

High potency
Competitive mechanism of action
Selective blocking action upon acetylcholine receptors of the skeletal muscles

Rapid development of the effect
Short-term neuromuscular block (at single administration it should be no longer than 10–15 min)
No potentiation or cumulation on repeated administration
Little or no side effects
Low toxicity
Lack of physiologic and toxicologic action of probable metabolites and their rapid elimination from the organism
Effective antagonists
Stability in storage
Profitable synthesis on an industrial scale

The interest in such agents lies in that they could be used for controlled neuromuscular block by intravenous infusion. It is known that this is the principle of controlled hypotension by short-acting ganglion-blocking agents (hygronium, Arfonad) or myotropic vasodilators (e.g., sodium nitroprusside). Should a sufficiently adequate drug for controlled neuromuscular block be designed, it could be applied in the majority of surgical interventions and could virtually claim to be a universal drug.

Long-acting neuromuscular blocking agents which do not affect hemodynamics or induce any other side effects are also required. They would be applied in prolonged surgery and for the complex therapy of tetanus. The neuromuscular blocking agents with only slight or essentially no effect on spontaneous respiration are of special interest. They are indicated for operations in which apnea is not obligatory or in which it is undesired.

The quest for new potent nondepolarizing neuromuscular blocking agents for oral administration which cause minimal suppressing effect on respiration also deserves attention. Such drugs would be of interest for the treatment of neurologic diseases associated with increased tone of the skeletal muscles.

Thus, new, more selectively acting drugs with the desired properties would enrich the stock of practically useful neuromuscular blocking drugs.

Acknowledgment. The author wishes to express his appreciation to Mrs. MARIA LIPMAN for the translation of this chapter.

References

Agoston S, Salt P, Newton D, Bencini A, Boomsma P, Erdmann W (1980) The neuromuscular blocking action of Org NC 45, a new pancuronium derivative, in anaesthetized patients. A pilot study. Br. J Anaesth 52 [suppl I]:53S–59S

Ahnefeld FW, Bergmann H, Burri C, Dick W, Halmágyi M, Hossli G, Rügheimer E (eds) (1980) Muskelrelaxanzien. Klinische Anästhesiologie und Intensivtherapie vol 22. Springer, Berlin Heidelberg New York

Bali JM, Dundee JW, Doggart JR (1975) The source of increased plasma potassium following succinylcholine. Anesth Analg 54:680–686

Barsukov PYa (1970) Application of cyclobutonium under endotracheal ether-oxygen anaesthesia during gynecological operations (in Russian). In: Kharkevich DA (ed) Novye kurarepodobnye i ganglioblokiruyushchie sredstva (New curare-like and ganglion-blocking agents). Meditsina, Moscow, pp 164–168

Biró K, Kárpáti E (1981) The pharmacology of a new short-acting nondepolarizing muscle relaxant steroid (RGH-4201). Arzneimittelforsch 31:1918–1924
Blogg CE, Brittain RT, Simpson BR, Tyers MB (1975) AH 10407: a novel short-acting competitive neuromuscular blocking drug in animals and man. Br J Pharmacol 53:446P
Bowman WC (1980) Pharmacology of neuromuscular function. Wright, Bristol
Bowman WC (1982) Non-relaxant properties of neuromuscular blocking drugs. Br J Anaesth 54:147–160
Buzello W (ed) (1981) Muskelrelaxantien. Neuere Konzepte ihrer Pharmakologie und klinischen Anwendung. Thieme, Stuttgart
Cavallito CJ (1980) Quaternary ammonium salts – advances in chemistry and pharmacology since 1960. Prog Drug Res 24:268–373
Cheymol J (ed) (1972) Neuromuscular blocking and stimulating agents, vol I–II. Pergamon, Oxford
Collier C (1975) Suxamethonium pains and fasciculations. Proc R Soc Lond 68:105–108
Colquhoun D, Dreyer F, Sheridan RE (1979) The actions of tubocurarine at the frog neuromuscular junction. J Physiol (Lond) 293:247–284
Cronnelly R, Morris RB (1982) Antagonism of neuromuscular blockade. Br J Anaesth 54:183–194
Danilov AF, Malygin VV, Starshinova LA, Khromov-Borisov NV, Torf SF, Cherepanova VP (1979) Tercuronium – a new nondepolarizing neuromuscular blocking agent with high activity and selectivity of effect (in Russian). Farmakol Toksikol 5:478–481
Dreyer F (1982) Acetylcholine receptor. Br J Anaesth 54:115–129
Durant NN, Katz RL (1982) Suxamethonium. Br J Anaesth 54:195–208
Durant NN, Marshall IG, Savage DS, Nelson DN, Sleigh T, Carlyle IC (1979) The neuromuscular and autonomic blocking activities of pancuronium, Org NC 45 and other pancuronium analogues, in the cat. J Pharm Pharmacol 31:831–836
Feldman SA (1973) Muscle relaxants. Major Problems in Anaesthesia, vol I. Saunders, London
Foldes FF (1957) Muscle relaxants in anesthesiology. Thomas, Springfield
Ginsburg S, Kitz RJ, Savarese JJ (1971) Neuromuscular blocking activity of a new series of quaternary N-substituted choline esters. Br J Pharmacol 43:107–126
Glover EE, Yorke MM (1971) Cyclic quaternary ammonium salts. Part IX. 1,1′-azoimidazo[1,2-α]-pyridinium salts. J Chem Soc *C*:3280–3286
Kárpáti E, Biró K (1980) Pharmacological study of a new competitive neuromuscular blocking steroids, pipecuronium bromide. Arzneimittelforsch 30:346–354
Katz RL (ed) (1975) Muscle relaxants. Monographs in anesthesiology, vol 3. Amsterdam
Katz B, Miledi R (1978) A re-examination of curare action at the motor endplate. Proc R Soc (Lond) Ser B 203:119–133
Kharkevich DA (1969) Farmakologiya Kurarepodobnykh sredstv (Pharmacology of neuromuscular blocking drugs). Meditsina, Moscow
Kharkevich DA (ed) (1970a) Novye Kurarepodobnye i ganglioblokiruyushchie sredstva (New curare-like and ganglion-blocking agents) Meditsina, Moscow
Kharkevich DA (1970b) On pharmacological properties of a new antidepolarizing neuromuscular blocking agent diadonium diiodide (in Russian). Farmakol Toksikol 5:531–536
Kharkevich DA (1973) Curare-like agents (in Russian). In: Kharkevich DA (ed) Uspekhi V sozdanii novykh lekarstvennykh sredstv (Advances in drug research). Meditsina, Moscow, pp 138–187
Kharkevich DA (ed) (1983) Novye miorelaksanty (New muscle relaxants) Meditsina, Moscow
Kharkevich DA, Fisenko VP (1981) The effect of neuromuscular blocking agents on the acetylcholine receptors of different skeletal muscles. Arch Int Pharmacodyn Ther 251:255–269
Kharkevich DA, Skoldinov AP (1980) On some principles of interaction of curare-like agents with acetylcholine receptors of skeletal muscles. J Pharm Pharmacol 32:733–739

Kharkevich DA, Arendaruk AP, Gracheva EA, Skoldinov AP (1967) On curare-like properties of mono-quaternary ammonium derivatives of cinnamic acid (in Russian). Farmakol Toksikol 5:562–567
Khromov-Borisov NN, Gmiro VE, Magazanik LG (1969) Removal of curare-like effect by direct inactivation of the myorelaxant molecule by disruption of the disulfide bond (in Russian). Dokl Akad Nauk SSSR 196:236–239
Kimenis AA, Klusha VE, Ginters YaYa, Veveris MM (1976) Dioksony. Farmakologiya i anesteziologicheskoe primenenie (Dioxonium. Pharmacology and anesthesiological application) Zinatne, Riga
Kitz RJ, Karis JH, Ginsburg S (1969) A study in vitro of new short-acting nondepolarizing neuromuscular blocking agents. Biochem Pharmacol 18:871–881
Lepakhin VK, Fisenko VP (1970) On comparative sensitivity of neuromuscular junctions of different muscles to diadonium and decadonium (in Russian). Farmakol Toksikol 3:288–292
Marshall IG (1980) Actions of non-depolarizing neuromuscular blocking agents at cholinoceptors other than at the motor endplate. In: Curare and curarization. Excerpta Medica, Amsterdam, pp 257–274
Marshall IG, Agoston S, Booij LHDJ, Durant NN, Foldes FF (1980) Pharmacology of Org NC 45 compared with other non-depolarizing neuromuscular blocking drugs. Br J Anaesth 52 [Suppl I]:11S–19S
Marshall RJ, McGrath JC, Miller RD, Docherty JR, Lamar J-C (1980) Comparison of the cardiovascular actions of Org NC 45 with those produced by other non-depolarizing neuromuscular blocking agents in experimental animals. Br J Anaesth 52 [Suppl I]:21S–32S
Peper K, Bradley RJ, Dreyer F (1982) The acetylcholine receptor at the neuromuscular junction. Physiol Rev 62:1271–1340
Savage DS, Sleigh T, Carlyle I (1980) The emergence of Org NC 45, I-[(2β,3α,5α,16β,17β)-3,17-bis(acetyloxy)-2-(I-piperidinyl)-androstan-16-yl]-I-methylpiperidinium bromide from the pancuronium series. Br J Anaesth 52 [Suppl I]:3S–9S
Savarese JJ, Wastila WB (1979) Pharmacology of BW 78521: a short-acting nondepolarizing neuromuscular blocking agent. Anesthesiology 51:S277
Savarese JJ, Ginsburg S, Braswell L, Kitz RJ (1979) Actions at neuromuscular and esteratic cholinoceptive sites of some phenylene diacryloyl bis-cholinium esters. J Pharmacol Exp Ther 208:436–445
Savarese JJ, Ali HH, Basta SJ, Ramsey FM, Rosow CE, Lebowitz PW, Lineberry CG, Cloutier G (1980) Clinical neuromuscular pharmacology of BW 785U, an ultra-short-acting nondepolarizing ester neuromuscular blocking agent. Anesthesiology 53:S274
Shaker N, Eldefrawi AT, Aguayo LG, Warnick JE, Albuquerque EX, Eldefrawi MW (1982) Interactions of d-tubocurarine with the nicotinic acetylcholine receptor/channel molecule. J Pharmacol Exp Ther 220:172–177
Smith SE (1976) Neuromuscular blocking drugs in man. In: Zaimis E (ed) Neuromuscular junction. Springer, Berlin, Heidelberg, New York, pp 593–660 (Handbook of experimental pharmacology, vol 42)
Stenlake JB (1979) Molecular interactions at the cholinergic receptor in neuromuscular blockade. In: Ellis GP, West GB (eds) Progress in medicinal chemistry, vol 16. Butterworths, London, pp 257–286
Stenlake JB (1980) Neuromuscular blocking agents. In: Wolff ME (ed) Alfred Burger's medicinal chemistry, 4th edn. Wiley-Interscience, New York
Trautmann A (1983) Tubocurarine, a partial agonist for cholinergic receptors. J Neural Transm [Suppl] 18:353–361
Wastila WB, Savarese JJ (1979) Autonomic/neuromuscular dose-ratios and hemodynamic effects of BW 785U, a short-acting nondepolarizing ester neuromuscular blocking agent. Anesthesiology 51:S278
Whittaker M (1980) Plasma cholinesterase variants and anaesthetist. Anaesthesia 35:174–179
Zaimis E (ed) (1976) Neuromuscular junction. Springer, Berlin Heidelberg New York (Handbook of experimental pharmacology, vol 42)

CHAPTER 2

The End-Plate Acetylcholine Receptors: Structure and Function

L. P. WENNOGLE

A. Introduction

The acetylcholine receptor (AChR) is central to neuromuscular transmission and, in a broader sense, to all of biology. As a model system, the junctional acetylcholine receptor has added greatly to our appreciation of such diverse subjects as autoimmune diseases, neuromuscular development, the evaluation of multisubunit proteins, and the important area of drug–receptor interactions. The nicotinic acetylcholine receptor, in particular, is an extraordinary research tool.

All of molecular biology anxiously awaits the structural resolution of the acetylcholine receptor at the atomic level. With the remarkable pace of advancement and the advent of highly sophisticated methods of analysis, this accomplishment is not far in the future. A thorough appreciation of how these atoms interact functionally should not be far behind.

I. Scope of Review

A number of excellent reviews cover various aspects of the acetylcholine receptor. CARTAUD (1980) outlines the electron microscopy; LINDSTROM (1979a, b) and FUCHS (1979) treat receptor immunology; LESTER (1977), NEHER and STEVENS (1977), STEINBACH (1980), and COLQUHOUN (1979) deal with the physiology; FAMBROUGH (1979), and PATRICK and BERMAN (1980) review muscle acetylcholine receptor development; several authors discuss receptor function (CHANGEUX 1981; KARLIN 1980; KARLIN et al. 1979, 1976; CHANGEUX et al. 1976a; CHANGEUX et al. 1980, 1976a, b; GIRAUDAT and CHANGEUX 1980; HEIDMANN and CHANGEUX 1978; OSWALD et al. 1981; RAFTERY et al. 1979, 1976; ELDEFRAWI and ELDEFRAWI 1975, 1977). The general structure of membrane channels is reviewed by OVCHINNIKOW (1980). Rather than attempting to supersede these reviews, this chapter deals chiefly with literature from 1978 and concentrates mainly on receptor structure and function from a molecular point of view. The references cited are, in general, the most recent ones.

Most of the work described herein is derived from studies on receptors isolated from the electric organs of fish such as species of the electric ray *Torpedo* and the freshwater eel *Electrophorus electricus*. Fortunately, a number of parallels have been drawn to the mammalian neuromuscular junction. As these electric organs have an embryologic origin from striated muscle (FESSARD 1958), the receptor's structure, including amino acid sequence and aspects of immunologic cross-reactivity with skeletal muscle receptor (LINDSTROM 1979a, b; LINDSTROM et al. 1976, 1979a, b), has been highly conserved.

The aims of this review are: (a) to detail the molecular topology of the receptor, including its ion channel and transmembrane orientation; (b) to appreciate functional aspects of the receptor, including its multiple ligand binding sites and multiple conformational states; and (c) to reassess old ideas, while suggesting and stimulating new ones.

II. Perspective of the Junctional Acetylcholine Receptor

The process of neuromuscular transmission is conceptually simple. Acetylcholine (ACh) liberated by nerve terminals diffuses across a narrow cleft of extracellular substances (500–1,000 Å) and reaches the acetylcholine receptor situated in the postsynaptic membrane. The receptor responds by binding acetylcholine and opening its ion channel, allowing the membrane to become permeable to sodium and other cations. (Channel opening is, of necessity, both brief and subtle – see Sect. B.IV) As predicted by the Nernst and Goldman equations, the muscle cell becomes locally depolarized. This depolarization, or end-plate potential, if sufficient to exceed threshold, results in the propagation of the muscle action potential. (Action potential propagation is chiefly the work of another quite interesting membrane protein, the voltage-sensitive sodium ionophore.) Finally, bound ACh is released, and the receptor returns to its resting state.

The mammalian junctional acetylcholine receptor is found in skeletal muscle innervated by somatic motoneurons of the voluntary nervous system. Classification of the receptor as the nicotinic type differentiates it from a class of cholinergic receptors found ubiquitously in brain and smooth muscle responsive to acetylcholine, but of the muscarinic type. Putative nicotinic receptors in brain and sympathetic ganglion are not as well characterized.

Historically, the pioneers of physiology and neurochemistry have studied the nicotinic acetylcholine receptor. LANGLEY (1907) originally conceived of receptors. KUFFLER (1943) studied the specific chemical excitability of the end-plate region. NACHMANSOHN (1955) and his students further conceptualized the receptor and distinguished it from acetylcholinesterase. Katz and co-workers defined the process of receptor desensitization (KATZ and THESLEFF 1957, see Sect. D.2); they also introduced many innovative electrophysiologic techniques. CHANGEUX (1981) gives a detailed historical account of the subject.

The advent of a class of α-neurotoxins from elapid and sea snake venoms (LEE 1972), specific and high affinity antagonists of the nicotinic acetylcholine receptor, revolutionized this field of study. This accomplishment, as well as the use of rich sources of receptor from the electric eel *Electrophorus electricus* and electric rays of the genus *Torpedo* (NACHMANSOHN 1955; CHANGEUX 1975), enabled the biochemical characterization of the end-plate receptor. With the introduction of "in vitro" translation technologies and DNA recombination techniques (MENDEZ et al. 1980; SUMIKAWA et al. 1981; ANDERSON and BLOBEL 1981; MERLIE and Sebbane 1980; MERLIE et al. 1981), work on the acetylcholine receptor has been revolutionized; the successful synthesis of receptor by frog oocytes after injection of

Torpedo messenger RNA has recently been reported (SUMIKAWA et al. 1981). In addition, clonal probes of all four subunits of the receptor are now available (SUMIKAWA et al. 1982; CLAUDIO et al. 1983; NODA et al. 1983; DEVILLERS-THIERY et al. 1983; TANABE et al. 1984).

III. Some Particularly Interesting Questions

Many intriguing challenges remain; for example, what is the nature of the acetylcholine receptor ionophore? Is it a simple water-filled pore with a gate, or a more subtle molecular arrangement, such as an ion-exchanging matrix pathway? Do multiple subunit interactions form the pore, or is it proper to a single receptor subunit? Indeed, what are the nature and function of subunit interactions? What are the functions of individual subunits? Are local anesthetics and piperidine alkaloids, such as histrionicotoxin (HTX) (KATO and CHANGEUX 1976), actually plugging the ionophore, or is their interaction of an allosteric nature? In addition to these questions, it is not known whether the receptor is regulated during development by some metabolic process such as phosphorylation or methylation. Phosphorylation may, for instance, regulate the mean channel open time and serve to differentiate developmental forms (SAITOH and CHANGEUX 1981). How is the receptor immobilized at the mature, junctional end-plate? Are cytoskeletal contacts important, and what role do basal lamellar elements play? Finally, what progress has been made in understanding the molecular details of ligand binding sites? How many are there, do they preexist, are they flexible, etc.? As we shall see, partial and provocative answers have been presented for each of these questions.

B. Molecular Properties of the Acetylcholine Receptor

In order to relate structure to function, the aim has been to achieve homogeneous preparations of receptor and to correlate functional characteristics throughout the isolation process. This approach has been quite successful from both points of view. Among other achievements, the importance of subunit-subunit interactions, as well as the significance of the lipid environment, is now quite clear. First, we shall recount the progress made in receptor purification; in short, this work has led to homogeneous preparations of receptor, either in a detergent-solubilized form or in membrane preparations after selective extraction procedures. Next, we will explore the molecular components of this receptor. And finally, we will detail how these components work together to achieve the control over ion translocation at the neuromuscular junction.

I. Purification

The nicotinic acetylcholine receptor has been purified to homogeneity from a variety of tissue sources, including *Torpedo (marmorata, ocelata,* and *californica), Electrophorus electricus,* and skeletal muscle. Standard membrane fractionation methods are quite useful with *Torpedo* electric organ. This organ is rich in receptor. The receptor is found in membranes with a distinctive density corresponding to 38% sucrose. With muscle sources, where receptor density is far less, affinity chromatography methods with detergent-solubilized receptor using immobilized cholinergic ligands or α-toxin are essential. These methods have been well reviewed (HEIDMANN and CHANGEUX 1978; KARLIN 1980; CHANGEUX 1981; see also: HESS et al. 1978; HARTIG and RAFTERY 1979; JOHANSSON et al. 1981; FLANAGAN et al. 1976). The need to inhibit endogenous calcium-activated proteases during purification cannot be overstressed (CHANG and BOCK 1977; VANDLEN et al. 1979; SAITOH et al. 1980).

It is a curious fact that *Torpedo* membrane vesicles, corresponding to pinched-off postsynpatic membranes (microsacs), reseal tenaciously after freezing, sonication (WENNOGLE and CHANGEUX 1980), or osmotic shock (WEST and HUANG 1980). Fresh microsac preparations are atypical in this respect. When compared, for instance, with red blood cell membranes, microsacs are extremely difficult to permeabilize. As receptor densities reach $10,000/\mu m^2$ and protein : lipid ratios exceed 1.5:1 certain properties of these microsacs resemble those of protein more than lipids. Interestingly, it is increasingly evident that the morphological and structural aspects of cell membranes, including their tendency to form sealed structures, are determined by a reticulum of protein associated with the inner surface of the cell membrane. Such a protein reticulum has been well characterized with the red blood cell membrane; as we shall see, evidence points to a similar situation in the junctional membrane.

1. Alkaline Extraction

Alkaline extraction is an effective way to open microsac preparations, particularly when performed at room temperature (WENNOGLE and CHANGEUX 1980; SOBEL et al. 1980). More importantly, as originally reported by NEUBIG et al. (1979), alkaline extraction removes essentially all nonreceptor proteins from microsac preparations. A peripheral protein of molecular weight 43,000 (see Sect. C) is eliminated by alkaline treatment.

After pH 11 treatment, if done at 0 °C, a fraction of microsacs reseal; the analysis of agonist-mediated efflux of cations is still possible (NEUBIG et al. 1979; WU et al. 1981). NEUBIG et al. (1979) originally reported that the internal volumes of microsac preparations were reduced to 15%–30% after pH 11 extraction. At any rate, it is clear that alkaline-extracted membranes are more fragile structures (BARRANTES 1982).

2. Summary

Starting with a dozen fresh *Torpedo marmorata,* we typically isolate highly purified microsac preparations within 24 h (SAITOH et al. 1980; SOBEL et al. 1977). About 40% of the protein in these microsacs is acetylcholine receptor, and yields of protein approach 0.25 g. Another day's work involving alkaline extraction and/or nondenaturing detergent solubilization followed by sucrose gradient sedimentation in detergent yields pure, functional receptor (Fig. 1). Specific activities approach 8×10^6 α-toxin sites per gram protein. Purification of the acetylcholine

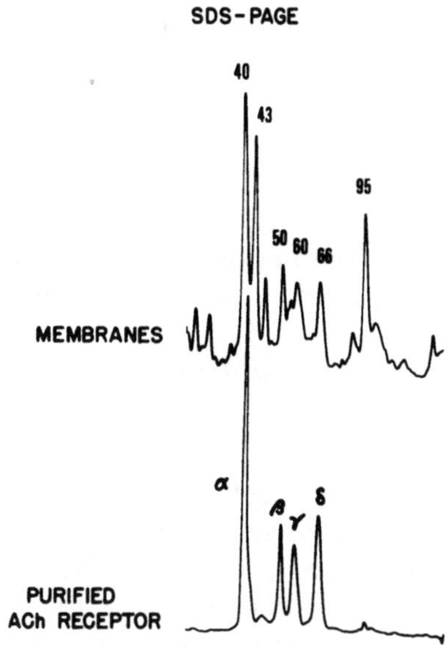

Fig. 1. Optical tracing of an SDS-PAGE rund after sample reduction of purified microsac preparations (*top*) and sucrose density gradient purified acetylcholine receptor (*bottom*) from *Torpedo marmorata*. Membrane and receptor purifications and electrophoresis conditions are outlined in WENNOGLE et al. (1981). Proteins were stained with Coomassie; molecular, weight is indicated $\times 10^{-3}$

receptor is one of the premier accomplishments of receptor molecular biology. The purified preparations have greatly facilitated biochemical characterization of the individual subunit components.

II. Subunits and Stoichiometry

The acetylcholine receptor of *Torpedo* contains five subunits with stoichiometry: $\alpha_2\beta\gamma\delta$ (REYNOLDS and KARLIN 1978). *Electrophorus electricus* and muscle receptors appear to be similar (LINDSTROM et al. 1979a, b, 1980a), attesting to their evolutionary conservation (see also DEUTSCH and RAFTERY 1979). Strong sequence homologies exist at the NH_2 terminus of the four chains from *Torpedo* (RAFTERY et al. 1980); these subunits are evidently derived from a common ancestor protein. Furthermore, in a series of elegant experiments, TZARTOS and LINDSTROM (1980) demonstrated antigenic homologies between α- and β- as well as between γ- and δ-subunits of *Torpedo* receptor. A five-subunit protein structure of this nature, by necessity, represents a curiously asymmetric molecule.

The purified receptor appears pure by polyacrylamide gel electrophoresis and NH_2 terminal analysis (DEVILLERS-THIERY et al. 1979; RAFTERY et al. 1980). At the level of its protein content, its homogeneity is unquestionable. However, during purification, certain types of artifacts may arise which could influence functional properties. Oxidation of susceptible amino acids may occur (WU and HUCHO 1977), particularly in detergents; it is unknown if the level of phosphorylation of individual chains (see VANDLEN et al. 1979) represents saturation or not. RU-

CHEL et al. (1981) claim that the presence of EDTA during purification adversely effects gating characteristics of the receptor. Other such modifications are easily imagined, but are yet to be studied.

III. Physical Properties

Native *Torpedo* acetylcholine receptor exists in situ as a dimeric structure with individual monomeric units connected by a disulfide bond or bonds between δ-subunits:

$$\begin{array}{r} \alpha_2\beta\gamma\delta-\mathrm{S} \\ | \\ \alpha_2\beta\gamma\delta-\mathrm{S} \end{array}$$

The situation with eel and muscle receptor is not as clear, perhaps because very mild proteolysis during isolation has split dimers into monomers (WENNOGLE et al. 1981). As an alternate hypothesis, mild proteolysis of the receptor (Sect. E) may be a normal cellular function which serves as a type of biologic clock. Perhaps proteolytic nicking serves to earmark receptor protein for metabolic turnover. The metabolic turnover time of junctional receptor from all species studied is on the order of weeks.

Receptor monomer and dimer are thought to be functionally equivalent (ANHOLT et al. 1980; RUCHEL et al. 1981; POPOT et al. 1981; but see CHANG and BOCK 1977). The monomeric unit ($\alpha_2\beta\gamma\delta$), produced by mild reduction, has a molecular weight of 250,000, a Stokes radius of 7.3 nm, and a $S_{20}w$ of 8.6 (REYNOLDS and KARLIN 1978). The amino acid composition and carbohydrate content is typical of membrane proteins (VANDLEN et al. 1979). However, it has so far been impossible to dissociate individual subunits from native receptor.

1. Subunit Association

Strong noncovalent forces keep subunits tightly associated with one another both in the membrane environment and after solubilization with nondenaturing detergents such as Triton X-100 and sodium cholate. Conditions as harsh as 6 *M* guanidine hydrochloride, 8 *M* urea, or >0.2% sodium dodecylsulfate (SDS) are needed to dissociate receptor subunits. The implications of this strong association to function and to the successful reconstitution of purified receptor will be discussed.

IV. Reconstitution, Flux, and Planar Lipid Bilayers

Reconstitution of membrane proteins has been somewhat loosely defined as effective solubilization from, then return to, a lipid environment with preservation of biologic activity. This feat was achieved from *Torpedo* by a number of laboratories (EPSTEIN and RACKER 1978; CHANGEUX et al. 1979; WILSON et al. 1979; ANHOLT et al. 1980; KILIAN et al. 1980; SOBEL et al. 1980; HEIDMANN et al. 1980; POPOT et al. 1981; RUCHEL et al. 1981; WU et al. 1981). Receptor subunits remain associated under these conditions. The key in maintaining function with solubilized receptor has been to extract with nonionic detergents (i.e., sodium cholate) supplemented with crude phospholipid mixtures (i.e., asolectin), ensuring that critical lipid-receptor interactions are maintained.

A wealth of information, including the reconstitution, the effect of phospholipases, and the influence of general anesthetics on receptor function, indicates that a very delicate interaction exists between receptor and lipids (reviewed by CHANGEUX 1981). The receptor is surrounded by a particular lipid annulus (MARSH and BARRANTES 1978; CHANG and BOCK 1979), which is necessary for correct functional activity; its exact composition is unknown. HEIDMANN et al. (1981) have shown that allosteric properties of the receptor are maintained in Tween 80 solution, without supplemental lipids.

In a stricter sense, it will be desirable to reconstitute receptor function after complete dissociation of its polypeptide components. This will ultimately reveal the function of individual subunits. Recently, HAGGERTY and FROEHNER (1981) successfully reconstituted α-toxin binding activity to SDS-purified α-subunits. Affinity of the isolated subunit was quite low and, therefore, altered. However, this important observation is consistent with the hypothesis that the α-subunit alone carries the α-toxin binding site. It may further indicate that interactions with other polypeptide chains are necessary to express correct conformations. With this level of expertise, both in purification and reconstitution, great progress has been made in understanding receptor function.

1. Flux Experiments

The end result of junctional receptor function is membrane permeability increase, resulting in the flux of cations down their respective ion gradients. The work done on agonist-mediated flux of radioactive cations (sodium, thallium, and rubidium) from microsacs and from reconstituted receptor will be reviewed briefly (see also KASAI and CHANGEUX 1971; POPOT et al. 1976; HESS and ANDREWS 1977; HESS et al. 1978; NEUBIG and COHEN 1980; HESS et al. 1981, 1982). Basically, by this analysis the receptor in isolated microsacs, as well as receptor reconstituted into artificial membranes, responds normally to agonist. This is to say that appropriate conformational changes, ionophore opening, and receptor desensitization [1] all occur (HEIDMANN et al. 1980). Quantitating the fraction of active receptor in these preparations has been a problem (HUGANIR et al. 1980; NEUBIG and COHEN 1980; WU et al. 1981; HESS et al. 1982). To circumvent this and other such problems, methods of measuring flux in the millisecond time range have recently been developed (NEUBIG and COHEN 1980; CASH and HESS 1980; CASH et al. 1981; HESS et al. 1981; WU et al. 1981). The results of these analyses are summarized in Sect. E.2.

The most esoteric application of reconstitution studies has been the successful incorporation of receptor microsacs (SCHINDLER and QUAST 1980, BOHEIM et al. 1981) and detergent-solubilized receptor (NELSON et al. 1980; BOHEIM et al. 1981)

1 The term desensitization has many connotations, most referring to the process whereby the response to neurotransmitter is somehow dampened. In the context used here, a much more defined conformational state of the acetylcholine receptor, an inactive one, is implied

into planer lipid bilayers. This method allows detailed electrophysiologic investigation of the receptor between two compartments with closely controlled ionic environments (see also HORN and PATLAK 1980). By most criteria, the receptor appears to function normally. NELSON et al. (1980), starting with purified receptor, report single-channel conductance of 16 pS and a mean channel open time of 35 ms after incorporation into vesicles of soybean lipids. This conductance agrees well with expected values; the open time is longer than predicted. The consequences of these technical developments are enormous; the experimentalist is given free access to both inner and outer faces of the membrane. One hypothesis will be worth testing, namely, does receptor phosphorylation (Sect. F) and dephosphorylation (SAITOH et al. 1979) by purified protein kinases and phosphatases influence mean channel open time?

An inescapable conclusion of reconstitution studies is that metabolic energy is unnecessary for function of the receptor, including agonist binding, channel opening, and desensitization. Energy of binding acetylcholine apparently suffices to drive these processes (MAELICKE et al. 1977).

C. The 43K Protein

The 43K protein is found in high concentrations in purified receptor-rich *Torpedo* microsacs (SOBEL et al. 1977). Although this protein is not part of the receptor itself, evidence suggests it is responsible for the immobility of the junctional receptor in the plane of the membrane, and possibly for the interaction of the receptor with cytoskeletal elements. This data will be reviewed.

There are approximately equal quantities of 43K protein to α-subunits in *Torpedo* microsac preparations. This protein is not a glycoprotein (WENNOGLE and CHANGEUX 1980) as it is not labeled by lectins such as Con A, phytohemagglutinin, or wheat germ agglutinin. Strong associations exist between the 43K protein and receptor-rich membranes; sonication, which is effective in opening these membranes, is unable to remove it from microsacs (WENNOGLE and CHANGEUX 1980). On the other hand, pH 11 extraction (NEUBIG et al. 1979) or nonionic detergent solubilization (SOBEL et al. 1977) will dissociate the 43K protein. Contact with membranes is evidently via ionic interactions (disrupted at pH 11) or via intimate lipid contact (disrupted by sodium cholate), or both. Specific reassociation with receptor-rich microsacs was demonstrated (SAITOH et al. 1979). In the absence of membranes, neutralization of the supernatant from pH 11 extracted membranes results in precipitation of the 43K protein. Perhaps this indicates a tendency to self-associate. GYSIN et al. (1981) have claimed the 43K protein is a heterogeneous family of similar proteins. However, their ν_1 may be the membrane-associated 43K protein; ν_2 and ν_3 are likely to represent trapped cytoplasmic contaminants in the preparation.

WENNOGLE and CHANGEUX (1980) located the 43K protein bound to the inner, cytoplasmic face of the receptor-rich microsacs. This location was confirmed by a number of groups employing labeling (ST. JOHN et al. 1982) as well as electron micrographic techniques (CARTAUD et al. 1981; SOBEL et al. 1980; SEALOCK 1980, 1981; BARRANTES 1982). The protein appears to be confined to the innervated face of *Torpedo* electric organ and to synaptic areas of rat diaphragm muscle (FROEHNER et al. 1981). CONTI-TRONCONI et al. (1982), on the other hand, claim this protein may be both intracellular and extracellular. This hypothesis appears unlikely, since pH 11 treated membranes were used. Treatment at pH 11 is a drastic treatment. It is known to fragment membranes and has a tendency to scramble the orientation of receptor molecules relative to the lipid bilayer (BARRANTES et al. 1980).

I. Receptor Mobility

The junctional receptor is normally immobile in the membrane; it is a prototype of an immobile membrane protein. ROUSSELET et al. (1979, 1981), LO et al. (1980), and BARRANTES et al. (1980) convincingly demonstrated that removal of the 43K protein is paralleled by a dramatic increase in receptor mobility. This effect was reversed by reassociation of a fraction highly enriched in the 43K protein (ROUSSELET et al. 1981). Likewise, the specific reassociation of the 43K protein restores a measure of thermal stability to the receptor (SAITOH et al. 1979). The 43K protein does not play a role in the agonist-mediated flux response (NEUBIG et al. 1979). The weight of this evidence suggests a direct association between this 43K protein and the receptor (see also BARRANTES 1982). A link between receptor (and/or 43K protein) and cytoskeletal elements is thought possible (STRADER et al. 1980; CARTAUD et al. 1981). These details remain to be elucidated.

In summary, the 43K protein most likely immobilizes the junctional receptor. It is a classic peripheral protein bound to the cytoplasmic surface of the postsynaptic membrane. This protein should be useful in studying the role played by postsynaptic density elements in neuromuscular development. In the same context, the role played in such processes by extracellular substances, such as basal lamina structures, represents another interesting avenue of research (e.g., BURDEN et al. 1979).

D. Multiple Binding Sites and Multiple Affinity States

The understanding of the ion flux response and the availability of purified receptor preparations have aided research into the molecular topology of the acetylcholine receptor. This study has proceeded at many levels. Some of the most rewarding insight has come from studies concerning (a) the distribution of ligand binding sites and (b) conformational responses of the molecule. The nicotinic acetylcholine receptor displays certain complex facets. It is an allosteric protein, containing multiple sites for ligand interactions (both homotropic and heterotropic effectors) and multiple conformational states. This complicates the analysis of dose-response characteristics. As we shall see, equilibrium techniques of ligand binding relate primarily to nonfunctional receptor conformations. To study the native form of the receptor firsthand, more sophisticated methods have been developed.

I. Multiple Sites for Ligand Interaction

The end-plate nicotinic acetylcholine receptor (Fig. 2) conceptually contains at least three distinct functional components.

1. One component binds agonists (acetylcholine, carbamylcholine, etc.). In ddition, this component also interacts with antagonists like gallamine, tubocurarine, etc., all acting competitively. They act by changing the apparent K_m for acetylcholine without changing the maximum response. This locus is referred to as the *agonist binding site* (see also the following discussion and ELDEFRAWI and ELDEFRAWI 1977). α-Bungarotoxin-like toxins are essentially irreversible inhibitors which act at this locus.

2. The second component is a site or locus responsible for the observed effects of certain quaternary aromatic amine local anesthetics (trimethisoquin, meproadifen) and possibly also the site of interaction of alkaloids of the HTX type. These agents, as well as others (see Sect. D.IV), all act noncompetitively to inhibit agonist-induced flux responses; this locus is referred to as the site for *noncompetitive blocking agents* (NCB).

3. The third component is the *ionophore,* responsible for ion translocation.

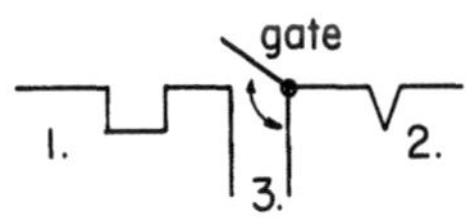

Fig. 2. End-plate nicotinic acetylcholine receptor

This admittedly oversimplified model will be expanded.

The site for noncompetitive blocking agents is sometimes loosely referred to as the *local anesthetic binding site*. Selective local anesthetics, typified by trimethisoquin, are capable of interacting stereospecifically, selectively, and with relatively high affinity for this site. In fact, at concentrations that are effective, it is likely that many local anesthetics exert an effect here, but affinities are low. Most classic local anesthetics also act nonspecifically on postsynaptic membranes, disturbing lipid bilayer properties (YOUNG et al. 1981; KOBLIN and LESTER 1979; TURNER and OLDFIELD 1979; HEIDMANN and CHANGEUX 1981). It is a healthy outlook not to limit our conceptualization of the receptor by simple models with a fixed number of sites. Indeed, it is not certain that all noncompetitive blocking agents act at the same site (see Sect. D.IV, and also COHEN et al. 1980; HEIDMANN and CHANGEUX 1981). On the contrary, as this receptor has five subunits and has a mass of 250,000 daltons, multiple sites can be anticipated.

II. The Three-State Model

In addition to a multiplicity of sites, the receptor can exist in a number of different conformational states. Three discrete states are well studied, while a fourth state has been proposed based upon recent evidence (reviewed by OSWALD et al. 1981). The three-state model (Fig. 3) indicates the receptor may exist in a resting, active,

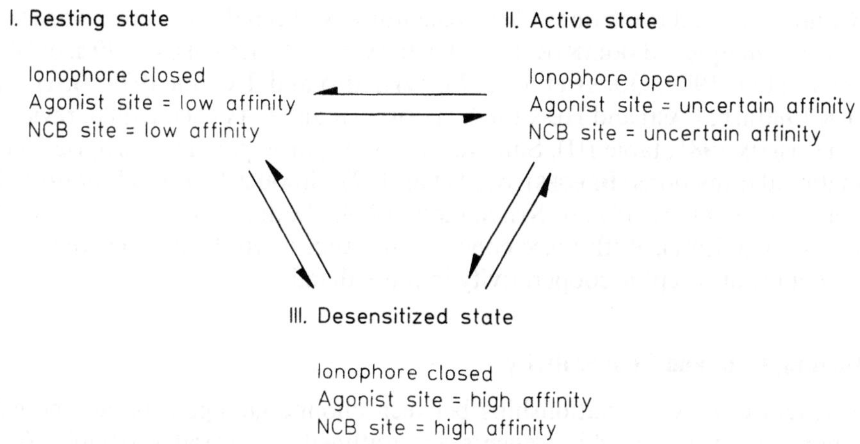

Fig. 3. Three-state model of AChR

or desensitized configuration. *Resting receptor* has a low affinity for agonist and local anesthetics, possibly as low as $K_D \sim 1$ m*M* for acetylcholine. *Active receptor* corresponds to one with an open ionophore. The *desensitized state* corresponds to receptor with bound agonist (or certain antagonists such as tubocurarine), but with a closed ionophore. Transition from resting to desensitized receptor is favored by high agonist concentrations and by local anesthetics. The affinity of both classes of receptor sites for agonists and local anesthetics increases in proceeding from resting to desensitized states.

As a word of clarification, proteins are not restricted in the same sense as an automobile's transmission, which is limited generally to four gears. Receptors, as all proteins, are flexible, dynamic entities which vibrate, conform, and interact in a complex manner. They are outside the restrictions imposed by nuts and gears. As receptors are embedded in a lipid matrix, the situation is further complicated. On the contrary, a model is an attempt to simplify this complex behavior and limit it to a minimum number of easily characterized conformations. The inadequacy of such an approach is inherent.

1. Desensitization

The shift of the receptor into high affinity, inactive conformations is well documented (Katz and Thesleffs 1952; Weber et al. 1975; Barrantes 1978; Weiland and Taylor 1979; Boyd and Cohen 1980a, b; Weiland et al. 1977, 1979; Walker et al. 1981; Grunhagen et al. 1977; Quast et al. 1978). Rate of desensitization is agonist dependent and accelerated by divalent cations (Cohen et al. 1974).

Desensitization serves to "shut down" the receptor when it is bombarded by high agonist concentrations. In a sense, it is a safety valve that prevents overstimulation. The phenomenon is a general one, also seen with other neurotransmitter receptor types such as serotonin postsynaptic receptors, muscarinic and glutamate receptors.

Conformational transitions of the receptor have been detected by a variety of sensitive techniques (BARRANTES 1978; HEIDMANN and CHANGEUX 1979a, b; WASSERMANN et al. 1979; QUAST et al. 1979; WEILAND and TAYLOR 1979; BOYD and COHEN 1980a, b; CASH and HESS 1980; HEIDMANN and CHANGEUX 1980; reviewed by CHANGEUX 1981, table III). Sulfhydryl groups participate in appropriate conformational transitions (BREGESTOVSKI et al. 1977; SUAREZ-ISLA and HUCHO 1977; MOORE and RAFTERY 1979b; STEINACKER 1979; BARRANTES 1980; STEINACKER and ZUAZUGA 1981). With these concepts at hand, it will be instructive to look at the details of receptor cooperativity in more detail.

2. Binding Data and Cooperativity

The structure–activity relationships between cholinergic agonists, antagonists, and noncompetitive blocking agents are outlined in several textbooks (e.g., chapter by Taylor in GILMAN et al. 1980; see also WAKSMAN et al. 1980; WASSERMANN et al. 1979). A variety of techniques have been employed to probe the nature and multiplicity of the binding sites for cholinergic agents. These techniques include: (a) radioactive α-toxin binding, and its inhibition by cholinergic effectors (WEBER et al. 1975; HESS et al. 1975; KATO and CHANGEUX 1976; BULGER et al. 1977; WEILAND et al. 1977; BLANCHARD et al. 1979; WEILAND and TAYLOR 1979; SINE and TAYLOR 1979, 1981; LEPRINCE et al. 1981; YOUNG et al. 1981); (b) kinetic or equilibrium analysis of the binding of radiolabeled agonists and antagonists (FU et al. 1977; NEUBIG and COHEN 1979; BOYD and COHEN 1980a, b); (c) binding of fluorescent probes, or the use of intrinsic protein fluoresence quenching (COHEN et al. 1974; GRUNHAGEN et al. 1977; BARRANTES 1978; HEIDMANN and CHANGEUX 1979a, b; QUAST et al. 1979; BODE et al. 1979; DUNN et al. 1980); (d) studying affinity labeling reagents (see Table 1), and (e) the use of radioactive local anesthetic derivatives, including HTX (ELDEFRAWI et al. 1978, 1980a, b; ELLIOTT and RAFTERY 1979; KRODEL et al. 1979; OSWALD et al. 1980; SAITOH et al. 1980; ALBUQUERQUE et al. 1980; SOBEL et al. 1980; SHAKER et al. 1981). There is general agreement that two agonist sites and most likely one noncompetitive blocking agent site exist per receptor monomer ($\alpha_2\beta\gamma\delta$); one ion channel per monomer seems most reasonable. From these results, what type of general picture has emerged that relates agonist binding to functional response?

3. Positive Cooperativity

As shown in a variety of species by electrophysiology (LAND et al. 1981) and by flux analysis (SINE and TAYLOR 1979, 1981; NEUBIG and COHEN 1980; CASH et al. 1981; HESS et al. 1981; CASH and HESS 1980), occupancy of two agonist sites is necessary for channel opening (see model of CASH et al. 1981, fig. 1).[2] These data are consistent with Hill coefficients which approach 2 for physiologic response measurements (COLQUHOUN 1979, table 1). Occupancy of a single site by antagonists is sufficient to block the response in eels (CASH and HESS 1980). Sigmoid binding curves are seen in *Torpedo* microsacs for the association of acetylcholine,

2 Under certain conditions the receptor may, however, respond slowly to a single agonist

indicating positive cooperativity (COHEN et al. 1974; GIBSON 1976; NEUBIG and COHEN 1979). The cooperative nature of the response of this receptor system is geared to allow rapid response, but over a limited agonist concentration range.

4. Nonequivalent Binding Sites

Another recent conclusion deals with the cooperative or heterogeneous nature of agonist binding sites and involves their individual nature. From an experimental point of view, the two agonist binding sites are nonidentical. This has been shown by: (a) direct binding of tubocurarine to *Torpedo* microsacs (NEUBIG and COHEN 1979); (b) inhibition of the binding of radiolabeled agonists or α-bungarotoxin by antagonists, as seen both in muscle cell lines and in *Torpedo* (NEUBIG and COHEN 1979; SINE and TAYLOR 1979, 1981); and (c) affinity labeling of agonist binding sites, again in *Torpedo* (DAMLE and KARLIN 1978). The result is consistent with the apparent negative cooperativity seen for the binding of a variety of antagonists to their respective binding sites (e.g., WEILAND and TAYLOR 1979), and with the biphasic α-toxin binding curves seen in certain cases (BULGER et al. 1977; YOUNG et al. 1981), but not in all (QUAST et al. 1978).

Whether this heterogeneity precedes (DAMLE and KARLIN 1978; SINE and TAYLOR 1981) or is an allosteric conformational response after the arrival of the first ligand is as yet unclear. The selectivity of antagonist for one agonist site versus the other varies considerably. Acetylcholine and α-toxins appear in most cases to be nonselective (NEUBIG and COHEN 1979).

5. Dose-Response Curves

Historically, measurements of the pharmacologic binding of agonists and of the flux response have been inconsistent. To correlate these data more closely, it was necessary to refine analysis by working on the millisecond time scale (HEIDMANN and CHANGEUX 1979a, b; BOYD and COHEN 1980b). Using stopped-flow techniques, an ED_{50} value for the flux response to carbamylcholine approaches 1 mM (CASH and HESS 1980; NEUBIG and COHEN 1980; BARRANTES 1980). This correlates with IC_{50} values for binding close to 1 μM for this compound. Indeed, lower values are very difficult to measure. Owing to the rapid conformational transition of the receptor, leading ultimately to a high affinity, nonfunctional state or states, rapid methods of analysis are necessary. As expected, the effectiveness of a ligand to induce desensitization does not correlate with its effectiveness in causing channel opening (WEILAND and TAYLOR 1979; and others). In general, all agonists, at equilibrium, produce receptor desensitization. With several notable exceptions (tubocurarine, nicotine), α-toxins and antagonists do not (see also WEBER et al. 1975). Once again, to obtain meaningful dose-response curves, response must be measured rapidly, before receptor desensitization has complicated the analysis.

There is, in general, agreement between different laboratories concerning these experimental results. However, several inconsistencies remain unresolved. The idea that antagonists like tubocurarine are strictly competitive antagonists has been challenged. Some evidence indicates that this compound may: (a) interact at high concentration with the ion channel (SHAKER et al. 1982); (b) interfere noncompetitively with agonist binding; and (c) show half-of-the-site reactivity in eel (BULGER et al. 1977; HESS et al. 1975; FU et al. 1977) and *Torpedo* (GIBSON 1976). Indeed, the possibility that the tubocurarine (antagonist?) site may be distinct from, but tightly coupled allosterically to, the agonist site has not been eliminated. There are also conflicting data as to the homogeneity of α-toxin binding sites (heterogeneous in eel and *Torpedo,* LEPRINCE et al. 1981; but homogeneous in *Torpedo* according to QUAST et al. 1978; BLANCHARD et al. 1979; see YOUNG et al. 1981). Of course, species differences, experimental protocol, steric hindrance, etc., may be sufficient to explain some of these inconsistencies.

III. The Agonist Binding Site

Primarily owing to the work of Karlin and co-workers (KARLIN 1969; KARLIN et al. 1976; DAMLE et al. 1978; DAMLE and KARLIN 1978, 1980), it is clear that the α-chain participates in the agonist binding site. Affinity labels specific for the acetylcholine receptor mark this chain, and the effect is blocked by α-toxins and the appropriate cholinergic ligands. Labeling is remarkably specific. The idea that all components of the agonist site are exclusively on the α-chain is provocative, but this has not been proven rigorously. Indeed, the other receptor subunits appear to be cross-linked by α-toxin affinity probes (see Table 1), indicating their proximity to the agonist site. On the other hand, assuming each α-chain binds one acetylcholine molecule, the pentameric structure $\alpha_2\beta\gamma\delta$ predicts that the two agonist binding sites are nonequivalent.

An important disulfide bond appears to be in the vicinity of the agonist active site. This bond is instrumental in coupling agonist binding to the opening of the AChR ionophore (BREGESTOVSKI et al. 1977; and others). Reduction of this bond: (a) is a prerequisite for affinity labeling with affinity labels employing sulfhydryl reagents (see Table 1); and (b) interferes with correct ion channel response to agonist (STEINACKER and ZUAZAGA 1981 and references therein; STEINACKER 1979). We do not now know what other amino acids participate in the active site. The agonist site exists by necessity (KATZ and THESLEFFS 1957) at the exterior surface of the postsynaptic membrane.

IV. Noncompetitive Blocking Agents

The δ-chain is specifically labeled (see Table 1) by a variety of local anesthetic affinity labels in the presence of agonist (OSWALD et al. 1980; SAITOH et al. 1980; OSWALD and CHANGEUX 1981 a, b). Labeling is enhanced by cholinergic agonists and blocked by HTX. By analogy to the α-chain, at least part of the binding site for noncompetitive blocking agents exists on the δ-chain. Using other allosteric enzymes as models, it is likely that the α- and δ-subunits are distinct entities, each containing the entire binding site for agonists and noncompetitive blockers, respectively.

The action of certain local anesthetics is intimately linked to the receptor ionophore. From the analysis of single-channel recordings, several researchers conclude that certain local anesthetics physically block the channel in an open conformation (reviewed by COLQUHOUN 1979; NEHER and STEVENS 1977; see also COLQUHOUN and SAKMANN 1982). The voltage dependency of their action supports this idea. However, it is equally possible that the effect of local anesthetics is at a site which is distinct from, but allosterically coupled to, the ionophore.

A number of laboratories are intensively studying the noncompetitive blocking agent site in order to characterize it pharmacologically. Binding of radioactive local anesthetics of the aromatic amine type such as ^{14}C-meproadifen or ^{3}H-trimethisoquin (KRODEL et al. 1979; SOBEL et al. 1980; OSWALD and CHANGEUX 1981 a) to nicotinic receptors has recently been demonstrated. In all cases, binding is enhanced by agonist and inhibited by cold HTX. The sites are detected when receptor is in a resting as well as a presumably desensitized form (HEIDMANN and CHANGEUX 1979 a, b). Affinity in the latter form is markedly higher. In the resting state, where affinities are quite low, equilibrium binding techniques are difficult to perform. Since these binding sites are seen while the receptor ionophore is closed, the idea of local anesthetics as pore blockers appears naive.

The localization of the ion channel and its relation to the noncompetitive blockers' site is, however, not yet clear. Evidence from kinetics of association of HTX in the presence or absence of carbamylcholine (OSWALD et al. 1982) indicates steric hindrance to HTX association. This hindrance may, for lack of a better explanation, involve slow occlusion by an ion channel. Selective binding to a transitory conformation is also possible.

An aesthetically pleasing aspect of the complex binding and conformational properties of the acetylcholine receptor is the intimate coupling between agonist site and local anesthetic site. In short, the two sites are totally interactive. It appears that any perturbation which induces a change in one site also changes the other. So, for instance, local anesthetics as well as agonists induce desensitization. Many ligands have an affinity for both sites, although with different potencies. A rather farfetched analogy would be to compare this situation to the conflict between the id and superego of Freudian psychoanalysis. At any rate, the interrelationship between agonist and the noncompetitive blockers' sites may tell us something about the spring-like mechanism of channel opening, and the inevitable event, subsequent channel closing.

Another approach has been the use of labeled HTX, a toxin from the Colombian frog *Dendrobates histrionicus* (ELDEFRAWI et al. 1978, 1980 b; ELLIOT and RAFTERY 1979). Although not a local anesthetic, this reagent is a noncompetitive blocker of the ACh response. It is possible that this probe binds to the noncompetitive blocking agent's binding site also; affinity is not markedly affected by the presence of agonist (see also ARONSTAM and WITKOP 1981; HEIDMANN and CHANGEUX 1981).

Phencyclidine (ELDEFRAWI et al. 1980 a; ALBUQUERQUE et al. 1980) is another ligand with a similar mode of action. Pharmacologic characteristics resemble those of the aforementioned ligands. Affinity is allosterically augmented by cholinergic agonists. However, ELDEFRAWI et al. (1980 a) claim HTX and phencyclidine bind to different sites, both compounds have characteristics of channel blockers (ELDEFRAWI et al. 1978, 1980 a; ALBUQUERQUE et al. 1980; see also LAUFFER and HUCHO 1982).

When analyzing this data, consider that biologic systems are often complex. Bear in mind the multiplicity of binding sites for another well-characterized receptor, the benzodiazepine receptor–effector complex (GABA, β-carbolines, Cl^-, picrotoxin, avermectin, presumably have distinct sites). Remember also, ligands which interact with competitive pharmacology need not bind to the same site. Tightly coupled allosteric sites can display competitive interactions.

Tricyclic antidepressants, typified by imipramine, bind to AChR-rich membranes and influence HTX binding (SHAKER et al. 1981; SCHOFIELD et al. 1981); binding is stimulated by agonists via affinity alterations. In addition to this effect, tricyclics are potent inhibitors of serotonin and norepinephrine uptake in platelets and brain tissue.

Several reports indicate that a component of the local anesthetic binding site is accessible at the exterior of the postsynaptic membrane (HORN et al. 1980; AGUAYO et al. 1981), although the interaction of local anesthetics from within the cell is postulated (e.g., AGUAYO et al. 1981). LESTER (1977), NEHER and STEVENS (1977), STEINBACH (1980), and COLQUHOUN (1979) have reviewed the complexity of local anesthetic interaction with the acetylcholine receptor in more detail.

These studies strengthen the idea that certain local anesthetics: (a) act as heterotropic effectors of the acetylcholine receptor, (b) bind to sites that are coupled allosterically with the agonist site, and (c) are somehow intimately associated with the ion channel. To integrate the binding data for such a number of noncompetitive blocking agents, many showing no structural similarity, one must conclude this locus shows a low level of discrimination.

E. Three-Dimensional Structure

I. Biochemical Characterization

The classical approach to three-dimensional structural elucidation entails both *vectorial probes*, such as chemical labeling reagents, antibodies, and proteolytic cleavages, and *physical techniques*, including diffraction methods, electron microscopy, and spectroscopy. With an advanced peptide mapping capability and some sequence data, structural resolution can be very revealing. At the present time, crystals of the acetylcholine receptor, amenable to X-ray diffraction analysis, are unavailable. Therefore, these classical methods must suffice.

1. Peptide Mapping

Peptide maps of individual receptor subunits have been published (FROEHNER and RAFTO 1979; NATHANSON and HALL 1979; LINDSTROM et al. 1979b). With one exception (LINDSTROM et al. 1979b), these maps are of low resolution, resolving only several peptides, and useful only for comparative purposes. Significantly more work is needed in this area.

2. Electron Microscopy and Diffraction Analysis

The gross morphology of the receptor situated in receptor-rich membranes is well understood (CARTAUD 1980). After negative staining, the protein appears as rosette structures in *Torpedo* microsac preparations. The rosettes, corresponding to receptor monomers, have a central pit which retains stain, surrounded by a five-subunit pattern which does not. Viewed normal to the membrane, the edge structure has grommet-like characteristics (KLYMKOWSKY and STROUD 1979). Rosettes viewed face on are 80 Å in diameter and are quite characteristic of nicotinic receptors. Viewed from a side projection, the rosette is about 110 Å long (see also WISE et al. 1981 a). The rosettes protrude 55 Å from one surface of the membrane, probably the extracellular one (KLYMKOWSKY and STROUD 1979; ROSENBLUTH 1975), and much less (15 Å) from the other (ROSS et al. 1977; KLYMKOWSKY and STROUD 1979). Diffraction analysis and image averaging techniques yield improved resolution (ROSS et al. 1977; ZINGSHEIN et al. 1980; WISE et al. 1981 a; KLYMKOWSKY and STROUD 1979). In addition to negative staining, deep etching (HEUSER and SALPETER 1979) and selective staining of cytoplasmic elements (e.g., 43K protein) with tannic acid (CARTAUD 1980; SEALOCK 1980, 1981) have been useful in defining receptor fine structure. Under certain conditions, aggregates (BARRANTES 1982) and crystalline arrays of rosettes are detected (KISTLER and STROUD 1981; see also CARTAUD et al. 1980, 1981; RASH et al. 1978). Recent work has focused on distinguishing the relative spatial arrangements of individual subunits (HOLTZMAN et al. 1982; WISE et al. 1981 a). The region of δ-δ chain contact was resolved by this method (WISE et al. 1981 a, b).

3. Immunology

Immunologic techniques have been used to probe a variety of structural properties. The most significant recent results involve the identification of the "main immunogenic region" of the receptor and the achievements made with monoclonal antibodies (GOMEZ et al. 1979; WEINBERG and HALL 1979; TZARTOS and LINDSTROM 1980; MOSHLY-ROSEN et al. 1979; RICHMAN et al. 1980; KAMO et al. 1982). TZARTOS and Lindstrom (1980) and TZARTOS et al. (1981) claim the main immunogenic region is (a) highly conserved between species, (b) distinct from the agonist binding site, (c) extracellular, and (d) specific to the α-chain. This region was shown to be a three-dimensional, conformational determinant (BARTFELD and FUCHS 1977, 1979a; LINDSTROM et al. 1979). Its biochemical identity is not known. This dominant antigen (or antigens) is evidently the antigen capable of inducing autoimmune reactions in a variety of species, and this phenomenon is referred to as *experimentally induced myasthenia gravis* (FUCHS 1979). In general, monoclonal antibodies produced against the receptor are not interacting at the agonist binding site (see however, GOMEZ et al. 1979). The methodology of selecting clones through α-toxin binding screens, at least in part, dictates this result.

FROEHNER (1981) and BARTFELD and FUCHS (1977, 1979) have distinguished conformational determinants of the receptor from those derived from primary sequence determinants. FROEHNER (1981) substantiated the idea that α, β-, and δ-chains are exposed extracellularly. Clearly, the utility of these approaches is only beginning to be realized.

4. Chemical Labeling and Cross-Linking

A comprehensive list of chemical modification reagents used on the acetylcholine receptor is presented in Table 1 (see also KARLIN 1980). Rather than repeat the

Table 1. Covalent modification of AChR[a]

	Comment	Reference
I. Agonist Site		
NPTMB	Activator, requires previous reduction	Cox et al. (1979a, b)
MBTA	α-Chain exclusively labeled, 0.5 sites per α-toxin site antagonist, requires reduction	Damle and Karlin (1978)
Bromoacetylcholine	α-Chain exclusively labeled, activator desensitizes, requires reduction 0.5 sites per toxin site	Damle et al. (1978) Karlin (1969) Moore and Raftery (1979a)
4-Azido-α-nitro benzyltrimethylammonium fluoroborate	α-Chain protected by toxin	Hucho et al. (1976)
DAPA	α-Chain preferentially, β and γ nearby	Witzemann and Raftery (1977)
TDF	α-Chain specific, 1 per α-toxin site, antagonist	Weiland et al. (1979)
	Agonist after DTT	Mautner and Bartels (1970)
NDF	Antagonist	Mautner and Bartels (1970)
p-Acetoxybenzenediafluoroborate	Antagonist	Mautner and Bartels (1970)
3-(α-Bromomethyl), 3′-(α(trimethylammonium)-methyl azobenzene	Activator, requires previous reduction	Bartels et al. (1971)
1,5-Difluoro-2,4-dinitrobenzene-α-neurotoxin	α- and δ-Chains labeled	Hamilton et al. (1978)
Azido-*S*-*S*-α-bungarotoxin	α- and δ-Chains labeled	Witzemann and Raftery (1978)
Azido-*R*-*S*-*S*-α-bungarotoxin	δ-Chain labeled	Witzemann et al. (1979)
NAP-α-bungarotoxin	α- and δ-Chains labeled	Hucho (1979)
NAP-α-*Naja naja toxin*	δ-Chain labeled	Hucho (1979)
Bromoacetylcholamine	Antagonist	Kalderon and Silman (1971)
Dithiobischoline	Antagonist	Bartels et al. (1971)
α-Bungarotoxin	α-, γ, and δ-Chains	Oswald and Changeux (1982)
II. Noncompetitive blocking agents site		
5-AT	δ-Chain exclusively labeled in presence of agonists	Oswald et al. (1980) Saitoh et al. (1980)
PCP Trimethisoquin H_{12}-HTX	δ-Chain preferentially labeled in presence of carbamylcholine	Oswald and Changeux (1981b)

[a] A compendium of ligands used to covalently modify the junctional acetylcholine receptor is presented. Most studies deal with *Torpedo* species.

Table 1 (continued)

	Comment	Reference
Quinacrine mustard	All chains labeled, blocks channel and conformational changes	LAUFFER et al. (1979)
Procaine amide azide	43K Protein labeled, not protected by HTX	BLANCHARD and RAFTERY (1979)
Chlorpromazine	α-, β-, γ-, and δ-Chains labeled	OSWALD and CHANGEUX (1982)
III. Nonspecific labels		
A. Sulfhydryl group reagents		
p-Chloromercuribenzoate (PCMB)	Inhibits ACh response, blocks transitions $R \rightleftharpoons D$	LUKAS and BENNETT (1980) KARLIN and BARTELS (1966) MOORE and RAFTERY (1979b)
1,4-Dithiothreitol (DDT)	Decreases ACh response, increases decamethonium response, Hill coefficient lowered	KARLIN and BARTELS (1966) BREGESTOVSKI et al. (1977)
	Blocks transition $R \rightleftharpoons D$	BARRANTES (1980)
	Reduces mean channel conductance and shortens open time	MOORE and RAFTERY (1979) BEN-HAIM et al. (1975)
N-Ethylmaleimide (NEM)	Inhibits ACh response after DTT	KARLIN and BARTELS (1966) SUAREZ-ISLA and HUCHO (1977) BARRANTES (1980)
	α-Chain not labeled unless reduced first	HAMILTON et al. (1979)
	Receptor dimer is natural form	CHANG and BOCK (1977)
Sodium bisulfite	Increases ACh sensitivity	STEINACKER (1979)
Dithionitrobenzoic acid	Carbamylcholine reduces reactivity	SUAREZ-ISLA and HUCHO (1977)
	Solubilization increases reactivity	
Diamine	Induces receptor dimers by cross-linking β-chains	HAMILTON et al. (1979)
	Affinity changes	STEINACKER and ZUAZAGA (1981)
B. General		
Acetic anhydride	Nonspecific label	PATRICK et al. (1973)
Lactoperoxidase	α, β, γ Labeled, δ "shielded"	HARTING and RAFTERY (1977)
Carbodiimides	Inhibit receptor function	NACHSEN and LANDAU (1977)
C. Cross-linking reagents		
Suberimidate	$\alpha_2, \beta, \gamma, \delta$	HUCHO et al. (1978)
Diamine	Induces β-*S*-*S*-β cross-links	HAMILTON et al. (1979)
Copper phenanthroline	Induces β-*S*-*S*-β and δ-*S*-*S*-δ cross-links after reduction	HAMILTON et al. (1979)

Table 1 (continued)

	Comment	Reference
D. Lipid bilayer probes		
5′-Iodonaphthyl-1-azide	α- and 90K Chains 13K α-Fragment reported[b]	TARRAB-HAZDAI et al. (1980)
Pyrenesulfonylazide	β and γ Labeled in membrane, α and β in detergent	SATOR et al. (1979)

[b] See Table 2 footnote b

details of Table 1, it will be more instructive to reexamine several theoretical considerations and to reflect upon several examples.

In principle, a chemical labeling agent should: (a) be well characterized chemically, (b) be chemically homogeneous, (c) react via well-known mechanisms and specificity, (d) have mild effects on protein structure, and (e) react in a way that reflects important spatial, temporal, or other aspects of the substrate. These requirements are rarely met. Furthermore, the homogeneity and intactness of substrate are equally important. In general, the time of reaction should be short (relative to protein conformational changes), yields should be indicative of specificity, and some independent test of specificity must be reported. The number of potential artifacts and misinterpretations is enormous. Reagents which react excessively will denature proteins readily (see, for instance, BARRANTES 1980). If yields are low, minor impurities in the reagent may be the reactive species; conversely, the probe may be reacting with a minor population of protein, for instance, with denatured material present. It is wise to assess critically each of these complications in analyzing the results of chemical labeling experiments. With vectorial probes, i.e., those which are designed to react only at one side of the membrane permeability barrier, these potential complications are markedly amplified. A small population of leaky or open membranes may be sufficient to nullify meaningful results.

Fortunately, a number of probes used with the receptor have been well characterized. The elegant experiments with MBTA, 5 AT, TDF, and bromoacetylcholine, for example, are clear and unambiguous. The work of Karlin and coworkers (HAMILTON et al. 1979) with diamine and copper phenanthroline is likewise very informative. These studies show that particular sulfhydryl groups of the δ-subunit normally participate in cross-linking the receptor, forming receptor dimers. Furthermore, reactive sulfhydryl groups are present on the β-subunit. Under certain conditions, these reactive sulfhydryl groups are easily oxidized. Diamine, for instance, induces an unnatural cross-linking of receptor monomers via β-subunits.

5. Proteolysis

Proteolytic enzymes can provide an indispensable understanding of native protein structure. When used in concert with primary sequence data and physical methods, such as X-ray diffraction, structural resolution is often greatly facilitated. These principles have been applied successfully to numerous membrane proteins such as bacterial rhodopsin, HLA, and glycophorin. In general, proteolytic digestion has served to tag a particular area or domain of the molecule in question, thereby orienting the data obtained from other methods.

The attack of native acetylcholine receptor by proteolytic enzymes has likewise been an extremely useful probe of three-dimensional structure. With proteases, the strategy is different from that adopted with covalent modification; a different type of information is obtained. In general, these probes enable one to locate accessible parts of the protein. Accessibility is determined by (a) steric factors, (b) protein conformation (generally a loosely held polypeptide chain or an unusual protrusion is susceptible), (c) the primary sequence compatability, and (e) overall affinity for the proteases. An important consideration is the relative concentration of protease to substrate. At high protease concentrations, steric factors predominate; at low concentrations, affinity factors do. Finally, the flexibility of the protein determines its resistance to proteases. As denatured proteins are many thousandfold more susceptible to proteases, conditions which relax protein conformations, even momentarily, are crucial. Fortunately, the hydrophobic forces which maintain protein structure are quite strong at physiologic temperatures.

Early work on the native acetylcholine receptor from eel and *Torpedo* indicate that the toxin binding activity is incredibly resistant to proteolytic attack (KALDERON and SILMAN 1971; RUBSAMEN et al. 1976; SHAMOO and ELDEFRAWI 1975; PATRICK et al. 1975; LINDSTROM et al. 1976).

Since this early work, emphasis has been on the characterization of stable or semistable fragments of each receptor subunit. Before listing these fragments, several important considerations are pertinent:

1. When dealing with proteolytic attack on intact receptor, care must be taken before denaturation and analysis (i.e., by SDS-PAGE) to stop the activity of the protease. Indeed, even after using soybean trypsin inhibitor, which stops trypsin completely in physiologic buffers, trypsin activity is unmasked after solubilization by SDS at room temperature (L. P. WENNOGLE and A. SOBEL 1979, unpublished work). Proper controls are essential. Trypsin is instantaneously inactivated by SDS at 100 °C.

2. When dealing particularly with very low protease concentrations and long incubations, a spreading of the molecular weight of subunits may occur. This is a characteristic of random attack. At the extreme, a subunit may essentially disappear from vision on SDS-PAGE, while it is present albeit as a very broad band! To circumvent these problems, we have generally employed a range of protease concentrations, including very high levels. It is clear that higher protease concentrations accentuate stable peptide fragments. Short incubation times are preferred because this limits the possibility of receptor unfolding.

3. To see effects of proteases, their actions need to be quantitative; the possibility of structural alterations subsequent to cleavage must be considered. Curiously, however, no major functional abnormality of the acetylcholine receptor has so far been reported after proteolytic degradation (see BARTFELD and FUCHS 1979b).

Table 2. Trypsin cleavage fragments of individual AChR subunits[a]

Chain	Approximate molecular weight		Relative stability to trypsin	Reference[d]
	Subunit	Fragment		
α	40,000		++	
		38,000	+++	WENNOGLE and CHANGEUX (1980)
		35,000	++++	WENNOGLE and CHANGEUX (1980)
		32,000	++++++	WENNOGLE and CHANGEUX (1980) BARTFELD and FUCHS (1979b)[b] KLYMKOWSKI et al. (1980)[b]
		13,000	N.A.	TARRAB-HAZDAI et al. (1980)[c]
β	50,000		+	
		38,000	++++	This chapter
γ	60,000		+	
		57,000	+	This chapter
		45,000	++	SAITOH et al. (1980)
δ	66,000		+	
		50,000 bis	++	WENNOGLE et al. (1981)
		49,000	+++	WENNOGLE et al. (1981)
		47,000	++++++	WENNOGLE et al. (1981)
		16,000	++	WENNOGLE et al. (1981)

[a] The stable fragments produced by treating native AChR with trypsin are listed, by subunit, in order of decreasing molecular weight (see text). N.A. not assayed.
[b] Fragments of slightly lower molecular weight reported, presumably from α-chain, most likely correspond to the 32K fragment reported here.
[c] Results obtained by the author indicate: (a) this peptide must have substantially lower molecular weight, as low as 2000 daltons; and (b) this peptide, labeled by lipophilic probes, is situated at the COOH terminal portion of the protein (see text and Fig. 5).
[d] ANDERSON and BLOBEL (1981) have found essentially the same patterns using in vitro translated receptor message (see text).

A series of stable and semistable fragments of each receptor subunit to trypsin are listed in Table 2. Qualitatively similar results are obtained with pronase, papain, chymotrypsin, and V8 protease. Results are similar with native receptor either in alkaline extracted membranes or the Triton X-100 solubilized form (WENNOGLE and CHANGEUX 1980).

a) Differential Subunit Susceptibility

There is a differential subunit vulnerability to attack by trypsin (WENNOGLE and CHANGEUX 1980; LINDSTROM et al. 1980b; BARTFELS and FUCHS 1979; STRADER and RAFTERY 1980). The δ-, γ-, and β-chains are particularly vulnerable (Table 2). The α-chain is noticeably more resistant. Similar results are obtained with both low concentration of protease and longer time and with high concentrations (as high as equimolar with receptor) and shorter times; steric factors rather then affinity factors are most likely responsible. With each subunit, trypsin attacks first at a site closer to one terminus, then moves progressively toward the other terminus (see Fig. 5). For the α- and δ-chains, the first attack is initially nearer to the

COOH terminus (WENNOGLE and CHANGEUX 1980; WENNOGLE et al. 1981). Clearly, the protease digests away existing and exposes or unmasks new protein domains. As functional properties are retained (at least those which were measured), even after extensive digestion, receptor denaturation is not likely to be responsible for the effect.

For the α- and δ-chains, "hot spots" of multiple protease attack are seen in particular regions; the α-chain is cut many times between approximate molecular weight 35 and 32,000; the δ-chain is cut repeatedly between 50 and 47,000. This evidence suggests that structural domains exist in the receptor which could be further characterized.

b) Stable Configurations of the Receptor

Three particularly stable configurations of the receptor, relevant to proteolytic digestion, can be singled out (Fig. 4). These configurations correspond to (I) native receptor, (II) receptor digested with low concentrations of trypsin, and (III) receptor after extensive digestion by trypsin. Apparently, the receptor cannot be degraded past configuration III unless it is denatured. By the limits of detection of standard 10% SDS-PAGE (i.e. ≧25,000 daltons), configuration II contains large polypeptide elements of the α-, β-, and δ-chains; configuration III contains elements of, at least, the α- and δ-chains. The ramifications of this very stable core structure will be discussed subsequently.

The question of whether other peptide fragments produced by trypsin remain receptor associated is complex and unresolved. LINDSTROM et al. (1980b) showed that major antigenic determinants remain receptor associated. However, it is clear that several fragments do not. The sedimentation behavior of the three configurations is indicated in Fig. 4 (see also BARTFELD and FUCHS 1979b; HUANG 1979; LINDSTROM et al. 1976). Trypsin clearly changes the sedimentation behavior as it changes the polypeptide profile. The shift from the H or 13 *S* form (corresponding to native receptor dimers) to a form of approximately 9 *S* is due to cleavage of a 16K fragment from the δ-chain, a fragment which carries an intermolecular disulfide bond (WENNOGLE et al. 1981). Within the limits of this analysis,mildly proteolyzed receptor is indistinguishable from reduced, native receptor monomer. With further digestion yielding configuration III, there is a slight shift in the *S* value to approximately 8.3. There are also data concerning another small peptide. A small peptide carrying the site for in vitro phosphorylation of the δ-chain is produced by mild trypsin digestion. This peptide is smaller than 1,500 daltons; it is exposed and accessible to the cytoplasmic side of the membrane (WENNOGLE et al. 1981). Neither the 16K nor 1.5K fragments remain associated with the core structure of the receptor after digestion and subsequent sedimentation through sucrose gradients (WENNOGLE et al. 1981).

On the contrary, elements of at least three subunits (α′, β′, and δ′) remain associated in configuration II, and at least two subunits in configuration III (α″ and δ″, see Fig. 4). The strong association of these fragments is reminiscent of the tenacity of interaction between subunits of native receptor. We are not certain of the exact fate of the other subunits. Curiously, the most stable configuration (III) retains carbohydrate (Con A, phytohemaglutinin, and wheat germ agglutinin sites) as well as components of agonist and noncompetitive blocking agents sites; the

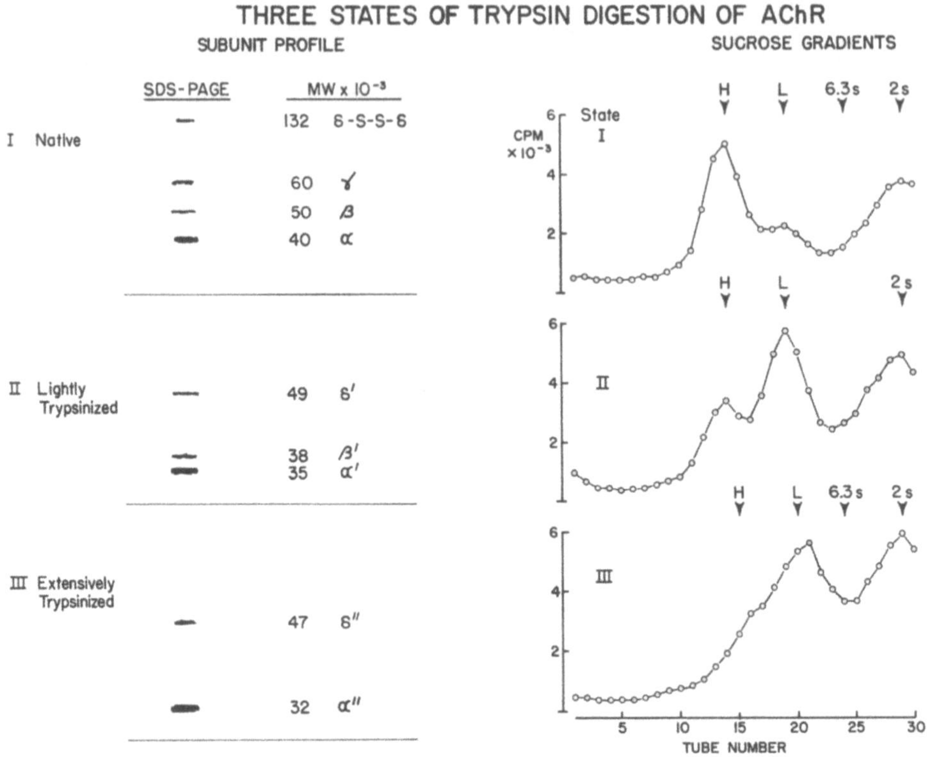

Fig. 4. When alkaline extracted microsacs or nonionic detergent extracted acetylcholine receptor from *Torpedo* are treated with trypsin several stable configurations are detected. Three such states are depicted (*left*) as an artist's conception of SDS-PAGE analysis, run without sample reduction. Exact conditions depend on trypsin and receptor concentration, pH, and temperature (see text). Depending upon conditions, the fragment derived from the δ-chain in configuration II has approximate molecular weight 49–47,000. Trypsin is inactivated prior to electrophoresis either with DFP, or by injecting samples into preheated (100 °C) SDS sample buffer. The 38 K chain was shown to be derived from β by peptide mapping; the origin of other fragments has been published (Table 2). (*Right*) membranes purified in the presence of NEM (CHANG and BOCK 1977) and extracted with pH 11 buffer (NEUBIG et al. 1979) were solubilized in 1% sodium cholate, 0.4 *M* NaCl, 3 m*M* EDTA, 10 m*M* sodium phosphate pH 7.5, and dialyzed. Trypsin was added to a final concentration of 0 (*top*), 8.0×10^{-4} (*middle*), or 0.0125 mg/ml (*bottom*), and reacted at 23 °C for 15 min; trypsin was then stopped with DFP (WENNOGLE et al. 1981). Excess quantities of α-bungarotoxin ^{125}I were incubated with the receptor mixture. Akaline phosphatase (*Escherichia coli*, 6.3 *S*) and native receptor labeled as above, but with excess ^{3}H-α-bungarotoxin to mark receptor dimer (H ~ 13 *S*), and monomer (L ~ 9 *S*) were included as internal standards. These sucrose gradients in 1% sodium cholate were described (WENNOGLE et al. 1981), and run 13 h at 36,000 rpm in a Beckman SW 56 swinging bucket rotor

MPTA and 5-AT label, respectively (WENNOGLE 1980; WENNOGLE et al. 1981). These fundamental ingredients for the proper functional response of the system are most highly preserved.

The next obvious question involves receptor function. As already mentioned, the α-toxin binding activity withstands even the most drastic digestion of the re-

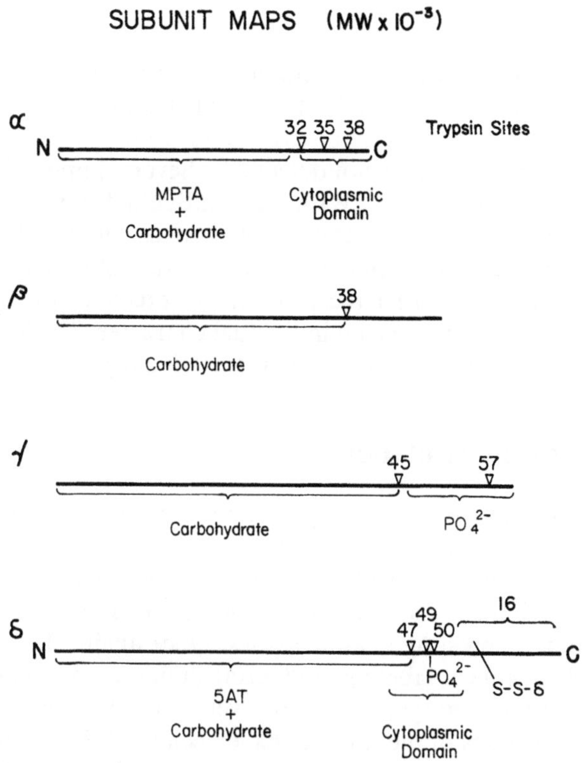

Fig. 5. Stick diagrams of the four subunits of *Torpedo* acetylcholine receptor are represented. Sites of trypsin attack are noted by *triangles* with molecular weight indicated $\times 10^{-3}$. Various domains of the individual subunits are indicated by *braces*. Precise location of agonist and carbohydrate sites is not yet known. The disulfide bond cross-linking receptor δ-subunits is indicated, as are sites for in vitro phosphorylation. The site for in vitro phosphorylation of the γ-chain is to the "left" of the 45 K clip; the 45 K fragment does not retain ^{32}P label

ceptor, indicating that the agonist binding site survives. By a variety of measurements of functional integrity (L. P. WENNOGLE, R. OSWALD, T. HEIDMANN and J. P. CHANGEUX, unpublished work), configuration II is normal. The situation with configuration III is less clear. Preliminary results suggest that this state, although expressing toxin binding activity, may be somehow altered. This important point must be reanalyzed. Electron microscopy using negative staining of membrane-bound receptor after extensive trypsin treatment indicates that receptor rosettes, observed from the external face of the membrane, are morphologically unaltered (KLYMKOWSKY et al. 1980).

In summary, the receptor was dissected by proteases into: (a) a domain of the receptor that is accessible to proteases, owing probably to a more flexible or loose configuration (elements of this domain are evidently dispensable for facets of receptor function); and (b) a stable core structure of reduced size with strong subunit interactions (a structure which has preserved many functional properties; Fig. 5).

6. In Vitro Translation

ANDERSON and BLOBEL (1981) have studied in vitro translation of receptor mRNA and the incorporation of these protein transcripts into microsomes. Under these conditions, receptor is ot assembled from transcripts of individual subunits; toxin binding activity is not detectable. Several important observations were made using this novel system. With these individual subunits, an essentially identical pattern of trypsin fragments was obtained, as compared with those with native receptor (Table 2). As the pattern of degradation of individual subunits remarkably resembles that of native receptor, one is forced to conclude that, in the native receptor, the subunits have an individual character. The idea that individual subunits have some level of autonomy is further supported.

II. Transmembrane Orientation

The transmembrane nature of the acetylcholine receptor is well established (DEL CASTILLO and KATZ 1955; TARRAB-HOZDAI et al. 1978; HUANG 1979; KLYMKOWSKY et al. 1980; STRADER et al. 1979). The specific molecular details remain to be worked out. On a subunit-by-subunit basis, the transmembrane nature of the receptor reveals some interesting characteristics. Once again, the combination of proteolytic attack, chemical labeling, and electron microscopy has been very instructive. Work with in vivo systems has not yet been fully exploited. In general, two different preparations of membranes have been used. Either *closed* microsac preparations, isolated from electric organs by sucrose density sedimentation, or membranes which have been *permeabilized* by a number of nondenaturing techniques to allow access to the cytoplasmic face (i. e., by sonication, freeze-thaw, alkaline extraction, nonionic detergent treatment, or saponin treatment). Unfortunately, sealed inside-out vesicles are not available. It is crucial for this type of analysis that the orientation and permeability characteristics of membranes be well characterized. When quantitative techniques such as protease digestion are used, a small percentage of fragmented or permeabilized membranes is of no concern. If trace labeling techniques are employed, such contamination can seriously bias the results (Sect. E.I.4).

The orientation and permeability characteristics of *Torpedo* microsac preparations are well known. These membranes form sealed vesicles with their orientation "right-side-out", referring to the organization of the postsynaptic membrane (see the discussion in WENNOGLE and CHANGEUX 1980). In the best preparations, greater than 90% of membranes appear sealed either by biochemical or electron micrographic criteria (WENNOGLE and CHANGEUX 1980; ST. JOHN et al. 1982; CARTAUD 1980). Occasionally, this value may be somewhat lower, but generally the majority of membranes are sealed.

The first question to be asked is: what subunits contact the external environment? All four (α, β, γ, and δ) are glycoproteins (VANDLEN et al. 1979; WENNOGLE and CHANGEUX 1980). The α-chain contains the externally located (DEL CASTILLO and KATZ 1955) agonist binding site. Both α- and β-chains are labeled by lactoperoxidase-catalyzed iodination (HARTIG and RAFTERY 1977). The α-, γ-, and δ-chains are cross-linked by toxin affinity probes (see Table 1). Antibodies to α, β-

and δ-chains label the exterior surface (FROEHNER 1981). By these criteria, the α-, β-, γ-, and δ-chains are accessible to the synaptic cleft.

Curiously, in uniformly sealed membranes, it was reported that subunit chains are resistant to proteolytic digestion by trypsin and pronase (WENNOGLE and CHANGEUX 1980). This resistance was overcome by the use of pepsin at pH 4, or by heating membranes briefly above 60 °C. Neither technique caused a breakdown in the permeability barrier. (The implications of this resistance are discussed by WENNOGLE and CHANGEUX 1980.) STRADER and RAFTERY (1980) published data which conflicts with this point; they claim all subunits from *Torpedo californica* could be attacked by trypsin from the exterior of sealed membranes. This point, therefore, should be reconfirmed. STRADER and RAFTERY (1980) used pH 11 treated membranes in their study, and in addition, the extent of subunit degradation in all cases was nonquantitative. Furthermore, when *permeabilized* microsacs or solubilized receptor are extensively digested with proteases to yield the protease-resistant core structure, all known elements of the exterior part of the receptor are spared. Both the exterior morphology (assessed by electron microscopy) and biochemical characteristics (toxin binding sites, local anesthetic binding site, antibody reactivity, and carbohydrates) are clearly maintained (KLYMKOWSKI and STROUD 1979; LINDSTROM et al. 1980b; WENNOGLE and CHANGEUX 1980; WENNOGLE et al. 1981). The barrier responsible for the incredible protease resistance of the core structure of the receptor (Sect. E.I.5) appears therefore, to be a property also of its external surface. When the permeability barrier is removed, for instance by detergent solubilization, protease rapidly attacks the cytoplasmic side of the receptor.

Since the receptor itself is a transmembrane structure, the next logical question is: which subunits contact the interior surface of the membrane? Assuming that structural changes are not occurring as membranes are permeabilized, the answer is simple; all four subunits do. They are all attacked by trypsin in permeabilized vesicles (WENNOGLE and CHANGEUX 1979, 1980; STRADER and RAFTERY 1980). However, as judged by differential susceptibility to proteolytic digestion (Sect. E.I.5), they are not all protruding into the cytoplasm to the same extent (WENNOGLE and CHANGEUX 1980). A model of the transmembrane orientation of the receptor is presented in Fig. 6.

The fact that all four subunit components extend into the cytoplasm has been verified by at least two elegant and very different methods (ANDERSON and BLOBEL 1981; KLOOG et al. 1980). The former technique of in vitro translation in dog pancreas microsomes and subsequent protease digestion has already been discussed. The latter, which employed in vitro methylation of individual receptor subunits either in intact or opened vesicles, supports the same conclusions. In addition, the in vitro site of phosphorylation of the δ-subunit was located at the cytoplasmic side of the membrane (WENNOGLE et al. 1981).

As regards maintaining structure after permeabilizing, the mildest technique, that of repeated freeze–thawing in the presence of trypsin (in our hands, about five cycles were quantitative), gave the same results as detergent solubilization or sonication in the presence of trypsin. Moreover, functional capabilities are maintained after permeabilizing by alkaline extraction (NEUBIG et al. 1979), indicating that no major structural rearrangement occurs.

III. Model

The model of the transmembrane nature of the receptor summarizes the extensive data available (Fig. 6). Subunits are presented as having distinct properties and individual character; the subunits are assembled in juxtaposition to one another. As yet, we have no evidence that the subunits are interwoven, as in the case of ribosomal proteins, or stacked, as may be true with cytochrome oxidase. Neither is there evidence of subunits shuttling between faces of the membrane. The structure more closely resembles that for gap junctional proteins, where cylindrical subunits surround a central pore. Indeed, the idea that gap junctional proteins can assume either tight or loose configurations is appealing here also. However, despite the work already described, we are basically naive as to the molecular nature of the ionophore. Bacterial rhodopsin is a multisubunit protein composed of three oligomers, but ions (protons) are thought to pass through each subunit, not between subunits. The labeling of the δ-chain of the receptor with local anesthetic probes is interesting, but this tells us little more about the ionophore than labeling the α-chain by agonist probes. Still, the possibility that the ionophore is related to the central pit seen at the exterior of the membrane by negative staining is inviting.

An apparent contradiction exists between the study of receptor structure with proteases and with physical techniques. The former technique predicts that a large part of the protein is accessible from the cytoplasmic side of the membrane.

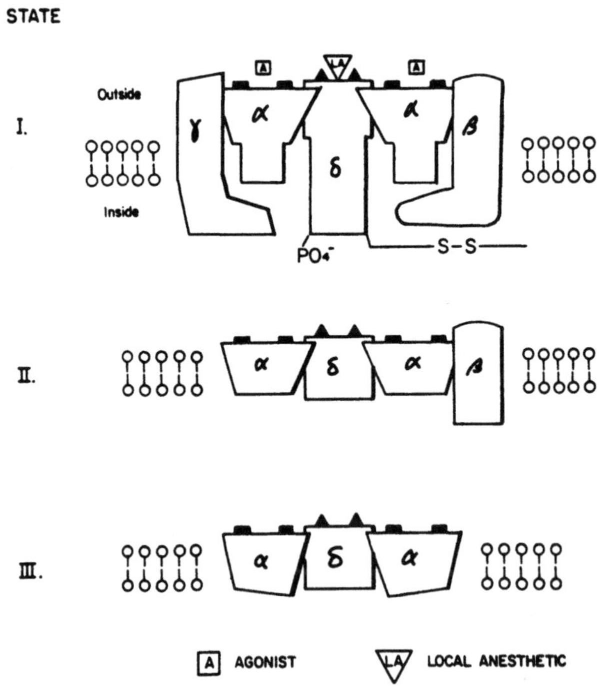

Fig. 6. A hypothetical model of the junctional acetylcholine receptor as a transmembrane protein composed of independent subunits tightly associated to one another. The location of the ion channel is assumed to be in the middle of the structure, between subunits

Different subunits are exposed to varying degrees. Physical techniques, for instance, neutron diffraction (WISE et al. 1981 a) and X-ray diffraction (KLYMKOWSKI et al. 1980), have been interpreted to indicate that the receptor protudes more toward the membrane exterior than toward the cytoplasm. In fact, these observations need not be contradictory. The sedimentation constant (of receptor monomers) is not markedly influenced by extensive trypsin treatment (Fig. 4). True, the subunit patterns (Fig. 4) are markedly different, but a limited number of cuts of a subunit which traverses the membrane many times would markedly influence the subunit pattern after denaturation by SDS. The alternate explanation, that the interpretations of the diffraction patterns have their membrane faces reversed, is not in keeping with the electron microscopy data.

F. Biochemical Control Over Receptor Activity: Phosphorylation, Methylation, and Glycosylation

It is not known if metabolic control is exerted over the acetylcholine receptor function. However, two means of chemical modification of the receptor by endogenous enzymes: phosphorylation and methylation, have been studied. Phosphorylation of receptor in microsac preparations (GORDON et al. 1979, 1980; SAITOH and CHANGEUX 1980, 1981; WENNOGLE et al. 1981) by ^{32}P-γ-ATP labels the δ- and γ-chains, as well as the 43K protein. The nature and control over the relevant kinases have been studied (GORDON et al. 1980; SMILOWITZ et al. 1981). Methylation of the four receptor subunits occurs (KLOOG et al. 1980). It is felt that both these processes occur at the cytoplasmic side of the membrane. More work will be required to assess the functional significance of these results. Little is known about the role of receptor glycosylation as a metabolic control mechanism. Progress is anticipated in this area as recently developed technologies are employed in investigating this question (ANDERSON and BLOBEL 1981; MERLIE et al. 1981).

G. Summary and Conclusions

The extensive research on the junctional acetylcholine receptor from a variety of sources is quite consistent. Both structurally and functionally, this protein is remarkably conserved. It will undoubtedly act, in reference to membrane receptors and ionophores, as a model analogous to that of hemoglobin as an allosteric protein.

The idea that the subunits of the receptor are individual components, each with a distinct function, is provocative. The function of α and δ in binding agonist and noncompetitive blockers, respectively, is clear. Unfortunately, the role played by γ and β is obscure. Could these chains function either in receptor desensitization or in channel opening or closing? More research (and innovative ideas) will be necessary. The subunits are not independent; their strong cohesion attests to this. Clearly the "whole" is greater than the sum of the parts; control over ion flux response is an emergent property of the macromolecular structure.

From a functional point of view, we are beginning to unravel the molecular details of ligand binding and conformational changes. Still, the events subsequent to the binding of a single agonist are unclear. The receptor's function is well documented as an allosteric enzyme displaying multiple binding sites with both heterotropic and homotropic cooperative effectors. The details of the molecular events of channel opening and the nature of the channel itself remain unsolved. It should be possible literally to titrate the co-ions which line the ion channel with radioactive cations whose affinity is not lost in SDS (e. g., Ca^{2+}), once detailed sequence and protein mapping techniques are available. As is so often the case, opening one door leads to many other closed ones. We have come far, but the curiosity of an eager scientific community persists, and we have, in the words of Robert Frost, "miles to go before (we) sleep."

Acknowledgments. I acknowledge the inspiration and help of Drs. CHANGEUX, SAITOH, POPOT, DEVILLERS-THIERY, OSWALD, SOBEL, PARHAM, and MEYERSON, the technical skills of M. GORDON, M. WILFRED, and B. CURTIS and the encouragement of my wife, BEVERLY, in preparing this text. The support of the Muscular Dystrophy Association, Foundation Philippe, and American Cyanamid Company is greatly appreciated.

References

Aguayo LG, Pazhenchevsky B, Daly JW, Albuquerque EX (1981) The ion channel of the acetylcholine receptor. Mol Pharmacol 20:345–355

Albuquerque EX, Tsai M-C, Aronstarn RS, Witkop B, Eldefrawi AT, Eldefrawi ME (1980) Phencyclidine interactions with the acetylcholine receptor. Proc Natl Acad Sci USA 77:1224–1228

Anderson DJ, Blobel G (1981) In vitro synthesis, glycosylation and membrane insertion of the four subunits of *Torpedo* acetylcholine receptor. Proc Natl Acad Sci USA 78:5598–5602

Anholt R, Lindstrom J, Montal M (1980) Functional equivalence of monomeric and dimeric forms of purified acetylcholine receptors. Eur J Biochem 109:481–487

Barrantes FJ (1978) Agonist-mediated changes of the acetylcholine receptor. J Mol Biol 124:1–26

Barrantes FJ (1980) Modulation of acetylcholine receptor states by thiol modification. Biochemistry 19:2957–2965

Barrantes FJ (1982) Oligomeric forms of the membrane-bound acetylcholine receptor. J Cell Biol 92:60–68

Barrantes FJ, Neugebauer DC, Zingsheim HP (1980) Peptide extraction by alkaline treatment is accompanied by rearrangement of membrane-bound acetylcholine receptor. FEBS Lett 112:73–78

Bartels E, Wassermann NH, Erlanger BF (1971) Photochromic activators of the acetylcholine receptor. Proc Natl Acad Sci USA 68:1820–1823

Bartfeld D, Fuchs S (1977) Immunological characterization of an irreversibly denatured acetylcholine receptor. FEBS Lett 77:214–218

Bartfeld D, Fuchs S (1979a) Fractionation of antibodies to acetylcholine receptor according to antigenic specificities. FEBS Lett 105:303–306

Bartfeld D, Fuchs S (1979b) Active acetylcholine receptor fragment obtained by tryptic digestion of acetylcholine receptor. Biochem Biophys Res Commun 89:512–519

Ben-Haim D, Dreyer F, Peper K (1975) Acetylcholine receptor: modification after treatment with a disulfide bond reducing agent. Pflugers Arch 355:19–26

Blanchard SG, Raftery MS (1979) Identification of polypeptide chains that interact with a local anesthetic analog. Proc Natl Acad Sci USA 76:81–85

Blanchard SG, Quast U, Reed K, Lee T, Schimerlik MI, Vandlen R, Claudio T, Strader CD, Moore H-PH, Raftery MA (1979) Interaction of (^{125}I)-α-bungarotoxin with acetylcholine receptor. Biochemistry 18:1975–1883

Bode J, Moody T, Schimerlik M, Raftery M (1979) Uses of fluorescent cholinergic analogues to study binding sites for cholinergic ligands in *Torpedo californica*. Biochemistry 18:1855–1861

Boheim G, Hanke W, Barrantes FJ, Eibl H, Sakmann B, Fels GS, Maelicke A (1981) Agonist-activated ionic channels in acetylcholine receptor reconstituted into planar lipid bilayers. Proc Natl Acad Sci USA 78:3586–3590

Boyd ND, and Cohen JB (1980a) Kinetics of binding of Acetylcholine to *Torpedo* postsynaptic membranes. Biochemistry 19:5344–5353

Boyd ND, Cohen JB (1980b) Kinetics of binding of acetylcholine to *Torpedo* membranes using rapid mixing techniques. Biochemistry 19:5353–5358

Bregestovski PD, Iljin VI, Jurchenko OP, Veprintsev BN, Vulfius CA (1977) Acetylcholine receptor conformational transition masks disulfide bonds against reduction. Nature 170:71–73

Bulger JE, Fu J-J, Hindy EF, Silverstein RL, Hess GP (1977) Allosteric interactions between the membrane-bound acetylcholine receptor and chemical mediators: Kinetic studies. Biochemistry 16:684–692

Burden SJ, Sargent PB, McMahan UJ (1979) Acetylcholine receptors in regenerating muscle. J Cell Biol 82:412–425

Cartaud J (1980) A critical re-evaluation of the structural organization of the excitable membrane in *Torpedo marmorata*. In: Taxi J (ed) Ontogenesis and functional mechanisms of peripheral synapses. INSERM symposium no 13. Elvevier, Amsterdam, pp 267–281

Cartaud J, Popot J-L, Changeux J-P (1980) Light and heavy forms of the acetylcholine receptor. FEBS Lett 121:324–332

Cartaud J, Sobel A, Rousselet A, Devaux PF, Changeux JP (1981) Consequences of alkaline treatment for the ultrastructure of acetylcholine-receptor-rich membranes. J Cell Biol 90:418–426

Cash DJ, Hess GP (1980) Molecular mechanism of acetylcholine receptor-controlled ion translocation across cell membranes. Proc Natl Acad Sci USA 77:842–846

Chang HW, Bock E (1977) Molecular forms of acetylcholine receptor. Biochemistry 16:4513–4520

Chang HW, Bock E (1979) Structural stabilization of isolated acetylcholine receptor: specific interaction with phosholipids. Biochemistry 18:172–179

Changeux JP (1975) The cholinergic receptor protein from fish electric organ. In: Iverson LL, Iverson SD, Snyder SH (eds) Handbook of psychopharmacology, vol 6. Plenum, New York, pp 235–301

Changeux JP (1981) The acetylcholine receptor: an "allosteric" membrane protein. Harvey Lect 75:85–253

Changeux JP, Podleski TR, Wofsy L (1967) Affinity labeling of the acetylcholine receptor. Proc Natl Acad Sci USA 58:2063–2070

Changeux JP, Benedotti L, Bourgeois JP, Brisson A, Cartaud J, Devaux P, Grunhagen H, Moreau M, Popot JL, Sobel A, Weber H (1976a) Some structural properties of the cholinergic receptor protein in its membrane environment relevant to its function as a pharmacological receptor. Cold Spring Harbor Symp Quant Biol 40:211–230

Changeux JP, Bon C, Briley MS, Grunhagen HH, Iwatsubo M, Sobel A, Teichberg VI (1976b) In Vitro studies of subsynaptic membrane fragments from *Torpedo marmorata* electric organ. International congress series no 404. Proceedings of the fifth international conference of the Muscular Dystrophy Association. Excerptia Media, Amsterdam

Changeux JP, Heidmann T, Popot JL, Sobel A (1979) Reconstitution of a functional acetylcholine regulator under defined conditions. FEBS Lett 105:181–187

Changeux JP, Giraudat J, Heidmann T, Popot J-L, Sobel A (1980) Functional properties of the acetylcholine receptor protein. Neurochemistry international vol 2. Pergamon, London, pp 219–231

Claudio T, Ballivet M, Patrick J, Heinemann, S (1983) Nucleotide and deduced amino acid sequences of *Torpedo californica* acetylcholine receptor γ subunit. Proc Natl Acad Sci USA 80:1111–1115
Cohen JB, Weber M, Changeux JP (1974) Effects of local anesthetics and calcium on the interaction of cholinergic ligands. Mol Pharmacol 10:904–932
Cohen JP, Boyd ND, Shera NS (1980) Interactions of anesthetics with nicotinic postsynaptic membranes. In: Fink BR (ed) Molecular mechanisms of anesthesia. Progress in anesthesiology vol 2. Raven, New York, pp 165–174
Colquhoun D (1979) The link between drug binding and response. In: O'Brien RD (ed) The receptors, vol 1, chapter 3. Plenum, New York, pp 93–142
Colquhoun D, Sakmann B (1982) Fluctuations in the microsecond time range through single acetylcholine channels. Nature 294:464–466
Conti-Tronconi BM, Dunn SMJ, Raftery MA (1982) Functional stability of *Torpedo* acetylcholine receptor. Biochemistry 21:893–899
Cox RN, Karlin A, Brandt PW (1979a) Activation of the frog sartorius acetylcholine receptor by a covalently attached group. J Membr Biol 51:133–144
Cox RN, Kawai M, Karlin A, Brandt PW (1979b) Voltage fluctuations at the frog sartorius motor endplate produced by a covalently attached activator. J Membr Biol 51:145–159
Damle VN, Karlin A (1978b) Affinity labeling of one of two α-neurotoxin binding sites in acetylcholine receptor from *Torpedo californica.* Biochemistry 17:2039–2045
Damle VN, Karlin A (1980) Effects of agonists and antagonists on the reactivity of the binding site disulfide in acetylcholine receptor from *Torpedo californica.* Biochemistry 19:3934–3932
Damle VN, McLaughlin M, Karlin A (1978) Bromoacetylcholine as an affinity label of the acetylcholine receptor from *Torpedo californica.* Biochem Biophys Res Commum 84(4):845–851
del Castillo J, Katz B (1955) On the localization of acetylcholine receptors. J Physiol 128:157–181
Deutsch JW, Raftery MA (1979) Polypeptide composition of acetylcholine receptor purified from teleost and elasmobranch electroplax membranes. Arch Biochem Biophys 197:503–515
Devillers-Thiery A, Changeux JP, Paroutaud P, Strosberg AD (1979) The amino acid sequence of the 40,000 molecular weight subunit of the acetylcholine receptor. FEBS Lett 104:99–105
Devillers-Thiery A, Giraudat J, Bentaboulet M, Changeux JP (1983) Complete mRNA coding sequence of the acetylcholine binding α-subunit of *Torpedo marmorata* acetylcholine receptor. Proc Natl Acad Sci USA 80:2067–2071
Dunn SMJ, Blanchard SG, Raftery MA (1980) Kinetics of binding to membrane-bound acetylcholine receptor. Biochemistry 19:5646–5651
Eldefrawi ME, Eldefrawi AT (1975) Molecular and functional properties of the acetylcholine-receptor. Ann NY Acad Sci 264:183–202
Eldefrawi ME, Eldefrawi AT (1977) Acetylcholine receptors. In: Cuatrecasas P, Greaves MF (eds) Receptors and recognition, vol 4, series A. Chapman and Hall, London, pp 199–258
Eldefrawi ME, Eldefrawi AT, Mansour NA, Daly JW, Witkap B, Albuquerque EX (1978) Acetylcholine receptor and ionic channel of *Torpedo.* Biochemistry 17:5474–5484
Eldefrawi ME, Eldefrawi AT, Aronstam RS, Maleque MA, Warnick JE, Albuquerque EX (1980a) 3H-Phencyclidine: a probe for the ionic channel of the nicotinic receptor. Proc Natl Acad Sci USA 77:7458–7462
Eldefrawi ME, Aronstam RS, Bakvy N, Eldefrawi AT, Albuquerque EX (1980b) Activation, inactivation and desensitization of acetylcholine receptors. Proc Natl Acad Sci USA 77:2309–2313
Elliott J, Raftery MA (1979) Binding of perhydrohistrionicotoxin to nicotinic acetylcholine receptor. Biochemistry 18:1868–1874
Epstein M, Racker E (1978) Reconstitution of carbamylcholine-dependent sodium ion flux and desensitization of the acetylcholine recepter from *Torpedo californica.* J Biol Chem 253:6660–6662

Fambrough DM (1979) Control of acetylcholine receptors in skeletal muscle. Physiol Rev 59(1):165–227
Fessard A (1958) Les organes electriques. In: Grasse PP (ed) Traite' de zoologie, vol 13A. Masson, Paris, pp 1143–1238
Flanagan SD, Barondos SH, Taylor P (1976) Affinity partitioning of membranes. J Biol Chem 251:858–865
Froehner SC (1981) Identification of exposed and buried determinants of the acetylcholine receptor. Biochemistry 20:4905–4915
Froehner SC, Rafto (1979) Comparison of the subunits of *Torpedo californica* acetylcholine receptor by peptide mapping. Biochemistry 18:301–307
Froehner SC, Gulbrandsen V, Hyman C, Jeng AY, Neubig RR, Cohen JB (1981) Immunofluorescence localization at the mammalian neuromuscular junction of the Mr 43,000 protein of *Torpedo* postsynaptic membranes. Proc Natl Acad Sci USA 78:5230–5234
Fu JL, Donner DB, Moore DE, Hess GP (1977) Allosteric interactions between acetylcholine receptor and chemical mediators: Equilibrium measurements. Biochemistry 16:678–684
Fuchs S (1979) Immunological analysis of acetylcholine receptor. In: Ceccarelli B, Clement F (eds) Advances of cytopharmacol, vol 3. Raven, New York, pp 279–286
Gibson RE (1976) Ligand interactions with the acetylcholine receptor from *Torpedo californica*. Biochemistry 15:3890–3901
Gilman AG, Goodman LS, Gilman A (1980) The pharmacological basis of therapeutics, 6th edn. Macmillian, New York
Giraudat SA, Changeux JP (1980) The acetylcholine receptor. Trends Pharmacol Sci (April): 198–202
Gomez CM, Richman DP, Berman PW, Burres SA, Arnason BGW, Fitch FW (1979) Monoclonal antibodies against purified nicotinic acetylcholine receptor. Biochem Biophys Res Commun 88:575–582
Gordon AS, Guillory RJ, Diamond I, Hucho F (1979) ATP-binding proteins in acetylcholine receptor-enriched membranes. FEBS Lett 108:37–39
Gordon AS, Davis CG, Milfay D, Kaur J, Diamond I (1980) Membrane-bound protein kinase activity in acetylcholine receptor-enriched membranes. Biochim Biophys Acta 600:421–431
Grunhagen H-H, Iwatsubo M, Changeux J-P (1977) Fast kinetic studies on the interaction of cholinergic agonists with acetylcholine receptor. Eur J Biochem 80:225–242
Gysin R, Wirth M, Flanagan SD (1981) Structural heterogeneity and subcellular distribution of nicotinic synapse-associated proteins. J Biol Chem 256:11373–11376
Haggerty JG, Froehner SC (1981) Restoration of (^{125}I)-α-bungarotoxin binding activity to the α subunit of *Torpedo* acetylcholine receptor. J Biol Chem 256:8294–8297
Hamilton S, McLaughlin M, Karlin A (1978) Crosslinking of the acetylcholine receptor. Fed Proc 37:529
Hamilton SL, McLaughlin M, Karlin A (1979) Formation of disulfide-linked oligomers of acetylcholine receptor in membrane from *Torpedo* electric tissue. Biochemistry 18:155–165
Hartig PR, Raftery MA (1977) Lactoperoxidase catalyzed membrane surface labeling of the acetylcholine receptor. Biochem Biophys Res Commun 78:16–22
Hartig PR, Raftery MA (1979) Right-side-out acetylcholine receptor enriched vesicles. Biochemistry 18:1146–1150
Heidmann T, Changeux JP (1978) Structural and functional properties of the acetylcholine receptor protein in its purified and membrane-bound states. Annu Rev Biochem 47:317–357
Heidmann T, Changeux JP (1979a) Fast kinetic studies on the interaction of a fluorescent agonist with membrane-bound acetylcholine receptor. Eur J Biochem 94:255–279
Heidmann T, Changeux JP (1979b) Fast kinetic studies on the allosteric interactions between acetylcholine receptor and local anesthetic binding sites. Eur J Biochem 94:281–296
Heidmann T, Changeux JP (1980) Interaction of a fluorescent agonist with the acetylcholine receptor. Biochem Biophys Res Commun 97:889–896

Heidmann T, Changeux JP (1981) Stabilization of the high affinity state of the acetylcholine receptor. FEBS Lett 131:239–244
Heidmann T, Sobel A, Popot JL, Changeux JP (1980) Reconstitution of a functional acetylcholine receptor. Eur J Biochem 110:35–55
Heidmann T, Cuisinier JB, Changeux JP (1981) Conservation des propritietes allosteriques de la proteine receptice en solution detergente. C R Seances Acad Sci 111:13–15
Hess GP, Andrews JP (1977) Functional acetylcholine receptor-electroplax membrane microsacs. Proc Natl Acad Sci USA 74:482–486
Hess GP, Aoshima H, Cash DJ, Lenchitz B (1981) Specific reaction rates of acetylcholine receptur–controlled ion translocation: a comparison of measurements with membrane vessicles and with muscle cells. Proc Natl Acad Sci USA 78:1361–1365
Hess GP, Bulger JE, Fu JL, Hindy EF, Silverstein RJ (1975) Allosteric interactions of the membrane bound acetylcholine receptor; Kinetic studies with α-bungarotoxin. Biochem Biophys Res Commun 64:1018–1027
Hess GP, Lipkowitz S, Struve GE (1978) Acetylcholine-receptor mediated ion flux in Electroplax membrane microsacs (vesicles): change in mechanism produced by assymetric distribution of sodium and potasium ions. Proc Natl Acad Sci USA: 75:1703–1707
Hess GP, Hoshima H, Cash DJ, Lenchitz B (1981) Specific reaction rate of acetylcholine receptor-controlled ion translocation. Proc Natl Acad Sci USA 78:1361–1365
Hess GP, Pasquale ED, Walker JW, McNamee MG (1982) Comparison of acetylcholine receptor-controlled cation flux in membrane vesicles from *Torpedo california* and *Electrophorus* electricus: Chemical Kinetic measurements in the millisecond region. Proc Natl Acad Sci USA 79:963–967
Heuser JE, Salpeter SR (1979) Organization of acetylcholine receptors. J Cell Biol 82:150–173
Holtzman E, Wise D, Wall J, Karlin A (1982) Electron microscopy of complexes of isolated acetylcholine receptor, biotinyl-toxin and avidin. Proc Natl Acad Sci USA 79:310–314
Horn R, Patlak J (1980) Single channel currents from excised patches of muscle membrane. Proc Natl Acad Sci USA 77:6930–6934
Horn R, Brodwick MS, Dickey WD (1980) Asymmetry of the acetylcholine channel revealed by quaternary anesthetics. Science 210:205–207
Huang L (1979) Transmembrane nature of acetylcholine receptor as evidenced by protease sensitivity. FEBS Lett 102:9–12
Hucho F (1979) Photoaffinity derivatives of α-bungarotoxin and α-*Naja Naja siamensis* toxin. FEBS Lett 103:27–32
Hucho F, Lager P, Kiefer HR, Bandini G (1976) Photoaffinity labeling and quaternary structure of the acetylcholine receptor from *Torpedo californica*. Proc Natl Acad Sci USA 73:2624–2628
Hucho F, Bandini G, Suarez-Isla BA (1978) The acetylcholine receptor as part of a protein complex. Eur J Biochem 83:335–340
Huganir RL, Racker E (1980) Endogenous and exogeneous proteolysis of the acetylcholine receptor from *Torpedo californica*. J Supramol Struct 14:215–221
Johansson G, Gysin R, Flanagan SD (1981) Affinity partitioning of membranes. J Biol Chem 256:9126–9135
Kalderon N, Silman I (1971) Water-soluble acetylcholine receptor from *Torpedo californica*. Biochim Biophys Acta 465:331–340
Kamo I, Furukawa S, Tada A, Mano Y, Iwasaki Y, Furuse T, Ito N, Hayashi K, Satoyoshi E (1982) Monodonal antibody to acetylcholine receptor. Science 215:995–997
Karlin A (1969) Chemical modification of the active site of the acetylcholine receptor. J Gen Physiol 54:245–264
Karlin A (1980) Molecular properties of nicotinic acetylcholine receptors. In: Cotman CW, Poste G, Nicolson GL (eds) The cell surface and neuronal function. Elsevier, New York, pp 191–260
Karlin A, Bartels E (1966) Effects of blocking sulfhydryl groups and of reducing disulfide bonds on the acetylcholine-activated permeability system of the electroplax. Biochim Biophys Acta 126:525–535

Karlin A, Weill CL, McNamee MG, Valderrama R (1976) Facets of the structures of acetylcholine receptors from *Electrophorus* and *Torpedo*. Cold Spring Harbor Symp Quant Biol 40:203–210
Karlin A, Damle V, Hamilton S, McLaughlin M, Valderrama R, Wise D (1979) Acetylcholine receptors in and out of membranes. Adv Cytopharmacol 3:163–188
Kasai M, Changeux JP (1971) In vitro exitation of purified membrane fragments by cholinergic agonists. J Membr Biol 6:1–23
Kato G, Changeux JP (1976) Effect of histrionicotoxin on the binding of (^{3}H) acetylcholine. Mol Pharmacol 12:92–100
Katz B, Thesleffs (1957) A study of "desensitization" produced by acetylcholine at the motor end-plate. J Physiol (Lond) 138:63–80
Kilian PL, Dunlap CR, Mueller P (1980) Reconstitution of acetylcholine receptor from Torpedo californica. Biochem Biophys Res Common 93:409–414
Kistler J, Stroud RM (1981) Crystaline arrays of acetylcholine receptor. Proc Natl Acad Sci USA 78:3678–3682
Kloog Y, Flynn D, Hoffman AR, Axelrod J (1980) Enzymatic carboxymethylation of the nicotinic acetylcholine receptor. Biochem Biophys Res Commun 97:1474–1480
Klymkowsky MW, Stroud RM (1979) Immunospecific identification of a membrane-bound acetylcholine receptor. J Mol Biol 128:319–334
Klymkowsky MW, Heuser JE, Stroud RM (1980) Protease effects on the structure of acetylcholine receptor membranes from *Torpedo*. J Cell Biol 85:823–838
Koblin DD, Lester HA (1979) Voltage-dependent and -independent blockade of acetylcholine receptors. Mol Pharmacol 15:559–580
Krodel EK, Beckman RA, Cohen JP (1979) Identification of a local anesthetic binding site in post-synaptic membrans. Mol Pharmacol 15:294–312
Kuffler SW (1943) Specific excitability of the end-plate region in normal and denervated muscle. J Neurophysiol 6:99–110
Land BR, Salpeter EE, Salpeter MM (1981) Kinetic parameters for acetylcholine interaction in intact neuromuscular junction. Proc Natl Acad Sci USA 78:7200–7204
Langley JN (1907) On the conctraction of muscle I. J Physiol 36:347–384
Lauffer L, Hucho F (1982) Triphenylmethylphosphonium is an ion channel ligand of the acetylcholine receptor. Proc Natl Acad Sci USA 79:2406–2409
Lauffer L, Weber KH, Hucho F (1979) Binding properties after covalent attachment of quinacrine. Biochim Biophys Acta 587:42–48
Lee CY (1972) Chemistry and pharmacology of polypeptide toxins in snake venoms. Annu Rev Pharmacol 12:265–286
Leprince P, Noble RL, Hess GP (1981) Comparison of the interactions of a specific neurotoxin with the acetylcholine receptor. Biochemistry 20:5565–5570
Lester HA (1977) The response to acetylcholine. Sci Am (February):107–117
Lindstrom J (1979a) Autoimmune response to acetylcholine receptors in myasthenia gravis. Adv Immunol 27:1–50
Lindstrom J (1979b) Antibodies to the acetylcholine receptor molecule. Adv Cytopharmacol 3:245–253
Lindstrom JM, Lennon VA, Seybold ME, Whillingham S (1976) Experimental autoimmune myasthenia gravis. Proc Natl Acad Sci USA 274:254–274
Lindstrom J, Walter (Nave) B, Einarson B (1979a) Immunochemical similarities between subunits of acetylcholine receptors from *Torpedo, Electrophoresis* and mammalian muscle. Biochemistry 18:4470–4480
Lindstrom J, Merlie J, Yogeeswaran G (1979b) Biochemical properties of acetylcholine receptor subunits from *Torpedo calificornia*. Biochemistry 18:4465–4470
Lindstrom J, Cooper J, Tzaros S (1980a) Acetylcholine receptors from *Torpedo* and *Electrophorus* have similar subunit structures. Biochemistry 19:1434–1438
Lindstrom J, Gullick W, Contri-Tronconi B, Ellisman M (1980b) Proteolytic nicking of the acetylcholine receptor. Biochemistry 19:4791–4795
Lo MMS, Garland PB, Lamprecht J, Barnard EA (1980) Rotational mobility of the acetylcholine receptor measured by phosphorescene depolarization. FEBS Lett 111:407–412

Lukas RJ, Bennett EC (1980) Chemical modification and reactivity of sulfhydryls and disulfides of acetylcholine receptors. J Biol Chem 265:5573–5577
Maelicke A, Fulpins BW, Klett RP, Reich E (1977) Acetylcholine receptor: responses to drug binding. J Biol Chem 252:4811–4830
Marsh D, Barantes FJ (1978) Immobilized lipid in acetylcholine receptor-rich membranes. Proc Natl Acad Sci USA 75:4329–4333
Mautner HG, Bartels E (1970) Interactions of p-nitrobenzene diazonium fluoroborate and analogues with the active sites of acetylcholine receptor and esterase. Proc Natl Acad Sci 67:74–78
Mendez B, Valenzuela P, Martial JA, Baxter JD (1980) Cell-free synthesis of acetylcholine receptor polypeptides. Science 209:695–697
Merlie JP, Sebbane R (1980) Acetylcholine receptor subunits transit a precursor pool before aquiring α-bungarotoxin binding activity. J Biol Chem 256:3605–3608
Merlie JP, Hofler JG, Sebbane R (1981) Acetylcholine receptor synthesis from membrane polysomes. J Biol Chem 256:6995–6999
Moore HH, Raftery MA (1979a) Studies of irreversible interactions of an alkylating agonist with the acetylcholine receptor. Biochemistry 18:1862–1867
Moore HPH, Raftery MA (1979b) Ligand-induced interconversion of acetylcholine receptor. Biochemistry 18:1907–1914
Moshly-Rosen D, Fuchs S, Eshhar Z (1979) Monoclonal antibodies against defined determinants of acetylcholine receptor. FEBS Lett 106:389–392
Nachmansohn D (1955) Metabolism and function of the nerve cell. Harvey Lect 49:57–99
Nachsen DA, Landau EM (1977) Re-examination of carbodiimide as a possible affinity label for the acetylcholine receptor of the frog neuromuscular junction. J Mol Biol 35:1–7
Nathanson N, Hall Z (1979) Subunit structure and peptide mapping of junctional and extrajunctional acetylcholine receptors. Biochemistry 18:3392–3401
Nelson N, Anholt R, Lindstrom J, Montal M (1980) Reconstitution of purified acetylcholine receptors with functional ion channels in planar lipid bilayers. Proc Natl Acad Sci USA 77:3057–3061
Neubig RR, Cohen JB (1979) Equilibrium binding of tubocurarine and acetylcholine by *Torpedo* postsynaptic membranes. Biochemistry 18:5464–5475
Neubig RR, Cohen JB (1980) Permeability control by cholinergic receptors in *Torpedo* post-synaptic membranes. Biochemistry 19:2770–2779
Neubig RR, Krodel EK, Boyd ND, Cohen JB (1979) Acetylcholine and local anesthetic binding to *Torpedo* nicotinic membranes after removal of nonreceptor peptides. Proc Natl. Acad Sci USA 76:690–694
Neher E, Stevens CF (1977) Conductance fluctuations and ionic pores in membranes. Annu Rev Biophys Bioeng 6:345–381
Noda M, Takahashi H, Tanabe T, Toyosata M, Kikyotani S, Hirose T, Asai M, Takashima H, Inayama S, Miyata T, Numa S (1983) Primary structures of β- and γ-subunit precursors of *Torpedo californica* acetylcholine receptors. Nature 301:751–755
Oswald R, Changeux JP (1981a) Ultraviolet light-induced labeling by noncompetitive blockers of the acetylcholine receptor. Proc Natl Acad Sci USA 78:3925–3929
Oswald RE, Changeux JP (1981b) Selective labeling of the δ-subunit of the acetylcholine receptor by a covalent local anesthetic. Biochemistry 20:7166–7174
Oswald RE, Changeux JP (1982) Crosslinking of alpha-bungarotoxin to the acetylcholine receptor from *Torpedo marmorata* by ultraviolet light irradiation. FEBS Lett 139:225–229
Oswald R, Sobel A, Waxsman G, Rogues B, Changeux JP (1980) Selective labeling by (^{3}H) trimethisoquin azide of polypeptide chains present in acetylcholine receptor-rich membranes form *Torpedo marmorata*. FEBS Lett 111:29–34
Oswald RE, Heidmann R, Changeux JP (1981) The acetylcholine receptor: an allosteric membrane protein. In: Peeters H (ed) Protides of the biological fluids. Pergamon, Oxford

Oswald RE, Heidmann T, Changeux JP (1983) Multiple affinity states for noncompetitive blockers revealed by ^{3}H-phencyclidine binding acetylcholine receptor rich fragments from *Torpedo marmorata*. Biochemistry 22:3128–3136

Ovchinnikow YA (1980) Ion channels: structure and function. Biochem Soc Symp 46:103–137

Patrick J, Berman PW (1980) Metabolism of nicotinic acetylcholine receptor. In: Cotman CW, Poste G, Nicolson GL (eds) The Cell Surface and Neuronal Fuction. North Holland, New York, pp 158–191

Patrick J, Lindstrom J, Culp B, McMillan J (1973) Studies on purified eel acetylcholine receptor and anti-acetylcholine receptor antibody. Proc Nat Acad Sci 70:3334–3338

Patrick J, Boulter J, O'Brien JC (1975) An acetylcholine receptor preparation lacking the 42,000 dalton component. Biochem Biophys Res Common 64:219–225

Popot JL, Sugiyama H, Changeux JP (1976) Studies on the electrogenic action of acetylcholine with *Torpedo* electric organ. J Mol Biol 106:469–483

Popot JL, Cartaud J, Changeux JP (1981) Reconstitution of a functional acetylcholine receptor. Eur J Biochem 118:203–214

Quast U, Schimerlik M, Lee T, Witzemann V, Blanchard S, Raftery MA (1978) Ligand-induced conformation changes in *Torpedo* acetylcholine receptor. Biochemistry 17:2405–2424

Quast U, Schimerlik MI, Raftery MA (1979) Ligand-induced changes in membrane-bound acetylcholine receptor. Biochemistry 18:1891–1901

Raftery MA, Vandlen RL, Reed KL, Lee T (1976) Characterization of *Torpedo californica* acetylcholine receptor: its subunit composition and ligand-binding properties. Cold Spring Harbor Symp Quant Biol 40:193–202

Raftery MA, Blanchard S, Elliott J, Hartig P, Moore HP, Quast U, Schimerlik MI, Witzemann V, Wu W (1979) Properties of the *Torpedo california* acetylcholine receptor. Adv Cytopharmacol 3:159–182

Raftery MA, Hunkapillar MW, Strader C, Hood LE (1980) Acetylcholine receptor: complex of homologous subunits. Science 208:1454–1457

Rash JE, Hudson CS, Ellisman MH (1978) Ultrastructure of acetylcholine receptors at the mammalian neuromuscular junction. In: Straub RW, Bolis L (eds) Cell membrane receptors for drugs and hormones. Raven, New York

Reynolds JA, Karlin A (1978) Molecular weight in detergent solution of acetylcholine receptor from *Torpedo californica*. Biochemistry 17:2035–2038

Richman DP, Gomez CM, Berman PW, Burres SA, Fitch FW, Arnason BGW (1980) Monoclonal anti-acetylcholine receptor antibodies can cause experimental myasthenia. Nature 286:738–739

Rosenbluth J (1975) Structure of the acetylcholine receptor. J Neurocytol 4:697–712

Ross MJ, Klymkowsky MW, Agard DA, Stroud RM (1977) Structural studies of a membrane-bound acetylcholine receptor form *Torpedo*. J Mol Biol 116:635–659

Rousselet A, Cartaud J, Devaux PF (1979) Importance des interactions proteine-proteine dans le maintien de la structure des fragments excitables de l'organe electrique de *Torpedo marmorata*. C R Seances Sci Acad 289D:461–463

Rousselet A, Cartaud J, Devaux PF (1981) Factors influencing the mobility of the acetylcholine receptor. Biochim Biophys Acta 648:169–185

Rubsamen H, Montgomery M, Hess GP Eldefrawi AT, Eldefrawi ME (1976) Characterization of the calcium-binding site if the purified acetylcholine receptor and identification of the calcium binding subunit. Biochem Biophys Res Common 70:1020–1027

Ruchel R, Watters D, Maelicke A (1981) Molecular forms and hydrodynamic properties of acetylcholine receptor. Eur J Biochem 119:215–223

Saitoh T, Changeux JP (1980) Phosphorylation in vitro of membrane fragments form *Torpedo marmorata* electric organ. Eur J Biochem 105:51–62

Saitoh T, Changeux JP (1981) Change in state of phosphorylation of acetylcholine receptor during maturation. Proc Natl Acad Sci USA 78:4430–4434

Saitoh T, Wennogle LP, Changeux JP (1979) Factors regulating the susceptibility of the acetylcholine receptor protein to heat inactivation. FEBS Lett 108:489–495

Saitoh T, Oswald R, Wennogle LP, Changeux JP (1980) Conditions for the selective labelling of the 66,000 dalton chain of the acetylcholine receptor by the covalent non-competitive blocker 5-azido-(^{3}H) trimethisoquin. FEBS Lett 116:30–36
Sator V, Gonzalez-Ros JM, Calvo-Fernandez P, Martinez-Carrion M (1979) Pyrenesulfonyl azide: a marker of acetylcholine receptor subunits in contact with membrane hydrophobic environment. Biochemistry 18:1200–1206
Schindler H, Quast U (1980) Functional acetylcholine receptor from *Torpedo marmorata* in planar membranes. Proc Natl Acad Sci USA 77:3052–3056
Schofield GG, Witkop B, Warnick JE, Albuquerque EX (1981) Differentiation of the open and closed states of the ionic channels of nicotinic acetylcholine receptors by tricyclic antidepressants. Proc Natl Acad Sci USA 78:5240–5244
Sealock R (1980) Identification of high acetylcholine receptor density with tannic acid. Brain Res 199:267–281
Sealock R (1981) Tannic acid fixation of the acetylcholine receptor and associated proteins. Trans Am Soc Neurochem 12:111
Shaker N, Eldefrawi AT, Miller ER, Eldefrawi HE (1981) Interaction of tricyclic antidepressants with the ion channel of the acetylcholine receptor. Mol Pharmacol 20:511–518
Shaker N, Eldefrawi AT, Aguayo LG, Warwick JE, Albuquerque EX, Eldefrawi ME (1982) Interactions of d-tubocurarine with nicotinic acetylcholine channel. J Pharmacol Exp Ther 220:172–177
Shamoo AE, Eldefrawi ME (1975) Carbamylcholine and acetylcholine-sensitive cation selective ionophore. J Membr Biol 25:47–63
Sine SM, Taylor P (1979) Functional consequences of agonist-mediated state transitions in the cholinergic receptor. J Biol Chem 254:3315–3325
Sine SM, Taylor P (1981) Relationship between reversible antagonistic occupancy and the functional capacity of the acetylcholine receptor. J Biol Chem 256:6692–6699
Smilowitz N, Hadjian RA, Dwyer J, Feinstein MB (1981) Regulation of acetylcholine receptor phosphorylation by calcium und calmodulin. Proc Natl Acad Sci USA 78:4708–4712
Sobel A, Weber M, Changeux JP (1977) Purification of the acetylcholine receptor from *Torpedo* electric organ. Eur J Biochem 80:215–224
Sobel A, Heidmann T, Cartaud J, Changeux JP (1980) Reconstitution of a functional acetylcholine receptor. Eur J Biochem 110:13–33
Steinacker A (1979) Sulphonation of cholingergic receptor disulfide bond. Nature 278:358–360
Steinacker A, Zuazaga DC (1981) Changes in neuromuscular junction endplate current time constants produced by sulfhydryl reagents. Proc Natl Acad Sci USA 78:7806–7809
Steinbach JH (1980) Activation of nicotinic acetylcholine receptors. In: Cotman CW, Poste G, Nicolson GL (eds) The cell surface and neuronal function. North Holland, New York, pp 120–157
St. John PA, Froener SC, Goodenough DA, Cohen JB (1982) Nicotinic postsynaptic membranes from *Torpedo*. J Cell Biol 92:333–342
Strader CD, Raftery MA (1980) Topographic studies of *Torpedo* acetylcholine receptor subunits. Proc Natl Acad Sci USA 77:5807–5811
Strader CD, Ruvel JP, Raftery MA (1979) Demonstration of the transmembrane nature of the acetylcholine receptor by labeling with anti-receptor antibodies. J Cell Biol 83:499–510
Strader CD, Lazarides E, Raftery MA (1980) The characterization of actin associated with postsynaptic membranes from *Torpedo californica*. Biochem Biophys Res Commun 92:365–373
Suarez-Isla BA, Hucho F (1977) Acetylcholine receptor: SH group reactivity as indicator of conformational changes and functional states. FEBS Lett 75:65–79
Sumikawa K, Houghton M, Emtage JS, Richards BM, Barnard EA (1981) Active multi-subunit ACh receptor assembled by translation of heterologeous in RNA in *Xenopos oocytes*. Nature 292:862–864

Sumikawa K, Houghton M, Smith JC, Bell L, Richards BM, Barnard EA (1982) The molecular cloning and characterization of cDNA coding for the α subunit of the acetylcholine receptor. Nucleic Acids Res 10:5809–5822

Tanabe T, Noda M, Furutani Y, Takai T, Takahashi H, Tanaka K, Hirose T, Inayama S, Numa S (1984) Primary structure of the β subunit precursor of calf muscle acetylcholine receptor. Eur J Biochem 144:11–17

Tarrab-Hazdai R, Geiger B, Fuchs S, Amsterdam A (1978) Localization of acetylcholine receptor in excitable membrane from the electric organ of *Torpedo*. Proc Natl Sci USA 75:2497–2501

Tarrab-Hazdai R, Bercovici T, Goldfarb V, Gitler C (1980) Identification of the acetylcholine receptor subunit in the lipid bilayer. J Biol Chem 155:1204–1209

Turner GL, Oldfield E (1979) Effect of local anesthetics on hydrocarbon chain order in membranes. Nature 277:669–670

Tzartos SJ, Lindstrom JM (1980) Monoclonal antibodies used to probe acetylcholine receptor structure. Proc Natl Acad Sci USA 77:755–759

Tzartos SJ, Rand DE, Einarson BL, Lindstrom JM (1981) Mapping of surface structures of electrophorus acetylcholine receptor using monoclonal antibodies. J Biol Chem 256:8635–8645

Tzartos SJ, Seybold ME, Lindstrom JM (1982) Specificities of antibodies to acetylcholine receptors in sera from myasthenia gravis patients. Proc Natl Acad Sci USA 79:188–192

Vandlen RL, Wu WCS, Eisenach JC, Raftery MA (1979) Studies of the composition of purified *Torpedo californica* acetylcholine receptor and of its subunits. Biochemistry 18:1845–1854

Waksman G, Changeux JP, Roques BP (1980) Structural requirements of agonist and noncompetitive blocking action of acetylcholine derivatives. Mol Pharmacol 18:20–27

Walker JW, McNamee MA, Pasquale E, Cash DH, Hess AP (1981) Acetylcholine receptor inactivation in *Torpedo californica* electroplax membrane vesicles. Detection of two processes in the millisecond and second time range. Biochem Biophys Res Commun 100:86–90

Wassermann NH, Bartels E, Erlanger BF (1979) Conformational properties of the acetylcholine receptor. Proc Natl Acad Sci USA 76:256–259

Weber M, David-Pfeuty T, Changeux JP (1975) Regulation of binding properties of the nicotinic receptor protein. Proc Natl Acad Sci USA 72:3443–3447

Weiland G, Taylor P (1979) Ligand specificity of state transitions in the cholinergic receptor. Mol Pharmacol 15:197–212

Weiland G, Georgia B, Lappi S, Chignell CF, Taylor P (1977) Kinetics of agonist-mediated transitions in state of the cholinergic receptor. J Biol Chem 252:7648–7656

Weiland G, Friesman D, Taylor P (1979) Affinity labeling of the subunits of the membrane associated cholinergic receptor. Mol Pharmacol 15:213–226

Weinberg CB, Hall EW (1979) Antibodies from patients with myasthenia gravis recognize determinants unique to extrajunctional acetylcholine receptors. Proc Natl Acad Sci USA 76:504–508

Wennogle LP, Changeux JP (1979) The three-dimensional strucutre of the acetylcholine receptor. Neurosci Lett [Suppl] 9:S255

Wennogle LP, Changeux (1980) Transmembrane orientation of proteins present in acetylcholine receptor-rich membranes from *Torpedo marmorata* studied by selective proteolysis. Eur J Biochem 106:381–391

Wennogle LP, Oswald R, Saitoh T, Changeux JP (1981) Dissection of the 66,000-dalton subunit of the acetylcholine receptor. Biochemistry 20:2492–2497

West LK, Huang L (1980) Transient permeabilization induced osmotically in membrane vesicles from *Torpedo* electroplax. Biochemistry 19:4418–4423

Wilson MHF, Wu WCS, Raftery M (1979) Rapid cation flux from *Torpedo californica* membrane vesicles. Biochem Biophys Res Commun 89:26–35

Wise DS, Schoenborn BP, Karlin A (1981 a) Structure of acetylcholine receptor dimer determined by neutron scattering and electron microscopy. J Biol Chem 256:4124–4126

Wise DS, Wall J, Karlin A (1981 b) Relative locations of the β&δ chains of the acetylcholine receptor. J Biol Chem 256:12624–12627
Witzemann V, Raftery MA (1977) Selective photoaffinity labeling of acetylcholine receptor using a cholinergic analogue. Biochemistry 16:5862–5868
Witzemann V, Raftery MA (1978) Affinity directed crosslinking of acetylcholine receptors. Biochem Biophys Res Commun 85:623–631
Witzemann V, Muchmore D, Raftery MA (1979) Affinity-directed crosslinking of membrane-bound acetylcholine receptor with photolabile α-bungarotoxin derivative. Biochemistry 18:5512–5518
Wu W, Moore HP, Raftery MA (1981) Quantitation of cation transport by reconstituted membrane vesicles. Proc Natl Acad Sci USA 78:775–779
Young AP, Oshiki JR, Sigmin DS (1981) Allosteric effects of volatile anesthetics on the acetylcholine receptor. Mol Pharmacol 20:506–510
Zingsheim HP, Neugebauer DC, Barrantes FJ, Frank J (1980) Structural details of membrane-bound acetylcholine receptor from *Torpedo*. Proc Natl Acad Sci USA 77:952–956

Pharmacodynamics and Pharmacokinetics of Neuromuscular Blocking Agents

CHAPTER 3

On the Principles of Postsynaptic Action of Neuromuscular Blocking Agents

D. COLQUHOUN

A. Introduction

In this chapter, the evidence concerning the mechanism of postsynaptic action of neuromuscular blocking agents will be discussed. Although it could certainly be argued that the important facts about tubocurarine were all known long before the voltage clamp was invented, it could not be argued that the reasons for its behaviour were understood. The emphasis in this chapter will be on the fundamental molecular effects of the drugs, rather than on the phenomena which they are empirically observed to produce. These limitations on the scope of this chapter reduce considerably the work that will be dealt with in any detail, because the amount of knowledge about molecular mechanisms of action is surprisingly small. This statement may seem odd in view of the vast amount of work that has been done on the neuromuscular junction, and on drugs that affect it. But inspection of the literature soon reveals that almost all of this work is done by methods that are not capable of giving rigorous information about mechanisms. For example, a blocking drug is often described as "competitive" for no better reason than that it fails to produce a depolarization; indeed, even membrane potential often is not directly observed, so perhaps one should say that it fails to behave *as though* it were producing a depolarization. This sort of statement can surely not be defended by any pharmacologist as an adequate definition of what is meant by "competitive". Similarly, the details of the mechanisms of action of those blockers that produce a depolarization have, with a few exceptions, yet to be investigated by modern electrophysiological methods.

There have been recently a number of books and reviews of the chemistry, effects, side effects and disposition of neuromuscular blocking drugs, which cover aspects of the subject that are not considered in detail here; see, for example, ZAIMIS (1976), STENLAKE (1979, 1980) and BOWMAN (1980). More fundamental aspects have been reviewed by KATZ (1966), NEHER and STEVENS (1977), STEINBACH (1980), WRAY (1980), ADAMS (1981), COLQUHOUN (1975, 1979, 1980, 1981 a, b) and PEPER et al. (1982).

In order to discuss critically the action of blocking agents, it is desirable first to review various possible mechanisms of action and the experimental methods that are needed to verify them (Sect. B) and then to discuss the present view of the receptor, the ion channel and the action of agonists (Sect. C). Then we shall discuss the actions of the classical, tubocurarine-like, nondepolarizing agents (Sect. D), and less specific (e.g. local anaesthetic-like) nondepolarizing agents (Sect. E). Finally, depolarizing agents will be considered (Sect. F).

B. Mechanisms of Action and Experimental Criteria for Them

I. Range of Conditions of Tests

It is, fortunately, true that many of the experimental methods that can be used to provide evidence for a particular mechanism of block (e.g. competitive block or ion channel block), are valid, even without detailed knowledge of how the agonist works. For example, if experiments are conducted with a low concentration of agonist (so that a small proportion of ion channels is open, and there is not much desensitization) the interpretation of the results may be greatly simplified. Similarly interpretation is usually simpler when response, or ligand binding, are measured at equilibrium. On the other hand, if a putative antagonist mechanism is tested *only* at a low agonist concentration, or *only* at equilibrium, the test is certainly less convincing than if it had been carried out over the whole range of agonist concentrations, or if rates as well as equilibria had been measured. It is common, for example, that two different antagonist mechanisms which would be easily distinguishable if tested over the whole range of agonist concentrations, are indistinguishable when tested only at low agonist concentrations. A depolarization of 10 or 20 mV at the end-plate, or an inward current of 50 nA, is considered a good sized response to an agonist. But it is important to remember that such a response is produced by opening only about 0.2% of the ion channels in the end-plate (see Sect. C. II and D.).

Furthermore, the conditions under which experiments are most easily conducted are quite different from those in the physiological process of neuromuscular transmission. In the latter, the agonist concentration is (transiently) very high, and the system is never at equilibrium (see Sect. C. III). This implies that, in order to understand fully the action of an antagonist on the physiological process, we need to know how the antagonist behaves when the agonist concentration is high, and that we need to know not only the equilibrium constant for the binding of the antagonist, but also the rate constant for its association and dissociation. It will emerge in the subsequent discussion that there is not, as yet, such complete information available for *any* neuromuscular blocking drug.

There are, in general, two experimental approaches: (a) measurement of binding of drugs to receptors, and (b) measurement of responses to agonists. In most cases these two approaches are employed by different workers, in different laboratories. The fact that those two groups frequently ignore, or misunderstand, each other's results is unfortunate for the progress of the subject. Clearly both approaches are essential, and neither, alone, is likely to provide conclusive results.

II. Measurement of Binding

The binding of drugs to receptors can be measured, either directly with labelled drug, or via inhibition of binding of a standard labelled drug by the (unlabelled) drug under investigation. The great virtue of binding measurements is that they give a more or less direct measurement of the actual receptor occupancy by a drug. The problems with binding measurements include the following.

1. Specificity of Binding

A major problem is that drugs are commonly bound at many sites in addition to the receptors of interest. In general (though not necessarily in every case) this problem is likely to be greater for ligands that have not got a very high affinity for the receptor itself; they will have to be used in a fairly high concentration so nonspecific binding, even if it has only a modest affinity may interfere seriously, with the results. Thus the problem of nonspecific binding tends to be greatest with agonists, which have a rather low affinity for the native receptor (see Sect. C II). But, at the nicotinic receptor, antagonists too have rather modest affinities (equilibrium constants of the order of μM) compared with, for example, muscarinic or histamine (H_1) receptors for which good antagonists have equilibrium constants around 1 nM. Results are normally presented in terms of "specific binding" of a labelled ligand, but the way in which this has been defined needs careful attention. Usually it is estimated by subtracting from the total binding a component of "nonspecific" binding, the latter being defined, fot example, as the binding observed when the experiment is carried out in the presence of a large concentration of an unlabelled ligand that combines with the receptor. This method supposes, of course, that the large concentration of unlabelled ligand suppresses all of the binding to the receptor to the receptor, but nothing else. But in so far as the labelled ligands have similar structures (or, even worse, the same structure) they are likely to share the same nonreceptor (as well as receptor) binding sites so the unlabelled compound will suppress binding to *any* saturable site, not only that to the receptor. Thus the "nonspecific" binding is likely to be underestimated and hence the "specific" binding is likely to be overestimated.

2. Time Resolution

Until recently, the relative slowness of methods for measuring binding has meant that only equilibrium studies have been possible. In earlier work, especially, problems arose even with the measurement of equilibrium binding of ligands (agonists or antagonists) when this was measured by inhibition of the binding of labelled α-bungarotoxin; because bungarotoxin binds essentially irreversibly, the receptor occupancy by the inhibitory ligand (which is assumed to equilibrate rapidly on the time scale of the experiment) must be inferred form the initial *rate* of bungarotoxin binding. Much earlier work was done using incubation times that were too long for the initial rate to be estimated properly, or on preparations in which the observed rate was diffusion controlled rather than being controlled, as required, by the rate of interaction with the receptor.

The restriction of measurement to equilibrium conditions meant that the binding of agonist was inevitably measured on desensitized preparations; in fact a rather slow increase in the binding affinity for agonists (but not for antagonists), which was presumed to reflect desensitization was detected, semiquantitatively, at a fairly early stage (e.g. WEBER et al. 1975; COLQUHOUN and RANG 1976).

Obviously what is needed (for agonists, especially) is a rapid method for measuring the binding rates. Recently a number of ingenious attempts have been made to do this. Nevertheless the problem is by no means solved, for two main reasons. (a) Although the fastest methods can resolve down to a few milliseconds, they are still not quite fast enough to resolve the binding that corresponds to the physiological channel opening process. They have, however, given useful information about the rate of desensitization (see Sect. C. IV). (b) The fast methods all involve measurements of fluorescence changes when agonists are added and there is no guarantee at all that these changes reflect *only* the binding stages. They could, and quite probably do, also reflect events subsequent to binding. The methods that have been used include the following: (a) measurement of the (rather small) changes in intrinsic fluorescence of membrane-bound receptor (BONNER et al. 1976); (b) use of fluorescent analogues of acethylcholine (ACh) (HEIDMANN and CHANGEUX 1979 a; JÜRSS et al. 1979) – these cannot, of course, give information about ACh itself; (c) use of reversibly bound fluorescent agents such as quinacrine (mepacrine), or ethidium (GRUNHAGEN and CHANGEUX 1977; QUAST et al. 1979) – these probably have side effects (e.g. local anaesthetic-like actions) of their own which complicate the results; and (d) covalent attachment of a fluorescent molecule to a site near the receptor (DUNN et al. 1980). The results with this method have not yet been correlated with physiological observations, and are inconsistent with those of HEIDMANN and CHANGEUX (1979 a).

III. Measurements of Response

The interpretation, in terms of underlying mechanism, of the inhibition of a response by a blocking agent depends on what sort of response is being measured. The types of response that have been measured cover a wide range. They can be classified as follows.

1. Measurements of Muscle Tension

The most direct information about what a neuromuscular blocking agent does to humans is obtained by experiments on humans, e.g. those in which the ulnar nerve is stimulated and the force of thumb adduction by the adductor pollicis muscle is measured (e.g. SUGAI et al. 1975). The most widely employed experimental method is measurement of the twitch tension evoked by nerve stimulation in whole animals (e.g. tibialis anterior muscle in anaesthetized cat) or isolated tissues (e.g. rat diaphram–phrenic nerve preparation). Such experiments can give valuable information to the anaesthetist and may give some indirect hints about the mode of action of blocking agents; however, no rigorous inferences about mechanisms can come from this sort of experiment because of the large number of uncontrolled variables (possible changes in ACh release, the ACh concentration at its site of action, the membrane potential, and so on). Probably the only sort of experiment in which muscle tension measurements can give good evidence about mode of action are those in which the equilibrium contracture tension of a slow muscle (e.g. frog rectus abdominis or chick biventer cervicis) is measured in response to a known concentration of agonist applied in bathing solution.

2. Measurements on Single Muscle Fibres

The most satisfactory way to investigate the mode of postsynaptic action of a neuromuscular blocking agent is clearly to look directly at its effect on end-plates of single muscle fibres. The simplest sort of experiment is to measure the agonist-induced depolarization of the end-plate by means of an intracellular voltage-recording microelectrode as in the classical study of JENKINSON (1960). This method is satisfactory when equilibrium responses to known (bath-applied) concentrations of agonist are measured (but see Sect. F. III). However, equilibrium studies alone can no longer be considered adequate. In order to establish a mechanism kinetic studies are also required, i.e. we must study the rate at which equilibrium is attained. For the investigation of kinetics, simple voltage recording is quite unsatisfactory; the reason for this is that as the membrane potential changes the membrane capacitance has to be charged or discharged and this process distorts the time course of the response which is the very thing that we are trying to measure. The sine qua non for measurement of kinetics is, therefore, to hold the membrane potential constant at a known value. The current that is observed to flow – the response – is then directly proportional to the number of ion channels that the agonist has opened in the end-plate, which is as near to the primary response as we can get. This can be done by inserting a second microelectrode into the muscle at the end-plate region (TAKEUCHI and TAKEUCHI 1960), or by inserting a dissected single muscle fibre into a Vaseline gap voltage clamp with the end-plate in the pool from which measurements are made (HILLE and CAMPBELL 1976) or by the patch clamp (single ion channel) method (NEHER and SAKMANN 1976; HAMILL et al. 1981; SAKMANN and NEHER 1983). The Vaseline gap method, and the isolated patch methods have the advantage that the "intracellular" composition can be controlled.

The main methods for investigation of kinetics are noise analysis (KATZ and MILEDI 1970, 1972), voltage jump relaxation analysis (ADAMS 1975a; NEHER and SAKMANN 1975; SHERIDAN and LESTER 1975) and single-channel analysis. Simple accounts of these methods have been given by COLQUHOUN (1981a) and CULL-CANDY (1981), and the more general theory underlying them has been presented in an introductory way by NEHER and STEVENS (1977), COLQUHOUN and HAWKES (1983) and more fully by COLQUHOUN and HAWKES 1977, 1981, 1982). These kinetic methods, together with equilibrium measurements (in so far as these can be achieved in the face of desensitization – see Sect. C. IV), constitute the methods from which mechanisms can be properly inferred. Information can also sometimes be obtained by measurement of nerve-evoked end-plate currents (EPC) in voltage-clamped fibres; however the method can be misleading if the decay phase of the EPC is partly controlled by the time course of ACh concentration in the synaptic cleft (see Sect. C. III). And, in addition, if the EPC decay is biphasic (see Sect. D. VII) it is not easy to relate the observed amplitudes of the two components to those predicted by any putative mechanism (because it is, at present, not possible to predict with any precision the initial state of the system at the moment at which measurement of the decay phase started, because of the complexity of the rising phase of the EPC). Experiments in which agonist is applied by iontophoresis will also, in general, be unsatisfactory for quantitative investigations of mechanism because the agonist concentration is unknown, and varies with both time and position along the end-plate. In some cases quantitative results *have* been inferred from iontophoretic experiments (DIONNE et al. 1978; DREYER et al. 1978) but the methods that have to be used to achieve this are both complex and, inevitably, somewhat indirect.

3. Measurements of Ion Flux In Vitro

It is not possible to make very satisfactory measurements of ion flux on intact muscle fibres, and the incentive to do so is small because electrophysiological measurements give similar information with much better time resolution. However, much useful information has come from studies of ion flux in membrane vesicles and cultured cells. The great advantage of such preparations is that both response and ligand binding can be measured on the *same* preparation. KASAI and CHANGEUX (1971) introduced a preparation of sealed vesicles made by differential centrifugation of homogenates of the electric organ of *Torpedo*. The influx or efflux of $^{22}Na^{+}$, in response to ACh-like agonists, could be measured in this preparation which contains a very high density of receptors in the vesicle membrane. Indeed the receptor density was *too* high for quantitative purposes. In all earlier work with such preparations the fluxes were so great relative to the vesicle volume, and the time scale so slow that ionic equilibration could take place before the measurement can be made (about 10 s after addition of agonist). As a result of this the flux response that was measured was not directly proportional to the number of open ion channels (the primary response). Therefore many sorts of quantitative assessment of putative mechanisms were impossible. More recently, faster methods of measuring the *initial* rate of ion flux (which should be directly proportional to the number of open channels) have been revised, which seem to work well, at least for moderate agonist concentrations. For example NEUBIG and COHEN (1980) use a rapid mix, quenched flow method with *Torpedo* vesicles which they assessed carefully both by measurements at short times (down to 24 ms after mixing with agonist) and also by reducing the ion flux with bungarotoxin. CASH et al. (1981) use a similar method based on vesicles from *Electrophorus* electric organ (which has a lower receptor density than *Torpedo*). SINE and TAYLOR (1979) have made elegant measurements of the initial rate of Na^{+} influx into monolayer cultures of a clonal muscle cell line (BC3H-1 cells) and compared them with ligand binding measurements in the same preparation. This method is not as fast as the quenched flow method (but the cells are larger, and less permeable, than *Torpedo* vesicles).

IV. Tests for Competitive Antagonism

A "competitive antagonist" will be defined as a molecule that can occupy the ACh receptor (without itself producing response) in a manner such that occupancy by the antagonist and by ACh (or its analogues) are mutally exclusive. It may be noted that this definition refers to a *mechanism*, not to directly observable predictions of the mechanism. Sometimes the observable phenomena (such as parallel shift of log concentration–response cures with unaltered maximum) are confused with, or even used as a definition of, a competitive mechanism. This definition takes no account of the rate at which the reactions take place; it clearly includes almost irreversible antagonists, like α-bungarotoxin which are indeed supposed to compete for the ACh receptor (although, in so far as the antagonist is genuinely irreversible, it will eventually win the competition). Clearly it is necessary to distinguish carefully measurements that are made at equilibrium form those that are not (e.g. those with α-bungarotoxin, in most cases).

1. Competitive Antagonists at Equilibrium

a) Binding Measurements

In most cases the equilibrium binding of the antagonist alone has been measured. The binding curve is then fitted with a hyperbolic Langmuir curve (HILL 1909; LANGMUIR 1918) or the sum of more than one such curve. Ideally, of course, mutual antagonism between agonist and antagonist binding should be demonstrated, and it should be shown that the number of binding sites is the same for both agonist and antagonist; this has been done only rarely (see Sect. D. IV).

b) Response Measurements

The discovery of the Schild method (SCHILD 1949; ARUNLAKSHANA and SCHILD 1959) produced something of a revolution in the study of competitive antagonism. This method provided a test of whether the mechanism of antagonism was competitive. And, if the test was passed, the method allows a simple calculation of the equilibrium constant for the binding of the antagonist to the receptor. This was, and still is, a remarkable advance because it meant that the characteristics of ligand *binding* could, for the first time be inferred from measurements of a biological *response*. Subsequent work (for example, PATON and RANG 1965) has frequently shown that the results obtained agree well with those obtained from direct binding measurements, even when the response that is measured is a complex quantity, far removed from the primary response to the agonist, such as the tension developed in a smooth muscle preparation.

The reason for the remarkable robustness of the method is esentially that it is a null method. The concentration of agonist (x_A, say) needed to produce a certain response in the absence of the antagonist is measured. Then the increased concentration (x'_A, say) that is needed to produce the same response in the presence of the antagonist (in concentration x_B, say) is determined. The ratio of these concentrations is defined as the *dose ratio*, denoted $r = x'_A/x_A$. The Schild equation states that, for a competitive antagonist at equilibrium, the dose ratio is given by

$$r = 1 + x_B/K_B\ , \tag{1}$$

where K_B ist the equilibrium dissociation constant for the binding of the antagonist to the receptor. For graphical purposes the Schild equation is commonly used in the logarithmic from

$$\log(r-1) = \log x_B - \log K_B\ . \tag{2}$$

A plot of $\log(r-1)$ against $\log x_B$ should be linear with a slope of unity. Interpolation of the value of x_B at $r=2$ gives an estimate of K_B. This result has been repeatedly observed to hold good, in many cases over a very wide range of antagonist concentrations.

It is noteworthy that the Schild equation does not involve the agonist concentration, or response size. This means that the dose ratio should be independent of which agonist is used and of the particular response level at which it is measured; in other words the shift of the log concentration–response curve to the right should be by a constant amount ($\log r$), so the curves with and without agonist are, in this sense, parallel (whether or not they are linear).

The reason for the robustness of the Schild method is, of course, that it makes very few assumptions about the relationship between receptor occupancy and the final measured response. It assumes only that a constant response size corresponds to the same agonist occupancy whether or not the antagonist is present and this is why it often works well even when the response measured is related to agonist occupancy by very complex pathways. The original derivation of the Schild equation was based on mutually exclusive Langmuir binding of agonist and antagonist to identical and independent receptor sites, and even now it is sometimes states that this condition is necessary for the Schild method to be valid. However, it has been shown (THRON 1973; COLQUHOUN 1973) that the Schild equation is valid for a wide range of mechanisms, including many with multiple binding sites and cooperative interactions.

c) Deviations from Competitive Predictions

If the Schild equation is not obeyed, or binding measurements do not follow competitive predictions, we can conclude that either the antagonist is not competitive (how it *does* work then being a question for investigation) or the antagonist is competitive, but one or more of the assumptions that have been made are not valid. For example, the experiment might be invalid because the system was not really at equilibrium, or because the concentration of agonist or antagonist at the receptor sites was not the same as that applied (e.g. because of uptake processes), or because the antagonist has other actions as well as being competitive. The Schild slope could be steeper than unity, as a result, for example, of concomitant open ion channel block (see Sect. D. VII), or as a result of desensitization at high agonist concentrations, or enhancement of desensitization by the antagonist (see Sect. E. IV).

Another possibility is that the mechanism, although competitive, is one of the few to which the Schild equation does not apply; for example the antagonist might bind competitively, but with unequal affinities for two nonequivalent subunits of the receptor channel complex. There is some evidence for this last hypothesis being true for tubocurarine, so the matter will be discussed in detail later, in Sect. C. V). This mechanism predicts Schild slope below unity (see Sect. D. V and Fig. 4). The Schild slope would also be less than unity for a mechanism in which ion channels could be blocked whether open or not (COLQUHOUN 1980).

2. Kinetics of Competitive Antagonists

Several methods have been used to try to measure the rate of action of competitive antagonists. None of them are entirely satisfactory. The results that have been obtained will be considered later (Sect. D. IV).

V. Tests for Noncompetitive Mechanisms

The term "noncompetitive" has been used in both the enzymological and the pharmacological literature to describe a variety of "mechanisms". These have usually been vaguely defined, and there is no good experimental evidence for them. The term "noncompetitive" will be used here merely to mean "not competitive".

Block by prolonged depolarization (see Sect. F. II), and block of open ion channels (see Sect. B. V, D. VII and F. IV) are both fairly well-defined mechanisms for which quite good evidence exists. There are also a number of other noncompetitive mechanisms which are rather less well defined, e.g. enhancement of desensitization, or block of shut ion channels. And there are some, such as "dual block" (Sect. F. IV) which are not defined as mechanisms at all. The evidence concerning these is rather weak. The sort of tests needed to provide evidence for block of ion channels will next be considered. The evidence for other noncompetitive mechanisms will be considered later in connection with particular drugs.

1. Block of Open Ion Channels

The simplest version of this mechanism can be written (see ARMSTRONG 1971; ADAMS 1976) as

$$\text{Shut} \underset{\alpha}{\overset{\beta'}{\rightleftharpoons}} \text{open} \underset{k_{-B}}{\overset{k_{+B}}{\rightleftharpoons}} \text{blocked.} \qquad (3)$$

A blocker molecule is supposed to prevent current flow through the open ion channel, quite possibly by simply plugging it. Notice that the blocker molecule is supposed to be able to enter and leave the ion channel only when it is in the open conformation. A number of predictions can be made from this mechanism. Some of these will be mentioned here (for more details see ADAMS 1976; COLQUHOUN and HAWKES 1983). The mechanism in Eq. (3) assumes that either the agonist binding or the conformation (open–shut) change is so fast that, in the absence of the blocking drug there are only two states in which the receptor ion channel can exist, open and shut (see Sect. C. II). This implies that kinetic experiments (e.g. EPC decay, voltage jump relaxations, noise) should show only a single exponential component, which is what is observed (see COLQUHOUN 1981 a for a basic account and examples). When the blocker is added we have, in Eq. (3), three distinct states for the receptor ion channel so we expect two kinetic components; for example we expect that the miniature end-plate current MEPC decay will, in the presence of blocker, be described by the sum of two exponential terms, one faster than the control (no blocker) and one slower. If the blocker dissociates rather slowly from the channel then we expect that only the faster component will be visible so the MEPC decay will appear simply to be speeded up by the blocker.

In fact it has become quite common for any agent that speeds up the decay of the MEPC, or makes it biphasic, to be described as a channel blocker. Such evidence is, of course, quite inadequate by itself. *Any* mechanism with three distinct states may be expected to give rise to two kinetic components. Another prediction of Eq. (3) is that the fractional inhibition by a fixed blocker concentration of the number of ion channels open at equilibrium should be greater at higher agonist concentrations than at low concentrations. This happens because there are more open ion channels available to be blocked at high agonist concentrations. But qualitatively similar behavior is expected for an antagonist that works by enhancing desensitization, an effect that is also expected to be bigger at high agonist concentrations (see Sect. F. IV).

In order to obtain evidence for Eq. (3) it is necessary to show quantitatively that the rate constants, and also the relative amplitudes of the kinetic components, vary with blocker concentration and with membrane potential in the way predicted by the mechanism.

C. The Mechanism of Action of Agonists

It has been pointed out (Sect. B. I and B. IV) that some tests for the mechanism of action of antagonists can be carried out without knowledge of how the agonist works. This is particularly true of equilibrium measurements at low agonist concentrations. However, a complete understanding of the antagonist can obviously be obtained only if the mechanism of action of the agonist alone is first understood. In this section we shall therefore review briefly the present view of acetylcholine action.

I. Structure of the Receptor Ion Channel

The most detailed studies have been done on receptors from *Torpedo*. In this species the receptor is a protein that consists of five subunits; there are two subunits (molecular weight about 40,000), and one each of heavier subunits denoted β, γ and δ (molecular weight about 50,000, 60,000 and 66,000, respectively). Thus the whole structure is denoted $\alpha_2\beta\gamma\delta$ and has a molecular weight of about 240–270,000 (see KARLIN 1980). The size of the receptor channel complex is very similar to this in other species that have been investigated, e.g. cat denervated muscle and chick muscle (innervated, denervated and embryonic); however, the subunit structure is less certain in these species (LO et al. 1981; SUMIKAWA et al. 1982). In receptor isolated from cat denervated muscle only one class of polypeptide (molecular weight about 43,000) was found by SHORR et al. (1981), despite careful avoidance of proteolysis during purification. Receptor from fetal calf muscle has been reported to be "a pentameric complex composed of two equivalent and three pseudoequivalent subunits" (CONTI-TRONCONI et al. 1982). The amino acid sequence has been determined for all four types of subunit of *Torpedo* receptor and those for the α and δ subunits have been published (SUMIKAWA et al. 1982; NODA et al. 1982; CLAUDIO et al. 1983). The subunits in *Torpedo*, and in mammalian receptor, have considerable similarities both in amino acid sequence and in immunological characteristics, which suggest a common genetic origin (RAFTERY et al. 1980; TZARTOS and LINDSTROM 1980).

Analysis of *Torpedo* membrane fragments by X-ray diffraction, electron microscopy and other methods, suggests that the polypeptide subunits are arranged around a central ion channel, the whole unit being about 11 nm long, i.e. considerably longer than the membrane thickness. The structure is thought to protrude, in a funnel shape, about 5.5 nm from the membrane on the extracellular side, and to protrude about 1.5 nm on the cytoplasmic side (KISTLER et al. 1982). The ion channel, while open, behaves like a rather large aqueous pore (ADAMS et al. 1980). The α-bungarotoxin binding sites, and hence probably the ACh binding sites, appear to be located near the outer ends of the unit. If this is correct, they would presumably be outside the electric field of the membrane, which is consistent with evidence that the binding of competitive antagonists (COLQUHOUN et al. 1979), and probably the binding of agonists also, is not dependent on membrane potential (see Sect. C. II, D. III).

The binding site for α-bungarotoxin and acetylcholine is located on the α-subunit, of which there are two (see KARLIN 1980). Thus the biochemical evidence now points strongly to there being two agonist binding sites per channel (see also Sect. C. II). Studies in which the purified receptor has been reconstituted into lipid membranes have shown clearly that the unit of molecular weight, 250–270,000 includes the ion channel structure as well as the agonist binding sites (SCHINDLER and QUAST 1980; NELSON et al. 1980; BOHEIM et al. 1981).

II. Opening of the Ion Channel by Agonists

The literature up to 1979 or 1980 has been reviewed by COLQUHOUN (1979), STEINBACH (1980) and ADAMS (1981). Therefore, for the sake of brevity, this discussion will concentrate mainly on recent developments.

1. Current Ideas About Mechanism

It is still true that binding measurements (see Sect. B. II) are too slow to cast much light on the rapid physiological process of ion channel inactivation. Ion flux studies (see Sect. B. III) have also been more informative about desensitization (see Sect. C. IV) than about the channel activation process. Nevertheless the results of these studies are, on the whole, compatible with the minimum mechanism that is required to account for electrophysiological results, viz. the sequential binding of two agonist molecules followed by isomerization from shut to open states, thus

$$\begin{array}{c} \mathrm{R} \\ k_{+1}\,\Big\uparrow\!\!\Big\downarrow\,k_{-1} \\ \mathrm{AR} \\ k_{+2}\,\Big\uparrow\!\!\Big\downarrow\,k_{-2} \\ \mathrm{A_2R} \underset{\alpha_2}{\overset{\beta_2}{\rightleftharpoons}} \mathrm{A_2R^*} \end{array} \tag{4}$$

In this scheme A denotes the agonist, R the shut, and R* the open, receptor channel. The arrows are labelled with the microscopic rate constants for each transition. The corresponding equilibrium constants are defined as

$$K_1 = k_{-1}/k_{+1}$$
$$K_2 = k_{-2}/k_{+2} \qquad (5)$$
$$L_2 = \beta_2/\alpha_2$$

The biochemical evidence is rather strongly in favour of there being two agonist binding sites (see Sect. C. I), so the Hill coefficient for agonist binding and for the response (fraction of channels that are open) should be between 1 and 2 (see, for example, COLQUHOUN 1973). This has been observed in most experiments though a few estimates lie between 2 and 3 (reviewed by COLQUHOUN 1979). These last values must now be suspect in view of the biochemical evidence.

According to Eq. (4) channels with only one agonist molecule bound cannot open. However, some electrophysiological evidence suggests that such complexes may open, though with a much lower probability than those with two agonist molecules bound (DIONNE et al. 1978; COLQUHOUN and SAKMANN 1981; COLQUHOUN and HAWKES 1982). Flux measurements by SINE and TAYLOR (1979; see Sect. B. III) are also consistent with this idea. Although none of the data are very compelling at the moment, they are consistent with the following slight extension of Eq. (4).

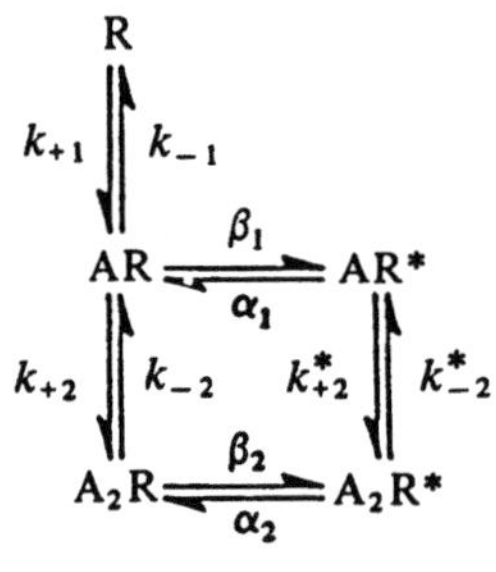

(6)

One more complication (apart from desensitization, see Sect. C. IV) may need to be added to this scheme as a result of evidence (SINE and TAYLOR (1979) which suggests that the binding of agonist may not be the same for the two binding sites. The ratio of affinities could be 15 or more (S.M. SINE 1982, personal communication) but the evidence (only on BC3H-1 cells so far) is suggestive rather than compelling at present (see, however, Sect. D. V).

2. Opening, Shutting and Dissociation Rates

In order to understand how the synapse, and blocking agents, work it would be nice to have some idea of the values for the rate constants in Eqs. (4)–(6). Most of the values are still rather vague. Until recently one would have said that α_2 in Eq. (4), the channel closing rate constant, was known because $1/\alpha_2$ is simply the mean lifetime of an open channel, and the latter can be found from noise analysis, or direct observation of single-channel currents. However, α_2 can be so found only under certain assumptions. In particular it must be assumed that the binding reactions in Eqs. (4) and (6) are very fast compared with the conformation change. The question of whether binding or conformation change (shut⇌open) is rate limiting has been a matter of controversy over the last 15 years (see reviews in COLQUHOUN 1979; STEINBACH 1980). It is not a trivial question because interpretation of kinetic experiments depends on the answer. Although both views have been advocated, most people have followed the hypothesis of ANDERSON and STEVENS (1973) that the binding step is the fast one (despite the fact that the authors made it quite clear that this was a working hypothesis for which there was only circumstantial evidence). The matter could be settled honourably if it turned out that neither step was rate limiting. For some time it was imagined that this was incompatible with the observation that noise and relaxations consist, as far as can be seen, of a single kinetic component. However, it was pointed out by COLQUHOUN and HAWKES (1977) and SAKMANN and ADAMS (1979) that this was not necessarily true. The easiest way to see why is to imagine what is happening to a single ion channel.

The hypothesis of rapid binding implies that as soon as an open channel shuts the bound ligand will (almost certainly) dissociate immediately, so the channel returns to its resting state. This, incidentally, also implies that there will be many brief occupanicies that do not lead to opening at all. Suppose, however, that the opening rate β is quite fast (as might be expected of a good transmitter at a fast synapse), and the dissociation rate (k_{-1}, k_{-2}) is not too fast. In this case it would be expected that, when an open channel shuts there would be quite a good chance that the channel might reopen before the agonist dissociated. The "opening" would therefore actually consist of a burst of several openings in quick succession, each opening being of length $1/\alpha$ on average. A noise analysis would give an estimate of, approximately, the whole burst length, not of the length of a single opening, so the noise analysis would underestimate α. COLQUHOUN and HAWKES (1977) gave a numerical illustration, a calculation based on values of β_2 and k_{-2} which seemed plausible at the time (there were no good experimental values available); this example showed bursts that consisted, on average, of about three openings (each of 1 ms mean length, i.e. $\alpha = 1{,}000\ s^{-1}$) occurring in quick succession (mean gap within a burst $= 34\ \mu s$). The mean burst length was therefore about 3 ms, and this was close to the time constant that a noise analysis would be expected go give. At about the same time in Göttingen, E. NEHER and B. SAKMANN (1979, personal communication), noticed that short gaps were quite often seen during ion channel openings. When a pair of openings occurred in quick succession, they referred to the second one as a *nachschlag* – a "second helping". The further analysis of this phenomenon had to await the development of higher resolution methods (see HAMILL et al. 1981). It was found by COLQUHOUN and SAKMANN (1981) that the 10 ms ion channel "opening" induced by the ACh analogue suberyldicholine (subCh) actually consisted, on average, of about four openings (each of about 2.5 ms. mean duration) separated by very short shut periods of 40–50 μs; this is illustrated in Fig. 1. More interestingly, this phenomenon did not appear to depend on the agonist concentration and so was unlikely to result from ion channel block by agonist molecules. Neither did it appear to be very voltage

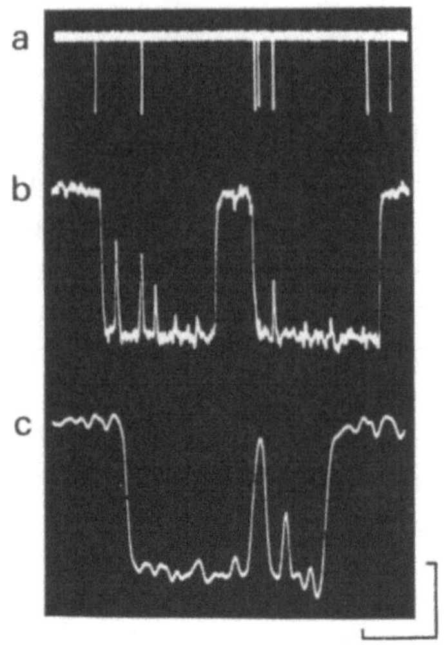

Fig. 1 a–c. Examples of single-channel currents, shown on three different time scales, recorded from frog neuromuscular junction in response to low concentrations of suberyldicholine (SubCh). Openings are shown as downward deflections. **a** illustration of the low overall frequency of elementary events (seven in 5 s in this case) with 100 n*M* SubCh (calibration bars 4 pA and 1 s), **b** several brief closures interrupting an open channel current, taken from the experiment illustrated in **a** (calibration bars, 2 pA and 10 ms), **c** another group of three openings at higher time resolutions than in **b**, elicited by 20 n*M* SubCh (calibration bars, 2 pA and 2 ms). (COLQUHOUN and SAKMANN 1981)

dependent which is consistent with other evidence which suggestes that β and k_{-2} are not very voltage dependent (see Sect. C.I and D.III). These facts do not, of course, prove that the phenomenon results from multiple ion channel openings during a single receptor occupancy, but this is, at least, a plausible working hypothesis. If this were the explanation then we could obtain from the results $\alpha = 1/2.5\ \text{ms} = 400\ \text{s}^{-1}$ (noise analysis would have given $1/10\ \text{ms} = 100\ \text{s}^{-1}$); the opening rate in Eq. (4), $\beta_2 \sim 15{,}000\ \text{s}^{-1}$; and the dissociation rate $k_{-2} \sim 2{,}000\ \text{s}^{-1}$. Such estimates are, of course, very tentative (see COLQUHOUN and SAKMANN 1983; COLQUHOUN and HAWKES 1982). If they are correct they are certainly too fast to be measured by any fast binding or ion flux measurement that has been devised up to now. Multiple openings are also seen with ACh (COLQUHOUN and SAKMANN 1983, and unpublished work).

There are of course many estimates of "mean open channel lifetime" in the literature from noise analysis and, more recently, from single-channel analysis. These measurements have often been interpreted as estimates of $1/\alpha$, but in the light of the *nachschlag* phenomenon this interpretation may not be correct.

There are very few estimates of the opening rate β available. LAND et al. (1981) interpret data on MEPC rise time according to Eq. (4). The argument is necessarily very indirect, but also suggests that β is quite large for ACh. SAKMANN and ADAMS (1979) found a more modest value ($1{,}560\ \text{s}^{-1}$) for carbachol (CCh) by noise analysis, but their data for ACh could not give a good estimate. The suggestion so far is that β/α is quite large for SubCh and ACh so they would be capable (if it were not for desensitization) of opening a large fraction of channel when present in a high concentration. This conclusion is reinforced by recording of single ion channels. These show that the channel, when it is not in the desensitized state, can be held open for a large fraction of the time by ACh, SubCh and CCh (SAKMANN et al. 1980; OGDEN and COLQUHOUN 1983). Again, this suggests that β is quite large. However, CASH et al. (1981), estimated from fast flux studies that $\beta/\alpha \sim 0.7$ for ACh, contrary to this suggestion. Electrophysiological studies at equilibrium have suggested rather lower values β/α for CCh than for ACh (DREYER 1978; DIONNE et al. 1978).

The value of β/α can, of course be taken as a measure of the efficacy of an agonist (STEPHENSON 1956; CASTILLO and KATZ 1957a; COLQUHOUN 1973). It seems that the long-term goal of the pharmacologist to measure separate physical parameters that correspond to affinity and efficacy is getting close.

3. The Affinity for Agonist Binding

A number of estimates have been made of the "apparent equilibrium constant for ACh binding" from electrophysiological studies; a value of $K_{eff} \sim 30\ \mu M$ is often cited for ACh and about $300\ \mu M$ for CCh (see SHERIDAN and LESTER 1977; DREYER et al. 1978; DIONNE et al. 1978). However, the interpretation of this number is dependent on the particular mechanism that is postulated; it is *not* an estimate of the microscopic binding constant K_1 (or K_2) in Eq. (4), and these cannot be inferred from it without knowledge of the mechanism and of values for other rate constants e.g. α and β. The nearest one can come to an interpretation common to these studies, of the meaning of the $K_{eff} = 30\ \mu M$ value is that it should be somewhere near to the ACh concentration that holds the ion channels open for about half the time that they are capable of being opened for (i.e. produce 50% of the maximum conductance response). A value of about $30\ \mu M$ for this quantity for ACh also has strong support from single-channel observations (SAKMANN et al. 1980). In the very simplest case (CASTILLO and KATZ 1957a) of a binding step (equilibrium constant K_A) followed by a conformation change (equilibrium constant α/β as described, the overall binding constant should also be K_{eff}, and in this case $K_{eff} = K_A/(1+\beta/\alpha)$ (see, for example, COLQUHOUN 1979). Since β is likely to be considerably greater than α (see earlier in this section and Sect. C.III), the microscopic equilibrium constant for ACh binding to the resting receptor is likely to be substantially greater than $30\ \mu M$. Qualitatively similar statements follow if the more realistic mechanisms of Eq. (5) or (6) are considered, rather than the Castillo–Katz mechanism.

On the other hand, even the "low affinity" binding detected in fast binding studies is usually much higher affinity than these values, being about $1\ \mu M$ for ACh (HEIDMANN and CHANGEAUX 1979a; NEUBIG and COHEN 1980; BOYD and COHEN 1980a, b). It seems likely that even the "low affinity" binding constant represents binding to some compound, partly desensitized state rather than to the resting state of the receptor (see discussion of desensitization; Sect. C.IV).

III. The End-Plate Current

The discussion of ACh action in the last section must now be related to physiological events. Further details are given by GAGE 1976; ROSENBERRY 1979; WATHEY et al. 1979; ADAMS 1980; LAND 1980, 1981.

Stimulation of the motor nerve leads to release of a few hundred quantal packets of ACh, each containing about 10,000 ACh molecules (KUFFLER and YOSHIKAMI 1975; FLETCHER and FORRESTER 1975). The postsynaptic membrane has about 20–25,000 α-bungarotoxin sites/μm^2 in the areas below release sites (MATHEWS-BELLINGER and SALPETER 1978; LAND et al. 1981) and probably about 10–12,000 ion channels/μm^2. Therefore a number of binding sites equal to the number of ACh molecules released is contained within an area of about 0.4 μm^2 (radius about 0.36 μm) of postsynaptic membrane. At first the ACh concentration will be huge (10^4 molecules, in 0.4 μm^2 of cleft would give over 1 m*M*), but it is expected to fall very rapidly, as a result of binding to receptors and of hydrolysis by acetylcholinesterase. The free ACh concentration probably falls to a very low value after 200–400 μs, and thereafter hydrolysis is fast enough to remove most ACh molecules as they dissociate (MAGLEBY and STEVENS 1972; ANDERSON and STEVENS 1973; and references already cited).

Initially, binding will occur very rapidly because of the high agonist concentration. For example if the concentration were $x_A = 2\,\mathrm{m}M$ and $k_{+1} = 5 \times 10^7\,M^{-1}\,s^{-1}$ in Eq. (4) then the mean waiting time before the first occupancy would be $1/(2k_{+1}x_A) = 5\,\mu s$ (see COLQUHOUN and HAWKES 1982, 1983). Diffusion over a 0.4 μm^2 area would be slower than this, taking perhaps 40 μs, but outside this area the receptor occupancy would fall rapidly, especially because the many vacant receptor sites will drastically slow down diffusion in the synaptic cleft, thus tending to restrict the ACh to a small area.

At the peak response to a quantum of ACh something like 1,500–2,000 ion channels are open, compared with a total number of channels of 4,000–5,000 in an area of 0.4 μm^2. Therefore the probability that in Eq. (4) a doubly occupied channel opens rather than losing an agonist molecule, viz. $\beta_2/(\beta_2 + 2k_{-2})$ must be substantial. Furthermore, once both binding sites are doubly occupied there will be, on average, a delay of $1/(\beta_2 + 2k_{-2})$, the mean lifetime of A_2R, before the channel opens (or loses one of its agonist molecules). The time to peak of the response is observed to be only 100–300 μs so this delay cannot be very large. The values suggested by *nachschlag* analysis (see Sect. C. II and COLQUHOUN and SAKMANN 1983) would suggest that this delay might be of the order of 20 μs. This is quite plausible, and is gratifyingly close to the value suggested by LAND et al. (1981); these authors conclude that both diffusion and channel opening control the time course of the onset of the response.

Clearly, if a third or half the channels are open at the peak response in the small area over which the ACh acts then the degree of receptor saturation over this area must be quite substantial at the time of the peak response. Various attempts (see references already cited) to simulate these events quantitatively agree roughly, but such calculations cannot be very precise because of the complex geometry of the synaptic cleft, and the considerable uncertainty as to the exact binding mechanism and rates which have a profound effect on the rate of ACh diffusion in the cleft. However, all agree that the effect of each quantum of ACh works over such a small area that these areas are unlikely to overlap when the motor nerve is stimulated, so each quantum works independently as long as acetylcholinesterase is normally active (HARTZELL et al. 1975); if the esterase is inhibited this is no longer true.

The classical work of ANDERSON and STEVENS (1973) showed that, at low temperature, the response to a quantum of ACh decays at a rate controlled by the lifetime of the open ion channels, rather than by the time course of ACh concentration in the synaptic cleft; the ACh concentration has already fallen to a very low level by, or shortly after, the peak response. (This general conclusion would not be altered even if the effective ACh-induced channel "opening" actually consisted of several openings in quick succession – see Sect. C.II.) At 8 °C the effective mean open lifetime was several milliseconds. At about 20 °C (and −80 mV) the effective mean lifetime is 1–1.5 ms (see reviews by COLQUHOUN 1979; WRAY 1980; and, for human muscle, CULL-CANDY et al. 1979). However, many reports (reviewed by WRAY 1980) have suggested that, at this higher temperature, the decay phase of the MEPC is somewhat (perhaps 40%) slower than this. It seems very likely that the time course of the ACh concentration in the synaptic cleft makes some contribution to the time course of MEPC decay at 20 °C. At 37 °C, when the effective mean channel open time is only around 0.3 ms in mouse and cat (DREYER et al. 1976; WRAY 1980; HEAD 1983) the time course of the ACh concentration is likely to be even more important.

The ion channel, while open, has a conductance of around 40 pS, i. e. it passes a current of about 3 pA at a membrane potential of −80 mV. In real life, of course, the membrane potential is not held constant by a voltage clamp (though electrophysiologists may sometimes appear to forget this inconvenient fact). At 37 °C in cat muscle the opening of a single ion channel causes a depolarization of about 0.1 μV (KATZ and MILEDI 1972; WRAY 1980) when the membrane potential is *not* clamped. At the peak about 300,000 channels are open so a large depolarization, the physiological end-plate potential, is produced as the membrane potential is pushed towards the equilibrium potential for the ACh-operated ion channels, i. e. about 0 mV. Since a depolarization of about 30 mV (i. e. to about −50 mV) is sufficient to produce a muscle action potential, and hence contraction, there is quite a large margin of safety in the transmission process (see Sect. D.VIII).

In summary it can be seen that the response to nerve stimulation is rather complex. The agonist concentration varies rapidly with both time and position and the system is far from equilibrium. The depolarization produced by the transmitter will tend to shorten the channel lifetime and hence hasten dissociation of transmitter; the binding characteristics of the receptors will therefore change rapidly with time and hence their important effect on the diffusion rate of transmitter in the cleft will also be changing. Furthermore the time course of the end-plate potential will be mainly determined not by the time course of ion channel opening, but by the passive membrane properties of the muscle fibre membrane; the decay rate of the end-plate potential will be close to the membrane time constant (around 2 ms in cat muscle at 37 °C; see WRAY 1980).

IV. Desensitization

It is doubtful whether desensitization is of any importance in the normal process of neuromuscular transmission (but see MAGLEBY and PALLOTTA 1981). However, it has been much discussed in the context of depolarizing blocking agents (see Sect. F) and some nondepolarizing agents have been said to enhance desensitization (see Sect. E.IV). Therefore the current state of knowledge will be discussed briefly.

The time course of desensitization is best studied in voltage-clamped preparations. If the membrane is allowed to depolarize the depolarization will cause a passive flux of chloride ion, and the time course will be distorted (JENKINSON and TERRAR 1973; see also Sect. F.III). The activation of an electrogenic sodium pump may, in some species, also distort the time course of membrane potential change (CREESE et al. 1976; see Sect. F.III).

It has been suspected for a long time that the time course of onset and recovery from desensitization needs at least two exponential components to describe it. The reason for this is that recovery from desensitization produced by relatively brief iontophoretic pulses of ACh was quite fast, the time constant being a few seconds (KATZ and THESLEFF 1957) whereas the time constant of recovery from desensitization after more prolonged exposure to ACh was several minutes (e.g. RANG and RITTER 1970; ADAMS 1975b). Two recent studies have shown this "biphasic" phenomenon clearly. FELTZ and TRAUTMANN (1982) measured ACh- and CCh-induced currents in voltage-clamped frog end-plates. They found that the onset of, and recovery from, desensitization were fitted by two exponential components. The onset rate constants depended on agonist concentration; the faster was around 1 s (probably overestimated) and the slower some tens of seconds. The faster time constant for recovery was about 10 s and the slower one 4–5 min, the former (and probably the latter) being the same for ACh and CCh. In the study by SAKMANN et al. (1980) the currents through single ion channels were measured with high concentrations of agonist in the pipette (so the behaviour that was observed was relevant to onset rather than offset of desensitization). The observed openings in rapid succession, as expected because of the high agonist concentration. But the rapid openings were separated by silent periods during which the channel remained shut (i.e. presumably desensitized). For example with 20 μM ACh the channel was open and shut for similar lengths of time, but after around 0.5 s of this behaviour, it would remain shut for a longer period averaging about 200 ms (in a short-lived desensitized state?). Furthermore after such bursts of openings had been going for 5 s or so, there would be a much longer silent period (on average about 0.5 min) until the next cluster of bursts appeared. It is not yet clear exactly how these short- and long-lived desensitized states are related to the other observations discussed here.

The mechanism originally postulated by KATZ and THESLEFF (1957) is

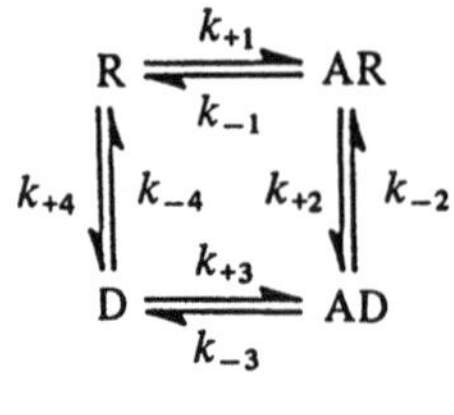

(7)

in which A represents agonist as before, and D represents a "desensitized" receptor channel. In the absence of any clear physical idea about the nature of the desensitized state D, this is more like an empirical scheme than a proper mechanism, but it is the best that can be done. In any case it is obviously very simplified because only one agonist molecule is bound, and because the channel opening step is not included explicitly. There is in fact clear evidence that many of the desensitized receptors have two agonist molecules bound (SINE and TAYLOR 1979; FELTZ and TRAUTMANN 1982). Clearly, to obtain a good representation of the data all of the states in the agonist mechanism of Eqs. (4) or (6) need to be connected to the analogous desensitized state, and perhaps to two different desensitized states – which would result in 15 different states in the case of Eq. (6). Not surprisingly, no one has yet attempted to fit such a scheme to data, though subsets of it have been considered (e.g. NEUBIG and COHEN 1980; CASH et al. 1981). For example, if we include two agonist binding sites, and activation, the obvious extension of Eq. (4) with a single sort of desensitized state would be

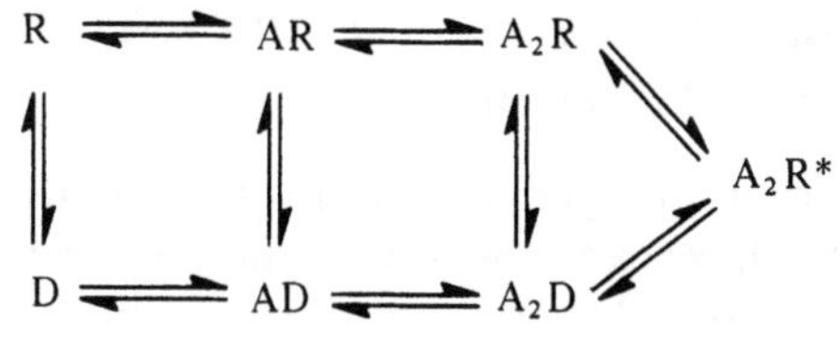

(8)

It is far from clear to what extent desensitization can proceed *without* channel opening. The results of RANG and RITTER (1970) with antagonists suggest that this route is not the primary one though WEILAND and TAYLOR (1979) suggest that opening is not necessary.

Much earlier physiological work (e.g. RANG and RITTER 1969, 1970) has supported a version of Eq. (7) in which both binding stages are assumed to be fast, so there are effectively only two states, and offset of desensitization should follow a simple exponential time course (the slow time constant mentioned previously). In these studies it has always been assumed that the R states were all activatable even though activation is not included explicitly in the mechanism. Several binding and ion flux studies also seem to be compatible with Eq. (7) though in most cases it is likely that the transition represented $R \rightleftharpoons AR$ is *not* directly related to activation, a fact which considerably complicates the comparison between physiological and binding studies.

A major contribution of binding studies has been to show, at first qualitatively (WEBER et al. 1975; COLQUHOUN and RANG 1976), and later quantitatively (SUGIYAMA et al. 1976; HEIDMANN and CHANGEUX 1979a; SINE and TAYLOR 1979; NEUBIG and COHEN 1980), that the time course of the slow increase in affinity for agonists is similar to the time course of the loss of the physiological response.

Direct binding studies with labelled ACh (BOYD and COHEN 1980ab, b), and "binding" (actually fluorescence change – see Sect. B.II) of the fluorescent ACh analogue DNS-C_6-Cho (HEIDMANN and CHANGEAUX 1979a) (both studies are on *Torpedo* membranes) could be fitted by Eq. (7) and the results were mainly in good agreement. The binding to the desensitized form was high affinity (K_3=

$k_{-3}/k_{+3} = 1\text{–}3\ \text{n}M$), and about 98% of the occupied forms were desensitized with ACh ($k_{-2}/k_{+2} \sim 70$). The "recovery time constant", $1/(k_{+4}+k_{-4})$, was about 5 min in the Boyd and Cohen work, or about 20 s (HEIDMANN and CHANGEUX 1979a). Both studies concluded that about 20% of the channels were already in the desensitized form in the absence of agonist ($k_{+4}/k_{-4} = 0.2\text{–}0.25$). In studies with cultured BC3H-1 cells, SINE and TAYLOR (1979) concluded that only about 2% of channels were desensitized with no agonist present; there is really no evidence about this quantity in intact muscle cells. Both studies found that k_{-1}/k_{+1} was about 1 μM for ACh, and HEIDMANN and CHANGEUX found that $1/k_{-1}$ was about 3.4 s; clearly, the affinity is too high, and the dissociation too slow for the reaction denoted R⇌AR to be the primary binding or activation step (see Sect. C.II.3). The actual physical events underlying this formal step are in doubt; it is possible that they could involve not only the initial binding and activation, but also binding to the putative "fast" desensitized state discussed previously.

D. Nondepolarizing Neuromuscular Blocking Agents: Tubocurarine and Similar Drugs

Having discussed the criteria for judging mode of action (Sect. B), and the mode of action of ACh (Sect. C), we can now discuss the evidence concerning the mode of action of nondepolarizing blockers. By far the most investigated drug is (+)-tubocurarine (Tc) (which is still widely, but improperly, referred to as d-tubocurarine, or even D-tubocurarine).

I. Numbers of Binding Sites

Two important facts have recently been established by binding measurements (see Sect. B.II). There has been much discussion about the relative number of binding sites for ACh, (+)-tubocurarine (Tc), and for α-bungarotoxin or other similar α-toxins. NEUBIG and COHEN (1979) performed direct binding measurements with ACh ^{3}H, Tc ^{3}H and α-toxin ^{3}H (from *Naja nigricollis*) on *Torpedo* membranes. They found equal numbers of binding sites with all three agents, and each agent inhibited the *binding* of the others, thus providing direct experimental evidence for what pharmacologists had always supposed. The second important piece of evidence comes from binding and flux studies on cultured BC3H-1 cells. SINE and TAYLOR (1980, 1981) provided evidence that occupation of *either* (or both) of the two ACh binding sites by α-toxin or tubocurarine-like drugs was sufficient to prevent any substantial response of the cells. The two sites could react equally well with α-toxin (though not with other blockers). A more surprising finding, in both of these pieces of work, was that the binding of nondepolarizing agents showed considerable heterogeneity. This finding appears to contradict some other biochemical and physiological results so it will be discussed in some detail in Sect. D.V.

II. Depolarization by Tubocurarine

Before discussing nondepolarizing actions it should perhaps be pointed out that there are circumstances in which tubocurarine *can* produce a depolarization. McINTYRE and KING (1943) reported briefly that Tc can cause contracture of 10-day denervated dog gastrocnemius muscle. ZISKIND and DENNIS (1978) found that Tc (1–10 μ*M*) could depolarize embryonic rat intercostal muscle (but not denervated adult rat intercostal or diaphragm). A similar effect is seen in cultured rat myotubes, in which currents through single channels opened by Tc have been observed (TRAUTMANN 1982); these channels had a similar conductance to those opened by ACh, but their mean open lifetime was only about 15% of that of the ACh-induced channels. On the other hand the cultured cell line BC3H-1 shows no detectable permeability increase with Tc (SINE and TAYLOR 1979) and there is no evidence for any such agonist effect on any adult muscle end-plate. In fact a small (40 μV) *hyperpolarization* is produced by Tc in frog, when acetylcholinesterase is inhibited (KATZ and MILEDI 1977); this probably results from inhibition of the effect of nonquantal ACh release from the nerve terminal.

III. Inhibition of Equilibrium Responses

1. Tubocurarine

The first measurements by good methods (see Sect. B.IV) were probably those by VAN MAANEN (1950). He measured contractures of the frog rectus abdominis muscle in response to ACh, and used the Schild method to infer an equilibrium constant of 0.8 μ*M* for binding of Tc to the ACh receptor. This value was much the same (i.e., the Schild equation remained valid) up to Tc concentrations of over 100 μ*M*; however, it is likely that only a small proportion of channels need be opened to produce a maximum contracture, so the dose ratios were probably measured at a small response level which limits the test of validity of the mechanism (see Sect. B.I). JENKINSON (1960) measured depolarization of single muscle fibres and again found the Schild equation to be valid up about 100 μ*M* Tc, and he found similar equilibrium constants of 0.4–0.7 μ*M*. Again, however, all the dose ratios were measured at a response level at which only a very small proportion of ion channels would be open. Many experiments along similar lines (reviewed by COLQUHOUN 1975; GINSBORG and JENKINSON 1976) give similar results, viz. equilibrium constants for Tc of 0.3–0.7 μ*M* in frog, and often slightly lower values, around 0.1–0.4 μ*M*, in mammalian muscle. A similar value was found in denervated rat muscle by FREEMAN and TURNER (1972). All of these obeyed the Schild equation (linear Schild plot with unit slope) within experimental error over the range tested, but all suffer from the same limited range of test conditions (see Sect. B.I).

Dose ratios were measured in voltage-clamped frog muscle fibres by COLQUHOUN et al. (1979), over an even more limited range of agonist and Tc concentrations. The results are shown in Fig. 2. At a membrane potential of -70 mV the Schild plot has a unit slope, but at -120 mV the Schild plot becomes steeper. However, the plots seem to converge at low antagonist concentrations. This was

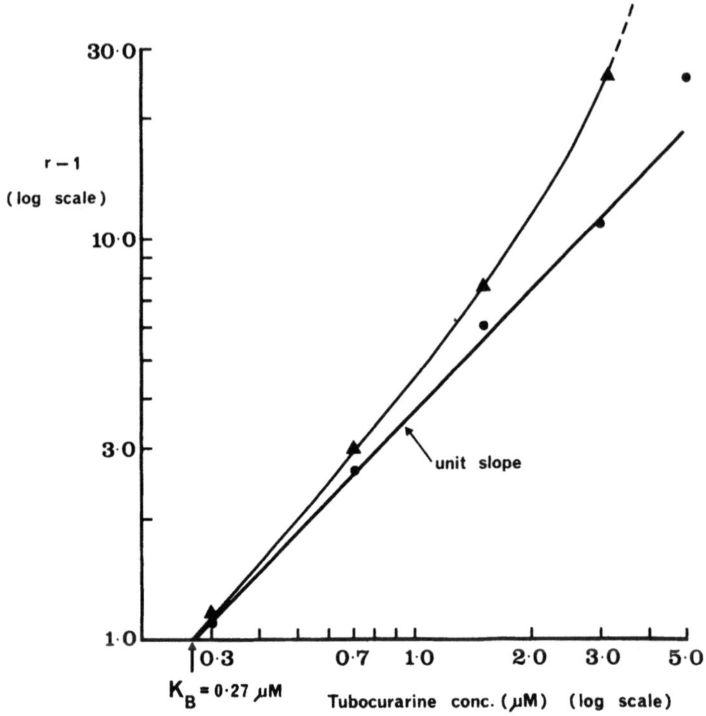

Fig. 2. Schild plot of dose ratios *r* against tubocurarine concentration. Response was the inward current elicited by carbachol in voltage-clamped frog end-plate. *Circles* dose ratios from equilibrium current at a membrane potential of −70 mV. *Triangles* dose ratios from equilibrium current at −120 mV. At both membrane potentials the equilibrium constant for tubucurarine, estimated as the concentration required to produce $r=2$, is 0.27 µ*M*. The slope is close to unity at −70 mV, but becomes steeper when the membrane is hyperpolarized. (COLQUHOUN et al. 1979)

interpreted to mean that the inibition by Tc was mainly competitive when the concentrations of agonist and antagonist were low and the membrane was not hyperpolarized. The equilibrium constant for competitive binding was 0.34 µ*M*, much as in earlier studies, but use of the voltage clamp showed, in addition, that this value was not dependent on the membrane potential. This is just what would be expected if the binding sites for agonist, and hence for competitive antagonist, are well away from the membrane electric field as suggested in Sect. C.I. It would also be inferred from this that the microscopic rate constants for agonist binding should not be dependent on membrane potential (see also Sect. C.II).

The upward curve in the Schild plot at −120 mV in Fig. 2 can most plausibly be attributed to an additional voltage-dependent channel-blocking action of Tc. This will be discussed later (Sect. D.VII). The most direct evidence for a genuine competitive mechanism comes from binding experiments in which mutual competition for binding of ACh, Tc and α-toxins is seen (see Sect. D.IV).

2. Other Tubocurarine-Like Drugs

Pancuronium (as well as Tc) gave unit Schild slope in guinea-pig muscle (WAUD et al. 1973). On the other hand, dose ratio experiments with gallamine have regularly yielded Schild plots with slopes slightly less than unity. For example RANG and RITTER (1969) in precise experiments found a slope of 0.88 ± 0.006 (chick muscle tension; CCh or succinylcholine as agonist) with no sign of nonlinearity with dose ratios between 1 and 50 or so. FREEMAN and TURNER (1972) found Schild slopes for gallamine of 0.63–0.88 (denervated rat diaphragm tension); and COLQUHOUN and SHERIDAN (1981) found 0.72 ± 0.04 at a membrane potential of -70 mV or 0.94 ± 0.07 at -130 mV (voltage-clamped frog muscle end-plate). It seems that, of the common nondepolarizing agents, gallamine is the only one for which shallow Schild plots have been reported regularly. Three of the possible reasons for this are as follows. (a) Gallamine could block ion channels when they are shut (as well as when they are open; see Sect. E.II), as discussed by COLQUHOUN (1980). (b) The two binding sites on the channel could show a greater degree of nonequivalence for gallamine than for other agents which give slopes near unity; this is discussed in Sect. D.V. (c) Gallamine is known to produce a multiplicity of effects other than neuromuscular block; for example it has actions like tetramethylammonium on potassium conductance (and hence on transmitter release) and also potent antimuscarinic effects in heart (discussed by COLQUHOUN and SHERIDAN 1981).

IV. Binding of Tubocurarine and Similar Agents

Until about 1979 everything seemed to be relatively simple for tubocurarine at least. COLQUHOUN (1979) reviewed 15 studies of tubocurarine binding. All were indirect (retardation of α-toxin binding or inhibition of labelled ACh or decamethonium binding). All values except one on *Electrophorus, Torpedo* and rat diaphragm (membranes, homogenate or purified) lay between 0.17 and 0.55 μ*M* exactly as found by the Schild method from response measurement. One value of 0.04 μ*M* for cat denervated leg muscle (BARNARD et al. 1977) was somewhat lower than usually seen, but these experiments were performed in 10 m*M* buffer, and the affinity of Tc is known to be increased by lowering the ionic strength (JENKINSON 1960). For the few other tubocurarine-like drugs that have been tested, the equilibrium constants obtained from binding studies were also quite similar to those found from response measurements.

In these studies the extent to which validity was checked, e. g. by determining the shape of the binding curve, varied considerably. Some estimates are from a single point only; others checked that the "equilibrium constant" was indeed independent of antagonist concentration. Thus, it appeared in 1979 that the competitive action, at least for Tc at equilibrium, was well characterized, despite the occasional small anomaly.

Addition of a competitive blocking agent (denoted B) to the agonist mechanism in Eq. (4) leads to the following view of the competitive mechanism (COLQUHOUN and SHERIDAN 1982).

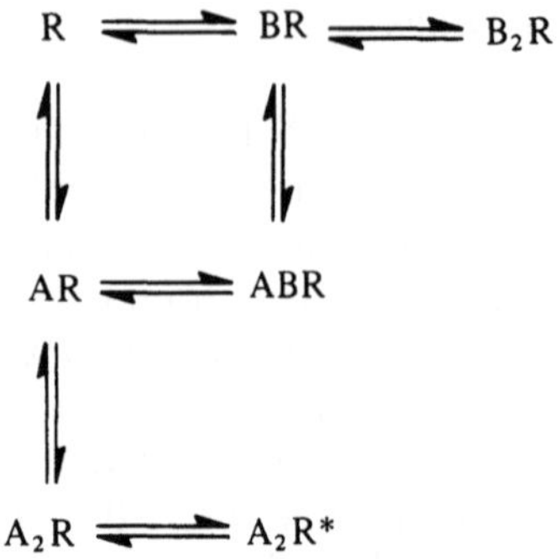

(9)

However, since 1979, some evidence has been found that, although Tc etc. do indeed compete with ACh, the affinity of these blockers is not the same for the two ACh binding sites (α-subunits) in the receptor channel complex. The evidence for and against this new complication will be discussed in the next section.

V. Evidence Concerning Nonequivalence of Binding Sites

In order to assess the evidence critically we need (a) to be able to interpret the "Hill coefficients" for antagonist binding as measured by retardation of α-toxin binding, and (b) to be able to predict the effects of nonequivalence on dose ratio experiments in which physiological responses are measured. The relevant theory for these two sorts of prediction must now be developed, in Sects. D.V.1 and D.V.2 before the experimental results are assessed in Sect. D.V.3. The predictions will be based on an extension of Eq. (9) in which the receptor channel is denoted $R_1 \cdot R_2$, rather than simply R, in order to represent the fact that it contains two nonequivalent sites, thus

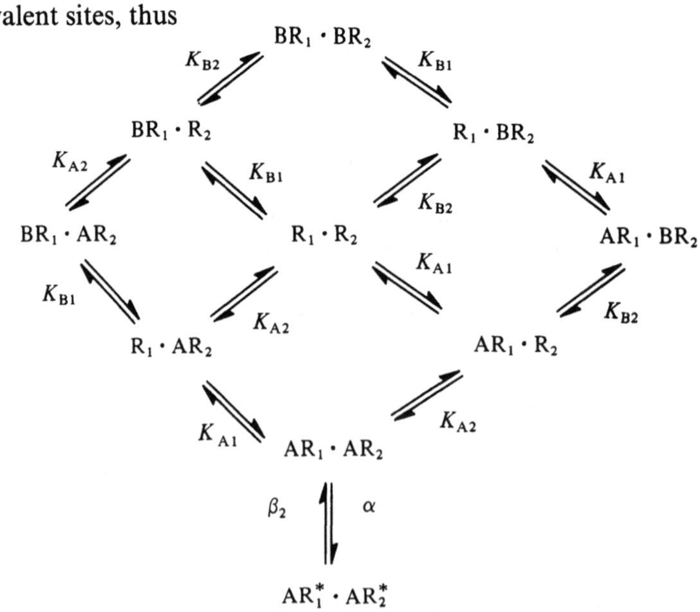

(10)

This is simplified to the extent that binding to one subunit is assumed to be independent of what is bound to the other.

1. Retardation of α-Toxin Binding: Theory

The usual assumption, that the antagonist equilibrates rapidly compared with the rate of toxin binding, implies that the time course of toxin binding will be exponential. Let us denote as k_T the association rate constant for toxin binding. When the concentration of toxin is x_T, the observed time constant for toxin binding will be $\tau_0 = (k_T x_T)^{-1}$ in the absence of antagonist, and

$$\tau = \tau_0 / [1 - p_B(\infty)]$$

in the presence of antagonist in a concentration sufficient to occupy a fraction $p_B(\infty)$ of receptors at equilibrium. In the case of simple Langmuir binding with antagonist concentration x_B, and equilibrium constant K_B, we have $p_B(\infty) = c_B / 1 + c_B)$ where the normalized antagonist concentration c_B is defined as x_B / K_B. Now define R as the ratio of rate constant for toxin binding in the absence of antagonist to that in its presence, i.e. $R \equiv \tau / \tau_0$. Thus

$$R \equiv \tau / \tau_0 = 1 / [1 - p_B(\infty)]. \tag{11}$$

In the case of homogeneous Langmuir binding we therefore have

$$R = 1 + c_B$$

so

$$R - 1 = (\tau - \tau_0) / \tau_0 = c_B \equiv x_B / K_B. \tag{12}$$

This form shows that a plot of log $(R-1)$ against log x_B should give a straight line with unit slope. This form of plot was used, for example, by COLQUHOUN and RANG (1976) and Lo et al. (1981); although it bears a formal analogy to the Schild plot (Sect. B.IV) its derivation and basis are quite different.

If, as in the studies to be discussed, this plot is found *not* to have unit slope, then in so far as the plot is linear (with slope n_R, say), the result can be described empirically by the modified equation

$$R - 1 = c_B^{n_R} \tag{13}$$

This is what would be expected if the antagonist binding were to be described by a Hill equation of the form

$$p_B(\infty) = c_B^{n_R} / [1 + c_B^{n_R}].$$

The "Hill coefficient" in this equation n_R is essentially an empirical constant with no simple physical significance (especially if $n_R < 1$).

No consider heterogeneous binding as described by Eq. (10). Suppose that toxin has the same association rate constant for both sorts of subunit (SINE and TAYLOR 1980). There are equal numbers of the two sorts of subunit in Eq. (10), so the fraction of free receptors will be

$$1 - p_B(\infty) = 0.5 \left(\frac{1}{1 + c_{B1}} + \frac{1}{1 + c_{B2}} \right), \tag{14}$$

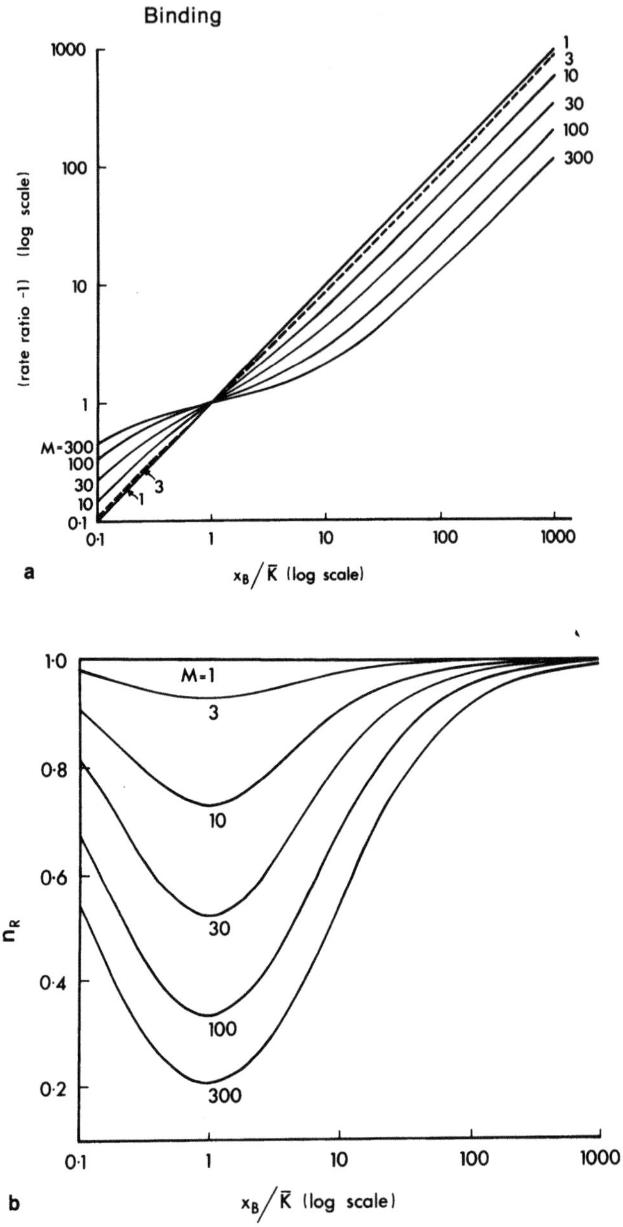

Fig. 3 a, b. Retardation of α-toxin binding with two nonequivalent binding sites, according to the mechanism of Eq. (10). The rate ratio R is defined as τ/τ_0 where τ and τ_0 are the time constants for toxin binding in the presence and absence, respectively, of the retarding drug. The concentration of the latter x_B is normalized with respect to the geometric mean $\bar{K}$ of the two equilibrium constants. The extent of heterogeneity is specified by the value of M, the ratio of the two equilibrium constants, which is marked on each curve. **a** plot of $\log(R-1)$ against $\log(x_B/\bar{K})$; calculated from Eq. (16), **b** the slope n_R of the curves shown in **a** plotted against the same abscissal value; calculated from Eq. (17)

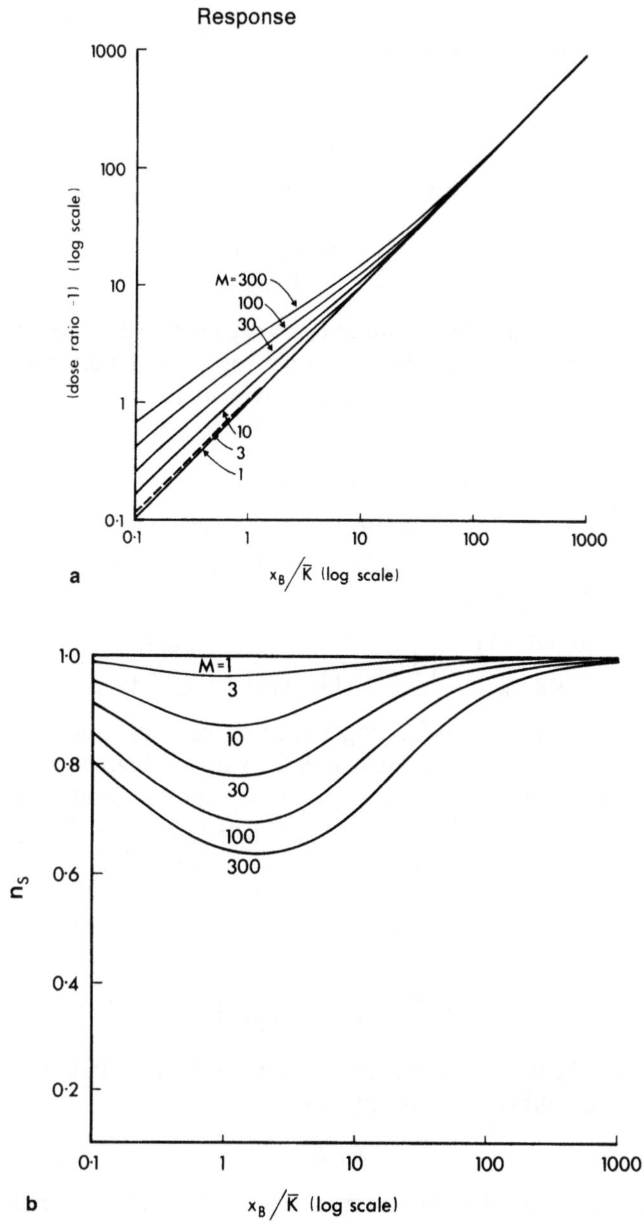

Fig. 4 a, b. Dose ratios r from response measurements in the case of two nonequivalent binding sites as specified in the mechanism of Eq. (10) **a** the Schild plot, $\log(r-1)$ against log antagonist concentration (normalized as in Fig. 3); calculated from Eq. (21), **b** the slope n_S of the curves shown in **a** plotted against the same abscissal values; calculated from Eq. (22)

where we normalize the concentration of agonist x_B with respect to the equilibrium constant for binding to each sort of subunit, i.e. we define

$$c_{B1} = x_B/K_{B1} \qquad (15)$$
$$c_{B2} = x_B/K_{B2}\,.$$

Thus, from Eq. (11), the relative rate of toxin binding is

$$R = \frac{2(1+c_{B1})(1+c_{B2})}{2+c_{B1}+c_{B2}}\,. \qquad (16)$$

In the case of homogeneous binding $c_{B1}=c_{B2}$ ($=c_B$, say) this reduces to Eq. (12), as expected. Plots of log $(R-1)$ against log x_B, calculated from Eq. (16) for various degrees of heterogeneity, are shown in Fig. 3a. The extent of the heterogeneity is defined by the relative affinities

$$M_B = K_{B1}/K_{B2} = c_{B2}/c_{B1}\,.$$

The graphs in Fig. 3a are not straight when binding is heterogeneous, i.e. their slopes are not constant, because Eq. (16) has not got the same form as Eq. (13). The slope follows from Eq. (16); it is

$$n_R \equiv \frac{\mathrm{d}\log(R-1)}{\mathrm{d}\log x_B} = \left(1+\frac{0.5(c_{B1}-c_{B2})^2}{c_{B1}(1+c_{B2})^2+c_{B2}(1+c_{B1})^2}\right)^{-1}. \qquad (17)$$

Clearly $n_R=1$ if binding is homogeneous so $c_{B1}=c_{B2}$, otherwise $n_R<1$ as expected. In Fig. 3b, n_R is plotted against R for various degrees of heterogeneity.

At low antagonist concentrations the slope tends to unity with, if $M \gg 1$, $K_{app}=K_{B1}/2$, where K_{app} is defined, from Eq. (1), as $x_B/(R-1)$. At high antagonist concentrations the slope also tends to unity with, if $M \gg 1$, $K_{app}=2k_{B2}$. In between these extremes the slope is shallower. It is easily shown that the minimum slope occurs at $R=2$ (i.e. at the point where antagonist halves the toxin binding rate) and is

$$n_{min} = \frac{2}{1+(1+M)/2\sqrt{M}}\,. \qquad (18)$$

At this point ($R=2$) the concentration of antagonist is equal to the geometric mean of the two equilibrium constants, i.e.

$$\bar{K} = (K_{B1}K_{B2})^{1/2}\,. \qquad (19)$$

Thus, if an estimate of the minimum slope is available, an estimate of the heterogeneity can be made from it, as

$$M = \left(\frac{2[1+(1-n_{min})^{1/2}]}{n_{min}} - 1\right)^2. \qquad (20)$$

2. Dose Ratio Experiments: Theory

Clearly the Schild method (Sect. B.IV) cannot work exactly for the mechanism of Eq. (10), if only because there are *two* equilibrium constants to be estimated. Furthermore Eq. (10) does not predict a parallel shift of the equilibrium log dose–response curves, so the dose ratio will depend not only on the antagonist concentration, but also on the agonist concentration, i.e. on the response level at which it is measured. However, when the fraction of open ion channels is small (relative to its maximum), as is the case in most experiments at the neuromuscular junction (see Sect. B.I), then Eq. (10) does predict an approximately parallel shift, with dose ratio given by

$$r \sim [(1+c_{B1})(1+c_{B2})]^{1/2}, \tag{21}$$

where the normalized blocker concentrations have been defined in Eq. (15). As long as the agonist concentration is sufficiently low, this approximation is valid whether or not the binding of *agonist* to the two sites is different.

The Schild plot, log $(r-1)$ against log x_B, for low responses (Eg. 21) is shown in Fig. 4a. The slope at low, and at high, antagonist concentrations tends to unity. Extrapolation of the high concentration asymptote gives

$$K_{app} = \bar{K} = (K_{B1}K_{B2})^{1/2}$$

as also found in Eq. (19). At intermediate antagonist concentrations the slope is less than unity though this effect is not so pronounced as in binding experiments (Fig. 3). In the extreme case that the affinity for one subunit is zero, the slope becomes 0.5, i.e. $r-1$ is proportional to the square root of the antagonist concentration. A heterogeneity of $M=3$ would probably be undetectable, and $M=10$ would be detectable only in a very precise experiment; but $M=100$ should be easily detectable. The Schild slope n_s corresponding to Eq. (21) is

$$n_s \equiv \frac{\mathrm{d}\log(r-1)}{\mathrm{d}\log x_B} \sim \frac{c_{B1}+c_{B2}+2c_{B1}c_{B2}}{2r(r-1)}. \tag{22}$$

This is plotted against r in Fig. 4b, for reference later. The Schild plot in Fig. 4a implies an antagonist concentration for $r=2$, K_{eff}, say, of

$$K_{eff} \sim 0.5(K_{B1}K_{B2})^{1/2}[(b^2+12)^{1/2}-b], \tag{23}$$

where we define

$$b = (1+M)/\sqrt{M}.$$

This concentration varies from $K_{B2}(=K_{B1})$ when $M=1$, to $3K_{B2}$ when $M \gg 1$. Notice that this is not the same as the analogous value for the binding experiment, given in Eq. (19).

3. Experimental Results

The main results of studies that show nonequivalence of binding sites for antagonists are summarized in Table 1. In some cases two-component binding curves have been fitted to the data, in others only Hill coefficients derived from α-toxin retardation experiments are given. The graphs in Fig. 3 show that the plots should be far from straight according to Eq. (10) so the latter sort of measurements are not easy to interpret. It seems likely that the Hill slopes given by the authors will be, if anything, steeper than the minimum Hill slope so the values for the hetero-

Table 1. Some studies in which heterogeneity of antagonist binding has been observed[a]

Drug		Reference (1)	(2)	(3)	(4)
Tubocurarine	M	233[b], 467[c], (22)[d]	(7.6)		(16.9)
	n_R	(0.23), (0.17), 0.58	0.78		0.63
Dimethyl-tubocurarine	M	20[c]	(8.5)	89 (32.1)	
	n_R	(0.60)	0.76	0.51 (0.35)	
Gallamine	M	500[c, e]	(13)	14.8 (11.7)	(14.4)
	n_R	(0.16)	0.68	0.7 (0.65)	0.66
Fazadinium (AH 8165)	M		(26)	10.3 (7.7)	
	n_R		0.55	0.78 (0.72)	
Pancuronium	M		(108)	7.6 (4.8)	
	n_R		0.32	0.86 (0.78)	

[a] Reference: (1) NEUBIG and COHEN (1979) (*Torpedo* membranes); (2) WEILAND and TAYLOR (1979) (*Torpedo* membranes); (3) SINE and TAYLOR (1981) (cultured BC3H–1 cells); (4) Lo et al. (1981) (cat denervated muscle; solubilized 9 *S* form in 50 m*M* Na phosphate buffer pH 8 with 0.2% Triton X–100). All values are obtained by measurement of retardation of labelled α-toxin binding except where noted. The table gives values of $M = K_{B1}/K_{B2}$, the ratio of high to low affinity, and n_H, the Hill slope for α-toxin retardation (see text). Values not in parentheses are those given by the authors. Values in parentheses are n_{min} calculated from the author's M by Eq. (18), or M calculate from the author's n_R by Eq. (20) (i.e. assuming the cited n_R to be the minimum Hill slope)
[b] From direct measurement of Tc ^3H binding
[c] From inhibition of Ach ^3H binding
[d] Toxin retardation results did not fit two-nonequivalent-site mechanism
[e] Poor fit to mechanism

geneity $M = K_{B1}/K_{B2}$ calculated from Eq. (20) which are given in Table 1 are likely to be underestimates. This is the case with the data of SINE and TAYLOR (1981) who give both Hill slopes and fitted values of K_{B1} and K_{B2}.

The results in Table 1 are picked because they show heterogeneity, but the results are not terribly consistent even on the same preparation. For example NEUBIG and COHEN (1979) find a very large heterogeneity for Tc ($M = 233$) by direct binding of Tc ^3H in *Torpedo* membranes, but they find that results from α-toxin retardation by Tc do not fit the two-nonequivalent-site mechanism at all. They conclude that "this points out the difficulty of using the α-neurotoxins as quantitative probes of small ligand binding". They do not suggest where the difficulty lies, however. This method is, nevertheless, that used by all the other authors in Table 2 and, at least in the case of SINE and TAYLOR (1981), it seems to give results that fit Eq. (10). Quantitatively the results for both tubocurarine and gallamine of NEUBIG and COHEN (1979) (on *Torpedo*) differ considerably from those of WEILAND and TAYLOR (1979) (also on *Torpedo*) or of Lo et al. (1981) (on cat denervated muscle). Also the results for pancuronium on *Torpedo* (WEILAND and TAYLOR 1979) differ considerably from those on cultured BC3H-1 cells (SINE and TAYLOR 1981).

Thus, although affinity ratios of up to 500 have been claimed, it is difficult to come to any firm conclusions about higher species at present. Apart from the inconsistencies in the data already mentioned, many other binding studies (see Sect. B.IV) fail to show evidence of heterogeneity; for example an α-toxin retardation study by COLQUHOUN and RANG (1976) on rat homogenized denervated diaphragm muscle showed a slope close to unity. In addition most dose ratio studies show a slope very close to unity. The most studied range of dose ratios, from 2 up to 10 or 20, is exactly that where deviations from unit slope are expected (Fig. 4) when there is heterogeneous binding. Certainly most results are incompatible with an affinity ratio of 100 or more which predicts (see Fig. 4) that the Schild plot slope should be below 0.8 for all dose ratios between 2 and 12, an effect that should be easily detectable. Small degrees of heterogeneity would be hard to detect in dose ratio measurements. In a precise study RANG and RITTER (1969) did find, for Tc in chick muscle, a slope of 0.95 ± 0.009 (significantly less than 1) for dose ratios up to 30 or so. This is compatible with $M = 10$ or so, but not with the large values found in Table 1. Furthermore the group (NEUBIG and COHEN 1980) who find the largest heterogeneity for Tc ($M = 467$ in Table 1) themselves found, when they measured dose ratios for the response of *Torpedo* vesicles, that 2 μM Tc produced a parallel shift of the dose-response curve with a dose ratio of about 5, corresponding to $K = 0.5$ μM, exactly in the range found by everybody else. If binding were heterogeneous, with $K_{B1} = 15$ nM and $K_{B2} = 7$ μM (which gave $M = 467$ in Table 1), a concentration of only about 0.4 μM should, from Eq. (21), have given this dose ratio. The authors attributed their entirely conventional results to errors resulting from (a) slow dissociation of Tc from the high affinity site, and (b) stabilization of the desensitized state by Tc (this itself is controversial, see Sect. E.IV).

In a kinetic study of the effect of blockers on miniature end-plate current amplitude, PENNEFATHER and QUASTEL (1981) found some results that were consistent with heterogenous binding of Tc. The argument, though indirect, provides the only electrophysiological evidence for heterogeneity.

In conclusion, different binding studies disagree quantitatively and qualitatively on the extent of heterogeneity of antagonist binding, though the evidence for heterogeneity in some preparations is certainly quite impressive. Most dose ratio measurements appear to be incompatible with the larger degrees of heterogeneity that have been claimed, unless the near unit slopes found in these experiments are attributed to cancellation of errors. For example a shallow slope resulting from heterogeneity (or from "block of shut channels" COLQUHOUN 1980) might be obscured by a steep slope resulting from desensitization or open channel block. This is possible but, if true, the cancellation is surprisingly precise in many different preparations. Another possibility is that binding studies give erroneous results because of binding to nonreceptor (e.g. ion channel) sites. This binding is α-toxin sensitive, but there is no real evidence about whether the "channel block site" is occluded by α-toxins, or whether it is the same as the "local anaesthetic site" (see Sect. E). Clearly, these problems need more work.

VI. Kinetics of Competitive Action

Even now very little is known of the rate at which antagonists associate with or dissociate from the ACh receptor, though a number of attempts have been made to investigate these rates of Tc (no other drugs have been investigated in any helpful way). One method is to measure the rate of recovery of the response to an agonist following a brief iontophoretic application for Tc (CASTILLO and KATZ 1957b; WAUD 1967; ARMSTRONG and LESTER 1979). However, it has been shown that rates measured in this way are likely to be limited by diffusion even when the nerve terminal is lifted away from the end-plate (WAUD 1967; ARMSTRONG and LESTER 1979; R. D. PURVES 1980, personal communication). Therefore the shortest time constant for recovery seen in such experiments (about 100 μs) provides only a lower limit (10 s^{-1}) for the value of the dissociation rate constant of Tc.

Another method that has been tried is to measure the extent to which a very rapidly equilibrating antagonist (such as hexamethonium) can relieve the block of the response to transient agonist application produced by a supposedly slower antagonist such as Tc. This method has been used by FERRY and MARSHALL (1973) and BLACKMAN et al. (1975). The latter authors interpreted their results in terms of receptor interaction rates according to the theory of GINSBORG and STEPHENSON (1974). They inferred that the dissociation rate constant for Tc war roughly 700 s^{-1} in rat diaphragm end-plates, but too fast to measure in amphibian end-plate.

Finally a voltage jump relaxation method has been used. It is expected that if an antagonist equilibrates slowly with receptors it will merely reduce the number of receptors, but will not alter the relationship between relaxation time constant and agonist concentration, α-bungarotoxin behaves like this. However, if the antagonist equilibrates very rapidly the relaxation time constant versus agonist concentration graph will be shifted by Tc according to the Schild equation (1). The latter result was found for Tc by SHERIDAN and LESTER (1977) in *Electrophorus,* and also by COLQUHOUN and SHERIDAN (1982) in frog end-plate. The results of COLQUHOUN and SHERIDAN (1982) when interpreted in terms of the mechanism of Eq. (9), seem to imply a dissociation rate constant for Tc of at least 1,000 s^{-1} ($\tau = 1$ ms). In conjunction with an equilibrium constant of 0.5 μM this would imply a very fast association rate constant of over $10^9 M^{-1} s^{-1}$ (which is barely possible; see COLQUHOUN and SHERIDAN 1982; GUTFREUND 1972). These fast rates render very implausible the hypothesis (TYERS 1978) that the rate of action of fazadinium (AH 8165) in vivo is controlled by its rate of dissociation from receptors.

It is quite likely that all of the attempts mentioned are foiled by the ubiquitous diffusion problem. Diffusion limitation of observed rates may mimic the Ginsborg–Stephenson phenomenon (GINSBORG and STEPHENSON 1974; D. R. WAUD 1963, personal communication; COLQUHOUN 1975). And the assumption of constant concentration that is made in the interpretation of voltage jump relaxations is also very dubious. For example, with 0.5 μM Tc present so that about half the receptors are occupied, there are roughly 200 bound Tc molecules present for every free, unbound, molecule in the synaptic cleft. Thus any perturbation that causes the antagonist occupancy to fall by 0.5% will *double* the free concentration of Tc. In Summary there are no really trustworthy results for the rates of competitive action of antagonists, even for Tc. If binding is heterogeneous, e.g. Eq. (10), this obviously complicates the question still further.

VII. Ion Channel Block by Nondepolarizing Antagonists

A very large number of agents have been found which appear to block open ion channels (see Sect. B. V), though in most cases this is probably not their only action (see Sect. E). However, of the classical neuromuscular blocking agents, only Tc and gallamine appear to have been investigated in any detail at the neuromuscular junction. In addition to their competitive action which is not noticeably dependent on membrane potential (COLQUHOUN et al. 1979), they seem to have an open ion channel-blocking action which is strongly voltage dependent (MANALIS 1977; KATZ and MILEDI 1978; COLQUHOUN et al. 1979; COLQUHOUN and SHERIDAN 1981; see also COLQUHOUN 1980, 1981 b).

The folly of referring indiscriminately to any nondepolarizing agent as "competitive" is nicely illustrated by the history of ganglion-blocking agents. These do not depolarize and have been widely supposed, for many years, to be competitive antagonists. But recent work by voltage-clamp methods has shown that most ganglion-blocking agents actually have very little effect on the receptor, but work almost entirely by blocking ion channels (ASCHER et al. 1979; RANG 1982; GURNEY and RANG 1984).

It has been suggested (MARTY and ASCHER 1980) that Tc might block ion channels when shut as well as when open, and that this might account for all of its actions without the need to invoke competition at all. The balance of evidence is not in favour of this hypothesis (COLQUHOUN 1980).

1. Channel Block by Tubocurarine and Gallamine

The association rate constant for block of open ion channels by Tc, at a membrane potential of -80 mV, is about $10^7 M^{-1}s^{-1}$ for Tc (COLQUHOUN et al. 1979), and a rather similar value of $3 \times 10^7 M^{-1}s^{-1}$ has been found for gallamine (COLQUHOUN and SHERIDAN 1981). The rate at which the blocker dissociates from the open channel is, however, very different for these two drugs. The dissociation rate constant (at -80 mV) is about 1,050 s^{-1} for gallamine, so a blockage lasts, on average for less than 1 ms (COLQUHOUN and SHERIDAN 1981); for Tc the dissociation rate constant is 1,000 times slower, about 1 s^{-1} so a blockage lasts, on average, for 1 s (COLQUHOUN et al. 1979). The affinity for the open ion channel is thus much greater for Tc than for gallamine, the equilibrium constant (at -80 mV) being about 0.1 μM for Tc, and 35 μM for gallamine. These values may by compared with equilibrium constants for competitive block, found in the same studies, of 0.3 μM for Tc and 2 μM for gallamine (COLQUHOUN et al. 1979; COLQUHOUN and SHERIDAN 1981). Tubocurarine has, at equilibrium, a much higher ratio of ion channel affinity to competitive affinity than has gallamine. This fact, however tells us very little about the effect of channel block in vivo. Even at equilibrium the amount of inhibition produced by Tc through its channel-blocking action will be slight at low agonist concentrations, despite the fact that (at -80 mV)

Tc has a higher affinity for the ion channel than for the ACh receptor. This is a result of the fact that the amount of equilibrium inhibition produced by open channel block increases with the agonist concentration, because channels must be opened before they can be blocked (see discussion in COLQUHOUN 1981). At high agonist concentrations, however, the channel-blocking effect would be greater than the competitive effect at equilibrium.

2. Channel Block in Vivo

In vivo we are never at equilibrium (Sect. C.III), consider, for example, Tc at a concentration of 2 μM, enough to produce a considerable competitive block. The forward channel-blocking rate would be $2\,\mu M \times 10^7\, M^{-1}\,s^{-1} = 20\,s^{-1} = 0.02\,ms^{-1}$ so only 2% of the available open ion channels would be blocked per millisecond, and even fewer during the rising phase of the response to a quantum of ACh which lasts only about 0.2 ms. Furthermore, during normal neuromuscular transmission the membrane depolarizes, so the rate of association with the ion channel is decreased and the rate of dissociation increased. At a membrane potential of 0 mV (which, incidentally, is the value relevant to many binding experiments) although the competitive action is more or less unchanged, the affinity for the open channel is much reduced (the equilibrium constant being about 1.2 μM for Tc, and about 775 μM for gallamine). In the light of these facts it is not surprising that the ion channel-blocking action of Tc seems to have little effect on neuromuscular transmission, even during a train of nerve-evoked end-plate currents. The decline in the amplitude of end-plate currents seen during a train in the presence of Tc is probably a presynaptic action of Tc, not a channel block effect (MAGLEBY et al. 1981).

3. Channel Block in Experiments

Although probably not important in vivo, channel-blocking effects are serious potential source of error in experimental work. Clearly, block by Tc is, per se, very poor evidence that a receptor ion channel is of the end-plate nicotinic type. The effects of open channel block look rather different for rapidly and slowly dissociating blockers. The rapid type, e.g. gallamine, produce a biphasic decay of nerve-evoked end-plate currents and a noise spectrum with two clear components, whereas the slow type, e.g. Tc, appear merely to speed up the end-plate current decay and noise spectrum (the slow component has too small an amplitude to be visible). In experiments in which the membrane is suddenly hyperpolarized, or the agonist concentration is suddenly increased, in the presence of a slow channel blocker like Tc (both perturbations lead to an increased extent of channel block) the current through ion channels will initially increase; but then slowly decline as channels become blocked – a slow "inverse relaxation" will be seen. This effect could easily be confused with desensitization (see Sect. C.IV). In the presence of the fast type of channel blocker, like gallamine, this is not seen because channels block rapidly. Further details are given in the references cited in this section.

VIII. Competitive Block Under Physiological Conditions

It seems (see Sect. D.VII) that the block of neuromuscular transmission produced in vivo by Tc, gallamine, and probably by a number of related drugs, is indeed primarily a result of competition with ACh for its receptor. It has been shown (PATON and WAUD 1967) that block of about 75% of the postsynaptic receptors is sufficient to overcome the normal margin of safety for transmission (in cat muscle) and to produce a just-detectable block of muscle twitch tension. To produce complete block of the muscle response about 92% of receptors had to be blocked. These values were found by the dose ratio method; suitable doses of agonist (e.g. succinylcholine or decamethonium) were given intra-arterially to determine the dose ratio r or various stages of block. From the Schild equation (1) the quantity $(r-1)/r = c_B/(1+c_B)$ was calculated as an estimate of the fraction of receptors occupied at equilibrium by the antagonist (in the absence of any agonist); these are the values already cited. If, of course, binding of antagonists were heterogeneous (see Sect. D.V) these conclusions would need some modification (thought the values given were very similar whichever agonist and antagonist were used).

One uncertainty concerns the question of whether or not the competitive antagonist is at equilibrium with ACh during block of nerve-evoked transmission. It is most unlikely that it is. Although the rate constants for association and dissociation of the competitive blocker may be quite fast (this is uncertain; see Sect. D.IV), it would be very surprising if they were fast enough for Tc and ACh to reach equilibrium during the brief rising phase of the evoked end-plate current (see Sect. C.III). In any case, even if the rate constants were fast enough, rapid equilibration would be prevented, in the case of any reasonably potent blocker, by the rapid rise in free antagonist concentration in the synaptic cleft that would follow any substantial fall in antagonist receptor occupancy provoked by ACh (this phenomenon has already been discussed at the end of Sect. D.VI). Thus receptor occupancy by a competitive antagonist is unlikely to change much during normal neuromuscular transmission despite the very high (but very transient) ACh concentration.

In order to understand the effect of receptor block under physiological conditions, the kinetic behaviour of the antagonist needs to be combined with the complex events with occur during normal transmission (Sect. C.III). Simple discussions of some aspects are given by KATZ and MILEDI (1973) and COLQUHOUN et al. (1977). But the most complete attack on the problem is in the interesting papers by PENNEFATHER and QUASTEL (1981, 1982). They observed, for example, that Tc was less effective at reducing the amplitude of MEPC than at reducing the effect of bath-applied agonists. This effect could be explained by supposing that normally most of the released ACh is taken up by receptors over a small area (see Sect. C.III), and that the MEPC amplitude is insensitive to receptor block because whenever ACh encounters a blocked receptor it can always diffuse a little further on until unblocked receptors are found. Thus, each quantum of ACh will act over a wider area in the presence of the blocker, but most ACh will still be taken up onto receptors, and will still open channels, until this area becomes so large that loss of ACh by hydrolysis during diffusion becomes important. The reduction of MEPC amplitudes by Tc was observed by Pennefather and Quastel to be even less effective when cholinesterase was inhibited, but more effective when receptor density was reduced. Some aspects of the results of PENNEFATHER and QUASTEL (1981) suggested heterogeneous binding of Tc to the subunits of the receptor (see Sect. D.V), but the authors concluded that no theory seems to be able to explain all the observations that we have. One can but agree.

E. Nondepolarizing Neuromuscular Blocking Agents: Miscellaneous Agents of Low Specificity

I. Some Drugs and Possible Mechanisms

A truly vast number of drugs has been tested for postsynaptic activity at the neuromuscular junction, though knowledge about how they work is very incomplete in most cases. Many of the results have been reviewed by ALBUQUERQUE et al. (1980). GAGE and HAMILL (1981) and by PEPER et al. (1982). The compounds tested include, among many others: local anaesthetics, e.g. procaine, and those particularly active on mucous membranes such as dibucaine (cinchocaine) and dimethisoquin; quaternary analogues of local anaesthetics, e.g. QX-222, QX-314; general anaesthetics; antimuscarinics (with local anaesthetic actions?) e.g. adiphenine, pipenzolate; drug metabolism inhibitors, e.g. proadifen (SKF-525A); barbiturates; mepacrine (quinacrine, an antimalarial and anthelmintic); amantadine (an antiviral and antiparkinsonism drug); and many others.

None of these compounds is thought to act primarily by competition with acetylcholine (though this question has only rarely been investigated quantitatively). The main mechanisms that have been postulated are: (a) selective block of open ion channels; (b) block of ion channels at a site that can be occupied whether the channel is open or shut; (c) reduction of single-channel conductance; (d) enhancement of desensitization; (e) fluidization of membrane lipid; and (f) change of membrane dielectric constant. These will now be considered. Roughly speaking, the earlier mechanisms in this list are most favoured for the most hydrophilic compounds (e.g. the quaternary lignocaine analogue, QX-222), whereas the later mechanisms are most favoured for the most hydrophobic compounds (e.g. volatile general anaesthetics). However the majority of the compounds to be discussed are weak bases, or in some cases weak acids, and they exist partially in the ionized form and partially in the nonionized form at physiological pH. These two forms may well act differently.

II. Block (Selective or Otherwise) of Ion Channels

The evidence for selective block of *open* ion channels, according to Eq. (3), is best for quaternary compounds such as QX-222 – and probably also gallamine and tubocurarine though these show competitive block too (see Sect. D.VII). Barbiturates (in their nonionized form) also appear to be fairly selective for open ion channels (ADAMS 1976). This action of QX-222 (STEINBACH 1968 a, b; RUFF 1977) appeared to be elegantly confirmed by a single ion channel recording study

(NEHER and STEINBACH 1978) which showed, for example, that although the blocker split up a channel opening into a burst of shorter openings as expected, the total charge carried per burst was unchanged in the presence of the blocker (in modest concentrations). However NEHER (1983) showed that when the concentration of QX-222 is higher (more than a few times the apparent equilibrium constant for binding to the open channel) the total charge that flows during a burst of openings is no longer constant, as predicted for simple block of open ion channels, but decreases with blocker concentration. Furthermore the distribution of the durations of shut periods within bursts (which are supposed to represent sojourns in a single blocked state) no longer show a simple exponential distribution. Evidently what had been considered the archetypal selective open channel-blocking drug is not in fact so simple. It appears that there must be some slow reaction (slow relative to the normal duration of openings) that leads from either the open or blocked states to another shut state that is not included in the simple blocking mechanism (Eq. 3). The obvious possibilities are that the blocked channel might shut with the blocker still bound (this is discussed next), or that a desensitized state may be entered (see Sect. E.IV) Many other agents that show signs of channel block also behave in a way that is not compatible with the simple scheme of Eq. (3): for example, procaine (ADAMS 1977), atropine (FELTZ et al. 1977) and benzocaine (OGDEN et al. 1981, 1986). The most obvious sign of this complex behaviour is that, in each case, the amount of inhibition of the equilibrium current induced by low agonist concentrations is considerably greater than that which is predicted from Eq. (3) (by means of the blocking and unblocking rate constants that are inferred from kinetic experiments). In addition, in some cases, where it has been checked (e.g. benzocaine; OGDEN et al. 1981, 1986), the relative amplitudes of the components of the noise spectrum are not as predicted from Eq. (3). Decamethonium (see Sect. F) can also block the ion channels that if opens, but voltage jump relaxations deviate from the predictions of Eq. (3) at higher concentrations (ADAMS and SAKMANN 1978; ADAMS and COLQUHOUN 1984).

The nature of the site or sites that can give rise to blockage even when the channel is shut are obscure at present. There are various possibilities. For example: (a) there may be one or more sites near the mouth of the channel that are different from the putative intrachannel blocking site, and that are accessible even when the channel is shut; (b) hydrophobic drugs may be able to reach an intrachannel site whether the channel is shut or not; and (c) it may be possible for the channel to shut with a blocking molecule trapped inside it. One of the most thorough investigations (block of acetylcholine-activated channels in submandibular ganglion cells) suggests that small quaternary molecules may be trapped, as in (c) (GURNEY and RANG 1984). The dissociation rate of trapped molecules is very slow, which may give rise to the artifactual appearance of a block as in (a) (RANG 1982).

III. Reduction of Single-Channel Conductance

In principle it is possible that blocking agents might reduce single-channel conductance. RUFF (1977) found, by noise analysis, that QX-222 reduced the apparent single-channel conductance in frog muscle. He sought to explain this observation by postulating that the drug produced only a partial blockage of the channel (if this were the case then QX-222 should increase rather than decrease the response to low agonist concentrations because the current passed through the partially blocked channel would be in addition to the unchanged charge per burst carried through open channels: this was not tested). In this case noise analysis would give a weighted average of the conductances of open and partially blocked channels and so would appear to be reduced. This view, on the evidence presented, appeared not to be adequate quantitatively (COLQUHOUN 1978), though new data has since been presented in its defence (RUFF 1982). Single-channel recording (albeit on a different tissue, cultured rat muscle) has shown that blockages by QX-222 are virtually complete and that the single-channel conductance while open is unchanged (NEHER 1983). Similarly it has been observed by OGDEN et al. (1986) that the single-channel conductance measured by noise analysis appeared to be reduced benzocaine, whereas single-channel recording on the same preparation (frog end-plate) showed little reduction. A similar phenomenon may have occurred in the case of two aromatic agonists (phenylpropyltrimethylammonium and its *meta*-hydroxy derivative) which appeared to have a substantially lower single-channel conductance than acetylcholine on frog end-plates (COLQUHOUN et al. 1975). These agonists showed two-component noise spectra that resembled qualitatively those seen with channel blockers. Reexamination of these, and other agonists, by single-channel recording both on outside-out membrane patches of cultured rat muscle and in frog end-plates (GARDNER et al. 1984) shows that all agonists have closely similar single-channel conductances. The reasons for these discrepancies between noise and single-channel measurements are not certain at present: in the studies mentioned the spectral density was extrapolated, on the basis of the fitted lorentzian components, to infinite frequency, so if variance was lost it must have been in extra spectral components that were entirely undetected.

Not all putative channel blockers appear to cause reduced single-channel conductance (at least in the concentration ranges tested): for example little effect was seen in noise experiments with atropine (FELTZ et al. 1977), tubocurarine (COLQUHOUN et al. 1979) or gallamine (COLQUHOUN and SHERIDAN 1981).

IV. Enhancement of Desensitization

RANG and RITTER (1969, 1970) found in chick muscle that certain agonists (e.g. decyltrimethylammonium and, in frog muscle, phenyltrimethylammonium) and antagonists (e.g. diphenyldecamethonium) behaved as though they had an unusually high affinity for a desensitized conformation of the receptor. They called these substances metaphilic drugs. Metaphilic agonists appeared to cause considerable desensitization without necessarily producing much activation. Metaphilic antagonists appeared to be more potent when applied with, or shortly after, an agonist. At the time this work was done, channel block, although it had been proposed by BLACKMAN (1959, 1970), was not yet a well-documented mechanism of action. The structure of metaphilic drugs is such as to suggest that they might well be channel blockers though it is unlikely that this can account for all of their actions; for example, the potency of antagonists was enhanced even when they were applied shortly *after* the agonist, at a time when the response to the agonist had disappeared and ion channels had presumably shut.

More recently it has been suggested that a number of drugs may "enhance desensitization", though this may not be their only effect. One of the problems with assessment of "enhancement of desensitization" as a mode of action is that there is still not really any firm knowledge of what desensitization is in the first place (see Sect. C.IV). This makes its enhancement difficult to define rigorously. Desensitization must, at present, be defined empirically in terms of observed phenomena. It involves, for example, the following features. (a) There is a relatively slow transition to one or more inactive (shut), and inactivatable, states in the prolonged presence of agonist; this is the only real definition. (b) The rate and eventual extent of this effect are increased as the agonist concentration is increased. (c) This transition occurs roughly in parallel with the development of a state or states with high affinity for agonist. But all of these phenomena are predicted to occur in the presence of a drug that blocks, and dissociates slowly from, open ion channels (this drug could be the agonist itself, or an added antagonist). The fractional block produced is predicted to be larger at high agonist concentrations, and the blocked channel in Eq. (3) is supposed to have an infinite affinity for the agonist. Qualitatively, therefore, the behaviour expected of an open-channel-blocking drug may bear many resemblances to that expected for a drug that, in some way, facilitated the entry of the system into the desensitized state or states. In order to distinguish these mechanisms careful quantitative arguments (and, preferably, a better understanding of desensitization) are needed; progress on both of these fronts has been slow.

It has been reported that SKF-525A (proadifen) and some structurally related compounds appeared to enhance desensitization in frog muscle, as judged by the decline in the depolarization response to iontophoretically applied agonist (see MAGAZANIK and VYSKOCIL 1973). The depolarization response to prolonged agonist application is, however, accompanied by a net entry of chloride ions (JENKINSON and TERRAR 1973), and this complicates the interpretation of such experiments. TERRAR (1974) therefore looked at frog muscle in solutions that contained the relatively impermeant anions isethionate or methylsulphate rather than chloride; he found that the rate of decline of the depolarization produced by low carbachol concentrations (10 or 20 μM) was greatly increased by adiphenine (which resembles proadifen in structure), and pipenzolate. Terrar points out that "although the rapid decline in carbachol depolarization which occurs in the presence of pipenzolate has been described as ‚desensitization' it is uncertain whether the mechanism is similar to that which operates in normal solution", and he goes on to discuss other possibilities, e.g. channel block; some more recent studies have not been so carefully interpreted.

It has been found at the frog end-plate that the fractional inhibition of the response to agonist produced by fixed concentration of benzocaine increases strongly with the agonist concentration that is used (OGDEN et al. 1986). The extrapolated inhibition at zero agonist concentration is small. This sort of behaviour is expected for the open channel block mechanism (Eq. 3). However independent measurement of the open channel-blocking action by noise or single-channel methods (OGDEN et al. 1981), showed that although such block does occur, it is far from being sufficient to explain quantitatively the strong agonist concentration dependence of the equilibrium block. This result, and the appearance of single-channel records in the presence of benzocaine, favour the view that benzocaine may increase the fraction of receptor channels that are in the desensitized state (or states) at equilibrium.

A number of biochemical studies have led to similar conclusions, although independent measurement of channel-blocking effects has not been possible by this approach. In most of these studies the slow increase in agonist binding affinity that is produced by exposure to agonist has been *defined* as desensitization; in the light of the previous discussion such a definition clearly cannot be regarded as infallible. It is now clear that a number of local anaesthetics and other drugs can accelerate the development of the state with high affinity for agonist, and that conversely agonists can increase the binding of local anaesthetics (to a site other than the receptor) (COHEN et al. 1974; WEBER and CHANGEUX 1974; WEILAND et al. 1977; HEIDMANN and CHANGEUX 1979 b; NEUBIG and COHEN 1980). Metaphilic antagonists (but not tubocurarine or gallamine) produce enhanced conversion to the high agonist affinity form in *Torpedo* membrane fragments (WEILAND and TAYLOR 1979). An elegant recent study is that of SINE and TAYLOR (1982) who used the clonal BC3H-1 cell line. They estimated that in these cells carbachol had a 300-fold greater affinity for desensitized receptors than for normal receptors. The results could be fitted by assuming that this ratio was little changed in the presence of local anaesthetics, but that the equilibrium constant between normal and desensitized receptors (0.01% desensitized in the absence of agonist) was greatly increased by dibucaine. QX-314 and histrionicotoxin. Further work on the mechanism of desensitization is obviously needed to clarify these observations.

V. Other Mechanisms: Correlations with Lipophilicity

There is quite a substantial body of knowledge concerning the effects of various drugs on cell membranes and on artifical lipid bilayers. For example, there are changes in lipid fluidity, membrane thickness and dielectric constant (see GAGE and HAMILL 1981 for further details). However, despite endless speculation, there is hardly any firm knowledge about the extent to which these undoubted effects are causally related to the observed effects of the drugs on cells. Correlations of potency with lipopholicity (e. g. oil/water partition coefficient) have been repeatedly noted since the turn of the century, but the question of causality remains unanswered: is it, for example, partition into membrane lipids, or partition into the hydrophobic areas of protein molecules that causes the pharmacological effect?

Many general anaesthetics cause an increase in the initial rate of decay of miniature end-plate currents, and so would be expected to reduce the amplitude of the physiological end-plate potential. The potency of inhalation and barbiturate anaesthetics in producing this effect is correlated with the lipophilicity, as is their potency in producing anaesthesia. Therefore anaesthetic potency is correlated with the ability to increase decay rate (see GAGE and HAMILL 1981).

The mechanisms are far from clear. All the mechanisms that have been discussed here have been invoked at one time or another. The reduction in mean channel lifetime has usually been ascribed to the production by the blocking agent of new discrete states (such as the blocked state) to which the channel can shut; the rate constant for shutting via this new pathway is added to the normal shutting rate constant α which is usually assumed to be unchanged in the presence of the blocking drug. It has been suggested that some blockers might act by increasing the normal channel shutting rate constant $\alpha\iota$ $\{\varepsilon\in\gamma\in$ GAGE et al. 1978). If one, or a few, molecules of the blocker interact with the channel without blocking it, but producing a species that shuts more rapidly than normal, there would be two types of open state present (those with the normal α and those with the increased α) so decay would be biexponential; however, if these states were in rapid equilibrium a simple exponential decay with increased rate would be seen. A similar effect would be observed if the drug were to exert a more or less continuously graded effect on each channel by dissolving in the lipid that surrounds the channel. These various possibilities are not easy to distinguish experimentally; there is still no clear consensus concerning the mechanisms of action of lipophilic blockers.

F. Depolarizing Blocking Agents

I. Some Possible Modes of Action

The clinical literature on the mode of action of depolarizing neuromuscular blocking agents, such as succinylcholine (suxamethonium) and decamethonium, is very confusing. Much of this literature has been reviewed by ZAIMIS (1976), ZAIMIS and HEAD (1976) and GINSBORG and JENKINSON (1976). It is clear that several different, but interacting, postsynaptic actions may contribute to block. Presynaptic effects have been reported, but are unlikely to be important for the blocking action. The reported postsynaptic actions include the following six.

1. The prolonged depolarization produced by activation of end-plate receptors will affect the adjacent electrically excitable muscle membrane, causing opening of potassium channels and inactivation of the membrane potential-operated sodium channels the opening of which normally underlies the rising phase of the muscle fibre action potential. Therefore the electrical threshold of the fibre will rise.
2. The prolonged action of depolarizing agents will produce desensitization of the acetylcholine receptors at the end-plate (see Sect. C.IV, E.IV).
3. The prolonged depolarization will cause changes in the ion concentrations in the fibre that result in slow extension of the depolarized and inexcitable area along the muscle fibre, in each direction away from the end-plate. An increase in internal sodium concentration may, in some species, activate an electrogenic sodium pump that causes slow repolarization.

4. The depolarizing agents can block the ion channels that they themselves have opened.
5. The increased conductance of the end-plate caused by the depolarizing agent will reduce the depolarization produced by acetylcholine release from the motor nerve ending.
6. Receptors will be occupied by the depolarizing agents, thus preventing the access of released acetylcholine in the synaptic cleft (this diffusion is strongly dependent on the presence of binding sites in the cleft) if the depolarizing agents were less efficacious than acetylcholine itself (which has *not* been clearly demonstrated for decamethonium or succinylcholine) then receptor occupancy by the former could inhibit the effect of the acetylcholine. These effects will now be considered in rather more detail.

II. Sodium Channel Inactivation

It is very likely that this is the principal mode of action of succinylcholine and decamethonium in humans, initially at least, though more prolonged application of these drugs may result in other factors becoming more important. In this respect succinylcholine and decamethonium appear to be no different from any other agonists; all the effects that they produce can be produced by other agonists including acetylcholine itself (as long as they are applied for a similar prolonged period). The essential features of the block were accurately described in the classical paper by BURNS and PATON (1951), and remarkably little new information about the physiological characteristics (as opposed to the underlying mechanisms) of the block has emerged in the 32 years since their work.

The receptors are, of course, almost entirely restricted to the end-plate so the depolarization induced by transmitter, or by succinylcholine, originates entirely from this small area. The depolarization is not, however, restricted to the end-plate. Depolarization at this point will produce a current that flows along the fibre through the cytoplasm, and out through the fibre membrane. The membrane near the end-plate will thus also be depolarized; the depolarization will spread further if the cytoplasmic resistance is low (e.g. large fibre diameter), or the membrane resistance is high. Depolarization of the end-plate that is prolonged for more than 10 or 20 ms will produce a steady state depolarization of the adjacent membrane that decays roughly exponentially, as $\exp(-x/\lambda)$, with distance x away from the end-plate (see JACK et al. 1975). The space constant, or length constant λ is of the order of a millimetre in mammalian muscle (see for example ADRIAN and MARSHALL 1977), so the depolarized region is far bigger than the end-plate itself. The fibre membrane contains sodium channels and potassium channels which are much like those found in axons – they are caused to open by membrane depolarization (i.e. they are electrically operated channels, as opposed to the chemically operated nonselective cation channel at the end-plate). They are responsible for the conduction of the action potential along the muscle fibre. The depolarization of the end-plate by agonist depolarizes the adjacent membrane; this causes the sodium channels (and, after a brief delay, the potassium channels) to open, and an action potential is initiated. If this depolarization is maintained however, a few more action potentials may be produced, but the sodium channels do not stay open for long; they enter an "inactivated" state in which they are not only shut (i.e. inactive), but are incapable of being opened by further depolarization (i.e. they are inactivatable). The physical nature of this inactivated state is no better understood than that of the rather analogous desensitized state or states of chemically operated channels. The potassium channels inactivate slowly if at all, so they stay open and thus contribute to the electrical inexcitability of the membrane near the end-plate.

The time constant for development of the inactivated state of the sodium channel is only a matter of milliseconds. Thus if the fibre membrane is held depolarized for longer than a few tens of milliseconds, an equilibrium level of inactivation is reached; the proportion of channels that are inactivated at equilibrium increases with the size of the depolarization (see for example ADRIAN and MARSHALL 1977; DUVAL and LEOTY 1978). Adrian and Marshall showed that in rat muscle, half the sodium channels were already inactivated at a membrane potential of −70 mV, a depolarization of only 10 or 15 mV from the resting potential. At −60 mV about 80% of sodium channels are inactivated, and at −50 mV virtually all are inactivated. Thus if the end-plate is held depolarized to −30 mV say (i.e. a 50 mV depolarization from the resting potential) sodium channels will be completely inactivated for a distance of 500 μm on either side of the end-plate, and 900 μm away they will be 80% inactivated. This area will be electrically inexcitable, and this is probably the main reason why BURNS and PATON (1951), who used decamethonium in cat muscle, made the following observations:

1. Decamethonium (40 μg/kg intravenously) produced a depolarization of the end-plate region that lasted for tens of minutes.
2. The depolarization (assessed by external electrodes) was initially restricted to a millimetre or two on either side of the end-plate zone (it spread to a wider region later: see Sect. F.III).
3. The time course of block of neuromuscular transmission paralleled that of the depolarization of the end-plate zone.
4. In normal muscle, or in the presence of tubocurarine, the ability of electrical stimulation to elicit a muscle action potential directly was the same along the whole fibre length. But after administration of decamethonium the end-plate zone, after an initial transient increase of excitability (associated with random fasciculations), became inexcitable to direct electrical stimulation, although the muscle remote from the end-plate region remained normally excitable.
5. In normal muscle an action potential elicited by electrical stimulation of one end of a muscle fibre will propagate along the whole length of the fibre. The same happens after tubocurarine treatment, but after decamethonium treatment the action potential was blocked when it reached the electrically inexcitable area around the end-plate.
6. The removal of decamethonium-induced depolarization by passing an anodal current at the end-plate region restored neuromuscular transmission.
7. Electrical depolarization of the muscle fibre by application of a cathode produced a block similar to that produced by decamethonium.
8. Acetylcholine, or rapid nerve stimulation, (in the presence of an anticholinesterase drug) produced a block similar to that of decamethonium.

Burns and Paton concluded that "the characteristic features of block by decamethonium and acetylcholine at the neuromuscular junction are simply those of any persistent cathode".

III. Changes in Intracellular Ion Concentration

Under resting conditions the muscle membrane is much more permeable to potassium than to sodium, but the permeability to chloride is considerably greater than that to potassium in both frog and mammalian muscle (HUTTER and PADSHA 1959; HODGKIN and HOROWICZ 1959; BRYANT and MORALES-AGUILERA 1971). Chloride, however, is not actively transported; the internal and external chloride concentrations are therefore dictated by the membrane potential, so normally the chloride equilibrium potential E_{Cl} is close to the resting potential. If the membrane adjacent to the end-plate is held depolarized for a prolonged period, chloride ions will redistribute according to the new membrane potential, i. e. chloride will enter the fibre and E_{Cl} will become less negative (JENKINSON and TERRAR 1973). Because the chloride permeability of the muscle membrane is relatively high the membrane potential is determined to a substantial extent by E_{Cl}, so this shift tends to maintain the membrane in a depolarized state. This means, for example, that when the agonist is removed so that the original cause of the depolarization has disappeared, the membrane may nevertheless remain depolarized (because of the raised internal chloride concentration) for some time (JENKINSON and TERRAR 1973). Furthermore, in the continuous presence of the agonist, the depolarized area will gradually spread further and further from the end-plate region as chloride enters under the influence of the depolarized area closer to the end-plate. This phenomenon was clearly described by BURNS and PATON (1951).

It is obvious that this phenomenon could cause serious errors if one attempted to assess the extent of receptor desensitization from measurements of depolarization; the membrane may stay depolarized even after agonist is removed (or when all end-plate channels are desensitized). These problems can be overcome by holding the membrane potential constant (with a voltage clamp) or by replacing external chloride ions by a less permeant anion such as isethionate (JENKINSON and TERRAR 1973). In addition to movement of chloride ions, prolonged depolarization will also cause a loss of potassium ions, and uptake of sodium. It is well known that treatment with decamethonium or succinylcholine causes a substantial rise in plasma potassium concentration in humans and other species.

A rise of intracellular sodium concentration might be expected to activate the electrogenic sodium pump. The operation of this pump would cause hyperpolarization during prolonged agonist application, and this might be interpreted, mistakenly, as repolarization caused by desensitization. This phenomenon has been observed in rat and guinea-pig muscle at 38 °C (CREESE et al. 1983); during prolonged application of agonist there is a slow recovery of membrane potential, but this recovery is inhibited by procedures that inhibit the sodium pump (application of ouabain or removal of external potassium). In the presence of ouabain a prolonged depolarization is maintained; in its absence the membrane potential recovers despite the fact that some, at least, of the end-plate channels continue to be active as judged by the continued entry of labelled sodium or decamethonium (CASEL et al. 1977; CREESE et al. 1977). In contrast, in frog muscle (at room temperature) inhibition of the sodium pump does not affect the recovery of membrane potential (TERRAR 1974).

IV. Desensitization, Channel Block and "Dual Block"

THESLEFF (1955) suggested that neuromuscular block by agonists was largely a result of desensitization. It is now clear that this is not true, at least for block of moderate duration in humans and cats. WRAY (1981) showed that depolarization in response to acetylcholine (1–2 μ*M*, enough to block transmission) is quite well maintained in cat muscle at 38 °C, and noise analysis showed that the channel opening frequency was also maintained so this plateau was not an artifact of ion shifts (see Sect. F.III). A similar conclusion is drawn from the fact that, in the same preparation, the depolarization in response to decamethonium is also well maintained even in chloride-free solution (ZAIMIS and HEAD 1976). This is in contrast with results in frog muscle at room temperature where the maintained depolarization in response to carbachol is partly an artifact of chloride entry: in chloride-free solution it was seen that desensitization was actually much faster than it appeared to be in normal solution (JENKINSON and TERRAR 1973). Thus in cats and humans the main mode of action appears to be depolarization as described in Sect. F.II.

There is much discussion in the literature of a phenomenon (or phenomena) known as "dual block" (see ZAIMIS 1976; ZAIMIS and HEAD 1976). This term refers to a block of neuromuscular transmission that initally has the characteristics of block by depolarization, but later changes to a block with the characteristics of "nondepolarizing", "curare-like" or "competitive" block. This change occurs quite rapidly in some species, such as rat or guinea-pig. A similar phenomenon occurs in humans, though more slowly; it can occur during prolonged exposure to the blocker, and is especially prominent when halogenated anaesthetics are used. The animal and human phenomena are quite similar, apart from the rate at which they appear, so both will be considered together in the absence of any good evidence that their mechanisms differ. It is often said that the "block changes with time from depolarizing block to competitive block". This statement makes no sense to me. The drug does not change, so it cannot be an agonist one minute and an antagonist (or partial agonist) the next. There is not any compelling evidence that depolarizing blockers are partial agonists though in some species, such as rat, decamethonium will not produce a large prolonged depolarization; this is more likely to result from channel block (ADAMS and SAKMANN 1978), from desensitization, or from an electrogenic sodium pump, than from genuine partial agonism. Much of the confusion seems to have arisen from the fact that in this field the term "competitive" has commonly been used in a very loose way (see Sect. A). Usually it implies not more than the observation that the inhibition of nerve-evoked muscle contraction is increased by tubocurarine and decreased by anticholinesterase agents, the opposite of what is found with the depolarizing block described in Sect. F.II.

In fact it seems to me to be quite possible that the whole phenomenon of "dual block" might be explicable simply on the basis of the gradual development of desensitization and/or ion channel block during prolonged application of the agonist. (A broadly similar argument could be used if repolarization during prolonged agonist application resulted from the activity of an electrogenic sodium pump, as discussed previously, rather than from desensitization. The essential feature is repolarization which, whatever causes it, will result in the recovery of electrical excitability.) This view has not been experimentally verified, or even discussed much, but it is perfectly plausible, and if correct would certainly simplify a very confusing section of the literature. The key to this explanation lies in the fact (which seems to have been rather neglected) that a small proportion only of ion channels need to be opened at the end-plate in order to produce a large depolarization. The input resistance of a muscle fibre is of the order of 0.5 MΩ, i.e. a conductance of 2 μS. Suppose that a sufficient concentration of succinylcholine is applied to increase the end-plate conductance by, say, 3 μS. This will produce a depolarization of about $80\ \mathrm{mV} \times 3/(2+3) \sim 50\ \mathrm{mV}$ which is enough to cause a high degree of transmission block (see Sect. F.II). The conductance of individual end-plate channels is about 30 pS, so a conductance increase of 3 μS corresponds to the opening of about 10^5 ion channels, i.e. about 1% of all the channels (about 10^7) at the end-plate. Suppose now that 95% of these channels become desensitized and/or blocked, so only 5% of channels remain functional. The succinylcholine will still open 1% of these so the conductance increase will be $0.05 \times 3 = 0.15$ μS, sufficient to depolarize by only 5 mV or so. Thus the membrane potential would largely recover, and therefore most of the sodium and potassium channels would return to their normal resting state (shut, but activatable). In this condition the conductance increase produced by nerve stimulation would be reduced to around 5% of normal so there would be a substantial block of transmission. Were tubocurarine now to be applied, it would occupy some of the available 5% of receptors and so would obviously aggravate the transmission block. On the other hand addition of an anticholinesterase agent would increase the concentration of acetylcholine in the cleft following nerve stimulation and so would ameliorate the transmission block (opening of 20% of the available 5% of channels would be equivalent to opening 1% of all channels, and would therefore suffice to produce a suprathreshold end-plate potential). The nerve-released acetylcholine would be present for only a short time, even in the presence of the anticholinesterase, so it would not be expected to increase the extent of desensitization much, at modest stimulation rates.

This argument shows that the mysterious "dual block" phenomenon might well reflect nothing more complicated than a slowly developing repolarization in the continued presence of the agonist, caused by progressive desensitization, channel block, or electrogenic sodium pump activity. Channel block by decamethonium has been demonstrated in frog muscle by ADAMS and SAKMANN (1978); the concentration needed is substantial in this species, but the channel-blocking potency is not known for mammals at physiological temperature so it is still not known whether channel block is a major factor.

The explanation of "dual block" that has been suggested is supported by the observation that the phenomenon is most prominent in humans when halogenated anaesthetics, such as halothane, are used (see ZAIMIS 1976; ZAIMIS and HEAD 1976). These anaesthetics are among those that appear to "enhance desensitization" (YOUNG and SIGMAN 1981, 1983; see Sect. E.IV). Proadifen (SKF-525A) which has also been reported to "enhance desensitization" (Sect. E.IV), has been bound to induce the "dual block" phenomenon in avian muscle (SUAREZ-KURTZ et al. 1969; ZAIMIS 1976); this observation also favours the explanation already suggested.

G. Conclusions

Very few neuromuscular blocking agents have been tested by methods that are adequate to establish rigorously their modes of action. Nevertheless some conclusions may be drawn from those agents that have been most thoroughly investigated, though these conclusions do not necessarily apply to all agents in clinical use.

All agents that have been tested, both depolarizing and nondepolarizing, have been found to block ion channels at the muscle end-plate in addition to their other actions. The simplest form of channel block is that in which the blocking molecule can enter and leave the ion channel only when it is open (the channel may be opened by acetylcholine, or by the blocking agent itself in the case of depolarizing blockers); while the molecule is in the channel ion permeation is prevented, but the channel is held in the open conformation. Tubocurarine, gallamine and decamethonium all behave in the way expected for this simple blocking mechanism, at relatively low concentrations anyway. At higher concentrations decamethonium behaves in a more complex way, and experiments with other blocking agents, as well as experiments on ganglia, suggest that this simple version of the channel block mechanism is at best only an approximation to the truth for most drugs. The importance of these recently discovered actions needs to be considered in two contexts, clinical use of the drugs to produce muscle relaxation, and their heuristic use for elucidation of physiological and pharmacological mechanisms.

The plethora of research in recent years has probably got little significance for the clinical effects of the drugs. Under clinical conditions it is probable that the *primary* mode of action of tubocurarine and gallamine is to compete with acetylcholine for the postsynaptic receptor in the manner which has long been assumed, and which was elegantly demonstrated by JENKINSON (1961). Similarly the *primary* mode of action of decamethonium and succinylcholine under clinical conditions is probably the production of an inexcitable area of muscle membrane surrounding the end-plate as a result of prolonged end-plate depolarization, as so clearly shown by BURNS and PATON (1951).

The situation is, however, quite different for the heuristic use of neuromuscular blocking agents. These recently discovered effects may complicate considerably the interpretation of experiments in which they are used to investigate mechanisms. The specificity of their actions can no longer be relied upon, even at the neuromuscular junction. A channel-blocking effect can strongly influence the results of kinetic experiments, even at concentrations so low that little inhibition is produced at equilibrium by the channel-blocking action. At other synapses the situation is even worse; for example in peripheral ganglia the block of the nicotinic response results almost entirely from channel block with little apparent action on the receptor, so the use of tubocurarine inhibition as a criterion for existence of nicotinic receptors in peripheral (and probably also central) ganglia is baseless.

References

Adams DJ, Colquhoun D (1984) Current relaxations with high agonist concentrations. Do acetylcholine and suberyldicholine block ion channels in frog muscle? J Physiol (Lond) 341:22–23 P

Adams DJ, Dwyer TM, Hille B (1980) The permeability of endplate channels to monovalent and divalent metal cations. J Gen Physiol 75:493–510

Adams PR (1975a) Kinetics of agonist conductance changes during hyperpolarization at frog endplates. Br J Pharmacol 53:308

Adams PR (1975b) A study of desensitization using voltage clamp. Pflugers Arch 360:135–144

Adams PR (1976) Drug blockade of open end-plate channels. J Physiol (Lond) 260:531–552

Adams PR (1977) Voltage jump analysis of procaine action at the frog end-plate. J Physiol (Lond) 268:291–318

Adams PR (1980) Aspects of synaptic potential generation. In: Pinsker HM (ed) Information processing in the nervous system. Raven New York

Adams PR (1981) Acetylcholine receptor kinetics. J Membr Biol 58:161–174

Adams PR, Sakmann B (1978) Decamethonium both opens and blocks endplate channels. Proc Natl Acad Sci USA 75:2994–2998

Adrian RH, Marshall MW (1977) Sodium current in mammalian muscle. J Physiol (Lond) 268:223–250

Albuquerque EX, Adler M, Spivak CE, Aguayo L (1980) Mechanism of nicotinic channel activation and blockade. Ann N Y Acad Sci 358:204–238

Anderson CR, Stevens CF (1973) Voltage clamp analysis of acetylcholine produced endplate current fluctuations at frog neuromuscular junction. J Physiol (Lond) 235:655–691

Armstrong CM (1971) Interaction of tetraethylammonium ion derivatives with the potassium channels of giant axons. J Gen Physiol 58:413–437

Armstrong DL, Lester HA (1979) The kinetics of tubocurarine action and restricted diffusion within the synaptic cleft. J Physiol (Lond) 294:365–386

Arunlakshana O, Schild HO (1959) Some quantitative uses of drug antagonists. Br J Pharmacol 14:48–58

Ascher P, Large WA, Rang HP (1979) Studies on the mechanism of action of acetylcholine antagonists on rat parasympathetic ganglion cells. J Physiol (Lond) 295:139–170

Barnard EA, Coates V, Dolly JO, Mallick B (1977) Binding of α-bungarotoxin and cholinergic ligands to acetylcholine receptors in the membrane of skeletal muscle. Cell Biol Int Rep 1:99–106

Blackman JG (1959) The pharmacology of depressor bases. PhD Thesis, University of New Zealand

Blackman JG (1970) Dependence on membrane potential of the blocking action of hexamethonium at a sympathetic ganglionic synapse. Proc University of Otago Med Sch 48:4–5

Blackman JG, Gauldie RW, Milne RJ (1975) Interaction of competitive antagonists: the anti-curare action of hexamethonium and other skeletal neuromuscular junction. Br J Pharmacol 54:91–100

Boheim G, Hanke W, Barrantes FJ, Eibl H, Sakmann B, Fels G, Maelicke A (1981) Agonist-activated ionic channels in acetylcholine receptor reconstituted into plainer lipid bilayers. Proc Natl Acad Sci USA 78:3586–3590

Bonner R, Barrantes FJ, Jovin TM (1976) Kinetics of agonist-induced intrinsic fluorescence changes in membrane-bound acetylcholine receptor. Nature 263:429–431

Bowman WC (1980) Pharmacology of neuromuscular function. Wright, Bristol

Boyd ND, Cohen JB (1980a) Kinetics of binding of [^{3}H]acetylcholine and [^{3}H]carbamylcholine to Torpedo postsynaptic membranes: slow conformational transition of the cholinergic receptor. Biochemistry 19:5344–5353

Boyd ND, Cohen JB (1980b) Kinetics of binding of [^{3}H]acetylcholine to Torpedo postsynaptic membranes: association and dissociation rate constants by rapid mixing and ultrafiltration. Biochemistry 19:5353–5358

Bryant SH, Morales-Aguilera A (1971) Chloride conductance in normal and myotonic muscle fibres and the action of monocarboxylic aromatic acids. J Physiol (Lond) 219:367–383

Burns BD, Paton WDM (1951) Depolarization of the motor end-plate by decamethonium and acetylcholine. J Physiol (Lond) 115:41–73

Case R, Creese R, Dixon WJ, Massey FJ, Taylor DB (1977) Movement of labelled decamethonium in muscle fibres of the rat. J Physiol (Lond) 272:283–294

Cash DJ, Aoshima H, Hess GP (1981) Acetylcholine-induced cation across cell membranes and inactivation of the acetylcholine receptor: chemical kinetic measurements in the millisecond time region. Proc Natl Acad Sci USA 78:3318–3322

Castillo J del, Katz B (1957a) Interaction at end-plate receptors between different choline derivatives. Proc R Soc Lond B Biol Sci 146:369–381

Castillo J del, Katz B (1957b) A study of curare action with an electrical micro-method. Proc R Soc Lond B Biol Sci 146:339–356

Claudio T, Ballivet M, Patrick J, Heinemann S (1983) Nucleotide and deduced aminoacid sequences of Torpedo californica acetylcholine receptor γ subunit. Proc Natl Acad Sci USA 80:1111–1115

Cohen JB, Weber M, Changeux J-P (1974) Effects of local anesthetics and calcium on the interaction of cholinergic ligands with the nicotinic receptor protein from Torpedo marmorata. Mol Pharmacol 10:904–932

Colquhoun D (1973) The relation between classical and cooperative models for drug action. In: Rang HP (ed) Drug receptors. Macmillan, London, pp 149–182

Colquhoun D (1975) Mechanisms of drug action at the voluntary muscle end-plate. Annu Rev Pharmacol 15:307–325

Colquhoun D (1978) Noise: a tool for drug receptor investigation. In: Bolis L, Straub RW (eds) Cell membrane receptors for drugs and hormones. Raven, New York

Colquhoun D (1979) The link between drug binding and response: theories and observations. In: O'Brien RD (ed) The receptors: a comprehensive treatise. Plenum, New York

Colquhoun D (1980) Competitive block and ion channel block as mechanisms of antagonist action on the skeletal muscle end-plate. Adv Biochem Psychopharmacol 21:67–80

Colquhoun D (1981a) How fast do drugs work? Trends Pharmacol Sci 2:212–217 (Reprinted in Lamble J (ed) Towards understanding receptors. Elsevier, Amsterdam, 1981)

Colquhoun D (1981b) The kinetics of conductance changes at nicotinic receptors of the muscle end-plate and of ganglia. In: Birdsall N (ed) Drug receptors and their effectors. Macmillan, London

Colquhoun D, Hawkes AG (1977) Relaxation and fluctuations of membrane currents that flow through drug-operated ion channels. Proc R Soc Lond [Biol] B199:231–262

Colquhoun D, Hawkes AG (1981) On the stochastic properties of single ion channels. Proc Roy Soc Lond [Biol] B211:205–235

Colquhoun D, Hawkes AG (1982) On the stochastic properties of bursts of single ion channel openings and of clusters of bursts. Philos Trans R Soc Lond [Biol] B300:1–59

Colquhoun D, Hawkes AG (1983) The principles of the stochastic interpretation of ion channel mechanisms. In: Sakmann B, Neher E (eds) Single channel recording. Plenum, New York

Colquhoun D, Rang HP (1976) Effects of inhibitors on the binding of iodinated α-bungarotoxin to acetylcholine receptors in rat muscle. Mol pharmacol 12:519–535

Colquhoun D, Sakmann B (1981) Fluctuations in the microsecond time range of the current through single acetylcholine receptor ion channels. Nature 294:464–466
Colquhoun D, Sakmann B (1983) Bursts of openings in transmitter-activated ion channels. In: Sakmann B, Neher E (eds) Single channel recording. Plenum, New York
Colquhoun D, Sheridan RE (1981) The modes of action of gallamine. Proc R Soc Lond [Biol] B211:181–203
Colquhoun D, Sheridan RE (1982) The effect of tubocurarine competition on the kinetics of agonist action on the nicotine receptor. Br J Pharmacol 75:77–86
Colquhoun D, Dionne VE, Steinbach JH, Stevens CF (1975) Conductance of channels openend by acetylcholine-like drugs in muscle end-plate. Nature 253:204–206
Colquhoun D, Large WA, Rang HP (1977) An analysis of the action of a false transmitter at the neuromuscular junction. J Physiol (Lond) 266:361–395
Colquhoun D, Dreyer F, Sheridan RE (1979) The actions of tubocurarine at the frog neuromuscular junction. J Physiol (Lond) 293:247–284
Conti-Tronconi BM, Gotti CM, Hunkapiller MW, Raftery MA (1982) Mammalian muscle acetylcholine receptor: a supramolecular structure formed by four related proteins. Science 218:1227–1229
Creese R, Franklin GI, Mitchell LD (1976) Two mechanisms for spontaneous recovery from depolarising drugs in rat muscle. Nature 261:416–417
Creese R, Franklin GI, Mitchell LD (1977) Sodium entry in rat diaphragm induced by depolarizing drugs. J Physiol (Lond) 272:295–316
Creese R, Humphrey PPA, Mitchell LD (1983) Recovery from decamethonium rat muscle and denervated guinea pig diaphragm. J Physiol (Lond) 334:365–377
Cull-Candy SG (1981) Synaptic noise and transmitter action at nerve muscle junctions. Trends Neurosci 4:1–3
Cull-Candy SG, Miledi R, Trautmann A (1979) End-plate currents and acetylcholine noise at normal and myasthenic human end-plates. J Physiol (Lond) 287:247–265
Dionne VE, Steinbach JH, Stevens CF (1978) An analysis of the dose-response relationship at voltage-clamped frog neuromuscular junctions. J Physiol (Lond) 281:421–444
Dreyer F, Muller K-D, Peper K, Sterz R (1976) The M omohyoideus of the mouse as a convenient mammalian muscle preparation. Pflugers Arch 367:115–122
Dreyer F, Peper K, Sterz R (1978) Determination of dose-responses curves by quantitative ionophoresis at the frog neuromuscular junction. J Physiol (Lond) 281:395–419
Dunn SMJ, Blanchard SG, Raftery MA (1980) Kinetics of carbamylcholine binding to membrane-bound acetylcholine receptor monitored by fluorescence changes of a covalently bound probe. Biochemistry 19:5645–5652
Duval A, Leoty C (1978) Ionic currents in mammalian fast skeletal muscle. J Physiol (Lond) 278:403–423
Feltz A, Trautmann A (1982) Desensitization at the frog neuromuscular junction: a biphasic process. J Physiol (Lond) 322:257–272
Feltz A, Large WA, Trautmann A (1977) Analysis of atropine action at the frog neuromuscular junction. J Physiol (Lond) 269:109–130
Ferry CB, Marshall AR (1973) Anti-curare effect of hexamethonium at the mammalian neuromuscular junction. Br J Pharmacol 47:353–362
Fletcher P, Forrester T (1975) The effect of curare on the release of acetylcholine from mammalian motor nerve terminals and an estimate of quantum content. J Physiol (Lond) 251:131–144
Freeman SE, Turner RJ (1972) Agonist-antagonist interaction at the cholinergic receptor of denervated diaphragm. Aust J Exp Biol Med Sci 50:21–34
Gage PW (1976) Generation of end-plate potentials. Physiol Rev 56:177–247
Gage PW, Hamill OP (1981) Effects of anesthetics on ion channels in synapses. In: Porter R (ed) Neurophysiology IV. University Park Press, Baltimore (International Review of Physiology vol 25)
Gage PW, McBurney RN, Van Helden D (1978) Octanol reduces end-plate channel lifetime. J Physiol (Lond) 274:279–298
Gardner P, Ogden DC, Colquhoun D (1984) Conductances of single ion channels opened by cholinominetic agonists are indistinguishable. Nature 309:160–162

Ginsborg BL, Jenkinson DH (1976) Transmission of impulses from nerve to muscle. In: Zaimis E (ed) Neuromuscular junction. Springer, Berlin Heidelberg New York, pp 229–364 (Handbuch der experimentellen Pharmakologie, vol 42)
Ginsborg BL, Stephenson RP (1974) On the simultaneous action of two competitive antagonists. Br J Pharmacol 51:287–300
Grunhagen H-H, Changeux J-P (1977) Fast kinetic studies on the interaction of cholinergic agonists with the membrane-bound acetylcholine receptor from Torpedo marmorata as revealed by quinacrine fluorescence. Eur J Biochem 80:225–242
Gurney AM, Rang HP (1984) The channel-blocking action of methonium compounds on rat submandibular ganglion cells. Br J Pharmacol 82:623–642
Gutfreund H (1972) Enzymes: physical principles. Wiley, London
Hamill OP, Marty A, Neher E, Sakmann B, Sigworth FJ (1981) Improved patch-clamp techniques for high-resolution current recording from cells and cell-free membrane patches. Pflugers Arch 391:85–100
Hartzell HC, Kuffler SW, Yoshikami D (1975) Post-synaptic potentiation: interaction between quanta of acetylcholine at the skeletal neuromuscular synapse. J Physiol (Lond) 251:427–463
Head SD (1983) Temperature and end-plate currents in rat diaphragm. J Physiol (Lond) 334:441–459
Heidmann T, Changeux JP (1979a) Fast kinetic studies on the interaction of fluorescent agonist with the membrane-bound acetylcholine receptor from Torpedo marmorata. Eur J Biochem 94:255–279
Heidmann T, Changeux J-P (1979b) Fast kinetic studies on the allosteric interactions between acetylcholine receptor and local anesthetic binding sites. Eur J Biochem 94:281–296
Hill AV (1909) The mode of action of nicotine and curari determined by the form of the contraction curve and the method of temperature coefficients. J Physiol (Lond) 39:361–373
Hille B, Campbell DT (1976) An improved vaseline gap voltage clamp for skeletal muscle fibre. J Gen Physiol 67:265–293
Hodgkin AL, Horowicz P (1959) The influence of potassium and chloride ions on the membrane potential of single muscle fibres. J Physiol (Lond) 148:127–160
Hutter OF, Padsha SM (1959) Effect of nitrate and other ions on the membrane resistance of frog skeletal muscle. J Physiol (Lond) 146:117–132
Jack JJB, Noble D, Tsien RW (1975) Electric current flow in excitable cells. Clarendon, Oxford
Jenkinson DH (1960) The antagonism between tubocurarine and substances which depolarize the motor end-plate. J Physiol (Lond) 152:309–324
Jenkinson DH, Terrar DA (1973) Influence of chloride ions on changes in membrane potential during prolonged application of carbachol to frog skeletal muscle. Br J Pharmacol 47:363–376
Jurss R, Prinz H, Maelicke A (1979) NBD-5-Acylcholine: Fluorescent analog of acetycholine and agonist at the neuromuscular junction. Proc Natl Acad Sci USA 76:1064–1068
Karlin A (1980) Molecular properties of nicotinic acetylcholine receptors. Cell Surf Rev 6:191–260
Kasai M, Changeux J-P (1971) In vitro excitation of purified membrane fragments by cholinergic agonists. I Pharmacological properties of the excitable membrane fragments. J Membrane Biol 6:1–23
Katz B (1966) Nerve muscle and synapse. McGraw-Hill, New York
Katz B, Miledi R (1970) Membrane noise produced by acetylcholine. Nature 226:962–963
Katz B, Miledi R (1972) The statistical nature of the acetylcholine potential and its molecular components. J Physiol (Lond) 224:665–699
Katz B, Miledi R (1973) The binding of acetylcholine to receptors and its removal from the synaptic cleft. J Physiol (Lond) 231:549–574
Katz B, Miledi R (1977) Transmitter leakage from motor nerve endings. Proc R Soc Lond [Biol] B196:59–72

Katz B, Miledi R (1978) A re-examination of curare action at the motor end-plate. Proc R Soc Lond [Biol] B203:119–133

Katz B, Thesleff S (1957) A study of the desensitization produced by acetylcholine at the motor end-plate. J Physiol [Lond] 138:63–80

Kistler J, Stroud RM, Klymkowsky MW, Lalancette RA, Fairclough RH (1982) Structure and function of an acetylcholine receptor. Biophys J 37:371–383

Kuffler SW, Yoshikami D (1975) The number of transmitter molecules in a quantum: an estimate from iontophoretic application of acetycholine at the neuromuscular synapse. J Physiol (Lond) 251:465–482

Land BR, Salpeter EE, Salpeter MM (1980) Acetylcholine receptor site density affects the rising phase of miniature end-plate currents. Proc Natl Acad Sci USA 77:3736–3740

Land BR, Salpeter EE, Salpeter MM (1981) Kinetic parameters for acetylcholine interaction in intact neuromuscular junction. Proc Natl Acad Sci USA 78:7200–7204

Langmuir I (1918) The adsorption of gases on plane surfaces of glass, mica and platinum. J Am Chem Soc 40:1361–1402

Lo MMS, Dolly JO, Barnard EA (1981) Molecular forms of the acetylcholine receptor from vertebrate muscles and Torpedo electric organ. Eur J Biochem 116:155–163

Magazanik LG, Vyskocil F (1973) Desensitization at the motor end-plate. In: Rang HP (ed) Drug receptors. Macmillan, London

Magleby KL, Pallotta BS (1981) A study of desensitization of acetylcholine receptors using nerve-released transmitter in the frog. J Physiol (Lond) 316:225–250

Magleby KL, Stevens CF (1972) A quantitative description of end-plate currents. J Physiol (Lond) 223:173–197

Magleby KL, Pallotta BS, Terrar DA (1981) The effect of (+)-tubocurarine on neuromuscular transmission during repetitive stimulation in the rat mouse and frog. J Physiol (Lond) 312:97–113

Manalis RS (1977) Voltage-dependent effect of curare at the frog neuromuscular junction. Nature 267:366–368

Marty A, Ascher P (1980) Les Modes d'action de la tubocurarine. In: La transmission neuromusculaire les mediateurs et le "milieu interieur". Fondation Singer-Polignac. Masson, Paris, pp 89–100

Matthews-Bellinger J, Salpeter MM (1978) Distribution of acetylcholine receptors at frog neuromuscular junctions with a discussion of some physiological implications. J Physiol (Lond) 279:197–213

McIntyre AR, King RE (1943) Contraction of denervated muscle produced by d-tubocurarine. Science 97:516

Neher E (1983) The charge carried by single channel currents of rat cultured muscle cells in the presence of local anaesthetics. J Physiol (Lond) 339:663–678

Neher E, Sakmann B (1975) Voltage-dependence of drug-induced conductance in frog neuromuscular junction. Proc Nat Acad Sci USA 72:2140–2144

Neher E, Sakmann B (1976) Single-channel currents recorded from membrane of denervated frog muscle fibres. Nature 260:799–802

Neher E, Steinbach JH (1978) Local anaesthetics transiently block currents through single acetylcholine-receptor channels. J Physiol (Lond) 277:153–176

Neher E, Stevens CF (1977) Conductance fluctuations and ionic pores in membranes. Annu Rev Biophys Bioeng 6:345–381

Nelson N, Anholt R, Lindstrom J, Montal M (1980) Reconstitution or purified acetylcholine receptors with functional ion channels in planar lipid bilayers. Proc Natl Acad Sci USA 77:3057–3061

Neubig RR, Cohen JB (1979) Equilibrium binding of [^{3}H]tubocurarine and [^{3}H]acetylcholine by Torpedo postsynaptic membranes: stoichiometry and ligand interactions. Biochemistry 18:5464–5475

Neubig RR, Cohen JB (1980) Permeability control by cholinergic receptors in Torpedo postsynaptic membranes: agonist dose-response relations measured at second and millisecond times. Biochemistry 19:2770–2779

Noda M, Takahashi H, Tanabe T, Toyosato M, Furutani Y, Hirose T, Asai M, Inayama S, Miyata T, Noma S (1982) Primary structure of α-subunit precursor of Torpedo californica acetylcholine receptor deduced from cDNA sequence. Nature 299:793–797

Ogden DC, Colquhoun D (1983) The efficacy of agonists at the frog neuromuscular junction studied with single channel recording. Pflügers Arch 399:246–248
Ogden DC, Siegelbaum SA, Colquhoun D (1981) Block of acetylcholine-activated ion channels by an uncharged local anaesthetic. Nature 289:596–598
Ogden DC, Siegelbaum SA, Colquhoun D (1986) Mechanisms of action of the uncharged local anaesthetic benzocaine (in preparation)
Paton WDM, Rang HP (1965) The uptake of atropine and related drugs by intestinal smooth muscle of the guinea-pig in relation to acetylcholine receptors. Proc R Soc Lond B Biol Sci 163:1–44
Paton WDM, Waud DR (1967) The margin of safety of neuromuscular transmission. J Physiol (Lond) 191:59–90
Pennefather P, Quastel DMJ (1981) Relation between subsynaptic receptor blockade and response to quantal transmitter at the mouse neumuscular junction. J Gen Physiol 78:313–344
Pennefather P, Quastel DMJ (1982) Modification of dose-response curves by effector blockade and uncompetitive antagonism. Mol Pharmacol 22:369–380
Peper K, Bradley RJ, Dreyer F (1982) The acetylcholine receptor at the neuromuscular junction. Physiol Rev 62:1271–1340
Quast U, Schimerlik MI, Raftery MA (1979) Ligand-induced changes in membrane-bound acetylcholine receptor observed by ethidium fluorescence. II Stopped flow studies with agonists and antagonists. Biochemistry 18:1891–1901
Raftery MA, Hunkapiller MW, Strader CD, Hood LE (1980) Acetylcholine receptor: complex of homologous subunits. Science 208:1454–1457
Rang HP (1982) The action of ganglion blocking drugs on the synaptic responses of rat submandibular ganglion cells. Br J Pharmacol 75:151–168
Rang HP, Ritter JM (1969) A new kind of drug antagonism: evidence that agonists cause a molecular change in acetylcholine receptors. Mol Pharmacol 5:394–411
Rang HP, Ritter JM (1970) On the mechanism of desensitization of cholinergic receptors. Mol Pharmacol 6:357–382
Rosenberry TL (1979) Quantitative simulation of endplate currents at neuromuscular junctions based on their reaction of acetylcholine with acetylcholine receptor and acetylcholinesterase. Biophys J 26:263–290
Ruff RL (1977) A quantitative analysis of local anaesthetic alteration of miniature endplate currents and end-plate current fluctuations. J Physiol (Lond) 264:89–124
Ruff RL (1982) The kinetics of local anaesthetic blockade of end-plate channels. Biophys J 37:625–631
Sakmann B, Adams PR (1979) Biophysical aspects of agonist action at frog end-plate. In: Jacob J (ed) Advances in pharmacology and therapeutics, vol 1: Receptors. Pergamon, Oxford, pp 81–90
Sakmann B, Neher E (1983) Single channel recording. Plenum, New York
Sakmann B, Patlak J, Neher E (1980) Single acetylcholine-activated channels show burst-kinetics in presence of desensitizing concentrations of agonist. Nature 286:71–73
Schindler H, Quast U (1980) Functional acetylcholine receptor from Torpedo marmorata in planar membranes. Proc Natl Acad Sci USA 77:3052–3056
Schild HO (1949) pA_x and competitive drug antagonism. Br J Pharmacol 4:277–280
Sheridan RE, Lester HA (1975) Relaxation measurements on the acetylcholine receptor. Proc Natl Acad Sci USA 72:3496–3500
Sheridan RE, Lester HA (1977) Rates and equilibria at the acetylcholine receptor of Electrophorus electroplaques. J Gen Physiol 70:187–219
Shorr RG, Lyddiatt A, Lo MMS, Dolly JO, Barnard EA (1981) Acetylcholine receptor from mammalian skeletal muscle. Oligomeric forms and their subunit structure. Eur J Biochem 116:143–153
Sine S, Taylor P (1979) Functional consequences of agonist-mediated state transitions in the cholinergic receptor. Studies in cultured muscle cells. J Biol Chem 254:3315–3325
Sine SM, Taylor P (1980) The relationship between agonist occupation and the permeability response of the cholinergic receptor revealed by bound cobra α-toxin. J Biol Chem 255:10144–10156

Sine SM, Taylor P (1981) Relationship between reversible antagonist occupancy and the functional capacity of the acetylcholine receptor. J Biol Chem 256:6692–6699

Sine SM, Taylor P (1982) Local anesthetics and histrionicotoxin are allosteric inhibitors of the acetylcholine receptor. J Biol Chem 257:8106–8114

Steinbach AB (1968a) Alteration by xylocaine (lidocaine) and its derivatives of the time course of the end-plate potential. J Gen Physiol 52:144–161

Steinbach AB (1968b) A kinetic model for the action of Xylocaine on receptors for acetylcholine. J Gen Physiol 52:162–180

Steinbach JH (1980) Activation of nicotinic acetylcholine receptors. Cell Surf Rev 6:119–156

Stenlake JB (1979) Molecular interactions at the cholinergic receptor in neuromuscular blockade. Prog Med Chem 16:257–286

Stenlake JB (1980) Neuromuscular blocking agents. In: Wolff ME (ed) Alfred Burger's medicinal chemistry, 4th edn. Wiley-Interscience. New York

Stephenson RP (1956) A modification of receptor theory. Br J Pharmacol 11:379–393

Suarez-Kurtz G, Paulo LG, Fonteles MC (1969) Further studies on the neuromuscular effects of β-diethylaminoethyl-diphenylpropylacetate hydrochloride (SKF-525-A). Arch Int Pharmacodyn 177:185–195

Sugal N, Hughes R, Payne JP (1975) The effect of suxamethonium alone and its interaction with gallamine on the indirectly elicited tetanic and single twitch contractions of skeletal muscle in man during anaesthesia. Br J Clin Pharmacol 2:391–402

Sugiyama H, Popot JL, Changeux JP (1976) Studies on the electrogenic action of acetylcholine with Torpedo marmorata electric organ. III Pharmacological desensitization in vitro of the receptor-rich membrane fragments by cholinergic agonists. J Mol Biol 106:485–496

Sumikawa K, Barnard EA, Dolly Jo (1982a) Similarity of acetylcholine receptors of denervated, innervated and embryonic chicken muscles. Subunit compositions. Eur J Biochem 126:473–479

Sumikawa K, Houghton M, Smith JC, Bell L, Richards BM, Barnard EA (1982b) The molecular cloning and characterization of cDNA coding for the α subunit of the acetylcholine receptor. Nucleic Acids Res 10:5809–5822

Takeuchi A, Takeuchi N (1960) The permeability of end-plate membrane during the action of transmitter. J Physiol (Lond) 154:52–67

Terrar DA (1974) Influence of SKF-525A congeners, strophanthidin and tissue-culture media on desensitization in frog skeletal muscle. Br J Pharmacol 51:259–268

Thesleff S (1955) The mode of neuromuscular block caused by acetylcholine, nicotine, decamethonium and succinylcholine. Acta Physiol Scand 34:218–231

Thron CD (1973) On the analysis of pharmacological experiments in terms of an allosteric receptor model. Mol Pharmacol 9:1–9

Trautmann A (1982) Curare can open and block ionic channels associated with cholinergic receptors. Nature 298:272–275

Tyer MB (1978) Factors limiting the rate of termination of the neuromuscular blocking action of fazadinium dibromide. Br J Pharmacol 63:287–293

Tzartos SJ, Lindstrom JM (1980) Monoclonal antibodies used to probe acetylcholine receptor structure: localization of the main immunogenic region and detection of similarities between subunits. Proc Natl Acad Sci USA 77:755–759

Van Maanen EF (1950) The antagonism between acetylcholine and the curare alkaloids D-tubocurarine, c-curarine-I, c-toxiferine-II and β-erythroidine in the rectus abdominis of the frog. J Pharmacol Exp Ther 99:255–264

Wathey JC, Nass WM, Lester HA (1979) Numerical reconstruction of the quantal event at nicotinic synapses. Biophys J 27:145–164

Waud BE, Cheng MC, Waud DR (1973) Comparison of drug-receptor dissociation constants at the mammalian neuromuscular junction in the presence and absence of halothane. J Pharmacol Exp Ther 187:40–46

Waud DR (1967) The rate of action of competitive neuromuscular blocking agents. J Pharmacol Exp Ther 158:99–114

Weber M, Changeux J-P (1974) Binding of Naja nigricollis ^{3}H-α-toxin to membrane fragments from Electrophorus and Torpedo electric organs. 2 Effect of cholinergic agonists and antagonists on the binding of the tritiated α-neurotoxin. Mol Pharmacol 10:15–34
Weber M, David-Pfeuty T, Changeux J-P (1975) Regulation of binding properties of the nicotinic receptor protein by cholinergic ligands in membrane fragments from Torpedo marmorata. Proc Natl Acad Sci USA 72:3443–3447
Weiland G, Taylor P (1979) Ligand specificity of state transitions in the cholinergic receptor: behaviour of agonists and antagonists. Mol Pharmacol 15:197–212
Weiland G, Georgia B, Lappi S, Chignell CF, Taylor P (1977) Kinetics of agonist-mediated transitions in state of the cholinergic receptor. J Biol Chem 25:7648–7656
Wray D (1980) Noise analysis and channels at the postsynaptic membrane of skeletal muscle. Prog Drug Res 24:9–56
Wray D (1981) Prolonged exposure to acetylcholine: noise analysis and channel inactivation in cat tenuissimus muscle. J Physiol (Lond) 310:37–56
Young AP, Sigman DS (1981) Allosteric effects of volatile anesthetics on the membrane-bound acetylcholine receptor protein. I Stabilization of the high affinity state. Mol Pharmacol 20:498–505
Young AP, Sigman DS (1983) Conformational effects of volatile anesthetics on the membrane-bound acetylcholine receptor protein: facilitation of the agonist-induced affinity conversion. Biochemistry 22:2155–2162
Zaimis E (1976) The neuromuscular junction: areas of uncertainty. In: Zaimis E (ed) Neuromuscular junction. Springer, Berlin Heidelberg New York, pp 1–21 (Handbuch der experimentellen Pharmakologie, vol 42)
Zaimis E, Head S (1976) Depolarising neuromuscular blocking agents. In: Zaimis E (ed) Neuromuscular junction. Springer, Berlin Heidelberg New York, pp 365–419 (Handbook of experimental pharmacology, vol 42)
Ziskind L, Dennis MJ (1978) Depolarising effect of curare on embryonic rat muscles. Nature 276:622–623

Weber M, Changeux J-P (1974) Binding of Naja nigricollis [illegible] to membrane fragments from Electrophorus and Torpedo electric organs. [illegible] Effect of cholinergic agonists and antagonists on the binding of the tritiated [illegible]. Mol Pharmacol [illegible]

Weber M, David-Pfeuty T, Changeux J-P (1975) Regulation of binding properties of the nicotinic receptor protein by cholinergic ligands in membrane fragments from Torpedo marmorata. Proc Natl Acad Sci USA 72:[illegible]

[illegible]

Weiland G, Georgia B, Lappi S, Chignell CF, Taylor P (1977) Kinetics of agonist-mediated transitions in state of the cholinergic receptor. J Biol Chem [illegible]

[illegible]

[illegible]

[illegible]

[illegible]

CHAPTER 4

On the Hydrophobic Interaction of Neuromuscular Blocking Agents with Acetylcholine Receptors of Skeletal Muscles

D. A. KHARKEVICH and A. P. SKOLDINOV

A. Introduction

One of the most important tasks of the pharmacology of neuromuscular transmission is to determine the principles of interaction of neuromuscular blocking agents with acetylcholine receptors of skeletal muscles. Such studies can be important for the specification of the structure of nicotinic receptors of the end-plate and for the purposeful synthesis of new neuromuscular blocking agents (CAVALLITO and GRAY 1960; STENLAKE 1963; BARLOW 1968; KHARKEVICH 1969, 1970c, 1973, 1974, 1975, 1983; KHARKEVICH and SKOLDINOV 1970, 1971, 1974, 1976, 1980; KIER 1971; TRIGGLE et al. 1971; ANICHKOV 1974; ALBERT 1979; CAVALLITO 1980, etc.).

The interaction of neuromuscular blocking agents with acetylcholine receptors involves various forces acting between molecules. Among these there are electrostatic, ion–dipole, dipole–dipole, hydrophobic, and van der Waals interactions and hydrogen bonds. This chapter mainly deals with hydrophobic interactions. Some aspects of this problem have been discussed in a number of publications (PATON 1956; CAVALLITO 1959a, b, 1962, 1967a, b; MIKHELSON and ZEIMAL 1970; ANICHKOV 1974; and others). The absence of sufficient data necessitated systematic research devoted to the relationship between the structure of the compounds, their hydrophobicity, and neuromuscular blocking action (KHARKEVICH 1970a, b; 1973, 1974; KHARKEVICH and SKOLDINOV 1970, 1971, 1974, 1975, 1980). The investigations were mainly carried out in the series of quaternary ammonium salts. The degree of hydrophobicity was changed by the introduction of hydrophobic radicals into the cationic head or the interonium part of the structure (for bisonium salts). In this chapter, the influence of radicals of different hydrophobicity and localization on the mechanism of action and the activity of several cholinergic agonists and antagonists are considered.

B. The Effect of Hydrophobic Radicals on the Mode of Action

I. *N*-1-Adamantyl Derivatives

In the series of monoquaternary ammonium salts, *N*-1-adamantyl (*N*-1-Ad) derivatives of the known cholinomimetics such as tetramethylammonium, choline, and acetylcholine were tested. It was shown that the replacement of an *N*-methyl group by the *N*-1-Ad radical was followed by changes in the mechanism of action. Thus, for instance, tetramethylammonium, an autonomic ganglion-stimulating agent, evokes spastic paralysis in pigeons and contracture of isolated rectus abdominis muscle of the frog. Choline and acetylcholine produced an analogous effect. Unlike their prototypes *N*-1-Ad derivatives of all the three compounds are nondepolarizing agents (Table 1, compounds 1–3). Their administration evokes flaccid paralysis in pigeons. They are devoid of cholinomimetic action on isolated frog rectus abdominis muscle, but antagonize carbachol (Table 1).

The *N*-1-Ad derivative of the monocholine ester of succinic acid (Table 1, compound 4) also differs from its trimethylammonium analog by a nondepolarizing mechanism of action.

These transformations were also observed in *N*-1-Ad derivatives of benzoic and cinnamic acid 4-dimethylaminobutyl esters (compounds 5–8). It was previously shown that these compounds in the form of trimethylammonium salts, possess a marked neuromuscular blocking activity. By their mechanism of action they are depolarizing agents (KHARKEVICH et al. 1967; ARENDARUK et al. 1967; KHARKEVICH 1973; see Chapt. 13). In all the experiments, the replacement of one *N*-CH_3 group by *N*-1-Ad affected the mechanism of action: depolarizing neuromuscular blocking agents were turned into nondepolarizing ones. This is confirmed by the following results. In experiments on cats, they did not cause muscle fasciculations, neostigmine was their antagonist; they induced flaccid paralysis in pigeons and chickens, and on the isolated rectus abdominis muscle of the frog they had only nicotinic blocking activity.

The introduction of *N*-1-Ad radicals affects not only the mechanism of action of the derivatives of benzoic and cinnamic acid esters, but also induces a pronounced decrease of their activity. Thus, the replacement of one methyl group in the cationic center of monoquaternary ammonium salts by the hydophobic adamantyl radical results in changes of the mechanism of action of the agents, turning the depolarizing agents into nondepolarizing agents (KHARKEVICH et al. 1973).

Bisquaternary ammonium salts (KHARKEVICH et al. 1974) provided further evidence for the established regularity. Succinylcholine and decamethonium, which are known to have a marked depolarizing action, and their *N*-1-Ad analogs, called diadonium and decadonium, respectively (Table 1, compounds 9, 10) were studied in most detail. Diadonium and decadonium are typical nondepolarizing neuromuscular blocking agents. Diadonium blocks the transmission from the sciatic nerve to the gastrocnemius muscle at doses of 0.25–0.3 mg/kg as does decadonium – at doses of 0.25–0.35 mg/kg i. v.

The mechanism of action and locus of action of both agents can be judged by the following characteristics:

1. Absence of muscle fasciculations prior to the neuromuscular block (experiments on cats and rabbits).
2. Enhanced pessimal inhibition (tetanic fade), marked posttetanic "decurarization", antagonism with neostigmine (experiments with the registration of the neuromuscular transmission in anesthetized cats).
3. Absence of changes of membrane potential, reduced amplitude of miniature potentials, the frequency being unchanged (experiments on phrenic nerve – diaphragm preparation of the rat).

Table 1. *N*-adamantyl derivatives of quaternary ammonium salts

Compound	Structure	Block of transmission from sciatic nerve to gastrocnemius muscle in cats (mg/kg i.v.)	Flaccid paralysis in pigeons (mg/kg i.v.)	Effect on frog isolated rectus abdominis muscle	
				Contracture	Antagonism with carbachol
	$R^1-\overset{+}{N}(CH_3)_2Ad\text{-}1 \cdot I^-$				
	R^1				
1	H_3C-	25.0 –30.0	~ 6.0	–	+
2	$HO-CH_2-CH_2-$	25.0 –30.0	10.0 –15.0	–	+
3	$H_3C-CO-O-CH_2-CH_2-$	25.0 –30.0	9.0 –10.0	–	+
4	$HOOC-CH_2-CH_2-COO-CH_2-CH_2-$	25.0 –30.0	1.5 – 2.0	–	+
	$R^1-CO-O-(CH_2)_4-\overset{+}{N}(CH_3)_2Ad\text{-}1 \cdot I^-$				
	R^1				
5	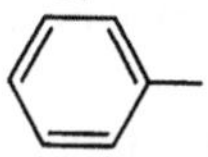	12.0 –14.0	8.0 –10.0	–	+
6	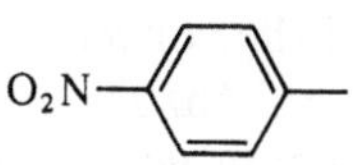	12.0 –14.0	2.0 – 3.0	–	+
7	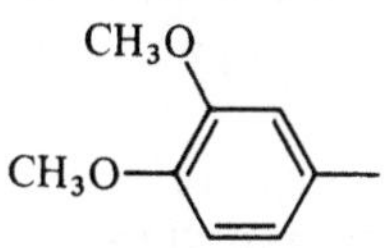	10.0 –12.0	2.0 – 4.0	–	+
8	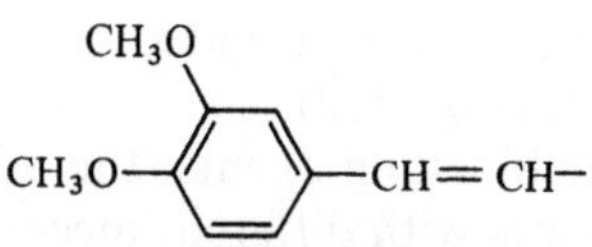	7.0 – 8.0	1.5 – 2.0	–	+
9	$1\text{-}Ad(CH_3)_2\overset{+}{N}-(CH_2)_{10}-\overset{+}{N}(CH_3)_2Ad\text{-}1 \cdot 2I^-$ [a]	0.25– 0.3	0.15– 0.16	–	+
10	$1\text{-}Ad(CH_3)_2\overset{+}{N}-(CH_2)_2-O-CO-(CH_2)_2-$ $-CO-O-(CH_2)_2-\overset{+}{N}(CH_3)_2Ad\text{-}1 \cdot 2I^-$ [b]	0.25– 0.35	0.03– 0.04	–	+
11	$1-Ad(CH_3)_2\overset{+}{N}-(CH_2)_3-O-$ 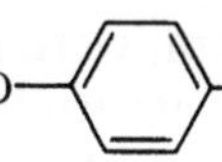 $-O-(CH_2)_3-$ $-\overset{+}{N}(CH_3)_2Ad\text{-}1 \cdot 2I^-$	1.3 – 1.5		–	+

(+) Effect; (–) no effect; 1-Ad= 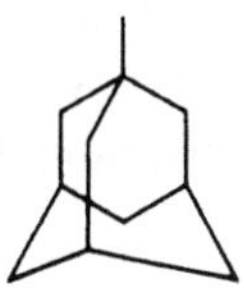

[a] Decadonium
[b] Diadonium

4. Antagonism with acetylcholine and carbachol, absence of stimulating effect, parallel shift of cumulative curves (carbachol as agonist), maximum amplitude remaining the same (experiments on isolated abdominis muscle of the frog).

The data presented indicate that diadonium and decadonium are competitive neuromuscular blocking agents with subsynaptic action.

The changes in the mechanism of action of succinylcholine and decamethonium due to the introduction of adamantyl radicals into both cationic heads were followed by decrease of their neuromuscular blocking activity. In the framework of more detailed analysis of the role of hydrophobic interactions in the pharmacodynamics of neuromuscular blocking agents, the relationships between chemical structure, hydrophobicity, and mode of action of the compounds were tested.

Physicochemical methods which allow one to estimate the hydrophobicity were chosen. Among the effects studied were the surface activity of aqueous solutions of neuromuscular blocking agents at interfaces, the ability of micelle formation in aqueous solutions, interaction with artificial bimolecular phosphatidylcholine membrane, and the affinity for the negatively charged macromolecules of the synthetic polyelectrolyte, polyacrylic acid, as the compounds tested can be regarded as diphilic hydrophobic–hydrophilic molecules (PRYANISHNIKOVA et al. 1974; DROZHZHIN 1975; DROZHZHIN and IBADOVA 1975, 1976).

Surface activity was determined by recording the isotherms of surface tension which expressed the relationship between the surface tension and the concentration of the neuromuscular blocking agents at the interface with air, benzole, and benzole solutions of lipoproteins of the nerve tissue. Surface tension was measured by a semistatic method of maximum drop pressure using the apparatus designed by Rebinder (see PRYANISHNIKOVA 1973).

The interface surface activity of neuromuscular blocking agents in relation to their structure was found to be different. The introduction of adamantyl radicals in the cationic groups of succinylcholine and decamethonium resulted in increased surface and interface activity of the compounds (Figs. 1, 2).

To estimate the hydrophobicity of neuromuscular blocking agents, the adsorption of surface-active neuromuscular blocking agents with different mechanisms of action was tested on the dropping mercury electrode. The hydrophobicity was estimated by the changes of the capacitance of the electric double layer formed by the directed adsorption of diphilic molecules on the surface of a negatively charged dropping mercury electrode.

Four neuromuscular blocking agents: succinylcholine, decamethonium, and their *N*-1-Ad analogs diadonium and decadonium, were found to have a significant ability to adsorb on the dropping mercury electrode, regardless of the potential of the latter. Since the formation of an electric double layer at the interface with mercury during the adsorption of diphilic neuromuscular blocking agents takes place in the aqueous phase, the increased adsorption of neuromuscular blocking agents following the introduction of *N*-1-Ad radicals into the molecule can be related only to the enhanced hydrophobic interaction of nonpolar groups in the adsorption layer.

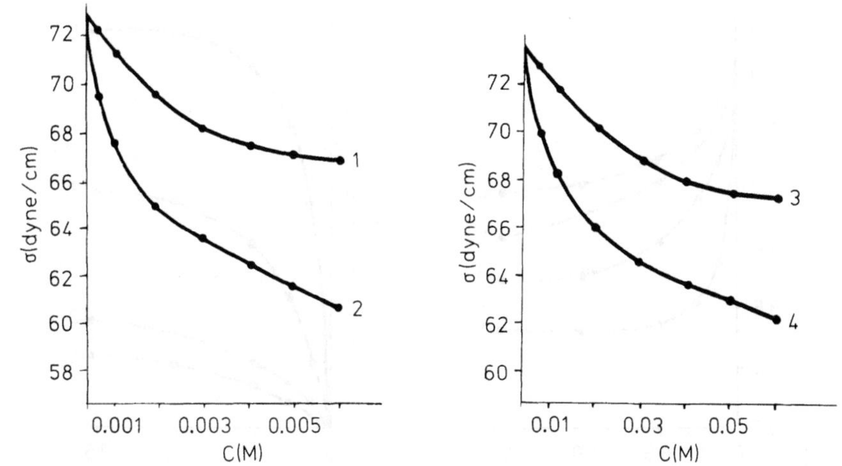

Fig. 1. Surface tension σ (water–air) of aqueous solutions of neuromuscular blocking agents. *1* diadonium; *2* decadonium; *3* succinylcholine; *4* decamethonium. Abscissa: concentration of substances (*M*); ordinate: surface tension (dyne/cm). Each point represents ten measurements. (DROZHZHIN 1975)

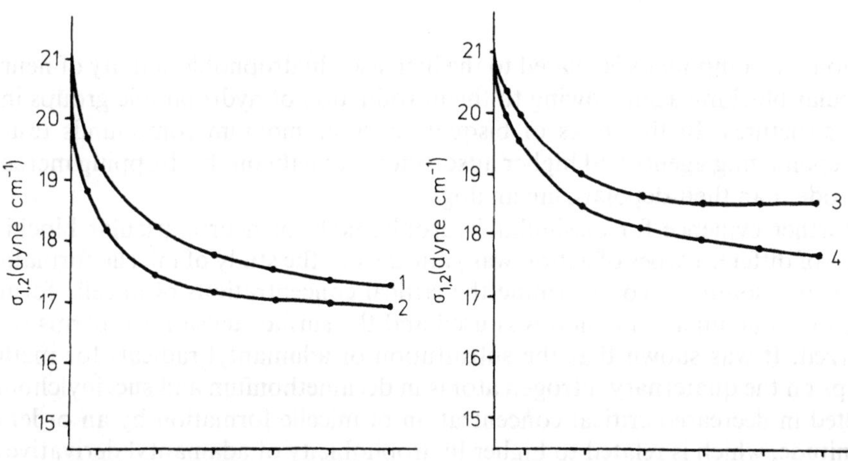

Fig. 2. Interface tension $\sigma_{1,2}$ (water–lipoproteins) of aqueous solutions of neuromuscular blocking agents. *1* decadonium; *2* diadonium; *3* decamethonium; *4* succinylcholine. (DROZHZHIN 1975)

There is a number of papers devoted to the adsorption of surface-active monoquaternary ammonium compounds on the dropping mercury electrode. Thus, for instance, ZUTRAUEN (1956) studied the series of alkyl (C_{12}–C_{18}) trimethylammonium compounds. It was shown that in this series the adsorption capability increased with increased length of the alkyl radicals, i.e., with increased hydrophobicity of the compound. Thus, the enhanced adsorption of quaternary

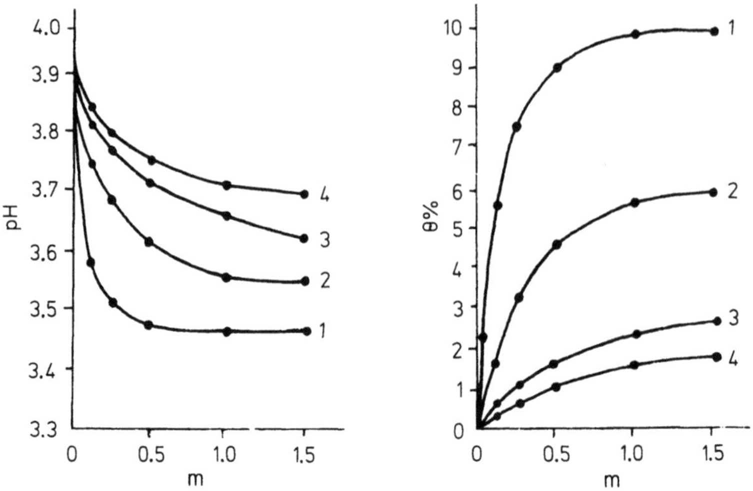

Fig. 3. The degree of interaction θ (%) of neuromuscular blocking agents with polyacrylic acid (3×10^{-3} *M*). *1* decadonium; *2* diadonium; *3* decamethonium; *4* succinylcholine. *m* Number of ions of neuromuscular blocking compounds corresponding to one link of polyacrylic acid. (DROZHZHIN 1975)

ammonium compounds is related to the increased hydrophobic activity of neuromuscular blocking agents owing to the introduction of hydrophobic groups into their structures. In the series of bisquaternary ammonium compounds tested, nondepolarizing agents had higher adsorption capacity on the dropping mercury electrode than their depolarizing analogs.

Further evidence for dissimilar hydrophobicity of neuromuscular blocking agents of different types of action was gained from the study of micelle formation in aqueous solution. To determine the critical concentrations of micelle formation, a conductimetric method was used and the surface tension isotherms were analyzed. It was shown that the substitution of adamantyl radicals for methyl groups on the quaternary nitrogen atoms in decamethonium and succinylcholine resulted in decreased critical concentration of micelle formation by an order of magnitude, which is related to higher hydrophobicity of adamantyl derivatives.

Electrostatic interaction of mono- and bisquaternary ammonium compounds in aqueous solution with negatively charged macromolecules of the synthetic polyelectrolyte, polyacrylic acid was studied by potentiometric titration (FELDSTEIN 1972). The experiments demonstrated that the introduction of *N*-1-Ad radicals into the molecules of decamethonium and succinylcholine resulted in enhanced binding with polyacrylic acid (Fig. 3). Electrostatic binding of surface-active compounds with negatively charged macromolecules of polyelectrolyte is stabilized by their hydrophobic interactions. Therefore, the enhancement of the binding should be regarded as related to the increased hydrophobicity of agents owing to adamantyl radicals introduced into their molecules. According to the degree of the binding with polyacrylic acid, the neuromuscular blocking agents tested can be ordered as follows: succinylcholine < decamethonium < diadonium < decadonium. Owing to stronger hydrophobic interactions, the binding of

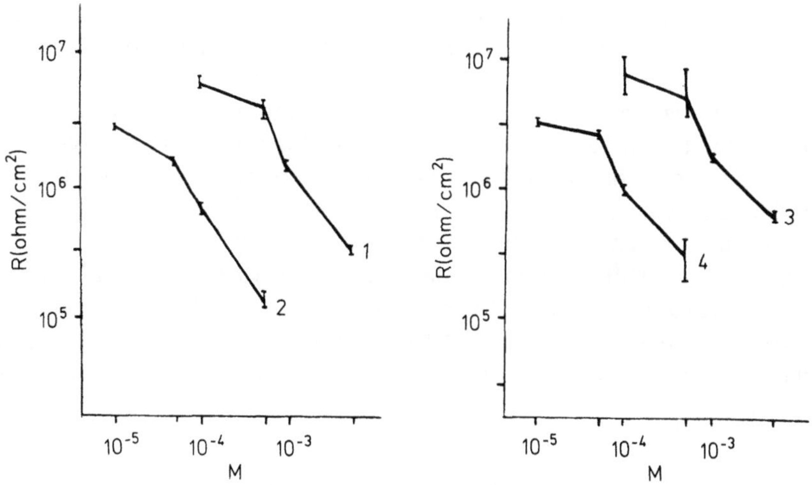

Fig. 4. The influence of neuromuscular blocking agents on the electrical resistance *R* of phosphatidylcholine bilayer membrane. *1* decamethonium; *2* decadonium; *3* succinylcholine; *4* diadonium. Agents were added from both sides. Resistance of control membrane $(2.1 \pm 0.1) \times 10^{-7}\ \Omega/cm^2$. Abscissa: concentration of agents (*M*). (DROZHZHIN 1975)

neuromuscular blocking agents with a polymethylene chain between quaternary nitrogen atoms (decamethonium and decadonium) with negatively charged macromolecules of polyacrylic acid is more marked than that of neuromuscular blocking agents with ester groups in the interonium chain (succinylcholine and diadonium).

The interaction of neuromuscular blocking agents with an artificial bimolecular membrane was studied by measuring the electric conductivity of the membrane. Phosphatidylcholine in *n*-heptane was used as the membrane-forming material. After neuromuscular blocking agents were introduced into the buffer solution from both sides of the membrane (symmetric action of the drugs on the bimolecular layer maintains membrane stability), a reduced electrical resistance of the artificial phospholipid membrane was observed. It was found that the agents with adamantyl radicals on quaternary nitrogen atoms (diadonium and decadonium) decreased the electrical resistance of the phosholipid bimolecular membrane more effectively than their *N*-methyl analogs (succinylcholine and decamethonium) and at significantly lower concentrations (Fig. 4).

The interaction of neuromuscular blocking agents tested with artificial phosphatidylcholine membrane, containing zwitterion lipids and electrostatically neutral at the given pH (7.4), suggests that the interaction of these neuromuscular blocking agents is most likely due to their hydrophobic groups which can be attached to the membrane. The addition of neuromuscular blocking agents with hydrophobic adamantyl radicals on quaternary nitrogen atoms (diadonium and decadonium) to the buffer solution results in a more effective decrease of electrical resistance of the artificial bimolecular phosphatidylcholine membrane than the addition of their *N*-methyl analogs (succinylcholine and decamethonium). This is due to higher hydrophobicity of the two former compounds.

Thus, the introduction of adamantyl radicals into the cationic groups of succinylcholine and decamethonium molecules causes a marked increase in the surface and interface activity at water–air and water–benzole solutions of lipoprotein interfaces, in the capability of adsorption on the dropping mercury electrode, and in micelle formation in aqueous solution; it also causes a more marked binding with the negatively charged macromolecules of the synthetic polyelectrolyte, polyacrylic acid. Moreover, *N*-1-Ad analogs of succinylcholine and decamethonium (diadonium and decadonium) are more effective in decreasing the electrical resistance of the artificial bimolecular phosphatidylcholine membrane. The changes are due to the increasing hydrophobicity of compounds following the introduction of adamantyl radicals into their cationic groups, which is associated with changes in their mechanism of action, leading to the development of competitive properties.

II. Quaternary Ammonium Compounds with Adamantyl Radicals in Various Parts of the Molecule

In the compounds tested, adamantyl radicals are part of their cationic groups. It is natural that the large size of adamantyl radicals should change the interaction of the "cationic head" of the compound with the anionic structure of the acetylcholine receptor. In particular, this is confirmed by the decreased neuromuscular blocking activity of the majority of the *N*-1-Ad compounds in comparison with their trimethylammonium analogs. To analyze the role of adamantyl radicals in the interaction of compounds with acetylcholine receptors of the skeletal muscles, mono- and bisquaternary ammonium compounds with adamantyl radicals in different loci were synthetized. Functionally active trimethylammonium groups were retained.

Among monoquaternary ammonium compounds, adamantyl derivatives of choline, acetylcholine, and its analogs – quaternary salts of aminoalkylesters 1-adamantanecarboxylic acids (Table 2, compounds 1–6) were studied. The introduction of the adamantyl radical at the β-position in choline and acetylcholine changed the mechanism of action of the agents. The adamantyl derivatives tested are nondepolarizing agents, in spite of the presence of the trimethylammonium group. Monoquaternary ammonium derivatives of 1-adamantanecarboxylic acid aminoalkylesters are also nondepolarizing agents; they differ by the length of the polymethylene chain, and consequently, by unequal distances between adamantyl and quaternary nitrogen atom.

In the series of bisquaternary ammonium salts, studies were made of the derivatives, analogs, and homologs of succinylcholine with an adamantyl radical in the interonium part of the molecule (bistrimethylammonium compounds) or in the cationic head, but separated from the quaternary nitrogen atom by one or two methylene groups (see also Chapt. 13, Sect. D.I.1). Moreover, bistrimethylammonium salts with two adamantyl radicals in the central part of the structure (3,3′-derivatives of 1,1′-diadamantyl) were studied (Table 2, compounds 7–12). The results were similar. All the bisquaternary ammonium salts tested are nondepolarizing agents, regardless of the locus of adamantyl radicals. They are effectively an-

Table 2. Some compounds with various sites of *N*-1-adamantyl radicals in the molecule

Compound	Structure	Block of transmission from sciatic nerve to gastrocnemius muscle in cats (mg/kg i.v.)	Flaccid paralysis in pigeons (mg/kg i.v.)	Effect on frog isolated rectus abdominis muscle	
				Contracture	Antagonism with carbachol
	$R^1-\overset{+}{N}(CH_3)_3 \cdot I^-$; R^1				
1	HO–CH(1-Ad)–CH_2–		17.0–20.0	–	+
2	H_3C–COO–CH(1-Ad)–CH_2–		5.0– 6.0	–	+
3	1-Ad–CH_2–COO–CH_2–CH_2–		2.5– 4.0	–	+
4	1-Ad–COO–CH_2–CH_2–		5.5– 6.0	–	+
5	1-Ad–COO–CH_2–CH_2–CH_2–		3.5– 5.0	–	+
6	1-Ad–COO–CH_2–CH_2–CH_2–CH_2–		2.0– 4.0	–	+
	$X\begin{cases} CO-O-CH(R)-CH_2-\overset{+}{N}(CH_3)_3 \\ CO-O-CH(R)-CH_2-\overset{+}{N}(CH_3)_3 \end{cases} \cdot 2I^-$; X / R				
7	X: –$(CH_2)_2$–; R: 1-Ad	12.0–15.0	7.0– 7.5	–	+
8	X: 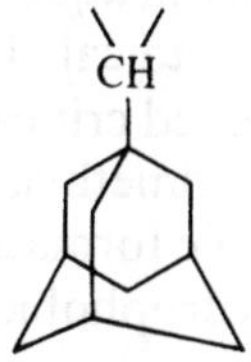 R: H	45.0–50.0	10.0–12.0	–	+
9	X: 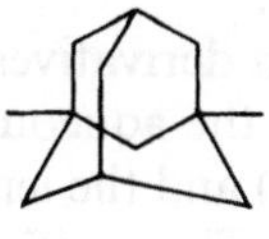 R: H	20.0–25.0	4.0– 4.5	–	+
10	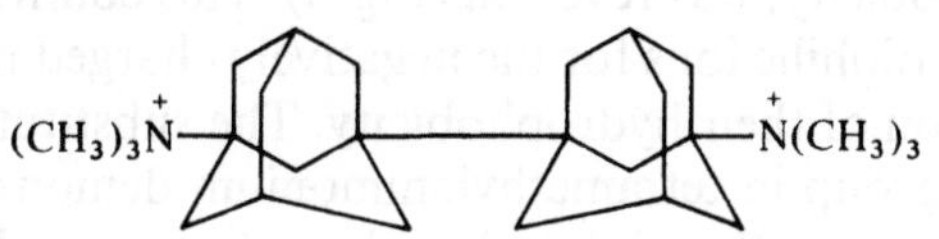	0.7– 0.9		–	+
11	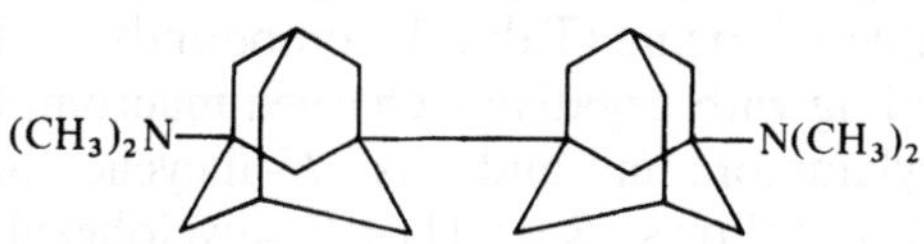	3.0– 4.0		–	+
12	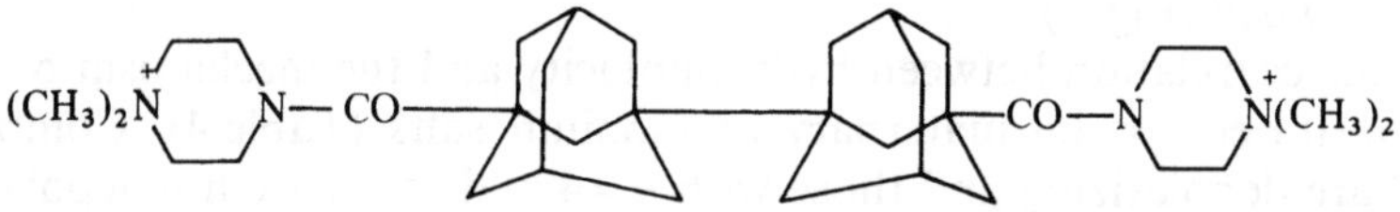	4.0– 5.0		–	+

tagonized by neostigmine. If the attachment of adamantyl radicals results in changes in the mechanism of action, this is associated, as a rule, with a decreased activity of the agents. Thus, the location of adamantyl radicals is not a decisive factor in the transformation of depolarizing neuromuscular blocking agents into nondepolarizing ones. The degree of hydrophobicity of a compound as a whole seems to be of major importance.

III. Alterations of the Mode of Action Evoked by Gradual Increase in Hydrophobicity

So far we have considered mono- and bisquaternary ammonium compounds with high hydrophobicity related to the presence of adamantyl radicals. To analyze the established regularities, a comparative study of the mechanism of action of structural analogs with gradually increasing hydrophobicity of radicals was carried out. For this purpose, a series of mono- and bisquaternary ammonium salts, containing aliphatic and alicyclic radicals with increasing hydrophobicity were synthetized and investigated.

In the series of monoquaternary alkyltrimethylammonium salts $C_nH_{2n+1}N^+(CH_3)_3 \cdot X^-$, it was found that the compounds with $n=2$–8 were depolarizing agents, whereas those with $n=10$ and more were nondepolarizing agents (Table 3, compounds 1–10). It was also shown that nondepolarizing compounds possessed a more pronounced surface and interface activity than depolarizing ones.

The alkyltrimethylammonium compounds tested form micelles in aqueous solutions. This ability increases with increasing length of the alkyl radical. The experimental data obtained are in good correlation with the calculated critical concentrations of micelle formation in a homologous series of alkyltrimethylammonium compounds. The latter indicated that the capability of micelle formation increased logarithmically. These data indicate the importance of hydrophobic interaction not only on the interfaces, but also in aqueous solution.

Furthermore, in this series of monoquaternary ammonium derivatives, a direct correlation between the degree of electrostatic binding in the aqueous solutions with negatively charged polyelectrolyte (polyacrylic acid) and the length of the alkyl radical, i.e., its hydrophobicity, was revealed (Fig. 5). This confirms the conclusion that the affinity of the diphilic ions for the negatively charged macromolecules of polyacids is a function of their hydrophobicity. The substitution of alicyclic radicals for one methyl group in tetramethylammonium demonstrated that depolarizing action persists only in *N*-cyclohexyl analogs. Compounds with *N*-bornyl or *N*-Ad radicals are nondepolarizing (Table 3, compounds 12–14).

According to the degree of binding with negatively charged macromolecules of polyacrylic acid, tetramethylammonium and its *N*-alicyclic analogs $RN^+(CH_3)_3 \cdot X^-$, can be ordered as follows: $R=CH_3 < R=$cyclohexyl$ < R=$bornyl$ < R=$1-Ad (Fig. 6).

The same correlation between hydrophobicity and the mechanism of action was demonstrated for bisquaternary ammonium salts (Table 4). Compounds with $n=2$ are depolarizing and those with $n=4$ and more are nondepolarizing.

Table 3. Monoquaternary ammonium compounds with different degrees of hydrophobicity of *N*-aliphatic and *N*-alicyclic radicals

Compound	Structure		Character of paralysis in pigeons (paralyzing doses mg/kg i.v.)	Effect on frog isolated rectus abdominis muscle	
				Contracture	Anatagonism with carbachol
	$R\overset{+}{N}(CH_3)_3 \cdot X^-$ $R=C_nH_{2n+1}$				
	n	X			
1	2	I	S (0.5– 1.0)	+	
2	4	I	S (0.04–0.05)	+	
3	6	I	S (0.3– 0.4)	+	
4	8	I	S (1.0– 1.5)	+	
5	10	I	F (3.0– 4.0)	–	+
6	12	Br	F (7.0– 7.5)	–	+
7	14	Br	F (16.0–16.5)	–	+
8	16	Br	F (21.0–22.0)	–	+
9	18	Br	F (10.0–11.0)	–	+
10	20	Br	F (31.0–33.0)	–	+
	R	X			
11	$-CH_3$	I	S (0.4– 0.5)	+	
12	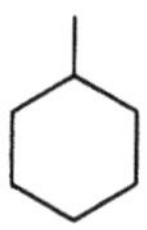	I	S (1.0– 1.5)	+	
13	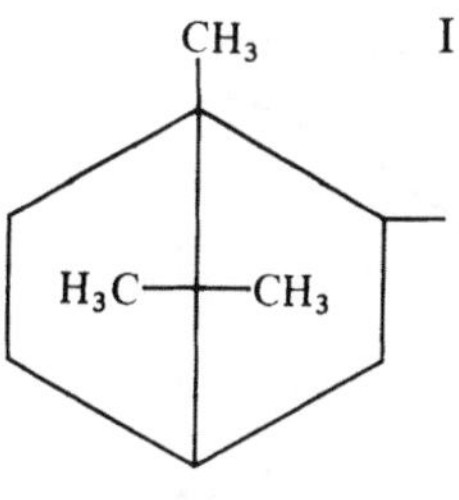	I	F (10.0–11.1)	–	+
14	1-Ad	I	F (12.0–15.0)	–	+

S, spastic paralysis; F, flaccid paralysis

The elongation of *N*-alkyl radicals, which is connected with the increase in their hydrophobicity, is followed by an increase in the interface and surface activity in this series of compounds, too.

It was shown in the study of compounds with $R=CH_3$, C_2H_5, and $n-C_4H_9$, that the ability to form micelles in aqueous solutions was enhanced with the elongation of alkyl radicals on quaternary nitrogen atoms. Consequently, nonde-

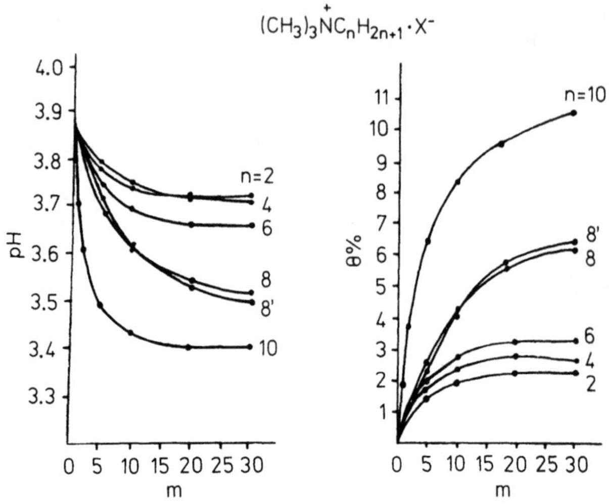

Fig. 5. The degree of interaction θ (%) of alkyltrimethylammonium compounds ($n = 2$–10) with polyacrylic acid (3×10^{-3} *M*). *m* Number of ions of compounds corresponding to one link of polyacrylic acid. For $n = 2$–8, $X = 1^-$; for $n = 8'$ and $n = 10$, $X = Br^-$. (DROZHZHIN 1975)

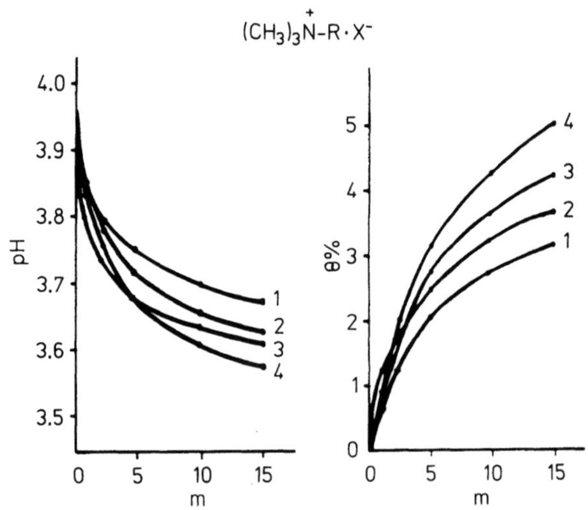

Fig. 6. The degree of interaction θ (%) of monoquaternary ammonium compounds with polyacrylic acid (3×10^{-3} M). *m* Number of ions corresponding to one link of polyacrylic acid. *1*, $R = CH_3$; *2*, R = cyclohexyl; *3*, R = bornyl; *4*, R = 1-adamantyl. (DROZHZHIN 1975)

Table 4. Bisquaternary ammonium compounds with different degrees of hydrophobicity of their *N*-aliphatic and *N*-alicyclic radicals

Compound	Structures	Character of paralysis in pigeons (Paralyzing doses, mg/kg i.v.)	Effect on frog isolated rectus abdominis muscle	
			Contracture	Antagonism with carbachol
	$R(CH_3)_2\overset{+}{N}-(CH_2)_{10}-\overset{+}{N}(CH_3)_2R \cdot 2I^-$ $R=C_nH_{2n+1}$ *n*			
1	2	S (0.2 – 0.3)	+	
2	4	F (0.4 – 0.5)	–	+
3	6	F (0.25– 0.3)	–	+
4	8	F (1.0 – 1.1)	–	+
5	10	F (10.0 –10.5)	–	+
	R			
6	$-CH_3$	S (0.05– 0.1)	+	
7	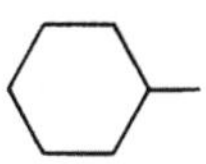	F (0.25– 0.3)	–	+
8	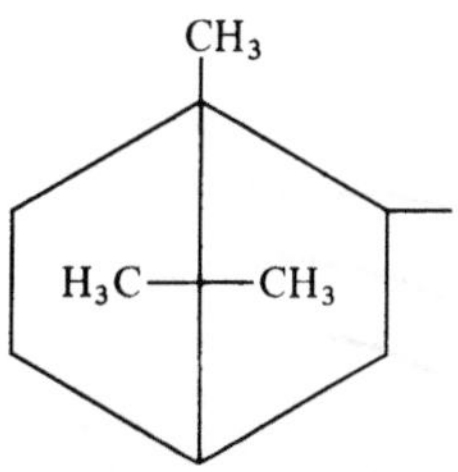	F (0.15– 0.2)	–	+
9	1-Ad	F (0.15–0.16)	–	+

polarizing agents have greater capability of micelle formation in aqueous solutions than the depolarizing ones. The degree of binding with polyacrylic acid is directly proportional to the elongation of *N*-alkyl radicals (from C_2H_5 up to $n-C_{10}H_{21}$; Fig. 7).

The attachment of alicyclic radicals (cyclohexyl, bornyl, 1-Ad) to quaternary nitrogen atoms transforms the compounds into acetylcholine blocking agents. In this series, only the unsubstituted bistrimethylammonium compound (decamethonium) acts as a depolarizing agent. According to the surface activity as related to R, the compounds mentioned can be ordered as follows: $R=CH_3$ (decamethonium) $<$ R = cyclohexyl $<$ R = bornyl $<$ R = 1-Ad (decadonium). The de-

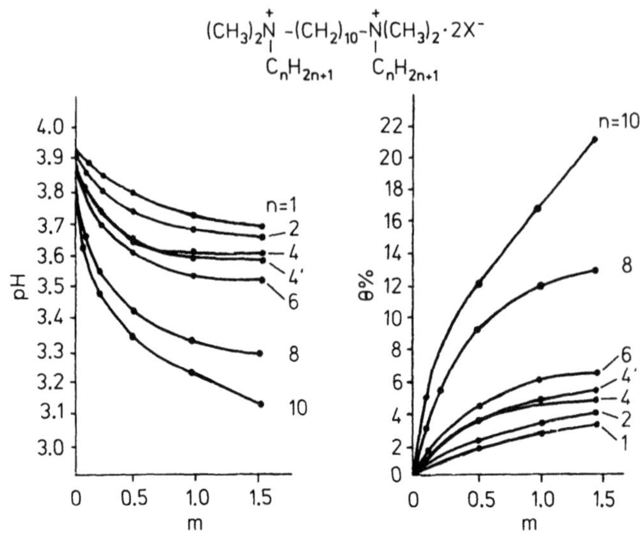

Fig. 7. The degree of interaction θ (%) of bisquaternary ammonium compounds with polyacrylic acid (3×10^{-3} *M*). Compounds with $n = 1, 2, 4, 6$ are iodides, while those with $n = 4'$, 8, 10 are chlorides. *m* Number of ions of bisquaternary ammonium compounds corresponding to one link of polyacrylic acid. (DROZHZHIN 1975)

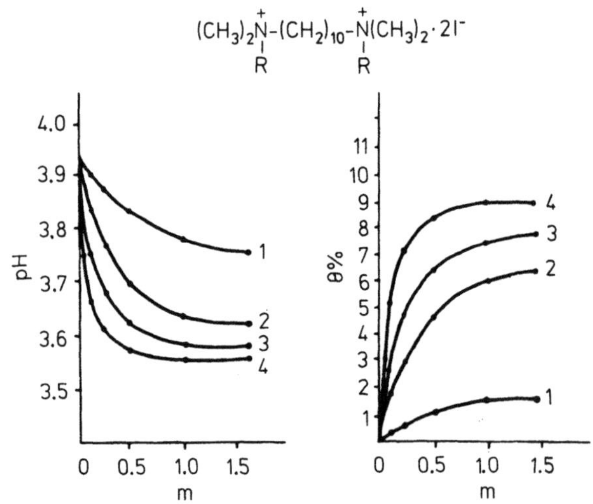

Fig. 8. The degree of interaction θ (%) of bisquaternary ammonium compounds with polyacrylic acid (3×10^{-3} *M*). *1*, $R = CH_3$; *2*, R = cyclohexyl; *3*, R = bornyl; *4*, R = 1-adamantyl. *m* Number of ions of bisquaternary ammonium compounds corresponding to one link of polyacrylic acid. (DROZHZHIN 1975)

gree of binding with polyacrylic acis is increased with increased hydrophobicity in the same order from decamethonium to decadonium (Fig. 8). Thus, the mono- and bisquaternary ammonium compounds tested display the same regularity: the insertion of radicals with a certain hydrophobicity into the structure of depolarizing agents transforms them into nondepolarizing ones.

C. The Effect of Hydrophobic Radicals on the Activity

I. The Role of the Initial Mechanism of Action

The data obtained reveal that the changes of the mechanism of action resulting from the introduction of hydrophobic radicals are usually associated with a decrease of neuromuscular blocking activity. Yet, there were a few compounds in which the latter was unaffected. At the same time, it was shown that the insertion of hydrophobic radicals might lead to the increase in neuromuscular blocking activity. It seems that this may be achieved if the hydrophobic structures of a neuromuscular blocking agent are highly complementary to hydrophobic sites of acetylcholine receptors. This was demonstrated for some derivatives of polymethylene bistrimethylammonium salts (Table 5). These compounds are particularly interesting, since a lot of them, decamethonium included, possess a marked neuromuscular blocking activity. The series represented in Table 5 attracts attention, since at $n=5$ and 6 the compounds are nondepolarizing, at $n=7$ and 8 they are mixed, and at $n=9$–11 they are depolarizing. After the *N*-1-Ad radical was introduced into both cationic heads (one methyl group was replaced on each quaternary nitrogen atom), all the compounds blocked neuromuscular transmission in a nondepolarizing manner (no muscle fasciculation; flaccid paralysis in pigeons; no stimulating effect on the frog rectus abdominis muscle, but antagonism with acetylcholine and carbachol; effective antagonism with neostigmine in cats). Thus, the introduction of *N*-1-Ad radicals into cationic groups abolished the depolarizing properties in polymethylene compounds, transforming them into nondepolarizing agents. If the initial agent was a nondepolarizing one, the attachment of the adamantyl radical affected only its activity.

The alterations of the activity induced by the 1-Ad radical were dissimilar. In experiments on cats, it was shown that the agents with $n=6$–8 were the most active. The shortening of the interonium distance up to $n=5$ or its elongation up to $n=9$ resulted in a decreased activity. At $n=10$, the activity was again somewhat increased and further elongation up to $n=11$ reduced the neuromuscular blocking activity which became about the same as at $n=9$. The same correlation was observed in experiments on rabbits. Compounds with $n=6$ and 8 caused head drop at doses of 40–45 µg/kg, those with $n=5$ at doses of about 300 µg/kg, and with $n=10$ at doses of 100–110 µg/kg i.v.

Thus, a rather definite correlation is revealed: the insertion of *N*-1-Ad radicals into cationic groups of depolarizing agents ($n=9$–11) is followed by changes in the mechanism of action and a decrease of the activity. The insertion of *N*-1-Ad radicals in ammonium groups of nondepolarizing compounds increases the activity of the latter by an order of magnitude at $n=5$ and by two orders of magnitude at $n=6$. Compounds with $n=7$ or 8 occupy an intermediate position; bistrimethylammonium salts with such an interonium distance have nicotinomimetic activity, though rather low (experiments on chickens and isolated frog rectus abdominis muscle). The activity of their *N*-1-Ad analogs was also increased, but to a smaller degree than at $n=5$ or $n=6$ (at $n=7$ it was 8–10 times higher and at $n=8$ about 1.5 times).

Table 5. Cholinotropic effects in bisquaternary ammonium compounds

$$(CH_3)_2\overset{+}{\underset{|\atop R}{N}}-(CH_2)_n-\overset{+}{\underset{|\atop R}{N}}(CH_3)_2\cdot 2I^-$$

R	Parameter	n							
		5	6	7	8	9	10	11	
$-CH_3$	Stimulation of frog isolated rectus abdominis muscle (conditional units of activity)[a,b]	0	0.1	1.3	7.1	36	71	91	Acetylcholine 1500
	95% Inhibition of transmission from sciatic nerve to tibialis muscle of the cat (mg/kg i.v.)[b]	40	40	1.9	0.16	0.036	0.03	0.06	Tubocurarine 0.3
−1-Ad	Block of transmission from sciatic nerve to gastrocnemius muscle of the cat (mg/kg i.v.)	0.8–1.2	0.12–0.2	0.2–0.25	0.09–0.12	0.45–0.55	0.25–0.3	0.45–0.55	Tubocurarine 0.18–0.23

[a] Activity at $n=12$ taken as 100
[b] According to PATON and ZAIMIS (1949)

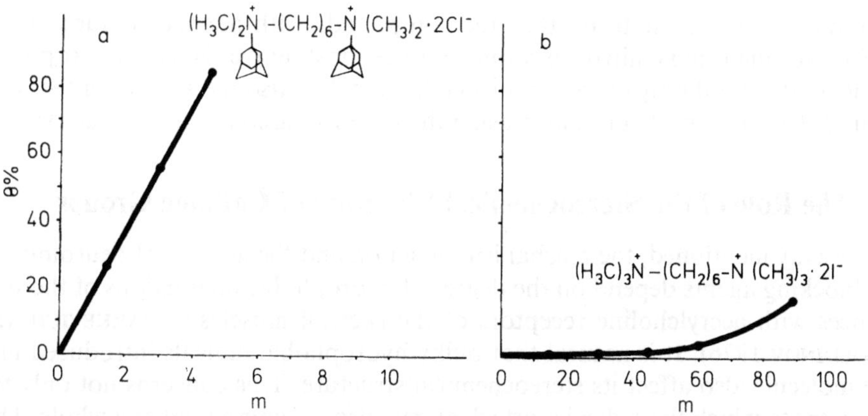

Fig. 9 a, b. The degree of interaction θ (%) of **b** hexamethonium and **a** its bis-*N*-1-adamantyl analog with polyacrylic acid (3×10^{-3} *M*). *m* Number of ions corresponding to one link of polyacrylic acid. (Drozhzhin 1975)

The substitution in compounds with $n=8$ of a carboxyl group for two methylene groups (structure I) decreased the activity about twofold.

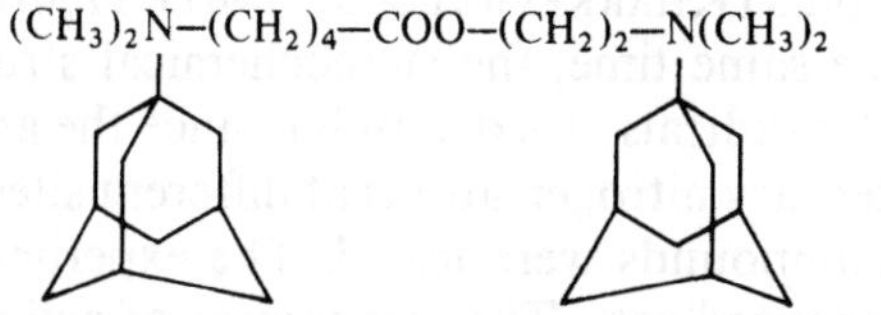

(I)

Compound I blocks neuromuscular transmission in cats at doses of 200–220 μg/kg and the corresponding polymethylene compound at $n=8$ (Table 5) at doses of 90–120 μg/kg i. v.

Consequently, the difference in activity between hexamethonium (Table 5, $n=6$) and its bis-*N*-1-Ad analog is especially marked. It is likely that adamantyl radicals provide additional points of fixation on hydrophobic sites of the subsynaptic membrane, thus stabilizing the ionic bonds formed at the expense of onium centers. Therefore, binding of the bis-*N*-1-Ad analog of hexamethonium with polyacrylic acid is much more solid than that of hexamethonium itself (Fig. 9). This indicates significant divergence in their hydrophobicity. Compound II, a (bisquaternary salt of 3,3′-bis(dimethylamino)-1,1′-diadamantyl) with two adamantyl radicals in the central part of the molecule and with an interonium distance close to that of hexamethonium, turned out to be a sufficiently active non-depolarizing neuromuscular blocking agent (it blocks neuromuscular transmission in cats at doses of 0.7–0.9 mg/kg). Neostigmine is its antagonist. It should be noted that in this case a bistertiary amine (structure II b) with adamantyl radicals is also a rather potent neuromuscular blocking agent (it blocks neuromuscular transmission in cats at doses of 3–4 mg/kg).

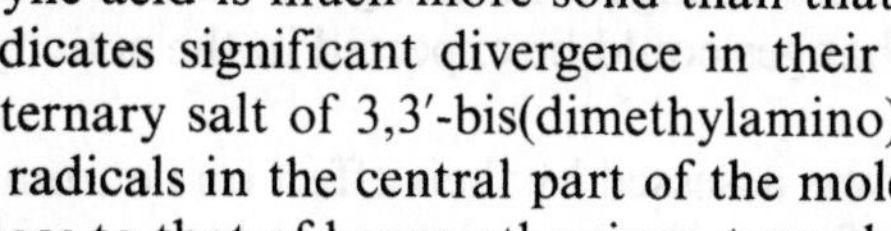

· 2RX a) RX = CH_3I (II)
b) RX = HCl

The data presented indicate that the compounds with adamantyl radicals attached to quaternary nitrogen atoms are the most active. Hence, hydrophobic radicals affect not only the mechanism of action, but also the activity of the compounds tested, in particular, increasing their neuromuscular blocking activity.

II. The Role of the Stereochemical Structure of Cationic Groups

As already mentioned, the mechanism of action and the activity of neuromuscular blocking agents depend on the degree of hydrophobic interactions of the substances with acetylcholine receptors of the skeletal muscles (KHARKEVICH and SKOLDINOV 1976). It is natural that bulky hydrophobic radicals introduced into the molecule also affect its stereochemical structure. This concerns not only the fragment to which the radical is attached, but also the compound as a whole. This affects the interrelationship of the compound with the acetylcholine receptor, i. e., their complementarity, and consequently, the conditions of intermolecular interactions. This section deals with the role of the stereochemical structure of cationic groups in the interaction of compounds with acetylcholine receptors of the skeletal muscles. 1-Adamantyl (1-Ad) or 2-adamantyl (2-Ad) radicals were used as hydrophobic radicals (Table 6); except for two compounds, they were attached to quaternary nitrogen atoms (KHARKEVICH et al. 1981). Hydropobicity of both radicals was equal. At the same time, the stereochemical structure of cationic groups with 1-Ad and 2-Ad radicals was dissimilar, since the adamantyl radicals used are attached to quaternary nitrogen atoms at different sites. Mono- and bisquaternary ammonium compounds were tested. The experiments were carried out on pigeons and anesthetized cats. The mechanism of action and the neuromuscular blocking activity of agents served as criteria.

All the agents tested with 1-Ad or 2-Ad were shown to cause flaccid paralysis in pigeons, i. e., to be nondepolarizing agents. This was also confirmed on recording the neuromuscular transmission when the antagonism of these agents with neostigmine was observed. It was previously reported (KHARKEVICH and SKOLDINOV 1980) that the hydrophobic-hydrophilic balance was of importance for the mechanism of action of the compounds. In the series tested, it was invariable for each pair of compounds, and it was only the spatial arrangement of adamantyl radicals and cationic centers that were changed. Although it failed to influence the mechanism of action, certain changes could be expected in the activity of the agents.

The activity of the compounds was assessed by their effect on neuromuscular transmission in cats. In the group of monoquaternary ammonium compounds – *N*-adamantyl derivatives of cholinomimetics: tetramethylammonium (Table 6, compound 1), choline (Table 6, compound 2), and acetylcholine (Table 6, compound 3), 1-Ad and 2-Ad derivatives had equal activity. They were all of very low activity.

1-Ad and 2-Ad derivatives of polymethylene bisquaternary ammonium compounds displayed a marked difference in activity. 2-Ad derivatives appeared to be less active than 1-Ad derivatives. This was exemplified by bis-*N*-Ad derivatives of hexamethonium (Table 6, compound 4), octamethonium (Table 6, compound 5), and decamethonium (Table 6, compound 6). Corresponding 2-Ad de-

Table 6. Neuromuscular blocking activity of quaternary ammonium compounds containing 1-adamantyl (1-Ad) or 2-adamantyl (2-Ad) radicals

Compound	Structures	R	Doses (mg/kg i.v.) blocking the transmission from sciatic nerve to gastrocnemius muscle in cats
1	$(CH_3)_3\overset{+}{N}-R \cdot I^-$	1-Ad	25 –30
		2-Ad	25 –30
2	$HO-CH_2-CH_2-\overset{+}{N}(CH_3)_2R \cdot I^-$	1-Ad	25 –30
		2-Ad	25 –30
3	$H_3C-COO-(CH_2)_2-\overset{+}{N}(CH_3)_2R \cdot I^-$	1-Ad	25 –30
		2-Ad	30 –35
4	$R(CH_3)_2\overset{+}{N}-(CH_2)_6-\overset{+}{N}(CH_3)_2R \cdot 2I^-$	1-Ad	0.12– 0.18
		2-Ad	2.0 – 3.0
5	$R(CH_3)_2\overset{+}{N}-(CH_2)_8-\overset{+}{N}(CH_3)_2R \cdot 2I^-$	1-Ad	0.09– 0.12
		2-Ad	0.6 – 0.8
6	$R(CH_3)_2\overset{+}{N}-(CH_2)_{10}-\overset{+}{N}(CH_3)_2R \cdot I^-$	1-Ad[a]	0.35– 0.4
		2-Ad	1 – 1.2
7	$[-CH_2-COO-(CH_2)_2-\overset{+}{N}(CH_3)_2]_2 \cdot 2I^-$	1-Ad[d]	0.35– 0.45
		2-Ad	2 – 2.5
8	$[-(CH_2)_2-COO-(CH_2)_2-\overset{+}{N}(CH_3)_2R]_2 \cdot 2I^-$	1-Ad	2.5 – 3
		2-Ad	4 – 6
9	$[-(CH_2)_4-COO-(CH_2)_2-\overset{+}{N}(CH_3)_2R]_2 \cdot 2I^-$	1-Ad	25 –30
		2-Ad	40
10	$R-CH[-COO-(CH_2)_2-\overset{+}{N}(CH_3)_3]_2 \cdot 2I^-$	1-Ad	45 –50
		2-Ad	40 –45

[a] Decadonium
[b] Diadonium

1—Ad 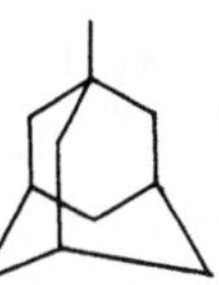 ; 2—Ad = 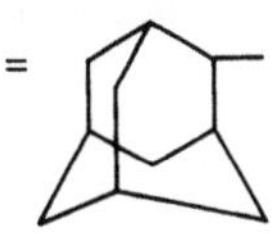

rivatives were 3–15 times less active than their 1-Ad analogs. This series is most significant, since, unlike other compounds containing ester groups (Table 6, compounds 3, 7–10), it cannot be hydrolyzed. Therefore, the activity of compounds 4–6 in Table 6 is best characterized by their neuromuscular blocking action.

In the series of bisquaternary ammonium derivatives of dicarboxylic acids, succinylcholine adamantyl analogs (Table 6, compound 7) showed the most marked difference. In this case, too, the 1-Ad derivative was more active than 2-Ad (about five times). The activity of the derivatives of adipinic (Table 6, compound 8) and sebacinic (Table 6, compound 9) acids differed insignificantly. Diiodomethylates of adamantylmalonic acid bis (*N*-dimethylaminoethyl)esters were found to be equally of very low activity. Unlike all the foregoing compounds, 1-Ad or 2-Ad radicals were attached to the carbonic atom in the central part of their molecule (Table 6, compound 10).

The comparison of the data obtained shows that the difference in activity between 1-Ad and 2-Ad derivatives is mainly relevant for the active neuromuscular blocking agents. It is likely that the stereochemical structure of the ammonium centers of compounds with 1-Ad radicals is responsible for their higher complementarity to the end-plate acetylcholine receptors than that of the compounds with 2-Ad radicals. This promotes a more effective hydrophobic interaction of 1-Ad derivatives with acetylcholine receptors of the skeletal muscles, thus stabilizing the electrostatic interaction of cationic centers of compounds with anionic centers of acetylcholine receptors. As far as the less active agents are concerned, owing to their low complementarity to acetylcholine receptors and the ability to bind with nonspecific receptors, a relatively insignificant difference in the stereochemical arrangement of adamantyl radicals on the quaternary nitrogen atom under given experimental conditions has essentially no effect on their neuromuscular blocking activity.

D. The Effect of Hydrophobic Radicals on the Main Pharmacologic Action

The role of hydrophobic radicals in the mechanism of neuromuscular blocking action of the agents and their activity has already been discussed. However, in the series of polymethylene bis-*N*-Ad derivatives still another possibility was observed. It was shown that the insertion of adamantyl radicals in certain cases can change the main site of action. Thus, for instance, hexamethonium (structure III a; $R=CH_3$) is an active ganglion-blocking agent, and it is practically devoid of neuromuscular blocking activity (see Table 5).

$$R(CH_3)_2\overset{+}{N}-(CH_2)_6-\overset{+}{N}(CH_3)_2R \cdot 2I^- \qquad \text{(III)}$$

a) $R=CH_3$
b) $R=1-Ad$

When two methyl groups are replaced by 1-Ad radicals (structure III b), ganglion-blocking activity of the compound is essentially lost and the neuromuscular activity predominates (Figs. 10, 11). Compound III b blocks neuromuscular transmission in cats at doses of 0.12–0.18 mg/kg i.v., whereas tubocurarine causes the same effect at doses of 0.18–0.23 mg/kg. Hence, the bis-*N*-1-Ad analog of hexamethonium has a marked neuromuscular blocking activity. Furthermore, it acquires antimuscarinic activity mainly directed toward the muscarinic receptors of the heart (see Chapt. 7, Sect. A). The importance of *N*-1-Ad radicals is also clearly exemplified by the bistertiary salt corresponding to hexamethonium.

$$RCH_3N-(CH_2)_6-NCH_3R \cdot 2HCl \qquad \text{(IV)}$$

a) $R=CH_3$
b) $R=1-Ad$

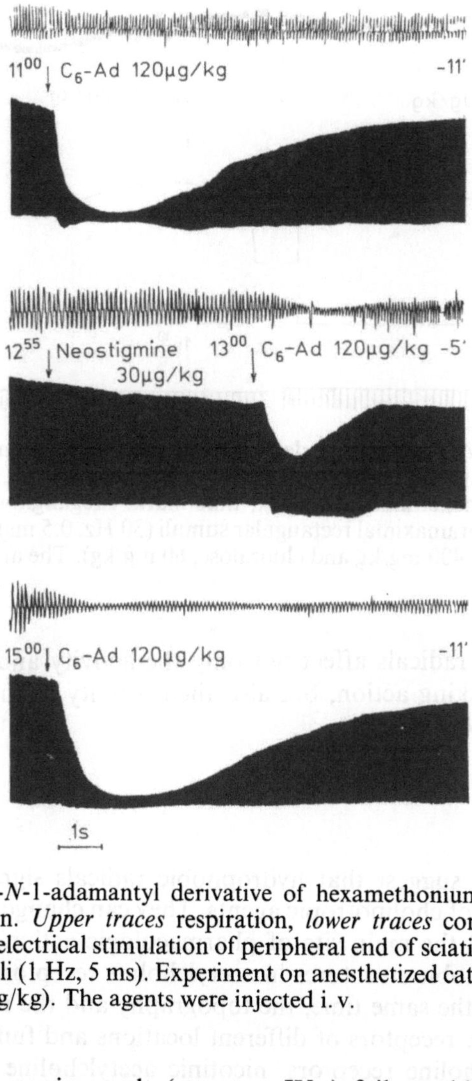

Fig. 10. Effect of bis-*N*-1-adamantyl derivative of hexamethonium (C_6–Ad) on neuromuscular transmission. *Upper traces* respiration, *lower traces* contractions of gastrocnemius muscle under electrical stimulation of peripheral end of sciatic nerve by supramaximal rectangular stimuli (1 Hz, 5 ms). Experiment on anesthetized cat (urethane, 400 mg/kg with chloralose, 60 mg/kg). The agents were injected i. v.

The bisdimethylammonium salt (structure IV a) fails to block neuromuscular transmission. The replacement of methyl groups by adamantyl radicals (structure IV b) on the nitrogen atoms results in the development of neuromuscular blocking activity (the compound blocks neuromuscular transmission in cats at doses of 10–12 mg/kg i. v.).

Analogous changes in the type of action were also observed in compounds with $n=5$. Pentamethonium, a bistrimethylammonium compound (structure V a), is also known as a ganglion-blocking agent, whereas its bis-*N*-Ad analog (structure V b) is characterized by neuromuscular blocking activity (Table 5).

$$R(CH_3)_2\overset{+}{N}{-}(CH_2)_5{-}\overset{+}{N}(CH_3)_2R \cdot 2I^- \qquad (V)$$

a) $R = CH_3$
b) $R = 1{-}Ad$

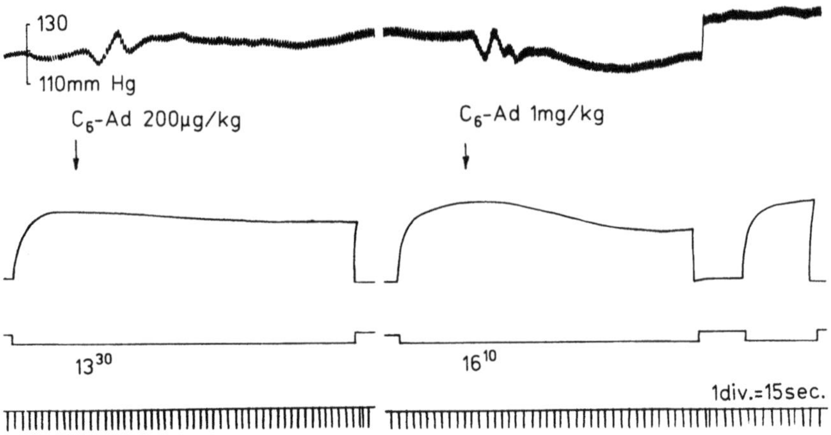

Fig. 11. Effect of bis-*N*-1-adamantyl derivative of hexamethonium (C_6–Ad) on arterial pressure and nictitating membrane tone. *Traces from top downward:* arterial pressure; nictitating membrane tone; stimulation mark; time mark. Preganglionic sympathetic fibers were stimulated by supramaximal rectangular stimuli (30 Hz, 0.5 ms). Experiment on anesthetized cat (urethane, 400 mg/kg and chloralose, 60 mg/kg). The agents were injected i. v.

Thus, hydrophobic radicals affect not only the activity and the mechanism of neuromuscular blocking action, but also their affinity to functionally different acetylcholine receptors.

E. Conclusions

The data presented suggest that hydrophobic radicals significantly affect the pharmacodynamics of cholinotropic agents. They can change their mechanism of action, activity, and the main site of pharmacologic action. This indicates the presence of hydrophobic structures in acetylcholine receptors and/or in the surrounding areas. At the same time, the topography and the role of hydrophobic sites in acetylcholine receptors of different locations and functional importance (muscarinic acetylcholine receptors, nicotinic acetylcholine receptors of autonomic ganglia and of the skeletal muscles) seem to vary significantly. This conclusion is suggested by the possibility of considerably altering the affinity of the agents for certain acetylcholine receptors by introducing the hydrophobic radicals. This is manifested by the changes in the main pharmacologic effect.

The insertion of hydrophobic radicals into cholinomimetics regularly transforms them into acetylcholine blocking agents. This is likely due to the fact that hydrophobic radicals provide additional sites of fixation of the agent on the acetylcholine receptor, thus increasing their interaction. This stabilizes the conformation of the acetylcholine receptor and prevents its depolarization. The interaction of agents with ionic channel sites of acetylcholine receptors also seems to play a certain role. This has already been mentioned for tubocurarine and gallamine (see Chapt. 1). It is also of interest that the blocking action of 1-adamantanamine hydrochloride on neuromuscular transmission is attributed to the interaction with ionic channel sites (ALBUQUERQUE et al. 1978; TSAI et al. 1978). This is so for N-alkyl-substituted analogs of 1-adamantanamine, too (WARNICK et al. 1982).

The introduction of hydrophobic radicals into the molecule of nondepolarizing neuromuscular blocking agents may sometimes significantly potentiate their activity. This seems to occur only if hydrophobic radicals are highly complementary to hydrophobic sites of the end-plate acetylcholine receptors.

In addition to their theoretical value, these data are important for the design of new acetylcholine blocking agents. The established regularities allowed the synthesis of the short-acting nondepolarizing neuromuscular blocking agent, diadonium which is approved in the USSR for practical use in patients (see Chap. 20 and 26). Some trends in the quest for atropine-like agents blocking muscarinic receptors of a given locus can be presently outlined, too. Hydrophobic interactions are also essential for anticholinesterase activity of agents and hydrolysis of esters by cholinesterases. Hence, hydrophobicity of cholinotropic compounds plays an important role in their effect on cholinergic processes.

Acknowledgment. The authors wish to express their appreciation to Mrs. MARIA LIPMAN for the translation of this chapter.

References

Albert A (1979) Selective toxicity, 6th edn. Methuen, London

Albuquerque EX, Eldefrawi AT, Eldefrawi ME, Mansour NA, Tsai M-C (1978) Amantadine: neuromuscular blockade by suppression of ionic conductance of the acetylcholine receptor. Science 199:788–790

Anichkov SV (1974) Izbiratelnoe deistvie mediatornykh sredstv (Selective action of transmitter agents). Meditsina, Leningrad

Arendaruk AP, Gracheva EA, Skoldinov AP, Kharkevich DA (1967) The basic esters and amides of substituted cinnamic acids and their analogs (in Russian). Khim-Farm Zh 4:20–25

Barlow RB (1968) Introduction to chemical pharmacology, 2nd edn. Methuen, London

Cavallito CJ (1959a) Influences of lipophilic chemical structures on curaremimetic and other activities of some quaternary ammonium salts. Atti XI Congr Soc Ital di Anestesiol 149–160

Cavallito CJ (1959b) Some interrelationships of chemical structure, physical properties and curaremimetic action. In: Bovet D, Bovet-Nitti F, Marini-Bettolo GB (eds) Curare and curare-like agents. Elsevier, Amsterdam, pp 288–303

Cavallito CJ (1962) Structure-action relations throwing light on the receptor. In: De Reuch AVS (ed) Curare and curare-like agents. Ciba Foundation Study Group No 12. Little, Brown and Co., Boston, pp 55–74

Cavallito CJ (1967a) Bonding characteristics of acetylcholine stimulants and antagonists and cholinergic receptors. Ann N Y Acad Sci 44:900–912

Cavalliti CJ (1967) Some speculations on the chemical nature of postjunctional membrane receptors. Fed Proc 20:1647–1654

Cavallito CJ (1980) Quaternary ammonium salts – advances in chemistry and pharmacology since 1960. Prog Drug Res 24:267–373

Cavallito CJ, Gray AP (1960) Chemical nature and pharmacological action of quaternary ammonium salts. Prog Drug Res 2:135–226

Drozhzhin AP (1975) O gidrofobnosti kurarepodobnykh veshchestv (On hydrophobicity of curare-like drugs). Cand Med Sci Thesis, Moscow, First Medical Institute

Drozhzhin AP, Ibadova DN (1975) On the role of hydrophobicity of mono-quaternary ammonium compounds in the mechanism of their curare-like action (in Russian). Farmakol Toksikol 3:289–294
Drozhzhin AP, Ibadova DN (1976) On the role of hydrophobicity of polymethylene-bisquaternary ammonium compounds in the mechanism of their curare-like action (in Russian). Farmakol Toksikol I:21–25
Feldstein MM (1972) Issledovanie vzaimodeistviya poverkhnostnoaktivnykh veshchestv s sinteticheskimi polipeptidami v vodnykh rastvorakh (Study of the interaction of ionogenic surface-active compounds with synthetic polypeptide in aqueous solutions). Cand Chem Sci Thesis, Moscow, State University
Kharkevich DA (1969) Farmakologiya kurarepodobnykh sredstv (Pharmacology of curare-like agents). Meditsina, Moscow
Kharkevich DA (1970a) On curare-like activity of decadonium diiodide (in Russian). Farmakol Toksikol 4:395–399
Kharkevich DA (1970b) On pharmacological properties of a new antidepolarizing curare-like drug diadonium diiodide (in Russian). Farmakol Toksikol 5:531–536
Kharkevich DA (ed) (1970c) Novye kurarepodobnye i ganglioblokiruyushchie sredstva (New curare-like and ganglion blocking agents) Meditsina, Moscow
Kharkevich DA (1973) Curare-like drugs (in Russian). In: Kharkevich DA (ed) Uspekhi v sozdanii novykh lekarstvennykh sredstv (Advances in drug research). Meditsina, Moscow, pp 138–187
Kharkevich DA (1974) New curare-like agents. J Pharm Pharmacol 26:153–165
Kharkevich DA (1975) Hydrophobic properties of the neuromuscular blocking agents. Proc Sixth Int Congr Pharmacol 1:33–47
Kharkevich DA (ed) (1983) Novye miorelaksanty (New muscle relaxants). Meditsina, Moscow
Kharkevich DA, Skoldinov AP (1970) New acetylcholine antagonists (in Russian). Zh Vses Khim O-va Im DI Mendeleeva 2:145–156
Kharkevich DA, Skoldinov AP (1971) On the effect of lipophilic radicals in the molecule of curare-like drugs on the mechanism of their action (in Russian). Dokl Akad Nauk SSSR 198:985–988
Kharkevich DA, Skoldinov AP (1974) On the significance of hydrophobic interactions for the mechanism of action of curare-like drugs (in Russian). Dokl Ak Nauk 219:762–765
Kharkevich DA, Skoldinov AP (1975) On the effect of 1-adamantyl radicals on curare-like activity (in Russian). Byull Eksp Biol Med 5:71–75
Kharkevich DA, Skoldinov AP (1976) On the significance of hydrophobic radicals for the interaction of curare-like agents with acetylcholine receptors (in Russian). Zh Vses Khim O-va Im DI Mendeleeva 2:124–129
Kharkevich DA, Skoldinov AP (1980) On some principles of interaction of curare-like agents with acetylcholine receptors of skeletal muscles. J Pharm Pharmacol 32:733–739
Kharkevich DA, Arendaruk AP, Gracheva EA, Skoldinov AP (1967) On neuromuscular blocking activity of mono-quaternary ammonium derivatives of cinnamic acid (in Russian). Farmakol Toksikol 5:562–567
Kharkevich DA, Skoldinov AP, Ibadova DN (1973) On the effect of adamantyl radicals on the mechanism of cholinergic action of monoquaternary ammonium compounds (in Russian). Farmakol Toksikol 2:201–205
Kharkevich DA, Skoldinov AP, Ibadova DN (1974) On the significance of adamantyl radicals for the mechanism of neuromuscular action of bis-quaternary ammonium compounds (in Russian). Farmakol Toksikol 3:166–171
Kharkevich DA, Skoldinov AP, Klimova NV, Lavrova LI, Agafonova VP (1981) On the curare-like activity of quaternary ammonium compounds with 2-adamantyl radicals (in Russian). Farmakol Toksikol 6:670–672
Kier LB (1971) Molecular orbital theory in drug research. Academic, London
Mikhelson MYa, Zeimal EB (1970) Atsetilkholin. O molekulyarnom mekhanizme deistviya (Acetylcholine. On the molecular mechanism of action). Leningrad, Nauka

Paton WDM (1956) Mode of action of neuromuscular blocking agents. Br J Anaesth 28:470–480
Paton WDM, Zaimis EI (1949) Pharmacological actions of polymethylene bis-trimethylammonium salts. Brit J Pharmacol 4:381–400
Pryanishnikova NT, Drozhzhin AP, Feldshtein MM (1974) On the relationship between some physicochemical properties of neuromuscular blocking agents and their mechanism of action (in Russian). Farmakol Toksikol 4:418–421
Pryanishnikova NT (1973) On the relationship between the analgetic effect and surface activity in some narcotic analgetics (in Russian). Farmakol Toksikol 2:195–197
Stenlake JB (1963) Some chemical aspects of neuromuscular block. In: Ellis GP, West GB (eds) Prog Med Chem 3:1–51
Triggle DJ, Moran JF, Barnard EA (eds) (1971) Cholinergic ligand interaction. Academic, New York
Tsai M-C, Mansour NA, Eldefrawi AT, Eldefrawi ME, Albuquerque EX (1978) Mechanism of action of amantadine on neuromuscular transmission. Mol Pharmacol 14:787–803
Warnick JE, Maleque MA, Bakry N, Eldefrawi AT, Albuquerque EX (1982) Structure-activity relationships of amantadine. I. Interaction of the N-alkyl analogues with ionic channels of the nicotinic acetylcholine receptor and electrically excitable membrane. Mol Pharmacol 22:82–93
Zutrauen HA (1956) Study of alkyltrimethylammonium salts of electrodes. J Chem Phys 53:54

CHAPTER 5

Prejunctional Actions of Cholinoceptor Agonists and Antagonists, and of Anticholinesterase Drugs

W. C. BOWMAN, A. J. GIBB, A. L. HARVEY, and I. G. MARSHALL

A. Introduction

In this chapter we discuss the vexed question as to whether prejunctional cholinoceptors exist at the neuromuscular junction, and if so, what physiological function they might serve. There is some controversy in the literature about several aspects of the subject that are best illustrated by posing a series of questions, as follows:

1. Does acetylcholine exert one or more actions on motor nerve endings, and are these mediated by nicotinic or muscarinic receptors, or both?
2. If acetylcholine can act on the motor nerve endings, has its action (or actions) at this site any physiological relevance in relation to the transmission process?
3. Do neuromuscular blocking drugs (e.g. tubocurarine, gallamine, decamethonium, snake α-toxins) act on motor nerve endings as well as on postjunctional cholinoceptors?
4. If they do act prejunctionally, is their action mediated through cholinoceptor sites in the nerve endings, or is it a quite separate action that is independent of their known ability to interact with nicotinic cholinoceptors?
5. Do anticholinesterase drugs (e.g. neostigmine, diisopropyl fluorophosphate) act on motor nerve endings as well as on junctional cholinesterase, and if they do so act, is their action direct or mediated by preserved transmitter acetylcholine?

Each of these questions has been, and continues to be, answered in contradictory ways by different workers. A number of reviews covering the subject have been published (WERNER and KUPERMAN 1963; STANDAERT and RIKER 1967; RIKER and OKAMOTO 1969; BOWMAN and WEBB 1972; GALINDO 1972; RIKER 1975; HOBBIGER 1976; MIYAMOTO 1978; BOWMAN 1980; FOLDES and VIZI 1980; BOWMAN and MARSHALL 1981), and therefore only the salient points, together with more recent papers, are dealt with here.

B. Cholinoceptor Agonists and Antagonists, and Anticholinesterase Drugs

I. Repetitive Antidromic Nerve Activity

The production of repetitive antidromic activity in the motor nerves by nicotinic agonists and anticholinesterase drugs, and block of this effect by antagonists that act at nicotinic receptors have been reviewed several times (BOWMAN and WEBB 1972; RIKER 1975; HOBBIGER 1976; MIYAMOTO 1978), and there is little new to add. The main references to the vast amount of original work will be found in these reviews.

MASLAND and WIGTON (1940) first showed that injection of small doses of acetylcholine into the popliteal artery of the anaesthetized cat, as well as stimulating the muscle, gave rise to action potentials in the motor nerve that could be recorded antidromically in the ventral root (backfiring). Masland and Wigton's observation with acetylcholine has frequently been confirmed in mammalian nerve–muscle preparation, not only with acetylcholine, but also with other nicotinic agonists including carbachol, suxamethonium and decamethonium. The muscarinic agonists metacholine and oxotremorine do not produce antidromic firing so that, whatever its origin, nicotinic rather than muscarinic receptors appear to be involved, and this is confirmed by the use of appropriate antagonist drugs. Thus, antagonists at nicotinic receptors abolish the backfiring produced by nicotinic agonists, whereas atropine in doses sufficient to block nuscarinic receptors is without effect. Abrupt close-arterial injection of a large dose of a nicotinic agonist causes a powerful brief contraction of the muscle. This is generally considered to arise from depolarization of the postjunctional motor end-plate membrane; neuromuscular transmission is subsequently temporarily blocked. With these large doses of nicotinic agonists, antidromic firing in the nerve commences and then abruptly ceases. That this is a self-blocking action is shown by the observation that subsequent smaller doses no longer produce antidromic firing during a period of several minutes.

Intravenous injection, or slow close-arterial injection of smaller doses, of nicotinic agonists produces muscle fasciculations (i.e. synchronous contractions of those muscle fibres comprising any one motor unit) and backfiring, without an initial powerful contraction of the whole muscle. Since, in this type of experiment, the ventral roots are severed, there is no question of the fasciculations arising reflexly from the muscle spindles, although in the intact organism such reflex activity augments the basic activity initiated at the neuromuscular junction. Twitches evoked by motor nerve stimulation during the period of fasciculations are augmented in size and associated with repetitive firing both in the muscle and in the nerve. During the self-blockade of the backdischarge, twitches are depressed or completely blocked.

MASLAND and WIGTON (1940) and FENG and LI (1941) also studied the actions of neostigmine and physostigmine respectively, in addition to those of acetylcholine. BROWN (1937) had previously shown that the twitch augmentation produced by physostigmine in the anaesthetized cat was associated with, and presumably a consequence of, repetitive firing of the muscle fibres. That is, physostigmine converted the muscle response to a single nerve shock into a brief asynchronous tetanus. MASLAND and WIGTON (1940) and FENG and LI (1941) went on to show that the nerve response also becomes repetitive. The initial evoked antidromic action potential in the ventral root was followed in the presence of neostigmine or physostigmine, by a burst of repetitive action potentials. In slightly larger doses both neostigmine and physostigmine, as well as other related drugs, produce antidromic nerve activity in the absence of electrically evoked nerve impulses. It thus seems that drug-induced repetitive firing, whether in the absence or in the presence of stimulus-evoked nerve impulses, is a consequence of the same basic mechanism, and that the nerve impulse merely serves to potentiate the drug effect.

Many other drugs classified as anticholinesterase agents including edrophonium, ambenonium, diisopropylfluorophosphate (for references, see reviews already cited), and more recently ecothiopate (MORRISON 1977), paraoxon (LASKOWSKI and DETTBAR 1979) and parathion (CLARK et al. 1979, 1983; CLARK and HOBBIGER 1983) have been shown to produce essentially the same effects as neostigmine at mammalian neuromuscular junctions both in vivo and in vitro, differing only in their time course of action. Treatments that increase or decrease acetylcholine release from the motor nerve endings, correspondingly modify the degree of anticholinesterase-induced repetitive firing in the nerve and muscle fibres after a single stimulus.

Antagonists of acetylcholine, classically thought to act at nicotinic receptors, including hexamethonium, tubocurarine, benzoquinonium, gallamine, pancuronium, and α-bungarotoxin, block repetitive firing in nerve and muscle, and the accompanying mechanical events (fasciculations and twitch augmentation), in doses which, because of the safety factor in transmission, are too small to affect normal twitches. The blocking action of these drugs is equally pronounced against the repetitive firing produced by nicotinic agonists or by anticholinesterase drugs.

The observations that nicotinic agonists (but not muscarinic agonists) and anticholinesterase drugs of diverse chemical types (carbamates, anilinium ions, oxamides, organophosphorous compounds) produce antidromic firing, together with the finding that nicotinic antagonists (but not muscarinic antagonists) block the actions of the agonists and of the anticholinesterase drugs, show conclusively that the effect is initiated by nicotinic cholinoceptors. The obvious explanation of the effect of anticholinesterase drugs is that they act by potentiating acetylcholine released from the nerve endings either spontaneously or by nerve impulses. The main argument against this idea has been that, in some series of compounds, the rank order of potency in producing repetitive firing may not be the same as the rank order of potency in inhibiting acetylcholinesterase in vitro. However, in vitro antienzyme studies do not take into account other features of molecules that may modify their ability to produce repetitive firing in an intact neuromuscular system. For example, ability to interfere with ion channel functions in excitable membranes, that is, actions such as those possessed by tetraethylammonium, 4-aminopyridine, guanidine or veratrum alkaloids, or an action to facilitate transmitter release such as that possessed by phenol, would potentiate the basic repetitive firing effect. On the other hand, a weak "curare-like" action, such as that possessed by methoxyambenonium (BLABER and KARCZMAR 1967), would oppose the repetitive firing effect. Actions of this sort would distort the rank order of potency determined from enzyme studies. The fact that drugs that block voltage-dependent ion channels (that is, those formerly known as membrane stabilizers) e.g. local anaesthetics, tetrodotoxin, phenytoin, meprobamate, barbiturates, general anaesthetics (Riker et al. 1959; van Poznak 1963; Riker and Standaert 1966; Raines and Standaert 1966; Usubiaga and Standaert 1968), but with no specific acetylcholine receptor antagonist action, also abolish repetitive firing does not detract from the view that cholinoceptors are involved in the initiation of the effect. Such drugs may act on the propagating mechanism at a stage after cholinoceptor stimulation.

The important question that arises is whether these cholinoceptors are located prejunctionally on the nerve endings, or postjunctionally on the motor end-plate membrane. The principal arguments supporting the involvement of a population of nerve ending nicotinic cholinoceptors that mediate the production of a nerve terminal depolarization are described in reviews (Bowman and Webb 1972; Riker 1975; Hobbiger 1976; Miyamoto 1978). It would be surprising if such receptors did not exist, since sensitivity to the nicotinic actions of acetylcholine seems to be a general property of nonmyelinated nerve membranes, having been demonstrated in nodes of Ranvier (Dettbarn 1960), sensory nerve terminals (Douglas and Ritchie 1960), crustacean motor nerve fibres (Dettbarn 1967), mammalian C fibres (Ritchie 1967), adrenergic nerve terminals (Kopin 1967), and the nerve endings at the locust neuromuscular junction (Fulton and Usherwood 1977). Consequently, it may be supposed that, in somatic motor nerves, cholinomimetic drugs depolarize the nonmyelinated terminal membrane (or some part of it) relative to the myelinated part of the axon, and thereby give rise to a standing potential resembling a generator potential at a sensory nerve ending. In a stimulated axon, the generator potential may sum with the negative afterpotential following the nerve impulse, so that the one potentiates the other.

Excessive agonist action may lead to block of repetitive firing and of fasciculations, because the cholinoceptors become desensitized, or because the excessive depolarization produced gives rise to a zone of conduction block in the affected terminal branches, or because the fall in terminal membrane potential reduces the quantal content of the end-plate potential as described by Hubbard et al. (1965).

The repetitive firing and fasciculations produced by anticholinesterase drugs in the absence of electrical stimulation of the nerve may have a similar basis to that described. The amount of acetylcholine released spontaneously far exceeds that formerly thought to be released in this way. In fact, as much as 99% of the spontaneous release of acetylcholine may be by continual leakage in a nonquantal manner (Katz and Miledi 1977). It may well be, therefore, that the total amount released spontaneously is adequate, when its destruction is prevented by cholinesterase inhibition, to diffuse to and excite prejuctional cholinoceptors and hence produce fasciculations. In the absence of electrical stimulation of the nerve the amount of acetylcholine released presumably remains below that necessary to cause block, since self-blockade of repetitive firing is not a feature of the response to even a large dose of an anticholinesterase drug under such conditions.

The main suggestion as to how postjunctional cholinoceptors could mediate antidromic firing in the nerve originates from experiments reported by KATZ (1962), who first proposed that a prolonged and increased postjunctional action of acetylcholine might result in sufficient efflux of K^+ from the muscle fibre into the junctional cleft to depolarize the nerve endings and give rise to antidromic repetitive firing. Although this "potassium hypothesis" was subsequently abandoned by Katz himself (KATZ 1969), it has recently been resurrected by HOHLFELD et al. (1981). These workers calculated postsynaptic potassium currents according to a simple mathematical model of the muscle membrane. Under conditions in which, according to the mathematical model, K^+ efflux from the rat diaphragm muscle would be enhanced (i.e. low extracellular $[K^+]$, exchange of chloride for nitrate, small muscle fibre diameter and consequent high input resistance as in young animals) antidromic repetitive firing following single stimuli was enhanced (although not markedly). The matter remains controversial, although on the whole the evidence in favour of a role for prejunctional cholinoceptors seems the most convincing.

In the absence of electrical stimulation of the nerve, the depolarization produced by a nicotinic agonist at any one neuromuscular junction may initially trigger propagating antidromic action potentials that will be conducted, not only into the corresponding ventral root, but also, by axon reflex, into all the branches of the motor unit. Hence the motor unit will respond as a whole, so that the mechanical response is fasciculation (synchronous contractions of fibres within a motor unit) rather than fibrillation (asynchronous contractions of individual fibres). Whether activity also spreads orthodromically in those axon branches postulated to respond by the production of "generator potentials" is not known. Conduction into mammalian nerve terminals even under normal conditions, depends on ionic mechanisms that differ from those in the preterminal axon (BRIGANT and MALLART 1982).

The question arises as to whether the repetitive activity in the nerve gives rise to the muscle repetition, or whether they are independent events. Transmitter acetylcholine, preserved in the presence of an anticholinesterase drug, may make multiple postjunctional receptor interactions (MAGLEBY and TERRAR 1975) and thereby give rise to a prolonged end-plate potential which would trigger repetitive muscle spikes, as first described by ECCLES et al. (1942). Likewise, evoked end-plate potentials superimposed on a background depolarization produced by a cholinesterase-stable nicotinic agonist might remain above threshold long enough to evoke repetitive action potentials in the muscle fibre membranes after each nerve shock. Repetitive muscle action potentials can be produced by anticholinesterases when antidromic nerve repetition has been blocked by a reduced $[Ca^{2+}]$:$[Mg^{2+}]$ ratio (CLARK 1982), by treatment with dithiothreitol (CLARK et al. 1979), or by treatment with certain neuronal depressant drugs, such as phenytoin or local anaesthetics (RIKER et al. 1959; RIKER and STANDAERT 1966). On the other hand, in some instances, end-plate potentials in response to single nerve shocks, rather than becoming prolonged, have been found to become repetitive in the presence of anticholinesterase drugs (edrophonium: BLABER 1972; paraoxon; CLARK 1982). It therefore seems that muscle repetition can be elicited solely by a postjunctional action, but it may be augmented or, rarely, actually initiated by nerve repetition.

It is difficult to see how antidromic repetitive firing, were it to be evoked by the transmitter, could play any physiological role in the transmission process, and so we are inclined to the view that the particular prejunctional cholinoceptors postulated to mediate the effect merely reflect a general tendency of nonmyelinated axonal membranes to respond to cholinomimetics; perhaps they are a vestigial remnant of a function long lost during phylogeny. Confirmation that repetitive firing of this type, both in the muscle and in the nerve, is irrelevant to the normal transmission process, comes from the observation that it does not occur at frequencies of nerve stimulation above 2 Hz (WERNER 1960; BLABER and BOWMAN 1963), and is therefore absent at frequencies within the physiological range. The effects of stimulation frequency on the responses to cholinoceptor agonists and anticholinesterase drugs are complex (for review see BOWMAN and WEBB 1972). At least two factors might contribute to absence of repetition after the second and subsequent nerve impulses of a high frequency train. (a) High frequency stimulation, by augmenting the positive afterpotentials, may hyperpolarize the site of the generator potential in the nerve endings to the extent that neuronal repetitive firing either is not initiated or is quickly abolished. (b) The high frequency nerve discharge may deplete the readily releasable transmitter to the extent that output falls below the level necessary, either to stimulate prejunctional cholinoceptors, or to stimulate the postjunctional cholinoceptors sufficiently repeatedly to prolong the end-plate potential beyond the refractory period of the muscle fibre membrane. Thus, twitch augmentation, fasciculation, and repetitive firing in nerve and muscle are characteristic of cholinoceptor agonist or anticholinesterase drug action only when the frequency of evoked nerve impulses is low, in fact below the physiological range. Fasciculations are produced in resting muscles, but repetitive firing would not accompany physiological muscle movements, apart, perhaps, from the first nerve impulse initiating the movement.

Higher doses of anticholinesterase drugs in the presence of high frequency stimulation produce a rapidly developing waning of tetanic tension. The effect is a consequence of neuromuscular block, and is generally regarded as a postjunctional block produced by accumulating acetylcholine (LI and TING 1941; DOUGLAS and PATON 1954), analogous to the block by depolarization produced by stable nicotinic agonists such as decamethonium (BURNS and PATON 1951). Alternatively, it may be the result of postjunctional receptor desensitization (THESLEFF 1959). It seems clear that there is a large component of postjunctional block in the waning tetanus. However, there may also be a component of prejunctional block at some end-plates, especially when the frequency of stimulation is high. In unpublished experiments on the isolated phrenic nerve–diaphragm of the rat in which the release of radiolabelled acetylcholine was measured, E.T. Abbs and D.N. Joseph (1981, personal communication) found that, during stimulation at 50 Hz, neostigmine produced failure of neuromuscular transmission associated with a marked diminution of acetylcholine output. (For method, see ABBS and JOSEPH 1981.) Probably the transmitter acetylcholine, preserved by neostigmine, depressed its own further release by depolarizing the nerve endings. Evidence that this was so was provided by the observation that tubocurarine restored the acetylcholine output at 50 Hz, presumably by blocking the depolarization of the nerve endings produced by the preserved transmitter. By restoring output towards normal, tubocurarine, under these conditions, thus appeared to cause an increase in acetylcholine release. When neostigmine was omitted from the experiment and the total ^3H outflow measured (presumably choline), stimulation at 50 Hz no longer caused block nor decreased output, and tubocurarine did not increase output. Abbs and Joseph were aware of the possible errors inherent in collecting choline rather than acetylcholine, but a not unlikely interpretation of their results is that when the released transmitter is hydrolysed with normal rapidity, it does not persist in the junctional cleft in high enough concentrations to reach and react with the prejunctional cholinoceptors postulated to mediate terminal depolarization. MILEDI et al. (1978) had previously described an increase in evoked acetylcholine release produced by α-bungarotoxin, again in the rat diaphragm preparation. Inevitably, cholinesterase had been inhibited, this time by diisopropylfluorophosphate. They used a lower frequency of stimulation (3 Hz), but it is possible that transmission was impaired at some junctions by terminal depolarization even at this lower frequency. α-Bungarotoxin may then have increased acetylcholine output by blocking the terminal depolarization produced by the preserved transmitter. There are other explanations of these results (MILEDI et al. 1978), and in fact the authors did not include the one suggested here.

An analogous nerve terminal block by depolarization may account for the reduction in the quantal content of the end-plate potential during phase II of the neuromuscular block produced by decamethonium in isolated guinea-pig muscle (CREESE et al. 1982). E. T. Abbs and D. N. Joseph (1981, personal communication) found that decamethonium, in a concentration of 10^{-7} *M* (lower concentrations had a different effect) produced neuromuscular block in the rat diaphragm preparation which was associated with a reduced output of ^3H (presumably choline, since no anticholinesterase was present in this experiment). They interpreted the result as an indication that decamethonium may impair transmission by depolarizing the nerve endings.

The prejunctional cholinoceptors postulated to mediate a localized nerve terminal depolarization are stimulated by stable or relatively stable depolarizing drugs (e.g. decamethonium, suxamethonium, carbachol) but are apparently not affected by transmitter acetylcholine unless the junctional cholinesterase is inhibited. It may therefore be, as suggested by HOBBIGER (1976), that the prejunctionally located acetylcholinesterase serves as a barrier to protect the nerve endings from depolarization by released transmitter. Although these prejunctional cholinoceptors apparently have no physiological function, they obviously have pharmacological importance, since they may play a part in the muscle fasciculations produced by anticholinesterase drugs or depolarizing drugs, such as suxamethonium; they may also contribute to the block of transmission produced by depolarizing drugs, and to the muscle weakness produced by anticholinesterase drugs during exercise of normal muscles.

Stimulation of these prejunctional receptors by accumulating acetylcholine is not likely to play a part in the "anticurare" actions of anticholinesterase drugs, since these receptors are the first to be blocked by curare-like drugs, and there is no antidromic repetitive firing in the nerve during the anticurare action of neostigmine and related drugs. Likewise, they are probably not involved in the improvement of muscle power produced by anticholinesterase drugs in myasthenia gravis, since repetitive firing and fasciculations are not associated with the enhanced muscle response in the disease. Possibly the circulating cholinoceptor antibody impairs the functioning of the prejunctional receptors as well as of the postjunctional receptors.

II. Tetanic Fade and Rundown of Trains of Nerve-Evoked Responses

1. Tension Depression and Fade are Separate Actions of Neuromuscular Blocking Agents

Tubocurarine, and drugs that act in a similar manner, not only depress peak tension, but also produce fade of responses as illustrated by the familiar phenomena of tetanic fade (HOFMANN 1903; WEDENSKI 1903; PATON and ZAIMIS 1951) and train-of-four fade (ALI et al. 1971). The electrophysiological counterparts of tetanic fade are the tetanic rundown of EPPS (LILEY 1956b) and EPCS (GLAVINOVIĆ 1979; and see Fig. 5). The apparently increased potency of tubocurarine that occurs with increased frequency of stimulation (PRESTON and VAN MAANEN 1953; BLACKMAN 1963) is a reflection of the same basic effect.

In the anaesthetized cat, differences in the action of tubocurarine on tetani are observed with different routes of drug administration (Fig. 1). When given intravenously tubocurarine produced the familiar depression of peak tetanic tension, associated with fade of tension during the tetanus. However, when the drug was administered by close-arterial bolus injection of a smaller dose, despite an equal degree of peak tension depression, virtually no fade was discernible. Intra-arterial infusion of tubocurarine produced a similar effect to that of intravenous injection. If the depression of peak tension and tetanic fade were a consequence of the same action of tubocurarine, then the same degree of tension depression would be expected to be accompanied by the same degree of fade. The different effects of close-arterial and intravenous bolus injections on fade therefore suggest that the two phenomena are due to separate actions of tubocurarine.

One explanation may be that tubocurarine binds more rapidly with the site of action involved in tension depression, presumably the recognition sites of postjunctional cholinoceptors, than it does with the site causing fade. After intra-arterial injection, the tubocurarine presumably combines with the postjunctional cholinoceptor recognition sites on its first passage through the muscle, hence reducing peak tension. However, if combination with the sites involved in fade is slower, there may not be sufficient time for it to occur in a single passage of the drug through the muscle. When the tubocurarine is injected by intra-arterial infusion, there would then be time for it to combine not only with the recognition sites of the postjunctional cholinoceptors, but also with the sites that produce tetanic fade.

A similar explanation probably underlies the observation (SUGAI et al. 1976) that during the course of a block in humans, tetanic fade is slower to develop than

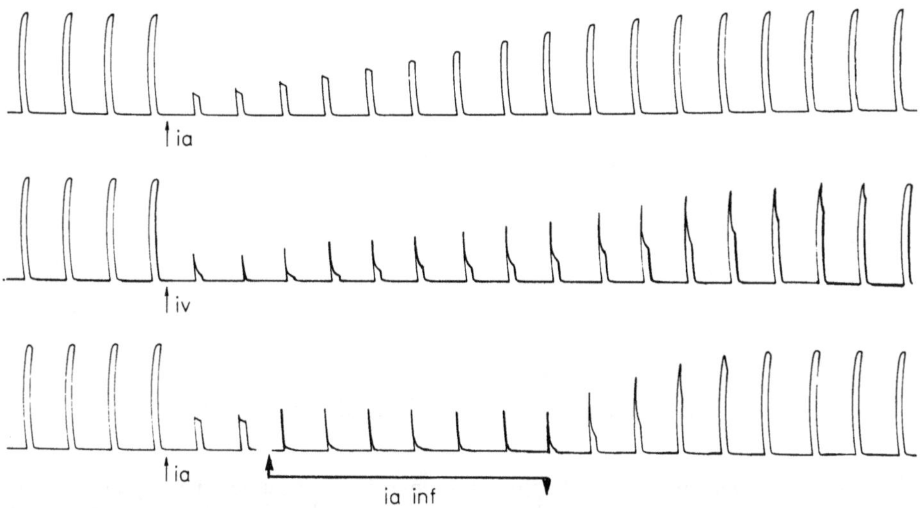

Fig. 1. Cats; chloralose anaesthesia. The *top* and *middle tracings* are from the same experiment. The *bottom tracing* is from a different experiment. Tetanic contractions of a tibialis anterior muscle were evoked by stimulating the motor nerve at a frequency of 50 Hz for 5 s in every 30 s. Tubocurarine was injected at the *arrows*, intra-arterially (*ia* 50-μg bolus *top* and *bottom records*), intravenously (*iv* 0.3-mg/kg bolus, *middle record*), or as an intra-arterial infusion (*ia inf* 9 μg-kg^{-1} min^{-1}). Note that while all injections depressed tension, only the intravenous bolus and intra-arterial infusion produced tetanic fade

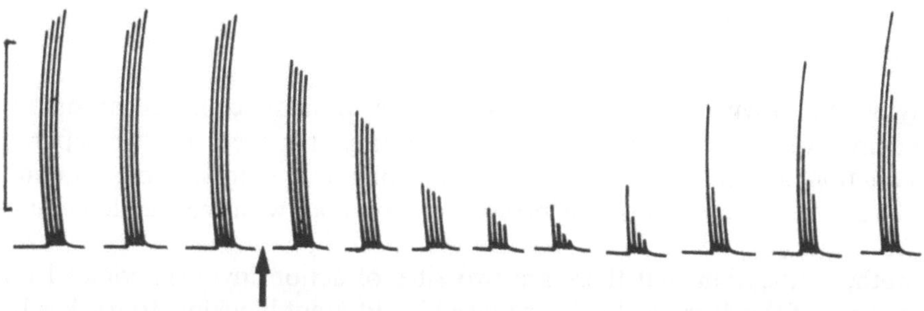

Fig. 2. Cat, chloralose anaesthesia. Maximal twitches of a tibialis muscle were evoked by stimulation of the motor nerve with train-of-four stimulation (2 Hz for 2 s) applied every 20 s. Vecuronium (30 μg/kg) was injected intravenously at the *arrow*. Selected groups of stimuli are illustrated during the onset of and the recovery from the block. Groups were selected in pairs to show matching degrees of block of the first twitches in each pair of groups during onset and recovery. Note that fade of trains is much more marked during recovery than during onset; that is, fade has a slower onset than has initial twitch depression. (Reprinted with permission from the International Anesthesia Research Society from "Prejunctional and postjunctional Cholinoceptors at the Neuromuscular Junction," by W. C. Bowman, Anesthesia and Analgesia, vol 59, pp 935–943)

is depression of peak tetanic tension. Tetanic fade is presumably an exaggeration of train-of-four fade, and it is a common observation, as illustrated in Fig. 2, that for equal degrees of block of the initial twitch in a group, train-of-four fade is more pronounced during recovery than during the onset of the block. This is clearly so with the first dose of blocking drug. It is important to emphasize that

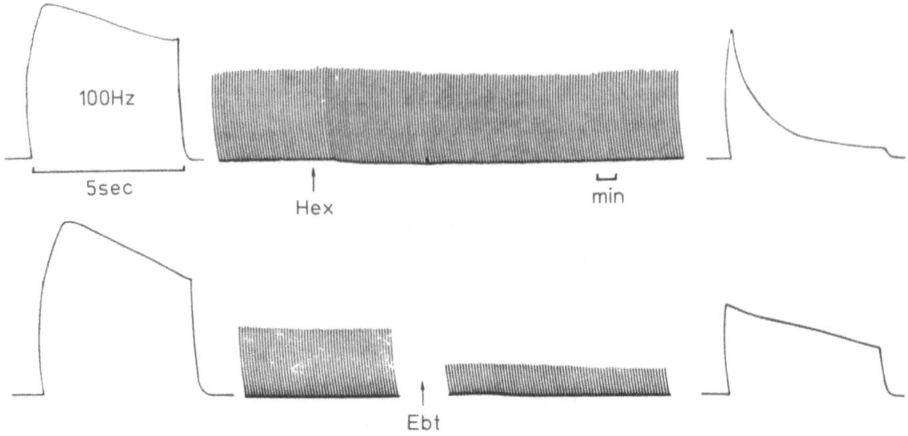

Fig. 3. Isolated phrenic nerve hemidiaphragm preparations of the rat. The *top* and *bottom traces* are from different experiments. Twitches were evoked by stimulating the motor nerve at a frequency of 0.1 Hz, and tetani (100 Hz for 5 s) were interposed. The tetani were recorded on faster moving paper and the gain for the tetani was halved. At *Hex*, 150 μg/ml hexamethonium was added to the bath. This dose was too small to depress the twitches, but produced marked tetanic fade in the subsequent tetanus. At *Ebt*, 0.32 μg/ml erabutoxin b was added to the bath. The twitches were slowly depressed by more than 50%. Tetanic tension was also depressed, but tetanic fade was no greater than that occurring in the control tetani at the beginning of the experiment

it may not be so with subsequent doses, because cumulative effects can modify the picture. Again, the most obvious explanation is that there are two separate sites of action, and that the first dose of a blocking drug combines more rapidly with the site that causes tension depression than it does with the site that causes fade.

Another indication that there are two sites of action involved comes from comparisons of the degrees of fade produced by different blocking drugs. The literature contains scattered references to the fact that, for equal degrees of tension depression, the degree of tetanic fade produced by some neuromuscular blocking drugs (e.g. toxiferine I, PATON and PERRY 1951; pancuronium, HUNTER 1970) is less than that produced by tubocurarine. BOWMAN and WEBB (1976) made a systematic comparison of the effects of a number of neuromuscular blocking drugs in the cat, and confirmed that tetanic fade is more pronounced with tubocurarine than with pancuronium. Similar observations have been made in humans from train-of-four stimulation experiments (WILLIAMS et al. 1980). Further observations in the cat were that the ganglion-blocking drug hexamethonium, which has only very weak neuromuscular blocking activity, produced complete tetanic fade to zero tension with doses that were too small to depress twitches evoked at 0.1 Hz and which caused only minimal depression of peak tetanic tension (BOWMAN and WEBB 1976). Conversely, the snake toxin, α-bungarotoxin, produced pronounced depression of both twitch and peak tetanic tension without producing any tetanic fade (LEE et al. 1977).

The difference between some blocking drugs in this respect can also be demonstrated in vitro. Figure 3 illustrates experiments on the isolated phrenic nerve–hemidiaphragm preparations of the rat. Only the two extreme effects (with hexamethonium and a snake toxin) are illustrated. With hexamethonium, a pronounced tetanic fade is produced even by concentrations that are too small to depress the twitches. On the other hand, the snake toxin erabutoxin b, which resembles α-bungarotoxin in its action (TAMIYA and ARAI 1966), produced pronounced twitch and peak tetanic tension depression, but virtually no tetanic fade. In isolated muscle, the clinically used neuromuscular blocking drugs tubocurarine, pancuronium, vecuronium, and atracurium all produced tetanic fade at lower concentrations than those producing single twitch depression, but no statistically significant difference between the drugs was detectable with respect to their relative ability to produce these two effects (GIBB and MARSHALL 1983). The results with hexamethonium and snake toxins both in vivo and in vitro are indicative of separate sites of action mediating depression of peak tension and tetanic fade. The apparent lack of selectivity for tension or fade of the clinically used neuromuscular blocking drugs when used in vitro might reflect a species difference (cat vs rat), or could be the result of the equilibrium conditions that pertain in vitro. If the differences in selectivity in vivo are determined by differences in the rates of combination with the two sites, as suggested, then they will be influenced by redistribution, metabolism and excretion; these variables play no part in vitro.

2. Possible Mechanisms Underlying Drug-Induced Fade and Rundown

It has long been known that in nerve–muscle preparations paralysed by tubocurarine, end-plate potentials run down in amplitude during a train of repetitive stimulation. This has been ascribed to a falloff in transmitter release during the train (HUTTER 1952; OTSUKO et al. 1962). Thus HUTTER (1952) observed that the mechanical response of a partially curarized cat muscle to close-arterial injection of acetylcholine remained at control level during tetanic fade. OTSUKA et al. (1962) obtained analogous results by electrophysiological techniques in frog muscle, and one of their experiments is illustrated in Fig. 4. They showed that when a train of evoked end-plate potentials, recorded in the presence of tubocurarine, had run down to about 50% of their initial amplitude, the response to iontophoretic application of acetylcholine did not differ from that recorded 10 s later when the evoked EPP had returned to control amplitude. Both HUTTER (1952) and OTSUKO et al. (1962) concluded that at the time of maximum tension fade or EPP rundown, there was no progressive reduction in sensitivity to acetylcholine during the train, and therefore the waning responses to nerve stimulation must reflect a falloff in acetylcholine release with successive nerve impulses evoked at a high frequency. It was generally supposed that in the absence of blocking drug, the waning transmitter output also occurred, but was not normally reflected in waning tetanic tension, because of the considerable safety margin in transmission (PATON and WAUD 1967; and see GINSBORG and JENKINSON 1976; and HOBBIGER 1976) in terms of both an excess of acetylcholine released and of an excess of cholinoceptors. It was further supposed that it was only when a pro-

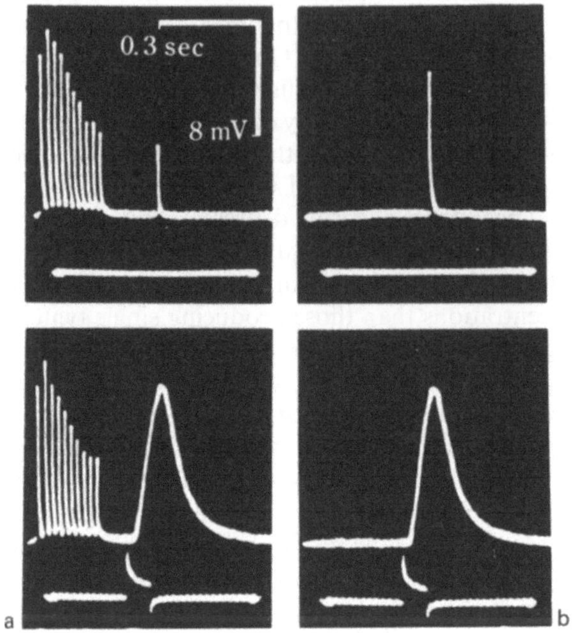

Fig. 4 a, b. End-plate potentials and acetylcholine potentials following a conditioning tetanic stimulation (50 Hz). Frog nerve–muscle preparation treated with tubocurarine 2×10^{-6} *M*. In each record, the *upper trace* shows the membrane potential recorded from the end-plate region. In the *lower trace* the strength of the current through the acetylcholine pipette was monitored. **a** end-plate potential (above) and acetylcholine potential (below) 0.18 s after repetitive stimulation, **b** end-plate potential (above) and acetylcholine potential (below) 10 s after repetitive stimulation. Conditioning stimuli were given every 20 s. Resting potential was 80 mV. Monitor calibration: 8 mV scale $= 1.9 \times 10^{-7}$ A. (OTSUKA et al. 1962)

portion of cholinoceptors was occluded by an acetylcholine antagonist, and the safety margin thereby reduced, that the waning transmitter output was unmasked, and was expressed as a fading tension response or a rundown of EPPS. Thus, according to the traditional view, drugs such as tubocurarine unmask the consequences of a spontaneously occurring waning in transmitter output, but they do not cause the waning. In order to study this problem it is necesary to use nerve–muscle preparations in which normal transmitter release may be estimated before and during treatment with tubocurarine. To this end, cut muscle preparations have been developed, and recently problems of nonlinear summation of end-plate potentials have been overcome by the use of voltage clamp techniques. When such techniques have been used to prevent contraction, little or no rundown of end-plate responses (potentials or currents) has been recorded in the absence of drugs, particularly when recordings are made near normal body temperature; a fall in temperature can itself lead to a rundown of responses. However, when tubocurarine is added to such a preparation, a striking rundown of responses becomes evident. Figure 5 illustrates an experiment of GLAVINOVIĆ (1979) which shows the effect of tubocurarine on trains of end-plate currents re-

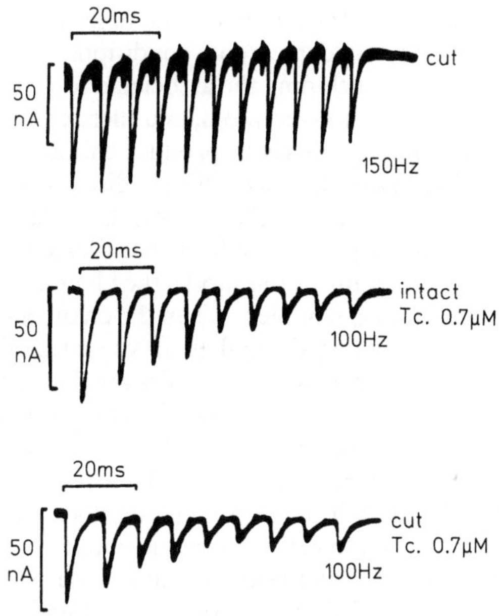

Fig. 5. Rat diaphragm fibres, voltage clamped at 60 mV. End-plate currents (EPC) generated by short tetanic trains of stimuli applied to the nerve. Note that in the absence of tubocurarine there is relatively little diminution of the responses in the cut fibre (*top*), but in the presence of tubocurarine, whether the fibre is intact (*middle*) or cut (*bottom*), tetanic rundown of responses is clear. (Compiled from Fig. 2 of GLAVINOVIĆ 1979)

corded from cut, voltage-clamped muscle fibres. Several experiments of this type now leave little doubt that tubocurarine and related drugs cause (rather than merely unmask) the rundown of end-plate responses during high frequency stimulation. Two possible mechanisms whereby this effect and the consequent tetanic fade might occur, may be considered.

a) Occlusion of Postjunctional Cholinoceptor-Operated Ion Channels

Despite the preexisting evidence that tetanic rundown in the presence of tubocurarine is a consequence of a falloff in transmitter release (HUTTER 1952; OTSUKA et al. 1962), and the more recent evidence that there is no rundown in the absence of tubocurarine, there has been, and to an extent still is, a reluctance to put these observations together and conclude that tubocurarine actually causes a progressive diminution in transmitter release during high frequency stimulation. Instead many physiologists and pharmacologists would prefer to invoke explanations based on postjunctional actions for both tension depression and rundown or fade.

One possibility that should be considered is that the blocking drugs, in addition to blocking the receptor recognition sites, also occlude the ion channels during tetanic stimulation, and that this type of "use-dependent" block is the basis of the rundown in end-plate responses and of tetanic fade. Work by MANALIS (1977), KATZ and MILEDI (1978) COLQUHOUN et al. (1979), LAMBERT et al. (1980),

Colquhoun and Sheridan (1981), and others described by Colquhoun in Chap. 3, shows that, at least under certain conditions, many neuromuscular blocking drugs are capable of occluding the associated ion channels as well as of blocking the recognition sites of the receptors, and there does seem to be a broad correlation between the ability of drugs to produce this action and their ability to cause tetanic fade. The matter has been clearly discussed by Dreyer (1982). The most widely accepted model for channel block is that first proposed by Adams (1975, 1976) and subsequently confirmed and extended by Ruff (1977) and Adler et al. (1978). According to this model the blocking drug interacts with the open form of the ion channel complex. When the channel is opened, the drug associates with the channel binding site and effectively plugs the channel, blocking the flow of ions and hence impairing transmission. This effect increases with membrane hyperpolarization, i.e. it is voltage dependent. Obviously if the unblocking rate is very slow the channel will remain blocked for a long time. During repetitive activation of the receptors, as evoked by a train of nerve-evoked impulse, a use-dependent block will develop if the unblocking rate is sufficiently slow that some channels are still blocked from one impulse when the second and subsequent impulses arrive; this will lead to tetanic fade. Evidence that such a mechanism can lead to tetanic fade and to rundown of end-plate responses was supplied by Gibb and Marshall (1982) with the ganglion-blocking drug trimethaphan. They found that the primary effect of trimethyphan in cut rat hemidiaphragm preparations was to reduce the time constant of decay of end-plate and miniature end-plate currents in a voltage-dependent manner, suggesting a channel-blocking action. It also produced a rundown of trains of end-plate currents that was voltage dependent. If this effect is mainly postjunctional in origin, a similar rundown of end-plate currents would be expected to occur if acetylcholine were applied to the end-plate by iontophoresis rather than by release from the nerve. During trains of iontophporetically evoked end-plate currents at a frequency of 50 Hz, trimethaphan produced a voltage-dependent rundown of the current trains (Fig. 6 a) in a manner similar to that which occurred during nerve stimulation at 50 Hz. These results confirm that channel block can result in tetanic fade. Channel block may at least contribute to the tetanic fade produced by hexamethonium. Hexamethonium has been shown to produce channel block in isolated rat parasympathetic ganglia (Ascher 1979) and also at the frog neuromuscular junction (Milne and Byrne 1981), although A. J. Gibb and I. G. Marshall (1981, unpublished work) were unable to confirm any channel-blocking action of hexamethonium in the rat diaphragm. In any case, a demonstrable channel-blocking action of some drugs does not preclude the possibility that other drugs that produce tetanic fade and rundown, e.g. tubocurarine and other clinically used neuromuscular blocking drugs, may act by a mechanism other than channel block.

Evidence that channel block is unlikely to be the cause of tetanic rundown produced by tubocurarine was provided by Magleby et al. (1981) who found that the rundown of end-plate current trains in the presence of the drug was not voltage dependent. Gibb and Marshall (1983; and unpublished work) have obtained similar findings with both tubocurarine and vecuronium. In addition, it has been shown that these drugs, in contrast to channel-blocking drugs, produce no run-

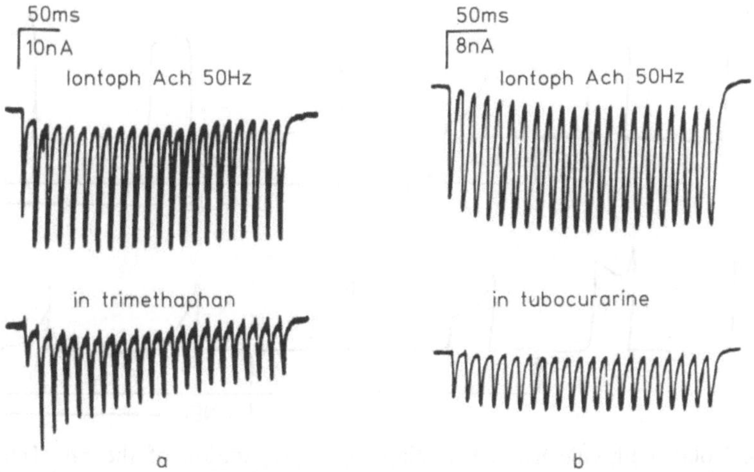

Fig. 6 a, b. Cut rat diaphragm fibres, voltage clamped at −60 mV. End-plate currents generated by iontophoretically applied jets of acetylcholine at a frequency of 50 Hz before and in the presence of either trimethaphan or tubocurarine. Note the rundown of responses in the presence of trimethaphan, but uniform depression with absence of rundown in the presence of tubocurarine. **a** and **b** are from different experiments. From A. J. GIBB and I. G. MARSHALL (1981, unpublished work)

down of trains of end-plate currents elicited by iontophoretically applied acetylcholine, although, as would be expected, the amplitude of the currents is reduced. Figure 6 b illustrates the effect of tubocurarine. It should be noted that ALBUQUERQUE et al. (1979) found a quantitative difference between the abilities of a channel blocker (perhydrohistrionicotoxin) to block "intrinsic" end-plate responses (i.e. those evoked by transmitter acetylcholine acting on receptors presumably near the middle of the end-plate region) compared with "extrinsic" end-plate responses (i.e. those evoked by inotophoretically applied acetylcholine and presumably located more peripherally). The authors considered that different populations of receptors could be involved, on the one hand, in neuromuscular transmission and, on the other hand, in responses to exogenously applied acetylcholine. However, the two populations of receptors did not behave differently towards tubocurarine, and so the absence of rundown of iontophoretically induced responses in the presence of this drug (Fig. 6 b) in the experiments of GIBB and MARSHALL (1983) seems to be another valid indication that the ability of tubocurarine to produce fade of responses evoked by repetitive nerve stimulation is not a consequence of ion channel occlusion.

A further observation that argues against an important role of channel occlusion in the production of tetanic fade by clinically used neuromuscular blocking drugs, is that anticholinesterases, such as neostigmine, not only increase the amplitude of tetani depressed by tubocurarine and related drugs, but also abolish tetanic fade (Fig. 7 a). Since the effects of channel occlusion are likely to increase with increased frequency of channel opening, as is produced by neostigmine (KATZ and MILEDI 1971), it would be expected, were channel occlusion impor-

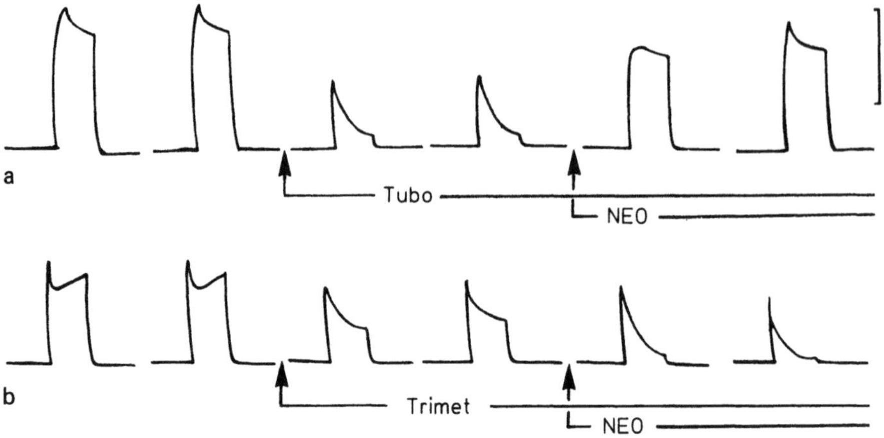

Fig. 7 a, b. Isolated phrenic nerve–hemidiaphragm preparation of the rat. Tetani were evoked by stimulating the motor nerve at a frequency of 50 Hz for 5 s every 30 s. At *Tubo, tubocurarine* (0.5 μg/ml), at *Trimet,* trimethaphan (50 μg/ml), and at *NEO,* neostigmine (0.5 μg/ml) were added to the bath. Once added, each drug was left in contact with the tissue throughout the parts of the experiments shown. The *upper* and *lower records* are from different experiments. Vertical calibration: 0.05 N. Both tubocurarine and trimethaphan depressed the contractions and produced tetanic fade. Neostigmine antagonized the effects of tubocurarine, but enhanced the fade produced by trimethaphan

tant, that neostigmine would enhance rather than relieve tetanic fade (DREYER 1982). In fact, when tetanic fade is produced by channel occlusion, as is the case with trimethaphan, neostigmine does enhance the falloff in tension (Fig. 7 b). It should be noted, however, that (as described in Sect. B.I) tetanic fade is produced by neostigmine in the absence of any other drugs, and the effect of neostigmine in conjunction with trimethaphan illustrated in Fig. 7 b should be examined by the appropriate electrophysiological techniques before definite conclusions regarding the underlying mechanism are made.

A study of the literature fails to show any instance in which a clinically used neuromuscular blocking drug has been found to produce channel occlusion at other than abnormally raised membrane potentials or at doses higher than those required to block receptors. Channel occlusion appears to make a relatively greater contribution to the effects of neuromuscular blocking drugs in muscles whose receptor recognition sites are relatively insensitive to them, e.g. those of the frog, so that high concentrations of the drug can be applied without extinguishing all measurable responses. The question therefore arises as to whether channel occlusion has any relevance to the clinical use of neuromuscular blocking drugs in humans. We are inclined to believe that, at the most, it may come into play only with abnormally large doses.

b) Prejunctional Action of Nicotinic Antagonists and Agonists

α) Antagonists. The experiments of GIBB and MARSHALL (1983), illustrated for tubocurarine in Fig. 6 b, exclude not only ion channel occlusion as a basis for fade phenomena produced by neuromuscular blocking drugs, but also any other type of postjunctional effect, including a so-called metaphilic effect. RANG and RITTER (1969) obtained evidence that agonists may cause a change in the molecular structure of the receptor (a metaphilic effect), such that the affinity of the receptor for certain antagonists is increased. However, the possibility that repetitive bombardment of the receptors by the transmitter during a tetanus may increase the binding of neuromuscular blocking drugs, and thereby give rise to fade, is excluded by the results of GIBB and MARSHALL (1983).

Since a postjunctional action underlying fade produced by neuromuscular blocking drugs is excluded by the results already referred by, a prejunctional action on the nerve endings must be responsible. In any case, an explanation depending upon an entirely postjunctional action ignores the earlier evidence (e.g. OTSUKA et al. 1962) that tetanic fade and rundown of EPPS are a consequence of a diminishing acetylcholine release per nerve impulse during each train of stimuli.

Thus, the overall impact of the evidence is to suggest that tubocurarine and related drugs act on the nerve endings to impair the release of transmitter, specifically during repetitive nerve activity above a certain minimal frequency (about 1 Hz). In fact several workers have proposed, on the basis of end-plate potential records, that tubocurarine impairs mobilization of acetylcholine within the nerve endings (LILLEHEIL and NAESS 1961; HUBBARD et al. 1969; MAENO and NOBE 1970; HUBBARD and WILSON 1973; BLABER 1973). This effect occurred with slower onset than the postjunctional effect. Such an action on transmitter mobilization may explain the reports of decreased quantal content recorded in the presence of pancuronium (SU et al. 1979) or vecuronium (TORDA and KILOH 1982a, b), as both drugs were studied at rates of nerve stimulation (1–2 Hz) that might be expected to result in rundown of recorded responses.

If it is accepted that curare-like drugs may exert this prejunctional action, it is necessary to consider the results of various workers who have attempted to measure directly the effect of tubocurarine on acetylcholine release. The original observations were made by DALE et al. (1936) who failed to detect any effect of tubocurarine on the release of acetylcholine evoked by nerve stimulation in the muscle of the tongue perfused by Locke's solution. Most other workers, using isolated or perfused muscles, have also reported an absence of any effect of tubocurarine on acetylcholine release (EMMELIN and MACINTOSH 1956; KRNJEVIĆ and MITCHELL 1961; CHEYMOL et al. 1962; CHANG et al. 1967; FLETCHER and FORRESTER 1975). The main exception has been BEANI et al. (1964) who found that tubocurarine did reduce acetylcholine release, but only under certain conditions. At normal body temperature, tubocurarine had no effect on acetylcholine output when the stimulation frequency was low, but it decreased the output when the stimulation frequency was raised to 50 Hz (i.e. a frequency at which transmitter mobilization is important). At low temperatures, tubocurarine reduced release even at low frequencies of stimulation. These results of BEANI et al. (1964) seem to match those deduced from electrophysiological recording, but it should be noted that CHANG et al. (1967) under apparently closely similar conditions, failed to confirm them. There are problems of interpretation of the results of experiments designed to assay released acetylcholine, and these arise for three main reasons.

1. Assay methods, expecially bioassay methods, may not be precise enough to detect what might be quite a small falloff in output, occurring only at certain stimulation frequencies and influenced by temperature.

2. For obvious reasons, collection of transmitter has almost invariably been made after inhibition of cholinesterase, and there are two reasons why this might confuse the issue. First anticholinesterase drugs prevent the production of tetanic fade, so that, in the presence of such drugs, the effect of tubocurarine, about which evidence is being sought, may have been prevented. Second, as explained in Sect. B.I of this chapter, anticholinesterase drugs may allow accumulating acetylcholine to depolarize some nerve terminals and thereby *reduce* acetylcholine release. Tubocurarine, by blocking this effect, restores the release, giving the appearance of an increased output. Such an increased output at some junctions coupled with a possible reduced output at others, may lead to little change in measured overall output.

3. It has long been suspected that some of the acetylcholine that is released spontaneously is nonquantal, but until more recently it was not realized that most of the spontaneous release, perhaps as much as 99%, is in this category (KATZ and MILEDI 1977). It is generally believed that only the acetylcholine capable of being released quantally gives rise to MEPPS and EPPS, and is immediately concerned in transmission. However, all of the acetylcholine released contributes to that measured by assay. In the face of a large spontaneous release, any falloff in the fraction concerned with transmission (though detectable by electrophysiological techniques) may be negligible in the total assayed release.

β) Agonists. The class of drugs already referred to that produce tetanic fade and associated phenomena apparently by a prejunctional action to impair transmitter mobilization, are all antagonists of acetylcholine at nicotinic receptors. This implies the existence of prejunctional nicotinic autoreceptors specifically concerned with mobilization of transmitter. If such receptors exist, agonists at nicotinic receptors, especially acetylcholine, should exert the opposite effect to antagonists, causing increased mobilization of transmitter. Some evidence in support of this possibility exists, but at the present time this is perhaps the weakest part of the story.

BLABER (1970) showed that, in appropriate low concentrations, decamethonium, which can be considered to have acetylcholine-like actions, increased the mobilization of transmitter in the nerve endings of the isolated cut tenuissimus muscle of the cat. This action is of course quite separate from its neuromuscular blocking action. BLABER (1970) also showed that this action of decamethonium was blocked by tubocurarine, suggesting that a nicotinic cholinoceptor was involved. E.T. Abbs and D.N. Joseph (1981, personal communication) obtained supporting results. They showed that decamethonium (10^{-8} *M*) facilitated the release of labelled choline from the isolated phrenic nerve–hemidiaphragm preparation of the rat at high frequencies of stimulation (50 Hz). As already mentioned, a ten times higher concentration of decamethonium (10^{-7} *M*) inhibited the release of choline, presumably by depolarizing the nerve terminals. Nicotine (10^{-5} *M*) or acetylcholine (10^{-5} *M*), like low concentrations of decamethonium, facilitated the release of choline in the experiments of Abbs and Joseph; tubocurarine blocked their effects.

In addition, anticholinesterase drugs, such as neostigmine, can abolish the tetanic fade and decrease the rundown of end-plate responses produced by tubocurarine; such drugs presumably work by potentiating the action of the transmitter. The evidence that is presently not fully available is whether it is possible to reverse tubocurarine-induced tetanic rundown of end-plate currents by the application of appropriate concentrations of acetylcholine. However, in preliminary unpublished experiments, Gibb and Marshall found that, provided the tubocurarine-induced rundown of EPC trains was not pronounced, it was possible to reverse it to a small, but statistically significant extent with iontophoretically applied acetylcholine. It is possible that the number of prejunctional cholinoceptors is relatively small, and their interaction with tubocurarine may be such that only limited competition with acetylcholine is possible. Furthermore a problem associated with experiments of this type is the rapid development of receptor desensitization to acetylcholine that occurs in the isolated muscle. The recent technique of BRIGANT and MALLART (1982) of applying drugs locally to specific regions of mammalian nerve terminals, while recording pre- and postjunctional activity with an extracellular microelectrode may solve this problem.

III. Is There Feedback Control of Transmitter Release?

Feedback control of transmitter release through presynaptic autoreceptors has been described for central and autonomic synapses (for reviews see LANGER 1977; STARKE 1977; VIZI 1979) and may be a general phenomenon at chemically transmitting junctions. However, the evidence for such a mechanism at the neuromuscular junction in skeletal muscle is speculative compared with that for many other sites.

1. Prejunctional Nicotinic Receptors

The main argument in favour of prejunctional nicotinic receptors mediating a stimulatory, or positive feedback, role on transmitter mobilization is that tubocurarine and other nicotinic antagonists cause a rundown in trains of neurally evoked EPPs or EPCs. The results of MAGLEBY et al. (1981) and GIBB and MARSHALL (1983) provide strong evidence against the hypothesis (DREYER 1982) that postjunctional channel block can totally explain train rundown. However, the concept of positive feedback control via nicotinic receptors has recently been challenged by the observations of WILSON (1982) who postulated a negative nicotinic feedback mechanism. In a study of the rundown of trains of end-plate potentials in the cut rat hemidiaphragm preparation, Wilson assessed postjunctional sensitivity during the train by measuring MEPP amplitude. By reducing the noise level of his recording system, and selecting a suitably low concentration of tubocurarine, Wilson was able to measure EPP quantal content in the absence and presence of tubocurarine by use of the direct method of quantal content assessment. Wilson found that the quantal content of the 50th EPP in a 50-Hz train was not significantly different in the presence of tubocurarine compared with control. However, he did find that the quantal content of the first EPP in the train was significantly elevated by tubocurarine, compared with control. The result was that the EPPs in the train appeared to run down in the same way as observed by other workers, but not to a lower plateau level of quantal content. WILSON (1982) also claimed that large doses of neostigmine and physostigmine reduced the quantal content during a train. He then concluded, in contrast to the tenor of the previous discussion, that nicotinic receptors are involved in a negative feedback pathway for the control of transmitter mobilization.

Two factors may allow a different interpretation of Wilson's conclusions. First, despite the fact that the membrane potential in his experiments was sufficiently high that the EPPs did not reach the reversal potential, there is an inherent difficulty in applying correction factors for nonlinear summation and membrane potential with large EPPs close to the reversal potential. The use of voltage clamp techniques would remove this difficulty. Second, the only nicotinic antagonist used in the study was tubocurarine. BLABER (1973) had previously shown similar facilitatory effects with tubocurarine attributable to an increase in the quantal content of the first EPP of a train, in the cut tenuissimus muscle preparation of the cat. However, BLABER (1973) did not ascribe this action to an effect on nicotinic receptors because the more potent nicotinic receptor-blocking agent dimethyltubocurarine (metocurine) did not possess this action. Instead BLABER (1973) ascribed the initial increase in release to an action of the phenolic groups of tubocurarine, which disappeared when these groups were methylated. Phenol and catechol increase acetylcholine release in a similar way; that is by increasing the fractional release without affecting the size of the available store (OTSUKA and NONOMURA 1963; GALLAGHER and BLABER 1973).

As described previously (Sect. B.II.2.b), the evidence for nicotinic receptor-controlled feedback based on results from experiments in which nicotinic agonists were used is more scant than that from experiments involving antagonists. However, taken together, the evidence from the use of agonists and antagonists seems sufficient to postulate the presence of nicotinic cholinoceptors on the membrane of the nerve terminals that in some way facilitate the mobilization of transmitter within the axoplasm, so that the demands of high frequency stimulation can be met. Since tubocurarine and related drugs cause the various "fade" phenomena apparently by blocking these prejunctional receptors, it seems likely that the receptors are normally activated during high frequency stimulation by the transmitter itself, which would then function in a positive feedback mechanism through nicotinic autoreceptors. Another observation that is compatible with this idea is that increasing the fractional release of transmitter, for example with 4-aminopyridine, greatly enhances the fade produced by nondepolarizing neuromuscular blocking drugs (GIBB et al. 1982). In the absence of a neuromuscular blocking drug, the enhanced transmitter release evoked by 4-aminopyridine may itself feed back to increase the mobilization of more transmitter so that the high demand is met. However, by blocking the prejunctional site, release soon fails to match the extra demand and enhanced fade ensues.

If the interpretation of the events underlying fade phenomena in terms of prejunctional nicotinic autoreceptors is accepted, then it has to be postulated that these cholinoceptors comprise a different population from those mediating antidromic repetitive firing described in Sect. B.I of this chapter. The reasons for postulating two different groups of prejunctional nicotinic receptors are threefold:

1. The receptor mediating antidromic repetitive firing (*Rrep* in Fig. 8) is readily blocked by small intra-arterial doses of tubocurarine and related drugs, as well as by intravenous doses, whereas the receptor postulated to mediate mobilization of transmitter (*Rmob* in Fig. 8), block of which gives rise to tetanic fade and related phenomena, interacts more slowly with tubocurarine and related drugs and is not readily blocked by intra-arterial injection of small doses.

2. The *Rrep* receptors (Fig. 8) are readily blocked by snake α-toxins, whereas the *Rmob* receptors are resistant; or are only very slowly blocked, possibly because of diffusion barriers to the large molecules. The absence of prejunctional binding of α-bungarotoxin noted by JONES and SALPETER (1983) is in line with the absence of fade phenomena produced by this toxin, although there is controversy as to whether α-bungarotoxin binds to nerve endings (LENTZ et al. 1977).

3. The *Rrep* receptors are thought to mediate depolarization at some site near the nerve terminal, thereby giving rise to a generator potential that triggers repetitive firing (or block with high concentrations of agonists). However, terminal depolarization, although it increases the frequency of miniature end-plate potentials, inhibits evoked transmitter release by reducing the quantal content of the end-plate potential (HUBBARD and WILLIS 1962), and so could not mediate enhanced mobilization.

It would appear, therefore, that not only do *Rrep* and *Rmob* represent different populations of prejunctional nicotinic cholinoceptors, but also their associated mechanisms differ; neuronal repetitive firing is triggered (through *Rrep*) by axonal depolarization giving rise to a generator potential, whereas transmitter mobilization is enhanced (through *Rmob*) by a different and as yet unknown mechanism, presumably involving a second messenger.

An alternative point of view has recently been put forward by STANDAERT (1982) who suggests that nondepolarizing neuromuscular blocking agents block sodium channels linked to the mobilization mechanism in nerve terminals. Part of this view is based on an earlier suggestion by BLABER (1970, 1973) that sodium ions are involved in mobilization, and that cholinoceptor agonists and antagonists may modify the process by affecting sodium ion movements. Standaert proposes that tetanic fade may arise because tubocurarine and related drugs plug open sodium channels in the nerve terminals in a manner analogous to their demonstrated effect (under certain conditions) on the postjunctional membrane. This is an interesting suggestion, although it lacks proof at the present time.

2. Muscarinic Receptors

Evidence for the presence of presynaptic muscarinic receptors that inhibit acetylcholine release and synthesis in both rodent (SZERB and SOMOGYI 1973; WALTERS and ROTH 1976; LANGER 1977) and human (MAREK et al. 1982) brain has been described. Similar inhibitory presynaptic muscarinic autoreceptors are present in the intestine (FOSBRAEY and JOHNSON 1980) and in autonomic ganglia (KOKETSU and YAMADA 1982). Ligand binding studies allied to subcellular fractionation studies in the *Torpedo* electroplax have shown that muscarinic receptors in this tissue are associated with the synaptosomal or nerve terminal fraction (KLOOG et al. 1980; STRANGE et al. 1980; DOWDALL et al. 1981). Oxotremorine, a muscarinic agonist, inhibits acetylcholine release from *Torpedo* synaptosomes, suggesting that the muscarinic receptors have an inhibitory role in transmission (MICHAELSON et al. 1979, 1980).

DAS et al. (1978) and GANGULY and DAS (1979) obtained evidence that suggested that the nerve terminal muscarinic receptors facilitated acetylcholine release from the rat phrenic nerve–diaphragm preparation. However, other workers were unable to repeat these findings (GUNDERSEN and JENDEN 1980). Some studies have shown the opposite. Thus, DUNCAN and PUBLICOVER (1979) found that carbachol and anticholinesterase agents produced reductions in MEPP frequency at the frog neuromuscular junction. This decrease in frequency was unaffected by tubocurarine, but was blocked by atropine. DUNCAN and PUBLICOVER (1979) considered that prejunctional muscarinic receptors mediated a negative feedback control of transmitter release, and that the muscarinic receptor was associated with prejunctional butyrylcholinesterase. ABBS and JOSEPH (1981) also found that muscarinic receptors mediated inhibition of acetylcholine release. They found that atropine enhanced the evoked release of labelled acetylcholine from the rat isolated phrenic nerve–hemidiaphragm, suggesting that transmitter release might normally be under continuous muscarinic receptor-mediated inhibitory modulation. The muscarinic agonist, oxotremorine, antagonized the facilitatory effects of atropine, although by itself it did not inhibit acetylcholine release. Possibly, under the conditions of the experiments (e.g. inhibited cholinesterase), the extent of muscarinic receptor-mediated inhibitory modulation by the transmitter itself was already maximal, so that oxotremorine could not produce any additional effect, except in the presence of atropine.

Overall, the weight of evidence suggests that prejunctional muscarinic receptors at the neuromuscular junction are inhibitory in nature. As yet, no mechanism has been proposed to explain how activation of the postulated muscarinic receptors depresses transmitter release.

Some neuromuscular blocking drugs (notably gallamine, pancuronium and fazadinium) are capable of blocking a subclass of muscarinic receptors present in the heart and on autonomic nerves (RIKER and WESCOE 1951; SAXENA and BONTA 1970; MARSHALL 1973; VERCRUYSSE et al. 1979; BOWMAN 1982). It might be that muscarinic receptors present on somatic motor nerve endings resemble this subclass, and that the small initial increase in transmitter release reported to be produced by small doses of gallamine and pancuronium (GERGIS et al. 1972; SOKOLL et al. 1973), is a consequence of block of prejunctional inhibitory muscarinic receptors.

3. Intra-axonal Cholinoceptors

TAUC and BAUX (1982) showed that presynaptic intra-axonal injection of acetylcholine, choline or carbachol into a cholinergic synapse of *Aplysia* increased the evoked quantal content, (but not the quantum size), whereas injection of acetylcholinesterase depressed the quantal content, at the same time as it reduced the cytoplasmic concentration of acetylcholine. Studies with various antagonists led them to believe that intracellular nicotinic cholinoceptors were involved.

KRNJEVIĆ et al. (1982) found, in anaesthetized cats, that intracellular injections of acetylcholine into lumbosacral motorneurons exerted a number of effects resembling those produced by tetraethylammonium, which blocks K^+ channels. The authors considered the most important effect to be prolongation of the action potential, an effect which is known to increase transmitter release by augmenting Ca^{2+} influx at the terminals. They proposed that fluctuations in free axoplasmic acetylcholine may form the basis of an autoregulatory mechanism for transmitter release of the type demonstrated at cholinergic *Aplysia* synapses by TAUC and BAUX (1982).

IV. Summary and Conclusions

Evidence is reviewed that may be interpreted to indicate the existence of two populations of nicotinic cholinoceptors, and a population of muscarinic cholinoceptors, on the membranes of somatic motor nerve terminals, as illustrated in Fig. 8.

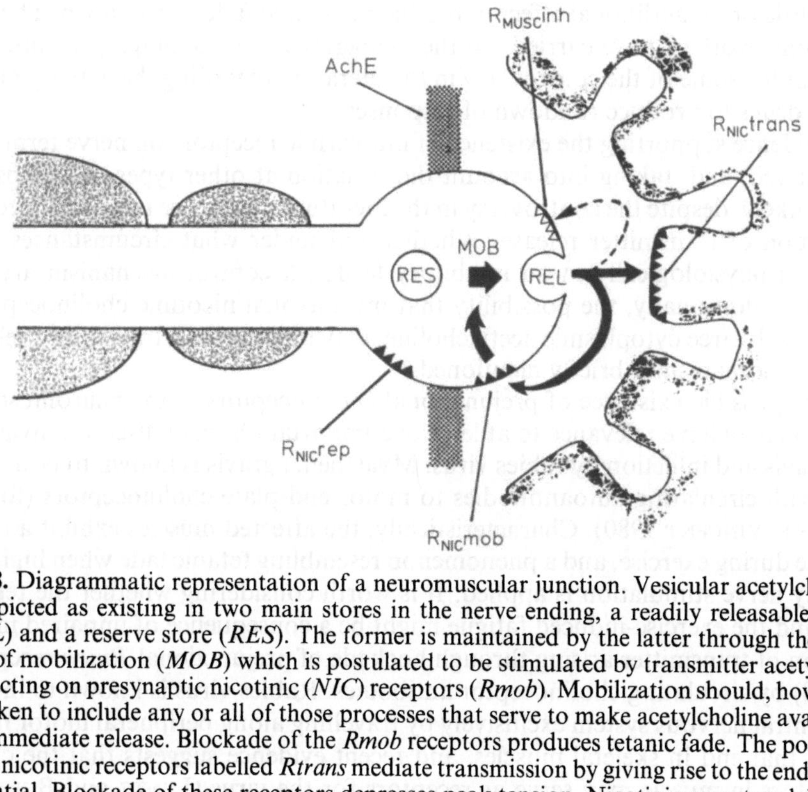

Fig. 8. Diagrammatic representation of a neuromuscular junction. Vesicular acetylcholine is depicted as existing in two main stores in the nerve ending, a readily releasable store (*REL*) and a reserve store (*RES*). The former is maintained by the latter through the process of mobilization (*MOB*) which is postulated to be stimulated by transmitter acetylcholine acting on presynaptic nicotinic (*NIC*) receptors (*Rmob*). Mobilization should, however, be taken to include any or all of those processes that serve to make acetylcholine available for immediate release. Blockade of the *Rmob* receptors produces tetanic fade. The postsynaptic nicotinic receptors labelled *Rtrans* mediate transmission by giving rise to the end-plate potential. Blockade of these receptors depresses peak tension. Nicotinic receptors labelled *Rrep* are depicted as being located nearer to the myelinated portion of the axon. They may mediate depolarization of the axon which gives rise either to repetitive firing in the axon, or to conduction block, depending on the magnitude of the depolarization. The acetylcholinesterase (*AchE*) of the nerve terminal membrane may act as a barrier that normally prevents transmitter acetylcholine from reaching the *Rrep* receptors. Muscarinic (*MUSC*) receptors that inhibit acetylcholine release (*Rinh*) may also be present on the nerve terminal

It is proposed that one of these populations of nicotinic cholinoceptors mediates antidromic repetitive firing and, after larger doses of nicotinic agonists, a prejunctional depolarization and block of conduction. Although these receptors may have pharmacological importance, no physiological importance has been ascribed to them.

The suggestion is made that a second population of nicotinic autoreceptors facilitates transmitter mobilization (i.e. any or all of those processes that make acetylcholine immediately available for release) during high frequency nerve activity. It is proposed that these cholinoceptors are involved in a feedback control mechanism, triggered by the transmitter itself, that maintains transmitter output during heavy traffic of nerve impulses. Blockade of these receptors gives rise to the well-known phenomena of tetanic fade and rundown of end-plate responses produced by curare-like drugs. Preliminary results by Gibb and Marshall indicate that the part of the mobilization process that is facilitated by stimulation of prejunctional nicotinic cholinoceptors is temperature dependent. It may be nonfunctional at low temperatures, with the result that acetylcholine antagonists then have little or no additional effect in producing tetanic fade or rundown. The fact that some workers have carried out their experiments at room temperature may account for some of the controversy in the literature regarding the activity of different drugs to produce rundown of responses.

Evidence supporting the existence of muscarinic receptors on nerve terminals is reviewed, and, taking into account the situation at other types of synapse, it seems likely, despite the controversy in the literature, that these receptors mediate inhibition of transmitter release. Whether and under what circumstances they function physiologically in an inhibitory feedback control mechanism are not known. Additionally, the possibility that intra-axonal nicotinic cholinoceptors, activated by free cytoplasmic acetylcholine, may function in a transmitter release control mechanism is briefly mentioned.

The possible existence of prejunctional cholinoceptors at the neuromuscular junction may have relevance to at least two important human diseases, myasthenia gravis and infection by rabies virus. Myasthenia gravis is known to be associated with circulating autoantibodies to motor end-plate cholinoceptors (for review, see VINCENT 1980). Characteristically, the affected muscles exhibit a rapid fatigue during exercise, and a phenomenon resembling tetanic fade when high frequency nerve stimulation is applied. It is worth considering whether the tetanic fade and the exercise-induced fatigue might be a consequence of impaired mobilization of transmitter arising through bockade of prejunctional *Rmob* receptors (Fig. 8) by circulating cholinoceptor antibodies. Rabies virus is known to invade the central nervous system exclusively by travelling along peripheral motor nerve axons that end in skeletal muscles, and recent evidence suggests that the cholinoceptors in muscle may serve as receptors for the virus (LENTZ and BURRAGE 1982). While the postjunctional cholinoceptors could serve to attract and concentrate the virus particles near the nerve terminals which they subsequently enter, it is difficult to see how such receptors could play a part in the penetration of the virus into the terminal axoplasm. However, it may be that the prejunctional cholinoceptors fulfil this role.

References

Abbs ET, Joseph DN (1981) The effects of atropine and oxotremorine on acetylcholine release in rat phrenic nerve-diaphragm preparations. Br J Pharmacol 73:481–483

Adams PR (1975) A model for the procaine end-plate current. J Physiol (Lond) 246:61–63P

Adams PR (1976) Drug blockade of open end-plate channels. J Physiol (Lond) 260:531–552

Adler ME, Albuquerque EX, Lebeda FJ (1978) Kinetic analysis of endplate current altered by atropine and scopolamine. Molec Pharmacol 14:514–529

Albuquerque EX, Gage PN, Oliveira AC (1979) Differential effect of perhydrohistrionicotoxin on 'intrinsic' and 'extrinsic' end-plate responses. J Physiol (Lond) 297:423–442

Ali HH, Utting JE, Gray TC (1971) Quantitative assessment of residual antidepolarizing block (Part II). Br J Anaesth 43:478–485

Ascher P, Large WA, Rang HP (1979) Studies on the mechanism of action of acetylcholine antagonists on rat parasympathetic ganglion cells. J Physiol (Lond) 295:139–170

Beani L, Bianchi C, Ledda F (1964) The effect of tubocurarine on acetylcholine release from motor nerve terminals. J Physiol (Lond) 174:172–183

Blaber LC (1970) The effect of facilitatory concentrations of decamethonium on the storage and release of transmitter at the neuromuscular junction of the cat. J Pharmacol Exp Ther 175:664–672

Blaber LC (1972) The mechanisms of the facilitatory action of edrophonium in cat skeletal muscle. Br J Pharmacol 46:498–507

Blaber LC (1973) The prejunctional actions of some non-depolarizing blocking drugs. Br J Pharmacol 47:109–116

Blaber LC, Bowman WC (1963) Studies on the repetitive discharges evoked in motor nerve and skeletal muscle after injection of anticholinesterase drugs. Br J Pharmacol 20:326–344

Blaber LC, Karczmar AG (1967) Multiple cholinoceptive and related sites at the neuromuscular junction. Ann NY Acad Sci 144:571–583

Blackman JG (1963) Stimulus frequency and neuromuscular block. Br J Pharmacol Chemother 20:5–16

Bowman WC (1980) Prejunctional and postjunctional cholinoceptors at the neuromuscular junction. Anesth Analg Cleve 59:935–943

Bowman WC (1982) Non-relaxant properties of neuromuscular blocking drugs. Br J Anaesth 54:147–160

Bowman WC, Marshall IG (1981) Die Rolle prä- und postsynaptischer cholinergische Rezeptoren bei der neuromuskulären Übertragung und deren Beeinflußbarkeit durch Muskelrelaxantien. In: Buzello W (ed) Muskelrelaxantien. Thieme, Stuttgart, pp 34–48

Bowman WC, Webb SN (1972) Acetylcholine and anticholinesterase drugs. In: Cheymol J (ed) International encyclopedia of pharmacology and therapeutics. Pergamon, Oxford, pp 427–502

Bowman WC, Webb SN (1976) Tetanic fade during partial transmission failure produced by non-depolarizing neuromuscular blocking drugs in the cat. Clin Exp Pharmacol Physiol 3:545–555

Brigant JL, Mallart A (1982) Presynaptic currents in mammalian motor endings. J Physiol (Lond) 333:619–636

Brown GL (1937) Action potentials of normal mammalian muscle. Effects of acetylcholine and eserine. J Physiol (Lond) 19:220–237

Burns BD, Paton WDM (1951) Depolarization of the motor endplate by decamethonium and acetylcholine. J Physiol (Lond) 115:41–73

Chang CC, Cheng HC, Chen TF (1967) Does d-tubocurarine inhibit the release of acetylcholine from motor nerve endings? Jpn J Physiol 17:505–515

Cheymol J, Bourillet F, Ogura Y (1962) Action de quelques paralysants neuromusculaires sur la libération de l'acétyl choline au niveau des terminaisons nerveuses motrices. Arch Int Pharmacodyn Thér 139:187–197

Clark AL (1982) Studies on the nature of twitch potentiation induced by organophosphate anticholinesterase at mammalian neuromuscular junctions. Ph.D. Thesis, University of London

Clark AL, Hobbiger F (1983) Twitch potentiation by organophosphate anticholinesterases in rat phrenic nerve diaphragm preparation. Br J Pharmacol 78:239–246

Clark AL, Hobbiger F, Terrar DA (1979) The effect of dithiothreitol on anticholinesterase induced antidromic firing and twitch potentiation. Br J Pharmacol 67:481–482P

Clark AL, Hobbiger F, Terrar DA (1983) The relationship between stimulus-induced antidromic firing and twitch potentiation produced by paraoxon in rat phrenic nerve-diaphragm preparations. Br J Pharmacol 80:17–25

Colquhoun D, Sheridan RE (1981) the modes of action of gallamine. Proc R Soc Lond B 211:181–203

Colquhoun D, Dreyer F, Sheridan RE (1979) the actions of tubo curarine at the frog neuromuscular junction. J Physiol (Lond) 293:247–284

Creese R, Head SD, Jenkinson D (1982) Prolonged action of depolarizing drugs in guinea-pig muscle. Br J Pharmacol 77:431P

Dale HH, Feldberg W, Vogt M (1936) Release of acetylcholine at voluntary motor nerve endings. J Physiol (Lond) 86:353–380

Das M, Ganguly DK, Vedasiromoni JR (1978) Enhancement by oxotremorine of acetylcholine release from the rat phrenic nerve. Br J Pharmacol 62:195–198

Dettbarn WD (1960) Effect of curare on conduction in myelinated isolated nerve fibres of the frog. Nature 186:891–892

Dettbarn WD (1967) The acetylcholine system in peripheral nerve. Ann NY Acad Sci 144:483–503

Douglas WW, Paton WDM (1954) The mechanism of motor endplate depolarization due to a cholinesterase inhibiting drug. J Physiol (Lond) 124:325–344

Douglas WW, Ritchie JM (1960) The excitatory action of acetylcholine on cutaneous non-myelinated nerve fibres. J Physiol (Lond) 150:501–514

Dowdall MJ, Golds PR, Strange PG (1981) Presynaptic muscarinic receptors on *Torpedo*. Biochem Soc Trans 9:412–413

Dreyer F (1982) Acetylcholine receptor. Br J Anaesth 54:115–130

Duncan CJ, Publicover SJ (1979) Inhibitory effects of cholinergic agents on the release of transmitter at the frog neuromuscular junction. J Physiol (Lond) 294:91–103

Eccles JC, Katz B, Kuffler SW (1942) Effect of eserine on neuromuscular transmission. J Neurophysiol 5:211–230

Emmelin NG, MacIntosh FC (1956) The release of acetylcholine from perfused sympathetic ganglia and skeletal muscles. J Physiol (Lond) 131:477–496

Feng TP, Li TH (1941) Studies on the neuromuscular junction XXIII. A new aspect of the phenomena of eserine potentiation and posttetanic facilitation in mammalian muscles. Chin J Physiol 16:37–54

Fletcher P, Forrester T (1975) The effect of curare on the release of acetylcholine from mammalian motor nerve terminals and an estimate of quantum content. J Physiol (Lond) 251:131–144

Foldes FF, Vizi ES (1980) Modulation of presynaptic acetylcholine release at the neuromuscular junction. In: Vizi ES (ed) Modulation of neurochemical transmission. Akadémiai Kiadó, Budapest, pp 355–382

Fosbraey P, Johnson ES (1980) Release-modulating acetylcholine receptors on cholinergic neurones of the guinea-pig ileum. Brit J Pharmacol 68:289–300

Fulton BP, Usherwood PNR (1977) Presynaptic acetylcholine action at the locust neuromuscular junction. Neuropharmacol 16:877–880

Galindo A (1972) The role of prejunctional effects in myoneural transmission. Anesthesiology 36:598–608

Gallagher JP, Blaber LC (1973) Catechol, a facilitatory drug that demonstrates only a prejunctional site of action. J Pharmacol Exp Ther 184:129–135

Ganguly DK, Das M (1979) Effects of oxotremorine demonstrate presynaptic and dopaminergic receptors on motor nerve terminals. Nature 278:645–646

Gergis SD, Dretchen KL, Sokoll MD, Long JP (1972) Effect of pancuronium bromide on acetylcholine release. Proc Soc Exp Biol Med 139:389–401
Gibb AJ, Marshall IG (1982) The effects of trimetaphan on tetanic fade and on endplate ion channels at the rat neuromuscular junction. Br J Pharmacol 76:187P
Gibb AJ, Marshall IG (1983) Pre- and postjunctional effects of tubocurarine and trimetaphan involved in tetanic fade at the rat neuromuscular junction. Br J Pharmacol 78:86P
Gibb AJ, Marshall IG, Bowman WC (1982) Increased tetanic fade produced by 3,4-diaminopyridine in the presence of neuromuscular blocking agents. In: Lechat P, Thesleff S, Bowman WC (eds) Aminopyridine and similarly acting drugs: effects on nerves, muscles, and synapses. Pergamon, Oxford, p 216
Ginsborg BL, Jenkinson DH (1976) Transmission of impulses from nerve to muscle. In: Zaimis E (ed) Neuromuscular junction. Handb Exp Pharmacol 42:228–364
Glavinović MI (1979) Presynaptic action of curare. J Physiol (Lond) 290:499–506
Gundersen CB, Jenden DJ (1980) Oxotremorine does not enhance acetylcholine release from rat diaphragm preparation. Br J Pharmacol 70:8–10
Hobbiger F (1976) Pharmacology of anticholinesterase drugs. In: Zaimis E (ed) Neuromuscular junction. Springer, Berlin Heidelberg New York, pp 486–581 (Handbook of experimental pharmacology, vol 42)
Hofmann FB (1903) Studies über Tetanus. Pflügers Arch Ges Physiol 95:484–532
Hohlfeld R, Sterz R, Peper K (1981) Prejunctional effects of anticholinesterase drugs at the endplate mediated by presynaptic acetylcholine receptors or by postsynaptic potassium efflux. Pflügers Arch Ges Physiol 391:213–218
Hubbard JI, Willis WW (1962) Mobilization of transmitter release by hyperpolarization. Nature 193:174–175
Hubbard JI, Wilson DF (1973) Neuromuscular transmission in a mammalian preparation in the absence of blocking drugs and the effect of d-tubocurarine. J Physiol (Lond) 228:307–325
Hubbard JI, Schmidt RF, Yokota T (1965) The effect of acetylcholine upon mammalian motor nerve terminals. J Physiol (Lond) 181:810–829
Hubbard JI, Wilson DF, Miyamoto M (1969) Reduction of transmitter release by d-tubocurarine. Nature 223:531–533
Hunter AR (1970) Recent advances in muscle relaxants – general discussion. Proc R Soc Med 63:699
Hutter OF (1952) Post-tetanic restoration of neuromuscular transmission blocked by d-tubocurarine. J Physiol (Lond) 118:216–227
Jones SW, Salpeter MM (1983) Absence of [^{125}I] α-bungarotoxin binding to motor nerve terminals of frog, lizard, and mouse muscle. J Neurosci 3:326–331
Katz B (1962) The transmission of impulses from nerve to muscle and the subcellular unit of synaptic action. Proc R Soc Lond B 155:455–477
Katz B (1969) the release of neural transmitter substances. Liverpool University Press, Liverpool
Katz B, Miledi R (1971) Further observations on acetylcholine noise. Nature 232:124–126
Katz B, Miledi R (1977) Transmitter leakage from motor nerve endings. Proc R Soc Lond B 196:59–72
Katz B, Miledi R (1978) A re-examination of curare action at the motor endplate. Proc R Soc Lond B 203:119–133
Kloog Y, Michaelson DM, Sokolovsky M (1980) Characterization of the presynaptic muscarinic receptor in synaptosomes of *Torpedo* electric organ by means of kinetic and equilibrium binding studies. Brain Res 194:97–115
Koketsu K, Yamada M (1982) Presynaptic muscarinic receptors inhibiting active acetylcholine release in the bullfrog sympathetic ganglion. Br J Pharmacol 77:75–82
Kopin IJ (1967) Acetylcholine, bretylium and release of norpinephrine from sympathetic nerve endings. Ann NY Acad Sci 144:558–570
Krnjević K, Mitchell JF (1961) The release of acetylcholine in the isolated rat diaphragm. J Physiol (Lond) 155:246–262

Krnjević K, Puil E, Werman R (1982) A possible mechanism of autoregulation of acetylcholine release. Abstracts: European Symposium on Cholinergic Transmission, Presynaptic Aspects, Strasbourg, p 69
Lambert JJ, Volle RL, Henderson EG (1980) An attempt to distinguish between the actions of neuromuscular blocking drugs on the acetylcholine receptor and on its associated ion channel. Proc Natl Acad Sci USA 77:5003–5007
Langer SZ (1977) Presynaptic receptors and their role in the regulation of transmitter release, Sixth Gaddum Memorial Lecture.Br J Pharmacol 60:481–497
Laskowski MB, Dettbarn WD (1979) An electrophysiological analysis of the effects of paraoxan at the neuromuscular junction. J Pharmacol Exp Ther 210:269–274
Lee C, Chen D, Katz RL (1977) Characteristics of nondepolarizing neuromuscular block I. Postjunctional block by alpha-bungarotoxin. Can Anaesth Soc J 24:212–219
Lentz TL, Burridge TG (1982) Is the acetylcholine receptor a rabies virus receptor? Sience 215:182–184
Lentz TL, Mazurkiewitz JE, Rosenthal J (1977) Cytochemical localization of acetylcholine receptors at the neuromuscular junction by means of horseradish peroxidase-labeled α-bungarotoxin. Brain Res 132:423–442
Li TH, Ting YC (1941) Studies on the neuromuscular junction XXI. Responses of cat muscles to acetylcholine during Wedensky inhibition and post-tetanic facilitation. Chin J Physiol 16:1–8
Liley AW (1956b) The quantal components of the mammalian end-plate potential. J Physiol (Lond) 133:571–587
Lilleheil G, Naess K (1961) A presynaptic effect of d-tubocurarine in the neuromuscular junction. Acta Physiol Scand 52:120–136
Maeno T, Nobe S (1970) Analysis of presynaptic effect of d-tubocurarine on the neuromuscular transmission. Proc Jpn Acad 46:750–754
Magleby KL, Terrar DA (1975) Factors affecting the time course of decay of end-plate currents: a possible co-operative action of acetylcholine on receptors at the frog neuromuscular junction. J Physiol (Lond) 244:467–495
Magleby KL, Pallotta BS, Terrar DA (1981) The effect of (+)-tubocurarine on neuromuscular transmission during repetitive stimulation in the rat, mouse, and frog. J Physiol (Lond) 312:97–113
Manalis R (1977) Voltage-dependent effects of curare at the frog neuromuscular junction. Nature 267:366–368
Marek KL, Bowen DM, Sims NR, Davison AN (1982) Stimulation of acetylcholine synthesis by blockade of presynaptic muscarinic inhibitory autoreceptors: observations in rat and human brain preparations and comparison with the effects of choline. Life Sci 30:1517–1524
Marshall IG (1973) The ganglion blocking and vagolytic actions of three short-acting neuromuscular blocking drugs in the cat. J Pharm Pharmacol 25:530–536
Masland RL, Wigton RS (1940) Nerve activity accompanying fasciculation produced by prostigmin. J Neurophysiol 3:269–275
Michaelson DM, Arissar S, Kloog Y, Sokolovsky M (1979) Mechanism of acetylcholione release: possible involvement of presynaptic muscarinic receptors in regulation of acetylcholine release and protein phosphorylation. Proc Nat Acad Sci USA 76:6336–6340
Michaelson DM, Arissar S, Ophir I, Pinchasi I, Angel I, Kloog Y, Sokolovsky M (1980) On the regulation of acetylcholine release. A study utilizing *Torpedo* synaptosomes and synaptic vesicles. J Physiol (Paris) 76:505–511
Miledi R, Molenaar PC, Polak RL (1978) α-Bungarotoxin enhances transmitter released at the neuromuscular junction. Nature 272:641–643
Milne RJ, Byrne JM (1981) Effect of hexamethonium and decamethonium on end-plate current parameters. Molec Pharmacol 19:276–281
Miyamoto MD (1978) the actions of cholinergic drugs on motor nerve terminals. Pharmacol Rev 29:221–247
Morrison JD (1977) The generation of nerve and muscle repetitive activity in the rat phrenic nerve-diaphragm preparation following inhibition of cholinesterase by ecothiopate. Br J Pharmacol 60:45–53

Otsuka M, Nonamura Y (1963) The action of phenolic substances on motor nerve endings. J Pharmacol Exp Ther 140:41–45
Otsuka M, Endo M, Nonomura Y (1962) Presynaptic nature of neuromuscular depression. Jpn J Pharmacol 12:573–584
Paton WDM, Perry WLM (1951) The pharmacology of the toxiferines. Br J Pharmacol 6:299–310
Paton WDM, Waud DR (1967) The margin of safety of neuromuscular transmission. J Physiol (Lond) 191:59–60
Paton WDM, Zaimis EJ (1951) The action of d-tubocurarine and of decamethonium on respiratory and other muscles in the cat. J Physiol (Lond) 112:311–331
Preston JB, van Maanen EF (1953) Effect of frequency of stimulation on the paralyzing dose of neuromuscular blocking agents. J Pharmacol Exp Ther 107:165–171
Raines A, Standaert FG (1966) Pre- and post-junctional effects of diphenylhydantoin at the cat soleus neuromuscular junction. J Pharmacol Exp Ther 153:361–371
Rang H, Ritter JM (1969) A new kind of drug antagonism: evidence that agonists cause a molecular change in acetylcholine receptors. Molec Pharmacol 5:394–411
Riker WF (1975) Prejunctional effects of neromuscular blocking and facilitatory drugs. In: Katz RK (ed) Muscle relaxants. American Elsevier, New York, pp 60–102
Riker WF, Okamoto M (1969) Pharmacology of motor nerve terminals. Ann Rev Pharmacol 9:173–208
Riker WF Jr, Standaert FG (1966) The action of facilitatory drugs and acetylcholine on neuromuscular transmission. Ann NY Acad Sci 135:163–176
Riker WF, Wescoe WC (1951) The pharmacology of Flaxedil with observations on certain analogs. Ann NY Acad Sci 54:373–394
Riker WF Jr, Werner G, Roberts J, Kuperman A (1959) Pharmacologic evidence for the existence of a presynaptic event in neuromuscular transmission. J Pharmacol Exp Ther 125:150–158
Ritchie JM (1967) On the role of acetylcholine in conduction in mammalian non myelinated nerve fibres. Ann NY Acad Sci 144:504–516
Ruff RL (1977) A quantitative analysis of local anaesthetic alteration of miniature end-plate currents and end-plate current fluctuations. J Physiol (Lond) 264:89–124
Saxena PR, Bonta IL (1970) Mechanism of selective cardiac vagolytic action of pancuronium bromide. Specific blockade of cardiac muscarinic receptors. Eur J Pharmacol 11:332–341
Sokoll MD, Dretchen KL, Gergis SD, Long JP (1973) The effects of gallamine on nerve terminals and endplates. Anesthesiology 38:157–165
Standaert FG (1982) Release of transmitter at the neuromuscular junction. Br J Anaesth 54:131–145
Standaert FG, Riker WF (1967) The consequences of cholinergic drug action on motor nerve terminals. Ann NY Acad Sci 144:517–533
Starke K (1977) Regulation of noradrenaline release by presynaptic receptor systems. Rev Physiol Biochem Pharmacol 77:3–124
Strange PG, Dowdall MJ, Golds PR, Pickard MR (1980) Ligand-binding properties of a muscarinic acetylcholine receptor from *Torpedo* electric organ. FEBS Lett 122:293–296
Su PC, Su WL, Rosen AD (1979) Pre- and postsynaptic effects of pancuronium at the neuromuscular junction of the mouse. Anesthesiology 30:199–204
Sugai N, Hughes R, Payne JP (1976) Sequential changes in the fade of tetanic tension after the administration of tubocurarine in anaesthetized man. Br J Anaesth 48:535–539
Szerb JC, Somogyi GT (1973) Depression of acetylcholine release from cerebral cortical slices by cholinesterase inhibition and by oxotremorine. Nature 241:121–122
Tamiya N, Arai H (1966) Studies on sea snake venoms: crystallization of erabutoxins a and b from *Laticauda semifasciata* venom. Biochem J 99:624–630
Tauc L, Baux G (1982) Intraneuronal pharmacology points to a possible presence of intracellular acetylcholine receptors in the cholinergic synaptic nerve terminal. Eur Symp on Cholinergic Transmission, Presynaptic Aspects. Strasbourg, p 93
Thesleff S (1959) Motor and-plate "desensitization" by repetitive nerve stimulation. J Physiol (Lond) 148:659–664

Torda TA, Kiloh N (1982a) Org NC 45 reduces quantal release of acetylcholine (a preliminary communication). Anaesth Intensive Care 10:127–129
Torda TA, Kiloh N (1982b) Myoneural actions of Org NC 45. Br J Anaesth 54:1217–1221
Usubiaga JE, Standaert F (1968) The effects of local anaesthetics on motor nerve terminals. J Pharmacol Exp Ther 159:353–361
Van Poznak A (1963) Inhalatation suppression of repetitive activity generated in motor nerve ending. Fed Proc 22:390
Vercruysse P, Bossuyt P, Hanegreefs G, Verbeuren TJ, Vanhoutte PM (1979) Gallamine and pancuronium inhibit prejunctional and postjunctional muscarinic receptors in canine saphenous vein. J Pharmacol Exp Ther 209:225–230
Vincent A (1980) Immunology of acetylcholine receptors in relation to myasthenmia gravis. Physiol Rev 60(3):756–824
Vizi ES (1979) Presynaptic modulation of neurochemical transmission. Prog Neurobiol 12:181–290
Walters JT, Roth RH (1976) Dopaminergic neurones: an *in vivo* system for measuring drug interaction with presynaptic receptors. Naunyn Schmiedebergs Arch Pharmacol 296:5–14
Wedensky NE (1903) Die Erregung, Hemmung und Narkose. Pflügers Arch Ges Physiol 100:1–144
Werner G (1960) Neuromuscular facilitation and antidromic discharges in motor nerves: their relation to activity in motor nerve terminals. J Neurophysiol 23:171–187
Werner G, Kuperman AS (1963) Actions at the neuromuscular junction. In: Koelle GB (ed) Cholinesterases and anticholinesterase agents. Springer, Berlin, pp 570–678 (Handbook of experimental pharmacology, vol 15)
Williams NE, Webb SN, Calvey TN (1980) Differential effects of myoneural blocking drugs on neuromuscular transmission. Br J Anaesth 52:1111–1115
Wilson DF (1982) Influence of presynaptic receptors on neuromuscular transmission in rat. Am J Physiol 242 (Cell Physiol II):C366–372

CHAPTER 6

On the Comparative Sensitivity of Acetylcholine Receptors of Various Groups of Skeletal Muscles to Neuromuscular Blocking Agents

D. A. KHARKEVICH and V. P. FISENKO

A. Introduction

It is well known that the comparative sensitivity of acetylcholine receptors to neuromuscular blocking agents is different for various groups of muscles. Thus, for instance, tubocurarine causes relaxation, first, of the cervical and facial muscles, then, of the muscles of the torso and limbs, and last, of respiratory muscles (intercostal and diaphragm). Often, this is considered to be true of other neuromuscular blocking agents, although the data on comparative sensitivity of end-plate acetylcholine receptors to other drugs are rather discrepant. The latter is sometimes believed to be related to diverse influences of different factors on the sensitivity of acetylcholine receptors to neuromuscular blocking agents, including the type of anesthesia, structure of skeletal muscles, their temperature, blood circulation, volume of hemorrhage during the operation, acid–base equilibrium, etc.

Taking into account that during the last two decades there have been designed a large number of neuromuscular blocking agents of different chemical structure, it seems of interest to compare the order of skeletal muscle relaxation caused by different drugs. These data, as well as knowledge of the main reasons for unequal sensitivity of end-plate receptors at different locations to neuromuscular blocking agents are not only of basic, but also of applied value, e.g. for their rational clinical use and the search for new drugs with predominantly paralyzing actions on definite groups of skeletal muscles.

B. On the Order of Skeletal Muscle Relaxation in Humans Under the Influence of Neuromuscular Blocking Agents

I. Investigations on Anesthetized Patients

The safety margin of the neuromuscular blocking action may be used to estimate the difference in sensitivity of acetylcholine receptors of various groups of skeletal muscles under clinical conditions. It is expressed by the ratio of the doses causing relaxation of the torso and limb muscles to the doses blocking respiration. For tubocurarine, this value is 1 : 1.7. The observation that limb muscles are more sensitive to tubocurarine than respiratory muscles has been confirmed by many authors. Under the tubocurarine analog, metocurine, the sequence of skeletal muscle relaxation was found identical to that under tubocurarine (STOELING et al. 1950). However, according to numerous authors, metocurine's safety margin was wider than that of tubocurarine (ULLET et al. 1950; SADOVE et al. 1951).

The study of gallamine showed that it also had a wider safety margin than tubocurarine (WALTON 1950; COLLIER 1951). When gallamine and tubocurarine were used in doses providing an adequate relaxation of the muscles of the upper part of the abdominal wall, artificial respiration was required in 45% of cases under tubocurarine and in 10% under gallamine (DOUGHTY and WYLIE 1951). Moreover, gallamine produced a lower effect on intercostal muscles than tubocurarine did. FOLDES et al. (1952) observed apnea in only 13% of gallamine applications aiming at abdominal muscle relaxation. At the same time, according to PATTERSON and PITTSBURG (1952) gallamine and tubocurarine suppressed respiration equally. A similar conclusion was drawn by GRIGORIEV and ANICHKOV (1957) and by VINOGRADOV and DYACHENKO (1961) who demonstrated that the safety margins of gallamine and tubocurarine were about the same. While estimating the safety margin of the neuromuscular action of diplacinum (see the structure in Chap. 1, Table 1), GRIGORIEV and ANICHKOV (1957) observed that it caused respiratory arrest in doses 3–4 times as high as those needed to provide an adequate relaxation of the abdominal wall and limb muscles.

Data on comparative sensitivity of muscles to the depolarizing neuromuscular blocking agent, decamethonium, are rather controversial. Thus, for instance, DAVIES and LEWIS (1949), and ORGANE et al. (1949) believe that decamethonium has a wider safety margin than tubocurarine. The results obtained by GRAY (1950), HEWER et al. (1949), SADOVE et al. (1951), and SPENCER and COAKLEY (1955) indicate a smaller safety margin of decamethonium as compared with tubocurarine. Moreover, decamethonium in doses causing a satisfactory relaxation of abdominal muscles was shown to eliminate spontaneous respiration entirely (HARRIS and DRIPPS 1950; FOLDES and MACHAY 1951).

The data on the neuromuscular blocking action of another depolarizing agent, succinylcholine, are also controversial. Many authors believe that to use it in doses providing a surgically adequate relaxation of torso and limb muscles, artificial ventilation is required (RICHARDS and YOUNGMAN 1952; GILLIES et al. 1954; HOPPE 1955). On the other hand, DARDEL and THESLEFF (1952), THESLEFF (1952), and DANILOV et al. (1957) consider that the employment of succinylcholine creates adequate conditions for spontaneous respiration of patients. NGAI et al. (1959) report that the neuromuscular blocking action of the depolarizing agent, imbretil on limb muscles predominates over its action on respiratory muscles.

Thus, there are discrepancies in clinical results dealing with the sensitivity of different skeletal muscles to neuromuscular blocking agents of various chemical structures. These might be accounted for by the fact that the factors affecting the sensitivity to neuromuscular blocking agents are quite numerous. For instance, diethyl ether and halothane decrease the difference in sensitivity of respiratory and peripheral muscles to tubocurarine and anatruxonium. Maximum potentiating action of these anesthetics is observed in the skeletal muscles most resistant to tubocurarine and anatruxonium, i.e. respiratory and abdominal muscles, respectively (LEPAKHIN 1970). Disturbances in acid–base equilibrium, and electrolyte balance, hemorrhage volume, and hypothermia may interfere with the interaction of neuromuscular blocking agents with acetylcholine receptors and, therefore, may affect the sensitivity of the latter to neuromuscular blocking agents. It is also reported that the resistance of respiratory muscles to neuromuscular blocking agents may be increased under pulmonary emphysema (GERTLER 1981).

II. Investigations on Volunteers

The studies of sensitivity of various skeletal muscles to neuromuscular blocking agents on volunteers allow one to rule out the effect of many of the factors already listed on neuromuscular transmission. The action of tubocurarine on the skeletal muscles was studied in most detail. Thus, UNNA et al. (1950) and UNNA and PELIKAN (1951) showed that the muscles of the forearm and hand were more sensitive to tubocurarine than the respiratory muscles. The decrease of the hand grip strength by 95% was associated with only 20%–30% decrease of the vital capacity of lungs. Similar results were obtained by JOHNSON et al. (1964) who studied the effect of tubocurarine on respiration, the force of neck muscles, muscles of the forearm, and hand. They demonstrated that tubocurarine in doses of 0.1–0.15 mg/kg i.v. decreased the force of neck muscles by 92%, that of the muscles of forearm and hand by 79%, the inspiration volume by 40%, and the expiration volume by 17%. While estimating the sensitivity of the muscles of lower limbs to tubocurarine, SECHER et al. (1981, 1982) found that neuromuscular transmission in the gastrocnemius muscle was less sensitive to this drug than the soleus muscle. Under partial muscle paralysis induced by tubocurarine, minimal changes of the function of respiratory muscles were observed (GAL and GOLDBERG 1981). Metocurine also blocked to a lesser degree the neuromuscular transmission in the respiratory muscles. A 95% decrease of grip strength was followed by a 10%–16% reduction of the vital capacity of lungs (UNNA and PELIKAN 1951; PELIKAN et al. 1950).

Lower sensitivity of respiratory muscles to gallamine as compared with limb muscles was observed by MUSHIN et al. (1949). They showed that, at a dose of 1 mg/kg i.v. it entirely blocked neuromuscular transmission in the hand and forearm muscles, rectus abdominis muscles, and had no significant effect on respiratory muscles. According to UNNA and PELIKAN (1951), gallamine-induced decrease of grip strength by 95% was associated with only 20% decrease of the vital capacity of lungs.

Decamethonium suppressed respiration less than tubocurarine did (DAVIES and LEWIS 1949; GROB et al. 1949; POULSEN and HOUGS 1957). According to JORGENSEN et al. (1966), decamethonium (40–50 μg/kg i.v. decreased the force of neck muscles by 88%, of the upper limb muscles by 93%, the inspiration volume by 27%, and the expiration volume by 33%. GAL and GOLDBERG (1980) observed marked resistance of respiratory muscles to decamethonium. On the other hand, FOLDES et al. (1961) found that after the decrease of grip strength by 50%–70% induced by tubocurarine, vital capacity of lungs was essentially unchanged, while after decamethonium administration it was decreased by about 31%–51%. According to UNNA et al. (1950), and PELIKAN et al. (1950), when grip strength was decreased by 95%, tubocurarine caused a 30% decrease of the vital capacity of lungs, and decamethonium a 60% decrease.

The results obtained on both surgical patients and volunteers indicate that respiratory muscles are less sensitive to neuromuscular blocking agents than the limb muscles. The different degree of neuromuscular block in respiratory muscles evoked by tubocurarine and decamethonium, gallamine and succinylcholine is worthy of note. Furthermore, unequal sensitivity of the muscles of limbs, neck, and abdominal wall to neuromuscular blocking agents of different chemical structure is revealed. However, there are quite a lot of factors interfering with the interaction of neuromuscular blocking agents with end-plate receptors in surgical patients. There are also difficulties involved in the comparison of results, because different criteria are used for the estimation of function of various muscles when neuromuscular blocking agents are administered to surgical patients and to volunteers (CHURCHILL-DAVIDSON 1973).

The reasons for unequal sensitivity of end-plate acetylcholine receptors to neuromuscular blocking agents may be quite numerous; however, in clinics it is rather difficult to elucidate the role of each of these in the dissimilar order of muscle relaxation induced by neuromuscular blocking agents. Therefore, the data obtained in animals are more informative.

C. Factors Which May Affect the Sensitivity of End-plate Acetylcholine Receptors to Neuromuscular Blocking Agents

I. Structure of Skeletal Muscles

As early as 1950, PATON and ZAIMIS found that the order of skeletal muscle relaxation induced by decamethonium was related to whether "fast" (white) fibers or "slow" (red) ones predominated in them. In particular, acetylcholine receptors of anterior tibialis ("fast") muscle were more sensitive to decamethonium than those of soleus muscle, intercostal muscles, and the diaphragm, containing predominantly "slow" fibers. Tubocurarine showed another regularity: tibialis anterior muscle was more resistant to it than soleus muscle or the diaphragm (PATON and ZAIMIS 1951; JEWELL and ZAIMIS 1954). Furthermore, it was shown that the respiration remained inadequate when the neuromuscular transmission in the tibialis anterior muscle had recovered up to the initial level (PATON and ZAIMIS 1951). There is also clinical evidence for the predominance of "slow" fibers in the soleus muscle being responsible for the greater effect of tubocurarine on the neuromuscular transmission in it, than in the gastrocnemius muscle, containing less "slow" fibers (SECHER et al. 1981, 1982). It might seem that the data on the existence of different types of innervation and unequal functional activity of "fast" and "slow" muscles, as well as on dissimilar neuromuscular transmission in them (OLSON and SWETT 1966; MILEDI et al. 1968; FEDOROV 1969; BARNARD et al. 1971; GERTLER and ROBINS 1978) allow one to suggest that the content of "fast" and "slow" fibers may be responsible for unequal sensitivity of acetylcholine receptors to neuromuscular blocking agents of various modes of action. However, there is evidence for equal rates of contraction of respiratory muscle (diaphragm included) and tibialis anterior muscle, which contain unequal amounts of "fast" and "slow" fibers (GLEBOVSKY 1961; BISCOE 1962).

Numerous pharmacologic investigations fail to confirm the leading role of the content of different fibers in the skeletal muscles for their sensitivity to neuromuscular blocking agents. For instance, research by BONTA and GOORISSEN (1968), and MARSHALL (1973) demonstrated that, although soleus muscle was more sensitive to tubocurarine than tibialis anterior muscle, pancuronium and stercuronium blocked the neuromuscular transmission more in tibialis anterior muscle than in the soleus. Dacuronium caused equal relaxation of both muscles. Furthermore, it was shown by KHARKEVICH and FISENKO (1981) that there were considerable dissimilarities in the sensitivity of muscles (abdominal, respiratory, fore- and hindlimbs) among both nondepolarizing (tubocurarine, anatruxonium, pancuronium, decadonium, diadonium) and depolarizing (decamethonium, succinylcholine) neuromuscular blocking agents.

Different order of muscle relaxation was also observed after successive administration of tubocurarine, anatruxonium, pancuronium, or diadonium to the same animal (KHARKEVICH and FISENKO 1981). Evidently, the content of "fast" and "slow" fibers in the muscles is not decisive for the unequal sensitivity of endplates of various muscles to neuromuscular blocking agents.

II. Blood Circulation in the Skeletal Muscles

Attempts have been made to relate unequal sensitivity of neuromuscular junctions to neuromuscular blocking agents to different levels of blood circulation in skeletal muscles. For instance, CHURCHILL-DAVIDSON and RICHARDSON (1952) believe that unequal sensitivity of skeletal muscles to neuromuscular blocking agents is related to dissimilar blood circulation in them: working muscles are less sensitive to the relaxation than resting ones. Under general anesthesia, the relative resistance of respiratory muscles to neuromuscular blocking agents may be explained by significantly increased circulation in them (as in permanently working muscles) as compared with the limb or torso muscles. SECHER et al. (1982) consider that the significant density of capillaries in "slow" muscles (ANDERSEN 1975; INGJER 1979) can be responsible for the enhanced uptake of tubocurarine by them and for a more marked neuromuscular block, as compared with "fast" muscles. If the sensitivity of skeletal muscles to neuromuscular blocking agents is related to the circulation in them, then the administration of neuromuscular blocking drugs should be associated with smaller changes of neuromuscular transmission in the diaphragm than in the torso muscles. However, KHARKEVICH and FISENKO (1981) reported that the sensitivity of the torso muscles (oblique and transverse abdominal muscles) to nondepolarizing neuromuscular blocking agents (anatruxonium, cyclobutonium, decadonium, diadonium) and to depolarizing ones (decamethonium, succinylcholine) was lower than that of respiratory muscles. Intercostal muscles were more resistant to steroid neuromuscular blocking agents pancuronium and pipecuronium than the abdominal muscles or the diaphragm. Besides, tubocurarine administered after decadonium, diadonium, anatruxonium, or pipecuronium caused changes in the order of muscle relaxation.

Research by TAYLOR et al. (1964) in vitro also failed to confirm the suggestion of Churchill-Davidson and Richardson and Secher et al. They have shown that acetylcholine receptors of isolated neuromuscular guinea pig preparations (m. latissimus dorsi, m. serratus anterior, m. phrenicus) have unequal sensitivity to tubocurarine, the diaphragm being the most resistant.

III. Temperature of Skeletal Muscles

The influence of the temperature of skeletal muscles on the sensitivity of acetylcholine receptors to neuromuscular blocking agents has also been widely discussed. ALDERSEN and MACLAGAN (1964) estimated the degree of neuromuscular block in the diaphragm, tibialis anterior muscles, and intercostal muscles in anesthetized cats and they showed the diaphragm to be more sensitive to tubocurarine than the tibialis anterior muscle. On the other hand, decamethonium-induced block in respiratory muscles was substantially less marked in tibialis anterior muscle. According to Aldersen and Maclagan, the dissimilarities in the sensitivity to tubocurarine and decamethonium are related to different temperatures of the muscles. Under normal conditions in anesthetized animals, the temperature of limb muscles is much lower than that of the diaphragm, the difference in temperature being 2°–3 °C 1 h after the onset of the general anesthesia. CANNARD and ZAIMIS (1959) reported that in humans, too, temperature decrease potentiated and prolonged the action of depolarizing neuromuscular blocking agents and, vice versa, reduced the blocking effect of the nondepolarizing ones.

Repolarization of postsynaptic membrane following depolarization may be slowed down by cooling (BIGLAND et al. 1958). This accounts for the increase of paralytic effect for depolarizing agents and its decrease for nondepolarizing ones. After the elimination of the temperature difference between the respiratory and limb muscles, the effects of tubocurarine and decamethonium on various skeletal muscles become equal, too. It is worthy of note that in experiments on phrenic nerve–diaphragm preparation of rats in vitro, FARREL et al. (1981) observed temperature-related changes in potency of the nondepolarizing agents, tubocurarine and pancuronium. Other authors (HOLMES et al. 1951; HORROW and BARTKOWSKI 1981, 1983), while testing the same preparation, reported that the decrease of temperature of the bath medium from 40° to 26 °C, or from 37° to 25 °C caused a significant reduction of tubocurarine potency, pancuronium activity remaining unchanged. In spite of some controversial items in the papers mentioned, the relationships between the activity of neuromuscular blocking agents and muscle temperature seems established, but it is unlikely to play the main role in the unequal sensitivity of neuromuscular junctions to neuromuscular blocking agents. This suggestion is strengthened by the results obtained by LEPAKHIN and FISENKO (1970), and KHARKEVICH and FISENKO (1981) on anesthetized cats at different temperatures of respiratory and limb muscles. The comparison of the inhibition of neuromuscular transmission in these skeletal muscles induced by decamethonium and succinylcholine and their nondepolarizing analogs decadonium and diadonium, respectively, revealed essentially no difference in sensitivity of respiratory and limb muscles to the action of two pairs of compounds: decadonium–decamethonium and diadonium–succinylcholine. Moreover, diadonium administered 2 h after succinylcholine to the same animal failed to alter the sensitivity of acetylcholine receptors at the sites already mentioned, although the temperature difference between the limb muscles and the diaphragm was increased.

IV. Acid–Base Equilibrium

Variations of acid–base equilibrium may affect the action of neuromuscular blocking agents. It is usually considered that acidosis potentiates and alkalosis reduces tubocurarine-induced neuromuscular block (MACLAGAN 1976). At the same time, acidosis reduces and alkalosis potentiates the neuromuscular block induced by both nondepolarizing agents (metocurine, gallamine, alcuronium) and depolarizing ones (decamethonium, succinylcholine). FUNK et al. (1980) reported that during significant variations of pH in anesthetized cats, respiratory and metabolic acidosis caused an increase in the sensitivity of the tibialis anterior muscle, soleus muscle, and the diaphragm to the nondepolarizing agent vecuronium (Org-NC-45). However, the order of relaxation of these muscles was essentially unchanged under such conditions. On the other hand, metabolic alkalosis is followed by a reduction of vecuronium neuromuscular block, the difference in sensitivity of the muscles disappearing entirely. It is suggested that pH changes result in the alterations of the blood flow in the skeletal muscles, distribution of substances, ionization of receptors, and drugs (PAYNE and HUGHES 1981; HORROW and BARTKOWSKY 1981).

V. Lability of Neuromuscular Junctions

Several authors account for the unequal sensitivity of the skeletal muscles to neuromuscular blocking agents by their different lability (according to Wedensky) of the neuromuscular junctions (KHRUSTALEV 1962, 1964; DOBRYANSKY 1966). Thus, Khrustalev showed that the degree of lability of the synapses of gastrocnemius muscle, intercostal muscles, and diaphragm was completely consistent with their sensitivity to the nondepolarizing neuromuscular blocking agent, diplacinum. Neuromuscular block in less labile synapses of the gastrocnemius muscle after diplacinum administration developed earlier, at lower doses, than in more labile synapses of intercostal muscles or the diaphragm. However, the difference in lability cannot entirely account for the unequal sensitivity of acetylcholine receptors to neuromuscular blocking agents, since the lability level is rather constant for particular neuromuscular synapses, while the comparative sensitivity to structurally different neuromuscular blocking agents varies (KHARKEVICH and FISENKO 1981). In addition, increase in the rate of stimulation of the peripheral end of motor nerves from 0.1 to 50 Hz fails to abolish the difference in sensitivity of hindlimb muscles or diaphragm to nondepolarizing (tubocurarine, vecuronium, atracurium, Duador, pancuronium) or depolarizing (succinylcholine) drugs (TRAN et al. 1982). Thus, the data suggest that the structure of skeletal muscles, their temperature, pH variations, and circulation in the muscles do not entirely account for the unequal sensitivity of acetylcholine receptors to neuromuscular blocking agents.

VI. Site and Mode of Action of Neuromuscular Blocking Agents

According to LEE et al. (1982), the unequal sensitivity of the diaphragm and tibialis anterior muscle to succinylcholine, tubocurarine, and vecuronium in anesthetized cats can be related to dissimilar affinity for pre- and postsynaptic acetylcholine receptors. The importance of the latter is discussed by BOWMAN (1980). Furthermore, the presynaptic component in the action of neuromuscular blocking agents can be responsible for the nonidentical sequence of relaxation of respiratory and other skeletal muscles. Unfortunately, it is very difficult to estimate the role of presynaptic structures in the effect of neuromuscular blocking agents with respect to different muscles. The available data on the presynaptic influence of neuromuscular blocking agents of various modes of action are highly controversial. Thus, for instance, JENKINSON (1960) points out that the minimal quantal content of the end-plate potential (EPP) of "slow" skeletal muscles can account for the high affinity of tubocurarine for them. GERTLER and ROBINS (1978) have demonstrated that the quantal content of EPP in m. extensor digitorum longus ("fast" muscle) is higher than in soleus muscle ("slow" muscle), and the ratio of EPP amplitude to the value of threshold depolarization required for the generation of action potential in soleus muscle turned out about twofold lower than in the "fast" muscle. HUBBARD and WILSON (1973), BLABER (1973), and SU et al. (1979) believe that neuromuscular blocking agents are characterized by the ability to inhibit neuromuscular transmission even at the presynaptic level. At the same time, BERANEK and VYSCOČYL (1967), BAUER (1971), and AUERBACH and BETZ (1971) did not observe presynaptic blocking action for tubocurarine. KEMMOTSU et al. (1982), while studying in vitro the locus of atracurium action in the neuromuscular junction, observed its inhibitory action on the amplitude of miniature EPP and the sensitivity of the end-plate to microiontophoretically applied acetylcholine. These data and the absence of changes of miniature EPP frequency suggested a predominantly postsynaptic character of neuromuscular block induced by atracurium. The locus of action of neuromuscular blocking agents was also studied by FISENKO et al. (1973). It was found in phrenic nerve–diaphragm preparations of rats in vitro that in equieffective concentrations (with respect to EPP amplitude) tubocurarine, anatruxonium, cyclobutonium, decadonium, diadonium, decamethonium, and succinylcholine failed to affect miniature EPP frequency, but decreased the amplitude of the latter. The discrepancy of the data presented, dealing with a presynaptic component in the action of neuromuscular blocking agents, does not reveal explicitly its significance for the unequal sensitivity of skeletal muscles to neuromuscular blocking agents.

The relationship between the sequence of muscle relaxation and the mode of action of neuromuscular blocking agents was analyzed by LEPAKHIN and FISENKO (1970), and KHARKEVICH and FISENKO (1981). They estimated the sensitivity of skeletal muscles in anesthetized cats to two pairs of neuromuscular blocking drugs: decadonium–decamethonium (structure I) and diadonium–succinylcholine (structure II). As already mentioned, decadonium and diadonium are nondepolarizing *N*-adamantyl analogs of decamethonium and succinylcholine, respectively (KHARKEVICH 1970 a, b).

$$R(CH_3)_2N^+\text{–}(CH_2)_{10}\text{–}N^+(CH_3)_2R \cdot 2I^- \qquad (I)$$

a) $R = -CH_3$

b) R = 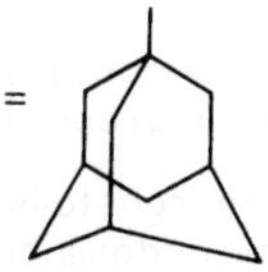

It was found that decadonium (I b) first of all caused a block of transmission to the masseter muscle, then to the gastrocnemius, the intercostal muscles, and the triceps brachii muscle. Abdominal muscles and the diaphragm appeared the most resistant to decadonium. The order of decamethonium-induced muscle relaxation was the same (Fig. 1). Decamethonium (I α) administered 1.5–2 h after decadonium to the same animal, caused no changes of the order of muscle relaxation. Tubocurarine administered 1.5–2 h after decadonium or decamethonium modified the order of the neuromuscular block and it developed in a tubocurarine-like manner.

$$R(CH_3)_2N^+\text{–}(CH_2)_2\text{–}OCO\text{–}(CH_2)_2\text{–}OCO\text{–}(CH_2)_2\text{–}N^+(CH_3)_2R \cdot 2X^- \qquad (II)$$

a) $R = -CH_3$

b) R = 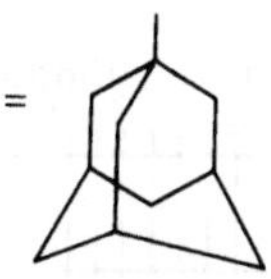

Diadonium (II b) blocked neuromuscular transmission most of all in the masseter muscle and in the triceps brachii and somewhat less in the respiratory

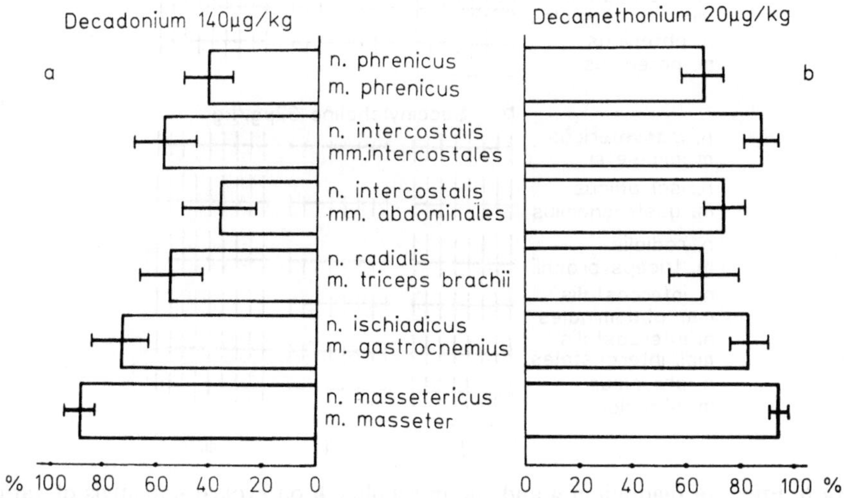

Fig. 1 a, b. Effect of decadonium **a** and decamethonium **b** on evoked potentials of various muscles. Mean values with confidence limits at $P = 0.05$ (10–12 experiments for each drug). Experiments on cats anesthetized with urethane (600 mg/kg) and chloralose (70 mg/kg) i.v.

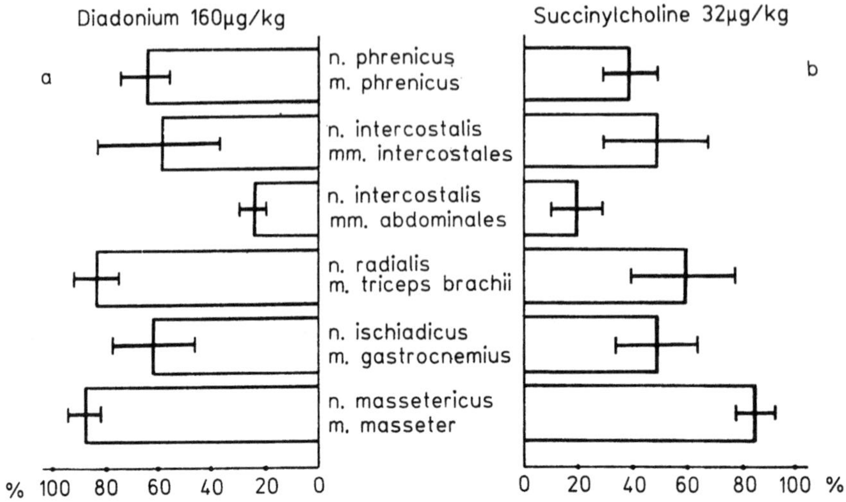

Fig. 2 a, b. Effect of diadonium **a** and succinylcholine **b** on evoked potentials of various muscles. Mean values with confidence limits at $P = 0.05$ (10–12 experiments for each drug). Experiments on cats anesthetized with urethane (600 mg/kg) and chloralose (70 mg/kg) i.v.

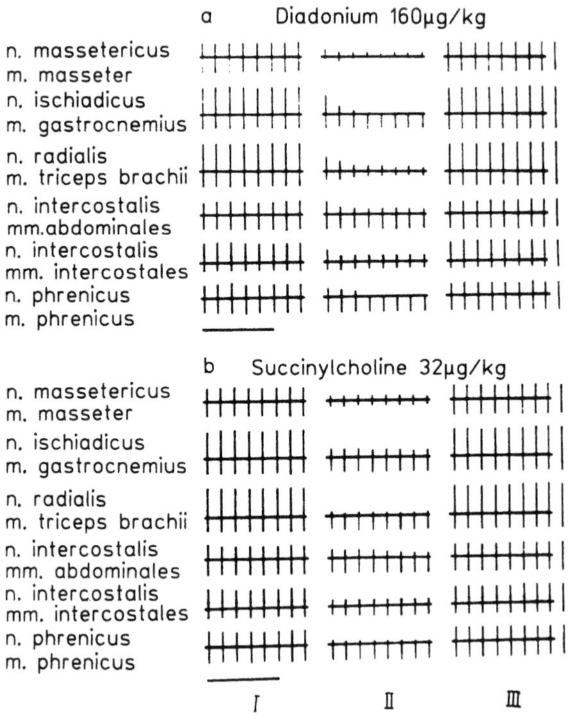

Fig. 3 a, b. Effect of diadonium **a** and succinylcholine **b** on evoked potentials of various muscles. Succinylcholine was administered 2 h after diadonium. I before drug administration; II 2 min after drug administration; III restoration within 40 min. *Vertical lines on the right* amplitude (1 mV). *Horizontal line* time (1 s). Experiment on cat anesthetized with urethane (600 mg/kg) and chloralose (70 mg/kg) i.v.

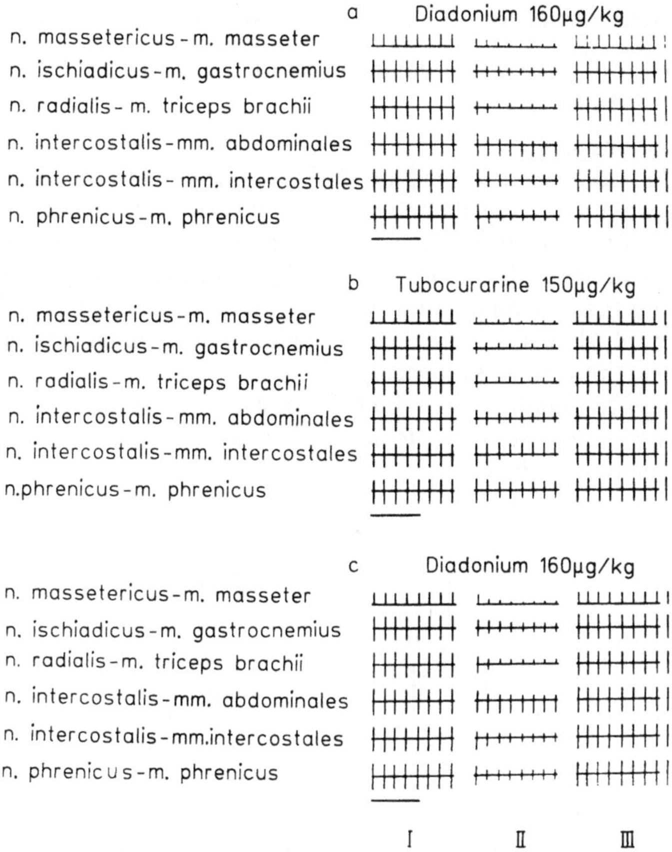

Fig. 4 a–c. Effect of diadonium **a**, tubocurarine **b**, and diadonium **c** on evoked potentials of various muscles. Tubocurarine was administered 2 h after diadonium. Second injection of diadonium 2 h after tubocurarine. I before drug administration; II 2 min after drug administration; III restoration within 40 min. *Vertical lines on the right* amplitude (1 mV). *Horizontal line* time (1 s). Experiment on cat anesthetized with urethane (600 mg/kg) and chloralose (70 mg/kg) i.v.

and abdominal muscles. Abdominal muscles turned out the most resistant. After administration of succinylcholine (II a), the neuromuscular block developed in the same order (Fig. 2). Succinylcholine administered 1.5–2 h after diadonium to the same animal failed to change the order of relaxation of the skeletal muscles (Fig. 3). On the other hand, tubocurarine administered 1.5–2 h after diadonium or succinylcholine changed the order of the neuromuscular block of various groups of skeletal muscles and it developed in a tubocurarine-like manner. Diadonium, when administered 2 h after tubocurarine, changed the order of the neuromuscular block again and readjusted it to its own pattern (Fig. 4). Thus, the comparative sensitivity of acetylcholine receptors of the skeletal muscles to structurally similar neuromuscular blocking agents, but having different mechanisms of neuromuscular block (decadonium–decamethonium, diadonium–succinylcholine) was identical.

VII. Chemical Structure of Neuromuscular Blocking Agents

1. Compounds with Similar Interonium Structure and Different Cationic Groups

The investigations carried out by LAPAKHIN (1967), LEPAKHIN and FISENKO (1970), and KHARKEVICH and FISENKO (1981) are devoted to the relationship between the sensitivity of acetylcholine receptors of various skeletal muscles and the chemical structure of neuromuscular blocking agents. They demonstrated that the relaxation of skeletal muscles induced by nondepolarizing neuromuscular blocking agents (anatruxonium and cyclobutonium, pancuronium and pipecuronium, respectively) having different cationic groups, but similar interonium structure, developed in the same sequence.

Thus, masseter muscle, then limb muscles, intercostal muscles, and the diaphragm were most sensitive to α-truxillic acid derivatives: anatruxonium (III a) and cyclobutonium (III b).

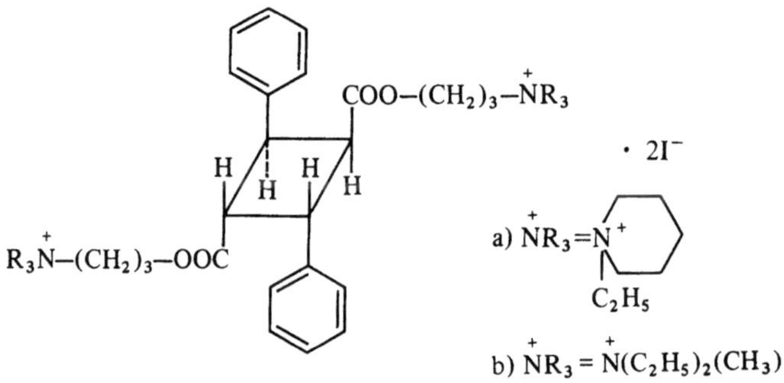

(III)

Abdominal muscles were the least sensitive to these agents (Fig. 5).

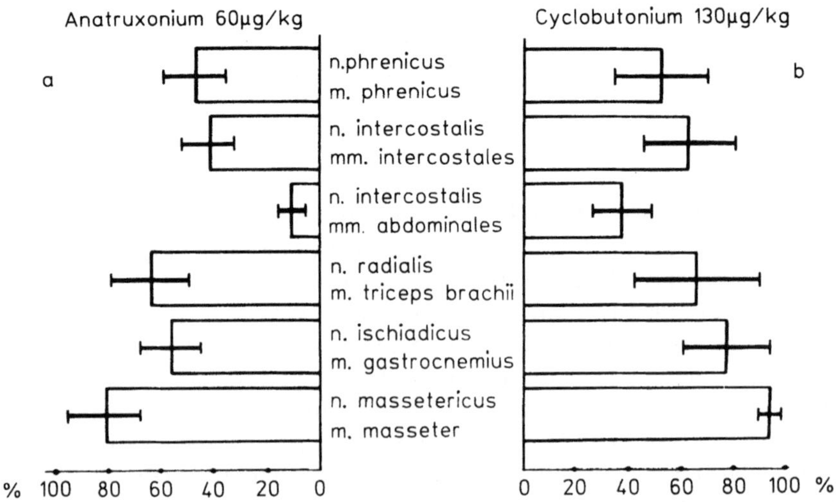

Fig. 5 a, b. Effect of anatruxonium **a** and cyclobutonium **b** on evoked potentials of various muscles. Mean values with confidence limits at $P = 0.05$ (10–12 experiments for each drug). Experiments on cats anesthetized with urethane (600 mg/kg) and chloralose (70 mg/kg) i.v.

Steroid neuromuscular agents pancuronium (IV a) and pipecuronium (IV b) first induced relaxation of the masseter muscles, then of the limb muscles, the diaphragm, abdominal muscles,

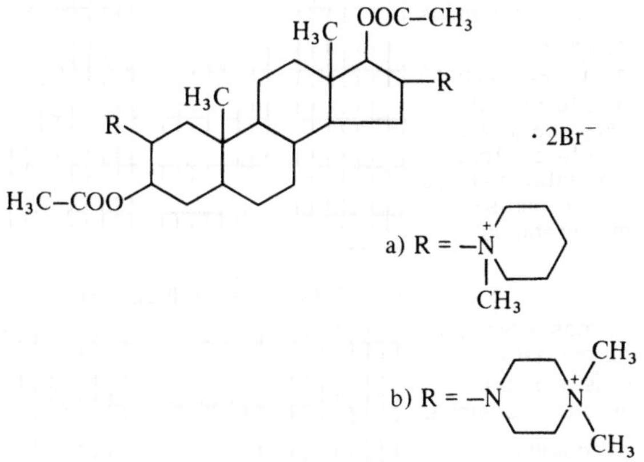

(IV)

and, finally, intercostal muscles (Fig. 6). It is interesting to note that tubocurarine administered after anatruxonium or cyclobutonium, pancuronium or pipecuronium changed the order of muscle relaxation, and it developed in a tubocurarine-like manner (Fig. 7).

2. Compounds with Identical Cationic Groups and Different Interonium Structure

The available data indicate that neuromuscular blocking agents with identical cationic groups, but different interonium structures can cause relaxation of skel-

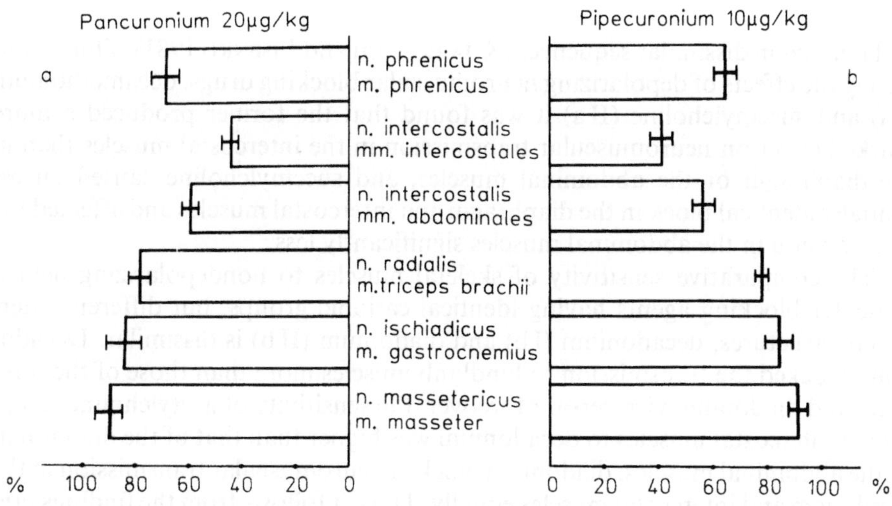

Fig. 6a, b. Effect of pancuronium **a** and pipecuronium **b** on evoked potentials of various muscles. Mean values with confidence limits at $P=0.05$ (10–12 experiments for each drug). Experiments on cats anesthetized with urethane (600 mg/kg) and chloralose (70 mg/kg) i.v.

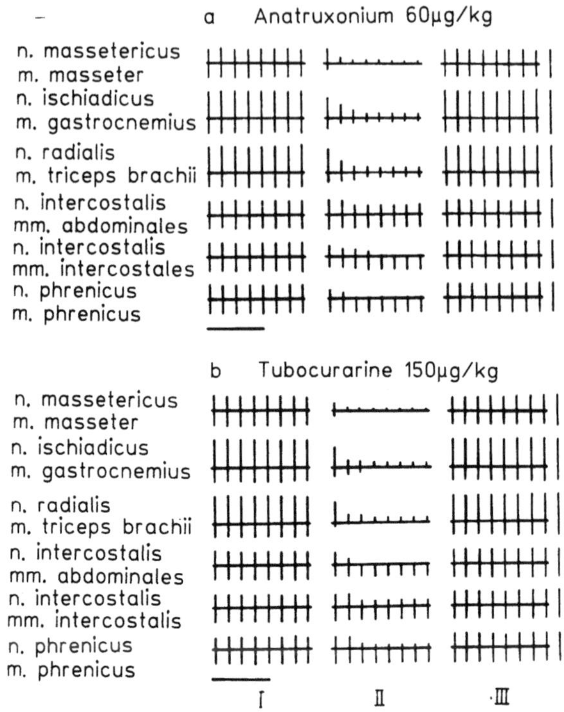

Fig. 7 a, b. Effect of anatruxonium **a** and tubocurarine **b** on evoked potentials of various muscles. Tubocurarine was administered 2 h after anatruxonium. I before drug administration; II 2 min after drug administration; III restoration within 40 min. *Vertical lines on the right* amplitude (1 mV). *Horizontal line* time (1 s). Experiment on cat anesthetized with urethane (600 mg/kg) and chloralose (70 mg/kg) i.v.

etal muscles in dissimilar sequences (KHARKEVICH and FISENKO 1981). Thus, comparing the effects of depolarizing neuromuscular blocking drugs, decamethonium (I a) and succinylcholine (II a) it was found that the former produced a more marked effect on neuromuscular transmission in the intercostal muscles than in the diaphragm or the abdominal muscles, and succinylcholine caused an essentially identical block in the diaphragm and intercostal muscles and affected the transmission in the abdominal muscles significantly less.

The comparative sensitivity of skeletal muscles to nondepolarizing neuromuscular blocking agents having identical cationic groups, but different interonium structures, decadonium (I b) and diadonium (II b) is dissimilar. Decadonium blocked the transmission to hindlimb muscles more than those of the forelimb, and diadonium vice versa. Moreover, the sensitivity of acetylcholine receptors of intercostal muscles to decadonium was higher than that of the diaphragm or the abdominal muscles; diadonium blocked neuromuscular transmission in the diaphragm and intercostal muscles equally. Thus, it follows from the findings presented that the structure of the interonium part of a neuromuscular blocking agent molecule is more significant for the sequence of its neuromuscular blocking action than the structure of cationic groups.

D. Conclusions

The data presented indicate a different order of blockade of acetylcholine receptors of various groups of muscles by chemically different neuromuscular blocking agents. Structural dissimilarities of agents causing relaxation of the skeletal muscles in different sequences have to do, mainly, with the interonium part of their molecule. For instance, succinylcholine and decamethonium, diadonium and decadonium have identical cationic groups and equal numbers of atoms between quaternary nitrogen atoms, but different interonium structure of the molecule. The sensitivity of skeletal muscles at different locations to these pairs of neuromuscular blocking drugs is not the same.

The suggestion of the significance of the interonium structure is corroborated by the examples of anatruxonium–cyclobutonium and pancuronium–pipecuronium, agents with different cationic groups, but identical or similar structure of the interonium fragment. The sequence of action of anatruxonium is analogous to that of cyclobutonium and that of pancuronium to the pipecuronium-induced sequence. Structural differences between decadonium–decamethonium and diadonium–succinylcholine also concern only their cationic groups. As was mentioned, the comparative sensitivity of acetylcholine receptors of skeletal muscles to decadonium is close to that of decamethonium, and that of diadonium is analogous to succinylcholine.

It follows from the data presented that the interonium structure of neuromuscular blocking agents affects the order of their neuromuscular blocking action, which may be related to unequal flexibility of the molecules and to the appearance of additional fixation points of the drugs on end-plate receptors, e.g., via hydrophobic, dipole–dipole, and other interactions.

It should be noted that the revealed differences in the sequences of muscle relaxation depending on the chemical structure of neuromuscular blocking agents are maintained under unequal content of "fast" and "slow" fibers in skeletal muscles, differences in temperature, values of acid–base equilibrium, and circulation level in the skeletal muscles.

The unequal sensitivity of acetylcholine receptors of skeletal muscles to neuromuscular blocking agents seems to be determined by the pecularities of the arrangement of both acetylcholine receptor sites and the other related functionally significant formations of the subsynaptic membrane (i.e., nonspecific sites of binding). The factors mentioned (temperature, acid–base equilibrium, intensity of blood supply, etc.) are of more or less universal importance for the sensitivity of acetylcholine receptors of various groups of muscles. However, it should be taken into account that the factors listed can produce diverse actions on the kinetics of interaction of structurally different neuromuscular blocking agents with acetylcholine receptors at various locations. It cannot be excluded that the peculiarities of interaction of neuromuscular blocking drugs with ionic channel sites, as well as with allosteric receptor sites of different groups of muscles, are of certain significance, too. However, special investigations are needed for the elucidation of all these questions. The detected dissimilarities in muscle relaxation suggest that it cannot be excluded that agents with predominant action on definite groups of skeletal muscles might be designed.

Acknowledgement. The authors wish to express their appreciation to Mrs. MARIA LIPMAN for the translation of this chapter.

References

Aldersen AM, Maclagan J (1964) The action of decamethonium and tubocurarine on respiratory and limb muscles of the cat. J Physiol (Lond) 173:38–56

Andersen P (1975) Capillary density in skeletal muscle of man. Acta Physiol Scand 95:203–205

Auerbach A, Betz W (1971) Does curare affect transmitter release? J Physiol (Lond) 213:691–705

Barnard RJ, Edgerton VR, Furukawa T, Peter JB (1971) Histochemical, biochemical and contractile properties of red, white and intermediate fibers. Am J Physiol 220:410–414

Bauer H (1971) Die Freisetzung von Acetylcholin an der motorischen Nervenendigung unter dem Einfluß von d-Tubocurarin. Pfluegers Arch 326:162–183

Beranek R, Vyskočyl F (1967) The action of d-tubocurarine on the normal and denervated rat diaphragm. J Physiol (Lond) 188:53–66

Bigland B, Goetzee B, Maclagan J, Zaimis E (1958) The effect of lowered muscle temperature on the action of neuromuscular blocking drugs. J Physiol (Lond) 141:425–434

Biscoe TJ (1962) The isometric contraction characteristics of cat intercostal muscle. J Physiol (Lond) 164:189–199

Blaber LC (1973) The prejunctional actions of some non-depolarizing blocking drugs. Br J Pharmacol 47:109–116

Bonta IL, Goorissen EM (1968) Different potency of pancuronium bromide in two types of skeletal muscle. Eur J Pharmacol 4:303–308

Bowman WC (1980) Prejunctional and postjunctional cholinoceptors at the neuromuscular junction. Anesth Analg 59:935–943

Cannard TH, Zaimis E (1959) The effect of lowered muscle temperature on the action of neuromuscular blocking drugs in man. J Physiol (Lond) 149:112–119

Churchill-Davidson HC (1973) A philosophy of relaxation. Anesth Analg 52:495–501

Churchill-Davidson HC, Richardson AT (1952) Decamethonium iodide: some observations on its action using electromyography. Proc R Soc Lond (Med) 45:179–185

Collier HOJ (1951) Problems of developing muscle relaxants in the laboratory. Proc R Soc Lond (Med) 44:627–636

Danilov AF, Mikhelson MYa, Rybolovlev RS (1957) Pharmacological properties of succinylcholine and its application in clinics (in Russian). In: Mndzhoyan AL (ed) Ditilin i opyt ego klinicheskogo primeneniya (Dithylinum and clinical experience with it). Armenian SSR Academy of Sciences Publishers, Yerevan, pp 92–108

Dardel O, Thesleff S (1952) Clinical experience with succinylcholine iodide, a new muscular relaxant. Anesth Analg 31:250–257

Davies DL, Lewis A (1949) Effects of decamethonium iodide on respiration and on induced convulsion in man. Lancet 1:775–777

Dobryansky VS (1966) Comparative lability of m. tibialis and diaphragm and their sensitivity to myorelaxants (in Russian). In: Second conf on anestesiol and reanimatol, pp 7–9

Doughty AG, Wylie WD (1951) An assessment of flaxedil. Proc R Soc Lond (Med) 44:375–388

Farrell L, Dempsey MJ, Waud BE, Waud DR (1981) Temperature and potency of d-tubocurarine and pancuronium in vitro. Anesth Analg 60:18–20

Fedorov VV (1969) Specific features of biopotentials of the fibers of slow and fast skeletal muscles of the rat (in Russian). Fiziol Zh SSSR 5:588–596

Fisenko VP, Polgar AA, Smirnova VS (1973) Microelectrophysiological investigation into the mechanism and localization of the effect produced by new curareform agents (in Russian). Farmakol Toksikol 36:206–209

Foldes FF, Machay TS (1951) Syncurine (decamethonium bromide). Its use with pentothal-sodium and nitrous oxide – oxygen anesthesia in abdominal surgery. Anesthesioly 12:366–375
Foldes FF, Machay TS, Hunt RD, McNall PG, Carberry PC (1952) Synthetic muscle relaxants in anesthesia. JAMA 150:1559–1566
Foldes FF, Monte AP, Brunn HM Jr, Wolfson B (1961) Studies with muscle relaxants in unanaesthetized subjects. Anesthesiology 22:230–236
Funk DI, Crul JF, Pol FM (1980) Effects of changes in acid-base balance on neuromuscular blockade produced by ORG NC 45. Acta Anaesth Scand 24:119–124
Gal TJ, Goldberg SK (1980) Diaphragmatic function in healthy subjects during partial curarization. J Appl Physiol 48:921–926
Gal TJ, Goldberg SK (1981) Relationship between respiratory muscle strength and vital capacity during partial curarization in awake subjects. Anesthesiology 54:141–147
Gertler RA (1981) Effect of emphysema on the action of curare. Anesthesiology 55:A211
Gertler RA, Robins N (1978) Differences in neuromuscular transmission on red and white muscles. Brain Res 142:160–164
Gillies DM, Cullen WG, Griffith HR (1954) Succinylcholine as a relaxant in abdominal surgery. Curr Res Anesth Analg 33:251–258
Glebovsky VD (1961) On contractile properties of respiratory muscles in adult and newborn animals (in Russian). Fiziol Zh SSSR 4:427–435
Gray AJ (1950) Decamethonium iodide as a muscle relaxant in abdominal surgery. Lancet 1:253–255
Grigoriev MS, Anichkov MN (1957) Kurare i kurarepodobnye sredstva v khirurgii. (Curare and curare-like drugs in surgery). Leningrad
Grob D, Holaday DA, Harvey AM (1949) The effect of bis-trimethylammonium decane on neuromuscular function on induced convulsions in man. New Engl Med J 241:812–816
Harris LC, Dripps RD (1950) The use of decamethonium bromide for the production of muscular relaxation. Anaesthesiology 11:215–223
Hewer AJH, Lucas BGB, Prescott F, Rowbotham ES (1949) Decamethonium iodide as a muscle relaxant in anesthesia. Lancet 1:817–818
Holmes PEB, Jenden DJ, Taylor DB (1951) The analysis of the mode of action of curare on neuromuscular transmission; the effects of temperature changes. J Pharmacol Exp Ther 103:382–402
Hoppe JO (1955) Observation on the potency of neuromuscular blocking agents with particular reference to succinylcholine. Anesthesiology 16:92–124
Horrow J, Bartkowski RR (1981) Cold reduces potency of curare; not pancuronium. Anesthesiology 55:A217
Horrow J, Bartkowski RR (1983) Pancuronium, unlike other nondepolarizing relaxants, retains potency at hypothermia. Anesthesiology 58:357–361
Hubbard JI, Wilson DF (1973) Neuromuscular transmission in a mammalian preparation in the absence of blocking drugs and the effect of d-tubocurarine. J Physiol (Lond) 228:307–325
Ingjer F (1979) Effects of endurance training on muscle fibre ATP-ase activity, capillary supply and mitochondrial content in man. J Physiol (Lond) 294:419–432
Jenkinson DM (1960) The antagonism between tubocurarine and substances which depolarize the motor end-plate. J Physiol (Lond) 152:309–324
Jewell PA, Zaimis E (1954) A differentiation between red and white muscle in the cat based on responses to neuromuscular blocking agent. J Physiol (Lond) 124:417–428
Johansen SH, Jørgensen M, Molbech S (1964) Effect of tubocurarine on respiratory and nonrespiratory muscle power in man. J Appl Physiol 19:990–994
Jørgensen M, Molbech S, Johansen SH (1966) Effect of decamethonium on head lift, hand grip and respiratory muscle power in man. J Appl Physiol 21:509–512
Kemmotsu O, Sokoll MD, Gergis SD (1982) Site action of BW3 3A at the rat neuromuscular junction. Anesthesiology 57:A289
Kharkevich DA (1970a) On curare-like activity of decadonium diiodide (in Russian). Farmakol Toksikol 33:395–399

Kharkevich DA (1970b) Pharmacological properties of a new antidepolarizing curare-like drug diadonium diiodide (in Russian). Farmakol Toksikol 33:531–536
Kharkevich DA, Fisenko VP (1981) The effect of neuromuscular blocking agents on the acetylcholine receptors of different skeletal muscles. Arch Int Pharmacodyn Ther 251:255–269
Khrustalev SI (1962) On comparative lability of neuromuscular synapses (in Russian). Farmakol Toksikol 25:88–93
Khrustalev SI (1964) Comparative sensitivity of neuromuscular synapses in different animals (in Russian). Fiziol Zh SSSR 50:1364–1372
Lee C, Durant N, Nguyen N, Tran B, Katz R (1982) Comparative neuromuscular pharmacology of the diaphragm and the tibialis anterior. Anesthesiology 57:A282
Lepakhin VK (1967) Comparative sensitivity of neuromuscular synapses of different muscles to the derivatives of truxillic acid and to tubocurarine chloride (in Russian). Farmakol Toksikol 30:152–156
Lepakhin VK (1970) The ether and fluorothane action on the comparative sensitivity of muscles to anatruxonium and d-tubocurarine chloride (in Russian). Farmakol Toksikol 33:538–542
Lepakhin VK, Fisenko VP (1970) Comparative sensitivity of neuromuscular synapses of different muscles to diadonium and decadonium (in Russian). Farmakol Toksikol 33:288–292
Maclagan J (1976) Competitive neuromuscular drugs. In: Zaimis E (ed) Neuromuscular junction. Springer, Berlin Heidelberg, pp 420–486 (Handbook of experimental pharmacology, vol 42)
Marshall IG (1973) The effects of three short-acting neuromuscular blocking agents on fast and slow-contracting muscles of the cat. Eur J Pharmacol 21:299–304
Miledi R, Stefani E, Zelena J (1968) Neural control of acetylcholine-sensitivity in rat muscle fibers. Nature 220:497–498
Mushin WW, Wein R, Mason DFJ, Langston GT (1949) Curare-like actions of tri(diethylaminoethoxy) benzene triethyliodide. Lancet 1:726–728
Ngai SH, Hanks EC, Fink BR, Holaday DA, Papper EM (1959) Quantitative study of action of imbretil and its modification in man. Anesthesiology 20:653–658
Olson CB, Swett CP (1966) A functional and histochemical characterization of motor unit in a heterogeneous muscle (flexor digitorum longus) of the cat. J Comp Neurol 128:475–491
Organe G, Paton WDM, Zaimis EJ (1949) Preliminary trials of bistrimethyl ammonium decane and pentane diiodide (C.10 and C.5) in man. Lancet 1:21–23
Paton WDM, Zaimis EJ (1950) Actions and clinical assessment of drugs which produce neuromuscular block. Lancet II:568–570
Paton WDM, Zaimis EJ (1951) The action of d-tubocurarine and of decamethonium on respiratory and other muscles in the cat. J Physiol (Lond) 112:311–331
Patterson RL, Pittsburg P (1952) The use of muscle relaxing drugs in anesthesia. Anesth Analg 31:65–68
Payne JP, Hughes R (1981) Clinical assessment of neuromuscular transmission. Br J Clin Pharmacol 11:537–548
Pelikan EW, Unna KR, Macfarlane DW, Cazort RJ, Sadove MS, Nelson JT (1950) Evaluation of curarizing drugs in man. Analysis of response curves and effects of repeated doses of d-tubocurarine, dimethyl-d-tubocurarine and decamethonium. J Pharmacol Exp Ther 99:215–225
Poulsen H, Hougs W (1957) The effect of some curarizing drugs in unanesthetized man. Acta Anaesth Scand 1:15–39
Richards H, Youngman HR (1952) The ultra-short acting relaxants. Br Med J 1:1334–1335
Sadove MS, Nelson JT, Unna KR (1951) Comparative evaluation of curare-like drugs. Anesth Analg 30:221–227
Secher NH, Rube N, Secher O (1981) Effect of tubocurarine on human soleus and gastrocnemius muscles. Acta Anaesth Scand [Suppl] 72:25–31

Secher NH, Rube N, Secher O (1982) Effect of tubocurarine on human soleus and gastrocnemius muscles. Acta Anaesth Scand 26:231–234
Spencer CH, Coakley CS (1955) Clinical evaluation of syncurine – a two year study. Anesthesiology 16:125–132
Stoeling VK, Graf JP, Theyl RA (1950) The use of metubine iodide in anesthesiology. Curr Res Anesth Analg 29:282–287
Su PC, Su Wen-huey L, Rosen AD (1979) Pre- and postsynaptic effects of pancuronium at the neuromuscular junction of the mouse. Anesthesiology 50:199–204
Taylor DB, Prior RD, Bevan JA (1964) The relative sensitivities of diaphragm and other muscles of the guinea pig to neuromuscular blocking agents. J Pharmacol Exp Ther 143:187–191
Thesleff S (1952) Pharmacological properties of succinylcholine iodide. Acta Physiol Scand 26:103–130
Tran DO, Amaki Y, Ohta Y, Nagashima H, Duncalf D, Foldes FF (1982) Simultaneous *in vivo* measurement of NM block on three muscles. Anesthesiology 57:A276
Ullet GA, Counts RM, Chapman AH, Parsons EH (1950) Clinical report on use of dimethyl ether of d-tubocurarine iodide in electroshock therapy. Am J Psychiat 107:184–189
Unna KR, Pelikan EW (1951) Evaluation of curarizing drugs in man. Ann NY Acad Sci 54:480–492
Unna KR, Pelikan EW, MacFarlane P, Cazort RG, Sadove MS, Nelson JT (1950) Evaluation of curarizing agents in man. JAMA 144:448–451
Vinogradov VM, Dyachenko PK (1961) Comparative clinical-pharmacological characterization of domestic curare substitutes. The action of curare-like agents on respiratory apparatus (in Russian). Transactions of Leningrad Chemical Pharmaceutical Institute N 13, pp 180–192
Walton FA (1950) Flaxedil, a new curarizing agent. Can Med Assoc J 63:123–129

CHAPTER 7

Antimuscarinic and Ganglion-Blocking Activity of Neuromuscular Blocking Agents

D. A. KHARKEVICH and V. A. SHORR

A. Antimuscarinic Activity

I. Introduction

Among the requirements set by anesthesiologists for new neuromuscular blocking agents, high selectivity of the action on end-plate acetylcholine receptors is one of the most important. Unfortunately, the majority of the known neuromuscular blocking agents, in addition to their main action, may also affect other biologic substrates adapted to the interaction with acetylcholine, such as nicotinic and muscarinic receptors of the autonomic nervous system as well as cholinesterases. Neuromuscular blocking agents with such activity can cause various side effects (changes of hemodynamics, heart rate, etc.).

The antimuscarinic properties of neuromuscular blocking agents attracted attention soon after the first clinical trials of gallamine, which was found to provoke a rather marked sinus tachycardia in patients and, in some of them, alterations of the cardiac rhythm (UNNA et al. 1950b; MURBURY et al. 1951; DOUGHTY and WYLIE 1951). Experimental analysis of the clinical data (RIKER and WESCOE 1951) has shown that gallamine in neuromuscular blocking doses has a vagolytic action due to blockade of the cardiac muscarinic receptors.

Further, similar action has been found in many nondepolarizing neuromuscular blocking agents belonging to various chemical classes and, therefore, today it can be stated that the presence of antimuscarinic activity in neuromuscular blocking agents is a frequent component of their pharmacologic spectrum.

This section presents a summary of the data on antimuscarinic activity of the majority of the known neuromuscular blocking agents and the specific features of muscarinic blocking action of some of them, as compared with that of classical atropine-like agents. Some other possible mechanisms which can be responsible for the appearance of autonomic side effects, characteristic of neuromuscular blocking agents with antimuscarinic activity, are also discussed.

II. Comparative Characteristics of Antimuscarinic Activity of Neuromuscular Blocking Agents

Antimuscarinic activity of neuromuscular blocking agents is estimated in experiments on the whole animal (mostly cats) and isolated organs by standard pharmacologic methods. From the practical point of view, the estimation of cardiac vagolytic activity of these agents, as related to their neuromuscular blocking activity, seems especially useful. Relative indices, such as ED_{50} vagal block/ED_{50} (or ED_{95}) neuromuscular block can be regarded as indices of cardiac antimuscarinic safety of neuromuscular blocking agents bearing a certain prognostic value concerning probable cardiac clinical side effects.

The analysis of clinical–experimental correlations shows that tachycardia, related probably to the vagolytic action of neuromuscular blocking agents, might develop in humans even when in experiments on cats the neuromuscular blocker causes a 50% block of the cardiac vagal transmission at doses 3–5 times as high as the 50% neuromuscular blocking dose (MARSHALL 1980). Therefore, it seems reasonable to agree with the suggestion of BOWMAN (1982), who considers neuromuscular blocking agents safe for humans in terms of probable development of antimuscarinic cardiac side effects, if in experiments on cats they have a ratio of ED_{50} for antimuscarinic vagal blockade to ED_{50} for neuromuscular blockade equal to, or higher than 10:1.

If, while calculating the experimental index of antimuscarinic safety, the doses causing a 95% (or complete) blockade of neuromuscular transmission are chosen as the criterion of neuromuscular blocking activity of the agents, then the neuromuscular blockers having a ratio between the cardiac vagolytic ED_{50} and neuromuscular blocking ED_{95} (or neuromuscular blocking dose) more than 5:1, can be considered as potentially safe for humans.

Using both these experimental criteria, the known neuromuscular blocking agents can be classified into two groups. To the first there belong the agents with a high index of cardiac antimuscarinic safety and thus devoid of antimuscarinic side effects; the second group consists of neuromuscular blocking agents with marked antimuscarinic activity which is manifested within or near the neuromuscular blocking dose range. These neuromuscular blockers can cause typical cardiac side effects in patients.

1. Neuromuscular Blocking Agents Without Antimuscarinic Activity

Tubocurarine is a typical representative of this group. In neuromuscular blocking doses (about 0.3 mg/kg i.v.) it produces a marked cardiac vagolytic effect in anesthetized cats and dogs, which, however, is not related to the block of muscarinic receptors of the neuroeffector junction, but is due to blockade of transmission at the ganglionic level (MAUTNER and LUISADA 1941; HUGHES 1972; DURANT et al. 1977; SAVARESE 1979; HUGHES and CHAPPLE 1981). At doses 2–4 times higher than the neuromuscular blocking dose, tubocurarine fails to affect the negative chronotropic effect of muscarinic agonists on the heart (SHORR 1972; KHARKEVICH 1974), and abolishes it only at doses 10–20 times as high as the neuromuscular blocking dose (MAUTNER and LUISADA 1941).

On isolated heart and atria (rabbits, guinea pigs) tubocurarine antagonizes the negative inotropic or chronotropic action of muscarinic agonists in concentrations more than ten times higher than those which can be reached in the whole animal after the administration of the agent in the effective neuromuscular blocking dose (GOAT and FELDMAN 1971; MARSHALL and OJEWOLE 1979; R. J. MARSHALL et al. 1980).

According to LEE SON and WAUD (1977, 1980) the affinity of tubocurarine for muscarinic receptors of the isolated guinea pig atrium (antagonism against negative chronotropic action of carbachol) is 264 times lower than its affinity for nicotinic receptors of the isolated lumbrical muscle of the guinea pig. It may be pointed out for comparison, that for neuromuscular blocking agents with marked cardiac muscarinic blocking activity (gallamine and pancuronium), these ratios are considerably lower: 2.4 times for gallamine and 5.3 times for pancuronium (LEE SON and WAUD 1977).

Experimental and certain clinical data indicate that tubocurarine in neuromuscular blocking and higher doses fails to produce a noticeable block of the muscarinic receptors of the vessels, small intestine, urinary bladder, bronchi, or salivary glands (UNNA et al. 1950a; PATON 1959; ELWELL 1960; SHORR 1972, 1975; KHARKEVICH 1974; MARSHALL and OJEWOLE 1979).

In addition to tubocurarine, neuromuscular blocking agents devoid of muscarinic blocking activity include nondepolarizing neuromuscular blockers: C-toxiferine (WASER and HARBECK 1959; FREY and SEEGER 1961), metocurine (UNNA et al. 1950a; HUGHES and CHAPPLE 1976a, b; CHAPPLE et al. 1976; DURANT et al. 1977; SAVARESE 1979), diplacinum (MASHKOVSKY and BRISKIN 1952), paramionum (BUTAEV 1953), qualidilum (MASHKOVSKY and SADRITDINOV 1962), tercuronium (DANILOV et al. 1979), pipecuronium (KÁRPÁTI and BIRÓ 1980; ALYAUTDIN et al. 1980), vecuronium (DURANT et al. 1979; MARSHALL and OJEWOLE 1979; I.G. MARSHALL et al. 1980), and atracurium (HUGHES and CHAPPLE 1981; PAYNE and HUGHES 1981); depolarizing neuromuscular blocking agents: succinylcholine (FOLDES et al. 1956; HUGHES 1970; SHORR 1972, 1975; KHARKEVICH 1974; DRANE and EVANS 1979; BIRÓ and KÁRPÁTI 1981) and decamethonium (UNNA et al. 1950a; ABRACHAMS and HILTON 1962; SHORR 1972, 1975; KHARKEVICH 1974; DRANE and EVANS 1979); and also neuromuscular blocking agents with mixed action: mytolon (HOPPE et al. 1955), mebutanum, nobutanum (MEDVEDEV 1962), and dioxonium (KLUSHA et al. 1970; SALMENPERÄ and TAMMISTO 1980).

2. Neuromuscular Blocking Agents With Antimuscarinic Activity

a) Peculiarities of the Spectrum of Antimuscarinic Activity of Neuromuscular Blocking Agents

In the first detailed experimental study of the antimuscarinic action of gallamine (RIKER and WESCOE 1951), it was shown that it differed substantially from that of typical atropine-like agents. The peculiarity of antimuscarinic action was that, unlike atropine and atropine-like agents, whose action on muscarinic receptors at various locations was essentially nonselective, gallamine in neuromuscular blocking and higher doses effectively blocked only cardiac muscarinic receptors, without inhibiting the muscarinic receptors of other organs. Further, the same prevailing blocking action on cardiac muscarinic receptors was observed in many nondepolarizing neuromuscular blocking agents belonging to various chemical classes (Table 1), including trisonium compounds related to gallamine (PELIKAN and UNNA 1952; KENSLER et al. 1954), truxillic acid derivatives (KHARKEVICH

Table 1. Neuromuscular blocking, ganglion-blocking (superior cervical ganglion) and cardiac vagolytic activities of neuromuscular blocking agents

Drug	Neuromuscular blocking ED_{50}* or ED_{95}** (mg/kg i.v.) *A*	Ganglion-blocking ED_{50} (mg/kg i.v.) *B*	Vagolytic ED_{50} (mg/kg i.v.) *C*	Index of ganglion-blocking safety (*B*/*A*)	Index of vagolytic safety (*C*/*A*)	Reference
I. Nondepolarizing						
a) Curare alkaloids and their derivatives						
Tubocurarine	0.2 (sol)*	0.45	0.25	2.2	1.2	DURANT et al. (1977)
N-Methyltubocurarine	0.08 (sol)*	0.42	0.47	5.3	5.9	DURANT et al. (1977)
Metocurine	0.03 (sol)*	> 1.3[g]	0.43[f]	> 44	14.4	DURANT et al. (1977)
	0.02 (gastr)*		0.50[f]		25	HUGHES and CHAPPLE (1981)
C-Toxiferine	0.005 (compl)		>0.3		> 60	WASER and HARBECK (1959)
Alloferin (alcuronium)	0.05 (gastr)*	> 0.32	0.25[f]	> 6.4	5	HUGHES and CHAPPLE (1976a)
b) Derivatives of α-truxillic acid						
Anatruxonium	0.12 (gastr)**	1.65	0.021[f]	13.7	0.2	SAMOILOV (1971a, b)
Cyclobutonium	0.15 (gastr)**	1.23	0.049[f]	8.2	0.3	SAMOILOV (1971a, b)
Dipyronium	0.02 (gastr)**	> 8	>0.1	> 400	> 5	LEMINA (1978)
Amidonium	0.02 (gastr)**	> 8	>0.1	> 400	> 5	LEMINA (1978)
c) Steroid compounds						
Pancuronium	0.018 (sol)*	4.3	0.062[f]	239	3.4	DURANT et al. (1977)
Stercuronium	0.2 (tib)*	> 4	0.1[f]	> 20	0.5	MARSHALL (1973)
Dacuronium	0.6 (tib)*		0.55[f]	8.5[b]	0.9	MARSHALL (1973)
Org-6368	0.128 (sol)*	8	0.125[f]	62.5	1	DURANT et al. (1979)
Pipecuronium (RHG-1106)	0.002 (tib)*	> 2	>1[f]	>1000	>500	KÁRPATI and BIRÓ (1980)
Vecuronium (Org-NC-45)	0.034 (sol)*	18.1	2.14[f]	532.3	63.1	DURANT et al. (1979)
Chandonium	0.047 (sol)*	> 1	0.104[f]	> 21.3	2.2	TEERAPONG et al. (1979)
HS-342	0.3 (sol)*	1.5	0.1[f]	5	0.3	GANDIHA et al. (1974)
Duador (RHG-4201)	0.113 (sol)*	> 5	0.06[f]	> 44.2	0.5	BIRÓ and KÁRPATI (1981)

d) *N*-Adamantyl compounds						
Diadonium	0.3 (gastr)**	> 3	0.026[a,f]	> 10	0.09	KHARKEVICH (1970b), SHORR (1972)
	0.18 (sol)*	> 5	0.06[f]	> 27.8	0.3	BIRÓ and KÁRPATI (1981)
Decadonium	0.25 (gastr)**	> 1	0.018[a,f]	> 4	0.07	KHARKEVICH (1970a), SHORR (1972)
e) Trisonium compounds						
Gallamine	0.8 (gastr)*		0.56[f]		0.7	HUGHES and CHAPPLE (1981)
	2 (tot par)	>30		> 15		MUSHIN (1949), RIKER and WESCOE (1951)
f) Miscellaneous compounds						
Fazadinium	0.7 (gastr)*	> 4	0.3[f]	> 5.7	0.4	HUGHES and CHAPPLE (1976a)
AH 10407	0.7 (tib)*	4	~1.5	5.7	~ 2	BRITTAIN et al. (1977)
Tercuronium	0.08 (tib)**	0.8	0.8	10	10	DANILOV et al. (1979)
Atracurium	0.13 (gastr)*	> 2–4	3.1	> 8–16	24.4	HUGHES and CHAPPLE (1981)
Qualidilum	0.25 (gastr)**	2 (compl)	3 (compl)	8	12	MASHKOVSKY and SADRITDINOV (1962)
Diplacinum[c]	1 (gastr)**	20 (compl)	1 (compl)	20	1	MASHKOVSKY and BRISKIN (1952)
II. Depolarizing						
Succinylcholine	0.052 (sol)*	> 1	>10	> 19.2	>192	BIRÓ and KÁRPATI (1981), DANILOV et al. (1957)
Decamethonium	0.017 (gastr)*	14.2		835		HOPPE et al. (1955)
	0.42 (gastr)*[§]		>1.6[§]		> 38[§]	HOPPE et al. (1955)
III. Mixed action						
Mytolon	0.055 (gastr)*	1.9		34.5		HOPPE et al. (1955)
	0.042 (gastr)*[§]		0.55[§]		14[§]	HOPPE et al. (1955)
Nobutanum[d]	0.02 (gastr)**	2 (compl)	1 (compl)	100	50	MEDVEDEV (1962)
Mebutanum[e]	0.02 (gastr)**	2 (compl)	1 (compl)	100	50	MEDVEDEV (1962)
Dioxonium	0.005 (gastr)*	> 1	>0.5	> 200	>100	KLUSHA et al. (1970)

All values (except for those marked with §) are obtained from experiments on anesthetized cats; § experiments on anesthetized dogs; sol soleus muscle; tib tibialis anterior muscle; gastr gastrocnemius muscle; compl complete block; tot par total paralyzing dose

[a] ED_{50} with respect to antagonism against negative chronotropic effect of acetylcholine

[b] Ratio ED_{30} (ganglion block)/ED_{30} (neuromuscular block)

[c] 1,3-Diethoxybenzol-ω, ω'-bisplatynecium dichloride

[d] 1,4-Bis[9-methyl-3,9-diazabicyclo[3,3,1]nonane-3]-butane diiodomethylate

[e] 1,4-Bis[9-methyl-3,9-diazabicyclo[3,3,1]nonane-3]-butane dimethylsulfomethylate

[f] Vagolytic dose for these substances characterizes the antimuscarinic effect

[g] The > sign in columns *B* and *C* is used when the substance was studied within a limited dose range and the maximal dose caused no effect or less than 50% effect.

1965, 1966, 1974; SHORR 1972, 1975), bisquaternary *N*-adamantyl compounds (KHARKEVICH 1970a, b, 1974; SHORR 1972, 1975), steroid compounds (SAXENA and BONTA 1970; MARSHALL 1973; MARSHALL et al. 1973; GANDIHA et al. 1975; SUGRUE et al. 1975; BIRÓ and KÁRPÁTI 1981), and some others.

This suggests that a subgroup of compounds with cardiotropic antimuscarinic action can be singled out from the neuromuscular blocking agents with marked antimuscarinic activity.

In addition to neuromuscular blocking agents with a cardiotropic muscarinic blocking effect, there are other neuromuscular blockers, which in paralytic doses cause a marked antimuscarinic effect, but unlike the agents of the first group, act nonselectively on the muscarinic receptors at various locations, i.e., according to the spectrum of antimuscarinic activity, they are similar to typical atropine-like agents. The agents of this subgroup can be called neuromuscular blocking agents with nonselective (atropine-like) antimuscarinic action.

b) Neuromuscular Blocking Agents With Cardiotropic Antimuscarinic Activity

A common experimentally detected feature of all the neuromuscular blocking agents of this group is their ability to block effectively the muscarinic receptors of the heart (cardiac vagolytic action, inhibition of chronotropic and inotropic effects of muscarinic agonists) in neuromuscular blocking or similar doses (lower or somewhat higher). At the same time, the sensitivity of muscarinic receptors at other locations to endogenous acetylcholine, or exogenous muscarinic agonists, remains essentially unaffected.

In patients treated with neuromuscular blocking agents having cardiotropic antimuscarinic action, as a rule, a moderate and sometimes a rather significant sinus tachycardia is observed, which is most marked within 5 min of the administration and sometimes persists throughout the period of curarization. The degree of tachycardia is in large part dependent on the initial heart rate, premedication, and also on the agents used for induction and maintenance of the anesthesia. When atropine or some other atropine-like agents are included in premedication, tachycardia to neuromuscular blocking agents is usually less marked or it may be altogether absent. On the other hand, the drugs producing vagotonia (morphine and other narcotic analgesics, general anesthetics such as thiopentone, halothane, and cyclopropane) can markedly enhance tachycardia.

In addition to a more or less marked increase in the heart rate, neuromuscular blocking agents with pronounced cardiac antimuscarinic activity can provoke in a number of patients transient disturbances of the cardiac rhythm (more often ventricular extrasystoles), which usually disappear independently of the decrease of tachycardia. Moreover, these agents can elevate cardiac output and blood pressure which results in increased myocardial oxygen consumption.

However, it should be emphasized that some investigators do not consider these hemodynamic changes as related exclusively to antimuscarinic vagolytic properties of the neuromuscular blocking agents. As will be shown, several agents with cardiotropic antimuscarinic action have a number of additional properties which, taken together, can be responsible for their sympathomimetic effects.

α) Trisonium Compounds. It has already been mentioned that the first neuromuscular blocking agent in which cardiotropic antimuscarinic action was detected was a trisonium compound, gallamine. In one of the first papers devoted to the pharmacology of this drug (BOVET et al. 1949), it was pointed out that, in cats and rabbits, gallamine in neuromuscular blocking doses (1–2 mg/kg i.v.) caused a marked cardiac vagus inhibition. Further, RIKER and WESCOE (1951) carried out a more detailed study of vagolytic action of gallamine.

The experiments in anesthetized cats have shown that cardiac vagolytic action of gallamine appears at an i.v. dose of 0.1 mg/kg, i.e., 20 times lower than the dose required for total paralysis, and complete vagal blockade develops after a dose of 1 mg/kg. In parallel to the inhibition of vagus–heart transmission, gallamine initially diminishes and then abolishes the negative chronotropic action of metacholine.

The most interesting finding of these investigations has been that gallamine fails to produce any inhibiting influence on the muscarinic receptors at other locations. Thus, it did not affect the depressor effect of metacholine or modify the action of diisopropylfluorophosphate on sweating, salivation, or intestinal motility. On the isolated ileum, gallamine at very high concentration (10^{-3} *M*) only slightly antagonizes the effect of acetylcholine. On the basis of these data, Riker and Wescoe have concluded that, unlike atropine-like compounds, the antimuscarinic action of gallamine within a wide dose range is limited to cardiac muscarinic receptors.

Further investigations have provided additional evidence for the predominant action of gallamine on the cardiac muscarinic receptors. In experiments on the isolated guinea pig atria and ileum, BROWN and CROUT (1970) demonstrated that in both tissues gallamine behaved as a competitive antagonist of acetylcholine. However, pA_2 of the neuromuscular blocking agent calculated for the atria was 5.73, whereas for the ileum it was less than 3. Thus, the affinity of gallamine for the cardiac muscarinic receptors was about 1,000 times higher than that for muscarinic receptors of the ileum.

A similar difference between the affinities of gallamine for the muscarinic receptors of the isolated ileum and the heart in situ (spinal cats and rats) has been observed by RATHBUN and HAMILTON (1970). These authors have shown that, unlike atropine, gallamine at doses which completely block the negative chronotropic action of acetylcholine, fails to affect the stimulating action of the latter on salivation and mucus secretion.

The general picture showing the peculiarities of the spectrum of antimuscarinic action of gallamine is added to by the findings of CLARK and MITCHELSON (1976). They have reported that this compound is a significantly less effective antagonist of carbachol and acetylcholine in the isolated urinary bladder and ileal longitudinal muscle of the guinea pig than in the isolated guinea pig atria (the ratios of dissociation constants urinary bladder : atrium and ileum : atrium are 43.7 : 1 and 10.2 : 1, respectively).

Experimental data indicate that other trisonium compounds, in particular, the gallamine analog SKF-2015 studied by PELIKAN and UNNA (1952), and also aliphatic trisonium compounds whose spectrum of antimuscarinic action has been investigated in detail by KENSLER et al. (1954) also have cardiotropic antimuscarinic activity. Like gallamine, these neuromuscular blocking agents at doses lower than the neuromuscular blocking doses, completely block the negative chronotropic action of acetylcholine and metacholine without affecting hypotension caused by these agonists. In addition, while studying aliphatic trisonium compounds in anesthetized dogs, it was found that these agents at vagolytic doses failed to affect salivation caused by chorda tympani stimulation.

Sinus tachycardia, observed in the majority of patients and persisting, as a rule, throughout the period of curarization, is the most common cardiovascular side effect of gallamine in humans (ARTUSIO et al. 1951; FOLDES et al. 1964; WALTS and PRESCOTT 1965; KENNEDY and FARMAN 1968; EISELE et al. 1971). It was reported that gallamine-induced tachycardia was dose-dependent and reached its maximum at approximately 1 mg/kg (EISELE et al. 1971; STOELTING 1973). It is interesting that gallamine at this dose and at higher doses (4–5 mg/kg) does not produce a complete inhibition of cardiac vagal control, since additional administration of atropine after the maximum gallamine effect was achieved, and resulted in further increase in heart rate (EISELE et al. 1971). According to EISELE et al. (1971) this fact may indicate that gallamine-induced tachycardia in humans may be determined not only by the vagolytic action, but also by some other mechanisms.

Other side effects of gallamine which may be due to its cardiac vagolytic action are moderate elevation of the blood pressure (STOELTING 1973; LONGNECKER et al. 1973; SHETH and IABUIS 1980) and of the cardiac output (SMITH and WHITCHER 1967; KENNEDY and FARMAN 1968; STOELTING 1973), as well as short-term arrhythmias, such as premature ventricular contractions or ventricular tachycardia (WALTS and PRESCOTT 1965). The latter are relatively rare and usually occur under halothane or cyclopropane anesthesia, whereas moderate elevation of the blood pressure associated with increased cardiac output is a characteristic side effect of gallamine, especially frequent when nitrous oxide is used as the sole anesthetic.

β) Bisquaternary Ammonium Derivatives of Diphenylcyclobutane Dicarboxylic (Truxillic) Acids. A great number of nondepolarizing neuromuscular blocking agents belonging to the bisquaternary ammonium derivatives of diphenylcyclobutane dicarboxylic acids have cardiotropic antimuscarinic action similar to that of gallamine (KHARKEVICH and KRAVCHUK 1961; KHARKEVICH 1965, 1966, 1974; SAMOILOV 1971a; SHORR 1972, 1975; KHARKEVICH et al. 1981; KHARKEVICH and SHORR 1980; see also Chap. 13). In anesthetized cats, the esters of this series in doses lower than the neuromuscular blocking doses completely block vagus bradycardia and negative chronotropic action of acetylcholine without affecting the hypotensive effect of the latter.

While studying the structure–activity relations with respect to the cardiac antimuscarinic effect in this series, it was found that for this kind of activity the steric configuration of truxillic acids (α-, β-, γ-), the distance between the cationic centers, the character of radicals at the quaternary nitrogen atoms, and the structure of the interonium moiety are of great importance. Taking into account that antimuscarinic action is undesired for neuromuscular blocking agents, those structural modifications which result in this series of compounds in a significant decrease of cardiac antimuscarinic activity with a parallel increase in the neuromuscular blocking activity deserve attention. In this respect the substitution of an amide group for the ester one gave the optimal effect. Thus obtained, the amide analogs of the parent bisesters, unlike the latter, have a high index of antimuscarinic safety, which allows one to assign these compounds to the neuromuscular blocking agents without essential cardiac antimuscarinic action (Chap. 13). This principle was successfully used for the design of two new active nondepolarizing neuromuscular blockers on the basis of the known bisesters of α-truxillic acid, dipyronium and amidonium (KHARKEVICH 1983) with sufficiently high indices of antimuscarinic safety (LEMINA 1978) (see Chap. 19).

Two bisesters of α-truxillic acid: anatruxonium and cyclobutonium, were comprehensively investigated with respect to their spectrum of antimuscarinic action. This investigation included the assessment in anesthetized cats of cardiac vagolytic activity of these compounds and also their interactions with the responses caused by intravenous administration of acetylcholine: bradycardia, hypotension, salivation, bronchospasm, and ileum and urinary bladder contractions. In addition, the nature and degree of anatruxonium antagonism against carbachol were studied in isolated spontaneously beating atria and ileum of the rat.

It was found that both agents, in doses lower than the neuromuscular blocking doses (ED_{95} of anatruxonium and cyclobutonium 0.12 and 0.15 mg/kg i.v., respectively), effectively inhibited vagal bradycardia (ED_{50} 0.021 and 0.049 mg/kg i.v., respectively) simultaneously diminishing chronotropic action of acetylcholine (ED_{50} 0.020 and 0.043 mg/kg i.v., respectively). Complete block of acetylcholine-induced bradycardia was observed after administration of 0.04–0.05 mg/kg anatruxonium and 0.1–0.12 mg/kg cyclobutonium. At doses providing abolition of the negative chronotropic effect of acetylcholine as well as after repeated administration in increasing doses (0.5 and 1 mg/kg), the compounds failed to produce inhibition of acetylcholine-induced hypotension, salivation, and contractile responses of the ileum and the urinary bladder.

On the other hand, among muscarinic receptors there were found those having a lower sensitivity than cardiac receptors, but distinct sensitivity to anatruxonium and cyclobutonium, namely, bronchial muscarinic receptors. At doses close to neuromuscular blocking ones, both anatruxonium (0.15 mg/kg i.v.) and cyclobutonium (0.2 mg/kg i.v.) decreased acetylcholine-induced bronchospasm by about 50%. Increasing the dose resulted in dose-dependent inhibition of this reaction, but even at 1–2 mg/kg the agents failed to abolish the constricting effect of acetylcholine on bronchi. Thus, in the whole animal anatruxonium and cyclobutonium had a prevailing, but not selective, action on the cardiac muscarinic receptors.

Predominant affinity of anatruxonium for cardiac muscarinic receptors was revealed also in isolated organs. Like gallamine, anatruxonium decreased the action of carbachol concentration-dependently in spontaneously beating atria and ileum of the rat, producing a parallel rightward shift of the dose–response curves to the agonist, without depression of the maximal response. However, the antagonism of the agent with carbachol in atria was significantly more marked ($pA_2 = 7.50$) than in ileum ($pA_2 = 6.28$). For atropine, there was no significant difference between pA_2 values in the two tissues (8.76 and 8.85, respectively).

In patients, cardiac antimuscarinic action of α-truxillic acid derivatives (truxilonium, anatruxonium, cyclobutonium) is manifested by only a moderate sinus tachycardia (KOTOMINA 1970; KOTOMINA and KHARKEVICH 1968; FIRSOV et al. 1970). The increase in heart rate produced by these agents is similar (10–40 beats/min), however, after repeated administration of the neuromuscular blocking agents during a long operation, tachycardia can be more marked which is likely due to the cumulation of the compounds and the corresponding potentiation of their vagolytic action (KOTOMINA 1970). Neuromuscular blocking agents of this group do not produce elevation of the blood pressure or disturbances of the heart rhythm.

γ) Bisquaternary N-Adamantyl Compounds. Diadonium and decadonium, N-adamantyl analogs of succinylcholine and decamethonium, respectively, belong to the neuromuscular blocking agents with a marked cardiotropic antimuscarinic activity (KHARKEVICH 1970a, b, 1974; SHORR 1972, 1975; KHARKEVICH and SHORR 1980; see also Chap. 20). The spectrum of their antimuscarinic action was studied in similar experiments to those for anatruxonium and cyclobutonium. In anesthetized cats, the cardiac vagolytic effect of diadonium and decadonium is developed in parallel with the decrease of negative chronotropic response of the heart to acetylcholine. A 50% block of acetylcholine-induced bradycardia is observed after administration of 0.026 mg/kg i.v. diadonium (neuromuscular blocking ED_{95} 0.3 mg/kg i.v.) and 0.018 mg/kg i.v. decadonium (neuromuscular blocking ED_{95} 0.25 mg/kg i.v.). At doses of 0.08–0.09 and 0.05–0.06 mg/kg, respectively, the drugs completely eliminate the bradycardiac effect of acetylcholine.

In these and higher doses (up to 2 mg/kg for diadonium and 1 mg/kg for decadonium) both agents either do not affect, or somewhat potentiate and prolong acetylcholine-induced hypotension. Decadonium at neuromuscular blocking and higher doses increases the responses of the bronchi, salivary glands, ileum, and urinary bladder to acetylcholine which, as well as the potentiation of the hypotensive effect of acetylcholine, may be due to considerable anticholinesterase activity of the agent.

Unlike decadonium, the effect of diadonium at neuromuscular blocking and higher doses on acetylcholine-induced bronchospasm is variable. In some cases, diadonium fails to affect it, in others it decreases or increases this response. It should be noted that the blocking action of diadonium on bronchial muscarinic receptors is always lower than the analogous effect of anatruxonium or cyclobutonium and the ability to increase the bronchospasm is significantly less marked than with decadonium. It is quite likely that this variability of diadonium with respect to acetylcholine-induced bronchospasm results from the competition of anticholinesterase and muscarinic blocking action of the agent. Diadonium to a lesser degree than decadonium enhances acetylcholine-induced salivation. The effect of acetylcholine on the ileum and urinary bladder during the action of diadonium is either unaffected, or somewhat increased.

Thus, in the whole animal diadonium and decadonium at neuromuscular blocking doses produce practically selective blockade of cardiac muscarinic receptors. A certain affinity for bronchial muscarinic receptors on the part of diadonium is so low that it is easily masked by its moderate anticholinesterase activity. In patients, diadonium evoked only a slight tachycardia (average heart rate increase by about 7%) associated with a certain decrease of the stroke index and stroke volume (OSTROVSKY and KISELEV 1980; DARBINYAN et al. 1980).

δ) Steroid Derivatives. More or less marked cardiac vagolytic action is typical of the majority of the known nondepolarizing steroid neuromuscular blocking agents. Many of them, namely pancuronium, dacuronium, stercuronium, chandonium, HS-342, Org-6368, Duador (RGH-4201) and some others belong to the compounds with high cardiac vagolytic activity (BONTA et al. 1968; BUCKETT et al. 1968; BUCKETT and SAXENA 1969; SAXENA and BONTA 1970; MARSHALL et al. 1973; MARSHALL 1973; GANDIHA et al. 1974, 1975; SUGRUE et al. 1975; TEERAPONG et al. 1979; BIRÓ and KÁRPÁTI 1981; MARSHALL et al. 1981). Of the neuromuscular blocking agents mentioned, only pancuronium and stercuronium were studied in detail with respect to their spectrum of antimuscarinic action and their assignment to the cardiotropic antimuscarinic compounds is quite evident.

The following experimental findings indicate the predominant action of pancuronium on cardiac muscarinic receptors. In anesthetized cats and dogs, pancuronium at doses somewhat higher than the neuromuscular blocking one, effectively inhibits vagus–heart transmission, does not alter ganglionic transmission in the superior cervical ganglion, insignificantly changes the responses of intestine and urinary bladder elicited by the stimulation of the corresponding parasympathetic nerves or the administration of dimethylphenylpiperazinium (DMPP), abolishes the negative chronotropic effect of muscarinic agonists, and does not essentially affect their hypotensive action (BONTA et al. 1968; BUCKETT et al. 1968; SAXENA and BONTA 1970). The affinity of pancuronium for muscarinic receptors of the isolated guinea pig ileum is significantly lower than for those of the isolated atria (SAXENA and BONTA 1970; MARSHALL and OJEWOLE 1979; R. J. MARSHALL et al. 1980) and in perfused guinea pig heart it inhibits the negative chronotropic and inotropic actions of muscarinic agonists without diminishing their coronary vasodilator effect (SAXENA and BONTA 1970).

Thus, experimental findings indicate that by the prevailing localization of its muscarinic blocking action, pancuronium is similar to gallamine, α-truxillic acid derivatives, diadonium and decadonium. Pancuronium differs from these compounds by having a larger index of antimuscarinic safety which, according to different data varies from 3.4 to 5.2. For gallamine, anatruxonium, cyclobutonium, diadonium, and decadonium this index is below 1 (see Table 1).

While having less marked antimuscarinic vagolytic properties in neuromuscular blocking doses than gallamine, pancuronium in humans causes as a rule smaller changes of the heart rate and hemodynamics than gallamine (KELMAN and KENNEDY 1971; STOELTING 1972; COLEMAN et al. 1972). However, in some cases tachycardia after pancuronium administration can be quite marked. Usually, a more pronounced tachycardia is observed in patients with low baseline heart rate or bradycardia induced by the agents causing vagotonia, such as cyclopropane, halothane, morphine, or thiopentone (GROSSMAN and JACOBI 1974; MILLER et al. 1975; QVIST et al. 1980). It is interesting that, unlike the positive chronotropic effect of gallamine, the analogous action of pancuronium is not dose dependent and is easily prevented by atropine premedication (COLEMAN et al. 1972; MILLER et al.

1975). The latter is in favor of the primary role of cardiac muscarinic receptor blockade in the genesis of pancuronium-induced tachycardia.

Like gallamine, pancuronium often evokes a moderate hypertension in patients which is likely due to the increase of the cardiac output at unchanged systemic vascular resistance (LOH 1970; STOELTING 1972; COLEMAN et al. 1972; SHETH and IABUIS 1980). If preceded by hypotension, the hypertensive effect of pancuronium can increase sharply (GROSSMAN and JACOBI 1974). The changes of the cardiac output and blood pressure induced by pancuronium are prevented by atropine (COLEMAN et al. 1972), which, as in the case of the positive chronotropic effect, indicates their dependence on the vagolytic activity of the agent. In a number of patients, especially in those operated under halothane or cyclopropane anesthesia, pancuronium can provoke ventricular premature beats, disappearing spontaneously within 3–8 min (STOELTING 1972; MILLER et al. 1975; QVIST et al. 1980).

While studying the antimuscarinic activity of stercuronium in anesthetized cats (MARSHALL 1973), it has been found that, unlike pancuronium, it produces a marked vagolytic action and blocks the negative chronotropic action of metacholine at doses lower than the neuromuscular blocking dose (low index of antimuscarinic safety). Like other neuromuscular blocking agents with a cardiotropic antimuscarinic action, stercuronium abolishes the negative chronotropic response to metacholine without significantly affecting its vasodepressor effect.

The study of comparative sensitivity of postsynaptic muscarinic receptors at various locations to stercuronium was carried out by LI and MITCHELSON (1980) in the whole animal and in the isolated organs (guinea pig). By the data of these authors, stercuronium is 2–5.8 times more effective in blocking the action of carbachol on the cardiac muscarinic receptors than on the vascular ones and its affinity for the muscarinic receptors of isolated atria is 16–17 times higher than its affinity for the muscarinic receptors of the isolated ileal longitudinal muscle and the urinary bladder.

The spectrum of muscarinic blocking action of dacuronium, chandonium as well as the compounds HS-342, Org-6368, and Duador, was studied in less detail than that of pancuronium and stercuronium. In anesthetized cats, these neuromuscular blocking agents at neuromuscular blocking doses markedly or completely block vagal cardiac response, inhibit or abolish the negative chronotropic effect of acetylcholine or metacholine, and, like all cardiotropic antimuscarinic neuromuscular blocking agents, fail to alter the vasodepressor effect of muscarinic agonists (MARSHALL 1973; MARSHALL et al. 1973; GANDIHA et al. 1974, 75; DURANT et al. 1979; TEERAPONG et al. 1979; BIRÓ and KÁRPÁTI 1981).

For chandonium and some of its analogs it was also shown that on the isolated guinea pig ileum these agents exhibited much less antagonism with muscarinic agonists than on isolated atria (MARSHALL and OJEWOLE 1979; TEERAPONG et al. 1979; MARSHALL et al. 1981). Indices of antimuscarinic safety of dacuronium, Org-6368, HS-342, and Duador are significantly lower than the index for pancuronium.

It should be stressed that marked cardiac muscarinic blocking activity is not an inevitable feature of steroid neuromuscular blocking agents. In particular, pipecuronium (RGH-1106, Arduan) according to experimental findings, produced an initial blockade of cardiac muscarinic receptors only at doses 7–10 times higher than its neuromuscular blocking dose (ALYAUTDIN et al. 1980). In the series of 26 pancuronium derivatives (both bis- and monoquaternary compounds) screened by DURANT et al. (1979), there were found 7 compounds having a higher index of antimuscarinic safety than pancuronium; in 4 compounds this index was equal or higher than 10 which allows one to consider these neuromuscular blocking agents as essentially devoid of antimuscarinic action.

One of these agents, vecuronium (Org-NC-45), a monoquaternary pancuronium analog having a very high index of antimuscarinic safety (63.1) as well as pipecuronium underwent clinical trials (BUNATYAN and MIKHEEV 1980; BOROS et al. 1980; CRUL and BOOIJ 1980; BOOIJ et al. 1982) which showed that neither of these neuromuscular blocking agents produced cardiovascular side effects typical of pancuronium or other neuromuscular blocking agents with cardiotropic antimuscarinic action.

ε) Miscellaneous Neuromuscular Blocking Agents. The azobisarylimidazo-1,2-α-pyridinium derivative, fazadinium (AH 8165, Fazadon) can be assigned to the neuromuscular blocking agents with predominant action at the cardiac muscarinic receptors. By the data of MARSHALL (1973), corroborated later by HUGHES and CHAPPLE (1976a), in anesthetized cats this agent at doses close to the neuromuscular blocking one, abolishes bradycardia produced by vagus nerve stimulation, and also the negative chronotropic response to metacholine, without altering its hypotensive effect. At doses eliminating vagal slowing of the heart, fazadinium does not alter vagally induced increase in stomach motility in cats (BRITTAIN and TYERS 1973). The index of antimuscarinic safety of fazadinium is 0.4–0.6 (MARSHALL 1973; HUGHES and CHAPPLE 1976a) and it is comparable to that of gallamine. As well as the latter, fazadinium's affinity for muscarinic receptors of the isolated guinea pig ileum is much lower than that for muscarinic receptors of isolated atria (MARSHALL and OJEWOLE 1979).

In patients, fazadinium causes a moderate to marked sinus tachycardia (COLEMAN et al. 1973; HUGHES et al. 1976b; FAMEVO 1981) sometimes associated with a slight increase in blood pressure and cardiac output. As in gallamine, the positive chronotropic action of fazadinium is dose dependent (HUGHES et al. 1976b), and it is significantly reduced after atropine premedication (FAMEVO 1981).

The experimental data concerning cardiac antimuscarinic action of Alloferin (alcuronium, Ro-4-3816) are rather limited and controversial. KAJIMOTO (1970), who tested its neuromuscular blocking activity and cardiovascular effects in anesthetized cats did not observe even an initial alteration of vagus nerve–heart transmission after 0.5 mg/kg i.v. Alloferin that was 20 times higher than its neuromuscular blocking ED_{50}.

However, HUGHES and CHAPPLE (1976a) have shown that, at the neuromuscular blocking dose, Alloferin initially depresses the transmission from the vagus to the heart in anesthetized cats, diminishing it by about 50% when the neuromuscular ED_{50} is increased fivefold. Since within the vagolytic dose range Alloferin inhibits the chronotropic action of metacholine and fails to affect the ganglionic transmission in the superior cervical ganglion, Hughes and Chapple have explained its cardiac vagolytic effect by muscarinic blocking properties.

Clinical data concerning cardiac effects of Alloferin suggest that moderate Alloferin-induced tachycardia in patients is a complex response, and the reduced vagal control of the heart rate is but one, and probably not the main component in its genesis (TAMMISTO and WELLING 1969; KENNEDY and KELMAN 1970; COLEMAN et al. 1972).

Marked cardiac vagolytic activity comparable or higher than the neuromuscular blocking activity has been revealed in many bisquaternary bis- and tetraisoquinolinium compounds by HUGHES (1972) and STENLAKE et al. (1981 a, b, c). Though the mechanism of vagal blockade by most of these agents was not a subject of special investigation, some data concerning selected members of this series allow one to suppose that high vagolytic activity of isoquinolinium compounds (in relation to neuromuscular blocking activity), determined in cats and monkeys, reflects their blocking action at cardiac muscarinic receptors.

In favor of this suggestion are the findings of HUGHES (1972) who has reported that two bisquaternary octamethylene bisisoquinolinium compounds (BW 252C64 and BW 403C65) at neuromuscular blocking doses markedly inhibit vagally induced bradycardia without essentially affecting the transmission in the superior cervical ganglion. Furthermore, BW 403C65 in preliminary clinical trials provoked such dramatic increases in heart rate in humans that the trials were ceased.

It is of considerable interest that according to STENLAKE et al. (1981 c) in the series of ω,ω-bisbenzyltetrahydroisoquinolinium polyalkylene diesters formed by certain modifications of 1-benzyl substituents, it is possible to obtain agents with extremely high ratios between vagal and neuromuscular blocking doses (17–42 in experiments on anesthetized cats). One of them, atracurium besylate, was recommended for clinical trials (HUGHES and CHAPPLE 1981) and, according to the preliminary results (PAYNE and HUGHES 1981) is devoid of any influence on the heart rate and blood pressure in humans.

3. Neuromuscular Blocking Agents With Nonselective (Atropine-like) Antimuscarinic Action

Only a limited number of neuromuscular blocking agents with a marked nonselective atropine-like activity are known. As typical representatives of this group, the polymethylene bisatropinium compounds tested by KIMURA and UNNA (1950) and ECKFELD (1959) can be named. These agents are rather active nondepolarizing neuromuscular blocking agents whose neuromuscular blocking activity corresponds to, or is higher than that of tubocurarine. Unlike neuromuscular blocking agents with cardiotropic antimuscarinic action, these compounds in anesthetized dogs at neuromuscular blocking or lower doses not only abolish vagal slowing of the heart, but also markedly suppress the chronotropic and hypotensive action of muscarinic agonists, i.e., cause a simultaneous blockade of cardiac and vascular muscarinic receptors. According to both (cardiac and vascular) antimuscarinic effects, hexa-, octa-, and decamethylene bisatropinium derivatives, and also hexamethylene biscopolaminium compounds, are more active than atropine.

In addition to the marked inhibition of cardiac and vascular muscarinic receptors, these neuromuscular blocking agents have a pronounced blocking action on the muscarinic receptors at other locations. They cause a mydriatic effect (cats) and also effectively antagonize the contractile responses of the isolated ileum and uterus to acetylcholine (rabbits, rats).

Imbretil, a neuromuscular blocking agent with a depolarizing action, has a relatively low and a likely nonselective antimuscarinic activity. Although the spectrum of its antimuscarinic action was not studied in detail, it was reported to antagonize acetylcholine in isolated guinea pig ileum and isolated frog heart and also to decrease the depressor action of acetylcholine in anesthetized dogs (KLUPP et al. 1953; KRAUPP et al. 1953; KRAUPP et al. 1954; BRÜCKE 1956).

Atropine-like action of Imbretil is usually masked by its muscarinic effects due to its anticholinesterase activity (KRAUPP et al. 1954). This is likely to be the reason why, both in experiments and in humans, Imbretil failed to produce any significant cardiovascular effects (LEONG et al. 1962; SAITO and AVIADO 1963).

III. Mechanism of Antimuscarinic Action of Neuromuscular Blocking Agents

The mechanism of interaction with muscarinic receptors has mainly been studied for neuromuscular blocking agents with marked cardiotropic antimuscarinic action. As a rule, the test objects for such investigations were isolated atria (spontaneously beating right or electrically driven left), and isolated ileum (or ileal longitudinal muscle). The antagonism of the compounds against the effects of various muscarinic agonists was evaluated by construction of the dose–response curves and graphic analysis of the data according to ARUNLAKSHANA and SHILD (1959).

In a number of studies dealing with gallamine (BROWN and CROUT 1970), pancuronium (SAXENA and BONTA 1970; MARSHALL and OJEWOLE 1979), chandonium and its analogs (TEERAPONG et al. 1979; MARSHALL and OJEWOLE 1979), anatruxonium (KHARKEVICH 1974), stercuronium (LI and MITCHELSON 1978, 1980), and fazadinium and vecuronium (MARSHALL and OJEWOLE 1979) it was shown that all these compounds within a certain concentration range behave as competitive antagonists at atrial or ileal muscarinic receptors.

However, Mitchelson and co-workers (CLARK and MITCHELSON 1976; LI and MITCHELSON 1978, 1980; LEUNG and MITCHELSON 1982), who studied in detail the mechanism of muscarinic blocking action of gallamine, stercuronium, and pancuronium found that in isolated atria they behaved as competitive antagonists only within a limited concentration range, and at higher concentrations the degree of their antagonism with muscarinic agonists tended toward a limiting value. Moreover, gallamine and stercuronium in combination with homatropine produced dose ratios (acetylcholine as the agonist) which were significantly smaller than those expected for the combination of two competitive antagonists (CLARK and MITCHELSON 1976; LI and MITCHELSON 1978).

On the basis of these data, the authors suggested that cardiac antimuscarinic activity of gallamine, stercuronium, and pancuronium was due to noncompetitive antagonism of the so-called metaffinoid type (a kind of allosteric antagonism) whereby the interaction of the antagonist at an allosteric site may modify the conformation of the receptor and its affinity for the agonist. The noncompetitive nature of interaction with cardiac muscarinic receptors, suggestive of allosteric antagonism, was revealed for gallamine (BIRDSALL et al. 1981) and also for diadonium and cyclobutonium (N. J. M. Birdsall 1982, personal communication) using the radioligand binding technique.

It is interesting that allosteric antagonism at atrial or ileal muscarinic receptors has been supposed for many bisquaternary ammonium compounds, such as toxogonin (obidoxime), hemicholinium-3, TMB-4, and a series of alkanebisammonium compounds (BIEGER et al. 1968; LÜLLMANN et al. 1969; KUHNEN-CLAUSEN 1970; MITCHELSON 1971; MADDEN and MITCHELSON 1975). The same mechanism of interaction with atrial muscarinic receptors was also suggested for symmetric tetramine compounds with disulfide or polymethylene groups in the central part of the molecule (BENFEY et al. 1979), whose secondary amino groups were highly protonated at physiologic pH.

These data permit one to propose that binding at allosteric sites can be a universal mechanism which provides the interaction of bisquaternary ammonium compounds (including neuromuscular blocking agents) and also several related compounds not only with cardiac muscarinic receptors, but also with muscarinic receptors at other locations (ileum, vessels, bronchi, etc.).

Although this supposition made by LI and MITCHELSON (1978) is not based on direct experimental evidence, several indirect data indicate that the interaction of some neuromuscular blocking agents with the muscarinic receptors of the ileum (antagonism of pancuronium and stercuronium with acetylcholine; R. J. MARSHALL et al. 1980; LI and MITCHELSON 1980) and vessels (antagonism of stercuronium with carbachol; LI and MITCHELSON 1980) is not compatible with the criteria of competitive antagonism.

It can be noted that the hypothesis of allosteric interaction seems rather convenient for the explanation of the fact that many neuromuscular blocking agents have relatively selective action on cardiac muscarinic receptors whereas atropine-like compounds do not have such selectivity.

Indeed, assuming in the framework of this hypothesis that muscarinic receptors at different locations (e.g., heart and ileum) have dissimilar allosteric sites, but identical sites for competitive antagonists, the dissimilarities between these receptors will be "recognized" only by allosteric antagonists, but not by competitive antagonists for which the receptors at both locations will be identical (indistinguishable).

IV. Some Other Possible Mechanisms of Cardiovascular Side Effects of Neuromuscular Blocking Agents

In experimental and clinical investigations of neuromuscular blocking agents with cardiotropic antimuscarinic activity (gallamine, pancuronium, and some others), it has been shown that in addition to a blocking action on the cardiac muscarinic receptors, these compounds have some other properties which can partially contribute to the cardiovascular side effects characteristic of the drugs of this group (tachycardia, hypertension, cardiac arrhythmias). These properties include the ability to block cardiac vagal transmission at the presynaptic level, as well as to activate sympathetic influences on the heart and vessels (sympathomimetic action).

1. Presynaptic Component of Cardiac Vagolytic Action

Presynaptic localization of cardiac vagolytic action of pancuronium and gallamine was assumed on the basis of a series of investigations of LEE SON and WAUD (1977, 1978, 1980), carried out on the isolated spontaneously beating guinea pig atrium. The data obtained indicated that, like atropine, pancuronium and gallamine altered vagal transmission not at the ganglia, but at the postganglionic level. However, the calculation of fractional occupation of receptors revealed an essential difference between neuromuscular blocking agents on one hand, and atropine, on the other. While atropine in concentrations producing a 50% inhibition of atrial response to postganglionic vagal stimulation occupied 86% of atrial receptors, gallamine and pancuronium in concentrations required for the same effect occluded only 43% and 69% of muscarinic receptors, respectively.

On the basis of these calculations it was concluded that the cardiac vagolytic effect of these neuromuscular blocking agents must depend on some other mechanism than muscarinic receptor occupation. This mechanism, as the authors suggest, is most likely the blockade of vagal nerve endings, resulting in the decrease of acetylcholine release from them (LEE SON and WAUD 1978, 1980; LEE SON et al. 1981). It is obvious that this hypothesis, offering a new explanation for the mechanism of cardiac vagolytic action of several neuromuscular blocking agents, requires additional verification in other experiments and using more direct methods for the assessment of the postulated presynaptic effect.

2. Sympathomimetic Action

An activating influence of gallamine on sympathetic innervation was suggested by clinicians (THOMAS 1963; MORGENSTERN and SPLINTH 1965; SMITH and WHITCHER 1967) and was then supported by experimental data (BROWN and CROUT 1966, 1968, 1970). The latter showed that in isolated myocardial tissues gallamine produced rather marked positive inotropic and chronotropic actions which were progressively diminished on repeated administrations, and could be antagonized by propranolol or pretreatment with reserpine, but not by atropine, hexamethonium, or cocaine. Tachyphylaxis in response to gallamine was eliminated after norepinephrine loading.

Furthermore, the sympathomimetic effects of gallamine were also demonstrated in whole animals. In cats pretreated with atropine, gallamine at doses somewhat higher than the neuromuscular blocking dose increased the contractility of the right ventricle by 50% and, as in isolated tissues, this inotropic effect was suppressed by propranolol. On the basis of these data, Brown and Crout concluded that gallamine had a tyramine-like action on sympathetic nerve endings and that this action, together with muscarinic blocking activity, can participate in the cardiac side effects of this drug in humans.

Moderate sympathomimetic properties have been experimentally revealed in pancuronium. It has been shown that this agent not only exerts a tyramine-like action (BONTA et al. 1966; SMITH et al. 1970; SEGARA DOMENECH et al. 1976), but also inhibits Uptake$_1$, thus increasing the effect of norepinephrine on various organs (IVANKOVICH et al. 1975; QUINTANA 1977; DOCHERTY and MCGRATH 1978, 1980; TOMLINSON 1979). It is interesting that similar action on Uptake$_1$ is also characteristic of chandonium, fazadinium, and pancuronium analogs Org-6368, Org-7268, and vecuronium (TOMLINSON 1979; SALT et al. 1980), but is not observed in gallamine (DOCHERTY and MCGRATH 1980). Of all the steroid neuromuscular blocking agents tested by SALT et al. (1980), pancuronium was the most active inhibitor of Uptake$_1$ in the perfused rat heart.

In addition to the tyramine-like effect (gallamine, pancuronium) and the inhibitory action on Uptake$_1$ (pancuronium and other steroid compounds, fazadinium), there is evidence indicating that a number of neuromuscular blocking drugs with cardiotropic antimuscarinic activity (gallamine, stercuronium, pancuronium) can in principle induce sympathetic activation via blockade of presynaptic inhibitory muscarinic receptors at adrenergic nerve endings (VERCRUYSSE et al. 1979; LI and MITCHELSON 1980).

According to VERCRUYSSE et al. (1979), gallamine and pancuronium in concentrations comparable to those blocking the agonist action on cardiac postsynaptic muscarinic receptors suppress the inhibitory effect of acetylcholine on the contractions of saphenous vein strips induced by transmural stimulation. LI and MITCHELSON (1980) found that the affinity of stercuronium for presynaptic muscarinic receptors of the isolated rabbit ear artery and postsynaptic muscarinic receptors of the isolated rabbit atria did not differ significantly. These data, together with the findings of FUDER et al. (1981) showing that gallamine fails to "differentiate" between presynaptic (adrenergic nerve endings) and postsynaptic muscarinic receptors in the rabbit heart, indicate that, among the mechanisms of sympathetic activation by gallamine, pancuronium, and stercuronium, there can be the facilitation of adrenergic transmission owing to the elimination of inhibitory cholinergic (muscarinic) control by parasympathetic nerves of norepinephrine release.

Finally, yet another probable mechanism of the enhancement of sympathetic influences induced by neuromuscular blocking agents with antimuscarinic properties is their suppressing action on the presumed inhibitory modulation of transmission at the sympathetic ganglia. According to some authors, these modulating influences can be produced by ganglionic dopaminergic interneurons called SIF cells (LIBERT and TOSAKA 1970; BJÖRKLUND et al. 1970; GREENGARD and KEBABIAN 1974). It is assumed that SIF cells are activated by acetylcholine released from preganglionic fibers via muscarinic receptors. Excitation of these interneurons results in dopamine-mediated hyperpolarization of ganglionic neurons (slow IPSP).

This slow hyperpolarization is preceded by rapid depolarization (rapid EPSP). The latter is easily inhibited by ganglion-blocking agents and reflects the transmission promoted by nicotinic receptors. After slow IPSP, slow depolarization (slow EPSP) is developed, which, as well as slow IPSP, is abolished by atropine, but not by nicotinic blocking agents, and is a manifestation of the excitation of muscarinic receptors of ganglionic neurons.

It has been shown that, in the cat superior cervical ganglion, gallamine and pancuronium in neuromuscular blocking doses evoke a marked decrease of slow IPSP, but fail to affect slow EPSP and that this may facilitate the transmission in the ganglia performed with the participation of muscarinic receptors (GARDIER et al. 1974, 1978).

On the basis of these data, it has been concluded that muscarinic receptors of SIF cells are not identical to those of ganglionic neurons (GARDIER et al. 1978). The former are similar to muscarinic receptors mediating the vagal effect on the heart (BROWN 1980; MARSHALL 1980). Moreover, these data suggest that if the inhibitory modulation of ganglionic transmission by SIF cells operates under physiologic conditions, then sympathomimetic effects of gallamine and pancuronium can be at least in some part dependent on the facilitation of sympathetic ganglionic transmission.

Some indirect clinical data indicate that neuromuscular blocking agents with antimuscarinic action can also cause sympathetic activation in humans. Thus, WALTS and PRESCOTT (1965), as well as MATSUKI et al. (1973), observed an increase of plasma catecholamine levels in gallamine-treated patients. It was also mentioned that the incidence of ventricular arrhythmias after gallamine administration was significantly higher under halothane or cyclopropane (WALTS and PRESCOTT 1965), which are known to increase the myocardial sensitivity to catecholamines. The clinical findings of EISELE et al. (1971) that gallamine in doses provoking maximal tachycardia fails to eliminate completely the vagal influence on the heart also seem to testify in favor of sympathomimetic effect of this agent.

NANA et al. (1973) revealed increased catecholamine levels in the blood of patients 5 min after administration of pancuronium at a dose of 0.05 mg/kg. They believe that this indicates the activation of the sympathetic nervous system, which could be the reason for pancuronium-induced tachycardia and hypertension. LEE et al. (1980) observed the contriction of capacitance vessels after pancuronium administration in patients with cardiopulmonary bypass undergoing open cardiac surgery and also explained this phenomenon by a sympathomimetic effect of the agent.

V. Conclusions

The data presented indicate that antimuscarinic, and especially, cardiotropic antimuscarinic action is characteristic of a large number of nondepolarizing neuromuscular blocking agents of various chemical classes. This should be considered both in their preclinical study and clinical use. The presence of marked muscarinic blocking activity in neuromuscular blocking agents should be regarded as a negative phenomenon, since these drugs can cause a number of undesirable cardiovascular side effects in patients. In the majority of cases, moderate hemodynamic changes provoked by the agents with cardiotropic antimuscarinic activity are of no danger. However, in some patients, in particular those suffering from coronary heart disease or heart failure, the use of these drugs may cause serious complications and is, therefore, inexpedient.

It is also not recommended to combine neuromuscular blocking agents which have antimuscarinic vagolytic activity with drugs causing vagotonia (cyclopropane, halothane, morphine, thiopentone) or increasing the myocardial sensitivity to catecholamines (cyclopropane, halothane). In the former case, cardioaccelerating and pressor actions of such neuromuscular blockers may be markedly potentiated, and in the latter their arrhythmogenic effect may develop.

Whereas from the practical point of view, the antimuscarinic properties of neuromuscular blocking agents can be regarded only as undesirable, theoretically, they are of great interest. The marked selectivity of numerous neuromuscular blockers with respect to cardiac muscarinic receptors, uncharacteristic of classical atropine-like agents, is convincing evidence for the nonidentical arrangement of the receptors at different locations. This evidence, together with the data of some other investigations, devoted to the heterogeneity of muscarinic receptors, is an important theoretical prerequisite for a successful quest for antimuscarinic agents selectively abolishing parasympathetic influences on certain organs.

B. Ganglion-Blocking Activity

I. Introduction

From the practical point of view, ganglion-blocking action should only be considered in those neuromuscular blocking compounds in which it is manifested at doses that are within or near the neuromuscular blocking dose range. Such compounds, when used in patients for muscle relaxation, can provoke several side effects related to the inhibition of autonomic ganglia. Some of these effects should be taken as undesirable. Among these, there are hypotension due to decreased sympathetic influences on the vessels and heart, and tachycardia related to the blockade of cardiac parasympathetic ganglia. In addition, it can be noted that blockade of ganglionic transmission results in inhibition of several homeostatic autonomic reflexes (e.g., baroreflexes, cardiocardiac reflexes), which are of importance for maintenance of adequate hemodynamics under normal conditions and stress.

There are certain clinically less important undesirable consequences of ganglionic blockade by neuromuscular blocking agents (mydriasis, a slight and short-term reduction of contractile activity of abdominal smooth muscle organs) and also some effects that can be regarded as beneficial (e.g., hyposalivation, inhibition of negative viscerovisceral reflexes provoked by anesthesiologic or surgical manipulations). Yet, on the whole, the presence of ganglion-blocking properties is considered as a disadvantage of neuromuscular blocking agents, hampering their clinical use.

As a rule, the action of neuromuscular blocking agents on ganglionic transmission is studied in experiments on anesthetized (more rarely, decerebrated) cats. To estimate the degree of the ganglion-blocking effect in neuromuscular blocking agents, usually a relative index is used – a ratio between the dose inducing a 50% ganglionic block (ED_{50}GB) and the dose required for a 50% inhibition of neuromuscular transmission (ED_{50}NMB). Several other, more rarely used indices (ED_{30}GB/ED_{30}NMB, ED_{50}GB/ED_{95}NMB, etc.) are essentially similar to ED_{50}GB/ED_{50}NMB. All of them, analogously to various indices of antimuscarinic safety, can be called indices of the ganglion-blocking safety of neuromuscular blocking agents.

Analysis of experimental and clinical data concerning a number of known neuromuscular blocking agents has suggested that ganglion-blocking action is clinically insignificant in those compounds whose experimentally obtained ED_{50}GB/ED_{50}NMB is over 10 (or ED_{50}GB/ED_{95}NMB is over 5). This section presents the experimental and some clinical data on ganglion-blocking activity of the majority of known neuromuscular blocking agents of different chemical structures and different modes of action.

II. Comparative Characteristics of Ganglion-Blocking Activity of Neuromuscular Blocking Agents

1. Nondepolarizing Neuromuscular Blocking Agents

a) Curare Alkaloids and Their Derivatives

Among neuromuscular blocking agents used in patients, tubocurarine is the one which has rather marked ganglion-blocking activity (MAUTNER and LUISADA 1941; BÜLBRING and DEPIERRE 1949; GUYTON and REEDER 1950; HOPPE et al. 1955; KHARKEVICH 1957, 1962; PATON 1959; HUGHES 1970; HUGHES and CHAPPLE 1976a; DURANT et al. 1977; SAVARESE 1979; and others). In many respects, the influence of tubocurarine on ganglionic transmission is similar to that of typical ganglion-blocking agents and can be accounted for by competitive antagonism against acetylcholine at the level of nicotinic receptors of ganglionic neurons (BROWN 1980).

From a practical point of view, it is most important that the inhibitory action of tubocurarine on numerous parasympathetic and sympathetic ganglia is manifested within the neuromuscular blocking dose range (see Table 1), i.e., the agent has a low index of ganglion-blocking safety. For instance, in doses providing a partial or complete block of transmission from motor nerves to the hindlimb muscles of cats and dogs (0.2–0.4 mg/kg i.v.), tubocurarine significantly inhibits ganglionic transmission in superior cervical ganglion (HOPPE et al. 1955; KHARKEVICH 1957, 1962; HUGHES 1970), ciliary ganglion (GUYTON and REEDER 1950), cardiac intramural ganglia (GUYTON and REEDER 1950; HUGHES and CHAPPLE 1976a), stellate ganglion (FLACKE and GILLIS 1968), celiac plexus ganglia (MCCULLOUGH et al. 1970), and some others.

After administration of tubocurarine in paralyzing doses to animals, its inhibitory action on parasympathetic ganglia is manifested by such effects as mydriasis, decreased salivation, reduced intestinal and gastric motility, and diminished contractility of the urinary bladder (GROSS and CULLEN 1945; GUYTON and REEDER 1950; PATON 1959; HALD and MYGIND 1967). Inhibition of the function of sympathetic ganglia by tubocurarine is considered to be responsible for characteristic hemodynamic alterations: hypotension, decrease of phasic aortic blood flow, stroke volume, cardiac contractility, heart rate, and systemic vascular resistance (HUGHES 1970; MCCULLOUGH et al. 1970, 1972).

It should be noted, however, that some of the hemodynamic changes can be due not only to ganglion-blocking activity of tubocurarine, but also to its histamine-liberating effect (MCINTOSH and PATON 1949; MCCULLOUGH et al. 1970) and also to its direct inhibitory action on the myocardium (BROWN and CROUT 1966; KAJIMOTO 1970).

Inhibition of several viscerovisceral reflexes from the receptors of larynx, roots of the lungs, carotid sinus, and celiac plexus to the heart and vessels revealed in experimental animals after administration of tubocurarine in neuromuscular blocking doses (BURSTEIN et al. 1950; HOPPE et al. 1955; WONG et al. 1971; BRITTAIN and TYERS 1973), is another manifestation of marked ganglion-blocking action of the drug.

Among the side effects of tubocurarine in humans, which may be associated with inhibition of the autonomic ganglia, hypotension is clinically the most important. After administration of tubocurarine, a decrease in blood pressure occurs in the majority of patients, reaches its maximum 1–3 min after administration, and persists for 15–30 min (BARAKA 1967; HARRISON 1972). During tubocurarine-induced hypotension, a reduction of stroke volume and cardiac output, and in many cases also a lowering of systemic vascular resistance have been observed (STOELTING 1972; COLEMAN et al. 1972). The degree of the fall in blood pressure is in large part dependent on the rate of tubocurarine administration and it is significantly increased if the agent is combined with anesthetics depressing the cardiovascular system (halothane) (SMITH and WHITCHER 1967; TAMMISTO and WELLING 1969; MUNGER et al. 1974).

It should be noted, however, that for the present the relative importance of the ganglion-blocking action of tubocurarine for its hypotensive effect in humans is still not quite clear. A number of clinical observations suggest that the histamine-liberating action of the drug (PATON 1957; STOELTING and LONGNECKER 1972), as well as its direct cardiodepressant effect (JOHNSTON et al. 1978) can play an important role in the genesis of hypotension.

Metocurine, according to experimental findings, unlike tubocurarine, has a high index of ganglion-blocking safety (COLLIER and HALL 1950; MCCULLOUGH et al. 1972; HUGHES and CHAPPLE 1976a, b, 1981; DURANT et al. 1977; SAVARESE 1979; see Table 1). In anesthetized cats the neuromuscular blocking activity of the drug is 9–14 times higher than that of tubocurarine, while its inhibitory potency with respect to superior cervical ganglion is 3–6 times lower than that of the latter.

In patients, metocurine as a rule does not cause hemodynamic alterations (CHAPPLE et al. 1976; HUGHES et al. 1976a). Moderate transient changes of the heart rate (tachycardia) and blood pressure (hypotension), observed in some patients after a fast administration of a paralyzing dose of metocurine (SAVARESE et al. 1977) should be considered as entirely due to the weak histamine-liberating effect of the drug (MCCULLOUGH et al. 1972; SAVARESE 1979).

A bisquaternary analog of tubocurarine, studied by DURANT et al. (1977) judging by the separation of neuromuscular blocking action from autonomic effects (blockade of superior cervical ganglion in cats), occupies an intermediate position between tubocurarine and metocurine (see Table 1).

Ganglion-blocking properties in another curare alkaloid C-toxiferine (WASER and HARBECK 1959) are essentially absent. In anesthetized cats, the agent causes a neuromuscular block at 5.4 μg/kg i.v., but even at 300 μg/kg i.v. it fails to affect the transmission in intramural cardiac ganglia. In humans, no hemodynamic changes were observed on the use of C-toxiferine for neuromuscular blockade (WASER and HARBECK 1959; FREY and SEEGER 1961; FOLDES et al. 1961).

The C-toxiferine derivative Alloferin (alcuronium) has essentially no ganglion-blocking activity. Although in doses 4–5 times as high as the neuromuscular blocking ED_{50} it markedly inhibits cardiac chronotropic response to vagal stimulation in cats, this effect is entirely related to the block of cardiac muscarinic receptors (HUGHES and CHAPPLE 1976a). Alloferin fails to induce a significant change of transmission in the superior cervical ganglion after four- to fivefold neuromuscular blocking doses. According to KAJIMOTO (1970), moderate hypotension develops in cats only after a tenfold neuromuscular blocking dose of the drug. In many cases, however, after Alloferin administration to humans, a moderate and short-term fall in blood pressure is observed (BARAKA 1967; TAMMISTO and WELLING 1969; KENNEDY and KELMAN 1970; COLEMAN et al. 1972), its causes being unclear.

b) α-Truxillic Acid Derivatives

Compounds of this series, in particular truxilonium, anatruxonium, and cyclobutonium have moderate ganglion-blocking activity (KHARKEVICH and KRAVCHUK 1961; KHARKEVICH 1965, 1966, 1970c; SAMOILOV 1971b). This is manifested by the inhibition of transmission in the superior cervical ganglion in cats after the drugs are administered in doses several times higher than the neuromuscular blocking doses (see Table 1). The compounds mentioned produce a marked cardiac vagolytic action, which, however, is primarily due to the blockade of cardiac muscarinic receptors (see Sect. A). Clinically moderate ganglion-blocking activity of truxilonium, anatruxonium, and cyclobutonium is displayed by a not too marked and short-term hypotension and also mydriasis (KOTOMINA 1970; KHARKEVICH 1973).

A striking separation of neuromuscular blocking action from autonomic effects in this series of compounds is associated with the substitution of an imino group for ester oxygen (SAMOILOV 1971b; Chap. 19). Thus, bisquaternary amides dipyronium and amidonium (see Chap. 19) possess extremely high indices of ganglion-blocking safety (more than 200 by their action on superior cervical ganglion).

c) Steroid Compounds

Neuromuscular blocking agents of this group, including pancuronium, stercuronium, dacuronium, Org-6368, pipecuronium, vecuronium, chandonium, HS-342, and Duador (RGH-4201), have a rather slight inhibitory action on transmission in superior cervical ganglion, i.e., their index of ganglion-blocking safety is high (BUCKETT et al. 1968; MARSHALL 1973; GANDIHA et al. 1974; SUGRUE et al. 1975; DURANT et al. 1979; TEERAPONG et al. 1979; ALYAUTDIN et al. 1980; KÁRPÁTI and BIRÓ 1980; BIRÓ and KÁRPÁTI 1981; MARSHALL et al. 1981). All the steroid compounds mentioned have a more or less high cardiac vagolytic activity, which, however, is entirely dependent on the blocking action on muscarinic receptors of the heart (see Sect. A).

d) *N*-Adamantyl Compounds

According to KHARKEVICH (1970a, b), in cats, bis *N*-adamantyl analogs of succinylcholine and decamethonium (diadonium and decadonium) in neuromuscular blocking doses are devoid of ganglion-blocking action (superior cervical ganglion). Moderate inhibition of ganglionic transmission (by 25%–30%) is observed after a 10- to 20-fold neuromuscular blocking dose of diadonium (KHARKEVICH 1970b; BIRÓ and KÁRPÁTI 1981) and a 3- to 4-fold dose of decadonium (KHARKEVICH 1970a). Both drugs have a pronounced cardiac vagolytic activity, which is entirely due to the blockade of cardiac muscarinic receptors (see Sect. A).

Adequate clinical relaxation requires diadonium doses 250–300 times higher than those in cats (DARBINYAN et al. 1980; OSTROVSKY and KISELEV 1980). In these doses, diadonium provokes in patients a short-term (3–5 min) hypotension (fall of blood pressure by 10%–15%) and tachycardia (heart rate increase by 15–20 beats/min) and also induces mydriasis, a certain increase in viscosity of bronchial gland excretion and a slight decrease of salivation. The effects mentioned may be related to the blockade of sympathetic and parasympathetic ganglia. It cannot be ruled out that some of them may result from a blocking action of diadonium on muscarinic receptors.

e) Miscellaneous Bisquaternary Ammonium Compounds

Fazadinium (AH 8165) at neuromuscular blocking doses (1.5–2 mg/kg, i.v.) effectively blocks cardiac muscarinic receptors in cats (cardiac vagolytic action) and causes initial inhibition of transmission in the superior cervical ganglion (MARSHALL 1973; HUGHES and CHAPPLE 1976a; HUGHES et al. 1976a). Pronounced blockade of this ganglion develops after a fazadinium dose 3–4 times as high as the neuromuscular blocking dose. A parallel fall in blood pressure occurs. In humans, after fazadinium administration, its cardiotropic antimuscarinic effect usually predominates: tachycardia and moderate hypertension appear (COLEMAN et al. 1973; HUGHES et al. 1976b; FAMEWO 1981).

A neuromuscular blocking agent of ultra-short action AH 10407 (1,1-azobis[3-methyl-2-phenylbenzimidazolinium]dimethanesulfonate) in experiments on anesthetized cats displays a rather marked ganglion-blocking action (BRITTAIN et al. 1977). Within the dose range evoking initial to complete blockade of neuromuscular transmission (0.5–2 mg/kg i.v.), AH 10407 decreases vagal depression of the heart rate by 8%–75%, and in doses of 0.5–4 mg/kg i.v. inhibits the transmission in superior cervical ganglion by 20%–50%. A parallel fall in blood pressure by 5–30 mmHg has been observed. High ganglion-blocking activity of AH 10407 has also been revealed in dogs, by the antagonism against the pressor effect of DMPP.

Tercuronium, studied by DANILOV et al. (1979) at neuromuscular blocking doses fails to inhibit the transmission in superior cervical ganglion and cardiac parasympathetic ganglia in anesthetized cats (see Chap. 21). To produce a 50% blockade of these ganglia, ED_{50} of tercuronium for neuromuscular blocking action (tibialis anterior muscle) should be increased tenfold, which indicates its rather high index of ganglion-blocking safety.

Atracurium, studied by STENLAKE et al. (1981 c) is essentially devoid of ganglion-blocking action (see Chap. 24). In anesthetized cats, it causes initial inhibition of transmission in superior cervical ganglion and cardiac intramural ganglia only in doses 8–16 times higher than the neuromuscular blocking dose (HUGHES and CHAPPLE 1981). The first clinical trials of atracurium have shown that it is devoid of any effect on the human autonomic nervous system (PAYNE and HUGHES 1981; MACMILLAN et al. 1983).

Moderate ganglion-blocking action has been found in the nondepolarizing neuromuscular blocking agent, qualidilum (MASHKOVSKY and SADRITDINOV 1962; see Chap. 14). In doses causing complete block of neuromuscular transmission in cats (0.3–0.5 mg/kg i.v.), it inhibits cardiac parasympathetic ganglia (heart response to acetylcholine remains unchanged), but does not exert inhibitory action on the transmission in superior cervical ganglion. In doses of 1–3 mg/kg i.v., qualidilum entirely blocks sympathetic and parasympathetic ganglia, causing a fall of blood pressure by 20–40 mmHg within 10–30 min. In patients, qualidilum does not affect blood pressure. A frequent side effect is tachycardia, which might be due to the blockade of cardiac parasympathetic ganglia (DEGTYAREVA and MIKHELSON 1964).

Diplacinum, according to MASHKOVSKY and BRISKIN (1952), at the neuromuscular blocking dose (1 mg/kg i.v.) produces a marked blocking action on transmission from the vagus to the heart (experiments on cats), whereas a short-term blockade of superior cervical ganglion is achieved only after 20 mg/kg. However, in studies of functional lability (according to Wedensky) of superior cervical ganglion (ability to reproduce high frequencies of stimuli), its decrease can be revealed even at low doses of diplacinum (0.1–0.25 mg/kg i.v.) (VALDMAN et al. 1955; KHARKEVICH 1956). Administration of diplacinum to patients usually is not associated with cardiovascular side effects (ZYABLINA 1959). A certain decrease of small intestine peristalsis and tone can be observed (VOLIKOV 1954).

f) Trisonium Compounds

In gallamine, belonging to this group of neuromuscular blocking agents, ganglion-blocking activity is low (BÜLBRING and DEPIERRE 1949; MUSHIN et al. 1949; RIKER and WESCOE 1951; HUGHES and CHAPPLE 1976 a). In doses 10–15 times higher than the neuromuscular blocking dose, it only causes initial inhibition of transmission in superior cervical ganglion in cats and dogs. Gallamine possesses a marked cardiac vagolytic activity, which is, however, related to the blockade of cardiac muscarinic receptors (RIKER and WESCOE 1951; HUGHES and CHAPPLE 1976 a).

2. Depolarizing Neuromuscular Blocking Agents

Succinylcholine, decamethonium, and Imbretil can be considered as neuromuscular blocking agents essentially devoid of ganglion-blocking activity. Slight inhibition of transmission in superior cervical ganglion in cats and dogs can be observed after a 40- to 50-fold neuromuscular blocking dose of succinylcholine (THESLEFF 1952; KHARKEVICH 1956; DANILOV et al. 1957) and more than 400-fold paralyzing dose of decamethonium (PATON 1949; HOPPE et al. 1955). Imbretil at doses 20–30 times higher than the neuromuscular blocking dose, only activates ganglionic neurons (KLUPP et al. 1953). The drugs mentioned fail to exert a significant influence on blood pressure in humans (BOURNE et al. 1952; ARDAMATSKAYA 1957; LEONG et al. 1962).

3. Neuromuscular Blocking Agents of Mixed Mode of Action

Mytolon (benzoquinonium), nobutanum, and mebutanum possess a low ganglion-blocking activity (HOPPE et al. 1955; MASHKOVSKY and MEDVEDEV 1960; MEDVEDEV 1962). According to HOPPE et al. (1955) Mytolon in cats and dogs inhibits cardiac parasympathetic ganglia and superior cervical ganglion in doses 10 and 20–30 times higher, respectively, than those needed for a 50% reduction of transmission in the gastrocnemius muscle. Ganglion-blocking properties of nobutanum and mebutanum are still less marked. For a pronounced blockade of cardiac parasympathetic and superior cervical ganglia in cats, neuromuscular blocking doses of these agents should be increased 25–50 and 50–100 times, respectively (MEDVEDEV 1962). Dioxonium has essentially no ganglion-blocking activity (KLUSHA et al. 1970).

III. Conclusions

The data given in this section indicate that among the neuromuscular blocking drugs most widely used in medical practice, only tubocurarine has marked ganglion-blocking activity. Other clinically employed agents in doses causing muscle relaxation, produce essentially little or no inhibitory action on autonomic ganglia and their pronounced ganglion-blocking effects can be revealed only after exceeding (sometimes manifold) their neuromuscular blocking doses.

Acknowledgment. The authors wish to express their appreciation to Mrs. MARIA LIPMAN for the translation of this chapter.

References

Abrachams VC, Hilton SM (1962) Blocking action of decamethonium at different sites in the autonomic nervous system of the cat. Br J Pharmacol 18:194–203

Alyautdin RN, Buyanov VV, Fisenko VP, Lemina EYu, Muratov VK, Samoilov DN, Shorr VA (1980) On some properties of a new steroid curare-like compound pipecurium bromide. Arzneimittelforsch 30(I), 2a:355–357

Ardamatskaya AN (1957) Surgical experience with succinylcholine (in Russian). In: Ditilin i opyt ego klinicheskogo primeneniya. Armenian SSR Academy of Sciences Publishers, Yerevan, pp 178–181

Artusio JF, Murbury BE, Grews MA (1951) A quantitative study of d-tubocurarine, tri-(ethylaminoethoxy)-1,2,3-benzene (Flaxedil) and a series of tri-methyl and dimethylethylammonium compounds in anaesthetized man. Ann NY Acad Sci 54,3:512–527

Arunlakshana O, Shild HO (1959) Some quantitative uses of drug antagonists. Br J Pharmacol 14:48–58

Baraka A (1967) A comparative study between diallylnortoxiferine and tubocurarine. Br J Anaesth 39:624–629

Benfey BG, Yong MS, Belleau B, Melchiorre C (1979) Cardiac muscarinic blocking and atropine blocking effects of tetramine disulfide with α-adrenoreceptor blocking activity. Can J Physiol Pharmacol 57:41–47

Bieger D, Lüllmann H, Wassermann O (1968) Über eine akute Anti-Acetylcholin-Wirkung von Hemicholinium No. 3 am Vorhof des Meerschweinschens. Naunyn Schmidebergs Arch Pharmacol 259:386–393

Birdsall NJM, Burgen ASV, Hulme EC, Stockton JM (1981) Gallamine regulates muscarinic receptors in the heart and cerebral cortex. Br J Pharmacol 74:798P

Biró K, Kárpáti E (1981) The pharmacology of new short-acting non-depolarizing muscle relaxant steroid (RGH-4201). Arzneimittelforsch 31:1918–1924

Björklund A, Cegrell L, Flack B, Ritzen M, Rosengren E (1970) Dopamin-containing cells in sympathetic ganglia. Acta Physiol Scand 78:334–338

Bonta IL, Buckett WR, Lewis JJ, Vargaftig BB (1966) 2β,16β-dipiperidino-5-androstane-3,17β-diol-diacetatedimethobromide (NA-97) a potent neuromuscular blocking steroid. In: Second International Congress of Steroids, Milan. Excerpta Medica, Amsterdam. Abstracts, p 344

Bonta IL, Goorissen EM, Derkx FH (1968) Pharmacological interaction between pancuronium bromide and anaesthetics. Eur J Pharmacol 4:83–90

Booij LHDJ, Vree TB, Crul JF (1982) Org NC 45: a new steroidal non-depolarizing muscle relaxant. Pharm Weekbl [Sci] 4:1–4

Boros M, Szenohradszky J, Marosi Gy, Toth J (1980) Comparative clinical study of pipecurium bromide and pancuronium bromide. Arzneimittelforsch 30, 2a:389–393

Bourne JC, Collier HOS, Somers GF (1952) Succinylcholine (Succinoylcholine): muscle relaxant of short action. Lancet 1:1225–1229

Bovet D, DEpierre F, Courvoisier S, Lestrange Y (1949) Recherches sur les poisons curarisants de synthèse. IIme Partie. Ethers phénoliques à fonction ammonium quaternaire. Action du tri-iodoéthylate de tri(diéthylaminoéthoxy)benzène (2559 F). Arch Int Pharmacodyn Ther 80:172–188

Bowman WC (1982) Non-relaxant properties of neuromuscular blocking drugs. Br J Anaesth 54:147–160

Brittain RT, Tyers MB (1973) The pharmacology of AH 8165: a rapid-acting, short-lasting, competitive neuromuscular blocking drug. Br J Anaesth 45:837–843

Brittain RT, Jack D, Tyers MB (1977) Pharmacological and certain chemical properties of AH 10407. An unusually short-acting, competitive neuromuscular blocking drug, and some related compounds. Br J Pharmacol 61:47–55

Brown BR, Crout JR (1966) Mechanism of the positive inotropic effect of gallamine on guinea pig heart muscle. Fed Proc 25:472

Brown BR, Crout JR (1968) The sympathomimetic effect of gallamine on the heart. Anesthesiology 29:179–180

Brown BR, Crout JR (1970) The sympathomimetic effect of gallamine on the heart. J Pharmacol Exp Ther 172:266–273

Brown DA (1980) Locus and mechanism of action of ganglion-blocking agents. In: Kharkevich DA (ed) Pharmacology of ganglionic transmission. Springer, Berlin Heidelberg New York, pp 185–235

Brücke F (1956) Dicholinesters of α,ω-dicarboxylic acids and related substances. Pharmacol Rev 8:265–335

Buckett WR, Saxena PR (1969) The pharmacology of dacuronium bromide – a new short-acting neuromuscular blocking drug of non-depolarizing type. In: Proceedings of the 4th International congress of pharmacology, Basel, 14–18 July, p 420

Buckett WR, Marjoribanks CEB, Marwick FA, Morton MB (1968) The pharmacology of pancuronium bromide (Org NA 97), a new potent steroidal neuromuscular blocking agent. Br J Pharmacol 32:671–682

Bülbring E, Depierre F (1949) The action of synthetic curarizing compounds on skeletal muscle and sympathetic ganglia, both normal and denervated. Br J Pharmacol 4:22–32

Bunatyan AA, Mikheev VI (1980) Clinical experience with new steroid muscle relaxant: pipecurium bromide. Arzneimittelforsch 30, 2a:383–385
Burstein CL, Jackson A, Bishop HF, Rovenstine EA (1950) Curare in the management of autonomic reflexes. Anesthesiology 11:409–421
Butaev BM (1953) Pharmacological study of paramionum, a new synthetic curare-like agent (in Russian). Cand Med Sci Thesis, Leningrad
Chapple DJ, Enderby DH, Hughes R, Payne JP (1976) Neuromuscular and cardiovascular studies with dimethyl tubocurarine in anaesthetized cats, rhesus monkeys and man. Br J Pharmacol 56:354P–355P
Clark AL, Mitchelson F (1976) The inhibitory effect of gallamine on muscarinic receptors. Br J Pharmacol 58:323–331
Coleman AJ, Downing JW, Leary WP, Moyes DG, Styles M (1972) The immediate cardiovascular effects of pancuronium, alcuronium, and tubocurarine in man. Anaesthesia 27:415–422
Coleman AJ, O'Brien A, Downing JW, Jeal DE, Moyes DG, Leary WP (1973) AH 8165: a new non-depolarizing muscle relaxant. Anaesthesia 28:262–267
Collier HOJ, Hall RA (1950) Pharmacology of d-OO-dimethyl tubocurarine iodide in relation to its clinical use. Br Med J I:1293–1295
Crul JF, Booij LHDJ (1980) First clinical experiences with Org NC45. Br J Anaesth 52, Suppl 1:49S–52S
Danilov AF, Mikhelson MYa, Rybolovlev RS (1957) Pharmacological properties of succinylcholine and clinical experience with it (in Russian). In: Ditilin i opyt ego klinicheskogo primeneniya. Armenian SSR Academy of Sciences Publishers, Yerevan, pp 92–107
Danilov AF, Malygin VV, Starshinova LA, Khromov-Borisov NV, Torf SF, Cherepanova VP (1979) Tercuronium – a new nondepolarizing muscle relaxant with high activity and selectivity of the effect (in Russian). Farmakol Toksikol 5:478–481
Darbinyan TM, Saliev RSh, Shloznikov BM, Saakyan ES (1980) Haemodynamics during various kinds of combined induction anaesthesia using antidepolarizing muscle relaxants for tracheal intubation (in Russian). Anesteziol Reanimatol 5:7–11
Degtyareva LG, Mikhelson VA (1964) Clinical experience with domestic neuromuscular blocking agent qualidil (in Russian). Eksp Khir Anesteziol 6:88–91
Docherty JR, McGrath JC (1978) Sympathomimetic effects of pancuronium bromide on the cardiovascular system of the pithed rat: a comparison with the effects of drugs blocking the neuronal uptake of noradrenaline. Br J Pharmacol 64:589–599
Docherty JR, McGrath JC (1980) A comparison of the effects of pancuronium bromide and its monoquaternary analogue, Org NC45, on autonomic and somatic neurotransmission in the rat. Br J Pharmacol 71:225–233
Doughty AG, Wylie WD (1951) An assessment of flaxedil (Gallamine triethiodide; B.P.). Proc R Soc Med 44:375–386
Drane SE, Evans MH (1979) The relative vagolytic potencies of six muscle relaxants in the rabbit. J Pharm Pharmacol 31:864–866
Durant NN, Bowman WC, Marshall IG (1977) A comparison of the neuromuscular and autonomic blocking activities of (+)-tubocurarine and its *N*-methyl and *O,O,N*-trimethyl analogues. Eur J Pharmacol 46:297–302
Durant NN, Marshall IG, Savage DS, Nelson DN, Sleigh T, Carlyle IC (1979) The neuromuscular and autonomic blocking activities of pancuronium, Org NC45 and other pancuronium analogues. J Pharm Pharmacol 31, 12:831–836
Eckfeld DK (1959) Curarizing and atropine-like properties of bis-atropinium and bis-scopolaminium compounds. J Pharmacol Exp Ther 126:21–23
Eisele JH, Marta JA, Davis HS (1971) Quantitative aspects of the chronotropic and neuromuscular effects of gallamine in anesthetized man. Anesthesiology 35:630–633
Elwell LH (1960) Effects of d-tubocurarine on the submandibular salivary gland. Am J Physiol 198:621–624
Famewo CE (1981) Clinical trial of fazadinium bromide (fazadon). Can Anaesth Soc J 28:149–152

Firsov AA, Zhilis BG, Stazhadze LL (1970) Anesthesiological experience with a neuromuscular blocking agent cyclobutonium (in Russian). In: Kharkevich DA (ed) Novye kurarepodobnye i ganglioblokiruyushchie sredstva (New curare-like and ganglion-blocking agents). Meditsina, Moscow, pp 152–159

Flacke W, Gillis RA (1968) Impulse transmission via nicotinic and muscarinic pathways in the stellate ganglion of the dog. J Pharmacol Exp Ther 163:266–276

Foldes FF, Machaj TS, Carberry PC (1954) The use of gallamine triethiodide (Flaxedil) with pentothal-sodium nitrous oxide-oxygen anesthesia in abdominal surgery. Anaesth Analg 33:122–128

Foldes FF, Swerdlow M, Liphschitz E, Hees GR, Shanor SP (1956) Comparison of the respiratory effects of suxamethonium and suxaethonium in man. Anesthesiology 17:559–568

Foldes FF, Wolfson B, Sokoll M (1961) The use of toxiferine for the production of surgical relaxation. Anesthesiology 22:93–99

Frey R, Seeger R (1961) Experimental and clinical experience with toxiferine (alkaloid of calabash curare). Can Anaesth Soc J 8:99–117

Fuder H, Meiser C, Wormstall H, Muscholl E (1981) The effects of several muscarinic antagonists on pre- and postsynaptic receptors in the isolated rabbit heart. Naunyn Schmiedebergs Arch Pharmacol 316:31–37

Gandiha A, Marshall IG, Paul D, Singh H (1974) Neuromuscular and other blocking actions of a new series of mono and bis quaternary aza steroids. J Pharm Pharmacol 26:871–877

Gandiha A, Marshall IG, Paul D, Rodger IW, Scott W, Singh H (1975) Some actions of chandonium iodide, a new shrot-acting muscle relaxant, in anaesthetized cats and on isolated muscle preparations. Clin Exp Pharmacol Physiol 2:159–170

Gardier RW, Ganansia M-F, Delaunois AL, Hamelberg W (1974) Enhancement of ganglionic muscarinic activity by gallamine. Anesthesiology 40:494–497

Gardier RV, Tsevdos EJ, Jackson DB (1978) The effect of pancuronium and gallamine on muscarinic transmission in the superior cervical ganglion. J Pharmacol Exp Ther 204:46–53

Goat VA, Feldman SA (1971) Action of muscle relaxant drugs on cardiac cholinergic receptors in the isolated rabbit heart. Br J Anaesthesiol 43:203–204

Greengard P, Kebabian JW (1974) Role of cyclic AMP in synaptic transmission in the mammalian peripheral nervous system. Fed Proc 33:1059–1067

Gross EG, Cullen SC (1945) The action of curare on the smooth muscle of the small intestine and on the blood pressure. Anesthesiology 6:231–238

Grossman E, Jacobi AM (1974) Hemodynamic interaction between pancuronium and morphine. Anesthesiology 40:299–301

Guyton AC, Reeder RC (1950) Quantitative studies on the autonomic actions of curare. J Pharmacol Exp Ther 98:188–194

Hald T, Mygind T (1967) Effect of curare on canine micturition. J Urol 97:101–104

Harrison GA (1972) The cardiovascular effects and some relaxant properties of four relaxants in patients about to undergo cardiac surgery. Br J Anaesth 44:485–494

Hoppe JO, Funnel JE, Lape H (1955) The effects of structural variation in the quaternary nitrogen centers of benzoquinonium chloride upon neuromuscular blocking activity. J Pharmacol Exp Ther 115:106–119

Hughes R (1970) Haemodynamic effects of tubocurarine, gallamine, and suxamethonium in dogs. Br J Anaesth 42:928–934

Hughes R (1972) Evaluation of the neuromuscular blocking properties and side-effects of the two new isoquinolinium bisquaternary compounds (BW 252C64 and BW 403C65). Br J Anaesth 44:27–41

Hughes R, Chapple DJ (1976a) Effects of non-depolarizing neuromuscular blocking agents on peripheral autonomic mechanisms in cats. Br J Anaesth 48:59–68

Hughes R, Chapple DJ (1976b) Cardiovascular and neuromuscular effects of dimethyl tubocurarine in anaesthetized cats and rhesus monkeys. Br J Anaesth 48:847–852

Hughes R, Chapple DJ (1981) The pharmacology of atracurium: a new competitive neuromuscular blocking agent. Br J Anaesth 53:31–44

Hughes R, Ingram GS, Payne JP (1976a) Studies on dimethyltubocurarine in anaesthetized man. Br J Anaesth 48:969–974
Hughes R, Payne JP, Sugai N (1976b) Studies on fazadinium bromide. (AH 8165): a new non-depolarizing neuromuscular blocking agent. Can Anaesth Soc J 23:36–47
Ivankovich AD, Miletich DJ, Albrecht RF, Zahed B (1975) The effect of pancuronium on myocardial contraction and catecholamine metabolism. J Pharm Pharmacol 27:837–841
Johnstone M, Mahmoud AA, Mrozinski RA (1978) Cardiovascular effects of tubocurarine in man. Anaesthesia 33:587–593
Kajimoto N (1970) Pharmacology of alloferin (diallyl-bis-nortoxiferine). Acta Sch Med Univ Kyoto 40:291–304
Kárpáti E, Biró K (1980) Pharmacological study of a new competitive neuromuscular blocking steroid, pipecurium bromide. Arzneimittelforsch 30, 2a:346–354
Kelman GR, Kennedy BR (1971) Cardiovascular effects of pancuronium in man. Br J Anaesth 43:335–338
Kennedy BR, Farman JV (1968) Cardiovascular effects of gallamine triethiodide in man. Br J Anaesth 40:773–780
Kennedy BR, Kelman GR (1970) Cardiovascular effects of alcuronium in man. Br J Anaesth 42:625–630
Kensler CJ, Zirkle CL, Matallana A, Condouris GJ (1954) The selective anticholinergic activity of aliphatic tris-quaternry ammonium compounds. J Pharmacol Exp Ther 112:210–219
Kharkevich DA (1956) Alteration of lability of the superior cervical ganglion under the influence of quaternary ammonium bases (in Russian). Byull Eksp Biol Med 10:34–38
Kharkevich DA (1957) The effect of cholinolytic agents on the functional lability of the superior cervical ganglion (in Russian). Byull Eksp Biol Med 12:70–76
Kharkevich DA (1962) Ganglionarnye sredstva (Ganglionic agents). Medgiz, Moscow
Kharkevich DA (1965) On the pharmacological properties of a new curare-like agent anatruxonium (in Russian). Farmakol Toksikol 3:305–309 or (1966) Fed Proc 25:T521–T523
Kharkevich DA (1966) On the pharmacology of a new nondepolarizing muscle relaxant cyclobutonium (in Russian). Farmakol Toksikol 1:47–53
Kharkevich DA (1970a) On curare-like activity of decadonium diiodide (in Russian). Farmakol Toksikol 4:395–399
Kharkevich DA (1970b) On the pharmacological properties of a new antidepolarizing curare-like agent diadonium diiodide (in Russian). Farmakol Toksikol 5:531–536
Kharkevich DA (1970c) Pharmacology of new antidepolarizing agents – anatruxonium, truxilonium, cyclobutonium and pyrocyclonium (in Russian). In: Kharkevich DA (ed) Novye kurarepodobnye i ganglioblokiruyushchie sredstva (New curare-like and ganglion-blocking agents). Meditsina, Moscow, pp 41–47
Kharkevich DA (1973) Curare-like agents (in Russian). In Kharkevich DA (ed) Uspekhi v sozdanii novykh lekarstvennykh sredstv (Progress in the design of new drugs). Meditsina, Moscow, pp 138–187
Kharkevich DA (1974) On predominant action of a number of pharmacological agents on cardiac m-cholinoreceptors (in Russian). Farmakol Toksikol 9:94–102
Kharkevich DA (ed) (1983) New neuromuscular blocking agents. Meditsina, Moscow
Kharkevich DA, Kravchuk LA (1961) On the pharmacology of a new curare-like agent truxilonium (in Russian). Farmakol Toksikol 3:318–324
Kharkevich DA, Shorr VA (1980) Cardiotropic antimuscarinic action of some curare-like agents. Arch Int Pharmacodyn Ther 248:238–250
Kharkevich DA, Skoldinov AP, Samoilov DN, Shorr VA (1981) The effects of anticholinergic agents on muscarinic receptors of different localization. In: Pepeu G, Ladinsky H (eds) Cholinergic mechanisms. Phylogenetic aspects, central and peripheral synapses, and clinical significance. Plenum, New York, pp 351–365
Kimura KK, Unna K (1950) Curare-like action of decamethylene-bis(atropinium iodide). J Pharmacol Exp Ther 98:286–292

Klupp H, Kraupp O, Stormann H, Stumpf Ch (1953) Über die pharmakologischen Eigenschaften einiger Polymethylen-Dicarbaminsäure-bischolinester. Arch Int Pharmacodyn Ther 96:161–182

Klusha VE, Kimenis AA, Sokolov GP (1970) The results of pharmacological and clinical study of a new domestic neuromuscular blocking agent dioxonium (in Russian). In: Eksperimentalnaya i klinicheskaya farmakoterapiya, 2 (Experimental and clinical pharmacotherapy). Zinatne, Riga, pp 55–74

Kotomina GL (1970) Comparative clinical evaluation of new neuromuscular blocking agents of the group of diphenylcyclobutanedicarboxylic acids' derivatives (in Russian). In: Kharkevich DA (ed) Novye kurarepodobnye i ganglioblokiruyushchie sredstva (New curare-like and ganglion-blocking agents). Meditsina, Moscow, pp 83–90

Kotomina GL, Kharkevich DA (1968) Clinical pharmacological characteristics of a domestic neuromuscular blocking agent anatruxonium (in Russian). Khirurgiya 4:23–28

Kraupp O, Klupp H, Stormann H, Stumpf Ch (1954) Cholinesterasehemmungswirkung und neuromuskuläre Wirksamkeit von Bischolin-Polymethylendicarbaminsäureestern. Arch Exp Pathol Pharmakol 222:180–182

Kuhnen-Clausen D (1970) Investigations on the parasympatholytic effect of toxogonin on the guinea-pig isolated ileum. Eur J Pharmacol 9:85–92

Lee C, Yang E, Lippmann M (1980) Constrictive effect of pancuronium on capacitance vessels. Br J Anaesth 52:261–263

Lee Son S, Waud BE (1977) Potencies of neuromuscular blocking agents at the receptors of atrial pacemaker and the motor end-plate of the guinea-pig. Anestesiology 47:34–36

Lee Son S, Waud DR (1978) A vagolytic action of neuromuscular blocking agents at the pacemaker of the isolated guinea pig atrium. Anesthesiology 48:191–194

Lee Son S, Waud DR (1980) Effects of non-depolarizing neuromuscular blocking agents on the cardiac vagus nerve in the guinea pig. Br J Anaesth 52:981–987

Lee Son S, Waud BE, Waud DR, Phil D (1981) A comparison of neuromuscular blocking and vagolytic effects of Org NC45 and pancuronium. Anesthesiology 55:12–18

Lemina EYu (1978) Farmakologiya novykh kurarepodobnykh sredstv dipironiya i amidoniya (Pharmacology of new curare-like agents dipyronium and amidonium). Cand Med Sci Thesis, Moscow, 1st Medical Institute

Leong W, Given JB, Little DM (1962) The clinical use of imbretil. Can Anaesth Soc J 9:312–318

Leung E, Mitchelson F (1982) The interaction of pancuronium with cardiac and ileal muscarinic receptors. Eur J Pharmacol 80:1–9

Li CK, Mitchelson F (1978) The effect of stercuronium on cardiac muscarinic receptors. Eur J Pharmacol 51:251–259

Li CK, Mitchelson F (1980) The selective antimuscarinic action of stercuronium. Br J Pharmacol 70:313–321

Libert B, Tosaka T (1970) Dopamine as a synaptic transmitter and modulator in sympathetic ganglia. A different mode of synaptic action. Proc Natl Acad Sci USA 67:667–673

Loh L (1970) The cardiovascular effects of pancuronium bromide. Anaesthesia 25:356–363

Longnecker DE, Stoleting RK, Morrow AG (1973) Cardiac and peripheral vascular effects of gallamine. Anaesth Analg 52:931–935

Lüllmann H, Ohnesorge FK, Shauwecker G-C, Wassermann O (1969) Inhibition of the actions of carbachol and DFP on guinea pig isolated atria by alkane-bis-ammonium compounds. Eur J Pharmacol 6:241–247

Macmillan RR, West DM, Williams NE, Calvey TN (1983) Effects of atracurium on neuromuscular transmission. Br J Clin Pharmacol 15:P154–P155

Madden J, Mitchelson F (1975) The interaction of hemicholinium-3 (HC-3) with cholinomimetics and atropine. Eur J Pharmacol 32:17–29

Marshall IG (1973) The ganglion-blocking and vagolytic actions of three short-acting neuromuscular blocking drugs in the cat. J Pharm Pharmacol 25:530–536

Marshall IG (1980) Actions of non-depolatizing neuromuscular blocking agents at cholinoceptors other than at the motor endplate. In: Curares and curarization. Excerpta Medica, Amsterdam, pp 257–274
Marshall IG, Paul D, Singh H (1973) The neuromuscular and other blocking actions of 4,17a-dimethyl-4,17a-diaza-d-homo-5a-androstane dimethiodide (HS-342) in the anaesthetized cat. Eur J Pharmacol 22:129–134
Marshall IG, Agoston S, Booij LHDJ, Durant NN, Foldes FF (1980) Pharmacology of Org NC 45 compared with other non-depolarizing neuromuscular blocking drugs. Br J Anaesth 52, Suppl I:11S–19S
Marshall IG, Harvey AL, Singh H, Bhardwaj TR, Paul D (1981) The neuromuscular and autonomic blocking effects of azosteroids containing choline or acetylcholine fragments. J Pharm Pharmacol 33:451–457
Marshall RJ, Ojewole JAO (1979) Comparison of the autonomic effects of some currently used neuromuscular blocking agents. Br J Pharmacol 66:77–78
Marshall RJ, McGrath JC, Miller RD, Docherty JR, Lamar J-C (1980) Comparison of the cardiovascular actions of Org NC 45 with those produced by other non-depolarizing neuromuscular blocking agents in experimental animals. Br J Anaesth 52, Suppl I:21S–31S
Mashkovsky MD, Briskin AI (1952) Pharmacological properties of a new curare-like drug "diplacine" (in Russian). Farmakol Toksikol 5:24–32
Mashkovsky MD, Medvedev BA (1960) On the pharmacology of symmetrical bis-quaternary derivatives of 9-methyl-3,9-diazobicyclo-(3,3,1)-nonane (in Russian). Farmakol Toksikol 6:493–499
Mashkovsky MD, Sadritdinov F (1962) Curare-like properties of 1,6-di-(3,3-benzylquinuclidyl-1,1)-hexane dichloride (Qualidil) (in Russian). Farmakol Toksikol 6:685–691
Matsuki A, Kothary SP, Zsigmond EK (1973) Effects of gallamine on plasma cortisol and catecholamine levels in man. Can Anaesth Soc J 20:539–545
Mautner H, Luisada A (1941) Antagonistic effect of asphyxia to curare paralysis of the vagus nerve. J Pharmacol Exp Ther 72:386–393
McCullough LS, Rier CE, Delaunois AL, Gardier RW, Hamelberg W (1970) The effect of d-tubocurarine on spontaneous postganglionic sympathetic activity and histamine release. Anesthesiology 33:328–334
McCullough LS, Stone WA, Delaunois AL, Rier CE, Hamelberg W (1972) The effect of dimethyl tubocurarine iodide on cardiovascular parameters, postganglionic sympathetic activity and histamine release. Anaesth Analg 51:554–559
McIntosh FC, Paton WDM (1949) The liberation of histamine by certain organic bases. J Physiol (Lond) 109:190–219
Medvedev BA (1962) Comparative study of neuromuscular blocking and other pharmacological properties of 1,4-bis[9-methyl-3,9-diazabicyclo-(3,3,1)-nonano-3]butane diiodomethylate and dimethylsulphomethylate (in Russian). Farmakol Toksikol 3:320–326
Miller RD, Eger EI II, Stevens WC, Gibbons K (1975) Pancuronium-induced tachycardia in relation to alveolar halothane, dose of pancuronium, and prior atropine. Anesthesiology 42:352–355
Mitchelson F (1971) Differentiation between the actions of acetylcholine and tetramethylammonium on the isolated taenia of the guinea-pig caecum by hemicholinium-3. Br J Pharmacol 42:43–55
Morgenstern C, Splinth G (1965) Untersuchungen über die Ursachen der Gallamintachykardie und ihre antagonistische Beeinflussung durch Beta-Adrenolytica. Anaesthesist 14:298–301
Munger WL, Miller RD, Stevens WC (1974) The dependence of d-tubocurarine-induced hypotension on alveolar concentration of halothane, dose of d-tubocurarine, and nitrous oxide. Anesthesiology 40:442–448
Murbury BE, Artusio JE, Wescoe WC, Riker WF (1951) The effects of synthetic curare-like compound tri-iodo salt of tris-(triethylaminoethoxy)-1,2,3-benzene (Flaxedil) on the anaesthetized surgical patients. J Pharmacol Exp Ther 103:280–287

Mushin WW, Wien R, Mason DFJ, Langston GT (1949) Curare-like action of tri-(diethylaminoethoxy)-benzene triethiodide. Lancet 1:726–728
Nana A, Cardan E, Domokos M (1973) Blood catecholamine changes after pancuronium. Acta Anaesth Scand 17:83–87
Ostrovsky VYu, Kiselev SO (1980) Anesthesiological experience with diadonium (in Russian). Anesteziol Reanimatol 4:7–10
Paton WDM (1949) The pharmacology of curare and curarizing substances. J Pharm Pharmacol 1:273–286
Paton WDM (1957) Histamine release by compounds of simple chemical structure. Pharmacol Rev 9:269–328
Paton WDM (1959) The effects of muscle relaxants other than muscular relaxation. Anesthesiology 20:453–463
Payne JP, Hughes R (1981) Evaluation of atracurium in anaesthetized man. Br J Anaesth 53:45–54
Pelikan EW, Unna KR (1952) The functional importance of the quaternary nitrogen atoms in Flaxedil. J Pharmacol Exp Ther 104:354–362
Quintana A (1977) Effect of pancuronium bromide on the adrenergic reactivity of the isolated rat vas deferens. Eur J Pharmacol 46:275–277
Qvist TF, Lang-Jensen T, Korshin JD, Skovsted P (1980) The cardiovascular effects of pancuronium during halothane and cyclopropane anaesthesia. Acta Anaesth Scand 24:415–418
Rathbun FJ, Hamilton JT (1970) Effect of gallamine on cholinergic receptors. Can Anaesth Soc J 17:574–590
Riker WF, Wescoe WC (1951) The pharmacology of flaxedil, with observations on certain analogs. Ann NY Acad Sci 54:373–394
Saito S, Aviado DM (1963) Pharmacological studies of hexamethylene-bis-carbaminoylcholine (Imbretil) on the cardiovascular and respiratory systems in dogs. Jpn Circ J 27:791–796
Salmenperä M, Tammisto T (1980) The use of dioxonium as a neuromuscular blocking agent. Acta Anaesth Scand 24:395–398
Salt PJ, Barnes PK, Conway CM (1980) Inhibition of neuronal uptake of noradrenaline in the isolated perfused rat heart by pancuronium and its homologues, Org 6368, Org 7268, and NC 45. Br J Anaesth 52:313–317
Samoilov DN (1971 a) Chemical structure and cholinolytic activity of bis-quaternary ammonium derivatives of truxillic acids (in Russian). Farmakol Toksikol 4:413–420
Samoilov DN (1971 b) On the chemical structure and ganglion blocking activity of truxillic acid derivatives (in Russian). Farmakol Toksikol 5:544–549
Savarese JJ (1979) The autonomic margins of safety of metocurine and d-tubocurarine in the cat. Anesthesiology 50:40–46
Savarese JJ, Ali HH, Antonio RP (1977) The clinical pharmacology of metocurine: Dimethyltubocurarine revisited. Anesthesiology 47:277–284
Saxena PR, Bonta IL (1970) Mechanism of selective cardiac vagolytic action of pancuronium bromide. Specific blockade of cardiac muscarinic receptors. Eur J Pharmacol 11:332–341
Segarra Domenech J, Santafe Oroz R, Rodriguez Sasiain JM, Quintana Loyola A, Santafe Oroz J (1976) Pancuronium bromide: An indirect sympathomimetic agent. Br J Anaesth 48:1143–1148
Sheth V, Iabuis II (1980) Comparison of cardiovascular effect of pancuronium bromide with d-tubocurarine and gallamine. Indian J Anaesth 28:42–50
Shorr VA (1972) On the action of curare-like agents on some acetylcholine effects (in Russian). Farmakol Toksikol 4:426–431
Shorr VA (1975) On the peculiarities of action of neuromuscular blocking agents on m-cholinoreceptors of various localizations (in Russian). Farmakol Toksikol 1:38–43
Smith G, Proctor DW, Spence AA (1970) A comparison of some cardiovascular effects of tubocurarine and pancuronium in dogs. Br J Anaesth 42:923–927
Smith NT, Whitcher CE (1967) Hemodynamic effects of gallamine and tubocurarine administered during halothane anesthesia. JAMA 199:704–708

Stenlake JB, Waigh RD, Urwin J, Dewar GH, Hughes R, Chapple DJ (1981 a) Bis-3,4-dihydroisoquinolinium salts as potential neuromuscular blocking agents. Eur J Med Chem 16:503–507

Stenlake JB, Waigh RD, Urwin J, Dewar GH, Hughes R, Chapple DJ (1981 b) Biodegradable neuromuscular blocking agents. Eur J Med Chem 16:508–514

Stenlake JB, Waigh RD, Dewar GH, Hughes R, Chapple DJ, Coker GG (1981 c) Biodegradable neuromuscular blocking agents. Part 4. Atracurium Besylate and related polyalkylene di-esters. Eur J Med Chem 16:515–524

Stoelting RK (1972) The hemodynamic effects of pancuronium and d-tubocurarine in anaesthetized patients. Anesthesiology 36:612–615

Stoelting RK (1973) Hemodynamic effects of gallamine during halothane-nitrous oxide anaesthesia. Anesthesiology 39:645–647

Stoelting RK, Longnecker DE (1972) Effect of promethazine on hypotension following d-tubocurarine use in anaesthetized patients. Anesth Analg 51:509–516

Sugrue MF, Duff N, McIndewar I (1975) On the pharmacology of Org 6368 (2β,16β-dipiperidino-5α-androstan-3α-ol acetate dimethobromide), a new steroidal neuromuscular blocking agent. J Pharm Pharmacol 27:721–727

Tammisto T, Welling I (1969) The effect of alcuronium and tubocurarine on blood pressure and heart rate: a clinical comparison. Br J Anaesth 41:317–322

Teerapong P, Marshall IG, Harvey AL, Singh H, Paul D, Bhardwaj TR, Ahuja NK (1979) The effects of dihydrochandonium and other chandonium analogues on neuromuscular and autonomic transmission. J Pharm Pharmacol 31:521–528

Thesleff E (1952) Succinylcholine iodide. Studies on its pharmacological properties and clinical use. Acta Physiol Scand 27, Suppl 99:103–129

Thomas ET (1963) The effect of gallamine triethiodide on blood pressure. Anaesthesia 18:316–323

Tomlinson DR (1979) On the mechanism of pancuronium-induced supersensitivity to noradrenaline in rat smooth muscle. Br J Pharmacol 65:473–478

Unna KR, Pelikan EW, Macfarlane DW, Cazort RJ, Sadove MS, Nelson JT, Drucker AP (1950 a) Evaluation of curarizing drugs in man. I. Potency, duration of action and effects on vital capacity of d-tubocurarine, dimethyl-d-tubocurarine and decamethylene-bis(trimethylammonium bromide). J Pharmacol Exp Ther 98:318–329

Unna KR, Pelikan EW, Macfarlane DW, Sadove MS (1950 b) Evaluation of curarizing drugs in man. IV. Tri-(diethylaminoethoxy)-1,2,3-benzene (Flaxedil). J Pharmacol Exp Ther 100:201–209

Valdman AV, Ivanova ZN, Kharkevich DA (1955) The effect of diplacine on the transmission in various links of reflex arch (in Russian). Farmakol Toksikol 2:3–11

Vercruysse P, Bossuyt P, Hanegreefs G, Verbeuren TJ, Vanhoutte PM (1979) Gallamine and pancuronium inhibit pre- and post-junctional muscarinic receptors in canine saphenous veins. J Pharmacol Exp Ther 209:225–230

Volikov AA (1954) Clinical use of domestic curare-like agents during intratracheal gas and ether anaesthesia (in Russian). Vestn Khir 8:10–18

Walts LE, Prescott FS (1965) The effects of gallamine on cardiac rhythm during general anesthesia. Anaesth Analg 44:265–269

Waser P, Harbeck P (1959) Erste klinische Anwendung der Calebassenalkaloide Toxiferin I und Curarin I. Anaesthetist 8:193–198

Wong KC, Wyte SR, Martin WE, Crawford EW (1971) Antiarrhythmic effects of skeletal muscle relaxants. Anesthesiology 34:458–462

Zyablina TF (1959) The use of diplacine during intrathoracic operations in pulmonary tuberculosis patients (in Russian). Vestn Khir 11:112–116

CHAPTER 8

The Interaction of Neuromuscular Blocking Agents With Human Cholinesterases and Their Binding to Plasma Proteins

F. F. FOLDES and A. DEERY

A. Introduction

The interactions of neuromuscular (NM) blocking agents (NMBA) with cholinesterases (ChE) have inportant pharmacologic and clinical implications. Some of the ester type NMBA, e.g., succinylcholine chloride (suxamethonium; SCh), suxethonium bromide, are hydrolyzed by human plasma butyrylcholinesterase (EC 3.1.1.8; BuChE) (TSUJI and FOLDES 1953; FOLDES et al. 1956). Their hydrolysis rate in most other mammalian plasmas is much lower than in human plasma (TSUJI and FOLDES 1953). In vivo inhibition of plasma BuChE will increse the intensity and duration of action of hydrolyzable NMBA.

Most hydrolyzable, and nonhydrolyzable NMBA have an inhibitory effect on both BuChE and acetylcholinesterase (EC 3.1.1.7; AChE) (DEERY et al. 1982) present in muscle (SMITH et al. 1963), brain (FOLDES et al. 1962), red cells, and most other tissues. Inhibition of muscle AChE will increse the concentration of ACh at the NM junction. This will augment the NM blocking effect of depolarizing NMBA and antagonize those of the nondepolarizing NMBA. From the point of view of their NM effect, the inhibitory effect of NMBA on muscle AChE is important. Obtaining a large enough pool of human muscle AChE, however, would be very difficult. Fortunately the kinetic characteristics (e.g., K_m, turnover rate, sensitivity to inhibitors) of human muscle and red cell AChE are very similar (F. F. FOLDES and A. DEERY, 1968, unpublished work). Therefore, the interaction of NMBA with AChE was studied on the easily obtainable red cell enzyme.

Only the unbound (free; F) form of a compound can interact with an enzyme. Therefore, the inhibitory effect of NMBA on plasma BuChE is dependent on the ratio of F to its total concentration T in plasma. Since only F of a compound is available for distribution to other tissues the inhibitory effect of NMBA on muscle AChE, indirectly, is also dependent on the value of F in plasma. The relationship between T, F and the bound concentration B depends primarily on binding to plasma proteins (protein binding). In other words, the inhibition of plasma BuChE or muscle AChE after the i.v. administration of a NMBA will be greatly influenced by its protein binding. Consequently, information on the protein binding of a NMBA is essential for the prediction of the inhibitory effect of a given dose of NMBA on plasma BuChE and muscle AChE.

It may be assumed that the protein binding of a compound in an assay system is proportional to the plasma concentration and that, for the same degree of inhibition of BuChE, F of the inhibitor must be constant. On the basis of these two assumptions, determination of the I_{50} of a NMBA for plasma BuChE in the presence of two different plasma concentrations (e.g., 5% and 50%) in the assay system, will make it possible to calculate T in undiluted plasma. It also follows that, in order to be able to compare the inhibitory effect of different NMBA, the plasma concentration of the assay system should be constant and, if possible, the same human plasma pool should be used for all the determinations.

B. Determination of Cholinesterase Activity

AChE and BuChE activity can be determined by a null-point potentiometric titration method using a pH-stat. ACh substrate can be used for both enzymes, or specific substrates, such as acetyl-β-methylcoline (MeCh) for AChE and butyrylcholine (BuCh) or benzoylcholine (BeCh) for BuChE, may be employed. The optimal ACh concentrations are $3 \times 10^{-3}\,M$ for AChE and $2.2 \times 10^{-2}\,M$ for BuChE (FOLDES 1978). The assay of enzyme activity is carried out at pH 7.4 and 37 °C with 0.1 M NaOH titrant. The concentration of plasma or hemolyzed red cells in the system is 5% (v/v). At first, the hydrolysis rate of ACh is determined with active and heat-inactivated enzymes and correction is made for nonenzymatic hydrolysis. The hydrolysis rates are expressed in μmol per milliliter plasma or red cell per hour. The inhibitory effect of NMBA is determined from the log concentration – inhibition curves obtained with five different concentrations of NMBA. To be able to compare the inhibitory effect of NMBA, experimental conditions, such as dilution of plasma or red cell or time of incubation of the inhibitor (NMBA) with the source of enzyme, must be the same. Decreasing the concentration of BuChE in the assay system increases activity when the source of enzyme is plasma of purified human plasma BuChE (FOLDES and SMITH 1966). Because of the binding of NMBA by plasma proteins when the source of BuChE is plasma, decreasing the plasma concentration in the assay system increases and increasing its concentration decreases the inhibitory effect (DEERY et al. 1982).

C. Determination of Protein Binding

The protein binding of a NMBA can be calculated from its I_{50} values determined at two different plasma concentrations (i.e., 5% and 50% or 5% and 90%) in the assay system. Based on calculations described elsewhere (FOLDES et al. 1982) when the I_{50} of a NMBA in an assay system containing X% or Y% plasma is designated as I_{50x} or I_{50y} then the amount of NMBA bound to proteins in undiluted plasma is

$$B = \frac{I_{50x} - I_{50y}}{X - Y} \tag{1}$$

and the concentrations of free (unbound) NMBA, required to produce 50% inhibition in undiluted plasma is

$$F = I_{50} - X \times \frac{I_{50x} - I_{50y}}{X - Y} = I_{50x} - X \times B \tag{2}$$

and the total concentration of NMBA required to cause 50% inhibition in undiluted plasma

$$T = B + F = B + I_{50x} - X \times B \tag{3}$$

and the percentage NMBA bound to proteins

$$P = \frac{B}{T} \times 100 \,. \tag{4}$$

For example, the I_{50} of tubocurarine for the hydrolysis of ACh by BuChE in assay systems containing 90% and 5% plasma were found to be 5.6×10^{-4} and $2.2 \times 10^{-4}\ M$ respectively. Substituting these values into Eq. (1).

$$B = \frac{5.6 \times 10^{-4} - 2.2 \times 10^{-4}\ M}{0.85} = 4.0 \times 10^{-4}\ M \,.$$

Substituting into Eq. (2)

$$F = 5.6 \times 10^{-4} - 0.9 \times 4.0 \times 10^{-4}\ M = 2.0 \times 10^{-4}\ M$$

Substituting into Eq. (3)

$$T = 4.0 \times 10^{-4} + 2.0 \times 10^{-4}\ M = 6.0 \times 10^{-4}\ M$$

Substituting into Eq. (4), the percentage of NMBA bound to proteins

$$P = \frac{4.0}{6.0} \times 100 = 66.7\% \,.$$

This method of the determination of protein binding has the advantages that it is simple, does not require dialysis, and that the concentration of the NMBA remains constant throughout the assay. The applicability of the method has two limitations. One of these is that the NMBA should not cause saturation of all the protein binding sites in concentrations that are lower than its I_{50} in undiluted plasma. Should this be the case, then the calculated percentage of bound NMBA will be lower than the percentage actually bound in the presence of lower, clinical concentrations of the NMBA (see Table 2). This possibility can be eliminated by calculating the protein binding at both the I_{30} and I_{50} levels. If the calculated protein binding, at these two levels of inhibition, are close to one another the data obtained at the I_{50} level may be considered valid. If the calculated binding at the I_{50} level is lower than at the I_{30} level then it must be assumed that lower than I_{50} concentrations of the NMBA caused complete saturation of all binding sites. In the case, protein binding should be calculated at the I_{20} level. If the calculated protein binding data at I_{20} and I_{30} levels are similar, the protein binding can be calculated at the I_{30} level. Otherwise, the method should be considered unsuitable for the calculation of the protein binding of the NMBA in question. The second limitation of the method is that the protein binding data obtained with it are only relevant clinically when the highest plasma concentration of NMBA, obtained immediately after i.v. injection of the initial dose, is lower than the value of I_{50} in undiluted plasma. The ranges of the highest plasma concentration of the 13 NMBA studied were all below their I_{50} values in undiluted plasma. The I_{50} value metocurine, however, was higher than its concentration that caused saturation of all binding sites. Consequently, its protein binding had to be determined at the I_{30} level.

D. Hydrolysis of Neuromuscular Blocking Agents by Cholinesterases

Of the NMBA in clinical use, only SCh is hydrolyzed by BuChE at a rate of about 6 and 4 μM per milliliter plasma per hour in males and females, respectively (FOLDES et al. 1963). Plasmas of most other mammals (TSUJI and FOLDES 1953) hydrolyze SCh more slowly than human plasma. Some mammalian plasmas (i.e., cat plasma), do not hydrolyze SCh at all (FOLDES and FOLDES 1965). There is a close relationship between the intensity and duration of action of ester type NMBA and their rate of hydrolysis by BuChE. Thus, for exampler, the NM potency and duration of action of suxethonium bromide, which is hydrolyzed by BuChE about 60% faster than SCh, is about half of that of SCh (FOLDES et al. 1956). Similarly, the NM potency and duration of action of several ω-amino fatty acid esters of choline, in humans, are inversely related to their enzymatic hydrolysis rate by BuChE (FOLDES and FOLDES 1965). Cat plasma BuChE does not hydrolyze these compounds and consequently in contrast to humans, the NM potency of the various ω-amino fatty acid esters of choline is about the same in cats (FOLDES and FOLDES 1965). Inhibiting the hydrolysis of SCh by BuChE by hexafluorenium bromide (Mylaxen) (FOLDES et al. 1960), or by other compounds that inhibit this enzyme, significantly increases the potency and duration of action of SCh. For example, pancuronium used to prevent rapid depolarization of the endplate by SCh can prolong the duration of action of SCh. In patients whose BuChE is inhibited by the prolonged topical application of eothiophate iodide (Phospholine), the duration of action of SCh may also be prolonged (PANTUCK 1966). The duration of action of conventional doses of SCh may be excessively prolonged in patients who have atypical or other abnormal types (e.g., fluoride-resistant, silent) of BuChE (FOLDES et al. 1963; FOLDES 1978).

E. Inhibition of Human Acetyl- and Butyrylcholinesterase by Neuromuscular Blocking Agents

All 14 NMBA investigated (Table 1) had a variable inhibitory effect on human plasma BuChE and, except for toxiferine chloride, pipecuronium bromide (Arduan), decamethonium, and suxethonium they also inhibited human red cell AChE. With the exception of benzoquinonium chloride (Mytolon) ($I_{50} = 2.2 \times 10^{-7}\ M$), they all were relatively weak inhibitors ($I_{50} > 10^{-5}\ M$) of AChE. The inhibitory effect of NMBA, except for benzoquinonium, was greater for BuChE than for AChE. The BuChE-inhibitory effect of the steroid base inhibitors, pancuronium, vecuronium, Duador (RGH-4201; 3α-pyrrolidino-17α-methyl-17α-aza-D-homo-5α-androstane dimethobromide), and pipecuronim bro-

Table 1. The inhibitory effect of neuromuscular blocking agents on human cholinesterases

Neuromuscular blocking agent	I_{50} (M) for Red cell AChE	I_{50} (M) for Plasma BuChE	(I_{50} AChE/ I_{50} BuChE)
Tubocurarine	7.2×10^{-4}	2.2×10^{-4}	3.3
Metocurine	3.2×10^{-3}	6.1×10^{-4}	5.2
Toxiferine	17% at 10^{-3}	1.3×10^{-5}	
Alcuronium	6.7×10^{-4}	6.4×10^{-5}	10.5
Gallamine	4.5×10^{-4}	4.5×10^{-5}	10.0
Benzoquinonium	2.2×10^{-7}	2.1×10^{-5}	0.01
Pancuronium	3.0×10^{-4}	6.1×10^{-8}	4,916.0
Vecuronium	6.6×10^{-5}	6.9×10^{-7}	95.6
Duador	6.4×10^{-4}	1.5×10^{-6}	426.6
Pipecuronium	4.3% at $\times 10^{-4}$	4.8×10^{-6}	
Atracurium	3.4×10^{-4}	4.2×10^{-4}	2.8
Decamethonium		1.4×10^{-5}	
Succinylcholine	2.4×10^{-3}	6.2×10^{-4}	3.9
Suxethonium		6.4×10^{-4}	

[a] Determined in assay systems containing 5% (v/v) hemolyzed red cell or plasma

mide (Arduan) is the greatest. Pancuronium is one of the most specific inhibitors of BuChE known (FOLDES 1978).

Inhibition of AChE antagonizes the NM effect of nondepolarizing NMBA and facilitates the reversal of the residual NM block at the end of surgery. Because of this, a weak inhibitory effect on AChE is a desirable property of NMBA. Since NMBA are distributed not only to the NM junction, but also to other parts of the extracellular compartment, they may also inhibit AChE at these sites. Inhibition of AChE at the parasympathetic nerve endings in the heart, tracheobronchial tree, salivary glands and gastrointestinal tract leads to the accumulation of ACh and causes unwanted muscarinic side effects, such as bradyarrhythmia, bronchiolar constriction, increased tracheobronchial and salivary secretion and gastrointestinal motility and abdominal cramps. The potent inhibitory effect of benzoquinonium on AChE was the reason for the discontinuation of its clinical use.

Plasma BuChE has no known physiologic function (FOLDES 1978) and its inhibition by nondepolarizing NMBA will only increase the potency and duration of action of ester type NMBA and the systemic toxicity of local anesthetic agents hydrolyzed by BuChE.

F. Binding of Neuromuscular Blocking Agents by Plasma Proteins

Most NMBA are strongly bound to plasma proteins (Table 2), primarily to serum albumin (F.F. Foldes and A. Deery 1981, unpublished work). By adding 4.5% human serum albumin to assay systems containing purified human plasma

Table 2. Binding of neuromuscular blocking agents to plasma proteins

Neuromuscular blocking agent	Range of plasma levels	I_{50} (M) with plasma concentrations of		Protein binding
	(10^{-6} M)[a]	5%	90%	(%)
Tubocurarine	5–10	2.2×10^{-4}	5.6×10^{-4}	66.7
Metocurine	2–5	1.8×10^{-4} [b]	2.6×10^{-4}	37.0
Toxiferine	0.5–1.0	1.3×10^{-5}	5.0×10^{-5}	80.1
Alcuronium	2–6	5.4×10^{-5}	1.9×10^{-4}	77.7
Pancuronium	2–4	6.1×10^{-8}	3.3×10^{-7}	87.5
Vecuronium	2–4	6.9×10^{-7}	4.5×10^{-6}	90.6
Duador	12–16	6.0×10^{-7}	1.3×10^{-6} [c]	79.8
Pipercuronium	2–3	2.2×10^{-6}	4.8×10^{-6} [c]	75.8
Atracurium	6–8	1.2×10^{-4}	3.3×10^{-4} [c]	81.9
Gallamine	30–50	4.5×10^{-4}	1.4×10^{-4}	73.9
Decamethonium	2–4	1.4×10^{-5}	3.0×10^{-5}	59.0
Succinylcholine	30–60	6.2×10^{-4}	2.3×10^{-3}	79.1
Suxethonium	50–100	6.4×10^{-4}	2.7×10^{-3}	82.4

[a] Highest levels immediately after injection of initial dose
[b] Because of relatively weak binding to plasma proteins, binding determined at the I_{30} level. For explanation see text
[c] Protein binding calculated from I_{50} determined in 5% and 50% plasma

BuChE, the I_{50} values of NMBA increase close to their I_{50} values in undiluted plasma. NMBA are only moderately bound to serum globulins.

G. Summary and Conclusions

Of the NMBA in clinical use, only SCh is hydrolyzed by human plasma BuChE. Most NMBA inhibit both human red cell and also muscle AChE and human plasma BuChE. With the exception of benzoquinonium (I_{50} for red cell AChE = 2.2×10^{-7} M), all NMBA have a greater inhibitory effect on BuChE than on AChE. Of all the NMBA, pancuronium is the most potent (I_{50} in assay systems containing 5% plasma = 6.1×10^{-8} M) and selective inhibitor of BuChE. Inhibition of plasma BuChE increases the potency and duration of action of SCh. Except for metocurine (protein binding 37%) 59%–90% of NMBA are bound to plasma proteins, primarily to serum albumin, when administered i.v. in the clinical dose range.

References

Deery A, Foldes FF, Benad G, McCloskey MA (1982) Interaction of neuromuscular blocking agents with human cholinesterases. Anesthesiology 57:A275

Foldes FF (1978) Enzymes of acetylcholine metabolism. In: Foldes FF (ed) Enzymes in anesthesiology. Springer, New York Heidelberg Berlin, pp 91–168

Foldes FF, Foldes VM (1965) ω-Amino fatty acid esters of choline: interaction with cholinesterases and neuromuscular activity in man. J Pharmcol Exp Ther 150:220–230

Foldes FF, Smith JC (1966) The interaction of human cholinesterases with anticholinesterases used in the therapy of myasthenia gravis. Ann NY Acad Sci 135:287–301
Foldes FF, Swerdlow M, Lipschitz E, van Hees GR, Shanor SP (1956) Comparison of the respiratory effects of suxamethonium and suxethonium in man. Anesthesiology 17:559–568
Foldes FF, Molloy RE, Zsigmond EK, Zwartz JA (1960) Hexafluorenium: its anticholinesterase and neuromuscular activity. J Pharmacol Exp Ther 129:400–404
Foldes FF, Zsigmond EK, Foldes VM, Erdos EG (1962) The distribution of acetylcholinesterase and butyrylcholinesterase in the human brain. J Neurochem 9:559–572
Foldes FF, Foldes VM, Smith JC, Zsigmond EK (1963) The relation between plasma cholinesterase and prolonged apnea caused by succinylcholine. Anesthesiology 24:208–216
Pantuck EJ (1966) Ecothiopate iodide eye drops and prolonged response to suxamethonium. Br J Anaesth 38:406–407
Smith JC, Foldes VM, Foldes FF (1963) Distribution of cholinesterase in normal human muscle. Can J Biochem Physiol 41:1713–1720
Tsuji FJ, Foldes FF (1953) Hydrolysis of succinylcholine in human plasma. Fed Proc 12:374

CHAPTER 9

On the Effect of Neuromuscular Blocking Agents on the Central Nervous System

D. A. KHARKEVICH

A. Introduction

Although investigations have been numerous, for the present, there is no universal opinion on the effect of neuromuscular blocking agents on the functions of the central nervous system. This is largely due to the fact that, according to clinical data, neuromuscular blocking agents on systemic administration do not induce marked central action. Therefore, the apparent effects, if any, are rather difficult to detect during clinical evaluation of neuromuscular blocking agents. Thus, while testing these drugs during operations, it should be taken into account that the patient receives many other drugs for premedication and anesthesia. This creates a certain background that may influence or mask the central action of a neuromuscular blocking agent. The interpretation of the data obtained is significantly hindered by different degrees of trauma involved in surgical interventions, total muscle paralysis, changes of the functional state of the visceral organs and systems, disturbances of water and electrolyte balance, the volume of the blood loss, and many other factors which can modify the function of the central nervous system. Certain difficulties are involved in the evaluation of experimental data, too. These are related to the ability of neuromuscular blocking agents to induce not only direct, but also indirect central action. Thus, for instance, total muscle paralysis induced by neuromuscular blocking agents changes the afferent proprioceptive firing, which significantly affects the activity of cerebral neurons. For example, EEG desynchronization occurring 30–60 min after the administration of neuromuscular blocking agents to conscious rabbits in high doses is regarded by BOVET and LONGO (1953) as a peripheral effect resulting from changed intensity of afferent impulses.

In addition, it is known that depolarizing agents evoke intense afferent firing from the muscle spindles which is followed by the inhibition of monosynaptic spinal reflexes (FUJIMORI and ELDRED 1961; MURATOV 1962). Proprioceptive impulses also reach the cerebral cortex, cerebellum, and hypothalamus. It should be also taken into account that immobilization is a stress for conscious animals, which is also likely to affect the bioelectrical activity of the brain (SAAVEDRA et al. 1979).

ALI (1981) while discussing probable causes of central effects of neuromuscular blocking agents, suggests that they are indirect and related to the agents' ability to affect afferent impulses from skeletal muscles and thus to change the activity of the activating brain stem reticular formation. To strengthen his suggestion, the author gives, in particular, the following data. It was noticed that the administration of gallamine to cats in paralyzing doses is followed by EEG synchronization, analogous to that observed after section of the spinal cord or its block by procaine (OSTOW and GARCIA 1949). Subparalytic doses of decamethonium promote the attenuation of conditioned reflexes in animals (see Sect. B.I). In humans, depolarizing agents induce muscle fasciculations and arousal response in the EEG (MORI et al. 1973). On the other hand, on the basis of an autoexperiment by Smith (SMITH et al. 1947) and his own data, Ali points out that during the administration of nondepolarizing drugs in subparalytic doses to humans, drowsiness and inebriation-like sensations are observed. It is also mentioned that patients under neuroleptanalgesia are often awakened when the neuromuscular block is eliminated and normal muscle tone is restored by quaternary anticholinesterase agents. At the same time, it is also reported that pancuronium decreases the halothane need (see Sect. B.VII). All this, according to Ali, may be due to the peripheral effect of neuromuscular blocking agents at the muscle spindle level.

The adequacy of artificial ventilation is also of great importance for the study of the influence of probable central action of neuromuscular blocking agents. Artificial ventilation can lead to the development of hypoxia, hypercapnia, or hypocapnia of various degrees. These factors per se can modify the bioelectrical activity of the brain (GIRDEN 1948). Thus, for instance, at 5%–50% CO_2 in the inspired air, the EEG amplitude of the rabbit is decreased after 0.5–3 min (REPIN 1960). Inspiration of air containing 8.9% CO_2 is followed by a 60% decrease of amplitude of monosynaptic potentials of the spinal cord. Polysynaptic potentials under these conditions are decreased by 20% (KIRSTEIN 1951). Analogous data were obtained by ESPLIN and ROSENSTEIN (1963).

Hyperventilation can also affect the spontaneous and evoked bioelectrical activity. According to OLIVER and FUNDERBURK (1965), artificial hyperventilation of cats evoked changes of EEG and markedly reduced cerebral blood flow. Facilitation of monosynaptic and polysynaptic spinal reflexes is observed, too (KITAHATA et al. 1969). The principles of acid–base equilibrium during long-lasting artificial ventilation evoked by neuromuscular blocking agents are discussed by GALLETTI et al. (1979).

It is also very important to consider the significance of the arterial pressure. Thus, when its level is substantially decreased, disturbance of interneuronal transmission is observed, which, in particular, results in changes of evoked and spontaneous potentials. The latter should be taken into account in view of central effects of neuromuscular blocking agents causing marked hypotension (e.g., tubocurarine (OCHS 1959). Moreover, hypotension, especially the acute type, results in alterations of afferent impulses from vascular baroreceptors. Naturally, the state of cerebral blood flow which can be changed by neuromuscular blocking agents is also of great importance for the functional state of the central nervous system (OLIVER and FUNDERBURK 1965). In brief, there is quite a number of probable indirect effects of neuromuscular blocking agents at various levels of the brain. Environmental temperature is also of great significance. For instance, OKUMA et al. (1965) pointed out its influence on electrocortical activity of immobilized cats.

The main problem is the possibility of penetration of neuromuscular blocking agents through blood–brain barrier. While for secondary and tertiary amine salts it is certain that they penetrate into the brain on systemic administration, for quaternary ammonium compounds it is well known that they penetrate tissue barriers rather poorly. A number of authors did not observe neuromuscular blocking agents to penetrate the blood–brain barrier. Thus, COHEN (1963) did not detect tubocurarine in the cerebrospinal fluid after its intravenous administration at doses of 0.3–3 mg/kg to anesthetized cats and three patients (a relatively insensitive fluorimetric method was used). DAL SANTO (1972) showed, using a radiochromatographic technique, that gallamine ^{14}C did not cross the blood–brain barrier in anesthetized dogs. According to WASER (1973) pancuronium ^{14}C intravenously administrered to mice did not essentially penetrate into the brain or spinal cord, either (histo- and whole body autoradiographic methods were used). However, a number of papers report that several neuromuscular blocking agents of the group of quaternary ammonium salts can penetrate the blood–brain barrier. Thus, MAHFOUZ (1949) using a bioassay method on frog rectus abdominis muscle after intravenous administration of tubocurarine to patients at a dose of 0.2 mg/kg, detected 2.5 µg/ml in the cerebrospinal fluid. DEVASANKARIA et al. (1973) using the same technique, also found neuromuscular blocking activity of the cerebrospinal fluid of dogs and humans after intravenous administration of tubocurarine, although in lower quantities. In anesthetized patients after the administration of 30 mg tubocurarine, it was detected in the cerebrospinal fluid in concentrations of 0.05–0.33 µg/ml. In dogs during intravenous administration of 0.3–3 mg/kg tubocurarine, perfusion from lateral ventricle to cisterna was performed. The agent could be identified within the range of 20–60 ng/min. The ability of tubocurarine (0.3 mg/kg i.v.) to pass through the blood–brain barrier was also confirmed by MATTEO et al. (1977), who, using a radioimmunoassay, could detect tubocurarine in the cerebrospinal fluid of humans in low amounts (25 ng/ml at 6 h). This was associated with marked individual variations of the amount of tubocurarine related to the character of disease and the intensity of the blood flow. In anesthetized dogs, DAL SANTO (1964) found that after intravenous injection of a trace of dimethyl-(+)-tubocurarine ^{14}C, of the order of 10^{-5} of the injected amount could be detected in cisternal cerebrospinal fluid (by radiochromatographic scans). Thus, in a number of papers it was found that some neuromuscular blocking agents could pass through the blood–brain barrier, although in small amounts.

It should be also taken into account that the permeability of the blood–brain barrier is unequal in different regions of the central nervous system. BOWMAN and RAND (1980) point out the following regions in which the blood–brain barrier seems to be ineffective: pineal gland, posterior pituitary gland and adjacent part of the hypothalamus, as well as area postrema, supraoptic crest, and intercolumnar tubercle. From this publication and a number of others (BAKAY 1954, 1957; BRATTGARD and LINDQVIST 1955; BRADBURY 1979; and others), it follows that within the blood–brain barrier there are regions with increased permeability. This seems to hold for quaternary ammonium compounds, too.

It is also known that the permeability of the blood–brain barrier is changed during several homeostatic alterations, a number of pathologic states, infections, intoxications, etc. Thus, for instance, it is increased under fever, hypoxia, brain edema, brain edema and anoxia, closed craniocerebral trauma, and X-irradiation of the head (ZAIKO 1962; BAKAY 1968). While studying the distribution of dimethyl-(+)-tubocurarine ^{14}C in the tissues of the body, it was shown that prolonged hypoxia, hypercapnia, circulatory shock, and hypothermia increased the penetrability of the blood–brain barrier for the drug tested (DAL SANTO 1964). However, in all cases, the concentration of the drug in the cerebrospinal fluid was comparatively low.

Thus, it cannot be ruled out that neuromuscular blocking agents of the group of quaternary ammonium compounds can traverse the blood–brain barrier in measurable amounts. If they pass into the cerebral tissues, the substrate with which they interact, i.e., acetylcholine receptors, is rather well represented in the central nervous system. The direct central action of neuromuscular blocking agents is especially marked when they are administered into the brain past the blood–brain barrier. This can be achieved by intraventricular administration and also by local and microiontophoretic application of neuromuscular blocking agents to certain groups of neurons, as well as by revealing their antagonism against cholinomimetics. However, it cannot be excluded that noncholinergic systems can also be the site of action of neuromuscular blocking agents in the brain. In particular, this is supported by the findings of BELESLIN and SAMARDZIC (1976) who studied the causes of convulsive action of tubocurarine on its intraventricu-

Table 1. Neuromuscular blocking activity

Drugs	Neuromuscular block (sciatic nerve – gastronomic muscle) in cats (approximate ED_{95} µg/kg i.v.)	Head drop in rabbits (µg/kg i.v.)
Tubocuranine	180–230	120–150
Metocurine	25–30	
Toxiferine	5.4	4
Pancuronium	24 (8[a])	
Pipecuronium	2[a]	
Vecuronium	38	
Truxilonium	150–180	30–40
Cyclobutonium	130–180	32–43
Anatruxonium	100–130	17–27
Dipyronium	18–20	10.8
Gallamine	850–900	200–250
Diplacinum	1,800	520
Paramionum[b]	200–300	100–150
Diadonium	250–350	130–180
Decadonium	250–300	100–110
Succinylcholine	60–80	120
Decamethonium	30	78
Pyrocurinum	120–130	26–30
Mellictinum	2,000	2,000
Dihydro-β-erythroidine		1,500

[a] ED_{50}, tibialis anterior muscle
[b] Paramionum, *meso*-3,4-bis(p-dimethylaminophenyl)-hexane bismethyl-iodide
[c] See also Table 1 in Chap. 7

lar administration and showed nicotinic blocking agents to be ineffective in this respect. However, antimuscarinic, ganglion-blocking agents, α- and β-adrenoceptor-blocking agents, serotonin and H_1-antagonists, and some antiepileptic agents (phenobarbital sodium, trimethadione) did not abolish tubocurarine-induced convulsions, either. According to TSOUKARIS-KUPFER et al. (1980; 1981), the convulsions occurring during intracisternal administration of tubocurarine to dogs are abolished by diazepam.

In this chapter, the data dealing with the central action of neuromuscular blocking agents that can appear in animals and humans under conventional (mainly, intravenous) administration of the agents are discussed. The practical value of the data obtained can be estimated by the ratio of neuromuscular blocking doses (Table 1) and those causing central effects.

B. Intravascular Administration of Neuromuscular Blocking Agents

I. The Effect on Conditioned Reflexes

To study central effects of neuromuscular blocking agents, several authors have used the method of conditioned reflexes. FELDMAN (1960), for instance, established that decamethonium in subparalytic doses (0.4–0.8 mg/kg intraperitoneally) abolished certain conditioned reflexes (visual, auditory, tactile) in rats. Similar results were gained by BRISKIN and FLEROV (1961) in experiments on rats by the method of motor food-conditioned reflexes. It was found that after subcutaneous administration of subparalytic doses of diplacinum (2–5 mg/kg) and tubocurarine (0.1 mg/kg), conditioned reflexes in animals were decreased and in some cases they were even abolished. This was not associated with any noticeable behavioral changes. Discussing the results obtained, both groups of authors tend to think that the inhibition of conditioned reflexes induced by neuromuscular blocking agents is probably unrelated to their direct central action, but is mediated by the peripheral effects of the neuromuscular blocking agents (changed intensity of afferent proprioceptive impulses, partial disturbance of neuromuscular transmission).

II. The Effect on the Electroencephalogram

Numerous investigators failed to observe any EEG changes on intravenous administration of neuromuscular blocking agents to warm-blooded animals or humans (SMITH et al. 1947; GIRDEN 1948; EVERETT 1947, JONES et al. 1956; CHARDON 1956; ACHESON et al. 1956; DANILOV et al. 1957; MIKHELSON 1957; DYACHENKO 1961; SKOROBOGATOV 1961; AMIROV et al. 1962; and others).

BOVET and LONGO (1953) observed EEG desynchronization in some anesthetized rabbits only 30–60 min after intravenous administration of tubocurarine, gallamine, succinylcholine, and decamethonium in doses 10–100 times higher than those causing respiratory arrest. It was shown that neither decreased arterial pressure or body temperature, nor anoxia were responsible for the observed EEG changes. The authors believe that EEG changes are related not to the direct action of neuromuscular blocking agents on the cerebral cortex, but are mediated by the changed afferent discharges from the muscle spindles. This suggestion is well grounded, since afferent impulses reaching the central nervous system are known to be of great importance for the EEG pattern. An analogous supposition concerning the influence of muscle relaxation on the brain function was also made by GELLHORN (1958 a, b) and other authors. According to MCINTYRE et al. (1946), in anesthetized dogs EEG inhibition (with a preceding transient phase of activation) was observed after intravenous administration of tubocurarine. It was shown that the EEG depression in the frontal areas occurred before peripheral paralysis.

It was demonstrated that in low doses diplacinum (0.75 mg/kg i.v.), and tubocurarine (0.15 mg/kg i.v.) had a marked inhibitory effect on EEG of conscious rabbits. Changes of bioelectrical activity were observed not only in the cortex, but also in the subcortical structures: superior colliculi, lateral geniculate body, reticular formation of brain stem and thalamus. These changes have a biphasic pattern: a short-term phase of increased bioelectrical activity is followed by a prolonged inhibitory phase (BRISKIN 1961).

In experiments on freely moving rats, systemic injection of tubocurarine changes the depth profile of the θ rhythm in the hippocampus, which lasts longer than its paralytic effect (WINSON 1976). In experiments on waking rabbits, LONGO (1955) showed that several neuromuscular blocking agents on intracarotid injection caused an arousal reaction on the EEG (decamethonium 2–5 μg; succinylcholine 3–7 μg; tubocurarine 10–20 μg). Gallamine turned out to be ineffective in this respect.

In the paper by WILD (1981), the effect of neuromuscular blocking agents on the central nervous system was studied in experiments on cats under general anesthesia and on *encéphale isolé* preparations. A separate group of experiments was carried out on animals after a series of electric shocks which led to the disturbance of blood–brain barrier permeability. EEG and evoked potentials were recorded on stimulation of the sciatic nerve and photostimulation. The agents were administered intravenously. Pancuronium and succinylcholine failed to affect the EEG under general anesthesia. Gallamine (20 mg/kg) modified both the cortical and subcortical EEG. On the *encéphale isolé* preparation, succinylcholine (1–5 mg/kg) caused an effect on the spontaneous EEG similar to an arousal reaction. Pancuronium did not produce this effect. No changes of evoked potentials related to the drugs tested were observed. After a series of electric shocks, pancuronium (0.1–0.4 mg/kg), succinylcholine (1–2 mg/kg), and gallamine (1–2 mg/kg) induced an arousal reaction and significant changes in EEG intensity with α-rhythm peaking in cortical and hippocampal structures. On the other hand, tubocurarine (1 mg/kg) decreased the EEG amplitude and frequency. The authors suggest that the effects observed are related to the action of the agents tested on the activity of the mesencephalic reticular formation.

According to KNECHT et al. (1980), intravenous administration of succinylcholine (0.125 mg/kg) to conscious dogs was followed by an approximately fivefold decrease of the amplitude of the fast EEG activity. The occurrence of slow waves (6–8 Hz) was more marked. In experiments on conscious dogs, SAAKOV

(1957a, b) studied the action of succinylcholine on the EEG and the response of auditory cortex neurons to sound stimuli of increasing intensity. He found that intramuscular succinylcholine at 0.8–1.0 mg/kg after 10–15 min caused the suppression of the dominant EEG rhythm in all the cortical areas observed (frontal, parietal, temporal, and occipital) and decreased the intensity of responses to the sound stimuli. The degree of the changes was directly related to the succinylcholine dose applied. MORI et al. (1973), in anesthetized patients, have seen an arousal response in the EEG evoked by succinylcholine, carbolonium, and decamethonium (i.v.). Gallamine and alcuronium produced no arousal response, but prevented it in response to a subsequent injection of succinylcholine.

Thus, estimations of the central action of neuromuscular blocking agents according to their effects on the EEG are rather controversial. This is no wonder, since numerous factors are involved. If the neuromuscular blocking agents are tested during an operation, the EEG depends on the general anesthetics and other drugs applied, the character of surgical intervention, the state of visceral organs, and hemeostasis as a whole. When they are tested on nonanesthetized volunteers, the mental state of the person, the degree of muscle relaxation, etc., may interfere. Naturally, such a variety of conditions inevitably results in dissimilar findings. To some extent, this is relevant for the experiments on animals, too. Moreover, slight changes in the functional state of some structures of the brain, if any, might not show up in the EEG, since the latter reflects total bioelectrical activity of large areas of the cortex and subcortical structures. Furthermore, they can be masked by EEG changes unrelated to the central action of neuromuscular blocking agents.

Sometimes, while studying the central action of neuromuscular blocking agents, the spontaneous activity of cortical neurons is recorded. For instance, MAIER et al. (1970) studied the influence of gallamine (1 mg/kg) and succinylcholine (2 mg/kg) on the spontaneous activity of cortical neurons in conscious rabbits. They showed that in the majority of neurons the activity was changed, predominantly toward depressed activity. This can be due to both direct action of the agents on the cortex and the changed afferent proprioceptive discharges related to the immobilization of the animals.

Probable central action of neuromuscular blocking agents is often tested by their influence on the effects of nicotinomimetics. For example, ALYAUTDIN (1978) studied a number of neuromuscular blocking agents on their intravenous administration to gallamine-immobilized cats. To obtain seizure discharges, corconium (subecholinum) was mainly used. Corconium (corconic acid β-dimethylaminoethyl ester diiodomethylate), a short-acting nicotinic stimulant drug, is a bisquaternary ammonium salt not penetrating the blood–brain barrier. Administered in the lateral ventricle, corconium (1 mg) caused high amplitude seizure discharges and high frequency, low amplitude waves which were recorded over 15–20 min. After repeated corconium administration (60–90 min later), the same EEG changes were observed. Neuromuscular blocking agents were administered usually prior to corconium. In some of the experiments, the agents were administered after the effects of corconium had developed. The interaction of neuromuscular blocking agents with nicotinic acetylcholine receptors of the brain was estimated by the variations of the duration of seizures discharges.

Intravenous pyrocurinum (0.65 mg/kg) prevented and eliminated high frequency, low amplitude waves evoked by corconium, but it affected the seizure discharges less. Mellictinum (3.5 mg/kg) eliminated all the effects of corconium. Diadonium (0.36–1.65 mg/kg), decadonium (0.3–1.5 mg/kg), dipyronium (0.02–0.1 mg/kg), pipecuronium (0.02–0.1 mg/kg), pancuronium (0.02–0.1 mg/kg), and paramionum (0.2–1 mg/kg) failed to affect the duration of corconium-induced seizure discharges.

Thus, of the agents tested, only mellictinum and pyrocurinum, tertiary ammonium compounds, when injected intravenously, eliminated or decreased corconium-induced EEG changes. The observed antagonism with corconium is likely related to the blocking action of mellictinum and pyrocurinum on brain nicotinic receptors. Quaternary ammonium compounds tested did not change the effects of corconium on their intravenous administration. On the other hand, injected intraventricularly, they all were observed to have anticonvulsant activity, indicating their probable interaction with brain nicotinic receptors.

III. The Effect on Interneuronal Transmission in the Afferent Pathways

The effect of neuromuscular blocking agents on the transmission in the afferent pathways was mainly studied during stimulation of somatic and visceral nerves and also during light and sound stimulation. In a number of papers, no direct central action of neuromuscular blocking agents was observed. Thus, for instance, while studying the primary cortical response evoked by the stimulation of the sciatic nerve, SKOROBOGATOV (1963) did not find any changes after intravenous injection of tubocurarine (0.5 mg/kg), diplacinum (4–6 mg/kg), succinylcholine (0.12 mg/kg), decamethonium (0.06 mg/kg), or paramionum (1 mg/kg), either. These doses are equal to, or somewhat higher than the neuromuscular blocking doses. After administration of succinylcholine (1 mg/kg), decamethonium (0.45–0.6 mg/kg), or paramionum (0.5–1 mg/kg), there were observed no changes of evoked potentials in specific or nonspecific cortical areas of anesthetized cats under sound or light stimulation. Succinylcholine had no effect on the potentials evoked by the stimulation of the sciatic nerve, either (SINITSYN and KHARKEVICH 1967).

ALYAUTDIN (1978) studied the action of bisquaternary compounds diadonium, decadonium, dipyronium, pipecuronium, and pancuronium, and also the tertiary amine salts pyrocurinum and mellictinum on the central links of the afferent system in anesthetized cats. The responses to the stimulation of the sciatic nerve were recorded in somatosensory area 1 and in the associative, acoustic, and visual cortical areas. During the stimulation of the vagus nerve, the potentials in area 1 were recorded (between the end of the coronary sulcus and the anterior rhinal fissure), and in area 2 (in the posterior part of the orbital gyrus) (CHERNIGOVSKY and ZARAISKAYA 1962), and also in the associative, acoustic, and visual areas of the cerebral cortex. The action of neuromuscular blocking agents on single evoked responses and also on the responses to paired stimuli were studied. It follows from the data obtained that the neuromuscular blocking agents mentioned, in doses causing total muscle paralysis and higher, do not affect the responses recorded to the stimulation of the sciatic and vagus nerves. They could not change the latency of the evoked potentials or the recovery cycle of testing responses, either.

LÖSSNER et al. (1970) did not observe gallamine (1–10 mg/kg i.v.) to affect cortical potentials evoked by the stimulation of the sciatic nerve, by light or sound stimuli (nonanesthetized rabbits and cats). Gallamine and succinylcholine at doses of 2 mg/kg i.v. did not cause a statistically significant difference in the responses of visual cortical neurons to light stimulation in conscious rabbits (KAMMERER and KRUG 1972). DYACHENKO (1961), in experiments on cats, did not observe tubocurarine, diplacinum, or succinylcholine to modify the evoked potentials of the neurons of the reticular formation of the brain stem and hypothalamus evoked by the stimulation of the somatic nerves.

After intravenous administration of succinylcholine (2 mg/kg) and gallamine (2 mg/kg) to nonanesthetized cats with implanted electrodes, no changes were observed in the responses of the cortex (posterior sigmoid gyri), thalamus (posteroventral lateral nucleus), brain stem reticular formation, or lateral geniculate body to auditory, visual, or painful stimuli (MUNROE et al. 1966). DAVID et al. (1963), using the techniques of extracellular registration of potentials, did not observe any changes of discharges of the neurons of the dorsal nucleus of the lateral geniculate body in response to orthodromic stimulation of the optic nerve and antidromic stimulation of the optic radiation after tubocurarine was injected into the carotid artery (50–500 μg/kg). In adult subjects with normal hearing, it was shown that succinylcholine in paralyzing doses did not change the auditory evoked potentials in the cortex (HARKER et al. 1977).

At the same time, there are several publications which report more or less marked central action of neuromuscular blocking agents. Thus, for instance, according to SHEA et al. (1954), EEG changes usually occurring in patients in response to light stimulation were absent after tubocurarine administration. This suggested that tubocurarine blocks the transmission along the optic pathway.

Central action of neuromuscular blocking agents was also found while studying their effects on evoked potentials appearing in the cerebral cortex under electrical stimulation of the central ends of the vagus and inferior cardiac nerves (SINITSYN and KHARKEVICH 1967). The experiments were carried out on anesthetized cats. Evoked potentials were recorded from the lateral surface of the anterior areas of the contralateral and ipselateral hemispheres (in the coronary, orbital, anterior suprasylvian, lateral, and posterior sigmoid gyri). Nondepolarizing and depolarizing neuromuscular blocking agents were tested: tubocurarine, gallamine, paramionum, truxilonium (see the structure in Chap. 19, Table 1), succinylcholine, and decamethonium. The agents were injected intravenously.

The data obtained indicate that some neuromuscular blocking agents have a marked inhibitory action on the evoked potentials. A marked depressing effect was observed after tubocurarine administration. It was shown that at doses of 0.5–1 mg/kg tubocurarine had an inhibitory effect on the transmission in the afferent pathways from the visceral organs. Cortical potentials evoked by the stimulation of the inferior cardiac and vagus nerves after tubocurarine administration had a smaller amplitude, their latent period increasing. The level of arterial pres-

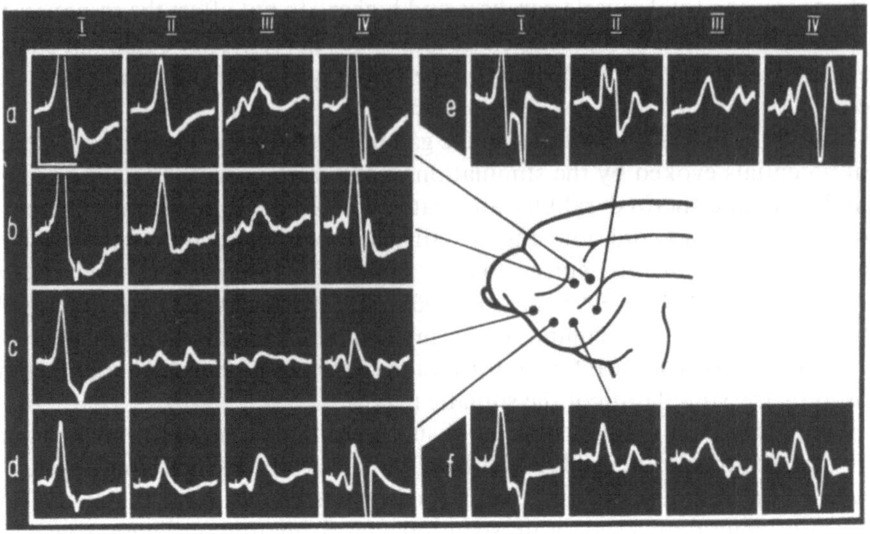

Fig. 1. The effect of succinylcholine on evoked potentials of the cerebral cortex under the stimulation of the inferior cardiac nerve. *a–f* positions of recording electrodes. *I*, before drug administration; *II*, 8 min after the first administration of succinylcholine (0.5 mg/kg i.v.); *III*, 5 min after the second administration (0.5 mg/kg, i.v.); *IV*, 1 hr after the second administration. Vertical line – amplitude (150 μV), horizontal line – time (200 msec). The nerve was stimulated by rectangular stimuli 1 msec, 7 V, 1 st per 3 sec. Experiment on anaesthetized cat (urethane 300 mg/kg and chloralose 60 mg/kg, i.v.)

sure after administration of tubocurarine in these doses varied, as a rule, within a relatively small range and, therefore, could not be responsible for the changes of the evoked potentials. Paramionum at doses of 0.5–1 mg/kg had a significant inhibitory effect on the cortical evoked potentials, too. Of the other nondepolarizing neuromuscular blocking agents, gallamine and truxilonium (0.2–2 mg/kg) either had no effect on the amplitude of the potentials or caused only insignificant changes.

Depolarizing neuromuscular blocking agents turned out quite effective. For instance, succinylcholine at doses of 0.5–1 mg/kg caused a marked decrease of the amplitude of the evoked potentials lasting about 50 min (Fig. 1). Decamethonium also had a marked activity. At doses of 0.1–0.2 mg/kg, it substantially decreased the amplitude of potentials and increased their latency. In the majority of tests, the tendency to the recovery of the potentials developed at 40–90 min. No drug-related changes of the arterial pressure were observed after succinylcholine or decamethonium administration (sometimes it was somewhat elevated).

The available data indicate that group I of the afferent fibers of stretch receptors of the anterior limb muscles of the cat is projected into the rostral part of the posterior sigmoid gyrus (OSCARSSON and ROSEN 1963, 1966). As already mentioned, in this cortical area the responses to the stimulation of both the inferior cardiac and the vagus nerves, are recorded. At the same time, it was also established (GRANIT et al. 1953) that depolarizing neuromuscular blocking agents, while activating the muscle spindles, lead to the inhibition of monosynaptic responses of the spinal cord (FUJIMORI et al. 1959; HENATSCH et al. 1959; SMITH 1963; MURATOV 1963a). It may be assumed that succinylcholine and decamethonium inhibit the potentials appearing in the cortex on stimulation of the inferior cardiac and the vagus nerves owing to the increased frequency of afferent impulses from muscle spindle receptors. However, there is experimental evidence against such interaction of the impulses from intero- and proprioceptors. This is confirmed by the following. At 1–3 min after succinylcholine administration (0.5–1 mg/kg) the discharges in the afferent fibers of the muscle spindles reach their maximum and after 5–10 min they are no longer recorded (GRANIT et al. 1953). In experiments with stimulation of the visceral nerves, complete block of the evoked potentials only occurred 5–10 min after succinylcholine administration and lasted 30–60 min. Furthermore, it was found that succinylcholine and decamethonium inhibited cortical responses to the stimulation of the vagus and the inferior cardiac nerves after gallamine premedication which decreases the excitation of the muscle spindles induced by depolarizing neuromuscular agents. Evoked potentials were also inhibited by the nondepolarizing neuromuscular blocking agents, tubocurarine and paramionum, which failed to increase the frequency of discharges in the afferent fibers of the muscle spindles. In addition, the absence of a marked depressive effect after gallamine or truxilonium administration suggests that total muscle relaxation is not responsible either for the inhibition of responses to the stimultion of the inferior cardiac and vagus nerves following administration of decamethonium, succinylcholine, tubocurarine, or paramionum. The data obtained could be regarded as evidence for a probable direct inhibitory action of neuromuscular blocking agents on the interneuronal transmission in the central nervous system.

OSTOW and GARCIA (1949) studied the effect of neuromuscular blocking agents on potentials on the sensorimotor area of the cerebral cortex evoked by the stimulation at the thoracic level of the spinal cord and the branches of the brachial plexus. They showed in anesthetized cats that intravenous administration of tubocurarine in doses 10–15 times as high as neuromuscular blocking doses was followed by a significant decrease of the amplitude of the evoked cortical potentials. However, it is known that the decrease of the amplitude of evoked cortical potentials may be due to the hypotensive action of tubocurarine (OCHS 1959).

The inhibitory action of neuromuscular blocking agents on the evoked cortical response was observed by LÖSSNER et al. (1970). Starting from doses which did not cause immobilization of the animals, gallamine (0.4–2 mg/kg), tubocurarine (0.05–0.2 mg/kg), succinylcholine (0.5–5 mg/kg), and decamethonium (0.1–3 mg/kg) injected intravenously to conscious cats and rabbits induced a suppression of cortical responses to the stimulation of the dental pulp. Intracarotid administration of the agents was followed by the same effect, but the inhibition developed after lower doses. Simultaneous registration of responses in the cortex, thalamus, and the sensory nucleus of the trigeminal nerve showed that neuromuscular agents affected the sensory nucleus of the trigeminal nerve.

PURPURA and GRUNDFEST (1956) while studying synaptic mechanisms in the cat cerebral cortex have shown that tubocurarine (3 mg/kg i.v.) blocks dendritic responses (evoked by direct, subcortical, or antidromic stimulation) and responses of the optic cortex to stimulation of the optic radiation. ANDY and AKERT (1955) observed increased duration of afterdischarges in the cerebral cortex induced by hippocampal stimulation after intravenous administration of tubocurarine in minimal neuromuscular blocking doses.

Intravenous gallamine (6.25 mg/kg) increased the duration of afterdischarges in the cat cortex during the stimulation of the adjacent cortical area by single stimuli or series of electrical stimuli (HALPERN and BLACK 1967, 1968). Gallamine also increased the duration of afterdischarges in cortical slices and on *encéphale isolé* preparations. On the basis of these observations, the authors suggested that gallamine had a central action.

However STRAW (1968) starting from the findings gained in analogous experiments arrived at a different conclusion. In cats, afterdischarges were recorded in the cerebral cortex during the stimulation of the medial ectosylvian gyrus. In gallamine-immobilized animals (5 mg/kg i.v.) the duration of afterdischarges was significantly longer than in the nonparalyzed state. An analogous increase in duration of afterdischarges was observed following spinal cord transection. Taking into account that on the *encéphale isolé* preparation gallamine failed to produce such an effect, the author concludes that the action of gallamine has a spinal or peripheral origin. Decamethonium (1 mg/kg i.v.) failed to affect the duration of afterdischarges.

DAVID et al. (1963) report that dihydro-β-erythroidine (100 µg/kg i.a. and higher) causes a significant decrease of the postsynaptic component of evoked potentials in the dorsal nucleus of the lateral geniculate body of cats during stimulation of the optic nerve. HELLNER and BAUMGARTEN (1961), using microelectrodes in anesthetized cats, studied the representation of afferent fibers of the vagus nerves in the sensory nucleus of the vagus in the medulla. They observed that succinylcholine (2 mg per animal i.v.) inhibited the discharges of the neurons of this region evoked by the stimulation of the central stump of the cervical vagus nerve.

Thus, there is evidence for the possible action of neuromuscular blocking agents on interneuronal transmission in some afferent pathways.

IV. The Effect on the Brain Stem

Several neuromuscular blocking agents have been shown to inhibit the linguomandibular reflex whose centers are located in the pons varolii and medulla oblongata (MURATOV 1963). This was tested for tubocurarine, paramionum, diplacinum, truxilonium, succinylcholine, and decamethonium. The agents were administered intravenously. Of the agents tested, only truxilonium had a marked action on the amplitude and latency of single discharges in the efferent part of the linguomandibular reflex induced by electrical stimulation of the lingual nerve. After administration of truxilonium (0.3–0.5 mg/kg) the amplitude of reflex discharges evoked by single supramaximal stimuli was decreased by 20%–25%. The latency after administration of 0.5–0.8 mg/kg truxilonium was increased by 1–1.5 ms.

The ability of neuromuscular blocking agents to change the lability (according to Wedensky) of the centers of the linguomandibular reflex was used as one of the criteria of their effect. Under rhythmic supramaximal stimulation of the lingual nerve, marked changes of the lability occurred only under truxilonium. After administration of 0.2–0.4 mg/kg truxilonium, the amplitude of some discharges was decreased. Impulse transformation occurred, especially marked at subpessimal frequency of stimulation (40–60 stimuli per second), and a shift of transformation threshold toward lower frequencies (decrease of maximal frequency at which stimuli produced a 1 : 1 response).

The application of series of submaximal stimuli to the lingual nerve allowed one to detect drug-related lability changes for other drugs tested. Under such experimental conditions, truxilonium also was the one to inhibit the lability of the centers of the linguomandibular reflex most markedly. Truxilonium administration at doses of 0.2–0.3 mg/kg usually resulted in an abrupt shift of the impulse transformation threshold toward lower frequencies. After it was administered at doses of 0.4–0.6 mg/kg at 20 stimuli per second, only single potentials of significantly smaller amplitude were recorded (Fig. 2).

Other nondepolarizing agents (tubocurarine, diplacinum, and paramionum) had a much less marked effect on the lability of the linguomandibular reflex. Thus, for instance, impulse transformation was only observed after high doses of diplacinum (8–10 mg/kg) and only at high frequencies of stimulation (60–80 stimuli per second). Tubocurarine and paramionum caused similar changes of lability at doses of 0.8–1 mg/kg and 6–8 mg/kg, respectively.

The depolarizing agents, decamethonium and succinylcholine, tested under rhythmic submaximal stimulation of the lingual nerve, had a marked inhibitory

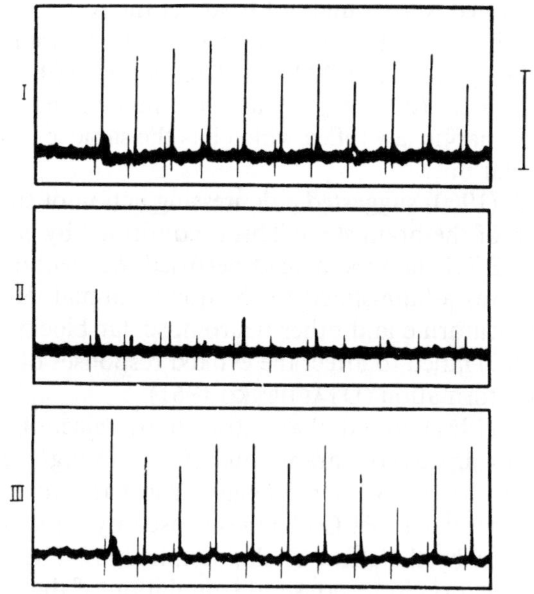

Fig. 2. The effect of truxilonium on the linguomandibular reflex. Evoked potentials of the motor branch of the inferior mandibular nerve: *I*, before drug administration; *II*, 3 min after truxilonium administration (0.5 mg/kg i.v.); *III*, after 35 min. *Vertical line* 200 μV. Lingual nerve was stimulated by submaximal stimuli 0.1 ms, 20 Hz. Experiment on decerebrated cat. (MURATOV 1963a)

effect on the lability. The administration of 0.25 mg/kg succinylcholine usually resulted in a shift of transformation threshold toward lower frequencies. After administration of 0.05–0.5 mg/kg decamethonium the changes of the lability were about the same. These changes of lability of the linguomandibular reflex (induced by depolarizing neuromuscular blocking agents) are unlikely to be influenced by proprioceptive afferent impulses from the muscle spindles. A number of authors report that the increased intensity of proprioceptive impulses due to depolarizing neuromuscular blocking agents resulted in the block of only monosynaptic reflexes, polysynaptic reflexes either remaining unchanged or being facilitated (ELDRED et al. 1957; HENATSCH and SCHULTE 1957 a, b; 1958 a, b; FUJIMORI et al. 1959; MURATOV 1963). It should be added that truxilonium, a nondepolarizing agent, was the most active with respect to the linguomandibular reflex. Considering these data and also the polysynaptic character of this reflex, it may be assumed that the changes observed after succinylcholine and decamethonium administration are likely due to their direct action on the central links of the linguomandibular reflex.

The effect of a number of other neuromuscular blocking agents: bisquaternary salts such as diadonium, decadonium, dipyronium, pancuronium, and pipecuronium, and also tertiary amines such as mellictinum and pyrocurinum, on that reflex was also studied (ALYAUTDIN 1978). The reflex responses were evoked by single stimuli, series of stimuli, and paired stimuli applied to the lingual nerve. The agents tested in neuromuscular blocking doses or at five times these doses, failed to affect the reflex responses. No drug-related changes of the duration of synaptic delay or the lability of the centers of the linguomandibular reflex were observed, either. The agents tested were found not to affect the recovery cycle of reflex responses or the amplitude of the response to the testing stimuli (the interval between the stimuli being up to 100–200 ms). Tubocurarine (0.6 mg/kg), gallamine (4 mg/kg), anatruxonium (0.35 mg/kg), and succinylcholine (0.75 mg/kg) failed to affect the monosynaptic jaw reflex which has the same level of central links as the linguomandibular one.

VASILESCU et al. (1960) suggested a depressing action of tubocurarine on the reticular formation of the brain stem. This is confirmed by cross-circulation experiments in which EEG changes in head-perfused recipient dogs were observed after tubocurarine was administered to the donor animal. However, in experiments on cats, tubocurarine and other neuromuscular blocking agents (diplacinum, succinylcholine) failed to affect the evoked responses of the neurons of the brain stem reticular formation (DYACHENKO 1961).

SKOROBOGATOV (1961) found that intravenous injection of one-tenth the neuromuscular blocking dose of paramionum (0.1–0.15 mg/kg) to conscious rabbits was followed by changes of spontaneous cortical activity (desynchronization of EEG). They are similar to EEG changes caused by direct stimulation of the reticular formation or by the effect of cholinomimetics. The ability of atropine to prevent EEG activation in response to stimulation of the reticular formation by paramionum or cholinomimetics suggests that paramionum produces an excitatory effect on acetylcholine receptors of the ascending activating system of the brain stem.

ELLIS et al. (1952), in experiments on anesthetized cats and dogs, demonstrated that after intravenous administration of tubocurarine (0.4 mg/kg), succinylcholine (0.5 mg/kg), and decamethonium (0.1 mg/kg), protracted respiratory arrest occurs which persists well after restoration of neuromuscular transmission in the diaphragm. The latter seems to be related to the central action of the drugs. KUMAGAI et al. (1954) reported that, when administered into the carotid artery, tubocurarine stimulated the respiratory center.

In cross-circulation experiments, tubocurarine, diplacinum, succinylcholine, and decamethonium were administered to dogs and cats in the circulating blood of the donor animal. In neuromuscular blocking doses, they did not cause alterations of respiration or arterial pressure in the recipient animal with vascularly isolated head (IRWIN and WELLS 1957; DYACHENKO 1961). Negative results were also obtained with head perfusion on the living dog and intracarotid injection of apneic doses of tubocurarine, gallamine, and succinylcholine (SAKUMA 1959). On the other hand, VASILESCU et al. (1960) observed, in cross-circulation experiments, apnea in the head-perfused recipient dog after 0.6 mg/kg intravenous tubocurarine administered to the donor animal.

TRIPPENBACH (1973), in experiments on anesthetized cats, also observed gallamine (7 mg/kg i.v.) and succinylcholine (0.2 mg/kg i.v.) to affect respiration. Under the influence of these two drugs, central respiratory rhythm was decreased and duration of the inspiratory phase increased. However, after vagotomy, only slight changes in the pattern and frequency of phrenic nerve discharges were observed. The author believes the observed changes of central respiratory activity to be of peripheral origin.

BOYD et al. (1969) investigated the effect of gallamine on the recovery cycle of the cuneate nucleus in unanesthetized cats. It was shown that gallamine in paralytic doses (3 mg/kg + 1.5 mg/kg every 30 min) increased the amount of recovery at all interstimulus intervals. They believe that although it penetrates poorly into the brain, gallamine has a direct central action. GALINDO et al. (1968) have found that, under intravenous administration of gallamine, the incidence of repetitive firing of cuneate neurons increased.

The injection of curare in the internal carotid artery of anesthetized cats was associated with short-term fall of arterial pressure and bradycardia which, according to EULER and WAHLUND (1941), indicates the excitatory action of curare on the centers of the vagus nerves. It was shown that intravenous dihydro-β-erythroidine did not depress the synaptic excitation of cerebellar neurons in the cat (CRAWFORD et al. 1966).

Thus, several neuromuscular blocking agents can affect the activity of certain neurons of the brain stem.

V. The Effect on the Spinal Cord

The majority of authors consider that neuromuscular blocking agents belonging to the quaternary ammonium salts do not directly affect the spinal cord in neuromuscular blocking doses. For instance, tubocurarine, diplacinum, truxilonium, and other nondepolarizing neuromuscular blocking agents failed to affect the amplitude of single mono- and polysynaptic discharges or the lability of the spinal neurons even after they were administered in doses 10–20 times higher than neuromuscular blocking doses (VALDMAN et al. 1955; SHAPOVALOV 1959; SKOROBOGATOV 1961; MURATOV 1963a; KHARKEVICH et al. 1969; MURATOV et al. 1970).

NAESS (1950) reported that tubocurarine (1.05–2.65 mg/kg i.v.) did not change the amplitude of evoked mono- and polysynaptic discharges of the spinal ventral roots in anesthetized cats. DE JONG et al. (1968), studying the effect of gallamine in anesthetized animals, found that it had no effect on reflex responses of the spinal cord (dorsal roots, ventral roots) at a dose of 6.25 mg/kg. The administration of 12.5 mg/kg gallamine was followed by a brief increase by 12% in the polysynaptic response which the authors consider to be related to the elevated arterial pressure. Monosynaptic potentials remained unchanged.

GINZEL et al. (1969) showed in cats that gallamine did not affect central nicotine inhibition of monosynaptic responses of the spinal cord. According to CURTIS and RYALL (1966), gallamine up to 8 mg/kg i.v. failed to reduce acetylcholine and early synaptic responses of Renshaw cells. Thus, systemically administered gallamine seems not to block nicotinic receptors of Renshaw cells. It was also shown that tubocurarine (1.2 mg/kg i.v.), metocurine (1 mg/kg i.v.), and gallamine (1 mg/kg i.v.) did not influence the transmission from the collaterals of motoneurons to Renshaw cells (ECCLES et al. 1954, 1956). Tubocurarine (i.v.) does not affect other types of inhibition of the spinal motoneurons either (CURTIS 1959).

ALYAUTDIN (1978) investigated the probable action on the spinal cord of bisquaternary salts: diadonium, decadonium, dipyronium, pipecuronium, and pancuronium. The drugs were administered intravenously to spinal cats at doses 1–5 times as high as the neuromuscular blocking doses. It was found that they had no effect on the amplitude, latency, configuration, or area of monosynaptic or polysynaptic potentials, either in deafferented animals or in those with unchanged afferent input (series of stimuli of 1–100 Hz frequency were applied to the dorsal roots and paired stimuli at varying intervals were applied to peripheral nerves). Furthermore, it was found that diadonium (0.35–1.65 mg/kg), decadonium (0.3–1.5 mg/kg), dipyronium (0.02–0.1 mg/kg), pipecuronium (0.02–0.1 mg/kg), and pancuronium (0.02–0.1 mg/kg) did not affect the inhibitory action of nicotine on monosynaptic potentials of the ventral roots of spinal cord evoked by stimulation of the gastrocnemius nerve.

However, several papers present data on the inhibitory action of some neuromuscular blocking agents on the spinal cord. Thus, according to MCCAWLEY (1949), in cats, tubocurarine at a dose of 0.9 mg kg^{-1} min^{-1} i.v. first blocked the polysynaptic reflex of the spinal cord, and in higher doses also blocked the monosynaptic ones. A number of authors (BERNHARD and TAVERNER 1951; BERNHARD et al. 1951) report that in anesthetized, decerebrated, decapitated, and spinal cats, tubocurarine (0.11–1.5 mg/kg i.v.) causes an increase in the monosynaptic extensor reflex induced by the stimulation of the gastrocnemius nerve, but has no effect on the polysynaptic reflex.

MALORNY (1953) has shown decamethonium to produce a depressive action on the flexor reflex. This conclusion was based on experiments indicating that decamethonium inhibited the flexor reflex in decerebrated and decapitated cats over

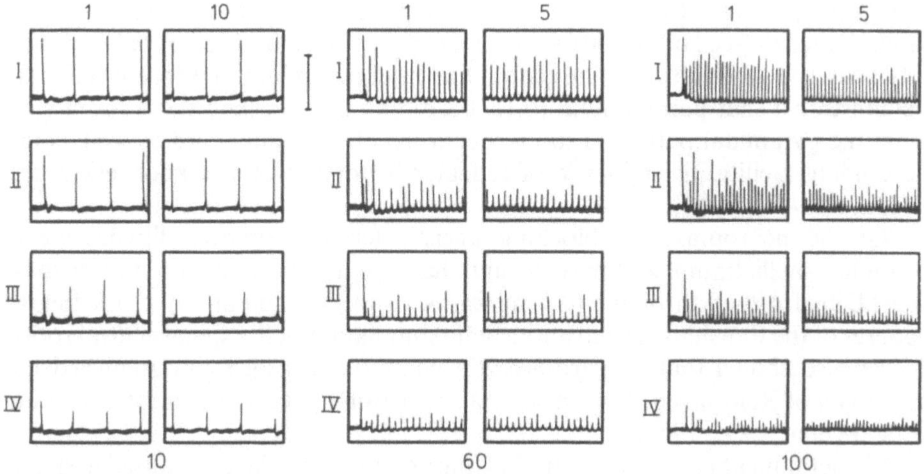

Fig. 3. The effect of paramionum on neuromuscular transmission in the spinal cord. *I* before drug administration; *II–IV* after administration of paramionum at 0.3 (*II*), 0.5 (*III*), and 1 mg/kg i.v. (*IV*). Figures above the potentials indicate the time (s) from the onset of stimulation. Dorsal root (S_1) was stimulated by series of supramaximal stimuli of 10, 60, and 100 Hz (indicated by figures under the potentials). Duration of each stimulus 0.1 ms. Potentials were recorded in the corresponding anterior root. *Vertical line* 500 μV. Experiment on decerebrated cat. (MURATOV 1963a)

a longer period than the transmission from the sciatic nerve to the gastrocnemius muscle.

Of the bisquaternary ammonium salts, paramionum has a marked inhibitory action on the spinal cord (SKOROBOGATOV 1961; MURATOV 1963). Thus, for instance, after paramionum was administered at subparalytic doses of 0.3–0.5 mg/kg i.v., the amplitude of monosynaptic discharges in ventral roots was decreased by 25%–30% and of the polysynaptic ones by about 1.5–2 times. Increasing the paramionum dose up to 0.8–1 mg/kg resulted in a complete block of polysynaptic discharges and an abrupt decrease of the amplitude of the monosynaptic ones. Paramionum also had a marked effect on the lability of spinal reflexes (Fig. 3). It causes especially marked changes of the lability when stimuli of subpessimal frequencies are applied. Paramionum-induced lability changes are manifested by progressive decrease of amplitude of evoked series of discharges and by marked transformation of impulses (MURATOV 1963).

Although at first, paramionum was classified as a nondepolarizing neuromuscular blocking agent, BUTAEV (1964) demonstrated that paramionum could induce contractions of rectus abdominis muscle of the frog. Its depolarizing component of action was also described in experiments on denervated lingual muscles (MAEVSKY 1964). Therefore, to study the mechanism of action of paramionum, control experiments with registration of proprioceptice activity in the peripheral nerve were carried out (ALYAUTDIN 1978). It turned out that after administration of 0.3 mg/kg i.v. paramionum, proprioceptive activity in the peripheral nerve was doubled and after 1 mg/kg it was increased fourfold. Apparently, the inhibitory action of paramionum on monosynaptic responses of the spinal cord is related to its depolarizing activity which is responsible for the increased intensity of afferent impulses (see later in this section). However, the inhibition of polysynaptic reflexes seems to result from the central action of paramionum. Moreover, it was demonstrated that paramionum (0.15–0.2 mg/kg i.v.) blocked the transmission from the pyramidal pathways to motoneurons of the spinal cord as well as the descending facilitating influence of reticular formation on the knee jerk reflex (SKOROBOGATOV 1961).

Of the neuromuscular blocking agents, tertiary amines: dihydro-β-erythroidine, mellictinum, and pyrocurinum have an inhibitory action on Renshaw cells. Dihydro-β-erythroidine (0.1 mg/kg i.v.) appeared to entirely block the responses of the Renshaw cells to antidromic stimulation of the spinal ventral roots. It was also shown that dihydro-β-erythroidine (0.4 mg/kg i.v.) diminished the sensitivity of Renshaw cells to the excitatory action of acetylcholine and nicotine (ECCLES et al. 1954, 1956).

Mellictinum at neuromuscular blocking doses (3.5 mg/kg i.v.) caused in spinal cats (deafferented or with intact afferent input) an increase in the amplitude of monosynaptic responses by 18%–21%, and also accelerated the recovery cycle of excitability of motoneurons. In experiments on cats, mellictinum (3.5 mg/kg i.v.) and pyrocurinum (130 μg/kg and 1 mg/kg i.v.) decreased the inhibitory action of nicotine on monosynaptic responses, mellictinum being more effective (ALYAUTDIN 1978). Thus, tertiary amines: dihydro-β-erythroidine, mellictinum, and pyrocurinum have a direct inhibitory action on nicotinic receptors of Renshaw cells.

GINZEL et al. (1952) observed in anesthetized and decapitated cats, that depolarizing neuromuscular blocking agents belonging to the group of dicholine esters, succinylcholine included, blocked the knee jerk reflex, when administered in doses which do not block neuromuscular transmission from the sciatic nerve to the gastrocnemius muscle. The same data were obtained for decamethonium (BRÜCKE 1953; GINZEL et al. 1953; KLUPP et al. 1953). The depolarizing agents, succinylcholine and decamethonium turned out to block the knee jerk reflex in doses in which they did not affect the flexor or contralateral extensor reflex or neuromuscular transmission.

It was demonstrated in experiments with isolated circulation of the lower limbs that the block of knee jerk reflex was not related either to the direct central action of the drugs or to the neuromuscular block (KLUPP et al. 1953). The findings of GRANIT et al. (1953) dealing with the activation by depolarizing agents of muscle spindles laid the basis for the suggestion that the block of knee jerk reflex observed after administration of succinylcholine and decamethonium was due to increase in the afferent impulses from the muscle spindles.

Further investigations led to the same conclusions (ELDRED et al. 1957; HENATSCH and SCHULTE 1958a, b; FUJIMORI et al. 1959; FUJIMORI and ELDRED 1961; MURATOV 1963; KATO and FUJIMORI 1965; GINZEL et al. 1969). The results obtained were confirmed by a number of clinical data (BRUNE et al. 1959, 1960; BRUNE and SCHENK 1960). Thus, the blocking action of the depolarizing agents, succinylcholine and decamethonium on monosynaptic reflexes of the spinal cord is related to their peripheral action. Several neuromuscular blocking agents can produce both direct and/or indirect (reflex) action on the spinal cord.

VI. The Effect on Reflex Responses of the Arterial Pressure

Neuromuscular blocking agents were shown to have a marked effect on the hypertensive responses of the arterial pressure observed during the stimulation of the central stump of peroneal or sciatic nerve (MURATOV 1963, 1967). It was found in decerebrated cats that tubocurarine was the most potent. Even at doses of 0.12–0.15 mg/kg i.v. it caused a decrease of pressor responses by 25%–30%. Similar changes occurred after 2 mg/kg i.v. diplacinum. A complete block of pressor responses was observed after 0.45–0.6 mg/kg tubocurarine. Diplacinum completely blocked pressor reflexes only at 8–10 mg/kg (4–5 times higher than the neuromuscular blocking doses). Other nondepolarizing agents, paramionum and truxilonium, produced a somewhat less marked action on pressor responses. Thus, for instance, truxilonium in doses 1.5–2 times higher than neuromuscular blocking doses (0.25–0.3 mg/kg i.v.), did not essentially affect them. However, increasing the truxilonium dose up to 0.6–0.75 mg/kg resulted in decreased pressor responses, by 40%–60%. A complete block of the latter only developed after 1.25–1.5 mg/kg i.v. truxilonium. The same effect was observed after 1.5 mg/kg paramionum.

The depolarizing agents, decamethonium and succinylcholine were also rather potent with respect to pressor responses. Even at low doses (decamethonium 0.03–0.04 mg/kg, succinylcholine 0.15–0.2 mg/kg i.v.) they decreased those responses by 40%–50% (Fig. 4). However, even at doses 10–12 times higher than the neuromuscular blocking doses they failed to cause a complete block of pressor reflexes.

Several nondepolarizing neuromuscular blocking agents have a marked ganglion-blocking activity (KHARKEVICH 1962, 1967; see also Chap. 7, Sect. B)

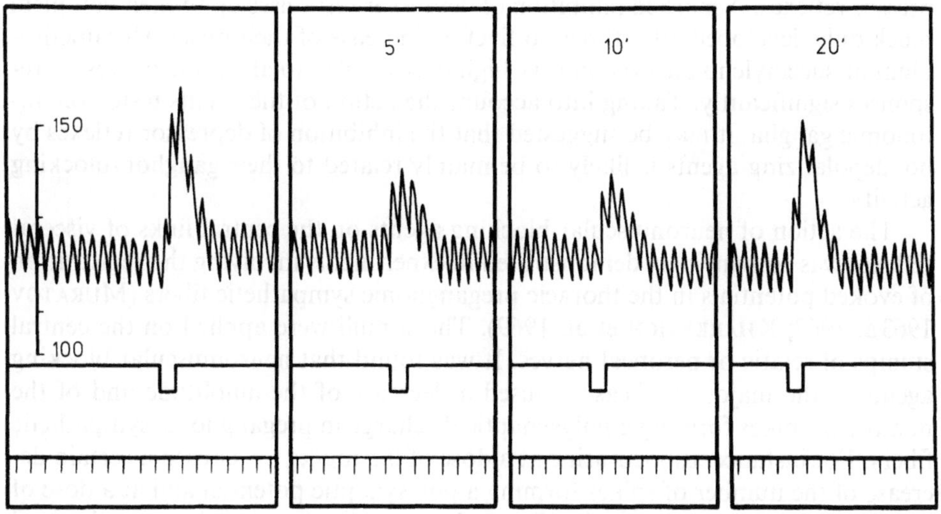

Fig. 4. The effect of succinylcholine on pressor reflexes. The figures indicate time (min) after i.v. administration of succinylcholine (0,2 mg/kg). First trace-control. Central section of the sciatic nerve was stimulated by rectangular stimuli (0.1 ms, 50 Hz). Experiment on decerebrated cat. (MURATOV 1963a)

which is likely to play a certain role in their diminishing effect on the pressor reflexes. However, it should be noted that even in high doses they fail to block sympathetic ganglia completely. Yet, they are highly potent with respect to pressor reflexes. Doses three times as high (tubocurarine) or five times as high (diplacinum, truxilonium, paramionum) as the neuromuscular blocking doses were enough to cause a complete block of pressor responses. Therefore, it cannot be excluded that the inhibition of pressor reflexes caused by nondepolarizing neuromuscular blocking agents might be related not only to their ganglion-blocking activity, but probably also to their influence on the central nervous system.

The depolarizing agents, succinylcholine and decamethonium also had a marked effect on pressor responses, though less marked than the nondepolarizing agents. Succinylcholine and decamethonium are known to have essentially no ganglion-blocking potency (KHARKEVICH 1962, 1967; Chap. 7, Sect. B). Taking the latter into account, it can be suggested that their inhibitory action on pressor reflexes is likely to be mainly due to their action on the central links of the corresponding reflexes.

Hypotension occurring on stimulation of the central stumps of the dissected vagus or depressor nerves was reduced only by nondepolarizing agents: tubocurarine, diplacinum, paramionum, or truxilonium. Tubocurarine, proved the most potent; its administration at doses of 0.15–0.2 mg/kg i.v. was followed by a 20%–25% decrease of depressor responses. A complete block of depressor responses required 0.45–0.6 mg/kg i.v. tubocurarine. Diplacinum in doses of 1.5–2 mg/kg did not significantly affect the value or duration of depressor responses. As much as 8–10 mg/kg diplacinum was needed to decrease the responses by 30%–60%. Diplacinum failed to cause a complete block of depressor reflexes even in doses of 15–25 mg/kg i.v. Truxilonium and paramionum in doses close to neuromuscular blocking doses (0.2–0.3 mg/kg i.v.) caused a 15%–20% decrease of depressor reflexes. A marked inhibition of depressor responses up to their complete block only developed after three- to fivefold increase of their doses. Decamethonium or succinylcholine even in very high doses, failed to affect the depressor responses significantly. Taking into account the action of the agents tested on autonomic ganglia, it may be suggested that the inhibition of depressor reflexes by nondepolarizing agents is likely to be mainly related to their ganglion-blocking activity.

The action of neuromuscular blocking agents on the central links of visceral reflexes was convincingly demonstrated in experiments involving the registration of evoked potentials in the thoracic preganglionic sympathetic fibers (MURATOV 1963a, 1967; KHARKEVICH et al. 1969). The stimuli were applied on the central stumps of sciatic or peroneal nerves. It was found that neuromuscular blocking agents in the majority of cases caused a decrease of the amplitude and of the number of spikes forming a polysynaptic discharge in preganglionic sympathetic fibers. For instance, tubocurarine at a dose of 0.5 mg/kg i.v. caused a certain decrease of the number of spikes forming a polysynaptic potential and at a dose of 1 mg/kg it caused even more significant changes. The same results were obtained after administration of 0.5–1.0 mg/kg paramionum (Fig. 5), 0.5 mg/kg truxilonium, 0.5–1.5 mg/kg succinylcholine, 0.25–0.3 mg/kg decamethonium, and 15–20 mg/kg diplacinum (i.v.).

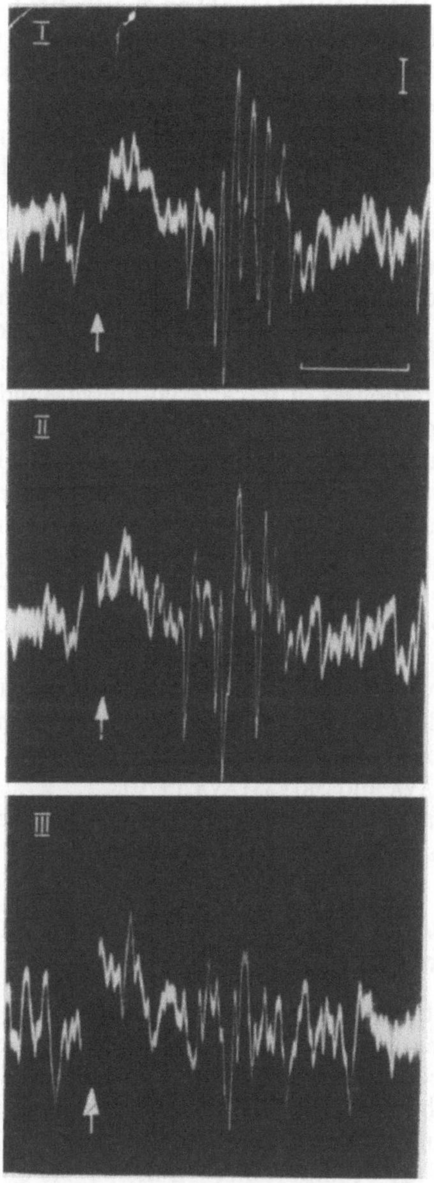

Fig. 5. The effect of paramionum on evoked potentials in the preganglionic fibers of the left sympathetic trunk (Th_8). *I*, before drug administration; *II*, 5 min after paramionum administration (0.5 mg/kg i.v.); *III*, 5 min after the second administration (1 mg/kg i.v.). Central end of the left sciatic nerve was stimulated by single supramaximal stimuli (0.5 ms). *Arrows* indicate stimulation. *Vertical line above* amplitude (10 μV). *Horizontal line* time (10 ms). Experiment on decerebrated cat. (MURATOV 1963a)

The experiments of MCLENNAN (1961) on anesthetized and conscious rabbits demonstrated that tubocurarine and metocurine (0.3–0.35 mg/kg) prevented hypotensive responses of the arterial pressure to slight low frequency stimulation of the auricular and saphenous nerves. In some cases, during the action of tubocurarine and metocurine, there occurred pressor responses instead of the depressor ones. This was associated with increase in the amplitude of discharges in the renal sympathetic nerves, which the author suggests are related to the direct action of tubocurarine on the vasomotor centers. However, no drug-related changes of either depressor responses or bioelectrical activity of the renal nerves were observed after administration of galamine or succinylcholine.

In the framework of the probable central action of tubocurarine, the data of PEISS and MANNING (1959) are of interest. They demonstrated its inhibitory action on hypertensive responses of the arterial pressure induced in vagotomized cats by the stimulation of medulla oblongata and hypothalamus. Intravenous administration of tubocurarine at doses of 0.1–0.15 mg/kg which does not block the transmission in the autonomic ganglia, almost entirely prevented pressor responses evoked by the stimulation of medulla oblongata and had little effect on pressor responses evoked by hypothalamic stimulation. A similar effect was observed after tubocurarine was injected into the carotid artery at doses of 0.1–0.15 mg/kg, or at 0.003–0.015 mg directly into the vasomotor area of medulla oblongata.

It was shown in experiments on dogs that intrathecal tubocurarine and gallamine facilitate, and decamethonium and succinylcholine inhibit spinal compression vasomotor response and urinary bladder response, i.e., have spinal autonomic effects (BHARGAVA and SRIVASTAVA 1970).

The findings indicate that a number of neuromuscular blocking agents might affect the central links of cardiovascular reflexes.

VII. The Interaction of Neuromuscular Blocking Agents with Other Neurotropic Drugs

This section deals with the interaction of neuromuscular blocking agents with centrally acting noncholinergic agents. The data on the interaction with cholinomimetics are presented in other sections of this chapter. Such works are rather scarce, but some of them deserve attention. Thus, for instance, the data on the ability of neuromuscular blocking agents to potentiate the action of other anesthetics and hypnotics are of importance. In the papers of HAZARD et al. (1954, 1956) it was shown that bisquaternary ammonium salts with a piperazine ring in the central part of the molecule (structure I) and having a neuromuscular blocking activity are capable of potentiating the hypnotic action of hexobarbital (Evipal) in mice

$$R^3R^2R^1\overset{+}{N}\text{—}(CH_2)_2\text{—}N\bigcirc N\text{—}(CH_2)_2\text{—}\overset{+}{N}R^1R^2R^3 \quad \cdot\ 2X^- \qquad \text{(I)}$$

where

$NR^1R^2 = N(C_2H_5)_2$, piperidyl

R^3 $= n\text{-}C_nH_{2n-1}$ $(n = 1–16)$; benzyl; o-, m-, p-chlorbenzyl

X $=$ I, Br

Compounds with $NR^1R^2 = N(C_2H_5)_2$, $R^3 = n\text{-}C_{12}H_{25}$, $X = Br$ and also with NR^1R^2 = piperidyl, $R^3 = C_2H_5$ or benzyl, $X = Br$ turned out the most potent in this respect. When used for premedication, they increased the duration of sleep induced by hexobarbital by 100%–125%. The authors see these results as a manifestation of the central inhibitory action of the neuromuscular blocking agents tested. However, to support this conclusion it would be desirable to have data on the influence of these agents on hexobarbital pharmacokinetics. BRISKIN (1959) reported that, in mice, diplacinum (10 mg/kg, s.c.) made the duration of hexobarbital sodium (i.v.) action twice as long.

It was shown that in humans pancuronium reduced halothane minimal alveolar concentration (MAC) by 25% (FORBES et al. 1979). It cannot be ruled out that this is due to central action of pancuronum. In particular, SAVARESE (1979) advances the opinion that one of the possibilities is the blocking action of pancuronium on muscarinic receptors of the ascending activating reticular formation. At the same time, total muscle paralysis significantly reduced the amount of afferent impulses entering the central nervous system, which results in alterations of the functional state of the central nervous system and can change the effectiveness of halothane. It should be also taken into account that pancuronium can influence nonspecific binding of halothane in the "sites of loss".

According to BOVET and LONGO (1953) the influence of the barbiturate numal and strychnine on the EEG after gallamine pretreatment was unaffected. At the same time, it was shown in spinal cats that tubocurarine intravenously administered after strychnine decreased the amplitude of the potentials of the spinal ventral roots, those changes being directly related to the dose ratio of strychnine and tubocurarine (TAVERNER 1953). On the basis of these data, the author suggests an inhibitory action of tubocurarine on the interneurons of the spinal cord.

Studying seizure discharges evoked in the cerebral cortex of monkeys by intravenous lidocaine, MUNSON and WAGMAN (1972) observed that gallamine increased the seizure threshold. The authors suggested that the effect was related to the direct central action of gallamine. However, DE JONG et al. (1981) did not observe metocurine or gallamine to affect seizure threshold by testing another local anesthetic, bupivacaine. Thus, several papers report interaction of neuromuscular blocking agents with centrally acting neurotropic agents. However, these data require further analysis, since the causes of the interaction mentioned are not quite clear.

C. Conclusions

Hence, opinions about the probable action of neuromuscular blocking agents on the central nervous system are rather controversial. The discrepancies mainly concern the experiments with intravenous administration of agents. Intraventricular injections, local, or microiontophoretic application give rather similar results. They indicate that there is in the brain a substrate sensitive to neuromuscular blocking drugs. It is chiefly represented by acetylcholine receptors. However, some noncholinergic sites of action seem to be of importance, too. Indeed, it has been demonstrated that excitatory and seizure actions of tubocurarine and of a number of other neuromuscular blocking drugs on their intraventricular administration are not related to their influence on acetylcholine receptors.

For practical medicine, the data obtained for common administration routes (as a rule, intravenous) are of greatest value. However, it is rather difficult to analyze and compare the majority of such papers, since they were carried out under different experimental conditions. Thus, for instance, a large number of experiments were conducted under general anesthesia, i.e., in a dramatically changed functional state of the central nervous system. Furthermore, these experiments essentially dealt with the combined action of general anesthetics with neuromuscular blocking agents. Also, general anesthetics with different pharmacodynamics and pharmacokinetics were used (for instance, chloralose, barbiturates, urethane) which should be taken into account, too. The adequacy of artificial ventilation, the state of acid–base equilibrium and electrolyte balance, variations of arterial pressure, the importance of immobilization stress, influence on proprioceptive afferent input, the significance of environmental temperature, etc., were not always considered. All the foregoing are of great importance for the estimation of the action of neuromuscular blocking agents on the central nervous system. Therefore, some papers probably contain artifacts and erroneous interpretations concerning the possibility of direct action of these agents on the central nervous system.

And yet, for tertiary amines (dihydro-β-erythroidine, mellictinum) their influence on the central nervous system is undoubted. They easily penetrate the blood–brain barrier and, passing into the central nervous system, interact with acetylcholine receptors. Of the quaternary ammonium compounds, paramionum can be singled out, which on intravenous administration produces marked central action (direct and indirect). Other nondepolarizing neuromuscular blocking agents injected intravenously may cause certain effects on several functions of the central nervous system. However, most often it is manifested after high doses, much higher than those needed for total muscle paralysis.

Depolarizing drugs are characterized by the ability to induce indirect action on the central nervous system at the expense of increasing the intensity of afferent discharges from muscle spindles. At the same time, it cannot be excluded that agents of the succinylcholine type on intravenous administration can also produce direct central action (probably, on some afferent pathways, some cardiovascular reflexes).

However, to provide further evidence for these conclusions it seems reasonable to study the content of neuromuscular blocking agents in the cerebrospinal fluid and cerebral tissues under different states of the organism which can alter the permeability of the blood–brain barrier. Obviously, such investigations will require highly sensitive quantitative techniques (like radioimmunoassay). Such data would allow one to differentiate more precisely between central and peripheral origin of central changes elicited by intravenous administration of neuromuscular blocking agents.

Acknowledgments. The author wishes to express his appreciation to V. K. MURATOV and R. N. ALYAUTDIN for their kind assistance in the preparation of the manuscript of this chapter and to Mrs. MARIA LIPMAN for the translation of this chapter.

References

Acheson F, Bull AB, Glees P (1956) Electroencephalogram of the cat after intravenous injection of lidocaine and succinylcholine. Anesthesiology 17:802–808

Ali HH (1981) Wirkungen von Muskelrelaxantien am Zentralnervensystem. In: Buzello W (ed) Muskelrelaxantien. Neuere Konzepte ihrer Pharmakologie und klinischen Anwendung. Thieme, Stuttgart, pp 57–62

Alyautdin RN (1978) On the effects of new neuromuscular blocking agents on the central nervous system (in Russian). Cand Med Sci Thesis, Moscow

Amirov RZ, Zolnikov SM, Ivanova VI, Strakhov SN (1962) Study of electroencephalographic data under up-to-date intubation anaesthesia (in Russian). Vestn Khir 6:57–61

Andy OJ, Akert K (1955) Seizure patterns induced by electrical stimulation of hippocampal formation in the cat. J Neuropathol Exp Neurol 14:198–213

Bakay L (1954) Studies on blood-brain barrier with radioactive phosphorus. AMA Arch Neurol Psychaitry 71:673–683

Bakay L (1957) Dynamic aspect of the blood-brain barrier. In: Richter D (ed) Metabolism of the nervous system. Pergamon, London, pp 136–150

Bakay L (1968) Changes in barrier effect in pathological states. Prog Brain Res 29:315–337

Beleslin DB, Samardzic K (1976) Failure of autonomic and central nervous system blocking agents to antagonise the gross behavioral effects of tubocurarine injected intraventricularly in conscious cats. J Pharm Pharmacol 28:519

Bernhard CG, Taverner D (1951) The action of d-tubocurarine on the monosynaptic extensor reflex. Br J Pharmacol 6:540–550

Bernhard CG, Taverner D, Widén L (1951) Differences in the action of tubocurarine and strychnine on the spinal reflex excitability of the cat. Br J Pharmacol 6:551–559

Bhargava KP, Srivastava PK (1970) Effects of d-tubocurarine and decamethonium on spinal autonomic loci. Eur J Pharmacol 9:220–226

Bovet D, Longo VG (1953) Action of natural and synthetic curares on the cortical activity of the rabbit. Electroencephalogr Clin Neurophysiol 5:225–234

Bowman WC, Rand MJ (1980) Textbook of Pharmacology, 2nd edn. Blackwell Scientific, Oxford

Boyd ES, Meritt DA, Aroesty S, Celso M (1969) Effects of gallamine and physostigmine on transmission through the cuneate nucleus. Am J Physiol 216:542–546

Bradbury MW (1979) The concept of a blood-brain barrier. Wiley Sons, New York

Brattgard S-O, Lindqvist T (1955) Differences in effectivity of the blood-brain barrier in different parts of central nervous system. Acta Psychol Neurol Scand 30:423–436

Briskin AI (1959) On the relationship of peripheral and central action of curare-like agents (in Russian). Conference on pathology and regeneration of circulation and respiration organs. Abstracts. Academy of Sciences of the USSR, Siberian Division, Novosibirsk, pp 72–74

Briskin AI (1961) On central effects of curare and curare-like agents (in Russian). Proceedings of the IX All-Union conference on pharmacology, Sverdlovsk, pp 33–34

Briskin AI, Flerov BA (1961) Influence of curare-like agents on conditioned reflex-activity of albino mice (in Russian). Farmakol Toksikol 5:523–529

Brücke VF (1953) Über den Wirkungsmechanismus peripher muskellähmende Stoffe. Arch Exp Pathol Pharmakol 218:70–81

Brune HF, Schenk E (1960) Spinale Wirkungen chemischer Muskelspindelaktivierung bei spastischen Zuständen. Dtsch Z Nervenheilkd 181:484–493

Brune HF, Schenk E, Voss H (1959) Der Einfluß chemisch aktivierter Muskelspindeln auf die monosynaptische Anregbarkeit spinaler Motoneurone des Menschen. Pfluegers Arch 269:555–569

Brune HF, Dammann R, Schenk E (1960) Chemische Aktivierung der Muskelspindeln beim Menschen. Die Bedeutung der Spindelmechanik für die spinalen Effekte. Pfluegers Arch 271:397–404

Butaev BM (1964) Cholinomimetic effects of several Soviet curare-like agents (in Russian). Byull Eksp Biol Med 10:63–65
Chardon G (1956) Action du 3.697 RP (flaxédil) sur l'electroencéphalogramme du Chien. CR Soc Biol (Paris) CL:525–527
Chernigovsky VN, Zaraiskaya SM (1962) On the representation of vagus nerve in the cortex of large hemispheres and limbic portion of the cat brain (in Russian). Dokl Akad Nauk SSSR 147:742–744
Cohen EN (1963) Blood-brain barrier to d-tubocurarine. J Pharmacol Exp Ther 141:356–362
Crawford JM, Curtis DR, Voorhoeve PE, Wilson VJ (1966) Acetylcholine sensitivity of cerebellar neurones in the cat. J Physiol (London) 186:139–165
Curtis DR (1959) Pharmacological investigations upon inhibition of spinal motoneurones. J Physiol (London) 145:175–192
Curtis DR, Ryall RW (1966) The acetylcholine receptors of Renshaw cells. Exp Brain Res 2:66–80
Dal Santo G (1964) Kinetics of distribution of radioactive-labelled muscle relaxants. I. Investigation with C^{14}-dimethyl-d-tubocurarine. Anaesthesiology 25:788–800
Dal Santo G (1972) Kinetics of distribution of radioactive-labelled muscle relaxants. IV. Urinary elimination of a single dose of ^{14}C-Gallamine. Br J Anaesth 44/4:321–329
Danilov AF, Mikhelson MYa, Rybolovlev RS (1957) Pharmacological properties of succinylcholine and its application in clinics (in Russian). In: Mndzhoyan AL (ed) Ditilin i opyt ego klinicheskogo primeneniya (Succinylcholine and data on its clinical application). Armenian Academy of Sciences, Yerevan, pp 92–107
David JP, Murayana S, Machne X, Unna KR (1963) Evidence supporting cholinergic transmission at the lateral geniculate body of the cat. Int J Neurophysiol 2:113–125
De Jong RN, Robles R, Morikawa K-I (1968) Gallamine (Flaxedil) and synaptic transmission in the spinal cord. Science 160:768–769
De Jong RH, Gamble CA, Bonin JD (1981) Neuromuscular blocking agents do not alter local anesthetic seizure thresholds. Exp Neurol 74:628–631
Devasancaraiah G, Haranath PSRK, Krishnamurty A (1973) Passage intravenously administered tubocurarine into the liquor space in man and dog. Br J Pharmacol 47:787–798
Dyachenko PK (1961) On the application of nitrous oxide for anaesthesia (in Russian). Vestn Khir 7:90–97
Eccles JC, Fatt P, Koketsu K (1954) Cholinergic and inhibitory synapses in a pathway from motor-axon collaterals to motoneurons. J Physiol (London) 126:524–562
Eccles JC, Eccles RM, Fatt P (1956) Pharmacological investigations on a central synapse operated by acetylcholine. J Physiol (London) 131:154–169
Eldred E, Fujimori B, Tokizane T (1957) Effects of muscle spindle discharge on monosynaptic reflex as revealed by syncurine administration. Fed Proc 16:34
Ellis CH, Morgan WV, De Beer EJ (1952) Central depressant actions of certain myoneural blocking agents. J Pharmacol Exp Ther 106:353–363
Esplin DV, Rosenstein R (1963) Analysis of spinal depressant actions of carbon dioxide and acetazolamide. Arch Int Pharmacodyn Ther 143:498–513
Euler US, Wahlund H (1941) Über zentrale Kurarewirkungen. Acta Physiol Scand 2:327–333
Everett GM (1947) The effect of d-tubocurarine on the central nervous system. Fed Proc 6:101
Feldman SA (1960) Effect of decamethonium upon conditional reflexes in rats. Anaesthesia 15:55–60
Forbes AR, Cohen NH, Eger EI (1979) Pancuronium reduces halothane requirement in man. Anesth Analg (Cleve) 58:497–499
Fujimori B, Eldred E (1961) Central effects of succinylcholine and decamethonium on monosynaptic reflexes. Am J Physiol 200:699–702
Fujimori B, Tokizani T, Eldred E (1959) Effects upon monosynaptic reflexes of decamethonium and succinylcholine. J Neurophysiol 22:165–176

Galindo A, Krnjević K, Schwartz S (1968) Patterns of firing in cuneate neurons and some effects of Flaxedil. Exp Brain Res 5:87–101
Galletti C, Maioli MG, Riva Sanseverino E (1979) Acid-base equilibrium during acute long-lasting experiments in artificially ventilated cats. Am J Physiol 236:R126–R131
Gellhorn E (1958a) The physiological basis of neuromuscular relaxation. AMA Arch Intern Med 102:392–399
Gellhorn E (1958b) The influence of curare on hypothalamic excitability and the electroencephalogram. Electroencephalogr Clin Neurophysiol X,4:697–703
Ginzel KH, Klupp H, Stormann H, Werner G (1952) Die Ausschaltung des Patellarsehnenreflexes durch muskellähmende Substanzen. Arzneim Forsch 2:271–274
Ginzel KH, Klupp H, Stormann H, Werner G (1953) Hemmung des Patellarsehnenreflexes durch zentral und peripher wirkende Stoffe. Arch Exp Pathol Pharmakol 218:308–312
Ginzel KH, Eldred E, Sasaki Y (1969) Comparative study of the actions of nicotine and succinylcholine on the monosynaptic reflex and spindle afferent activity. Int J Neuropharmacol 8:515–533
Girden E (1948) The electroencephalogram (EEG) in curarized mammals. J Neurophysiol 11:169–173
Granit R, Skoglund S, Thesleff S (1953) Activation of muscle spindles by succinylcholine and Decamethonium. The effect of Curare. Acta Physiol Scand 28:134–151
Halpern LM, Black RG (1967) Flaxedil (gallamine triethiodide): evidence for a central action. Science 155:1685–1687
Halpern LM, Black R (1968) Gallamine triethiodide facilitation of local cortical excitability compared with neuromuscular blocking agents. J Pharmacol Exp Ther 162:166–173
Harker LA, Hosick E, Voots RJ, Mendel MI (1977) Influence of succinylcholine on middle component auditory evoked potentials. Arch Otolaryngol 103:133–137
Hazard R, Cheymol J, Chabrier P, Gay Y, Muller P (1954) Recherche d'une action centrale chez certains curarisants de synthèse. Thérapie 9:314–324
Hazard R, Cheymol J, Chabrier P, Gay Y (1956) Recherche d'une action centrale chez certains curarisants de synthèse. Thérapie 11:684–692
Hellner K, von Baumgarten R (1961) Über ein Endigungsgebiet afferenter, kardiovasculärer Fasern des Nervus vagus in Rautenhirn der Katze. Pfluegers Arch 273:223–234
Henatsch H-D, Schulte F-J (1957a) Das Verhalten der Muskelspindeln des Frosches unter dem Einfluß von Muskel-Relaxantien. Pfluegers Arch 266:88–89
Henatsch H-D, Schulte FJ (1957b) Zur Aktivierung des intrafusalen Apparates von Kaltblüter-Muskel-Spindeln. Pfluegers Arch 266:89
Henatsch H-D, Schulte FJ (1958a) Einflüsse von Curare und Flaxedil auf die Muskelspindeln des Frosches. Naunyn Schmiedebergs Arch Pharmacol 234:247–263
Henatsch H-D, Schulte FJ (1958b) Reflexwirkungen chemisch erregter Muskelspindeln auf tonische und phasische Motoneurone der Katze. Pfluegers Arch 268:36
Henatsch H-D, Schulte FJ, Busch G (1959) Wandelbarkeit des tonisch-phasischen Reaktionstyps einzelner Extensor-Motoneurone bei Variation ihrer Antriebe. Pfluegers Arch 270:161–173
Irwin RL, Wells JB (1957) The respiratory activity of certain neuromuscular blocking compounds: a direct peripheral and central comparison. J Pharmacol Exp Ther 119:329–342
Jones CS, Meyerovitz LS, Sweeney GD (1956) The effect of curarizing doses of gallamine triethiodide ("Flaxedil") on the electroencephalogram in man. Cur Res Anesth Analg 35:425–428
Kammerer E, Krug M (1972) Einfluß von Muskelrelaxantien auf photisch evozierte kortikale Einzelzellantworten. Acta Biol Med Germ 28:363–369
Kato M, Fujimori B (1965) On the mechanism of fascicular twitching following administration of succinylcholine cloride. J Pharmacol Exp Ther 149:124–130
Kharkevich DA (1962) Ganglionarnye sredstva (ganglionic agents). Medgiz, Moscow
Kharkevich DA (1967) Ganglion-blocking and ganglion-stimulating agents. Pergamon, Oxford

Kharkevich DA, Muratov VK, Sinitsyn LN (1969) On the influence of neuromuscular blocking drugs on the central nervous system. Pharm Res Comm 1:353–362
Kirstein L (1951) Early effects of oxygen lack and carbon dioxide excess on spinal reflexes. Acta Physiol Scand [Suppl] 23 80:1–54
Kitahata LM, Taub A, Sato I (1969) Hyperventilation and spinal reflexes. Anesthesiology 31:321–326
Klupp H, Ginzel KH, Werner G (1953) Elective Hemmung des Patellarsehenreflexes durch peripher wirkende Stoffe. Arch Exp Pathol Pharmakol 218:141–142
Knecht CD, Kazmierczak K, Katherman AE (1980) Effects of succinylcholine on the electroencephalogram of dogs. Am J Vet Res 41:1435–1440
Kumagai H, Yui T, Ogawa K, Sabuma A (1954) Studies on the effect of d-tubocurarine chloride upon the central nervous system by means of the perfusion of dog's head in situ. Folia Pharmacol Jap 50:214
Longo VG (1955) Acetylcholine, cholinergic drugs and cortical electrical activity. Experientia 11:76–78
Lössner B, Schmidt J, Ruthrich H (1970) Der Einfluß von Muskelrelaxantien auf verschiedenartig ausgelöste Potentiale nichtnarkotisierter Kaninchen und Katzen. Acta Biol Med Germ 25:635–643
Maevsky VE (1964) On the mechanism of curare-like action of paramionum (in Russian). Farmakol Toksikol 2:184–186
Mahfouz M (1949) The fate of tubocurarine in the body. Br J Pharmacol Chemother 4:295–303
Maier R, Lössner B, Schmidt J (1970) Der Einfluß systemisch applizierter Muskelrelaxantien auf die Spontanaktivität kortikaler Neurone. Acta Biol Med Germ 25:869–874
Malorny G (1953) Versuche zur Aufhebung der Dekamethonium Lähmung am Reflex- und Nervmuskelpräparat der Katze. Arch Exp Pathol Pharmakol 218:147–148
Matteo RS, Pua EK, Khambatta HJ, Spector S (1977) Cerebrospinal fluid levels of d-Tubocurarine in man. Anesthesiology 46:396–399
McCawley EL (1949) Certain actions of curare on the central nervous system. J Pharmacol Exp Ther 97:129–139
McIntyre AR, Dunn AL, Tullar PE (1946) The effect of d-tubocurarine on the electrical activity of dogs' brains. Fed Proc 5:67
McLennan H (1961) On the response of the vasomotor system to somatic afferent nerve stimulation and the effects of anaesthesia and curare thereon. Pfluegers Arch 273:604–613
Mikhelson VA (1957) Application of curare-like drug succinylcholine in surgery (in Russian). In Mndzhoyan AL (ed) Ditilin i opyt ego klinicheskogo primeneniya (Succinylcholine and clinical data on it). Armenian SSR Academy of Sciences, Yerevan, pp 158–167
Mori K, Iwabuchii K, Fujita M (1973) The effects of depolarizing muscle relaxants on the electroencephalogram and the circulation during halothane anesthesia in man. Br J Anaesth 45:604–610
Munroe JP, Jenkins LC, Ling GM (1966) Experimental central nervous system studies related to anaesthesia: clinical implications III. Effects of muscle relaxants on sensory inflow. Can J Anaesth 13:109–118
Munson ES, Wagman JH (1972) Elevation of lidocaine seizure threshold by gallamine. Arch Neurol 28:324–329
Muratov VK (1962) On the effect of neuromuscular blocking agents on the central nervous system (in Russian). Farmakol Toksikol 6:758–759
Muratov VK (1963a) O vliyanii kurarepodobnykh veshchestv na techenie nekotorykh refleksov (On the effect of neuromuscular blocking agents on some reflexes). Cand Med Sci Thesis, Moscow, 1st Medical Institute
Muratov VK (1963b) On the effect of neuromuscular blocking agents on linguomandibular reflex (in Russian). Farmakol Toksikol 5:597–602
Muratov VK (1967) The effect of neuromuscular blocking agents on reflex alterations of arterial pressure (in Russian). Farmakol Toksikol 2:156–159

Muratov VK, Sinitsyn LN, Uspensky AE, Kharkevich DA (1970) On the effect of depolarizing and nondepolarizing neuromuscular blocking agents on central nervous system (in Russian). In: Kharkevich DA (ed) Novye kurarepodobnye i ganglioblokiruyushchie sredstva (New neuromuscular blocking and ganglion blocking agents). Meditsina, Moscow, pp 63–73
Naess K (1950) The effect of d-tubocurarine on the mono- and polysynaptic reflex of the spinal cord including a comparison with the effect of strychnine. Acta Physiol Scand 21:34–40
Ochs S (1959) Curare and low blood pressure effects on direct cortical responses. Am J Physiol 197:1136–1140
Okuma T, Fujimori M, Hayashi A (1965) The effect of environmental temperature on the electrocortical activity of cats immobilized by neuromuscular blocking agents. Electroencephalogr Clin Neurophysiol 18:392–400
Oliver KL, Funderburk WH (1965) Possible role of hyperventilation in the CNS effects attributed to tubocurarine. Electroencephalogr Clin Neurophysiol 19:501–508
Oscarsson O, Rosen I (1963) Projection to cerebral cortex of large muscle-spindle afferents in forelimb nerves of the cat. J Physiol (London) 169:924–945
Oscarsson O, Rosen I (1966) Short-latency projections to the cat's cerebral cortex from skin and muscle afferents in the contralateral forelimb. J Physiol (London) 182:164–184
Ostow M, Garcia F (1949) Effect of curare on cortical responses evoked by afferent stimulation. J Neurophysiol 12:225–229
Peiss CN, Manning JW (1959) Excitability changes in vasomotor areas of the brain stem following d-tubocurarine. Am J Physiol 197:149–152
Purpura DP, Grundfest H (1956) Nature of dendritic potentials and synaptic mechanisms in cerebral cortex of cat. J Neurophysiol 19:573–595
Repin IS (1960) On the profound and peculiar EEG inhibition during hypercapnia (on the role of CO_2 as a regulator of the tone of brain-stem reticular formation) (in Russian). Proceedings of the 1st Conference on physiology, morphology, pharmacology and clinics of reticular formation of the brain stem, Moscow, pp 93–94
Saakov BA (1957a) Changes of cortical biopotentials in big hemispheres caused by succinylcholine (in Russian). Byull Eksp Biol Med 6:44–48
Saakov BA (1957b) On pharmacodynamics and clinical application of succinylcholine (in Russian). In: Mndzhoyan AL (ed) Ditilin i opyt ego klinicheskogo primeneniya (Succinylcholine and clinical data on it). Armenian SSR Academy of Sciences, Yerevan, pp 56–66
Saavedra JM, Kvetnansky R, Kopin IJ (1979) Adrenaline, noradrenaline and dopamine levels in specific brainstem areas of acutely immobilized rats. Brain Res 160:271–280
Sakuma A (1959) Study on the central action of neuromuscular blocking agents. Jap J Pharmacol 9:30–40
Savarese JJ (1979) How may neuromuscular blocking drugs affect the state of general anaesthesia. Anaesth Analg (Cleve) 58:449–451
Shapovalov AI (1959) Transmission in the spinal cord under curarization (in Russian). Fiziol Zh SSSR 8:952–958
Shea LT, Davidson GM, Davis J (1954) Electroencephalographic studies of the curarized patient. Med J Aust 2:656–659
Sinitsyn LN, Kharkevich DA (1967) Effect of neuromuscular blocking drugs on cerebral cortex potentials evoked by the stimulation of the lower cardiac and vagus nerves (in Russian). Farmakol Toksikol 4:423–427
Skorobogatov VI (1961) Effect of paramionum on the central nervous system (in Russian). In: Valdman AV (ed) Issledovaniya po farmakologii retikulyarnoi formatsii i sinapticheskoi peredachi (Pharmacological studies on reticular formation and synaptic transmission). 1st Leningrad Medical Institute Publishers, Leningrad, pp 251–256
Skorobogatov VI (1963) Central action of neuromuscular blocking agents (in Russian). In: Valdman AV (ed) Aktualnye problemy farmakologii retikulyarnoi formatsii i sinapticheskoi peredachi (Actual problems of pharmacology of reticular formation and synaptic transmission). 1st Leningrad Midical Institute Publishers, Leningrad, pp 306–320

Smith CM (1963) Neuromuscular pharmacology: drugs and muscle spindles. Ann Rev Pharmacol 3:223–242
Smith SM, Brown HO, Toman JEP, Goodman LS (1947) The lack of cerebral effects of d-tubocurarine. Anesthesiology 8:1–14
Straw RN (1968) The effect of selected neuromuscular blocking agents and spinal cord transection on cortical after-discharge duration in the cat. Electroencephalogr Clin Neurophysiol 25:69–72
Taverner D (1953) The action of d-tubocurarine on slow potentials recorded from the dorsum of the spinal cord in the cat. Acta Physiol Scand [Suppl] 29 106:65–72
Trippenbach T (1973) The effect of peripheral action of muscle relaxants on the activity of phrenic motoneurones. Acta Physiol Pol 24:793–802
Tsoukaris-Kupfer D, Liblau L, Legrand M, Schmitt H (1980) Central cardiovascular actions of d-tubocurarine and inhibition of the hypotensive effect of clonidine. Eur J Pharmacol 65:301–304
Tsoukaris-Kupfer D, Liblau L, Legrand M, Schmitt H (1981) Effects centraux de quelques curarisants sur la pression artérielle et la fréquence cardiaque du chien. Leur influence sur les actions centrales de la clonidine. J Pharmacol 12:100
Valdman AV, Ivanova ZN, Kharkevich DA (1955) Effect of diplacine in the transmission in various links of reflex arch (in Russian). Farmakol Toksikol 2:3–10
Vasilescu V, Cinca J, Drocan J, Oproiua A, Suteanu S (1960) Data concerning the action of curare on the respiratory centre. Rumanian Med Rev 4:7–11
Waser PG (1973) Localization of ^{14}C-Pancuronium by Histo- and wholebody-Autoradiography in normal and pregnant mice. Naunyn-Schmiedebergs Arch Pharmacol 279:399–412
Wild K (1981) Der Einfluß systemisch verabreichter Muskelrelaxantien auf die spontanen und evozierten hirnelektrischen Aktivitäten der Katze. In: Buzello W (ed) Muskelrelaxantien. Neuere Konzepte ihrer Pharmakologie und klinischen Anwendung. Thieme, Stuttgart, pp 62–80
Winson J (1976) Hippocampal theta rhythm. Depth profiles in the curarized rat. Brain Res 103:57–70
Zaiko NN (1962) Changes in permeability of biological barriers under various pathogenic factors (in Russian). Patol Fiziol Eksp Ter 3:13–15

CHAPTER 10

Biodegradation and Elimination of Neuromuscular Blocking Agents

J. B. STENLAKE

A. Introduction

All neuromuscular blocking agents are characterised by the presence of quaternary ammonium groups. Such groups are essentially stable chemical functions. Thus, although quaternary ammonium compounds can be degraded to tertiary amines, in general this requires treatment with caustic alkali, i.e. approximately pH 14, at temperatures of around 100 °C (HOFMANN 1851). It is not surprising, therefore, that the tetraalkylammonium group is not normally subject to biotransformation by the usual metabolic pathways of oxidation, reduction, hydrolysis, and conjugation in mammalian species. Consequently simple quaternary ammonium salts are excreted unchanged.

Gallamine is not metabolised (DAL SANTO 1972), and in the dog 84% of the administered dose is excreted unchanged in the urine in 24 h. Tubocurarine has been recovered, 33% as unmetabolised drug from urine in 10–15 h after injection in humans (KALOW 1953). Similarly, alcuronium although much shorter acting, is eliminated almost entirely unchanged mainly in the urine (50% in 4 h), but with some 10% in the bile (WASER and LÜTHI 1966).

Quaternary ammonium salts are not only essentially stable chemical entities, but because of their ionic characteristics they are, typically, hydrophilic. These intrinsically salt-like hydrophilic properties predispose them towards excretion via the urinary system rather than by the biliary and faecal route. Additionally, such metabolism as may occur necessarily favours the formation of more water-soluble components which also tend to be excreted primarily via the kidneys and urine. It is not surprising, therefore, that those neuromuscular blocking agents which are extensively metabolised are excreted more rapidly, or that there is a high correlation between extent of metabolism and duration of action. It follows, too, from the dependence of quaternary ammonium compounds on the urinary excretion pathway that elimination is diminished in patients with renal failure or insufficiency, and in these circumstances blockade can be substantially prolonged.

Neuromuscular blocking agents which are biodegradable fall for the most part into two categories. Those which are essentially dependent on degradation by the normal drug-metabolising enzyme pathways of ester hydrolysis and azoreductase. These include suxamethonium, pancuronium, vecuronium, and fazadinium. The second group of which atracurium is the principal example, are degraded primarily by novel totally nonenzymic mechanisms such as the Hofmann elimination. The features, advantages, and disadvantages of these various in vivo pathways are discussed in the following sections.

B. Ester Hydrolysis

I. Suxamethonium

The influence of biodegradation upon the time course of action of neuromuscular blocking agents first emerged with the advent of suxamethonium (Fig. 1, compound 1). The importance of its close structural analogy to decamethonium with the same bistrimethylammonium group and similar, but not identical ten-atom unit interquaternary chain was evident in the close similarity of their neuromuscular blocking properties. However, in contrast to the prolonged block seen with decamethonium in all species, that observed with suxamethonium was relatively short.

This marked difference in duration of action between suxamethonium and decamethonium was ascribed to enzymatic hydrolysis of the two ester functions in the interquaternary chain (BOVET-NITTI 1949). In 1952, WHITTAKER and WIJESUNDERA showed that suxamethonium was hydrolysed rapidly in the plasma of normal human subjects. The molecule is split first to monosuccinylcholine (compound 2) and ultimately to succinic acid and choline (Fig. 1), all of which have negligible neuromuscular blocking potency. It was established that plasma pseudocholinesterase, and not the acetylcholinesterase of erythrocytes, was primarily responsible for this hydrolysis in vivo, and this was confirmed in a trial which showed the reversal of apnoea in human patients after administration of suxamethonium by cholase, a preparation of human pseudocholinesterase (EVANS et al. 1953).

Despite its rapid metabolism and short action in normal subjects, suxamethonium is not without problems, both pharmacological and metabolic. Serious metabolic malfunction due to genetically determined deficiencies in the biosynthesis of pseudocholinesterase (EVANS et al. 1953) is fortunately a rare occurrence amongst Caucasian populations. Enzyme biosynthesis is controlled by two allelic genes normal (E_1^u) and abnormal (E_1^a) (KALOW and STARON 1957). Individuals therefore fall into three types. The bulk of the population (approximately 96%) consists of homozygotes ($E_1^uE_1^u$) in which suxamethonium hydrolysis is normal (KALOW and GENEST 1957). Sensitive individuals are homozygotes ($E_1^aE_1^a$) and are

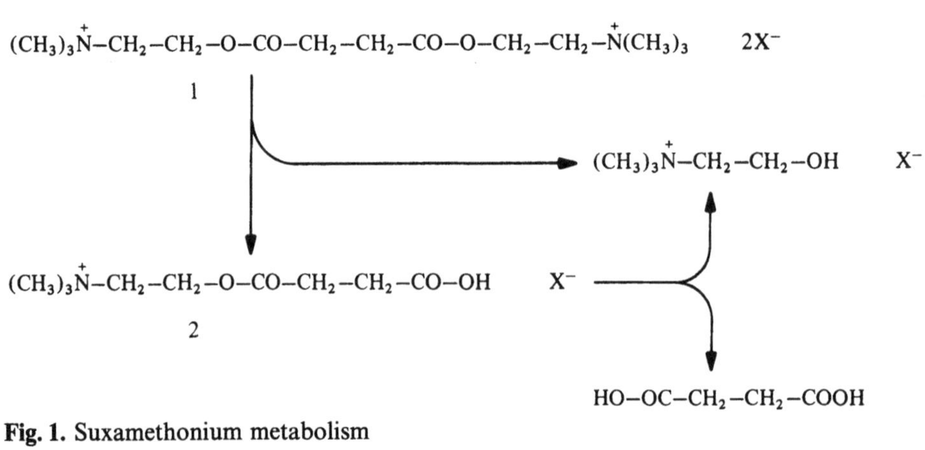

Fig. 1. Suxamethonium metabolism

relatively few in number (approximately 0.5% of the population). A third genotype consists of relatives (approximately 4% of the population) who are heterozygotes ($E_1^uE_1^a$), having some cholinesterase deficiency. The concept of a silent gene (E_1^s) that produces a nonfunctional enzyme protein has further led to the identification of heterozygotes ($E_1^uE_1^s$, approximately 0.5% of the population) with normal cholinesterase levels, heterozygotes ($E_1^aE_1^s$) with abnormally low cholinesterase levels and homozygotes ($E_1^sE_1^s$) who are totally anenzymatic (LIDDELL et al. 1962; HART and MITCHELL 1962). The incidence of silent genes is generally low in Caucasian populations (WHITTAKER 1980), but unusually high incidences have been reported in Caucasian South Africans (PANNALL et al. 1976) and in Eskimos (GUTSCHE et al. 1967).

In practical terms about one in 2,800 persons has an atypical plasma pseudocholinesterase that hydrolyses suxamethonium at slower than normal rates (KALOW and GUNN 1959). Such enzyme deficiences can increase the half-life of suxamethonium from 2 min in normal subjects to 4 h or more in cases of severe deficiency (MERRETT et al. 1983).

Plasma pseudocholinesterase levels are also reduced in liver disease (BOWEN 1960) myocardial infarction and various cancers (KANIARIS et al. 1979). Pseudocholinesterase levels also fall in otherwise normal female subjects during pregnancy. A recent study (EVANS and WROE 1980) confirms that suxamethonium sensitivity is greatest at full term and in the first 7 days following delivery. Some 11% of patients are considered to be at risk should this agent be required for rapid intubation prior to caesarean section.

II. Pancuronium

Pancuronium (compound 3) has a significantly shorter duration of action than tubocurarine and shows in humans a 50% recovery of the peak tetanic contraction of the adductor pollicis muscle after neuromuscular paralysis in about 45 min compared with about 75 min for tubocurarine (PAYNE and HUGHES 1981). It is pertinent that the total excretion (33%–44% in the urine and 11% in the bile over 30 h) of pancuronium and its metabolites in humans (AGOSTON et al. 1973) is substantially less than that for tubocurarine (60%–70% in the urine over 24 h, KALOW 1953). The more rapid recovery from pancuronium, therefore, is almost certainly largely due to the formation and more rapid excretion of metabolites. Hydrolysis of the acetoxy groups at positions 3 and 17 occurs (Fig. 2) with the formation of all three possible metabolites: compounds (4), (5) (dacuronium), and (6). Deacetylation occurs mainly at the less-hindered 3 position, with about 20% of the administered dose being recovered as 3-*O*-deacetylpancuronium in 30 h, and decreasing amounts of compounds (5) and (6) respectively.

Deacetylation leaves the steroid nucleus of pancuronium intact, so that in contrast to suxamethonium, all three pancuronium metabolites retain the bisquaternary structure and hence some neuromuscular blocking activity, the 3-*O*-deacetyl compound being about half as potent as pancuronium in humans (MILLER et al. 1978). Metabolites (5) and (6) have much lower potencies, 1/50th and 1/54th that of pancuronium respectively. Apparent elimination half-lives of 110, 68, 73, and 71 min are reported for pancuronium and the 3-hydroxy, 17-hydroxy,

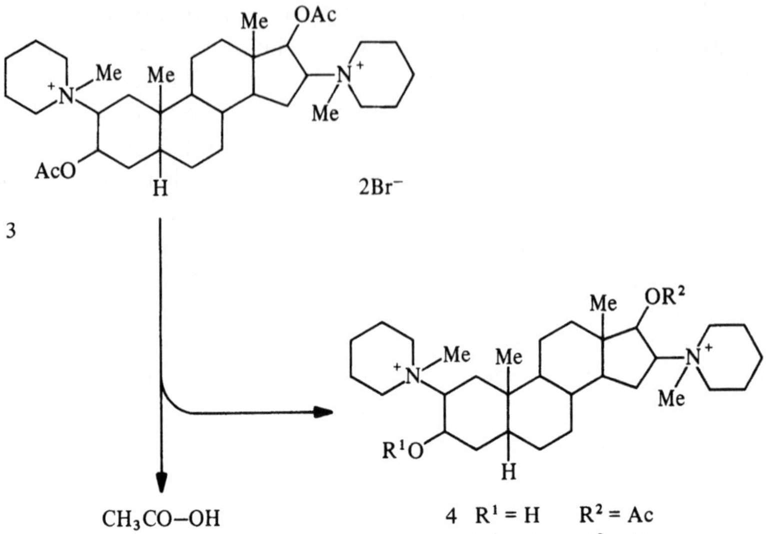

3

4 $R^1 = H$ $R^2 = Ac$
5 $R^1 = Ac$ $R^2 = H$
6 $R^1 = H$ $R^2 = H$

Fig. 2. Pancuronium metabolism

and 3,17-dihydroxy metabolites respectively. It follows from the metabolic pathway and excretion kinetics that clearance rates may be reduced some fivefold (McLEOD et al. 1976), and duration of blockade almost doubled (STOJANOV 1969) in patients with renal failure.

Metabolic deacetylation of pancuronium occurs primarily in the liver. Consequently, duration of action is also prolonged in patients with liver obstruction or damage (SOMOGYI et al. 1977). Hepatic drug-metabolising ability and renal function both diminish with advancing age. Circulation also slows in the elderly. The combination of these effects shows in the marked decrease of plasma clearance of pancuronium from approximately 6 l/h at age 20 years to approximately 2 l/h at age 85 years (McLEOD et al. 1979), and accounts for the increase in recovery time frequently seen in older patients (aged 60 years and over).

III. Vecuronium

Vecuronium (Fig. 3, compound 7), 2-*N*-demethylpancuronium, is a monoquaternary compound with the same potential as tubocurarine to equilibrate with its protonated 2,16-dication (compound 8) at physiological pH. That this dication is the principal active neuromuscular blocking species is evident from the direction of potency changes observed with changing plasma pH (FUNK et al. 1980).

The duration of vecuronium and recovery from full block are significantly shorter than those of pancuronium in the cat and dog (DURANT et al. 1979 a). Onset (4.5 min), duration (27 min), and recovery after continuous infusion (17 min) are also significantly shorter in humans (AGOSTON et al. 1980) than for pancuro-

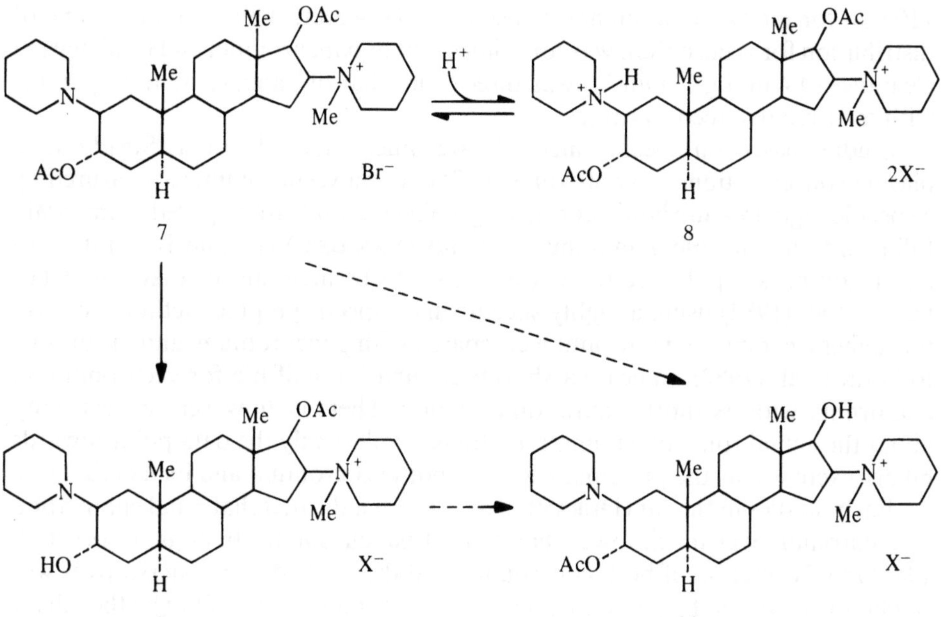

Fig. 3. Vecuronium metabolism

nium (7.0, 58, and 46 min respectively). These differences can be partly ascribed to the enhanced rate of metabolic *O*-deacetylation in vecuronium.

The rapid *O*-deacetylation of the ditertiaryamino diacetate precursor of pancuronium was ascribed to neighbouring group participation of the adjacent amino groups which assist hydrolysis. A similar effect accounts for the equally ready 3-*O*-deacetalation of vecuronium (Fig. 3, compound 7). Thus hydrolysis occurs in aqueous solution at 20 °C within 0.5 h (SAVAGE et al. 1980) in contrast to the relative stability of pancuronium (compound 3) and 3-*O*-deacetylpancuronium (compound 4) over 6 h under the same conditions. The reported equally rapid (0.5 h at 20 °C) in vitro 17-*O*-deacetylation of 3-*O*-deacetylvecuronium (compound 9; Org-7268), however, cannot be ascribed to neighbouring group participation, since this compound retains the same adjacent 16-*β*-quaternary ammonium group as pancuronium and 3-*O*-diacetylpancuronium, in which this deacetylation is relatively slow. The report (VAN DER VEEN and BENCINI 1980) that 17-*O*-deacetylvecuronium (compound 10) is not found in plasma samples after vecuronium administration lends support to this view.

There is now a substantial body of evidence to indicate that the shorter duration of action of vecuronium and recovery therefrom compared with pancuronium arise as a consequence of the different excretion patterns and kinetics of the two compounds. Thus, a study in young adults (SOHN et al. 1982) confirms that recovery from vecuronium is about three times as fast as from pancuronium, but the accompanying pharmacokinetics based on measurements using the nonspecific fluorimetric method (KERSTEN et al. 1973), which does not distinguish the drugs from their metabolites, show similar elimination half-lives of 103 and

110 min for the two compounds respectively. However, the apparent volume of distribution for vecuronium was four times that of pancuronium and total plasma clearance (4.4 ml kg^{-1} min^{-1}) was three times that of pancuronium, in parallel with their relative recovery rates.

Studies based on specific methods are much more clear-cut. Steady state plasma concentrations of approximately 250 ng/ml vecuronium, as measured by a specific high pressure liquid chromatographic method, are required to maintain full neuromuscular block in human patients (VAN DER VEEN and BENCINI 1980) and recovery is rapid once these levels cease to be maintained. Likewise, CRONELLY et al. (1982) using a highly specific mass spectrographic method, confirm the higher clearance of vecuronium compared with pancuronium, and in contrast to SOHN et al. (1982), indicate a shorter elimination half-life for vecuronium in conformity with its shorter duration of action. There is, however, less certainty about the actual route of elimination, though collectively the data point towards biliary excretion as the principal route for both vecuronium and its metabolites.

Early studies in the cat (DURANT et al. 1979 b) indicated that renal elimination of vecuronium was much lower than that of pancuronium. In dogs (BOOIJ et al. 1981) renal excretion of both compounds and their 3-*O*-deacetyl derivatives was similar (5%–20% in 6 h) with a similar level of excretion (7%–28%) by the biliary route. However, marked and contrasting differences in the excretion pattern of vecuronium and pancuronium were observed in the rat by a specific mass spectrometry assay (UPTON et al. 1982). Pancuronium was excreted primarily in the urine (85%–6% of the administered dose in 12 h), with biliary excretion accounting for barely 7.5% in 7 h. In contrast, only 3.5% of vecuronium appeared in the urine (12 h), but 46% was excreted in the bile (7 h). The much higher biliary clearance rate of vecuronium was ascribed to the greater lipophilicity of this monoquaternary compound compared with that of bisquaternary pancuronium. The importance of the biliary route in the rat is confirmed by the threefold increase in the duration of action of vecuronium in rats in both extrahepatic cholestasis and during intravenous infusion of taurocholate (WESTRA et al. 1981).

The report that duration of neuromuscular blockade following vecuronium administration was not prolonged by renal ligation in cats (DURANT et al. 1979 b) provides further support for the view that urinary excretion is not the principal route of excretion in this or indeed other species. As yet, however, there is little direct evidence that biliary excretion is the major pathway for the elimination of vecuronium in humans, though a number of studies seem to point in that direction, as for example that of CRONELLY et al. (1982). FAHEY et al. (1981) found no difference in the pharmacokinetics of vecuronium in patients with and without renal failure which also implies that urinary excretion is not the controlling factor in its elimination. However, the steady state dose requirement (SSDR) is age dependent and recovery rates are prolonged in older patients (D'HOLLANDER et al. 1982).

C. Azo Fission

Fazadinium

The early promise of fazadinium (AH 8165; Fig. 4, compound 11) has not been fully confirmed in humans particularly in the incidence of cardiovascular effects at less than full paralysing doses and in its more prolonged action and much slower recovery (HUGHES et al. 1976). Differences in duration of action in animals and humans appear to be due to differences in metabolism and excretion. In the rat and dog (BOLGAR et al. 1972) about 70% was excreted in the first 24 h after intravenous injection. Radioactivity was equally distributed in both bile and urine in both species. Two metabolites were observed after large doses. Metabolite 1 was identified as 3-methyl-2-phenylimidazo[1,2-*a*]-pyridine (compound 12). The second and major metabolite, which was not identified, was neutral in reaction and presumably an *N*-glucuronide or other conjugate of metabolite 1. It was suggested that enterohepatic circulation might account for the apparent persistence of the drug (and metabolites) after large doses. In vitro studies showed that fazadinium was rapidly metabolised by an NADPH-dependent azoreductase present in liver microsomes to metabolite 1 and nitrogen (Fig. 4).

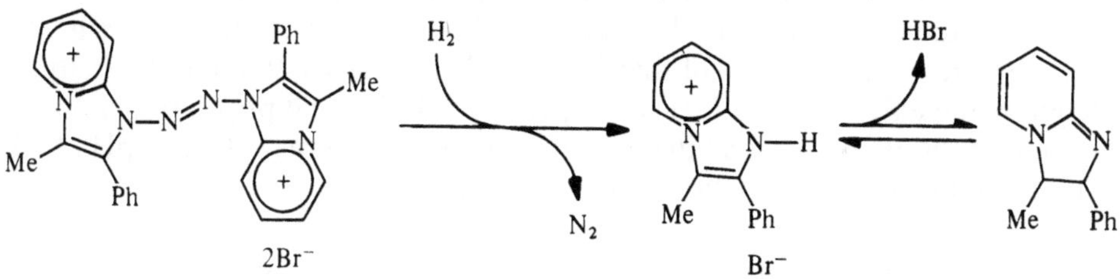

11

12

Fig. 4. Fazadinium metabolism

Metabolism of fazadinium in humans follows the same pathway (DUVALDESTIN et al. 1978 b), but is much slower so that it is excreted mainly unchanged in the urine with an elimination half-life $t_{1/2\beta}$ of approximately 85 min in patients with normal renal function (DUVALDESTIN et al. 1979). This readily accounts for its more prolonged action and longer recovery time in humans. Dependence on renal function for elimination is emphasised by the extension of $t_{1/2\beta}$ to approximately 140 min in patients with end-stage renal failure and the observation that even in such patients plasma concentration of metabolites accounted for no more than 7% of unchanged fazadinium (DUVALDESTIN et al. 1979).

The importance of the urine as the main elimination pathway for fazadinium is emphasised by the results of a pharmacokinetic study in patients with liver disease (DUVALDESTIN et al. 1980). A 90% increase in both $t_{1/2\alpha}$ from 10 to 19 min and $t_{1/2\beta}$ from 82 to 153 min was observed in patients with cirrhosis. Similarly, in total biliary obstruction $t_{1/2\beta}$ was increased, but only marginally to 103 min.

D. Hofmann Elimination

Atracurium

Atracurium (Tracrium, Fig. 5, compound 13) is the first compound to be introduced into clinical practice in which the principal degradation pathway is completely independent of enzymic function (COKER et al. 1981). It was designed (STENLAKE 1978; STENLAKE et al. 1981) to degrade to inactive and innocuous products at physiological pH and temperature primarily by Hofmann elimination. As first described by HOFMANN (1851) this reaction occurs at temperatures around 100 °C when quaternary ammonium salts are heated with sodium hydroxide (i.e. pH ~14). The positively charged nitrogen of the quaternary ammonium salt is electron attracting and functions as an electron sink. It weakens adjacent C–C and C–H bonds, facilitating attack by hydroxyl ions or other nucleophiles at the β-C–H bond, leading to bond breaking and formation of water, an olefin and a tertiary amine.

$$HO^- \; H{-}CH_2{-}CH_2{-}\overset{+}{N}R_3 \xrightarrow{pH\ 14/100\ ^\circ C} H_2O + CH_2{=}CH_2 + NR_3$$

The course and rate of this reaction are markedly influenced by steric and electronic effects of substituents on the quaternary nitrogen and adjacent carbon atoms (INGOLD 1962). In particular, the presence of a second electron-attracting group (X) attached to the β-carbon atom facilitates the reaction. The design of atracurium rests in the choice of that functional group, such that the elimination will proceed readily at physiological temperature (37 °C) and pH (7.4) to limit its duration of action, but be inhibited at still lower temperature (5 °C) and pH (3–4) so that a stable injection solution can be prepared.

$$HO^- \; H{-}\underset{\displaystyle X}{\underset{|}{CH}}{-}CH_2{-}\overset{+}{N}R_3 \rightarrow H_2O + \underset{\displaystyle X}{\underset{|}{CH}}{=}CH_2 + NR_3$$

This novel nonenzymic biodegradation mechanism combined with an optimum pharmacodynamic profile is achieved in atracurium (compound 13) by the introduction of ester carbonyl functions adjacent to the β-carbon atoms in the interquaternary chain. Atracurium is a competitive neuromuscular blocking agent readily reversible by neostigmine and virtually free of the major disadvantges of other neuromuscular blocking agents in current clinical practice (HUGHES and CHAPPLE 1981; PAYNE and HUGHES 1981). Its biodegradation is promoted by the combined electron-attracting properties of the two β-linked ester carbonyl groups and the positively charged nitrogens of the quaternary ammonium groups.

The resulting molecular fragmentation (Fig. 5) yields laudanosine and other products (STENLAKE et al. 1981; NEILL and CHAPPLE 1982) which are without neuromuscular cardiovascular or other effects at the concentrations present following clinical doses (CHAPPLE and CLARK 1983; SKARPA et al. 1983). Whole body autoradiography in the rat with ^{14}C-labelled atracurium also shows that neither atracurium nor the laudanosine produced from it enter the central nervous system (SKARPA et al. 1983). Likewise, administration to anaesthetised pregnant cats and

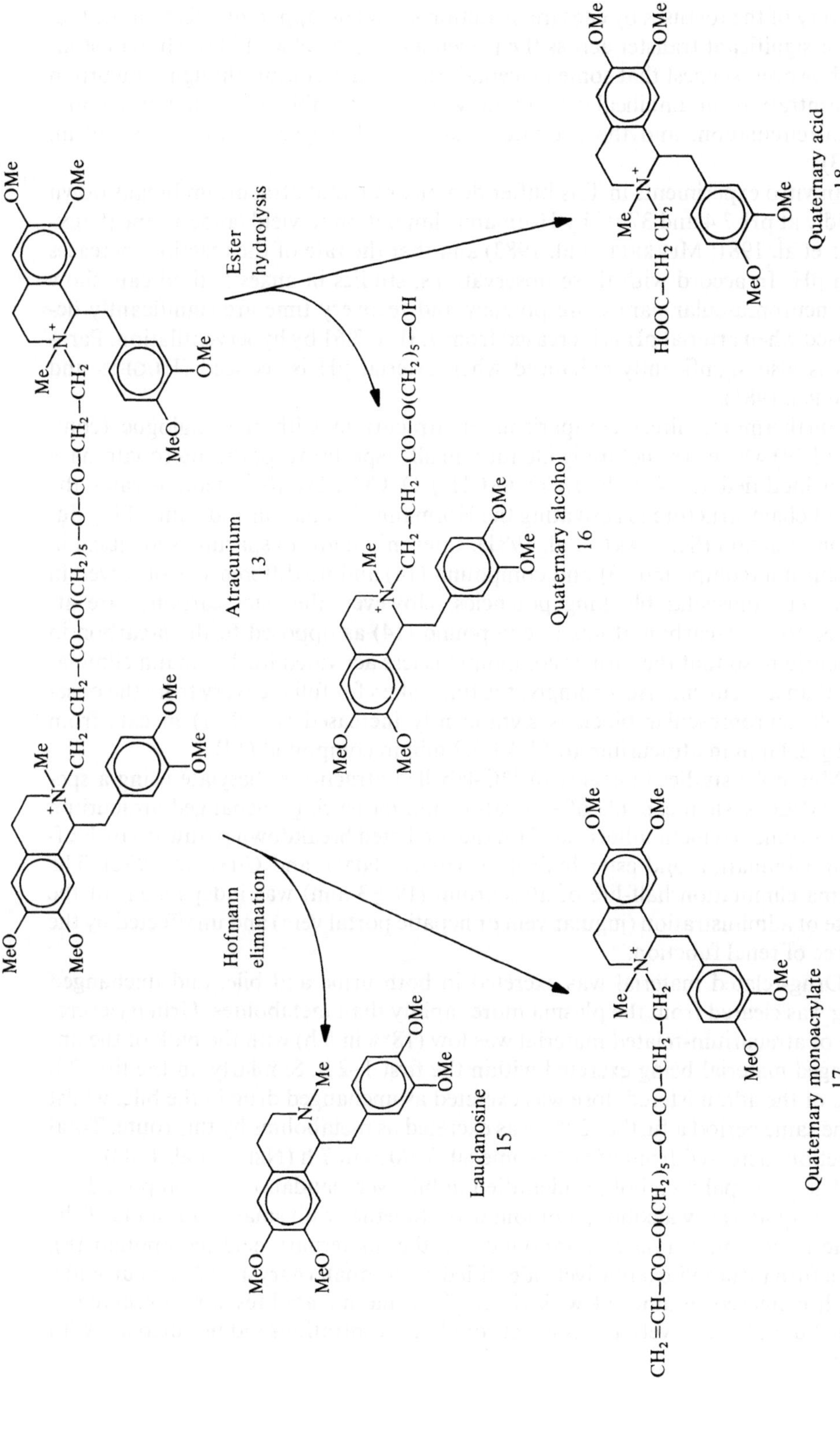

Fig. 5. Atracurium metabolism (STENLAKE et al. 1983)

delivery of the foetuses by caesarean section shows no apparent effect on the foetus or significant transfer across the placenta (SKARPA et al. 1983). Human studies, however, suggest that some placental transfer does occur, though atracurium concentrations in umbilical cord blood were considerably lower than in the maternal circulation, and the neonates had normal Apgar scores (FRANK et al. 1983).

In vitro experiments in Tris buffer demonstrate that atracurium breaks down rapidly at pH 7.4 and 37 °C by Hofmann elimination to yield laudanosine (STENLAKE et al. 1981; MERRETT et al. 1983) and that the rate of the reaction increases with pH. In accord with these observations, studies in anaesthetised cats show that neuromuscular paralysing potency and recovery time are significantly decreased when arterial pH is increased from 7.31 to 7.61 by hyperventilation. Paralysis is also significantly enhanced when arterial pH is reduced (HUGHES and CHAPPLE 1981).

Furthermore, direct comparison of atracurium with the analogue (compound 14) which is structurally identical in all respects except the interquaternary chain modified to $N^+(CH_2)_3CO \cdot O(CH_2)_3O \cdot CO(CH_2)_3N^+$ demonstrates the role of chain structure in activating the Hofmann elimination and controlling duration of action (STENLAKE et al. 1981). Interonium group spacing is identical in atracurium (compound 13) and compound (14) and no difference is observed in their neuromuscular blocking potencies. However, the ester carbonyl are attached to the γ-carbon atoms in compound (14) as opposed to the β-carbon in atracurium, so that the former compound is less activated for Hofmann elimination than atracurium. Accordingly, the time taken for full recovery from the onset of full neuromuscular block is significantly increased ($P<0.01$) in cats from 22.2 ± 2.9 min in atracurium to 51.4 ± 5.2 min in compound (14).

Metabolic studies in cats with ^{14}C-labelled atracurium besylate using a specific HPLC system capable of separating and measuring unchanged atracurium and its principal metabolites, confirm the predicted breakdown pathways by Hofmann elimination and ester hydrolysis (Fig. 5; NEILL and CHAPPLE 1982). The plasma elimination half-life of atracurium (19 ± 3 min) was independent of the route of administration (jugular vein or hepatic portal vein) and unaffected by the degree of renal function.

Drug-related material was excreted in both urine and bile, and unchanged drug was cleared from the plasma more rapidly than metabolites. Urinary excretion of atracurium-related material was low (18% in 5 h) with the bulk of the unchanged material being excreted within the first 1–2 h. Similarly, in the first 2 h 10% of the administered dose was excreted as unchanged drug in the bile, whilst in the same period a further 27% was excreted as metabolites by this route. Total excretion increased from 65% to some 80%–90% in 7 h (NEILL et al. 1983).

The principal metabolites identified in bile were laudanosine (compound 15) and the quaternary alcohol (compound 16) together with smaller amounts of the quaternary monoacrylate (compound 17), the quaternary acid (compound 18), and a fifth metabolite tentatively identified as the quaternary acid *N*-glucuronide, which increased in amount with time. The same metabolites were observed in urine but at lower levels and *N*-glucuronide concentrations did not increase with time.

Plasma elimination half-life of atracurium (19±2 min) and the area under the curve (AUC) were unaltered when the drug was administered via the hepatic portal vein. Total urinary excretion of drug and metabolites were higher (43% in 5 h) and excretion in bile lower (38% in 5 h) than when administered via the jugular vein (18% in urine and 47% in bile in 5 h). Similarly, the plasma elimination half-life of atracurium (23±5 min) was not significantly different after bilateral renal ligation and administration via the jugular vein, though the AUC representing both drug and metabolites was increased by about 50%. Excretion in bile was similar to that in animals with normal renal function (15% atracurium and 33% metabolites in 2 h).

MERRETT et al. (1983) measured the degradation of atracurium in vitro following incubation in buffer and blood plasma at 37 °C by determining the fall with time in whole body effects (loss of motor activity) after intravenous injection of measured volumes of the drug–plasma-incubated mixtures in mice. The resulting in vitro degradation half-lives of atracurium were shorter in mice ($t_{1/2}<4$ min; $n=2$) and rats ($t_{1/2}<14$ min; $n=2$) than in cats ($t_{1/2}<31$ min; $n=4$), dog ($t_{1/2}<33$ min; $n=1$), rabbit ($t_{1/2}<44$ min; $n=1$) of humans ($t_{1/2}<27$ min; $n=4$). These measurements do not distinguish between degradation by Hofmann elimination and ester hydrolysis both of which are base catalysed. The half-life in human plasma is, however, about half that found in Tris buffer at the same pH, indicating that ester hydrolysis is probably enzyme catalysed.

Similar experiments in human plasma from patients with normal and genetically determined deficiencies of pseudocholinesterase activity show that this enzyme is not involved in the breakdown of atracurium (MERRETT et al. 1983). These show that whereas the mean half-life of suxamethonium was extended from 2.6 min in plasma from patients with normal pseudocholinesterase activity to >4 h in plasma from homozygotes with virtually zero pseudocholinesterase activity, no significant difference was observed in the half-life of atracurium in atypical and normal plasma samples.

Human pharmacokinetic studies on atracurium in three groups of patients undergoing routine surgery with ages ranging from 24 to 69 years and average weight 75 kg (range 51–91 kg) receiving 0.3 mg/kg, 0.6 mg/kg, and 0.9 mg/kg respectively gave remarkably consistent data, showing an elimination half-life of 20 min and plasma clearance of 5.1 ml min^{-1} kg^{-1} body weight (WEATHERLEY et al. 1983). The linearity of the kinetics of bolus-injected and infused atracurium is also established, and an extended two-compartment open model to take account of the nonenzymic Hofmann elimination has been developed.

Inactivation of atracurium by both Hofmann elimination and ester hydrolysis is slowed by cooling. Its half-life in human plasma in vitro increased from 18 min at 37 °C to 15.5 h at 5 °C (MERRETT et al. 1983). Induced hypothermia in which body temperature was lowered to 25°–26 °C for 45–133 min in cardiopulmonary bypass operations permitted the maintenance of adequate surgical relaxation with significantly lower mean infusion rates of atracurium (0.0040±0.003 mg kg^{-1} h^{-1}) than during normothermia (mean infusion rate 0.0060±0.003 mg kg^{-1} h^{-1}).

References

Agoston S, Vermeer GA, Kersten UW, Meijer DKF (1973) The fate of pancuronium in man. Acta Anaesthesiol Scand 17:267–275

Agoston S, Salt P, Newton D, Bencini A, Boomsma P, Erdman W (1980) The neuromuscular blocking action of Org NC45, a new pancuronium derivative, in anaesthetized patients. Br J Anaesth 52:53S–59S

Bolgar L, Brittain RT, Jack D, Jackson MR, Martin LE, Mills J, Pointer D, Tyers MB (1972) Short-lasting, competitive neuromuscular blocking activity in a series of azobisarylimidazo[1,2,*a*]-pyridinium dihalides. Nature 238:354–355

Booij LHDJ, Vree TB, Hurkmans F, Reekers-Ketting JJ, Crul FF (1981) The pharmacokinetics and pharmacodynamics of the muscle relaxant drug Org NC45 and each of its hydroxy-metabolites in dogs. Anaesthesist 3077:329–333

Bovet-Nitti F (1949) Degradation of curare-like substances by the action of cholinesterases. Rend ist super sanita 12:138–157

Bowen RA (1960) Anaesthesia in operations for the relief of portal hypertension. Anaesthesia 15:3–10

Chapple DJ, Clark JS (1983) Pharmacological actions of breakdown products of atracurium and related substances. Br J Anaesth 55:11s–22s

Coker GG, Dewar GH, Hughes R, Hunt TM, Payne JP, Stenlake JB, Waigh RD (1981) A preliminary assessment of atracurium, a new competitive neuromuscular blocking agent. Acta Anaesthesiol Scand 25:67–69

Cronelly R, Gencorelli P, Miller RD, Fisher DM, Nguyen L (1982) Pharmacokinetics of pancuronium and Org NC45 (Norcuron). Anesth Analg (Cleve) 61:176–177

Dal Santo G (1972) Kinetics of distribution of radioactive labelled muscle relaxants. IV. Urinary elimination of a single dose of ^{14}C-gallamine. Br J Anaesth 44:321–329

D'Hollander A, Massaux F, Nevelsteen M, Agoston S (1982) Age-dependent dose-response relationship of Org NC45 in anaesthetised patients. Br J Anaesth 54:653–657

Durant NN, Marshall IG, Savage DS, Nelson DJ, Sleigh T, Carlyle IC (1979 a) The neuromuscular and autonomic blocking activities of pancuronium, Org NC45 and other pancuronium analogues, in the cat. J Pharm Pharmacol 32:831–836

Durant NN, Houwertjes MC, Agoston S (1979 b) Renal elimination of Org NC45 and pancuronium. Anesthesiology 51:S266

Duvaldestin P, Agoston S, Henzel D, Kersten UW, Desmonts JM (1978 a) Pancuronium pharmacokinetics in patients with liver cirrhosis. Br J Anaesth 50:1131–1136

Duvaldestin P, Henzel D, Demetrious M, Desmonts JM (1978 b) Pharmacokinetics of fazadinium in man. Br J Anaesth 50:773–777

Duvaldestin P, Bertrand JC, Concina D, Henzel D, Lareng L, Desmonts JM (1979) Pharmacokinetics of fazadinium in patients with renal failure. Br J Anaesth 51:943–947

Duvaldestin P, Saada J, Henzel D, Saumon G (1980) Fazidinium pharmacokinetics in patients with liver disease. Br J Anaesth 52:789–794

Evans FT, Gray PWS, Lehman H, Silk E (1953) Effect of pseudocholinesterase level on action of succinylcholine in man. Br Med J 1:136

Evans RT, Wroe JM (1980) Plasma cholinesterase changes during pregnancy. Anaesthesia 35:651–654

Fahey MR, Morris RB, Miller RD, Nguyen T-L, Upton RA (1981) Pharmacokinetics of Org NC45 (Norcuron) in patients with and without renal failure. Br J Anaesth 53:1049–1053

Frank M, Flynn PJ, Hughes R (1983) Atracurium in obstetric anaesthesia. Br J Anaesth 55:113s–114s

Funk DI, Crul JF, van der Pol FM (1980) Effects of changes in acid-base balance on neuromuscular blockade produced by Org NC45. Acta Anaesthesiol Scand 24:119–124

Gutsche BB, Scott EM, Wright RC (1967) Hereditary deficiency of pseudocholinesterase in Eskimos. Nature 215:322–323

Hart SM, Mitchell JV (1962) Suxamethonium in the absence of pseudocholinesterase. Br J Anaesth 34:207–209

Hofmann AW (1851) Beiträge zur Kenntnis der flüchtigen organischen Basen. Ann Chemie 78:253–286
Hughes R, Chapple DJ (1981) The pharmacology of atracurium: a new competitive neuromuscular blocking agent. Br J Anaesth 53:31–44
Hughes R, Payne JP, Sugai N (1976) Studies on fazadinium bromide (DH 8165): a new non-depolarizing neuromuscular blocking agent. Canad Anaesth Soc J 23:36–47
Ingold CK (1962) The mechanism of olefin elimination. Proc Chem Soc 265–274
Kalow W (1953) Urinary excretion of *d*-tubocurarine in man. J Pharmacol Exp Ther 109:74–82
Kalow W, Genest K (1957) A method for the detection of atypical forms of human serum cholinesterase: determination of dibucaine numbers. Can J Biochem Physiol 35:339–346
Kalow W, Gunn DR (1959) Some statistical data on atypical cholinesterase of human serum. Ann Hum Genet 23:239–250
Kalow W, Staron N (1957) On distribution and inheritance of atypical forms of human serum cholinesterase as indicated by dibucaine numbers. Can J Biochem Physiol 35:1305–1320
Kaniaris P, Fassoulaki A, Liarmakopoulou K, Dermitzakis E (1979) Serum cholinesterase levels in patients with cancer. Anesth Analg 58:82–84
Kerten VW, Meijer DFK, Agoston S (1973) Fluorimetric and chromatographic determination of pancuronium bromide and its metabolites in biological materials. Clin Chim Acta 44:59–66
Liddell J, Lehman H, Silk E (1962) A 'silent' pseudocholinesterase gene. Nature 193:561–562
McLeod K, Watson MJ, Rawlins MD (1976) Pharmacokinetics of pancuronium in patients with normal and impaired renal function. Br J Anaesth 48:341–345
McLeod K, Hull CJ, Watson MS (1979) Effects of aging on the pharmacokinetics of pancuronium. Br J Anaesth 51:435–438
Merrett RA, Thompson CW, Webb FW (1983) In vitro degradation of atracurium in human plasma. Br J Anaesth 55:61
Miller RD, Agoston S, Booij LHD, Kersten UW, Crul JF, Ham J (1978) The comparative potency and pharmacokinetics of pancuronium and its metabolites in anaesthetized man. J Pharmacol 207:539–543
Neill EAM, Chapple DJ (1982) Metabolic studies in the cat with atracurium: a neuromuscular blocking agent designed for non-enzymtic inactivation at physiological pH. Xenobiotica 12:203–210
Neill EAM, Chapple DJ, Thompson CW (1983) The metabolism and kinetics of atracurium: a overview. Br J Anaesth 55:23s–26s
Pannall PR, Potgeiter GM, Raubenheimer MM (1976) Plasma cholinesterase variants – an unexpectedly high incidence of the silent allele. S Afr Med J 50:304–306
Payne JP, Hughes R (1981) Evaluation of atracurium in anaesthetised man. Br J Anaesth 53:45–54
Savage DS, Sleigh T, Carlyle I (1980) The emergence of Org NC45, 1-[(2β,3α,5α,16β)-3, 17-bis(acetyloxy-2-(1-piperidinyl)-androstan-16-yl]-1-methylpiperidinium bromide, from the pancuronium series. Br J Anaesth 52:3S–9S
Skarpa M, Dayan AD, Follenfant M, James DA, Thomson PM, Lucke JN, Morgan M, Lovell R, Medd R (1983) Toxicity testing of atracurium. Br J Anaesth 55:27s–30s
Sohn YJ, Scaf AHJ, Bencini A, Gregoretti S, Agoston S (1982) Comparative pharmacokinetics of vecuronium (Org NC45) and pancuronium in man. Fed Proc 41:6229
Somogyi AA, Shanks CA, Triggs EJ (1977) Disposition kinetics of pancuronium bromide in patients with total biliary obstruction. Br J Anaesth 49:1103–1108
Stenlake JB (1978) Biodegradable neuromuscular blocking agents. In: Stoclet FC (ed) Advances in pharmacology and therapeutics, vol 3. Ions – cyclic nucleotides – cholinergy. Pergamon, Oxford, pp 303–312
Stenlake JB, Waigh RD, Dewar GH, Hughes R, Chapple DJ, Coker GG (1981) Biodegradable neuromuscular blocking agents. Part 4. Atracurium Besylate and related polyalkylene di-esters. Eur J Med Chem 16:515–524

Stenlake JB, Weigh RD, Urwin J, Dewar GH, Coker GG (1983) Atracurium. Conception and inception. Br J Anaesth 55:3S–10S
Stojanov E (1969) Possibilities for clinical use of the new steroid neuromuscular blocker pancuronium bromide in anaesthesiological practice. Arzneimittelforsch 19:1723–1725
Upton RA, Nguyen TL, Miller RD, Castagnoli N (1982) Renal and biliary elimination of vecuronium (Org NC45) Anesth Analg (Cleve) 61:313–316
Van der Veen F, Bencini A (1980) Pharmacokinetics and pharmacodynamics of Org NC45. Br J Anaesth 52:37S–41S
Waser PG, Lüthi U (1966) Verteilung, Metabolismus und Elimination von ^{3}H-Diallyl-nor-Toxiferin (Alloferin) bei Katzen. Helv Physiol Pharmacol Acta 24:259–273
Weatherley BC, Williams SG, Neill EAM (1983) The metabolism and kinetics of atracurium: an overview. Br J Anaesth 55:47S–52S
Westra P, Keulemans GTP, Houwertjes MC, Hardonk MJ, Meijer DKF (1981) Mechanisms underlying the prolonged duration of action of muscle relaxants caused by extrahepatic cholestasis. Br J Anaesth 53:217–227
Whittaker M (1980) Plasma cholinesterase variants and the anaesthetist. Anaesthesia 35:174–179
Whittaker VP, Wijesundera S (1952) Hydrolysis of succinyl-dicholine by cholinesterase. Biochem J 52:475–479

On the Relationship Between the Chemical Structure and the Neuromuscular Blocking Activity

CHAPTER 11

Methods for the Experimental Evaluation of Neuromuscular Blocking Agents

D. A. KHARKEVICH

A. Introduction

In the quest for more perfect neuromuscular blocking agents numerous techniques have been suggested (see reviews by LEWIS and MUIR 1959a, b, c; BOWMAN 1964; BARLOW 1968; KHARKEVICH 1969; CHEYMOL and BOURILLET 1972; MACLAGAN 1976). They are mainly based on the ability of the agents to block the neuromuscular transmission on the electrical stimulation of the corresponding motor nerve and to cause myorelaxation. Nondepolarizing agents can also be tested by their antagonism with cholinomimetics (by the elimination of the stimulating effect of the latter). At the same time, depolarizing neuromuscular blocking agents can stimulate the skeletal muscles, thus leading to more or less stable contractions (from twitch to contracture). Neuromuscular blocking agents are studied in various animal species and on their neuromuscular preparations both in vivo, and in vitro. The experiments are carried out in vivo on conscious animals and on animals immobilized by general anesthetics or decerebrated animals. In such experiments the agents tested are usually administered intravenously or intraperitoneally (rarely subcutaneously). In addition, close-arterial injection is sometimes used.

When the compounds are tested in vitro, they are introduced into the artificial bath medium into which the muscle is placed. Naturally, such artificial media cannot entirely substitute for natural extracellular fluids which inevitably affects the results of the tests. Nevertheless, the experiments on isolated neuromuscular preparations (muscles) are an important supplement to experiments on whole animals. Moreover, sometimes in vitro techniques can be applied in tests, especially those dealing with the mode and the site of action of the agents which in vivo are impeded or impossible.

This chapter is mainly devoted to methods of screening neuromuscular blocking agents on various animal species, and also to relatively simple techniques of the study of the mechanism and locus of their specific activity. Other methods of analysis of pre- and postsynaptic action of neuromuscular blocking agents are mentioned in Chaps. 3 and 5. Furthermore, in this chapter the main principles of the toxicologic study of neuromuscular blocking agents are presented.

B. Evaluation of Neuromuscular Blocking Activity

To study the neuromuscular blocking activity, various cold-blooded and warm-blooded animals are used. The history of the experimental study of curare begins with classical experiments on frogs carried out by Claude Bernard and his colleagues (BERNARD and PELOUZE 1850) and PELIKAN (1857a, b). Neuromuscular blocking activity was estimated by the block of transmission from the sciatic nerve (electrically stimulated) to the gastrocnemius muscle. Besides, the sciatic nerve–sartorius muscle preparation can be used. In both cases, the experiments are carried out both in vivo and in vitro (LEWIS and MUIR 1959a). The preparation of the isolated sartorius muscle of the frog is considered to be the most suitable. The amplitude of the muscle contractions under indirect stimulation is invariant for many hours, and the maximal effect develops rapidly, owing to the small size of the sartorius muscle (ING and WRIGHT 1931, 1933). The sartorius muscle preparation allows one to differentiate between pre- and postsynaptic action of the compounds. To do this, evoked potentials from the nerve endings are recorded (HUBBARD and SCHMIDT 1963; KATZ and MILEDI 1965) as well as evoked and miniature potentials from the end-plate (FATT and KATZ 1951, 1952; KITZ et al. 1969).

Some authors make comparative estimations of neuromuscular blocking agents on intact frogs. The righting reflex is the main criterion. The agents are usually injected into the lymphatic sac and at various times subsequently, it is determined whether the animal placed on its back can recover from this position. This allows one to establish the mean effective dose (ED_{50}). When necessary, the mean lethal dose can also be determined. However, frogs are not very suitable for the screening of neuromuscular blocking agents, since their sensitivity to these agents differs considerably from that of warm-blooded animals. Besides, the reaction of frogs to neuromuscular blocking agents changes, depending on the season.

The mechanism of action of the drugs has been widely studied on the frog rectus abdominis muscle (LANGLEY 1907, 1914; CHANG and GADDUM 1933; GARCIA DI JALON 1947; CHEYMOL 1957; BRITTAIN et al. 1959)[1]. Depolarizing neuromuscular blocking agents evoke contracture of the muscle. The effect of depolarizing agents, such as decamethonium, is usually observed 3–5 min after their first contact with the muscle. In such experiments EC_{50} (and consequently pD_2) are usually determined. The cumulative curves of logarithm of concentration vs effect provide a more precise characterization of depolarizing or mixed action. The effect of nondepolarizing agents is usually determined by their antagonism with acetylcholine or carbachol. The agent to be tested is administered 1–3 min prior to the cholinomimetic and the decrease of the amplitude of muscle contracture is recorded. Estimation of EC_{50} and pA_2 is the most suitable. The amplitude and shape of the cumulative curves of logarithm of concentration versus effect, obtained by using the combination of agonist and antagonist in different concentrations, allow one to determine whether the mechanism of action of the given nondepolarizing agent is competitive, noncompetitive, or mixed (VAN ROSSUM et al. 1958a, b).

1 Leech muscle can also be used for the same purpose (MINZ 1932)

Quite often, the mechanism of action of neuromuscular blocking agents is estimated in experiments on birds – chickens and pigeons. In birds, depolarizing agents evoke spastic paralysis and nondepolarizing ones flaccid paralysis (BUTTLE and ZAIMIS 1949; CHEYMOL 1957; DANILOV 1968, 1970). Danilov considers the experiments on pigeons more informative, since nondepolarizing neuromuscular blocking agents with marked antiacetylcholinesterase activity may cause spastic paralysis in chickens. In pigeons, nondepolarizing agents, even with high anticholinesterase activity, invariably cause flaccid paralysis. Phases in the action of agents of mixed type are quite obvious in birds. Thus, for instance, the administration of tridecamethonium to chickens during a contracture of the limbs is followed by a flaccid paralysis of the muscles of the neck which gradually affects the limbs and all other muscles (ZAIMIS 1953).

For the quantitative evaluation of neuromuscular blocking agents, it is suitable to use the gastrocnemius muscle of chickens (PELIKAN et al. 1954) and pigeons (GINZEL et al. 1951). The experiments are carried out on anesthetized (pentobarbital sodium, 25–30 mg/kg i.v.), artificially ventilated chickens.

PELIKAN et al. (1954) suggested two values for the estimation of the activity of agents such as decamethonium. One of them reflects the ability of the agent to evoke contracture. For this purpose, CD_{25} is established, i.e., the dose in which the drug induces contracture equal to 25% of the maximal one. The contracture evoked by decamethonium at doses of 50–60 µg/kg or acetylcholine at a dose of 1 mg/kg is taken as 100%. These "standard" drugs are usually administered at the beginning and end of the experiment. Besides, PD_{50} is determined – the dose in which the agents inhibit neuromuscular transmission by 50%. In this case, the initial amplitude of the muscle contractions evoked by indirect stimulation by supramaximal impulses is taken as 100%.

During the examination of decamethonium, it was established that successive doses could be administered at 15- to 20-min intervals. In this case, the previous injection of the drug does not affect the following one. CD_{25} and PD_{50} are estimated graphically by the ratio effect: log dose. In these experiments the drugs tested are injected intravenously or intra-arterially to the gastrocnemius muscle.

Indirectly stimulated biventer cervicis muscle of the chicken is a very useful preparation for the study of depolarizing agents in vitro (GINSBORG and WARRINER 1960). The sensitivity of this muscle to decamethonium is 100 times higher than that of the frog rectus abdominis muscle. In addition, neuromuscular preparations of pigeon biventer cervicis muscle are used (GRYGLEWSKI and MIKOS 1964). Also suggested for the test of depolarizing agents are: the semispinalis muscle of the back of the neck of chicken (CHILDE and ZAIMIS 1960), chicken sciatic nerve–tibialis anterior muscle preparation (VAN RIEZEN 1968), chicken gastrocnemius muscle (THESLEFF and UNNA 1954), and neuromuscular preparations of pigeon gastrocnemius muscle (GINZEL et al. 1951, 1953).

Preliminary studies of neuromuscular blocking agents are often carried out on mice. The estimation of myorelaxation by the inhibition of the righting reflex is the simplest variant (COLLIER et al. 1949 a, b). The response is recorded in the alternative form. The evaluation of the results can be somewhat extended. Thus, first of all, the doses in which the agents induce head drop (HDD_{50}) are determined. Then the paralyzing dose (PD_{50}) is determined by the inhibition of the

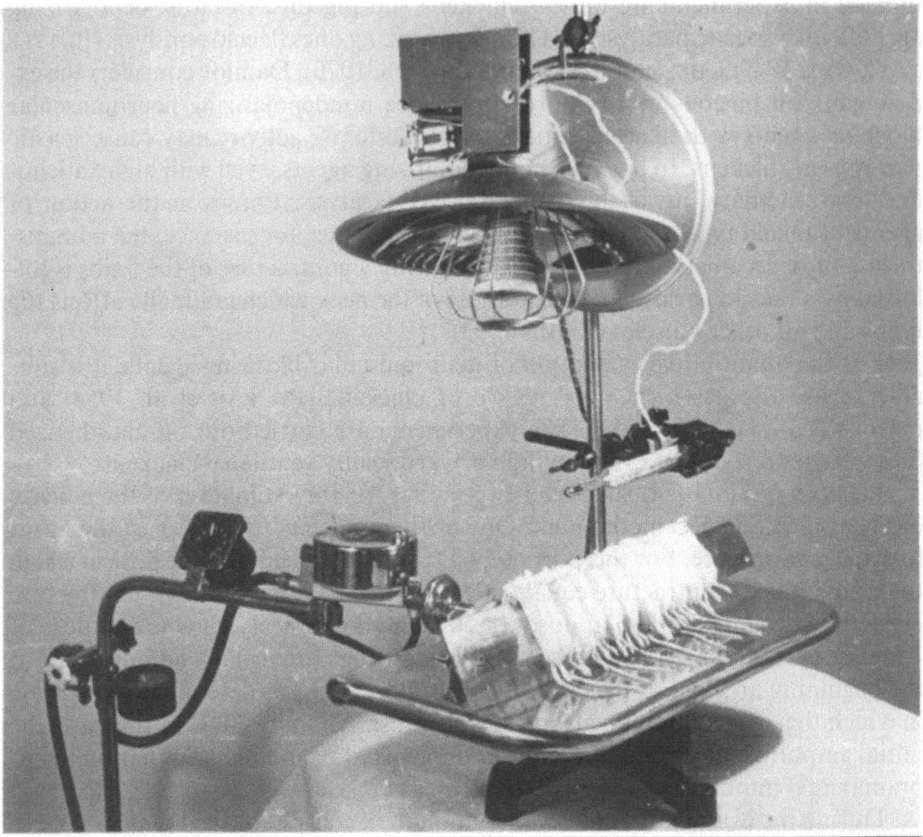

a

b

Fig. 1 a–e. General view **a** and fragments **b–e** of the equipment for artificial ventilation in mice. **c** fixation of mice by upper incisors using a loop connected with a movable coupling, **d** *1*, normal coposition of trachea and esophagus; *2*, placing the mouse on its back with neck bent provides the mechanical compression of the esophagus. The air flowing through the cannula gets only as far as the trachea, **e** respiration cannula: *1*, tube (inner diameter 1.2 mm, outer diameter 2.5 mm); *2*, movable coupling with a metallic loop and fixing screw; *3*, rubber pad on the conical part of the tube

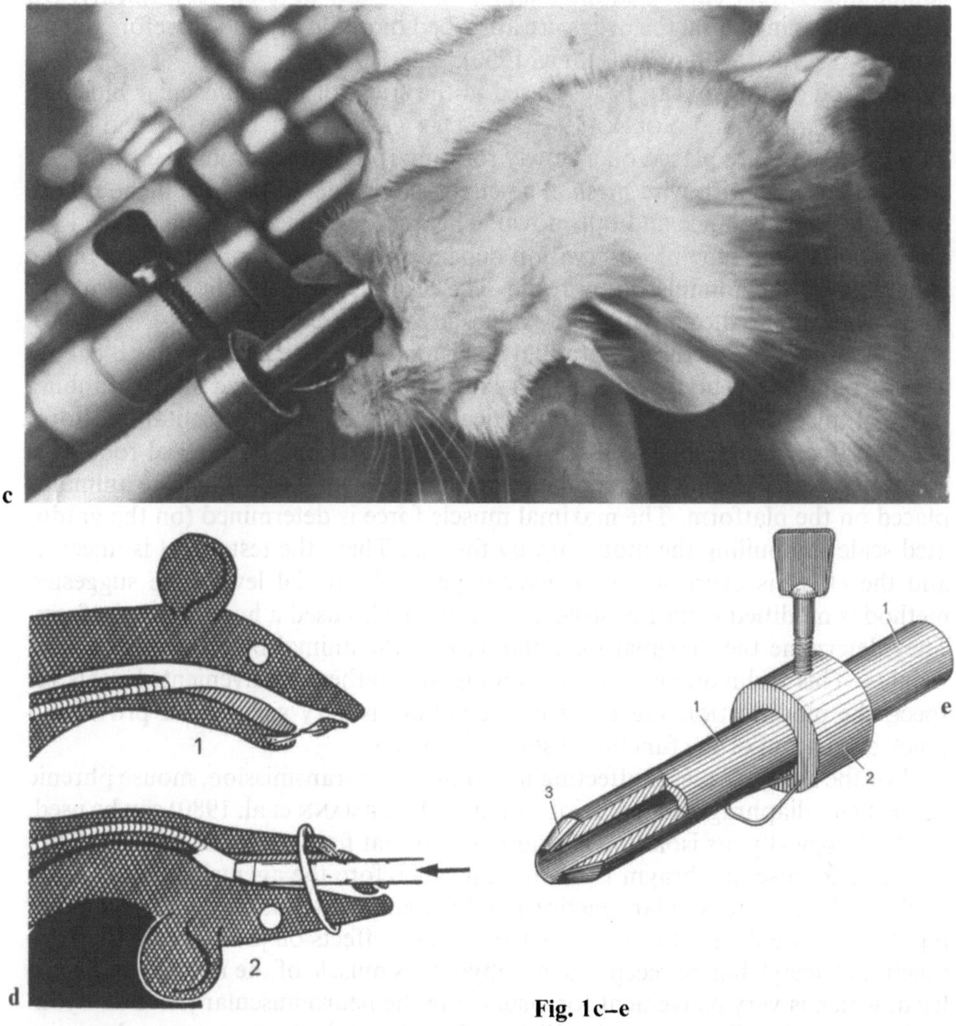

Fig. 1c–e

righting reflex. After that, the apnea-inducing dose (AD_{50}) is estimated. The ratios $AD_{50}:HDD_{50}$ or $AD_{50}:PD_{50}$ show the so-called safety margin. The latter characterizes the range between the doses inducing myorelaxation and those needed for apnea. The duration of action of drugs, in particular, under artificial ventilation can be also measured in mice (KRAVCHUK 1966, 1970; Fig. 1). In the latter case, it is convenient to administer the agents at doses equal to, or multiples of AD_{50} (3–5 AD_{50}).

The inclined plane method is sometimes used (THOMPSON 1946; PRADHAM and DE 1953; SMITH et al. 1953). The mice are put on a fine-meshed screen placed at an angle of 50°–60 °C. The effect is considered as positive when the mouse cannot maintain its balance and rolls down. The effect is recorded within the first 10–30 min following the injection of the test drug into the tail vein. However, the observations should be started from the first minute, since the latent period for some compounds is very short and the maximal effect may develop very quickly. On subcutaneous injection, the drugs are absorbed more slowly and therefore the results are recorded within 3 h (HOPPE 1950).

Rotarod and drum can be also used in the study of neuromuscular blocking activity (SKINNER and YOUNG 1947; COLLIER et al. 1949; DUNHAM and MIYA 1957). The mice are placed on a slowly rotating rod or on a drum, the surface of which is covered with wire mesh. The effect is considered positive if the mouse cannot keep its balance and falls down. The mice are placed on the rotarod for about 2 min. The time of observation depends on the duration of action of the agent and on the administration route. The result is estimated in the alternative form. To estimate the activity PD_{50} values are compared.

In addition, the method of gradual determination of neuromuscular blocking activity on mice has been suggested (TEREKHINA 1965). It is based on the grabbing reflex of mice. The device is a light vertical (it may be also horizontal) platform made of fine wire mesh. The platform is loosely fixed on two vertical rods. It is connected with a spring dynamometer graduated from 0 to 250 g. The animal is placed on the platform. The maximal muscle force is determined (on the graduated scale) by pulling the mouse up by the tail. Then, the test agent is injected, and the effect is expressed as a percentage of the initial level. The suggested method is modified from FROMMEL et al. (1957) who used a horizontal platform. They determine the maximal load endured by the animal by attaching various weights. This technique is time consuming and rather inconvenient. In this respect, the modification suggested by Terekhina is very simple and provides a quick evaluation of the functional state of muscles.

For the study of agents affecting neuromuscular transmission, mouse phrenic nerve–hemodiaphragm preparation in vitro (HABERMANN et al. 1980) can be used. Methodologically, its isolation is analogous to that for the rat (BÜLBRING 1946). However, mouse diaphragm is thinner and therefore the agents penetrate more easily to the neuromuscular junctions and are eliminated more rapidly on washing. For a more detailed assay of pharmacologic effects on junctional and extrajunctional acetylcholine receptors, omohyoideus muscle of the mouse can be utilized, which is very convenient for visualizing the neuromuscular junction by interference optics (DREYER et al. 1976). It should be taken into account that mice are much more sensitive to nondepolarizing agents than to depolarizing ones.

During the study of neuromuscular blocking agents, rats can be used in the same experimental conditions as mice. The activity of agents may be estimated by the inhibition of the righting reflex, by inclined plane or rotarod methods, and by apnea and death. However, it should be kept in mind that rats, like mice, have low sensitivity to depolarizing agents.

The isolated phrenic nerve–diaphragm preparation of the rat is quite often used in examination of nondepolarizing agents (BÜLBRING 1946). One end of the isolated diaphragm with a part of the rib is fixed by a platinum hook. A thread connected with a light isotonic level which records the muscle contractions is attached to the tendon. According to WEST (1947, 1949) and ZOHA (1958), the most stable responses and good restoration are observed at 20 °C for the bath medium. The temperature should be at the same level throughout the experiment since the slightest fluctuations significantly affect the functional state of the preparation.

Test drugs are placed in contact with the preparation for about 3 min (CHOU 1947). The compounds tested may be administered by perfusion via cannulated vena cava (GIBBERD 1966). It should be taken into account that the phrenic nerve–diaphragm preparation has low sensitivity to depolarizing agents. The activity of different agents can be compared by their EC_{50} (KLUPP et al. 1953a, b) or by the concentration of the agents causing equal neuromuscular block (BARLOW and ING 1949; see also MOGEY et al. 1949). The phrenic nerve–diaphragm preparation of the rat can be successfully used for microelectrode registration of membrane potential, end-plate potential, miniature potentials (LILLEY 1956; BARSTAD and LILLEHEIL 1968), and also for the examination of acetylcholine release from the nerve terminals (STRAUGHAN 1960; CHEYMOL et al. 1962; FISENKO et al. 1973). This allows one to estimate pre- and postsynaptic components in the action of neuromuscular blocking agents. For the study of neuromuscular blocking agents in vitro, the obturator nerve–gracilis muscle preparation has also been suggested (LAITY 1967).

The possibility of intracellular registration of potentials from the tail muscles of the rat in vivo is also of interest (STEG 1964; ROBERTS and THESLEFF 1965). This method is convenient because the fixation of the tail is rather simple. This method allows one to record the evoked potentials, the circulation being unaffected. Neuromuscular blocking agents are also studied in vivo on the motor nerve–masseter muscle of the rat (VANĚČEK et al. 1955).

Since the phrenic nerve–diaphragm preparation of the rat has low sensitivity to depolarizing agents, the latter may be tested on the same preparation of the guinea pig (JENDEN 1955; CREESE et al. 1959, 1963; LU 1970). The effect of agents on the neuromuscular junction of the guinea pig is also studied on the lumbrical muscle in vitro (WAUD and WAUD 1975).

One widely used test is the head drop symptom in rabbits (DUTTA and MCINTOSH 1949; VARNEY et al. 1949; PRADHAM and DE 1953), although the sensitivities of rabbits and humans to neuromuscular blocking agents differ significantly. The agents are injected intravenously. This administration route provides the most stable and reproducible results. The agents can be administered gradually or by bolus. The rate of administration is of importance during the infusion. This is in large part related to the latent period, inactivation rate, and elimination of the drugs. It is recommended that tubocurarine is injected at the rate of 0.1 ml every 15 s (1 ml contains 0.3 mg drug). The effect is supposed to develop within 3–8 min. The neuromuscular blocking agents analogous to tubocurarine should be tested according to the same experimental schedule. Only the concentration of the agent in the solution should be varied. A rabbit cannot be injected with the drug more than once daily. For each rabbit, the dose causing head drop is determined. Then, the mean dose is calculated. Relative activity of the agents may be estimated for each rabbit. In this case, on one day, the tested drug is injected, and on the next, tubocurarine or another standard agent.

It is more convenient to administer by bolus those drugs which have a short latent period, are rapidly inactivated, or quickly eliminated from the organism, i.e., short-acting neuromuscular blocking agents. In this case, the mean effective dose in which the agents induce head drop in 50% of the rabbits is determined. This is the way to estimate the activity of all neuromuscular blocking agents.

Moreover, this allows one to determine more precisely the onset of the block, its duration, and the rate of restoration of neuromuscular transmission for the agent. The disadvantage of the latter variant is that it takes a long time and a large number of animals. The head drop symptom in rabbits can be used to study the mechanism of action of agents. If neostigmine (0.1 mg/kg subcutaneously) pretreatment (15 min prior) decreases the activity of the neuromuscular blocking agents tested, it may be suggested that the latter is a nondepolarizing agent. In rabbits, it is possible to define the doses in which the agents induce myorelaxation (paralyzing dose, PD_{50}) and apnea (apneic dose, AD_{50}). This is the way to determine the safety margin ($AD_{50}:PD_{50}$).

In the paper by LOOMIS (1963), the method of registration of diaphragm contractions of the rabbit in situ in response to nerve stimulation is described. In anesthetized rabbits under artificial ventilation, the chest is opened, and a plastic spoon-like lever bent according to the outline of the right part of the diaphragm is introduced through the incision in the abdominal wall between the diaphragm and the liver. A light spring connected with the lever makes the latter press on the lower diaphragmatic surface. The lever is connected with transducer and recorder. This method is convenient since the blood supply of the diaphram is unaffected. The test agents are injected intravenously. A method for study of neuromuscular transmission on phrenicodiaphragmatic preparation in vivo has also been suggested by BOISSIER et al. (1966). Sometimes, the action of the agents on neuromuscular transmission is studied on flexor digitorum longus muscle (NAESS 1952a). Lumbricalis muscle of the rabbit is also used (JENDEN et al. 1954).

The sensitivity of cats to neuromuscular blocking agents is comparatively close to that of humans. This applies to the activity of agents, but not the duration of their action. In conscious cats, neuromuscular blocking avtivity of agents can be estimated by inhibition of the righting reflex. The experiments are often carried out under anesthesia. Urethane (1–1.4 g/kg), urethane (300–500 mg/kg) plus chloralose (60–80 mg/kg), or chloralose (60–80 mg/kg) plus pentobarbital sodium (4–6 mg/kg) are often used. They are administered intravenously or intraperitoneally. These agents have practically no effect on the activity of neuromuscular blocking agents and do not affect neuromuscular transmission. At the same time, some anesthetics may inhibit neuromuscular transmission (for example, diethyl ether; this is much less pronounced in barbiturates). The neuromuscular blocking agents are also tested on decerebrated cats. However, it should be taken into account that decerebration is usually performed under inhalation anesthesia, and therefore, the experiments should be started no less than 1.5–2 h after decerebration, since even the small concentrations of some anesthetics (e.g., diethyl ether) may significantly affect the activity of nondepolarizing agents.

The method of registration of tibialis anterior or gastrocnemius muscle contractions on stimulation of the sciatic nerve is widespread (BROWN 1938; BOWMAN 1964; KHARKEVICH 1969). The stimulation is applied by single impulses, low frequency impulses, or short series of high frequency impulses with registration of tetanic contractions. On low frequency stimulation, depolarizing and mixed neuromuscular blocking agents first increase the amplitude of muscle contractions and only after that the inhibition develops. Nondepolarizing agents do not cause facilitation; the inhibition of neuromuscular transmission occurs immediately after their administration.

The action of agents of different mechanisms of action is also typical at tetanus. Depolarizing agents uniformly decrease the amplitude of all the responses. Nondepolarizing agents act differently. From the very first, intense muscle contractions are observed (they may reach the initial amplitude) followed by a quick decrease of the tone in spite of continued stimulation (Wedensky or pessimal inhibition, tetanic fade).

Low frequency of stimulation makes the statistical processing of the data easier. The doses in which the agents injected intravenously evoke equal effects are determined. Mean effective doses can serve another criterion. However, defining ED_{50} can be hampered since at repeated administration of many agents the preceding injection affects the following one, even at long intervals (1.5–2 h). Thus, for instance, on repeated administration the action of nondepolarizing agents is potentiated, while that of agents of mixed type is reduced. Thus, only the first administration provides a reliable result. Therefore, it is possible to estimate ED_{50} in each experiment only for short-acting nondepolarizing agents which do not cumulate and whose effect remains unchanged on repeated administrations.

Agents with a short latent period and sufficiently long duration of action can be administered gradually with 1–2 min intervals between the injections (during continuous stimulation of the nerve by impulses of low frequency; one stimulus every 5–10 s). The point is to obtain a stepwise curve from the initial up to almost complete inhibition of neuromuscular transmission (an experimental schedule is described for superior cervical ganglion by KHARKEVICH 1967). This permits one to plot the effect against the logarithm of the dose and to determine ED_{50} graphically. At the same time, gradual increase of the dose allows one to determine the threshold dose and the dose needed for a complete neuromuscular block.

In order to locate the action of neuromuscular blocking agents, indirect stimulation is alternated with direct stimulation. For direct stimulation, a silver needle electrode introduced into the muscle serves as a cathode. A thin silver wire fixed on the muscle tendon can be used as the anode.

The method of evaluation of the lability level (according to Wedensky) of neuromuscular synapses is highly sensitive (VALDMAN and ZAKUSOV 1952; NAESS 1952a, b, 1953; VALDMAN et al. 1955; KHARKEVICH 1958). Muscle contractions are recorded either mechanographically or electrographically at increasing frequency of stimulation (5–300 Hz). The administration of nondepolarizing neuromuscular blocking agents decreases the lability level which is manifested by the shift of the level of impulse transformation and the pessimal inhibition toward lower frequencies.

In a simpler variant, motor nerves are stimulated by rhythmic impulses of relatively low frequency (2–10 Hz). The higher the frequency, the more marked the effect of nondepolarizing agents. It is considered that maximal decrease of the amplitude of the potentials is reached at potentials 6–8 (BLACKMAN 1963). The mean value of these three potentials and their initial value characterizes the activity of agents. Simultaneously, the mechanism of action can be determined. Progressive decrease of the amplitude of potentials, starting from the second potential, indicates nondepolarizing action of the drug.

In recent years, the estimation of neuromuscular blocking agents by a short series of stimuli (2 Hz, 2 s, train-of-four stimulation) became widespread in patients and, to some extent, in experimental pharmacology. The degree of neuromuscular blocking action is measured by the ratio between the amplitude of the fourth stimulus and the amplitude of the first one (ALI and SAVARESE 1976; BOWMAN 1980; ALI et al. 1981).

To analyze the mechanism of action of neuromuscular blocking agents, the so-called posttetanic decurarization is often used (HUTTER 1952; LILEY and NORTH 1953; CHURCHILL-DAVIDSON et al. 1960). First, the motor nerve is stimulated by single impulses, then a short tetanus (50 Hz, 5 s) occurs, followed once again by single testing stimuli (or low frequency stimuli of about 1–3 Hz). During the incomplete block caused by competitive neuromuscular blocking agents, after tetanus, a temporary increase in the amplitude of single potentials is observed. This increase is called posttetanic decurarization. After depolarizing agents, neither tetanic fade, nor posttetanic decurarization is observed. Under single stimuli and tetanic stimulation, the amplitude of potentials is uniformly decreased and is maintained at a constant level.

It is known that acetylcholine receptors of different groups of muscles have unequal sensitivity to neuromuscular blocking agents. This determines the order of muscle relaxation. This problem is discussed in more detail in Chap. 6.

While studying neuromuscular blocking activity in cats, it is desirable to know their safety margin. For this purpose, it is necessary to define the range between the doses evoking relaxation of the body muscles and the doses inhibiting respiration. In such experiments, the respiratory volume is registered simultaneously with neuromuscular transmission by any of the conventional methods. Moreover, there are methods which permit one to determine the function of the diaphragm in situ. For this, a rubber cylinder can be used. It is placed between the liver and the diaphragm and is connected with a Marey capsule via a rubber tube (LOOMIS 1963). Adequate results can be obtained from the recordings of diaphragm evoked potentials (LEPAKHIN 1970; LAPAKHIN and FISENKO 1970; KHARKEVICH and FISENKO 1981). Using special electrodes, diaphragm potentials can also be recorded in chronic experiments (SCHOOLMAN and FINK 1963). The registration of evoked muscle potentials provides a characterization of neuromuscular transmission in any muscle (see Chap. 6).

While studying the mechanism of action of neuromuscular blocking agents, BURNS and PATON (1951) recorded the potentials of gracilis muscle of the cat. The agents were injected into the artery toward the muscle. The tongue–hypoglossal nerve preparation of the cat in situ is suitable for the study of neuromuscular blocking agents (ERHARDT and SOINE 1975). The advantage of this technique is that the muscle is maintained in its natural medium. Tongue muscle is fixed by a thread to the recorder, hypoglossal nerve is stimulated by supramaximal impulses.

Earlier, DALE et al. (1936) during a study of the effect of curarine on acetylcholine production by motor fibers, performed the perfusion of the cat tongue. All the branches of common carotid artery and the external jugular veins, except for the tongue arteries and the corresponding veins, were ligated. The perfusion was performed by a pump. The outflow fluid was collected through one of the jugular veins. One of the hypoglossal nerves was stimulated. The contractions of the tongue muscle were registered. Sometimes, for in vitro studies of neuromuscular blocking agents, the isolated phrenic nerve–diaphragm preparation of kittens is used (WIEN 1948) or the tenuissimus muscle of the cat (MACLAGAN 1962).

In experiments on conscious dogs head drop (HOPPE 1950), general myorelaxation, and apnea are determined. If necessary, intubation and artificial ventilation are used. The duration of action is estimated by the restoration of spontaneous respiration and by the moment when the animal rises and makes its first steps (KLUPP et al. 1963b).

Neuromuscular blocking agents are also tested on anesthetized dogs. Pentobarbital sodium (35 mg/kg), barbital sodium (250 mg/kg) and other general anesthetics are injected intravenously or intraperitoneally. The action of the agents on neuromuscular transmission is estimated by the alterations of the amplitude of gastrocnemius muscle contractions on electrical stimulation of the sciatic nerve. The intervals between the repeated injections are about 3 h. However, this interval is not constant and depends on the duration of action of the agents tested. For each neuromuscular blocking agent, the minimal interval between the repeated administrations should be determined which provides unchanged effect at 3–4 injections of equal doses. The activity of the agents is compared by mean effective doses (ED_{50}) (MACRI 1954).

In some cases, neuromuscular blocking agents are tested on monkeys (MUSHIN and MAPLESON 1964; BIGGS et al. 1964; BAMFORD et al. 1967; HUGHES 1972; BRITTAIN and TYERS 1973; GYERMEK and WYMORE 1982). To evaluate the effect, conventional methods can be used: head drop test, relaxation of body muscles, inhibition of respiration, and transmission of impulses from the motor nerve to the skeletal muscle, including the technique employed in anesthetic practice with the stimulation of ulnar nerve and registration of contractions or potentials of the manus muscles. It is of great importance that several monkeys are very similar to humans with respect to the duration of action of neuromuscular blocking agents. According to GYERMEK and WYMORE (1982), a high correlation in this respect is observed in the rhesus macaque.

Thus, there exist a variety of methods for the study of neuromuscular blocking agents in animals. However, a final decision as to whether or not a neuromuscular blocking agent may be used in anesthetic practice can be taken only after its trial on nonanesthetized humans (volunteers) and on patients under anesthesia.

Detailed description of the principles of investigation of neuromuscular blocking agents in humans is presented in Chap. 25 (see also MIKHELSON 1967; KOVANEV et al. 1970; DRIPPS 1976; SMITH 1976; BELOYARTSEV 1980; VON CRUL 1980; HUGHES and PAYNE 1981; VIBY-MOGENSEN 1982). In this chapter, only some possibilities of using human isolated neuromuscular preparations will be presented. Thus, for instance, in the study of neuromuscular blocking agents, isolated phrenic nerve–diaphragm preparations of 16–24 week fetus have been used (BULLER and YOUNG 1949; HAINING et al. 1960); 1–2 h after the extraction of the fetus, the diaphragmatic section with the nerve and part of the rib is dissected and placed in oxygenated Tyrode's solution at 36°–37 °C. Maximal contractions appear when the nerve is stimulated by rectangular pulses of 1 ms duration at 5 stimuli per minute (HAINING et al. 1960).

Furthermore, an isolated preparation of human intercostal muscles is described (DILLON et al. 1955; CREESE et al. 1957; DILLON and SUBAWALE 1959). Muscles are isolated during opertions with thoracotomy. They should be dissected with a part of periosteum, so that both ends of muscular fibers may be retained. It was mentioned that muscles from the dorsal part of the thorax are more suitable for the experiment. The dissected part of the muscles is separated into fibers. A separated fiber (about 12 mm long, 0.5–0.8 mm diameter) is placed into the special artificial bath fluid. Temperature is maintained constant (38 °C). One end of the fiber is fixed on a plastic or glass hook, the other is connected with a recorder by a thread. Direct stimulation is applied by two electrodes located at both ends of the fiber (JENDEN et al. 1954). Since in this case it is impossible to isolate motor fibers, the impulses are applied under conditions that provide stimulation of the intramuscular nerve fibers. The electrodes are placed at both ends of the muscle fiber at a distance of 5 mm. To stimulate the nerve fibers, rectangular impulses of 500–1,000 mA amplitude and 2–5 µs duration are used. The selectivity of the action of stimulating impulses on the nerve fibers under these conditions is confirmed by the fact that tubocurarine blocks their effect without affecting the amplitude of directly stimulated muscle contractions (CREESE et al. 1957). The muscle and the nerves are stimulated alternately at 10 stimuli per minute. Intercostal muscle fibers are also used in microelectrode assays.

Thus, there exists a wide variety of techniques for the study of neuromuscular blocking activity.

C. Evaluation of Toxicity

Toxicologic study of neuromuscular blocking agents is rather specific, since they inhibit respiration. Acute toxicity is estimated under artificial ventilation (by intubation, tracheostomy, or by placing the animal in a respiratory box of the "iron lung" type). Regular estimation of LD_{50} in conscious animals is not suitable in this case since their death is not due to the toxicity of the drugs, but to paralysis of the respiratory muscles, i.e., the specific action of neuromuscular blocking agents. Usually, the so-called LD_{50} is equal to the dose in which the agents induce a stable apnea, i.e., AD_{50} is estimated in such experiments.

To test acute toxicity of potential neuromuscular blocking drugs, various animal species are used (e.g., mice, rats, rabbits, cats, dogs). Several authors examined drugs in conscious animals. However, for humane reasons and to rule out consequences of stress related to total muscle paralysis, the animals should be pretreated with general anesthetics which do not interfere with hemodynamics and do not interact with neuromuscular blocking agents (chloralose, chloralose with urethane, chloralose with a low dose of pentobarbital sodium, gaseous anesthetics). Generally, it is desired that these experiments be carried out under light anesthesia. The state of animals is evaluated by ECG and arterial pressure. Test drugs are usually administered intravenously in increasing doses. The doses of neuromuscular blocking agents which cause death of animals can be determined. However, for some agents these doses exceed the neuromuscular blocking doses in thousands of times. Therefore, if the agent is well tolerated, it is more suitable to use doses no higher than 50- to 100-fold apneic dose (AD) and, after that, if the animal does not die, to await the restoration of neuromuscular transmission. Then, the observations of the animal are conducted within 24–48 h. In the latter variant, it would be adequate to investigate the function of the most important organs and systems before and after administration of the neuromuscular blocking agent.

During the screening of large series of compounds, preliminary assessment of toxicity can be carried out in experiments on mice (KRAVCHUK 1966, 1970). The experiments are carried out on male mice weighing 18–25 g. The agents are administered intravenously at 3–5 AD. Some 10–15 s after the termination of drug administration, the animals with complete neuromuscular block and apnea are connected by a system of tubes to the respiratory pump. This device allows one to perform artificial ventilation of ten mice at a time (Fig. 1 a). Artificial ventilation is performed throughout the experiment. A metallic respiratory cannula is introduced into the mouth of each mouse. At the end of the respiratory cannula there is a rubber gasket with a conical nipple easily entering the peripharyngeal ring, where the cannula is tightly fixed. This method is much simpler than the introduction of a respiratory tube into the trachea of mice, and it is less time consuming. The head of the animal is fixed by the upper incisors in the necessary position by means of a metallic rope and a clamp loosely moving along the cannula and fixed by a screw. In order to prevent air entering the stomach, mice are laid on their backs, their heads bent backward, and the upper part of the body raised (Fig. 1 b–d). This position ensures that, at inspiration, the air from the pump fills a small part of the throat adjacent to the cannula, and enters the lungs via the trachea, since the esophageal foramen is compressed. The air from the respiratory pump to the mice at expiration is supplied by a system of glass tubes with one tap to the pump and ten taps connected with metallic cannulae. The recommended conditions are as follows: respiration rate 140 per minute; pressure at inspiration 15 mmHg; at expiration 10 mmHg; ratio of duration of inspiration and expiration 1 : 1. Body temperature in mice drops rather quickly under total muscle paralysis. To prevent hypothermia and to rule out changes of neuromuscular blocking activity, mice should be permanently and uniformly heated, to maintain normal temperature.

In experiments under artificial ventilation, the true value of LD_{50} of agents can be determined by injecting them in increasing doses. Cardiac arrest indicates the animal's death. However, it can be enough to estimate the tolerance to the agent within a certain dose range (e.g., up to 50 AD).

For a more detailed characteristization of the safety of neuromuscular blocking agents their chronic toxicity should also be studied. Such observations are also carried out on a few animal species (e.g., rats, guinea pigs, cats, dogs). In the USSR, it is considered that agents designed for single administration should be given to animals for 12–14 days. In experiments, the agents are used in subapneic doses and also in apneic doses and higher (5–10 times). It seems best to inject the drugs the way they are injected to patients, i.e., intravenously. Intraperitoneal and intramuscular routes are also acceptable.

The general principles of toxicologic study of neuromuscular blocking agents in subapneic doses are the same as for any other drugs. The task is more difficult when the agents are used in doses causing apnea. In this case, artificial ventilation is needed; it is performed mainly by intubation, tracheostomy, or by iron lungs.

The intubation is carried out using conventional techniques with the connection of the intubation tube to the respiratory pump. However, the intubation is suitable only in big laboratory animals (e.g., dogs). Besides, it requires general anesthetics. Long-term artificial ventilation with positive intrapulmonary pressure at inspiration can also result in unwanted cardiovascular effects (decrease of venous return, decrease of cardiac output and peripheral venous congestion, compression of pulmonary capillaries with decrease of blood saturation by oxygen, overload of the right ventricle, etc.). It should be taken into account that daily intubation can injure the upper respiratory tract. In this respect, chronic tracheal fistula may be preferable. It is natural that the latter should be carried out in advance and the experiments can only be started after the wound has healed. While working out an adequate pattern of artificial ventilation, PCO_2 in expired air or blood should be determined. Chronic toxicity in mice can also be studied according to the method described (see Fig. 1).

Respiratory apparatus of the iron lung type is more suitable, especially for smaller laboratory animals. Figure 2 shows this apparatus designed for rats (KOROBOV 1982). Immediately after the injection of neuromuscular blocking agents, the animals are placed in sealed chambers with the distal part of the head (nose, etc.) left outside. On rarefaction of the air inside (using the respiratory pump), the atmospheric air enters the respiratory tracts of the rat. Expiration takes place on equalization of pressure or increase in the pressure in the chamber. According to KOROBOV, in rats weighing 220–240 g, adequate respiration is reached at 95–100 respirations per minute, negative pressure in the chamber at inspiration being 55–60 mmH_2O and positive pressure at expiration 10–20 mmH_2O. Artificial ventilation is performed until spontaneous respiration is restored.

It is most suitable that during daily administration of neuromuscular blocking agents artificial ventilation lasts up to 30 min. The main physiologic values thus remain stable. Short-term experiments up to 20–30 min can be carried out on conscious animals, although before the test, neuromuscular blocking agents, small doses of noninhalation anesthetics which do not interact with the neuromuscular blocking agents, hypnotics, or tranquilizers can be administered to eliminate probable stress responses. At more prolonged artificial ventilation this premedication is necessary. Throughout the artificial ventilation the animals should be heated.

The functional state of the animals is examined prior to, during, and after the period of neuromuscular blocking agent administration. This examination consists of the usual set of physiologic, biochemical, hematologic, and other tests providing an adequate estimation of the functional state of the main systems and organs. At the end of the experiment, a detailed morphological examination is carried out.

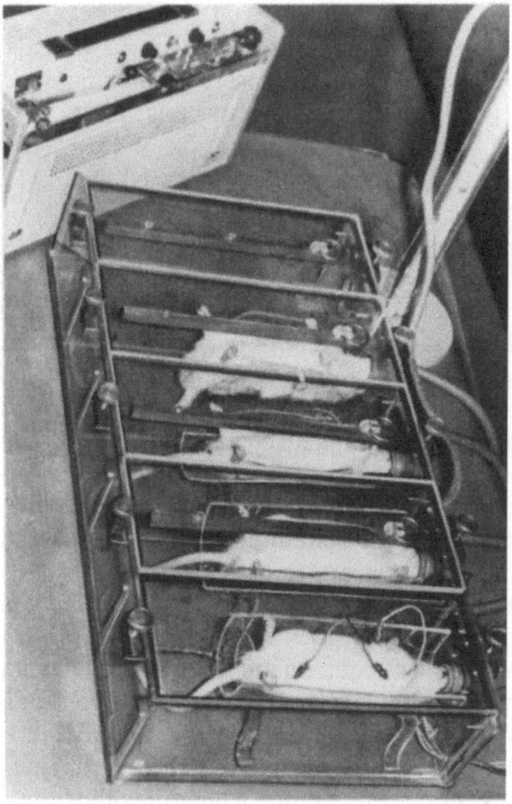

a

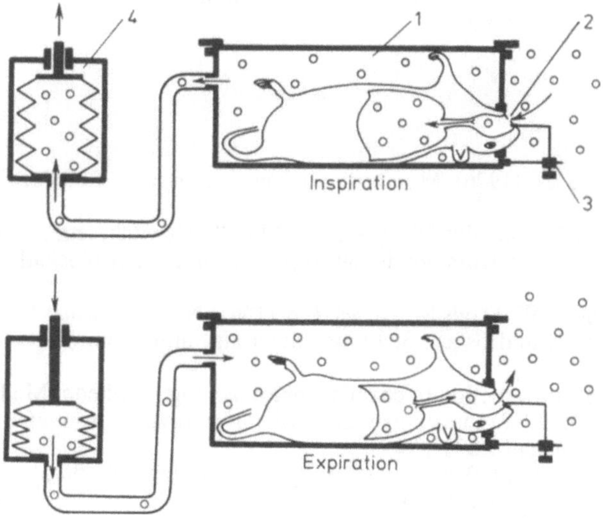

b

Fig. 2 a, b. Box respiratory apparatus for small laboratory animals **a** general view, **b** scheme of function: *1*, respiratory chamber; *2*, respiratory opening; *3*, fixation point; *4*, compressor

D. Conclusions

Thus, to test neuromuscular blocking agents, various animals can be used. Of the common laboratory animals, cats are closest to humans with respect to sensitivity to neuromuscular blocking agents. However, this applies to the estimation of potency of these agents, but not to the duration of their action. Nonhuman primates, e.g., rhesus macaques, are relatively close to humans according to duration of action of neuromuscular blocking agents.

Depending on the task, the experiments can be carried out on neuromuscular preparations, both in situ and in vitro. If it is known that the agents are neuromuscular blockers, the experiments can be carried out on conscious animals, recording the relaxation of certain groups of muscles or total muscle relaxation.

The most important characteristics of neuromuscular blocking agents related to their interaction with end-plate acetylcholine receptors are: activity, onset of blocking action, duration of complete block, rate of restoration of neuromuscular transmission after the block, and order of relaxation of various groups of muscles. It is also important to find out the mechanism of action of the test neuromuscular blocking agents. Naturally, for a comprehensive characterization of neuromuscular blocking agents, it is necessary to know their pharmacokinetics, side effects, and other properties. Some chapters of this volume are devoted to these aspects.

Toxicologic study of neuromuscular blocking agents has its specific features, the main of these being the requirement to carry out tests under artificial ventilation. The requirement of adequate artificial ventilation leads to certain difficulties in the study of chronic toxicity under daily administration of apneic doses of neuromuscular blocking agents over 14 days and more. Therefore, it is of the utmost importance to choose the optimal technique of artificial ventilation for the animal species used in the experiment.

Acknowledgment. The author wishes to express his appreciation to Mrs. MARIA LIPMAN for the translation of this chapter.

References

Ali HH, Savarese JJ (1976) Monitoring of neuromuscular function. Anesthesiology 45:216–249

Ali HH, Savarese JJ, Lebowitz PW, Ramsey FM (1981) Twitch, tetanus and train-of-four as indices of recovery from nondepolarizing neuromuscular blockade. Anesthesiology 54:294–297

Bamford DG, Biggs DF, Davis M, Parnell EW (1967) Neuromuscular blocking properties of stereoisomeric androstane-3, 17-bis-quaternary ammonium salts. Br J Pharmacol 30:194–202

Barlow RB (1968) Introduction to chemical pharmacology, 2nd edn. Methuen, London

Barlow RB, Ing HR (1948) Curare-like action of polymethylene bis-quaternary ammonium salts. Br J Pharmacol 3:298–304

Barstad JAB, Lilleheil G (1968) Transversally cut diaphragm preparation from rat. Arch Int Pharmacodyn Ther 175:373–390

Beloyartsev FF (1980) Electromiografiya v anesteziologii (Electromyography in anesthesiology). Meditsina, Moscow

Bernard C, Pelouze TJ (1850) Recherches sur le curare. CR Séances Acad Sci [D] 31:533–537

Biggs RS, Davis M, Wien R (1964) Muscle relaxant properties of a steroid bisquaternary ammonium salt. Experientia 20:119
Blackman JG (1963) Stimulus frequencies and neuromuscular block. Br J Pharmacol 20:5–16
Boissier J-R, Viars P, Simon P (1966) Préparation „diaphragmephrénique" in vivo. J Physiol (Paris) 58:115–118
Bowman WC (1964) Neuromuscular blocking agents. In: Laurence DR, Bacharach AL (eds) Evaluation of drug activities: pharmacometrics. Academic, London, pp 325–351
Bowman WC (1980) Pharmacology of neuromuscular function. Wright, Bristol
Brittain RT, Tyers MB (1973) The pharmacology of AH 8165; a rapid-acting short lasting competitive neuromuscular blocking drug. Br J Anaesth 45:837–843
Brittain RT, Chesher BG, Collier HOJ, Grimshaw JJ (1959) Assay of suxamethonium and laudexium on the frog rectus abdominis. Br J Pharmacol 1:158–163
Brown GL (1938) The preparation of the tibialis anterior (cat) for close arterial injection. J Physiol (Lond) 92:22–23
Bülbring E (1946) Observations on the isolated phrenic nerve diaphragm preparation of the rat. Br J Pharmacol 1:38–61
Buller AJ, Young HI (1949) The action of d-tubocurarine chloride on foetal neuromuscular transmission and the placental transfer of this drug in the rabbit. J Physiol (Lond) 109:412–420
Burns BD, Paton WDM (1951) Depolarization of the motor end-plate by decamethonium and acetylcholine. J Physiol (Lond) 115:41–73
Buttle GAH, Zaimis EJ (1949) The action of decamethonium iodide in birds. J Pharm Pharmacol 1:991–992
Chang HC, Gaddum JH (1933) Choline esters in tissue extracts. J Physiol (Lond) 79:255–285
Cheymol J (1957) Appréciation qualitative et quantitative d'une substance curarisante. Thérapie 12:321–356
Cheymol J, Bourillet F (1972) Inhibitors of postsynaptic receptors. In: Cheymol J (ed) Neuromuscular blocking and stimulating agents, vol 1. Pergamon, Oxford, pp 297–356
Cheymol J, Bourillet F, Ogura Y (1962) Action de quelques paralysants neuromusculaires sur la libération de l'acétylcholine au niveau des terminaisons nerveuses motrices. Arch Int Pharmacodyn Ther 139:187–197
Childe KJ, Zaimis E (1960) A new biological method for the assay of depolarizing substances using the isolated semispinalis muscle of the chick. Br J Pharmacol 15:412–416
Chou TC (1947) A method of estimating curare-like activity on the isolated phrenic nerve diaphragm preparation of the rat. Br J Pharmacol 2:1–7
Churchill-Davidson HC, Christie TH, Wise RP (1960) Dual neuromuscular block in man. Anesthesiology 21:144–149
Collier HOJ, Fieller EC, Hall RA (1949a) The assay of curarizing activity in the conscious mouse and rat. Analyst 74:583–588
Collier HOJ, Hall RA, Fieller EC (1949b) Use of a rotating drum in assessing the activities of paralysant, convulsant and anaesthetic drugs. Analyst 74:592–596
Creese R, Dillon JB, Marshall J, Sabawala PB, Shneider DJ, Taylor DB, Zinn DE (1957) The effect of neuromuscular blocking agents on isolated human intercostal muscles. J Pharmacol Exp Ther 119:485–494
Creese R, Taylor DB, Tilton B (1959) The effect of curare on the uptake and release of a depolarizing agent labelled with ^{131}I. In: Bovet B, Bovet-Nitti F, Marini-Bettolo GB (eds) Curare and curare-like agents. Elsevier, Amsterdam, pp 386–390
Creese R, Taylor DB, Tilton B (1963) The influence of curare on the uptake and release of a neuromuscular blocking agent labelled with radioactive iodine. J Pharmacol Exp Ther 139:8–17
Dale HH, Feldberg W, Vogt M (1936) Release of acetylcholine at voluntary motor nerve endings. J Physiol (Lond) 86:353–380

Danilov AF (1968) O stroenii kholinoretseptorov skeletnykh myshts i o mekhanizme deistviya miorelaksantov (On the structure of skeletal muscles acetylcholine receptors and the mechanism of action of neuromuscular blocking agents). Dr Med Sci Thesis, Leningrad

Danilov AF (1970) Blockade of neuromuscular transmission by decamethylene bispyridinium (in Russian). Farmakol Toksikol 5:536–537

Dillon JB, Sabawale PB (1959) The mode of action of depolarizing drugs. Acta Anaesthesiol Scand 3:83–100

Dillon J, Fields J, Gumas T, Jenden DJ, Taylor DB (1955) An isolated human voluntary muscle preparation. Proc Soc Exp Biol Med 90:409

Dreyer F, Müller K-D, Peper K, Sterz R (1976) The m. omohyoideus of the mouse as a convenient mammalian muscle preparation. A study of junctional and extrajunctional acetylcholine receptors by noise analysis and cooperativity. Pfluegers Arch 367:115–122

Dripps RD (1976) The clinician looks at neuromuscular blocking drugs. In: Zaimis E (ed) Neuromuscular function. Springer, Berlin Heidelberg New York, pp 583–592

Dunham NW, Miya TS (1957) A note on a simple apparatus for detecting neurological deficit in rats and mice. J Am Pharm Assoc Sci Ed 46:208–209

Dutta NK, Macintosh FC (1949) Assay of curare preparations by the rabbit head-drop method. Analyst 74:588–592

Erhardt PW, Soine TO (1975) Stereochemical preferences for curarimimetic neuromuscular junction blockade I: enantiometric monoquaternary amines as probes. J Pharm Sci 64:53–62

Fatt P, Katz B (1951) An analysis of the end-plate potential recorded with an intra-cellular electrode. J Physiol (Lond) 115:320–370

Fatt P, Katz B (1952) Spontaneous subthreshold activity at motor nerve endings. J Physiol (Lond) 117:109–128

Fisenko VP, Polgar AA, Smirnov VS (1973) Microelectrophysiological study of the mechanism and locus of action of several new neuromuscular blocking agents (in Russian). Farmakol Toksikol 2:206–209

Frommel E, Gold P, Fleury C (1957) Méprobamate ou phénobarbital ou méprobamate et phénobarbital. Schweiz Med Wochenschr 87:1480–1485

Garcia di Jalón PD (1947) A simple biological assay of curare preparations. Q J Pharm Pharmacol 20:28–30

Gibberd FB (1966) Action of decamethonium on rat diaphragm. Br J Pharmacol 28:128–136

Ginsborg BL, Warriner J (1960) The isolated chick biventer cervicis nerve-muscle preparation. Br J Pharmacol 15:410–411

Ginzel KH, Klupp H, Werner G (1951) Die Wirkung einiger aliphatischer α-ω-Bis-quaternärer Ammonium-Verbindungen auf die Skelettmusculatur. Arch Exp Path Pharmakol 213:453–466

Ginzel KH, Klupp H, Werner G (1953) Die Veränderung der Wirkungsweise von Decamethonium (CIO) durch eine Polymethylenbisphosphoniumverbindung. Naunyn Schmiedebergs Arch Pharmacol 218:140–141

Gryglewski R, Mikoś E (1964) The isolated pigeon biventer cervicis nerve-muscle preparation. Dissertat Pharmac 16:1–7, Kraków

Gyermek L, Wymore A (1982) Old and new nondepolarizing muscle relaxants. A reassessment on subhuman primates. Arch Int Pharmacodyn Ther 257:114–120

Habermann E, Dreyer F, Bigalke H (1980) Tetanus toxin blocks the neuromuscular transmission in vitro like botulinum A toxin. Naunyn Schmiedebergs Arch Pharmacol 311:33–40

Haining CG, Johnston RG, Smith JM (1960) The neuromuscular blocking properties of a series of bis-quaternary tropeines. Br J Pharmacol 15:71–81

Hoppe JO (1950) A pharmacological investigation of 2,5-bis-(3-diethyl-aminopropylamino)benzoquinone-bis-benzyl chloride (WIN 2747). A new curarimimetic drug. J Pharmacol Exp Ther 100:333–345

Hubbard JI, Schmidt RF (1963) An electrophysiological investigation on mammalian motor nerve terminals. J Physiol (Lond) 166:145–167
Hughes R (1972) Evaluation of the neuromuscular blocking properties and side effects of the two new isoquinolinium bisquaternary compounds BW 252C64 and BW 40365. Br J Anaesth 44:27–42
Hughes R, Payne JP (1981) Clinical assessment of neuromuscular transmission. Br J Clin Pharmacol 11:537–548
Hutter OF (1952) Post-tetanic restoration of neuromuscular transmission blocked by d-tubocurarine. J Physiol (Lond) 118:216–227
Jenden DJ (1955) The effect of drugs upon neuromuscular transmission in the isolated guinea-pig diaphragm. J Pharmacol Exp Ther 114:398–408
Jenden DJ, Kamijo K, Taylor DB (1954) The action of decamethonium on the isolated rabbit lumbrical muscle. J Pharmacol Exp Ther 111:229–240
Ing HR, Wright WM (1931) The curariform action of quaternary ammonium salts. Proc R Soc Lond [Biol] 109B:337–353
Ing HR, Wright WM (1933) Further studies on the pharmacological properties of onium salts. Proc R Soc Lond [Biol] 114B:48–63
Katz B, Miledi R (1965) Propagation of electric activity in motor nerve terminals. Proc R Soc Lond [Biol] 161:453–482
Kharkevich DA (1958) On combined action of diplacinum and narcotics (in Russian). Farmakol Toksikol 3:12–17
Kharkevich DA (1967) Ganglion-blocking and ganglion-stimulating agents. Pergamon, Oxford, pp 18–21
Kharkevich DA (1969) Farmakologiya kurarepodobnykh sredstv (Pharmacology of neuromuscular blocking agents). Meditsina, Moscow
Kharkevich DA, Fisenko VP (1981) The effect of neuromuscular blocking agents on the acetylcholine receptors of different skeletal muscles. Arch Int Pharmacodyn Ther 251:255–269
Kitz RJ, Harris JK, Ginsburg S (1969) A study in vitro of new short-acting nondepolarizing neuromuscular blocking agents. Biochem Pharmacol 18:871–881
Klupp H, Ginzel KH, Werner G (1953a) Elektive Hemmung des Patellarsehenreflexes durch peripher wirkende Stoffe. Naunyn Schmiedebergs Arch Pharmacol 218:141–142
Klupp H, Kraupp O, Stormann H, Stumpf C, Honetz H, Kobinger W (1953b) Über die pharmakologischen Eigenschaften einiger Polymethylen-Dicarbaminsäure-bis-cholinester. Arch Int Pharmacodyn Ther 96:161–182
Korobov NV (1982) On the assessment of toxicity of neuromuscular blocking agents (in Russian). Farmakol Toksikol 2:115–118
Kovanev VA, Khmelevsky YaM, Beloyartsev FF (1970) Myshechnye relaksanty (Muscle relaxants). Meditsina, Moscow
Kravchuk LA (1966) A method of artificial ventilation of curarized mice (in Russian). Bull Eksp Biol Med 1:53–60
Kravchuk LA (1970) On the duration of neuromuscular blocking action of bis-quaternary ammonium derivatives of truxillic acids (in Russian). In: Kharkevich DA (ed) Novye kurarepodobnye i ganglioblokiruyushchie sredstva (New curare-like and ganglion-blocking agents). Meditsina, Moscow, pp 35–41
Laity JLH (1967) A new nerve muscle preparation: the obturator nerve-anterior gracilis preparation of the rat. J Pharm Pharmacol 19:265–266
Langley JN (1907) On the contraction of muscle, chiefly in relation to the presence of "receptive" substance. J Physiol (Lond) 36:347–384
Langley JN (1914) The antagonism of curare and nicotine in skeletal muscle. J Physiol (Lond) 48:73–108
Lepakhin VK (1970) Sensitivity of different muscles to alpha-truxillic acid derivatives (in Russian). In: Kharkevich DA (ed) Novye kurarepodobnye i ganglioblokiruyushchie sredstva (New curare-like and ganglion-blocking agents). Meditsina, Moscow, pp 48–63

Lepakhin VK, Fisenko VP (1970) On the comparative sensitivity of neuromuscular junctions of different muscles to diadonium and decadonium (in Russian). Farmakol Toksikol 3:288–292
Lewis JJ, Muir TC (1959a) The laboratory estimation of curare-like activity in natural and synthetic products. Lab Pract 8:333–338
Lewis JJ, Muir TC (1959b) The laboratory estimation of curare-like activity in natural and synthetic products. Lab Pract 8:364–368
Lewis JJ, Muir TC (1959c) The laboratory estimation of curare-like activity in natural and synthetic products. Lab Pract 8:404–407
Liley AW (1956) The quantal components of the mammalian end-plate potential. J Physiol (Lond) 133:571–587
Liley AW, North KAR (1953) An electrical investigation of effects of repetitive stimulation on mammalian neuromuscular junction. J Neurophysiol 16:509–527
Loomis TA (1963) Neuromuscular effects of piperidylmethylandrostane-diol. Arch Int Pharmacodyn Ther 141:412–422
Lu TC (1970) Affinity of curare-like compounds and their potency in blocking neuromuscular transmission. J Pharmacol Exp Ther 174:560–566
Maclagan J (1962) A comparison of the responses of the tenuissimus muscle to neuromuscular blocking drugs in vivo and in vitro. Br J Pharmacol 18:204–216
Maclagan J (1976) Competitive neuromuscular blocking drugs. In: Zaimis E (ed) Neuromuscular junction. Springer, Berlin Heidelberg New York, pp 421–486 (Handbook of experimental pharmacology, vol 42)
Macri FJ (1954) Curare-like activity of some bis-fluorenyl-bis-quaternary ammonium compounds. Proc Soc Exp Biol Med 85:603–606
Mikhelson VA (1967) Myshechnye relaksanty kak komponent kombinirovannoi anestezii (Muscle relaxants as a component of combined anaesthesia). Meditsina, Moscow
Minz B (1932) Pharmakologische Untersuchungen am Blutegelpräparat zugleich eine Methode zum biologischen Nachweis von Azetylcholin bei Anwesenheit anderer pharmakologisch wirksamer körpereigener Stoffe. Naunyn Schmiedebergs Arch Pharmacol 168:292–304
Mogey GA, Trevan JW, Young PA (1949) The assay of d-tubocurarine chloride on the isolated rat diaphragm. A statistical examination. Analyst 74:577–582
Mushin WW, Mapleson WW (1964) Relaxant action in man of dipyrandinium chloride (M&B 9105A). Br J Anaesth 36:761–768
Naess K (1952a) The mechanism of action of curare. Acta Pharmacol Toxicol 8:149–163
Naess K (1952b) Effects of brief and protracted curarization. Acta Pharmacol Toxicol 8:400–408
Naess K (1953) Addendum to effects of brief and protracted curarization. Acta Pharmacol Toxicol 9:196–197
Pelikan E (1857a) Notice sur les propriétés physiologo-toxicologiques du curare. CR Acad Sci [D] 44:507–509
Pelikan E (1857b) Physiologische und toxikologische Untersuchungen über Curare. Arch Path Anat Physiol 11:401–410
Pelikan EW, Smith CM, Unna KR (1954) Mode of action of antagonists to curare. II. Anticurare action of hydroxyphenyltrialkylammonium compounds in avian muscle. J Pharmacol Exp Ther 3:30–42
Pradhan SN, De NN (1953) Hayatin methiodide: a new curariform drug. Br J Pharmacol 8:399–405
Roberts DV, Thesleff S (1965) Neuromuscular transmission in vivo and the actions of decamethomium: a micro-electrode study. Acta Anaesthesiol Scand 9:165–172
Schoolman A, Fink BR (1963) Permanently implanted electrode for electromyography of the diaphragm in the waking cat. Electroencephalogr Clin Neurophysiol 15:127–128
Skinner HG, Young DM (1947) A mouse assay for curare. J Pharmacol Exp Ther 91:144–146
Smith CM, Pelikan EW, Maramba LR, Unna KR (1953) Relationship between structure and activity in a series of bis-isoquinolinium compounds. J Pharmacol Exp Ther 108:317–329

Smith SE (1976) Neuromuscular blocking drugs in man. In: Zaimis E (ed) Neuromuscular junction. Springer, Berlin Heidelberg New York, pp 593–660 (Handbook of experimental pharmacology, vol 42)
Steg G (1964) Efferent muscle innervation and rigidity. Acta Physiol Scand 61 [Suppl 225]:1–53
Straughan DW (1960) The release of acetylcholine from mammalian motor nerve endings. Br J Pharmacol 15:417–424
Terekhina AI (1965) Izuchenie zakonomernostei tsirkulyatsii meprobamata v organisme v usloviyakh odnokratnogo i povtornogo vvedeniya (Study of regularities of circulation of meprobamate in the organism during single and repeated administration). Cand Med Sci Thesis, Moscow
Thesleff S, Unna KR (1954) Differences in mode of neuromuscular blockade in a series of symmetrical bis-quaternary ammonium salts. J Pharmacol Exp Ther 111:99–113
Thompson RE (1946) Biological assay of insulin. Objective determination of the quantal response of mice. Endocrinology 39:62
Valdman AV, Zakusov VV (1952) Effects of tubocurarine, acetylcholine and neostigmine on the transformation of impulses in neuromuscular junctions (in Russian). Farmakol Toksikol 15:4–9
Valdman AV, Ivanova ZN, Kharkevich DA (1955) Effect of diplacinum on the neurotransmission in various links of reflex arch (in Russian). Farmakol Toksikol 2:3–11
Vaněček M, Votava L, Šramková, Šragerová H (1955) Pharmacological properties of new, synthetic curarelike substances. Sulfonium analogs of methonium salts, derivatives of pyridinecarboxylic and piperidinecarboxylic acids. Physiol Bohemoslov 4:220–228
Van Riezen H (1968) Classification of neuromuscular blocking agents in a new neuromuscular preparation of the chick in vitro. Eur J Pharmacol 5:29–36
Van Rossum JM, Ariens EJ, Linssen GH (1958a) Three basic types of curariform drugs. Atti Soc Ital Anesthesiol XI:229–235
Van Rossum JM, Ariens EJ, Linsen GH (1958b) Basic types of curariform drugs. Biochem Pharmacol 1:193–199
Varney RF, Linegar CR, Holday HA (1949) Assay of curare by rabbit "head drop" method. J Pharmacol Exp Ther 97:72–83
Viby-Mogensen J (1982) Clinical assessment of neuromuscular transmission. Br J Pharmacol 54:209–223
Von Crul JF (1980) Erfassung und Registrierung der Impulsübertragung und ihrer Beeinflussung. In: Ahnefeld FW, Bergmann H, Burri C, Dick W, Halmágyi M, Hossli G, Rügheimer E (eds) Muskelrelaxanzien. Springer, Berlin Heidelberg New York, pp 51–66 (Klinische Anästhesiologie und Intensivtherapie, vol 22)
Waud BE, Waud DR (1975) The effects of diethyl ether, enflurane and isoflurane at the neuromuscular junction. Anesthesiology 42:275–280
West GB (1947) Note on the biological assay of tubocurarine. QJ Pharm Pharmacol 20:518–527
West GB (1949) The assay of curare and curare-like substances: a modification of the method using the rat's phrenic nerve-diaphragm preparation. Analyst 74:582
Wien R (1948) The phrenic nerve-diaphragm preparation of different species in the estimation of curare-like activity. J Physiol (Lond) 107:44P–45P
Zaimis EJ (1953) Motor end-plate differences as a determining factor in the mode of action of neuromuscular blocking substances. J Physiol (Lond) 122:238–251
Zoha MS (1958) Studies on the mode of action of pharmacologically active substances from natural sources with additional studies on some synthetic neuromuscular blocking agents. Thesis, University of Glasgow

CHAPTER 12

Steroid Derivatives

M. RIESZ, E. KÁRPÁTI, and L. SZPORNY

A. Introduction

The various hypotheses concerning the relationship between neuromuscular blocking activity and interonium structures make the study of almost rigid molecules of particular interest. The use of a fairly fixed steroid skeleton to which reactive side chains are attached (ALAUDDIN and MARTIN-SMITH 1962) was an important step forward in the search for useful curarizing drugs and resulted in a very powerful competitive acetylcholine antagonist, pancuronium (BUCKETT and BONTA 1966; BUCKETT et al. 1968; SAVAGE et al. 1971). Reports on its use in human practice started with BAIRD and REID (1967) and the drug is still the most widely utilized one. In spite of its general sucess, pancuronium is not yet the ideal curare-like agent because of its relatively long duration of action and of its multicausal cardiovascular stimulating property.

Our opinion on the necessity of developing curare-like substances of different characteristics and duration of action is in agreement with KHARKEVICH (1974). Relaxants with different duration of action (from 5–10 min to several hours) are needed in practice, depending on the duration of manipulation. For intubation and thoracic surgery, complete suppression of breathing is inevitable. On the other hand, the maintenance of spontaneous respiration would be preferable with relaxation of striated muscles of other areas (e.g., in childhood and old age).

Studies performed over the last 10 years resulted in the two most advantageous products: pipecuronium (Arduan) and vecuronium (Org-NC-45, Norcuron). Clinical and experimental data prove that both new substances are practically free of cardiovascular side effects. Pipecuronium has a clinical duration of action about 45 min and vecuronium about 20 min. Although these two drugs are almost ideal for surgery of intermediate and longer duration, further research is seen to be necessary for developing new molecules – eventually steroidal ones – for ultrashort nondepolarizing action, without side effects. In the following, we compare the published neuromuscular effects of steroids with curare-like activity according to chemical groups.

B. Neuromuscular Blocking Activity

I. Androstane Derivatives

The chemical structures and neuromuscular blocking potencies are given in Table 1.

Table 1. Structures and neuromuscular blocking potencies of androstane derivatives

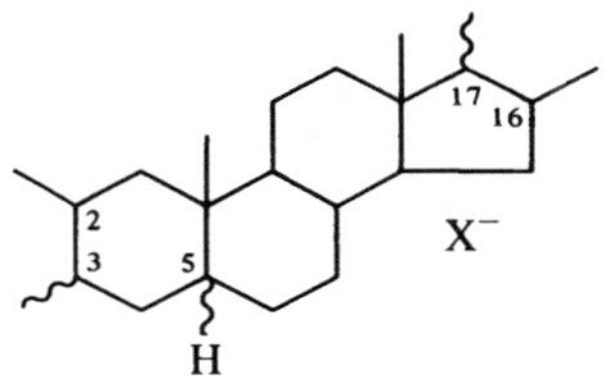

No. and name of compound	Positions and configurations of substitutions						Cat gastrocnemius Relative potency[a] (tubocurarine=1)	Cat tibialis ED_{50} (μg/kg)	References
	2	3	5	16	17	X			
1	MePi	αAcO	α	H	O[c]	Br	0.06[b]		[1]
2	AcO	αMePi	α	H	βAcO	Br	0.02[b]		[1]
3	AcO	αMePi	α	H	O[c]	Br	0.015[b]		[1]
4	Me_3N	αAcO	α	H	O[c]	OH	0.015[b]		[1]
5	AcO	αMePi	α	H	βAcO	OH	0.02[b]		[1]
6	AcO	αMePi	α	H	O[c]	OH	0.015[b]		[1]
7	MePy	αAcO	α	H	O[c]	Br	0.01[b]		[1]
8	MePi	αAcO	α	H	H	Br	<0.01		[2]
9	H	H	α	MePi	βAcO	Br	<0.01		[2]
10	H	αMe_3N	α	H	αMe_3N	2I	0.267[b]		[3]
11	H	αMe_2EtN	α	H	βMe_2EtN	2I	0.837[b]		[3]
12	H	αEt_2MeN	α	H	βEt_2MeN	2I	0.874[b]		[3]
13	H	αEt_3N	α	H	βEt_3N	2I	0.455[b]		[3]
14 Dipyrandium	H	βMePy	α	H	βMePy	2Cl		71	[4]
15	H	βMePy	α	H	αMePy	2I		97	[4]
16	H	αMePy	α	H	βMePy	2I		173	[4]
17	H	αMePy	α	H	αMePy	2I		226	[4]
18	H	βMePy	β	H	βMePy	2I		107	[4]
19	H	βMePy	β	H	αMePy	2I		145	[4]
20	H	αMePy	β	H	βMePy	2I		1,080	[4]
21	H	αMePy	β	H	αMePy	2I		1,080	[4]

22		H	βPy	α	H	βPy	2Cl		1,680	[5]
23		H	βMePy	α	H	βPy	2Cl		64	[5]
24		H	βPy	α	H	βMePy	2I		432	[5]
25		H	βMePy	α	H	O[c]	I		2,760	[5]
26		H	O[c]	α	H	βMePy	I		53,100	[5]
27		H	βMePy	α	H	βOH	I		10,500	[5]
28		H	βOH	α	H	βMePy	I		8,550	[5]
29		H	H	α	H	βMePy	Cl		> 7,710	[5]
30		H	H	α	H	αMePy	Cl		> 7,710	[5]
31		H	βMe$_2$AcOEtN	α	H	βMe$_2$AcOEtN	2Br	0.37		[2]
32		AcO	αMePy	α	H	βMePy	2Br	0.48		[2]
33		AcO	αMePi	α	H	βMePy	2Br	1.15		[2]
34		MePi	αAcO	α	H	βMePy	2Br	0.20		[2]
35		MePi	αAcO	α	H	βMePi	2Br	<0.10		[2]
36		MePi	αAcO	α	Aco	βMePi	2Br	0.59		[2]
37		MePi	αOH	α	OH	βMePi	2I	0.15		[2]
38		MePi	αAcO	α	AcO	βMePy	2I	0.70		[2]
39		MePi	αAcO	α	AcO	βMe$_3$N	2I	0.57		[2]
40		MePy	αAcO	α	AcO	βMePy	2I	0.13		[2]
41	Pancuronium	MePi	αAcO	α	MePi	βAcO	2Br	9.41	34	[2, 6]
42		MePi	αOH	α	MePi	βAcO	2Br	5.15	59	[2, 6]
43	Dacuronium Br	MePi	αAcO	α	MePi	βOH	2Br	1.72	230	[2, 6]
44		MePi	αOH	α	MePi	βOH	2Br	0.55		[2]
45		MePi	αAcO	α	MePi	βAcO	2I	15.0		[2]
46	Dacuronium I	MePi	αAcO	α	MePi	βOH	2I	1.29		[2]
47		MePi	αAcO	α	MePi	αAcO	2Br	6.15		[2]
48	Org-6368	MePi	αAcO	α	MePi	H	2Br	2.4	121	[6, 7]
49		MePi	αOH	α	MePi	O[c]	2Br	0.67		[2]
50		MePi	αOH	α	MePi	O[c]	2I		1,650	[6]
51		MePi	αAcO	α	MePi	O[c]	2Br	1.53		[2]
52		MePi	αAcO	α	MePi	O[c]	2I		240	[6]
53		MePi	αHCOO	α	MePi	βHCOO	2Br	1.30		[2]
54		MePi	αPrO	α	MePi	βPrO	2Br	7.10	87	[2, 6]
55		MePi	αOH	α	MePi	βPrO	2Br		544	[6]
56		MePi	αAcO	α	MePi	βPrO	2Br	5.60		[2]
57		MePi	αBuO	α	MePi	βBuO	2Br		50	[6]

Table 1 (continued)

No. and name of compound	Positions and configurations of substitutions						Cat gastrocnemius Relative potency[a] (tubocurarine=1)	Cat tibialis ED_{50} (μg/kg)	References
	2	3	5	16	17	X			
58	MePi	αOH	α	MePi	βBuO	2Br		250	[6]
59	MePi	αi-BuO	α	MePi	βi-BuO	2Br		161	[6]
60	MePi	αOH	α	MePi	βi-BuO	2Br		511	[6]
61	MePi	αMe_3CCOO	α	MePi	βMe_3CCOO	2I	3.80		[2]
62	MePi	αMe_3CCOO	α	MePi	βOH	2I	1.70		[2]
63	MePi	αPhCOO	α	MePi	βPhCOO	2Br	1.00		[2]
64	MePi	αPhCOO	α	MePi	βOH	2I	0.84		[2]
65	MePi	αAcO	α	AlPi	βAcO	2Br	7.23		[2]
66	AlPi	αOH	α	AlPi	βOH	2Br	2.50		[2]
67	PgPi	αAcO	α	PgPi	βAcO	2Br	6.31		[2]
68	PgPi	αOH	α	PgPi	βOH	2Br	0.77		[2]
69 Vecuronium	Pi	αAcO	α	MePi	βAcO	Br	6.0	38	[2, 6]
70 Org-7268	Pi	αOH	α	MePi	βAcO	Br		39	[6]
71	Pi	αAcO	α	MePi	βOH	Br	0.40		[2]
72 Org-7402	Pi	αOH	α	MePi	βOH	Br		1,850	[6]
73	Pi	αAcO	α	MePi	O[c]	Br	0,39	1,100	[2, 6]
74	Pi	αOH	α	MePi	O[c]	I		2,200	[6]
75	Pi	αPrO	α	MePi	βPrO	Br		520	[6]
76	Pi	αOH	α	MePi	βPrO	Br		544	[6]
77	Pi	αBuO	α	MePi	βBuO	Br		405	[6]
78	Pi	αi-BuO	α	MePi	βi-BuO	Br		2,000	[6]
79	Pi	αOH	α	EtPi	βAcO	Br		17	[6]
80	Pi	αOH	α	EtPi	βBuO	Br		260	[6]
81	Pi	αAcO	α	AlPi	βAcO	Br	3.44		[2]
82	Pi	αOH	α	AlPi	βAcO	Br		73	[6]
83	Pi	αOH	α	AlPi	βBuO	Br		670	[6]
84	Pi	αOH	α	PgPi	αAcO	Br		232[d]	[6]
85	Pi	αOH	α	PgPi	βBuO	Br		790	[6]

86		MeMo	αAcO	α	MeMo	βAcO	2I	3.44		[2]
87		MeMo[e]	αAcO	α	MeMo	βAcO	2I	0.95		[2]
88		MeMo	αOH	α	MeMo	βOH	2I	0.01		[2]
89		MeMo	αOH	α	MeMo	αOH	2I	<0.10		[2]
90		MeMo	αOH	α	MeMo	O[c]	2I	0.06		[2]
91		MePy	αAcO	α	MePy	αAcO	2I	5.90		[2]
92		MePy	αOH	α	MePy	βOH	2Br	0.17		[2]
93		MePi	αAcO	α	Me_3N	βAcO	2Br	0.40		[2]
94	Pipecuronium	Me_2Pa	αAcO	α	Me_2Pa	βAcO	2Br		2.04	[8]
95		Me_2Pa	αAcO	α	Me_2Pa	βOH	2Br		180	[9]
96		Me_2Pa	αOH	α	Me_2Pa	βOH	2Br		210	[9]
97		Me_2Pa	αAcO	α	Me_2Pa	αAcO	2Br		880	[9]
98		MePrPa	αAcO	α	MePrPa	βAcO	2Br		18	[9]
99		MeAlPa	αAcO	α	MeAlPa	βAcO	2Cl		25	[9]
100		MePi	αAcO	α	Me_2Pa	βAcO	2Br		27	[9]
101		Pi	αAcO	α	Me_2Pa	βAcO	Br		51	[9]
102		Pi	αOH	α	Me_2Pa	βOH	Br		590	[9]
103		Me_2Pa	αAcO	α	MePi	βAcO	2Br		18	[9]
104		Me_2Pa	αAcO	α	Pi	βAcO	Br		120	[9]
105		MePrPa	αAcO	α	Pi	βAcO	Br		31	[9]
106		Me_2Pa	αOH	α	Pi	βOH	Br		1,150	[9]

Abbreviations: Me methyl; Et ethyl; Pr Propyl; i-Pr isopropyl; Bu Butyl; Oc octyl; Al allyl; Cr crotyl; Pg propargyl; MeCl chlormethyl; OMe methoxy; CH_2OMe methoxymethyl; CH_2CH_2OMe β-methoxyethyl; EtOH β-hydroxyethyl; EtOAc β-acetoxyethyl; CH_2COOEt ethoxycarbonylmethyl; CH_2Ph benzyl; CH_2CH_2OPh β-phenoxyethyl; Me_2N dimethylamino; Me_3N trimethylamino; Me_2EtN dimethylethylamino; Et_2MeN diethylmethylamino; Et_3N triethylamino; Me_2AlN dimethylallylamino; $Me_2AcOEtN$ dimethyl-β-acetoxyethylamino; Py pyrrolidino; MePy *N*-methylpyrrolidino; EtPy *N*-ethylpyrrolidino; Pi piperidino; MePi *N*-methylpiperidino; EtPi *N*-ethylpiperidino; AlPi *N*-allylpiperidino; PgPi *N*-propargylpiperidino; MeMo *N*-methylmorpholino; Me_2Pa 4,4-dimethylpiperazino; MePrPa 4-methyl-4-propyl-piperazino; MeAlPa 4-methyl-4-allylpiperazino; AcO acetoxy; PrO propionyloxy; BuO butyryloxy; i-BuO isobutyryloxy; Me_3CCOO pivaloiloxy; PhCOO benzoiloxy

[a] Unless otherwise indicated, relative potencies are on a weight basis
[b] Relative molar potencies
[c] Oxo group
[d] Neuromuscular data assessed on soleus muscle instead of tibialis
[e] Configuration at position 2 is α instead of β

References: [1] LEWIS et al. (1967), [2] BUCKETT et al. (1973), [3] ALAUDDIN et al. (1965), [4] BAMFORD et al. (1967), [5] BAMFORD ET AL. (1971), [6] DURANT et al. (1979), [7] SUGRUE et al. (1975), [8] KÁRPÁTI and BIRÓ (1980), [9] M. RIESZ and E. KÁRPÁTI (1983, unpublished results)

1. One Amino Group

The compounds 1–9 and 25–30 contain a quaternary nitrogen atom at the C-2, C-3, C-16, or C-17 position. This type of molecule results in a very weak curare-like action (1%–10% compared with tubocurarine). The position of the nitrogen atom seems to be unimportant, but the compounds containing a cyclic nitrogen center are more potent (CAVALLITO and GRAY 1960; LEWIS et al. 1967).

2. 3,17-Diamino Compounds

In this group (10–24 and 31–33), cyclic nitrogen also brings about stronger action. Among *N*-alkyl substitutions, diethylmethyl is more preferable than trimethyl, triethyl (ALAUDDIN et al. 1965), or the extranuclear acetylcholine fragment (compound 31). The *trans* anellation of A and B rings is always more favorable than *cis* connection (14–17 and 18–21). In positions 3 and 17, the β configuration yields more active derivatives than the α one (14, 18 and 17, 21). According to BAMFORD et al. (1967), the steric position of the 3-amino substituent plays the decisive role, while isomerization of the basic center at 17 has little influence on potency. It is interesting to note that in the case of compounds 16 and 32, nonacetylcholine-like substitution in ring A resulted in a many times stronger curarizing effect. The ditertiary congener (compound 22) was comparatively inactive. It was pointed out (BAMFORD et al. 1971) that quaternization is more important at position 3 than at 17 (see 14, 23, and 24) and potency is retained even if the 17 nitrogen is tertiary. Among the 2β-acetoxy-3α, 17β-diamino androstanes, the 3-piperidino derivative was a more effective blocker than 3-pyrrolidino (compounds 33, 32).

3. 2,17-Diamino Compounds

Compounds 34–40 belong to this group. Exchange of the 2-acetoxy and 3-piperidino substituents resulted in a marked reduction of neuromuscular activity (compounds 33, 34). Incorporation of a second ester group at position 16 offered more pronounced curarizing action (compounds 34, 38 and 35, 36), but none of the compounds reached the effectivity of tubocurarine. Similarly to the observations in previous series, piperidino groups confer greater potency than pyrrolidino ones at position 2 (compounds 38, 40). On the other hand at position 17 pyrrolidino seems to be the preferable substituent (compounds 34 and 35).

4. 2,16 Diamino Compounds

This is the most widely studied and richest chemical group (compounds 41–106). Investigating the structure–activity relationship in this field, it becomes strikingly evident that the interrelation of substituents plays a decisive role in many cases. Therefore, only a few general remarks can be made, as follows. The optimum activity was reached by incorporation of 3α,17β-diacetoxy groups and lack of esterification (e.g., 3,17-diols) resulted in decreasing action. 17α-Acetoxy epimers of pancuronium (compound 47) and pipecuronium (compound 97) were weaker than the parent compounds (compounds 41, 94). The presence of two longer chain esters always reduced potency and this effect was more pronounced in the

16-monoquaternary series. The esterification of OH at position 17 is of greater importance, since in the case of monoquaternary compounds the 3-ol-derivatives can be as active as the diesters (see compounds 69, 70 and 75, 76). The fall in effectivity following identical modification in the bisquaternary series may be partly explained by its greater hydrophilic–lipophilic balance enhancing the binding to inactive acceptor sites in the body. It is interesting that highly active structures may be replaced with structures containing a tertiary nitrogen at ring A without greatly weakening the action, while compounds originally possessing lower paralyzing action become virtually inactive on similar modification (DURANT et al. 1979; and our personal data). The use of a morpholino ring instead of piperazino, and pyrrolidino seems to be less successful, probably owing to the charge-dissipating effect of the incorporated oxygen atom (BOVET 1972; BUCKETT et al. 1973).

II. D-Homoazaandrostanes and Androstenes

The members of this chemical group are given in Table 2. The activity of 5,6-unsaturated and the saturated derivatives do not differ significantly. Stereoisomerism at position 3 does not alter the potency (see compounds 117–118 and 120–121). At position 3 pyrrolidino substitution is more advantageous than piperazino (compounds 120, 122). Monoquaternary or compounds containing only one nitrogen atom are much less potent (see compounds 112–116). Choline- or acetylcholine-like structures do not improve the activity (compounds 107, 110, 111).

III. 4-Azaandrostanes

The chemical structures and activity are presented in Table 3. The only remark about this group is that two nitrogen centers are needed to provide sufficient curare-like action.

IV. Miscellaneous Azasteroids

Form this small group of compounds (Table 4) it is concluded that both quaternary nitrogens can be incorporated into the steroid nucleus (e.g., diazasteroid) while still retaining considerable curare-like action (see compound 129).

V. Pregnane Derivatives

Table 5 shows the structures and activities of pregnane derivatives. In general the pregnane skeleton seems to be less favorable than the androstane one. This can be explained by the enhanced lipophilicity (BUCKETT 1972). For neuromuscular blocking action, the presence of two nitrogens is mandatory. Considering the results obtained by the head drop method in rabbits (KHUONG HUU and PINTO-SCOGNAMIGLIO 1964; ROQUET et al. 1971 as detailed by BUCKETT 1975), 20-amino substitution confers greater potency than 16-amino substitution. The steric posi-

Table 2. Structures and neuromuscular blocking potencies of D-homoaza androstanes and 5-androstenes

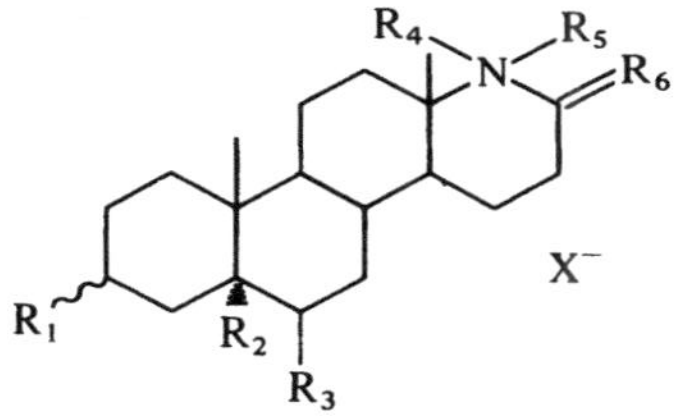

No. and name of compound	R_1	R_2	R_3	R_4	R_5	R_6	X	Anesthetized cat ED_{50} (μg/kg) Tibialis	Soleus	References
107 HS-310, Chandonium	βMePy			Me	Me	H	2I	70, 39	60, 47	[1, 2]
108 HS-705	βEtPy			Me	Et	H	2I	218	222	[2]
109 HS-704	βEtPy			Et	Et	H	2I	336	465	[2]
110 HS-627	βMePy			Me	EtOAc	H	2I	58	74	[3]
111 HS-626	βMePy			Me	EtOH	H	2I	57	46	[3]
112 HS-311	βMePy			H	[a]	O[b]	I	9,500	10,000	[1]
113 HS-433	βAcO			Me	Me	H	I	5,000	5,200	[1]
114 HS-308	βOH			Me	Me	H	I	1,500	2,200	[1]
115 HS-465	βOH			Me	EtOAc	H	I	>4,000	>4,000	[3]
116 HS-408	βOH			Me	EtOH	H	I	>4,000	>4,000	[3]
117 HS-692	βMePy	H	H	Me	Me	H	2I	80	86	[2]
118	βMePy	H	H	Me	Me	H	2Br	125		[4]
119 HS-693	βEtPy	H	H	Me	Et	H	2I	190	213	[2]
120 Duador	αMePy	H	H	Me	Me	H	2Br	105	113	[5]
121	αMePy	H	H	Me	Me	H	2I	120		[4]
122	αMe_2Pa	H	H	Me	Me	H	2Br	510		[6]
123	αMeAlPa	H	H	Al	Me	H	2Br	365		[6]

[a] Not quaternized at 17a position
[b] Oxo group
References: [1] GANDIHA et al. (1974), [2] TEERAPONG et al. (1979), [3] MARSHALL et al. (1981), [4] TUBA et al. (1981), [5] BIRÓ and KÁRPÁTI (1981), [6] M. RIESZ and E. KÁRPÁTI (1983, unpublished results)

Table 3. Structures and neuromuscular blocking potencies of some 4-azaandrostanes

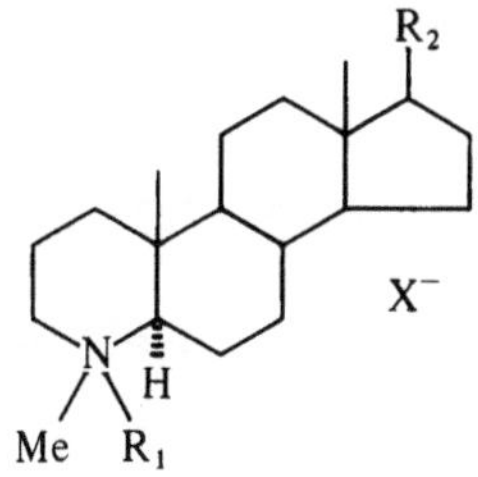

No. and name of compound	R_1	R_2	X	Anesthetized cat ED_{50} (μg/kg)		References
				Tibialis	Soleus	
124 HS-467	Me	Me_3N	2I	325	375	[1]
125 HS-435	Me	AcO	I	2,250	3,000	[1]
126 HS-419	Me	OH	I	2,500	3,000	[1]
127 HS-523	EtOAc	AcO	I	>4,000	>4,000	[2]
128 HS-522	EtOH	OH	I	>4,000	>4,000	[2]

References: [1] GANDIHA et al. (1974), [2] MARSHALL et al. (1981)

tion of 3, 16, and 20 substituents does not exert an important effect on activity. In cats, both lack of quaternization at the 3β nitrogen (compounds 153, 154) and 5,6 unsaturation (compounds 151–152) negatively influenced potency, however, similar differences were not observed in rabbits.

VI. Conessine Derivatives

These compounds are presented in Tables 6 and 7. Their relative potencies compared with tubocurarine are generally weaker and only some are equal or slightly stronger. The structure–activity relationships have been discussed in detail by BUSFIELD et al. (1968) and by BUCKETT (1975); no further comment is necessary.

VII. Cholane and Norcholane Derivatives

Members of this chemical group, given in Table 8, practically lack curarizing properties, presumably owing to a shortened interonium distance, and/or to stronger lipophilicity; only the facilitation of twitch responses points to some action on neuromuscular junctions.

VIII. Miscellaneous Steroids

These compounds are presented in Table 9 and Figs. 1 and 2. GARCIA et al. (1980) report the interesting fact that compounds 200–203, androstane derivatives, are depolarizing neuromuscular blockers. Piperidylmethyl androstane diol (compound 204) caused dose-dependent potentiation (at low doses) and blockade (at

Table 4. Structures and neuromuscular blocking activities of some miscellaneous azasteroids

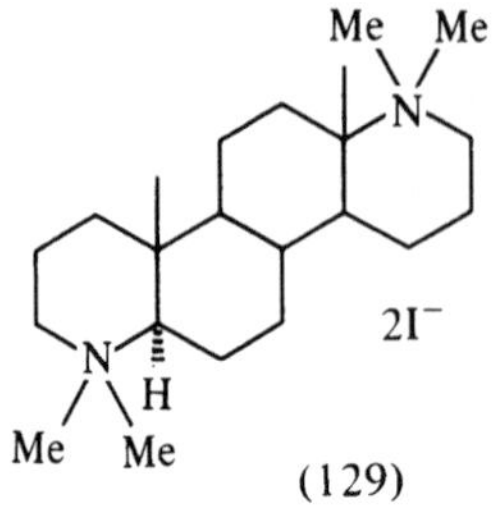

(129)

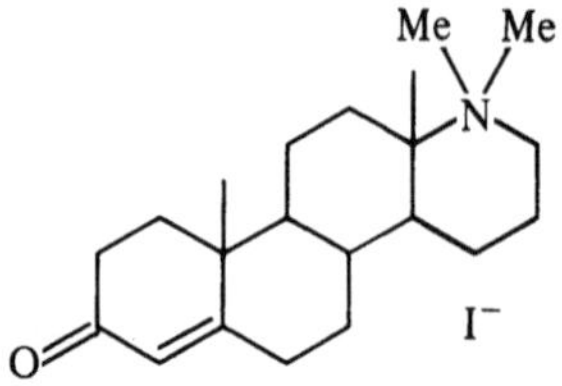

(130)

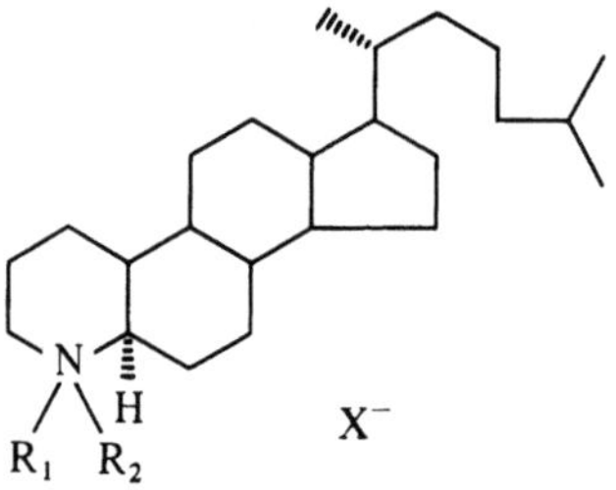

	R_1	R_2	X
131	EtOAc		
132	EtOAc	Me	I
133	EtOH	Me	I

No. and name of compound	Anesthetized cat ED_{50} (μg/kg)		References
	Tibialis	Soleus	
129 HS-342	290, 250	320, 300	[1, 2]
130 HS-316	1,500	2,000	[2]
131 HS-460[a]			[3]
132 HS-464[a]			[3]
133 HS-405[a]			[3]

[a] Small and inconsistent effects on neuromuscular transmission

References: [1] MARSHALL et al. (1973), [2] GANDIHA et al. (1974), [3] MARSHALL et al. (1981)

higher doses) of the indirectly elicited contractions of in situ rabbit diaphragm and isolated rat diaphragm (LOOMIS 1963). According to LOOMIS (1963), the block was located at the neuromuscular junction and the substance was not a potent relaxant, since the effective dose determined in rabbits was 10 mg/kg.

Quillaiate of choline iodide (compound 205), a steroid-related triterpenoid agent, at 10 mg/l partly blocked indirectly elicited twitches and tetani of the rat hemidiaphragm nerve–muscle preparation (HAMED and EL GHOLMY 1972). The transmission failure developed slowly and was enhanced by neostigmine, tetraethylammonium, and choline, probably owing to prejunctional inhibiting action.

Table 5. Structures and neuromuscular blocking potencies of pregnane derivatives

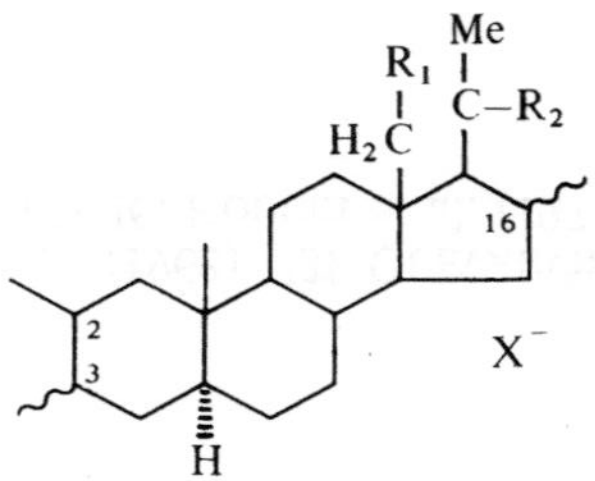

No. and name of compound	Positions and configurations of substitutions						Anesthetized cat			Rabbit head drop		References
	2	3	16	R_1	R_2	X	Relative potency[a] (tubocurarine = 1)		Tibialis ED_{50} (μg/kg)	Relative potency (tubocurarine = 1)	ED_{50} (μg/kg)	
							Gastrocnemius	Tibialis				
134	MePy	αAcO	H	H	O[c]	Br	0.02[b]					[1]
135	MePi	αOH	H	H	O[c]	Br	0.02[b]					[1]
136	Me_3N	αOH	αMe	H	O[c]	Br	0.15[b]					[1]
137	H	βOH	H	H	αMe_3N	I		<0.05				[2]
138	H	O[c]	H	H	αMe_3N	I		<0.05				[2]
139	H	βAcO	H	H	αMe_3N	I		<0.06				[2]
140 Malouetin	H	βMe_3N	H	H	αMe_3N	2I		0.82				[2]
141 Malouetin	H	βMe_3N	H	H	αMe_3N	2Cl				0.74, 1.20	150	[3, 4]
142	H	βMe_3N	H	H	βMe_3N	2Cl				1.20	150	[4]
143	H	αMe_3N	H	H	αMe_3N	2Cl				0.90	200	[4]
144	H	αMe_3N	H	H	βMe_3N	2Cl				0.90	200	[4]
145 Trimethyl-funtuminium	H	αMe_3N	H	H	O[c]	[d]				0.018		[5]
146 Trimethyl-funtumidinium	H	αMe_3N	H	H	αOH	[d]				0.013		[5]

Table 5 (continued)

No. and name of compound	Positions and configurations of substitutions						Anesthetized cat			Rabbit head drop		References
	2	3	16	R_1	R_2	X	Relative potency[a] (tubocurarine = 1)		Tibialis ED_{50} (μg/kg)	Relative potency (tubocurarine = 1)	ED_{50} (μg/kg)	
							Gastrocnemius	Tibialis				
147	H	βMe_3N	βMe_3N	AcO	βAcO				350		1,670	[6]
148	H	βMe_3N	βMe_3N	AcO	αAcO				270		1,470	[6]
149	H	βMe_3N	αMe_3N	AcO	βAcO				620		1,630	[6]
150	H	βMe_3N	αMe_3N	AcO	αAcO				280		1,800	[6]
151[e]	H	βMe_3N	βMe_3N	AcO	βAcO				550		1,440	[6]
152[e]	H	βMe_3N	αMe_3N	AcO	αAcO				750		1,670	[6]
153	H	βMe_2N	αMe_3N	AcO	βAcO				1,000		1,500	[6]
154	H	βMe_2N	αMe_3N	AcO	αAcO				460		1,800	[6]

[a] Unless otherwise indicated, relative potencies are on a weight basis
[b] Relative molar potencies
[c] Oxo group
[d] Methylsulfate derivatives
[e] 5-Pregnene skeleton instead of pregnane

References: [1] Lewis et al. (1967), [2] Busfield et al. (1968), [3] Quévauvillier and Lainé (1960), [4] Khoung-Huu and Pinto-Scognamiglio (1964), [5] Blanpin and Bretaudeau (1961), [6] Roquet et al. (1971) cited by Buckett (1975)

Table 6. Structures and neuromuscular blocking potencies of conessine derivatives assessed on anesthetized cat tibialis preparation (BUSFIELD et al. 1968)

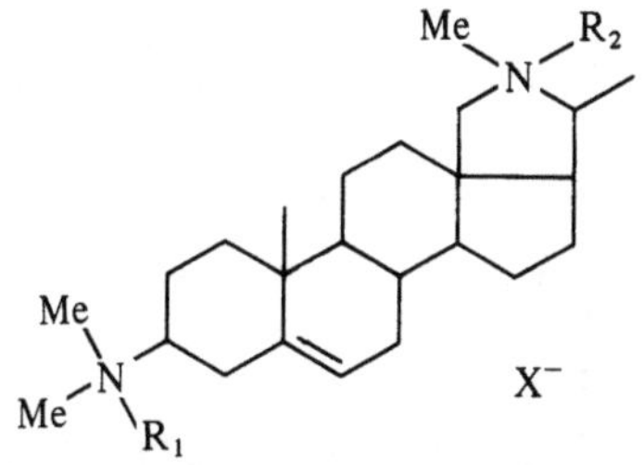

No. and name of compound	R_1	R_2	X	Relative potency (tubocurarine = 1)
155	Me		I	0.12
156	Et		I	0.33
157	Bu		I	0.64
158	MeCl		Br	0.22
159	CH_2COOEt		Br	0.10
160	CH_2Ph		Br	0.58
161	CH_2CH_2OPh		Br	0.45
162	O			0.02
163 *N,N'*-Dimethyl-conessine	Me	Me	2I	0.62
164	Et	Et	2I	0.79
165	Pr	Pr	2I	0.47
166	i-Pr	Me	2I	1.61
167	Bu	Me	2I	1.54
168	Oc	Me	2I	0.25
169	Al	Al	2Br	0.67
170	Pg	Pg	2Br	0.56
171	Cr	Cr	2Br	1.00
172	H	Me	2Cl	0.35
173	OMe	Me	2I	0.61
174	MeCl	Me	Br I	0.71
175	CH_2OMe	H	2Cl	0.21
176	CH_2CH_2OMe	CH_2CH_2OMe	2I	0.45
177	CH_2COOEt	CH_2COOEt	2Br	0.97
178	CH_2Ph	CH_2Ph	2I	0.51
179	CH_2Ph	Me	2Cl	1.09
180	CH_2CH_2OPh	Me	Br I	1.37

Table 7. Structures and neuromuscular blocking potencies of some saturated and unsaturated compounds related to conessine assessed on the anesthetized cat tibialis preparation

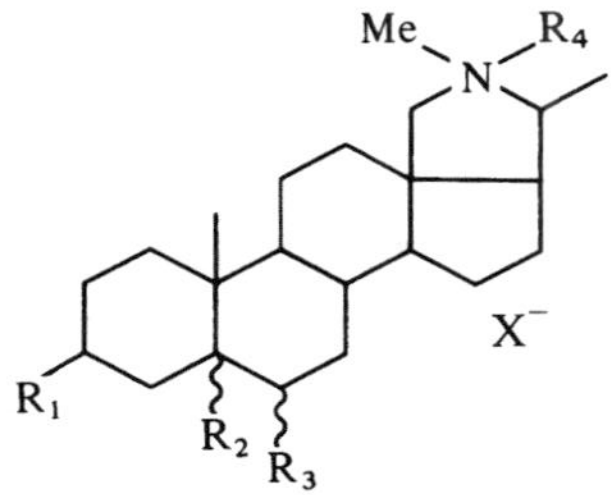

No. and name of compound	Position of double bonds in the ring system	R_1	R_2	R_3	R_4	X	Relative potency (tubocurarine = 1)	References
181	Δ^3, Δ^5				Me	I	<0.03	[1]
182		Me_3N	H	H	Me	2I	0.82	[1]
183[a]	Δ^5	Me_3N			Me	2I	0.62	[1]
184		Me_2AlN	H	H	Al	2Br	0.97	[1]
185	Δ^4, Δ^6	Me_3N			Me	2I	0.61	[1]
186 Stercuronium	Δ^4, Δ^6	Me_2EtN				I	1.62	[2]
187		Me_3N	5α, 6α	O[b]	Me	2I	0.47	[1]
188		Me_3N	5β, 6β	O[b]	Me	2I	0.60	[1]
189		Me_3N	H	O[c]	Me	2I	1.19	[1]
190		Me_3N	OH	OH	Me	2I	0.20	[1]
191		Me_3N	H	αOH	Me	2I	0.90	[1]
192	Δ^6	Me_2N	OH	AcO	Me	2I	0.39	[1]
193	Δ^6	Me_2N	OH		Me	2I	0.35	[1]
194	N-20	Me_3N	H	H		2Cl	1.16	[1]
195	N-20	Me_2N	H	H		2Cl	0.18	[1]

[a] Configuration of the methyl group at position 20 is α instead of β
[b] 5,6-Epoxides
[c] Oxo group
References: [1] BUSFIELD et al. (1968); [2] MARSHALL (1973)

Table 8. Structures of cholane and norcholane derivatives investigated by MARSHALL and MARTIN-SMITH (1972)

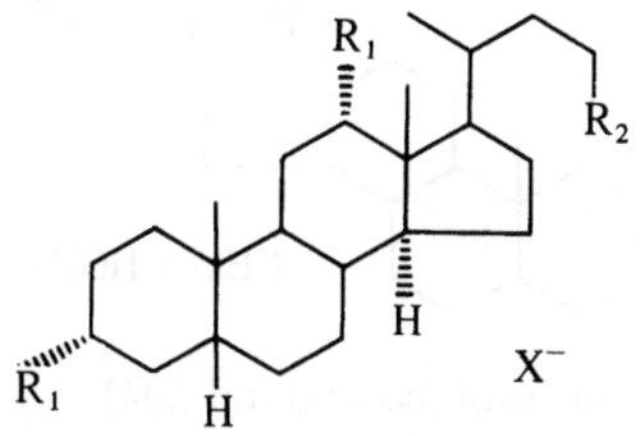

No. of compound	R_1	R_2	X
196	Me_3N	Me	2I
197	Me_2EtN	Me	2I
198	Me_3N	H	2I
199	Me_2EtN	H	2I

Table 9. The neuromuscular effects of some depolarizing steroid relaxants investigated by GARCIA et al. (1980)

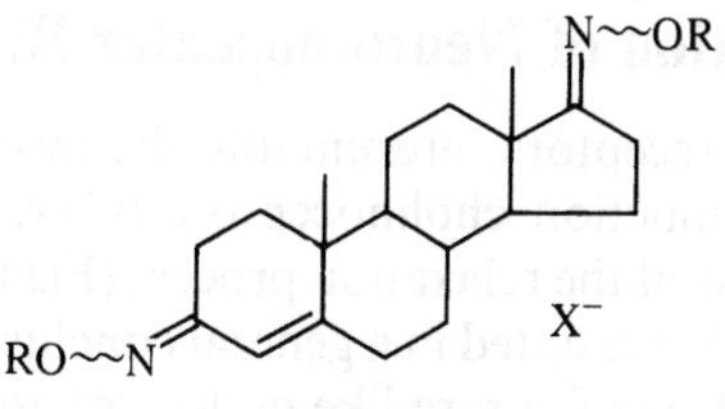

No. of compound	R	Configuration of OR substituents	X	Mouse treadmill test ED_{50} (μ*M*/kg) (intraperitoneal administration)
200	$Me_3N-CH_2CH_2$	E,E	2Br	10.0×10^{-2}
201	$Me_3N-CH_2CH_2$	Z,Z	2Br	7.05×10^{-2}
202	$Me_3N-CH_2CH_2CH_2$	E,E	2Br	9.80×10^{-2}
203	$Me_3N-CH_2CH_2CH_2$	Z,Z	2Br	5.10×10^{-2}

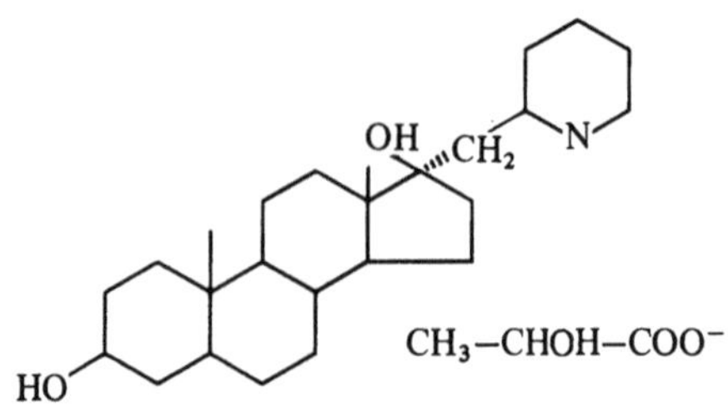

Fig. 1. Piperidylmethyl androstane diol (compound 204)

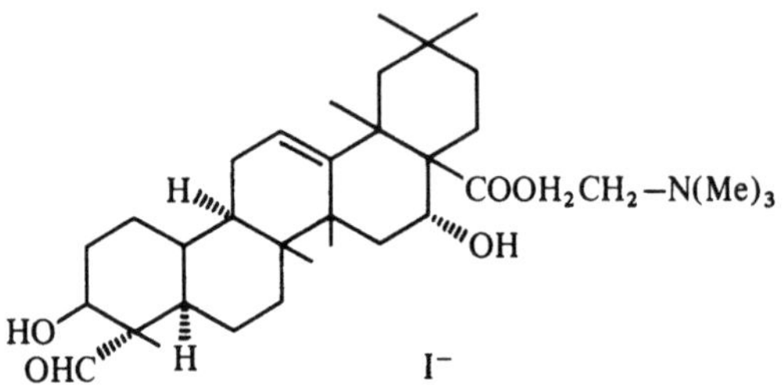

Fig. 2. Quillaiate of choline iodide (compound 205)

C. Onset and Duration of Neuromuscular Blocking Effect

Among the nicotinergic receptors, presumably the most readily accessible ones are skeletal myoneural junction cholinoceptors (CAVALLITO and GRAY 1960); however, detailed analysis of the relaxation process (FELDMAN and TYRRELL 1970; FELDMAN 1976; HULL 1982) resulted in a general conclusion that receptor binding and subsequent dissociation of curare-like molecules are not simple functions of its blood and extracellular fluid concentration dynamics. Data obtained by computer simulation of the time course of various relaxants (D'HOLLANDER and DELCROIX 1981), taking into account the plasma levels, the association and dissociation constants of the drug–receptor interaction, and the dependence of the twitch height on the fractional receptor occupation, were in good agreement with the experimental findings, indicating that the factors incorporated into the mathematical model are the most important determinants of the temporal pattern of paralysis.

Drugs having similar duration of action may show characteristic differences in the time required for development of blockade and for reactivation. Recent experiments in this laboratory (M. RIESZ and E. KÁRPÁTI 1983, unpublished work) suggested such a deviation in the effect of vecuronium (compound 69) and Duador (RGH-4201) (compound 120). In cats, at 75%–95% blocking dose, the onset time of vecuronium was slower, followed by a virtually constant depressed period and a short recovery in contrast to Duador which has two or three times shorter onset; the reactivation started immediately after the maximal twitch depression, but the recovery rate was longer.

Our present knowledge on the molecular features determining the time course of action of steroidal relaxants is very limited because up to the last decade structure–activity research was focused on the factors influencing potency. In the literature, four main statements were formulated: drug molecules (a) less strongly binding to the receptor (CRUL 1976), (b) of greater affinity for liver or inactive acceptor sites (AGOSTON and KERSTEN 1976; AGOSTON et al. 1977), (c) having a weak anticholinesterase action (GANDIHA et al. 1975), and (d) which may undergo rapid enzymatic or nonenzymatic breakdown (MARTIN-SMITH 1971), are promising for a short duration of action. Because paralysis which develops and terminates independent of the functional state of a given organ or enzyme is more desirable, the main direction of research should be concentrated on the investigation of factors determining the stability of the steroid–cholinergic receptor complex and a search for a rapidly and spontaneously eliminated drug.

The time course of action of steroidal curare-like agents are tabulated in Table 10. The following observations can be made: rapid onset (less than 2 min) shows a high coincidence with short duration (half that of tubocurarine or recovery rate less than 3 min; see compounds 43, 52, 55, 58, 59, 75, 76, 84, 85). Long duration of action (more than the duration of tubocurarine or recovery rate more than 6 min) frequently occurs in parallel with slow onset of block (more than 4 min; see compounds 41, 94, 98, 99, 101, 104, 105). All compounds of long duration are diester derivatives, most of them are diacetylesters (see compounds 38, 40, 41, 45, 65, 94, 97–99, 101, 104, 105). The majority of long-acting substances have high potency (at least fivefold that of tubocurarine; see compounds 41, 45, 65, 94, 97, 99, 101, 105) however, among derivatives of short duration, only compound 14 reaches this activity level.

D. Conclusions

The mode of action of the vast majority of compounds tested is competitive. The optimal supporting nucleus seems to be the androstane structure; among the series investigated so far, several derivatives belonging to the 2,16-diamino androstanes emerged as highly potent neuromuscular blockers.

Concerning chemistry, only some general conclusions can be drawn. For powerful curare-like action, at least two nitrogen centers are needed, preferably cyclic ones. One of them should be quaternized. The interonium distance can vary over a wider range than previously postulated. The incorporation of an acetylcholine-like part into the steroidal skeleton is not always necessary for high potency. The quaternary nitrogen atom should not be directly and rigidly connected to the steroidal nucleus. The last three remarks are derived from the high biologic activity of compounds which differ from the pancuronium series only in one main respect: the piperidino rings are replaced by piperazino substituents. These molecules lack the acetylcholine-like fragment. Inspection of a Dreiding stereo model of pipecuronium showed that, in the four probable conformations, the shortest distance between the quaternary (remote) nitrogens is greater than 12 Å; however, the maximum (in another conformer) is greater than 14 Å. The position of the quaternary nitrogens is less determined in the dipiperazino compounds than in the dipiperidino derivatives owing to the possible rotation of piperazino rings about the C-2–N and C-16–N bonds.

Table 10. Duration and time course of action in anesthesized cats

No. and name of compound	Relative duration (tubocurarine = 1)		Onset time [a] (min)		Recovery rate [b] (min)		Duration [c] (min)		References
	Tibialis	Gastrocnemius	Tibialis	Soleus	Tibialis	Soleus	Tibialis	Soleus	
14 Dipyrandium	0.24								[1]
31		0.44							[2]
32		0.65							[2]
33		0.75							[2]
34		0.70							[2]
36		0.73							[2]
37		0.62							[2]
38		1.1							[2]
39		1.0							[2]
40		1.5							[2]
41 Pancuronium		1.15	4.3	4.7	8.8	4.7			[2, 3]
42		1.0	1.7	2.3	5.1	6.8			[2, 3]
43 Dacuronium Br		0.49	1.5	1.5	4.7	4.7			[2, 3]
44		0.55							[2]
45		1.1							[2]
46 Dacuronium I		0.32							[2]
47		0.75							[2]
48 Org-6368			1.6	1.9	3.7	2.4			[3]
49		0.47							[2]
50			3.5	2.5					[3]
51		0.37							[2]
52			1.1	1.2	3.8	1.7			[3]
53		0.46							[2]
54		0.95	2.4	2.5	6.6	5.1			[2, 3]
55			1.3	1.4	2.5	2.8			[3]
56		0.85							[2]
57			2.0	2.6	4.3	2.9			[3]
58			1.4	1.5	1.4	1.9			[3]
59			0.9	1.4	1.9	3.6			[3]
60			2.6	2.9	3.1	4.9			[3]
61		1.47							[2]
62		0.5							[2]
63		>3.0							[2]
64		0.72							[2]
65		1.16							[2]
66		1.0							[2]
67		0.81							[2]
68		0.44							[2]
69 Vecuronium		0.77	3.5	3.8	3.9	2.9			[2, 3]
70			2.8	3.2	4.1	4.1			[3]
71		0.45							[2]
72 Org-7402			1.4	3.8	3.0	1.7			[3]
73		0.44	1.4	1.4	5.1	4.7			[2, 3]
74			1.0	1.3	4.4				[3]
75			1.1	1.6	2.8	2.8			[3]
76			0.5	1.1	2.7	2.7			[3]
77			1.5	1.5	3.6	3.7			[3]
79			4.6	4.5	4.7	3.9			[3]
80			1.3	1.7	3.5	4.4			[3]

Table 10 (continued)

No. and name of compound	Relative duration (tubocurarine = 1)		Onset time[a] (min)		Recovery rate[b] (min)		Duration[c] (min)		References
	Tibialis	Gastrocnemius	Tibialis	Soleus	Tibialis	Soleus	Tibialis	Soleus	
81		0.89							[2]
82			2.5	2.7	6.0	4.0			[3]
83			1.7	2.4	2.1	2.4			[3]
84			1.0	1.3	2.2				[3]
85			1.7	1.9	2.4	2.5			[3]
86		0.96							[2]
87		0.95							[2]
90		0.80							[2]
91		1.0							[2]
92		0.32							[2]
93		1.05							[2]
94 Pipecuronium			8.2		11.4				[4]
95			1.9		3.5		8.4		[4]
96			2.2		5.2		12.2		[4]
97			3.0		14.5				[4]
98			9.3		34.7				[4]
99			6.3		7.2		19.0		[4]
100			5.0		4.9		13.6		[4]
101			6.3		10.7		18.8		[4]
102			2.9		5.5		14.6		[4]
103			6.8				15.2		[4]
104			10.3		14.4		32.8		[4]
105			7.5		11.1		28.0		[4]
106			2.1		3.4		9.2		[4]
107 HS-310 Chandonium			2.1	3.9	3.2	3.8	12.3	17.7	[5]
108 HS-705			2.5	3.6	3.6	6.8	14.8	25.8	[5]
109 HS-704			2.1	3.6	3.9	6.6	12.4	22.9	]5]
110 HS-627			3.6	4.7	2.7	4.3	11.9	15.7	[6]
111 HS-626			2.7	3.0	3.5	3.4	11.0	12.5	[6]
117 HS-692			2.4	3.2	4.6	5.2	14.5	17.2	[5]
118			2.0		5.0				[4]
119 HS-693			2.6	3.8	4.3	6.7	13.5	23.9	[5]
120 Duador			1.2	1.5			12.2	10.9	[7]
121			1.6		5.9		16.2		[4]
122			2.3		8.9		25.0		[4]
123					12.2				[4]
129 HS-342	0.37								[8]
140 Malouetin							5–15	5–15	[9]
155							5–10	10–20	[9]
156							15–20	15–20	[9]
157							5–15	10–15	[9]
158							10–20	15–30	[9]
159							15–20	15–30	[9]
160							10–15	10–20	[9]
161							5–15	10–20	[9]
162							5–15	5–15	[9]
163 *N,N'*-Dimethylconessine							5–20	10–20	[9]

Table 10 (continued)

No. and name of compound	Relative duration (tubocurarine = 1)		Onset time[a] (min)		Recovery rate[b] (min)		Duration[c] (min)		References
	Tibialis	Gastrocnemius	Tibialis	Soleus	Tibialis	Soleus	Tibialis	Soleus	
164							10–20	10–20	[9]
165							10–25	10–30	[9]
166							15–20	15–25	[9]
167							10–15	10–15	[9]
168							10–20	10–15	[9]
169							10–20	10–20	[9]
170							20–30	25–35	[9]
171							15–30	20–30	[9]
172							10–15	15–20	[9]
173							10–15	15–25	[9]
174							15–30	20–30	[9]
175							5–15	10–15	[9]
176							10–20	15–20	[9]
177							20–30	20–30	[9]
178							20–30	25–35	[9]
179							10–20	10–20	[9]
180							5–15	5–20	[9]
182							10–20	10–20	[9]
183							20	20	[9]
184							10–20	15–25	[9]
185							5–20	10–20	[9]
187							30–60	30–60	[9]
188							5–30	10–35	[9]
189							15–40	20–40	[9]
190							15–30	15–30	[9]
191							15–20	15–20	[9]
192							5–15	10–20	[9]
193							15–25	15–30	[9]
194							10–20	15–25	[9]
195							5–25	5–25	[9]

[a] Time from injection to maximal effect
[b] Time elapsed between 25% and 75% reactivation
[c] Time to 90%–95% recovery
References: [1] BAMFORD et al. (1967), [2] BUCKETT et al. (1973), [3] DURANT et al. (1979), [4] M. RIESZ and E. KÁRPÁTI (1983, unpublished work), [5] TEERAPONG et al. (1979), [6] MARSHALL et al. (1981), [7] BIRÓ and KÁRPÁTI (1981), [8] MARSHALL et al. (1973), [9] BUSFIELD et al. (1968)

Regarding the onset time and duration of action of steroidal relaxants, further refinement of our knowledge seems to be necessary to give a decisive answer, whether the shortening of effect by chemical modifications is limited by the use of a steroidal skeleton or not.

References

Agoston S, Kersten UVW (1976) The pharmacokinetics of the steroid muscle relaxants. In: Spierdijk J, Feldman SA, Mattie H (eds) Anaesthesia and pharmacology. Williams and Wilkins, Baltimore

Agoston S, Crul EJ, Kersten UW, Houwertjes MC, Scaf AHJ (1977) The relationship between disposition and duration of action of a congeneric series of steroidal neuromuscular blocking agents. Acta Anaesthesiol Scand 21:24–30

Alauddin M, Martin-Smith M (1962) Biological activity in steroids possessing nitrogen atoms. Part I. Synthetic nitrogenous steroids. J Pharm Pharmacol 14:325–349

Alauddin M, Caddy B, Lewis JJ, Martin-Smith M, Sugrue MF (1965) Non-depolarising neuromuscular blockade by 3α,17α-bis/quaternary ammonium/5α-androstanes. J Pharm Pharmacol 17:55–59

Baird WLM, Reid AM (1967) The neuromuscular blocking properties of a new steroid compound, pancuronium bromide (a pilot study in man). Br J Anaesth 39:775–780

Bamford DG, Biggs DF, Davis M, Parnell EW (1967) Neuromuscular blocking properties of stereoisomeric androstane-3,17-bisquaternary ammonium salts. Br J Pharmacol Chemother 30:194–202

Bamford DG, Biggs DF, Davis M, Parnell EW (1971) The neuromuscular blocking activity of some monoquaternary androstane derivatives. J Pharm Pharmacol 23:595–599

Biró K, Kárpáti E (1981) The pharmacology of a new short acting nondepolarising muscle relaxant steroid (RGH-4201). Arzneimittelforsch 31:1918–1924

Blanpin O, Bretaudeau J (1961) Etude pharmacodynamique des sulfométhylates de triméthylfuntuminium et de triméthylfuntumidinium. CR Soc Biol (Paris) 155:878–883

Bovet D (1972) Synthetic inhibitors of neuromuscular transmission. In: Cheymol J (ed) International encyclopedia of pharmacology and therapeutics, sect 14, vol 1. Neuromuscular blocking and stimulating agents. Pergamon, Oxford

Buckett WR (1972) Aspects of the pharmacology of aminosteroids. In: Briggs MH, Christie G (eds) Advances in steroid biochemistry and pharmacology, vol 3. Academic, London

Buckett WR (1975) Steroidal neuromuscular blocking agents. In: Harper NJ, Simmonds AB (eds) Advances in drug research, vol 10. Academic, London

Buckett WR, Bonta IL (1966) Pharmacological studies with NA97 (2β,16β-dipiperidino-5α-androstane-3α17β-diol diacetate dimethobromide). Fed Proc 25:718

Buckett WR, Marjoribanks CEB, Marwick FA, Morton MB (1968) The pharmacology of pancuronium bromide (Org NA97), a new potent steroidal neuromuscular blocking agent. Br J Pharmacol Chemother 32:671–682

Buckett WR, Hewett CL, Savage DS (1973) Pancuronium bromide and other steroidal neuromuscular blocking agents containing acetylcholine fragments. J Med Chem 16:1116–1124

Busfield D, Child KJ, Clarke AJ, Davis B, Dodds MG (1968) Neuromuscular blocking activities of some steroidal mono and bis-quaternary ammonium compounds with special reference to NN′-dimethylconessine. Br J Pharmacol Chemother 32:609–623

Cavallito CJ, Gray AP (1960) Chemical nature and pharmacological actions of quaternary ammonium salts. Prog Drug Res 2:135–225

Crul JF (1976) New muscle relaxants. In: Spierdijk J, Feldman SA, Mattie H (eds) Anaesthesia and pharmacology. Williams and Wilkins, Baltimore

D'Hollander A, Delcroix C (1981) An analytical pharmacodynamic model for nondepolarising neuromuscular blocking agents. J Pharmacokinet Biopharm 9:27–40

Durant NN, Marshall IG, Savage DS, Nelson DJ, Sleigh T, Carlyle IC (1979) The neuromuscular and autonomic blocking activities of pancuronium, Org NC45, and other pancuronium analogues in the cat. J Pharm Pharmacol 31:831–836

Feldman SA (1976) Pharmacokinetics of the non-depolarizing muscle relaxants. In: Spierdijk J, Feldman SA, Mattie H (eds) Anaesthesia and pharmacology. Williams and Wilkins, Baltimore

Feldman SA, Tyrell MF (1970) A new theory of the action of muscle relaxants. Br J Anaesth 42:91–92

Gandiha A, Marshall IG, Paul D, Singh H (1974) Neuromuscular and other blocking actions of a new series of mono and bisquaternary aza steroids. J Pharm Pharmacol 26:871–877
Gandiha A, Marshall IG, Paul D, Rodger IW, Scott W, Singh H (1975) Some actions of chandonium iodide, a new short-acting muscle relaxant, in anaesthetized cats and on isolated muscle preparations. Clin Exp Pharmacol Physiol 2:159–170
Garcia DB, Brown RG, Delgado JN (1980) Synthesis, stereochemical analysis and neuromuscular blocking activity of selected E,E- and Z,Z-isomeric O-ethers of 4-androstene-3,17-dione dioximes. J Pharm Sci 69:995–999
Hamed MI, El-Gholmy Z (1972) Quillaiate of choline iodide: a new monoquaternary neuromuscular blocking agent. Arzneimittelforsch 22:2133–2136
Hull CJ (1982) Pharmacodynamics of non-depolarizing neuromuscular blocking agents. Br J Anaesth 54:169–182
Kárpáti E, Biró K (1980) Pharmacological study of a new competitive neuromuscular blocking steroid, pipecuronium bromide. Arzneimittelforsch 30:346–354
Kharkevich DA (1974) New curare-like agents. J Pharm Pharmacol 26:153–165
Khuong Huu F, Pinto-Scognamiglio W (1964) Activité curarisante du dichlorure de 3β-20αbistrimethylammonium 5α-prégnane (malouétine) et de ses stéréoisomeres. Arch Int Pharmacodyn 147:209–219
Lewis JJ, Martin-Smith M, Muir TC, Ross HH (1967) Steroidal monoquaternary ammonium salts with nondepolarising neuromuscular blocking activity. J Pharm Pharmacol 19:502–508
Loomis TA (1963) Neuromuscular effects of piperidylmethyl-androstane-diol. Arch Int Pharmacodyn 141:412–422
Marshall IG (1973) The effects of three short-acting neuromuscular blocking agents on fast- and slow-contracting muscles of the cat. Eur J Pharmacol 21:299-304
Marshall IG, Martin-Smith M (1972) Ganglion-blocking activity of bisquaternary ammonium salts derived from 3α,12α-diamino-5β-cholane and 3α,12α-diamino-24-nor-5β-cholane. Eur J Pharmacol 17:39–43
Marshall IG, Paul D, Singh H (1973) The neuromuscular and other blocking actions of 4,17a-dimethyl-4,17a-diaza-D-homo-5α-androstane dimethoiodide (HS 342) in the anaesthetized cat. Eur J Pharmacol 22:129–134
Marshall IG, Harvey AL, Singh H, Bhardway TR, Paul D (1981) The neuromuscular and autonomic blocking effects of azasteroids containing choline or acetylcholine fragments. J Pharm Pharmacol 33:451–457
Martin-Smith M (1971) Rational elements in the development of superior neuromuscular blocking agents. In: Ariens EJ (ed) Drug design, vol 2. Acedemic, New York
Quévauviller MA, Lainé MF (1960) Sur la toxicité et le pouvoir curarisant du chlorure de malouétine. Ann Pharm Fr 18:678–680
Roquet F, Harring J, Godard F, Aurousseau M (1971) Communication to the European medical chemistry congress, Lyon
Savage DS, Cameron AF, Ferguson G, Hannaway C, Mackay IR (1971) Molecular structure of pancuronium bromide (3α,17β-diacetoxy-2β,16β-dipiperidino-5α-androstane dimethobromide), a neuromuscular blocking agent. Crystal and molecular structure of the water: methylene chloride solvate. J Chem Soc C:410–415
Sugrue MF, Duff N, McIndewar I (1975) On the pharmacology of Org 6368 (2β,16β-dipiperidino-5α-androstan-3α-ol acetate dimethobromide), a new steroidal neuromuscular blocking agent. J Pharm Pharmacol 27:721–727
Teerapong P, Marshall IG, Harvey AL, Singh H, Paul D, Bhardwaj TR, Ahuja NK (1979) The effects of dihydrochandonium and other chandonium analogues on neuromuscular and autonomic transmission. J Pharm Pharmacol 31:521–528
Tuba Z, Marsai M, Görög S, Biró K, Kárpáti E, Szporny L (1981) Procedure for the synthesis of new 3-amino-17a-aza-D-homo androstan derivatives, its acid addition and quaternary salts (in Hungarian). Pat Hung 175 287

CHAPTER 13

The Derivatives of Carboxylic Acids

D. A. KHARKEVICH and A. P. SKOLDINOV

A. Introduction

In the design of new neuromuscular blocking agents, attention is attracted to the derivatives of the esters of various carboxylic and dicarboxylic acids. The reason for this is that they are hydrolyzed in the organism, the duration of their action being, therefore, limited. Since succinylcholine was found to possess neuromuscular blocking activity and was widely adopted into medical practice, a great number of quaternary salts of ester derivatives have been synthesized. The results of those studies have been summarized in numerous reviews and monographs (BOVET 1951, 1959; BRÜCKE 1956; KHARKEVICH 1969; MIKHELSON and ZEIMAL 1970; CHEYMOL 1972; ZAIMIS 1976; BOWMAN 1980; and many others). This chapter presents the data on some mono- and bisquaternary ammonium derivatives of a number of esters.

Truxillic acid derivatives are the most representative group discussed. Attention was drawn to them after the isolation of the alkaloid thesine from a central Asian plant *Thesium minkwitzianum* (ARENDARUK 1953). Pharmacologic study showed thesine [bis(α-1-oxymethylpyrrolyzidine ester) of *p,p'*-dihydroxy-α-truxillic acid] (structure I) and especially, its diiodomethylate to have neuromuscular blocking activity (MASHKOVSKY 1955). This laid the basis for the search for neuromuscular blocking agents among the derivatives and analogs of truxillic acid (stereoisomeric 1,3-diphenylcyclobutane-2,4-dicarboxylic acids).

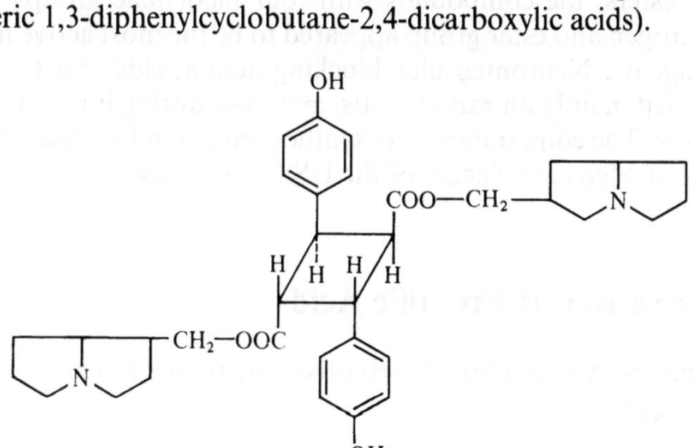

(I)

To obtain more effective neuromuscular blocking agents and to make them more easily available, the synthesis and pharmacologic study of a large number of bisquaternary and bistertiary ammonium salts of basic esters and amides of truxillic acids were carried out (ARENDARUK et al. 1963; KHARKEVICH 1966b, 1973, 1975; KHARKEVICH and SKOLDINOV 1970, 1971, 1976, 1980).

The high neuromuscular blocking activity of numerous bisquaternary derivatives of 1,3-diphenylcyclobutane-2,4-dicarboxylic acids encouraged the study of structural fragments of their molecules. Thus, in the second part of this chapter the data on neuromuscular blocking activity of monoquaternary salts of cinnamic acid aminoalkyl esters and some of their derivatives and analogs are described (ARENDARUK et al. 1967, 1970; KHARKEVICH et al. 1967; KHARKEVICH 1973). The latter (structure II) can be considered as one half of the molecule of bisquaternary ammonium salts of diphenylcyclobutanedicarboxylic acid aminoalkyl esters (structure III).

$$Ar{-}CH{=}CH{-}COO{-}(CH_2)_n{-}\overset{+}{N}R_3 \cdot X^- \qquad (II)$$

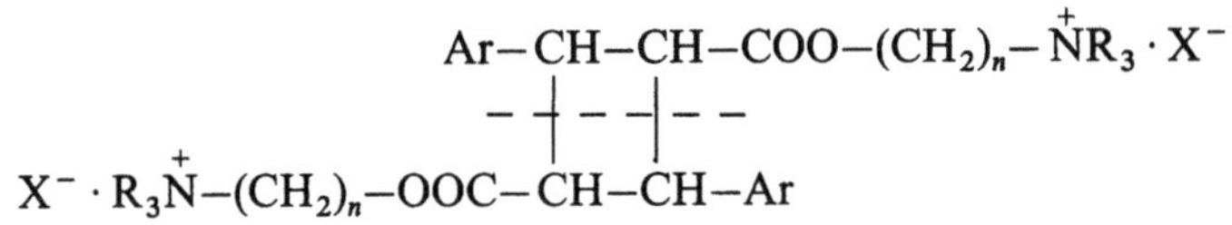

(III)

In addition, an analogous group of monoquaternary ammonium derivatives of benzoic acid aminoalkyl esters was studied (KHARKEVICH 1970; ARENDARUK et al. 1972a).

In the final part of this chapter the data on the bisquaternary ammonium derivatives of aliphatic dicarboxylic acid esters are given. One group includes the compounds containing adamantyl radicals. It was shown in Chap. 4 that the introduction of radicals with a marked degree of hydrophobicity turns the depolarizing agents into nondepolarizing ones. These regularities were used in the quest for active short-acting nondepolarizing neuromuscular agents.

Compounds of another group are the bisquaternary salts of δ-dimethylaminobutyl esters of different aliphatic dicarboxylic acids. This choice was determined by the fact that in the series of monoquaternary ammonium derivatives of benzoic and cinnamic esters, the compounds with four methylene groups between the quaternary nitrogen and ester group appeared to be the most active neuromuscular blocking agents. Neuromuscular blocking action, side effects, and toxicity have been studied mainly on rabbits, cats, and mice during intravenous administration of agents. The compounds were synthesized at the Institute of Pharmacology, Academy of Medical Sciences of the USSR, Moscow.

B. The Derivatives of Truxillic Acids

I. Bisquaternary Ammonium Derivatives of Basic Esters of Truxillic Acids

The data on bisquaternary ammonium salts of aminoalkyl esters of truxillic acids are summarized in a number of reviews (KHARKEVICH 1969, 1970, 1973). These investigations dealt with the struture–activity relationships of compounds with general chemical structure IV.

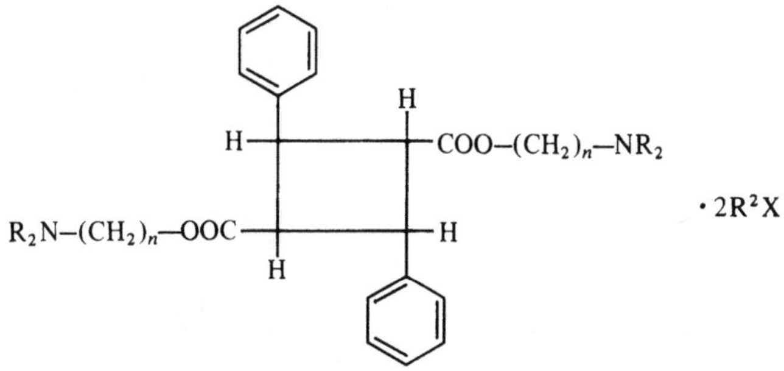

(IV)

where $n=2, 3, 4, 5, 7$
$NR_2 = N(CH_3)_2$; $N(CH_3)(C_2H_5)$; $N(C_2H_5)_2$; *N*-pyrrolidinyl; *N*-piperidinyl; *N*-morpholinyl
$R^2X = CH_3I$, C_2H_5I, $n\text{-}C_3H_7I$

The chemical modifications concerned the distance between the cationic centers, the interonium structure, and the nature of radicals on the quaternary nitrogen atoms. Bisquaternary ammonium salts of stereoisomeric α-, ε-, or γ-truxillic acids (structure V) and some stereoisomers with an asymmetric atom in the side chains were tested.

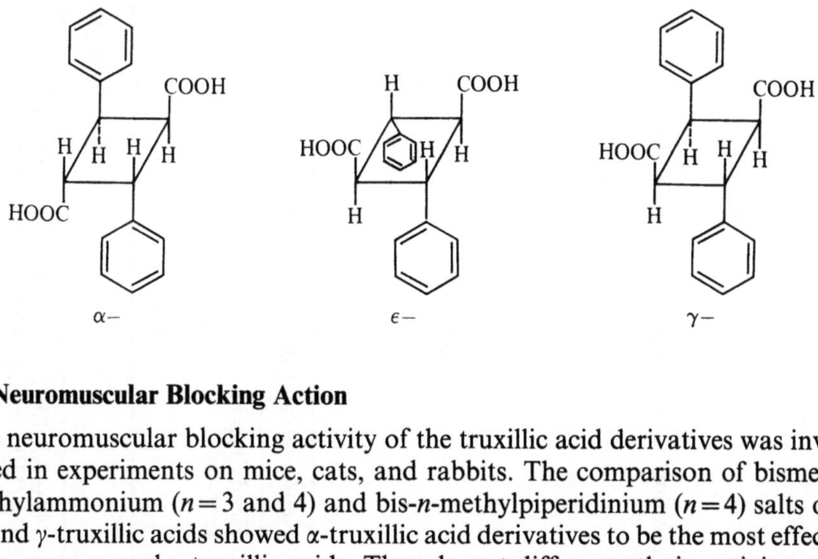

(V)

1. Neuromuscular Blocking Action

The neuromuscular blocking activity of the truxillic acid derivatives was investigated in experiments on mice, cats, and rabbits. The comparison of bismethyldiethylammonium ($n=3$ and 4) and bis-*n*-methylpiperidinium ($n=4$) salts of α-, ε-, and γ-truxillic acids showed α-truxillic acid derivatives to be the most effective. Then come ε- and γ-truxillic acids. They do not differ greatly in activity; about 3–5 times according to the head drop test in rabbits (Table 1 see also Table 8). Since the structure of α-truxillic acid seems to be optimal for the interaction with the end-plate acetylcholine receptors, the majority of the compounds synthesized and tested are derivatives of this acid.

One of the most important characteristics of bisonium salts is the distance between cationic centers. In the series of bis-*N*-methylpiperidinium and bis-*N*-methyldiethylammonium compounds in the experiments on cats and rabbits, the highest activity was observed at $n=3$, i.e., when the distance between the cationic centers was 13 atoms (11 carbon and 2 oxygen atoms). The neuromuscular block-

Table 1. Effect of stereoconfiguration of the central part of the molecule

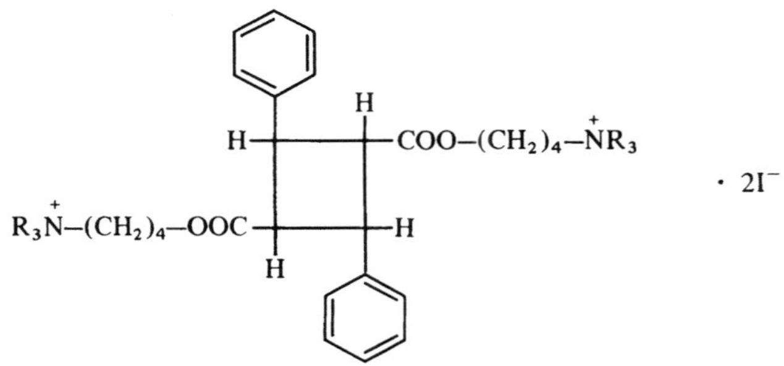

Compound	Stereoconfiguration of truxillic acid	Latency of neuromuscular block (s)[a]	Neuromuscular blocking activity (μg/kg)				Safety margin AD_{50}/HDD_{50} (mice)[f]	Duration of neuromuscular block in mice (min)[g] by restoration of	
			Head drop in rabbits[b]	Neuromuscular block in cats[c]	Head drop in mice (HDD_{50})[d]	Apnea in mice (AD_{50})[e]		Respiration	Righting reflex
			$-\overset{+}{N}R_3 = -\overset{+}{N}(CH_3)(C_2H_5)_2$						
1	α	30–40	48 (41.3–55.7)	250–350	125 (89–175)	700 (648–756)	5.6 (4–7.8)	30.2 (±3.4)	49.1 (±7.1)
2	ε	10–35	130 (110.1–153.4)	400–500	522 (462–587)	815 (751–884)	1.56 (1.2–1.9)	6.7 (±0.7)	8.5 (±0.9)
3	γ	20–30	250 (236.9–263.7)	800–1,000	360 (300–432)	1,040 (1,008–1,073)	2.89 (2.4–3.47)	30.7 (±2.5)	57.2 (±5.1)

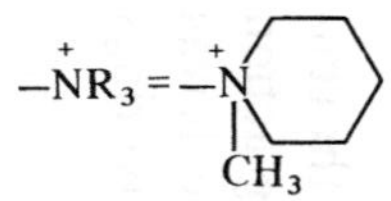

4	α	20–50	46 (43.3–48.7)	150–200	180 (153–212)	512 (427–614)	2.85 (2.24–3.6)	15.8 (±0.8)	24.9 (±2.2)
5	ε	10–35	110 (104.7–115.5)	200–250	385 (341–435)	500 (456.5–547)	1.29 (1.12–1.48)	10 (±0.9)	14.9 (±2.2)
6	γ	15–35	153 (136–171.4)	400–500	270 (218–334)	900 (770–1,080)	3.3 (2.6–4.2)	33 (±3.3)	41 (±6.2)
Tubocurarine chloride			137 (105–178)	180–230	92 (67–126)	190 (164–220)	2 (1.8–2.2)	18.5 (±1.9)	32 (±3.4)

[a] Time from intravenous injection to onset of neuromuscular block (experiments with neuromuscular transmission in anesthetized cats and intact rabbits with head drop registration)

[b] Mean effective doses (ED_{50}) with confidence limits ($P=0.05$). Agents administered by bolus within 3–5 s into auricular vein

[c] Transmission from sciatic nerve to gastrocnemius muscle in anesthetized cats (chloralose, 60 mg/kg + urethane, 300–400 mg/kg i.v.). Nerve stimulated by supramaximal rectangular pulses (0.5 ms) at 1 Hz; muscle contractions recorded mechanographically

[d] Mean effective doses in which the agents caused head drop (HDD_{50}) with confidence limits ($P=0.05$). Agents injected into tail vein

[e] Mean apneic doses (AD_{50}) with confidence limits ($P=0.05$)

[f] AD_{50}/HDD_{50} for mice with confidence limits ($P=0.05$)

[g] Compounds injected into tail vein at doses equal to 5 LD_{50}. Observations carried out under artificial ventilation, mean effective doses for 7–10 experiments with standard errors

ing activity was slightly lower at $n=4$ or 5. The activity of bismethyldiethylammonium derivatives at $n=5$ was about the same as at $n=3$. This slight difference in the neuromuscular blocking activity at $n=3$–5 might be due to flexibility of the side chains. When the chain was shortened to $n=2$ (11 atoms between quaternary nitrogen atoms) or elongated up to $n=7$ (21 atoms between quaternary nitrogen atoms) the neuromuscular blocking activity was much less marked (Table 2).

The importance of the structure of cationic centers was studied in compounds with the optimum interonium distance, i.e., when $n=3$ (Table 3). In the series of compounds tested, trimethylammonium derivatives were characterized by a relatively low neuromuscular blocking activity. Successive replacement of methyl by ethyl groups was followed by an increase in potency. Among trialkylammonium compounds, the triethylammonium salt was found to posses the highest neuromuscular blocking activity.

Bis-*N*-alkylpiperidinium and bis-*N*-alkylpyrrolidinium compounds are the most potent neuromuscular blocking agents. Thus, for instance bis-*N*-ethylpiperidinium salt (Table 3, compound 20) evokes head drop in rabbit at a dose of 21.7 μg/kg (ED_{50}) and neuromuscular block in cats at a dose of 100–130 μg/kg. Tubocurarine evoked the same effects at doses of 137 and 180–230 μg/kg, respectively. In experiments on mice, this derivative of α-truxillic acid (compound 20) was found to be one of the most active compounds, too. The bis-*N*-alkylmorpholinium salt is comparatively low in activity.

Using stereoisomeric truxillic acids, it was found that the structure of the central part of a molecule is of importance for its neuromuscular blocking action (Table 1); this was also confirmed by many other facts. Thus, for instance, it was shown that the introduction of substituents in *para* positions of the phenyl rings had a negative effect.

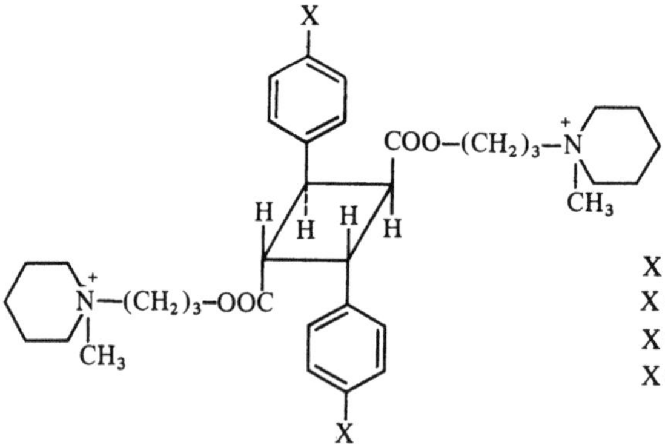

X = –H (VI)
X = $-NO_2$ (VII)
X = $-OCH_3$ (VIII)
X = –OH (IX)

The compound with X = H (Structure VI) is the most active. It blocks neuromuscular transmission in cats at doses of 70–90 μg/kg and at 20–25 μg/kg it evokes head drop in rabbits. When $X=NO_2$, the block of neuromuscular transmission is induced by doses of 120–140 μg/kg (structure VII). The activity of compounds VIII and IX is essentially equal. They block the transmission from the sciatic nerve to the gastrocnemius muscle in cats at doses of 150–180 μg/kg. It should be noted that in this case, the approximation to the structure of the natural alkaloid thesine (structure I) (containing the same acidic fragment as in IX), is not optimal.

Table 2. Effect of distance between quaternary nitrogen atoms

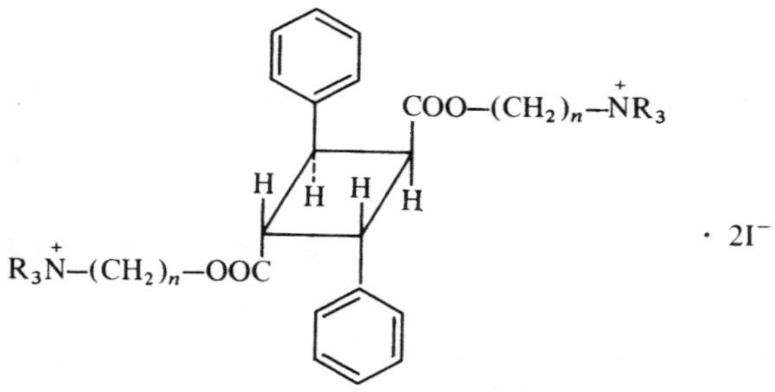

Compound	n	$-\overset{+}{N}R_3$	Latency of neuromuscular block (s)[a]	Neuromuscular blocking activity (μg/kg): Head drop in rabbits[a]	Neuromuscular block in cats[a]	Head drop in mice (HDD_{50})[a]	Apnea in mice (AD_{50})[a]	Safety margin AD_{50}/HDD_{50} (mice)[a]	Duration of neuromuscular block in mice (min)[a] by restoration of: Respiration	Righting reflex
7	2	$-\overset{+}{N}(C_2H_5)_2(CH_3)$	10–25	170 (147.8–195.5)	600–700	750 (664–847)	2,300 (2,000–2,600)	3.1 (2.6–3.7)	17 (±1.2)	31 (±2.8)
8	3		25–35	37.5 (32.4–43.3)	150–250	67 (58–77)	490 (408–588)	7.3 (5.8–9.1)	18.6 (±2.5)	39.1 (±4.5)
1	4		30–40	48 (41.3–55.7)	250–350	125 (89–175)	700 (648–756)	5.6 (4–7.8)	30.2 (±3.4)	49.1 (±7.1)
9	5		30–50	43.5 (37.2–49.6)	130–160	182 (157–211)	430 (377–490)	2.36 (1.98–2.88)	22.6 (±1.6)	29.4 (±1.8)
10	7		70–150	98 (87.5–109.7)	500–600	1,380 (1,160–1,640)	2,600 (2,260–2,990)	1.88 (1.5–2.35)	4.1 (±0.3)	4.4 (±0.3)
11	2	N-methylpiperidinium ($-\overset{+}{N}$ in ring, CH_3)	10–30	73 (62–84.7)	600–700	135 (104–175)	920 (836–1,048)	6.8 (5.1–9.1)	17.4 (±1.6)	23.2 (±0.6)
12	3		30–40	22.5 (15.6–32.4)	110–150	115 (103.5–127)	330 (300–363)	2.86 (2.49–3.28)	25.1 (±2.8)	33.5 (±3.6)
4	4		20–50	46 (43.3–48.7)	150–200	180 (153–212)	512 (427–614)	2.85 (2.24–3.6)	15.8 (±0.8)	24.9 (±2.2)
13	5		35–55	48 (42.8–53.7)	200–250	210 (179–245)	427 (381–478)	2 (1.65–2.42)	15.8 (±1.37)	17.7 (±2)
14	7		80–120	140 (107.6–182)	1,000–1,200	2,270 (1,960–2,630)	5,000 (4,540–5,500)	2.2 (1.36–2.56)	4.3 (±0.6)	4.4 (±0.8)
Tubocurarine chloride				137 (105–178)	180–230	92 (67–126)	190 (164–220)	2 (1.8–2.2)	18.5 (±1.9)	32 (±3.4)

[a] See Table 1

Table 3. Effect of radicals on the quaternary nitrogen atoms

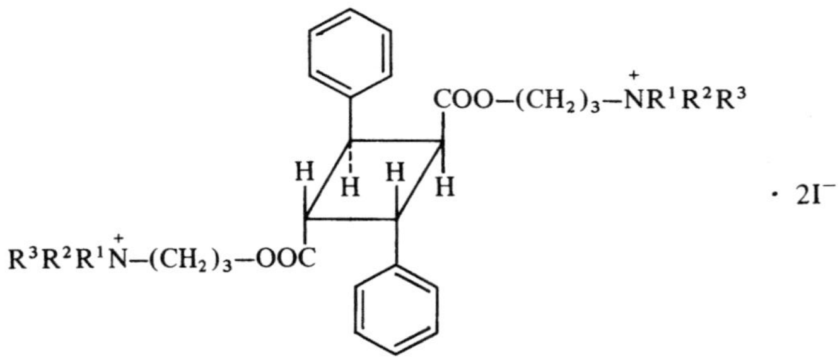

Compound	$-\overset{+}{N}R^1R^2R^3$	Latency of neuromuscular block (s)[a]	Neuromuscular blocking activity (μg/kg)				Safety margin AD_{50}/HDD_{50} (mice)[a]	Duration of neuromuscular block in mice (min)[a] by restoration of	
			Head drop in rabbits[a]	Neuromuscular block in cats[a]	Head drop in mice (HDD_{50})[a]	Apnea in mice (AD_{50})[a]		Respiration	Righting reflex
15	$-\overset{+}{N}(CH_3)_3$	15–30	255 (232.8–279.2)	600–800	500 (424–590)	2,200 (1,900–2,600)	4.4 (3.5–5.5)	10.5 (±1.5)	24.1 (±1.9)
16	$-\overset{+}{N}(CH_3)_2(C_2H_5)$		78 (55.7–109.2)	250–350					
8	$-\overset{+}{N}(CH_3)(C_2H_5)_2$ (cyclobutonium)	25–35	37.5 (32.4–43.3)	130–180	67 (58–77)	490 (408–588)	7.3 (5.8–9.1)	18.6 (±2.5)	39.1 (±4.5)
17	$-\overset{+}{N}(C_2H_5)_3$		34 (31.6–36.5)	120–150					
18	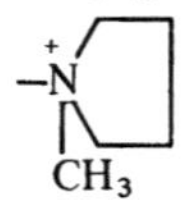	20–35	31.5 (28.8–34.3)	130–170	53 (43–65)	310 (267–359)	5.8 (5–6.7)	24.7 (±3)	43.2 (±3)

19	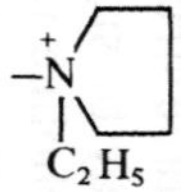 (pyrocyclonium)	20–30	24.2 (21.8–26.9)	120–140	80 (70–90)	332 (294–375)	4.15 (3.5–4.98)	25 (±2.7)	44.1 (±4.5)
12	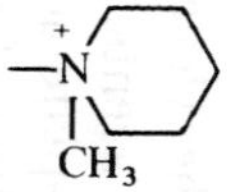	30–40	22.5 (15.6–32.4)	110–150	115 (103.5–127)	330 (300–363)	2.86 (2.49–3.28)	25.1 (±2.8)	33.5 (±3.6)
20	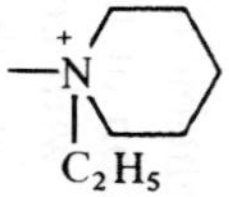 (anatruxonium)		21.7 (17.3–27.1)	100–130					
21	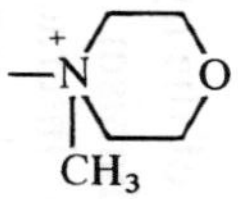		81 (74.5–88)	500–600					
	Tubocurarine chloride		137 (105–178)	180–230	92 (67–126)	190 (164–220)	2 (1.8–2.2)	18.5 (±1.9)	32 (±3.4)

[a] See Table 1

The relationship between the structure and the duration of action of truxillic acid derivatives was studied in detail in artificially ventilated mice (KRAVCHUK 1966a, b, 1970; see also Chap. 11). The agents tested were injected intravenously at doses equal to 5 LD_{50}. The duration of action was estimated by restoration of the righting reflex. Comparing the data on the derivatives of stereoisomeric truxillic acids, it was shown that the derivatives of ε-truxillic acid were the shortest acting. The neuromuscular blocking action of corresponding salts of α-truxillic acid was 1.7–5 times longer and γ-truxillic acid salts were the longest acting (Table 1).

The relationship between the duration of action and the interonium distance is rather regular. The maximal duration was observed for the derivatives of α-truxillic acid at $n=3$ or 4. Shortening the chain up to $n=2$ or lengthening it up to $n=5$ or 7 was followed by a decrease of the duration of the neuromuscular block (Table 2). The compounds with $n=7$ were especially short-acting.

The data on the structure of cationic groups is not very clear. Thus, among α- and γ-truxillic acid derivatives, bis-*N*-methyldiethylammonium salts act longer than bis-*N*-methylpiperidinium analogs, whereas for the derivatives of ε-truxillic acid, the situation is the opposite. The bis-*N*-methylmorpholinium salt ($n=4$) is particularly short-acting. Furthermore, the figures presented in Tables 1–3 provide comparative data on the duration of relaxation of respiratory and body muscles.

According to the mechanism of action, bisquaternary ammonium derivatives of truxillic acids belong to the nondepolarizing neuromuscular blocking agents. Their action is antagonized by neostigmine. It is worthy of note that even bistrimethylammonium salts (e.g., structure X) are nondepolarizing agents while their closest aliphatic analogs (structure XI) with $n=2$ (succinylcholine) and $n=3$ (glutaryldicholine) are typical depolarizing neuromuscular blocking agents.

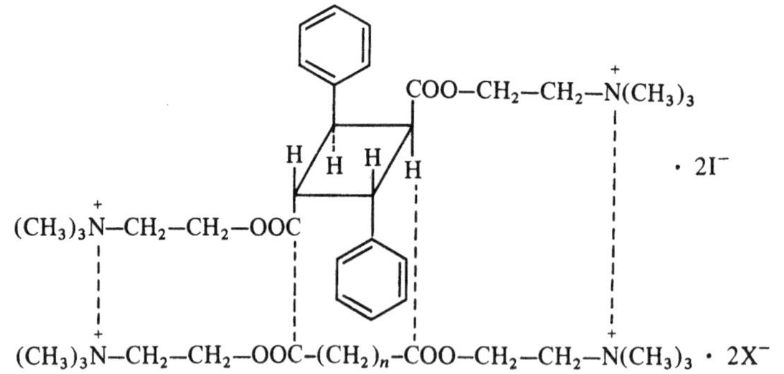

(X)

(XI)

$n = 2$ or 3

Thus, the presence of the hydrophobic diphenylcyclobutane structure significantly affects the mechanism of action of the agents. It may be suggested that, owing to hydrophobic interactions with end-plate acetylcholine receptors this fragment of the molecule promotes stabilization of the interaction of cationic heads with anionic centers of acetylcholine receptors.

2. Assessment of Side Effects and Toxicity

To estimate the clinical prospects of the truxillic acid derivatives tested, experiments were carried out with the aim of detecting probable side effects. The main attention was focused on the relationship between the structure of diphenylcyclobutanedicarboxylic acid derivatives and their antimuscarinic, ganglion-blocking, and anticholinesterase activity (SAMOILOV 1971 a, b, c).

It had been previously shown that compounds of this series had a certain antimuscarinic activity, manifested by the blockade of a cardiotropic effect evoked by the stimulation of the vagus nerve and acetylcholine injection, acetylcholine-induced hypotension being unaffected (KAHRKEVICH and KRAVCHUK 1961; KHARKEVICH 1965, 1966 a). In patients, the cardiotropic antimuscarinic action is manifested by sinus tachycardia. The antimuscarinic activity of the agents was judged by their ability to decrease heart response to acetylcholine and to the stimulation of the peripheral section of the vagus nerve by 50% (ED_{50}) on their intravenous administration to cats.

For all tested derivatives of truxillic acid esters, it was shown that the doses causing the antimuscarinic action were usually lower and sometimes significantly lower than those needed for neuromuscular block. The investigation of bis-*N*-methyldiethylammonium derivatives of esters ($n=4$) of stereoisomeric truxillic acids showed that the antimuscarinic activities of α- and ε-truxillic acid salts were equal, while that of γ-truxillic acid salts was about 3–4 times lower (Table 4).

The distance between the quaternary nitrogen atoms has a certain effect on the antimuscarinic activity (Table 5). Among bis-*N*-methyldiethylammonium and bis-*N*-methylpiperidinium derivatives, it is most pronounced at $n=4$ and $n=3$. Shortening of the interonium distance ($n=2$) or its extension ($n=5$ or 7) is followed by a decrease of antimuscarinic potency. For all the compounds tested, the antimuscarinic action was observed after lower doses than those required for the neuromuscular blocking effect. The bis-*N*-methyldiethylammonium compound is an exception. It blocks muscarinic receptors of the heart and nicotinic receptors of the gastrocnemius muscle in cats within approximately the same dose range (Tables 2 and 5).

The role of different radicals on the quaternary nitrogen atoms was investigated at the same interonium distance ($n=3$) (Table 6). It was shown that the successive replacement of *N*-methyl by *N*-ethyl radicals resulted in an increase of cardiotropic antimuscarinic activity. Among the bistrialkylammonium compounds tested, the bistriethylammonium salt was found to be the most effective. Its vagolytic action and antagonism with acetylcholine was observed at doses of 25 and 30 μg/kg (ED_{50}), respectively. At the same time, it blocks neuromuscular transmission in cats only at doses of 120–150 μg/kg. Bis-*N*-alkylpiperidinium and bis-*N*-alkylpyrrolidinium derivatives have pronounced antimuscarinic activity. The bis-*N*-alkylmorpholinium salt turned out less active.

These data indicate that, in the majority of cases, the changes of cardiotropic antimuscarinic activity and neuromuscular blocking activity resulting from the particular structural modifications discussed develop in the same direction.

Thus, these derivatives of truxillic acid esters are characterized by an unfavorable proportion of cardiotropic and neuromuscular blocking activities. First,

Table 4. Effect of stereoconfiguration of the central part of the molecule

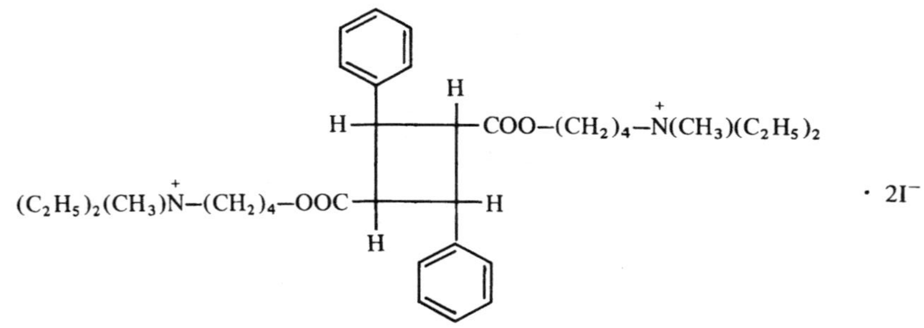

Compound	Stereoconfiguration of truxillic acids	Antimuscarinic activity (μg/kg i.v.) by inhibition of negative chronotropic effect of acetylcholine on cat heart[a]	Ganglion-blocking activity (μg/kg i.v.) by block of transmission in superior cervical ganglion in cats[a]	Anticholinesterase activity[b]		Toxicity LD_{50} (μg/kg i.v.) in artificially ventilated mice	Safety margin LD_{50}/HDD_{50}
				Acetylcholinesterase	Cholinesterase		
1	α	26 (19.1–32.9)	670 (420–920)	No effect	1.9×10^{-5}	330,000 (273,000–399,000)	2,640 (1,795–3,880)
22	ε	28 (21.7–34.3)	860 (460–1,260)	1.2×10^{-2}	0.6×10^{-5}	162,000 (134,000–196,000)	310 (256–375)
23	γ	99 (53.2–144.3)	1,570 (930–2,210)	4.7×10^{-4}	2.4×10^{-6}	30,000 (21,500–42,000)	83.3 (57–121.6)
Atropine sulfate		1.3 (0.5–2.1)					
Hexamethonium			740 (610–870)				
Neostigmine				1.5×10^{-7}	1.5×10^{-7}		
Tubocurarine chloride			250 (180–320)	No effect	2.0×10^{-6}		>2,000

[a] Mean effective doses (ED_{50}) with confidence limits, $P = 0.05$
[b] Molar concentrations in which the agents inhibit cholinesterase activity by 50%
[c] See HDD_{50} in Table 1

Table 5. Effect of distance between cationic centers

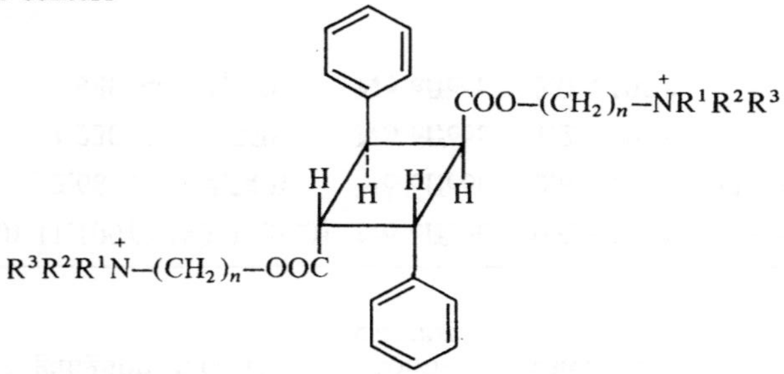

$\cdot\ 2I^-$

Compound	$-\overset{+}{N}R^1R^2R^4$	n	Antimuscarinic activity (μg/kg i.v.) by inhibition of negative chronotropic effect of acetylcholine on cat heart[a]	Ganglion-blocking activity (μg/kg i.v.) by block of transmission in superior cervical ganglion in cats[a]	Anticholinesterase activity[a]		Toxicity LD_{50} (μg/kg i.v.) in artificially ventilated mice	Safety margin LD_{50}/HDD_{50}
					Acetylcholinesterase	Cholinesterase		
7	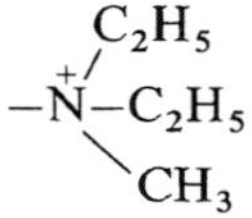	2	184 (143.0–225.0)	1,660 (1,360–1,960)	No effect	2.9×10^{-5}	695,000 (542,900–889,600)	927 (708–1,224)
8		3	43 (26.5–59.5)	1,230 (900–1,560)	No effect	1.2×10^{-5}	330,000 (266,100–409,200)	4,925 (3,788–6,402)
1		4	26 (19.1–32.9)	670 (420–920)	No effect	1.9×10^{-5}	330,000 (273,000–399,000)	2,640 (1,795–3,880)
9		5	130 (105.0–155.0)	1,250 (480–2,020)	No effect	1.4×10^{-5}	96,000 (71,600–128,600)	527.4 (381.8–727)
10		7	145.5 (98.4–192.6)	2,360 (2,060–2,660)	4.3×10^{-4}	1.0×10^{-6}	47,000 (32,400–681,000)	34 (22.7–51)
11	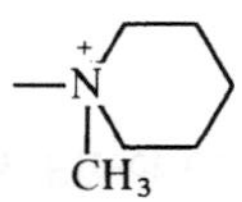	2	50 (37.1–62.9)	1,800 (620–2,980)	No effect	0.6×10^{-5}	100,000 (87,000–115,000)	740 (448–1,205)
12		3	30 (24.1–35.9)	1,930 (510–3,550)	No effect	1.2×10^{-6}		
4		4	24 (20.5–27.5)	1,210 (850–1,570)	1.2×10^{-3}	1.3×10^{-6}		
13		5	74 (53.7–94.3)	1,300 (940–1,660)	5.6×10^{-5}	0.6×10^{-6}	68,000 (46,900–98,600)	323 (220.2–476)
14		7	352 (326.6–377.4)	3,500 (3,250–3,750)	2.6×10^{-6}	3.5×10^{-7}		
Atropine sulfate			1.3 (0.5–2.1)					
Hexamethonium				740 (610–870)				
Neostigmine					1.5×10^{-7}	1.5×10^{-7}		
Tubocurarine chloride				250 (180–320)	No effect	2.0×10^{-6}		>2,000

[a] See Table 4
[b] See HDD_{50} in Table 2

Table 6. Effect of radicals on quaternary nitrogen atoms

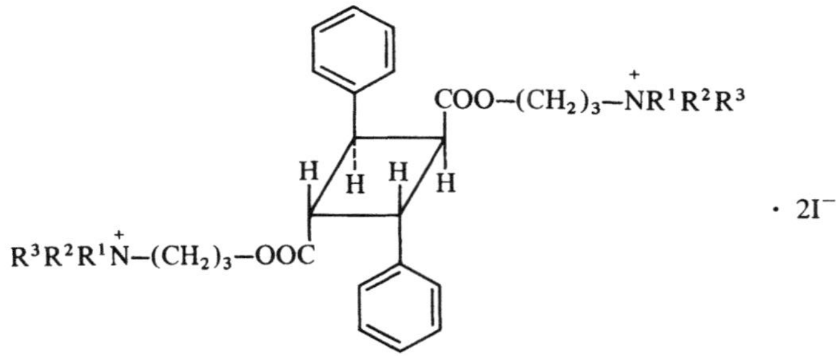

Compound	$-\overset{+}{N}R^1R^2R^3$	Antimuscarinic activity (μg/kg i.v.) by inhibition of negative chronotropic effect of acetylcholine on cat heart[a]	Ganglion-blocking activity (μg/kg i.v.) by block of transmission in superior cervical ganglion in cats[a]	Anticholinesterase activity[a]: Acetylcholinesterase	Anticholinesterase activity[a]: Cholinesterase	Toxicity LD_{50} (μg/kg i.v.) in artificially ventilated mice	Safety margin LD_{50}/HDD_{50}[b]
15	$-\overset{+}{N}(CH_3)_3$	155 (139.0–171.0)	11,100 (7,180–15,020)	No effect	0.6×10^{-4}	75,000 (50,000–122,000)	150 (96.5–232.5)
16	$-\overset{+}{N}(CH_3)_2(C_2H_5)$	56 (26.1–85.9)	2,680 (620–4,740)	No effect	0.6×10^{-5}	330,000 (266,000–409,200)	4,925 (3,788–6,402)
8	$-\overset{+}{N}(CH_3)(C_2H_5)_2$	43 (26.5–59.5)	1,230 (900–1,560)	No effect	1.2×10^{-5}		
17	$-\overset{+}{N}(C_2H_5)_3$	25 (16.4–33.6)	940 (440–1,440)	No effect	2.0×10^{-6}		

18	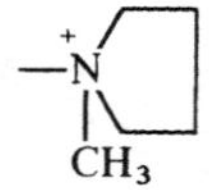	44 (34.0–54.0)	1,870 (1,550–2,190)	No effect	1.2×10^{-6}	86,000 (63,200–116,900)	1,623 (1,119–2,353)
12	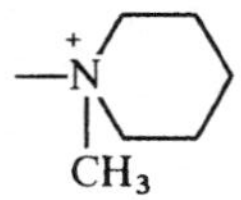	30 (24.1–35.9)	1,930 (510–3,550)	No effect	1.2×10^{-6}		
20	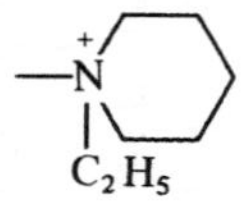	19.8 (12.4–27.2)	1,650 (1,210–2,090)	No effect	1.2×10^{-6}		
21	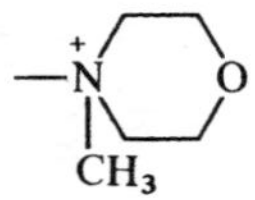	74 (48.3–99.7)	3,690 (700–6,680)	No effect	1.2×10^{-4}		
Atropine sulfate		1.3 (0.5–2.1)					
Hexamethonium			740 (610–870)				
Neostigmine				1.5×10^{-7}	1.5×10^{-7}		
Tubocurarine chloride			250 (180–320)	No effect	2.0×10^{-6}		>2,000

[a] See Table 4
[b] See HDD_{50} in Table 3

there develops an antimuscarinic effect upon the heart and then, after the dose is increased several times, the neuromuscular blocking effect appears. There is only one agent (compound 9, Table 5) which evokes both effects in similar doses.

The ganglion-blocking activity was studied on the superior cervical ganglion of cats. The agents were injected intravenously. In a number of compounds, the ability to block ganglionic transmission is substantial, but in all of them the ganglion-blocking doses are higher than the neuromuscular blocking doses (Tables 4–6).

In experiments with bis-*N*-methyldiethylammonium salts, it was shown that the derivatives of α- and ε-truxillic acids had practically equal blocking effect on the superior cervical ganglion. The analogous γ-truxillic acid derivative is less active (about half). The neuromuscular blocking action of these compounds is caused by smaller doses (Table 4).

We turn now to the importance of the interonium distance. In bis-*N*-methyldiethylammonium derivatives of α-truxillic acid, the ganglion-blocking activity was most marked at $n=4$. The shortening of the polymethylene chain ($n=3$ or 2) or its elongation ($n=5$ or 7) attenuated the ganglion-blocking activity. In the series of bis-*N*-methylpiperidinium salts, the compounds in which $n=4$ or 5 had the highest ganglion-blocking activity, whereas the agents with $n=2$ and 3 or $n=7$ were less effective (Table 5).

The substitution of *N*-ethyl for *N*-methyl radicals at $n=3$ resulted in the potentiation of ganglion-blocking activity. Bistriethylammonium compounds block the superior cervical ganglion in doses about 12 times lower than the bistrimethylammonium analog (Table 6).

The ganglion-blocking activity of bis-*N*-methylpiperidinium, bis-*N*-ethylpiperidinium, and bis-*N*-methylpyrrolidinium salts is essentially equal, while the bistriethylammonium derivative is twice as active, on average. The compound with *N*-methylmorpholinium radicals produces an insignificant effect on the ganglia. Furthermore, the anticholinesterase activity of the compounds was tested in vitro. Acetylcholinesterase was obtained from homogenates of the grey matter of rabbit cortex and cholinesterase from normal horse serum without preservative. It turned out that in physiologic concentrations, the compounds tested did not, as a rule, affect the activity of acetylcholinesterase or it was only slightly inhibited. However, the same compounds inhibited cholinesterase to a much greater degree.

Among the derivatives of stereoisomeric truxillic acids with *N*-methyldiethylammonium groups ($n=4$) the γ-truxillic acid derivative produced a greater effect on acetylcholinesterase, but nevertheless its action was 3,000 times lower than that of neostigmine. Antiacetylcholinesterase activity of the ε-truxillic acid derivative is even lower and the salts of α-truxillic acid are entirely devoid of such action (Table 4). According to the ability to block cholinesterase, bis-*N*-methyldiethylammonium derivatives of truxillic acids can be ordered $\alpha:\varepsilon:\gamma=1:3:8$.

On changing the distance between quaternary nitrogen atoms, it was found that, in the series of bis-*N*-methyldiethylammonium derivatives, only at $n=7$ was a hardly noticeable antiacetylcholinesterase activity observed. At shorter interonium distances ($n=2$–5), it was absent. Bis-*N*-methylpiperidinium derivatives with $n=2$ and 3 were ineffective. The antiacetylcholinesterase activity appears at $n=4$ and at $n=7$ it is increased. The last compound (Table 5) is the most active antiacetylcholinesterase agent of all the tested derivatives of truxillic acids, neostigmine being 18 times as active. Inhibition of cholinesterase was observed for all the compounds, but it was most marked at $n=7$, too. All the bis-*N*-methyldiethylammonium compounds with $n=2$–5 have about the same activity (0.5% –1.2% of the activity of neostigmine). In bis-*N*-methylpiperidinium salts, inhibition of cholinesterase increased with the increase of n (from 2 to 7).

The comparison of the derivatives of truxillic acids with different cationic heads demonstrated that they had no effect on acetylcholinesterase at $n=3$, but they inhibited cholinesterase (Table 6). The successive replacement of *N*-methyl by *N*-ethyl radicals was followed by an increase in the inhibitory effect of cholinesterase. It was shown that bis-*N*-methylpyrrolidinium, bis-*N*-methylpiperidinium, and bis-*N*-ethylpiperidinium compounds were more active and the bis-*N*-morpholinium salt was the least active.

Toxicity of compounds was studied under artificial ventilation. In experiments on mice, with the bis-*N*-methyldiethylammonium salt it was shown that the α-truxillic acid derivative was the least toxic and that of γ-truxillic acid was the most toxic, the ε-truxillic acid derivative occupying an intermediate position (Table 4). The agents tested can be ordered in the same manner according to the safety margin ($\alpha > \varepsilon > \gamma$).

As to the distance between the quaternary nitrogen atoms, in the same series of bis-*N*-ethyldiethylammonium salts it was shown that the safety margin was maximal at $n=3$, followed by the compound with $n=4$. Decrease ($n=2$) or increase ($n=5$ or 7) of the interonium distance results in a fall in the safety margin (Table 5). The toxicity of some bis-*N*-methylpiperidinium and other compounds was not determined because of their insufficient solubility in water. Of the compounds tested, a number of salts which seemed most promising for use in patients were selected and studied more comprehensively (Chap. 19). Of these, anatruxonium, truxilonium, and cyclobutonium (for structures see Table 3 and Chap. 19, Table 1) were tested in patients. All three agents proved to be highly active nondepolarizing neuromuscular blockers, more potent than tubocurarine in this respect. When used in average neuromuscular blocking doses, the agents act within 60 min. However, increase of the dose can prolong the effect considerably. Among the side effects of anatruxonium, cyclobutonium, and truxilonium there was observed moderate sinus tachycardia (resulting from cardiotropic antimuscarinic action) and a slight decrease of arterial pressure.

In addition, anatruxonium, cyclobutonium, and truxilonium are characterized by slow recovery from muscle relaxation, which can aggravate the postanesthetic period. In comparison with the three neuromuscular blocking agents, after tubocurarine administration the neuromuscular transmission is restored more rapidly. This might be due to the fact that in the structure of tubocurarine one cationic center is a quaternary nitrogen atom and the other a tertiary one (EVERETT et al. 1970). The tertiary nitrogen atom is known to have weaker bonds with the anionic center of the acetylcholine receptors of the skeletal muscles than the quaternary nitrogen atom. This is likely to account for the comparatively prompt restoration of neuromuscular transmission after tubocurarine administration. Thus, further search for neuromuscular blocking agents in this series was directed to the design of compounds with a higher selectivity of action and prompter restoration of neuromuscular transmission. The main attention was focussed on the derivatives of α-truxillic acid aminoalkylamides, since one such substance had been previously shown to be devoid of cardiotropic antimuscarinic action at neuromuscular blocking doses (SAMOILOV 1971 b; KHARKEVICH 1974 b; KHARKEVICH and SHORR 1980). Therefore, corresponding aminoalkylamides of α-truxillic acid in the form of bisquaternary ammonium salts and bistertiary salts were synthesized and investigated.

II. Bisquaternary Ammonium Salts of α-Truxillic Acid Aminoalkylamides

In the series of bisquaternary ammonium derivatives of α-truxillic acid basic esters the compounds with $n=3$ and 4, $NR_2=N(C_2H_5)_2$, *N*-piperidinyl, or *N*-pyrrolidinyl were found to be the most active neuromuscular blocking agents. Therefore, the main attention was given to the synthesis and pharmacologic study of amides designed on the pattern of the most active esters of general structure XII (KHARKEVICH 1983).

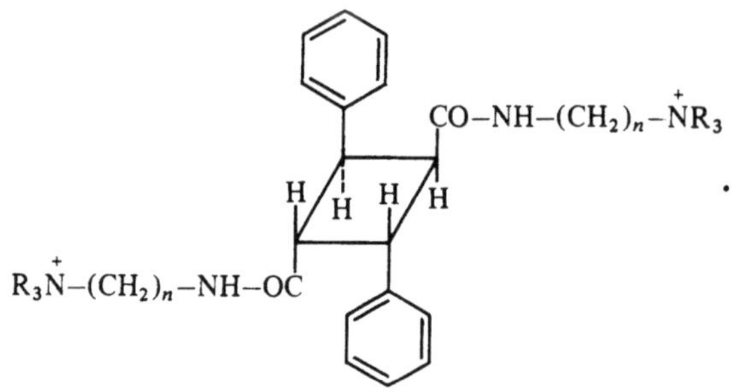

$\cdot 2X^-$ (XII)

The neuromuscular blocking activity was generally studied in experiments on anesthetized cats and intact rabbits (administration by intravenous bolus). In experiments on cats, the doses blocking transmission from the sciatic nerve to the gastrocnemius muscle were determined. The results obtained indicate high neuromuscular blocking activity of the synthesized agents (Table 7). It is most marked at $n=3$. Under these experimental conditions they are ten times as active as tubocurarine. Elongating the chain to $n=4$ was followed by a three- to fivefold decrease of neuromuscular blocking activity. The salts with $n=2$ are the least active. This relationship is identical to that previously observed for structurally similar esters.

As far as the substituents on the quaternary nitrogen atoms are concerned, in this group of compounds, one mostly uses those radicals which modify the action of the ester derivative in the desired manner. High neuromuscular blocking activity was observed in the agents in which both cationic centers contained *N*-alkylpyrrolidinium or *N*-alkylpiperidinium groups. The bistrimethylammonium group turned out somewhat less active. In those compounds, the replacement of one ethyl group by a methyl group with $n=2$ and 3 reduced the neuromuscular blocking activity (compounds 24–26 and 29–30, Table 7).

Table 7. Neuromuscular blocking activity of bisquaternary ammonium derivatives of α-truxillic acid aminoalkylamides

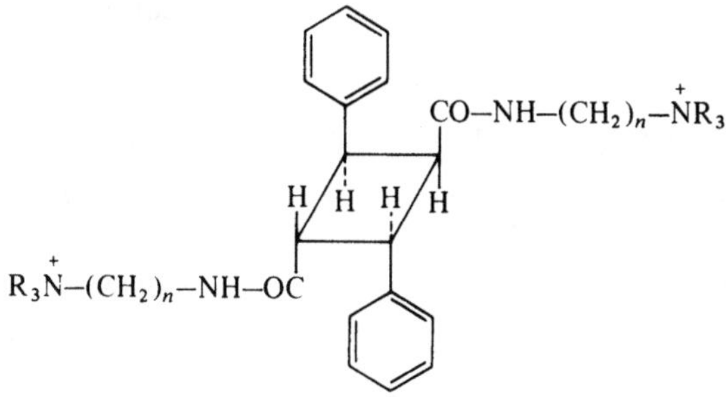

· $2X^-$

Compound	n	$-\overset{+}{N}R_3$	X^-	Neuromuscular block in anesthetized cats (mg/kg i.v.)[a]
24	2	$-\overset{+}{N}(CH_3)(C_2H_5)_2$	I^-	1,500–1,600
25	2	$-\overset{+}{N}(CH_3)(C_2H_5)_2$	$CH_3OSO_3^-$	1,400–1,500
26	2	$-\overset{+}{N}(C_2H_5)_3$	I^-	700–800
27	2	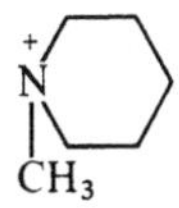	I^-	220–240
28	2	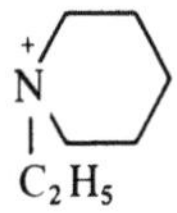	I^-	200–220
29	3	$-\overset{+}{N}(CH_3)(C_2H_5)_2$	I^-	45–50
30	3	$-\overset{+}{N}(C_2H_5)_3$	I^-	25–30
31	3		I^-	18–20
32	3		n-$CH_3C_6H_4SO_3^-$	18–20
33	3	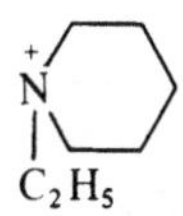	n-$CH_3C_6H_4SO_3^-$	18–20
34	3	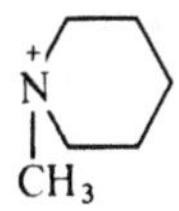	I^-	18–20
35	3	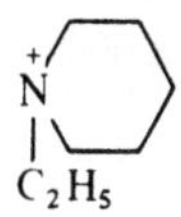	I^-	15–20

Table 7 (continued)

Compound	n	$-\overset{+}{N}R_3$	X^-	Neuromuscular block in anesthetized cats (mg/kg i.v.)[a]
36	4	$-\overset{+}{N}(CH_3)(C_2H_5)_2$	I^-	130–140
37	4	$-\overset{+}{N}(C_2H_5)_3$	I^-	110–130
38	4	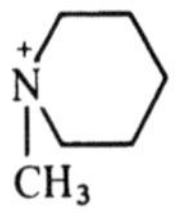	I^-	70–80
39	4	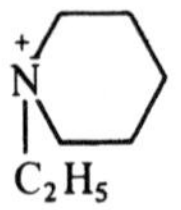	I^-	80–100
Tubocurarine chloride				180–230

[a] See Table 1

Table 8. Effect of chain between quaternary nitrogen atoms

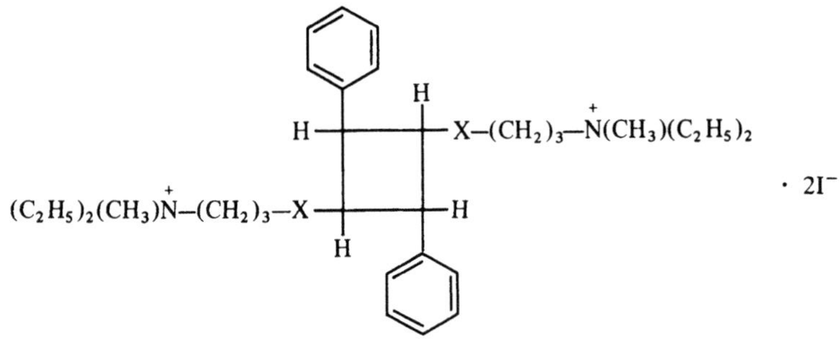

Compound	Stereo-configuration of truxillic acids	Neuromuscular blocking activity (μg/kg)[a]	
		Head drop in rabbits	Neuromuscular block in cats
		X = –COO–	
8	α	37.5 (32.4–43.3)	150–250
22	ε	126 (108.6–146.2)	400–450
23	γ	260 (218.4–309.4)	900–1,100
		X = –CONH–	
29	α	21.3 (16.8–25.8)	40–50
40	ε	1,200 (900–1,500)	6,000–8,000
41	γ	3,600 (2,600–4,900)	10,000–12,000

[a] See Table 1

A number of active agents were tested in experiments on rabbits. Their ability to cause head drop was studied. This effect was evoked by lower doses than those needed for neuromuscular block in cats. Thus, for instance, bis-*N*-methylpyrrolidinium (Table 7, compound 31) and bis-*N*-ethylpiperidinium salts (Table 7, compound 35) with $n=3$, caused head drop in rabbits at doses of about 10 µg/kg. As compared with esters, the corresponding amides turned out more active. The duration of action of the bisquaternary ammonium salts of α-truxillic acid aminoalkylamides was longer than that of tubocurarine.

The importance of the interonium fragment for neuromuscular blocking activity was shown by modifying the stereoconfiguration of the central part of the molecule and also by substituting methyl groups for the hydrogen atoms in the amide groups. The significance of stereoconfiguration of the central part of the molecule was exemplified by the derivatives of α-, ε-, and γ-truxillic acids (structure V). The bis-*N*-methyldiethylammonium salt of the substituted amide of α-truxillic acid ($n=3$) was considerably more active as a neuromuscular blocking agent than the salts of analogous derivatives of ε- and γ-truxillic acids (Table 8). The amide derivative of α-truxillic acid is about 1.75 times more active in experiments on rabbits and 3.75–5 times more active in experiments on cats than the analogous ester derivatives. On the other hand, for the esters of ε- and γ-truxillic acids, the neuromuscular blocking activity was more pronounced than in the corresponding amides, by 9.5 and 13.8 times, respectively (experiments on rabbits). Such dissimilarities are likely due to higher complementarity of α-truxillic acid derivatives to acetylcholine receptors of the skeletal muscles as compared with the derivatives of other stereoisomeric truxillic acids. Additional interactions involving hydrogen bonds of the amido groups and partial charges of the carbonyl groups seem to develop only when the neuromuscular agent approaches closer to the receptor which is more likely for α-truxillic acid derivatives.

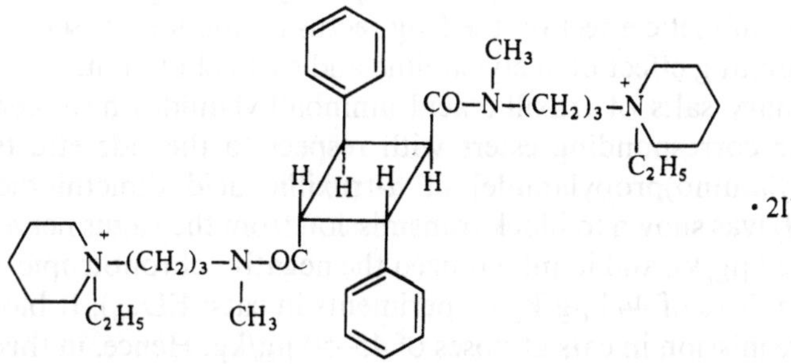

(XIII)

It was shown that the derivatives of bis-(*N*-methyl)amides were considerably less active than nonmethylated amides. Thus, structure XIII blocks neuromuscular transmission in cats at doses of 100–120 µg/kg, and the corresponding nonmethylated amide (compound 35, Table 7) does so at doses of 15–20 µg/kg.

The presence of an asymmetric carbon atom (structure XIV, asterisks) in the aliphatic chains of the interonium part of the molecule also contributes to the neuromuscular blocking activity of the compounds.

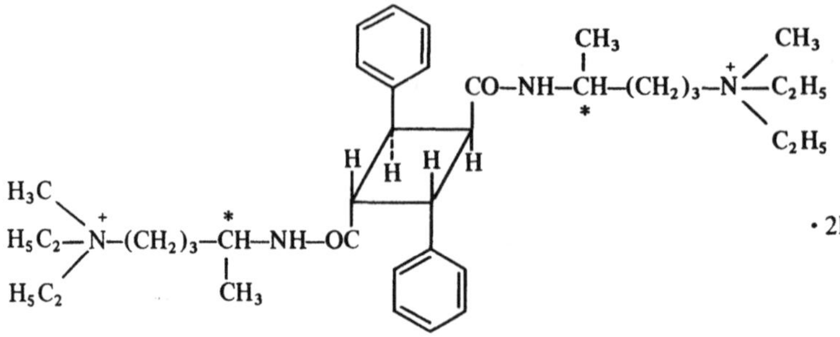

$\cdot 2I^-$ (XIV)

Compound XIV was obtained in two stereoisomeric forms: on the basis of the known regularities, the compound with higher melting point was assigned the structure of a *meso* form and the one with lower melting point that of a racemate. The *meso* form was found to be more active. The latter evoked head drop in rabbits at a dose of 29 μg/kg (ED_{50}), and blocked neuromuscular transmission in cats at doses of 60–70 μg/kg. The racemate was about ten times less active (it evoked the same effects at doses of 310 and 650–700 μg/kg, respectively).

Thus, the neuromuscular blocking activity of bisquaternary ammonium salts of substituted amides of truxillic acids is related to the structure of cationic centers, the distance between them, and the structure of the interonium part of the molecule.

The mechanism of action of the agents was studied in cats by the interaction with neostigmine, in pigeons by the type of paralysis, and also in the isolated frog rectus abdominis muscle by direct stimulating effect on nicotinic receptors or interaction with nicotinomimetics. Evidence was obtained that all the agents tested were nondepolarizing. This is confirmed by the following. Neostigmine antagonizes the effect of quaternary salts of truxillic acid aminoalkylamides on neuromuscular transmission; they evoke flaccid paralysis in pigeons; they fail to produce a nicotinomimetic effect on the frog rectus abdominis muscle, but they prevent the stimulating effect of acetylcholine and carbachol on it.

Bisquaternary salts of truxillic acid aminoalkylamides have certain advantages over the corresponding esters with respect to the side effects. Thus, the bis[1,3-(diethylamino)propylamide] of α-truxillic acid dimethiodide (Table 7, compound 29) was shown to block transmission from the vagus nerve to the heart at a dose of 527 μg/kg and it antagonized the negative chronotropic effect of acetylcholine at a dose of 444 μg/kg (experiments in cats: ED_{50}). It blocked neuromuscular transmission in cats at doses of 45–50 μg/kg. Hence, in this case the relations are other than for ester derivatives. First, there occurs the block of transmission from the sciatic nerve to the gastrocnemius muscle and only an increased dose (about tenfold) can induce a cardiotropic antimuscarinic effect. It is of importance that in neuromuscular blocking doses, this compound does not inhibit cardiac muscarinic receptors and hence it is unlikely to evoke tachycardia.

The lower ganglion-blocking activity of the amide derivatives under investigation is another advantage over the corresponding ester derivatives (cf. Table 7, compound 29 and Table 5, compound 8). The former blocks interneuronal transmission in the superior cervical ganglion at a dose of 8.76 (6.44–11.08) mg/kg and the corresponding ester derivative does so at a dose of 1.23 (0.90–1.56) mg/kg

($P = 0.05$). At the same time, the neuromuscular blocking activity of the amide (40–50 μg/kg) is higher than that of the ester (150–250 μg/kg). Thus, there is a significant difference between the neuromuscular blocking and ganglion-blocking doses of the amide (about 30-fold). Such a difference is undoubtedly of interest, since such neuromuscular blocking agents should not inhibit the tone of the autonomic innervation, vasoconstrictors in particular.

The amide derivative is completely devoid of inhibitory action on acetylcholinesterase and cholinesterase. It is thus characterized by a higher selectivity of neuromuscular blocking action. Being a highly potent neuromuscular blocking agent, it is devoid of antimuscarinic, ganglion-blocking, or anticholinesterase effects.

No side effects were detected in other bisquaternary salts of aminoalkylamides (Table 7) when they were used in neuromuscular blocking doses. They did not affect blood pressure or heart activity. They had no effect on autonomic ganglia or muscarinic receptors of the heart. No histamine release was observed. The toxicity of the agents is low and the safety margin sufficiently large.

Of the tested bisquaternary salts of α-truxillic acid aminoalkylamides, dipyronium and amidonium were selected (componds 31 and 35, Table 7) as the two most promising agents for use in patients. Detailed characteristics of these agents will be given in Chap. 19.

III. Bistertiary Ammonium Salts of α-Truxillic Acid Aminoalkylamides

In the series of bisquaternary ammonium salts of the derivatives of truxillic acid esters and amides, compounds with three methylene groups between the amide group and the quaternary nitrogen were shown to be the most potent neuromuscular blocking agents. Moreover, maximum activity was found in those bisquaternary salts in which the onium centers are represented by triethylammonium, *N*-methyl- or *N*-ethylpyrrolidinium, and *N*-methyl- or *N*-ethylpiperidinium groups. Therefore, pharmacologic studies of bistertiary salts of substituted aminoalkylamides have mainly dealt with the optimal structure of different fragments of the molecule (KHARKEVICH 1983).

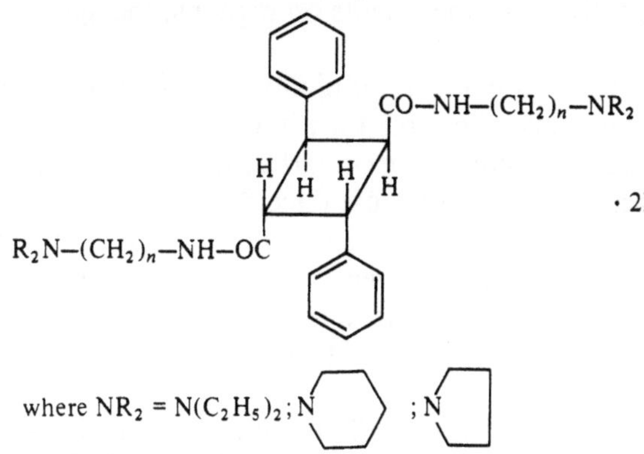

· 2HX (XV)

$n = 2–4$

X = anion of organic or inorganic acid

The pharmacologic study of the compounds was mainly devoted to the neuromuscular blocking activity and the detection of the side effects found in the available neuromuscular blocking agents among the bisquaternary ammonium salts.

The neuromuscular blocking activity was studied in experiments on cats, rabbits, and pigeons. In experiments on anesthetized cats, the amplitude of contractions of the gastrocnemius muscle evoked by electrical stimulation of the sciatic nerve was recorded. In experiments on rabbits, head drop was determined. In pigeons, the potency of the agents and the type of paralysis induced were analyzed. The effect on hemodynamics was estimated by the blood pressure level and the ECG. In addition, the action of the agents on the cardiac response to acetylcholine was tested in cats. The agents were administered intravenously in all the experiments.

The results obtained indicate that the optimal distance between tertiary nitrogen atoms is the same as it was for bisquaternary salts of ester and amide derivatives. In this case, too, the highest activity was observed at $n=3$ (Table 9). Decrease ($n=2$) or increase ($n=4$) of the interonium distance were followed by reduction of the neuromuscular blocking activity. As to the radicals attached to tertiary nitrogen atoms, the highest activity was observed in bis-*N*-pyrrolidinyl and bis-*N*-piperidinyl salts. Bisdiethylamine derivatives turned out to be about half as active. Therefore, the main attention was focused on the compounds with $n=3$ containing pyrrolidinyl or piperidinyl fragments.

In experiments on cats, compounds with $n=3$ block the transmission from the sciatic nerve to the gastrocnemius muscle at doses of 120–350 µg/kg. Tubocurarine caused an equal effect at doses of 180–230 µg/kg. No muscle fasciculation prior to the block was observed; complete block developed rapidly and lasted about 5–8 min, the restoration of the transmission was also prompt. Neostigmine is an effective antagonist of the compounds tested.

In experiments on rabbits, head drop occurred within the dose range of 26–88 µg/kg (tubocurarine caused the same effect at doses of 105–178 µg/kg). For two compounds with heterocyclic radicals, values of ED_{50} (with confidence limits) are presented in Table 10. An analogous bisdiethylamine derivative induced head drop at a dose of 80.5 (73.18–88.55) µg/kg ($P=0.05$). Hence, as in experiments on cats, compounds with bispyrrolidinyl and bispiperidinyl substituents were the most active. In experiments on pigeons, the agents induced flaccid paralysis.

The compounds tested are devoid of nicotinomimetic action on frog isolated rectus abdominis muscle. At the same time, they decrease or completely abolish the stimulating effect of acetylcholine or carbachol. These data, the antagonism with neostigmine, and the ability to cause flaccid paralysis in pigeons suggest that the compounds tested are nondepolarizing neuromuscular blocking agents. In neuromuscular blocking doses, the agents do not affect blood pressure or heart activity. They affect acetylcholine cardiotropic action only insignificantly and, therefore, they essentially do not induce tachycardia. No other side effects were observed.

Thus, the bistertiary salts of aminoalkylamides of α-truxillix acid have marked neuromuscular blocking activity. They are characterized by a rapid development of the block, its comparatively short duration, and quick restoration of neuro-

Table 9. Neuromuscular blocking activity of bistertiary ammonium salts of α-truxillic acid amides

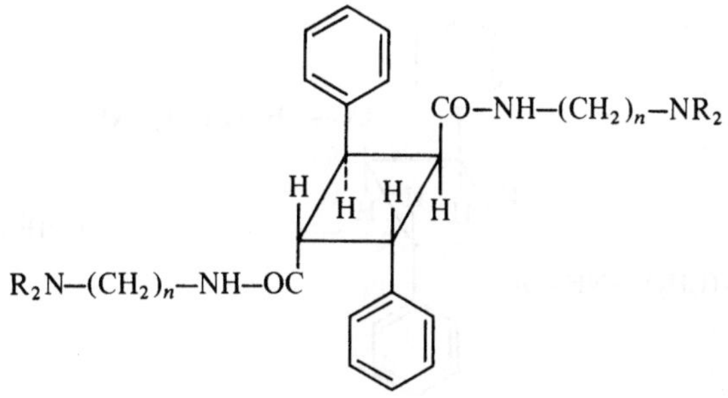

· 2HX

Compound	n	NR_2	HX	Neuromuscular block in cats (μg/kg i.v.)[a]
42	2	$N(C_2H_5)_2$	HCl	5,500–6,500
43	3	$N(C_2H_5)_2$	HCl	400–450
44	3		HCl	120–130
45	3		HOOC–C–H ‖ HOOC–C–H	150–160
46	3		HOOC–C–H ‖ H–C–COOH	160–170
47	3		H_2SO_4	150–160
48	3		HBr	170–180
49	3	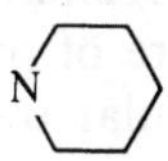	HCl	150–180
50	3	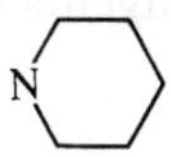	HBr	180–200
51	3	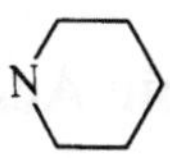	HOOC–C–H ‖ HOOC–C–H	180–200
52	4	$N(C_2H_5)_2$	HCl	750–800
53	4	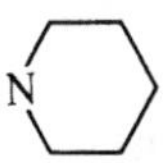	HCl	450–500
Tubocurarine chloride				180–230

[a] See Table 1

Table 10. Neuromuscular blocking agents from the bistertiary ammonium salts of α-truxillic acid amides

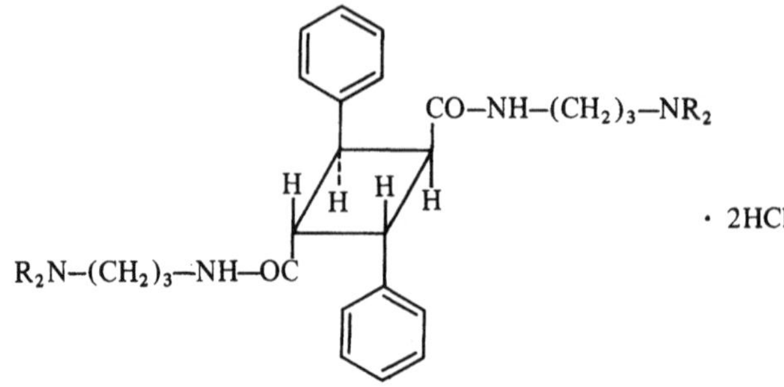

Compound	NR_2	Neuromuscular block in cats (μg/kg i.v.)	Head drop in rabbits ED_{50} (μg/kg i.v.) with confidence limits, $P=0.05$	Mode of action
Pyrocurinum		120–130	28.5 (26.1–30.9)	Nondepolarizing
Amidocurinum	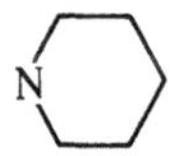	160–180	31.3 (27.2–36.0)	Nondepolarizing
Tubocurarine chloride		180–230	137 (105–178)	Nondepolarizing

muscular transmission. Their safety margin is large. As to the mechanism of action, they belong to the nondepolarizing neuromuscular blocking agents. Neostigmine is their antagonist. When used in neuromuscular blocking doses they cause no side effects.

Compounds with bis-*N*-piperidinyl and bis-*N*-pyrrolidinyl substituents with $n=3$ (pyrocurinum and amidocurinum, Table 10) are of particular interest for anesthesiologists. They selectively block neuromuscular transmission and they can be regarded as nondepolarizing neuromuscular agents of medium duration of action.

C. The Derivatives of Cinnamic and Benzoic Acids

I. Cinnamic Acid Derivatives

The compounds tested (see structures II, III) have the following general structure (XVI).

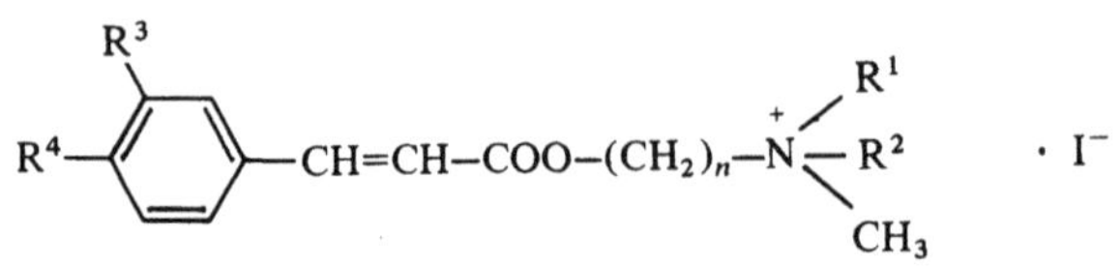

(XVI)

The importance of the length of the polymethylene chain was analyzed in the series of trimethylammonium derivatives (KHARKEVICH et al. 1967; ARENDARUK et al. 1970). It was found that $n=4$ was the optimum. This compound (Table 11, compound 56) blocked neuromuscular transmission in cats at doses of 0.25–0.3 mg/kg. The effects developed rapidly, the complete block lasted 2–4 min. There was a simultaneous depression of respiration. Immediately after the administration, muscular fasciculations were observed. Head drop in rabbits occurred after intravenous injection of compound 56 at doses of 0.45–0.5 mg/kg. Increasing the number of methylene groups to $n=7$ or decreasing it to $n=2$ resulted in a fall of neuromuscular blocking activity (Table 11).

Investigating the influence of alkyl substitution on the ammonium moiety of cinnamic acid derivatives (structure XVI; $R^3=R^4=H$), it was shown that trimethylammonium compounds ($R^1=R^2=CH_3$) were the most active neuromuscular blocking agents. Methyldiethylammonium ($n=2$–5) and *N*-methylpiperidinium ($n=3$ and 4) salts had low neuromuscular blocking activity (Table 11).

As to the mode of action of the compounds tested, trimethylammonium compounds were found to act as depolarizing agents. Thus, compound 56 at doses of 0.2–0.4 mg/kg caused spastic paralysis in intact chickens (within 1–2 min). Neostigmine (30–50 μg/kg) potentiated its effect in cats.

In further research in the series of esters of cinnamic acid derivatives and analogs, only the structure of the acidic fragment of the molecule was varied (Table 12). All the compounds were methiodides of δ-dimethylaminobutyl esters, i.e., the length of the polymethylene chain of the amino alcohol remained unchanged ($n=4$), as this value of n was optimal in the series of trimethylammonium

Table 11. Neuromuscular blocking activity of quaternary ammonium derivatives of cinnamic acid aminoalkyl esters

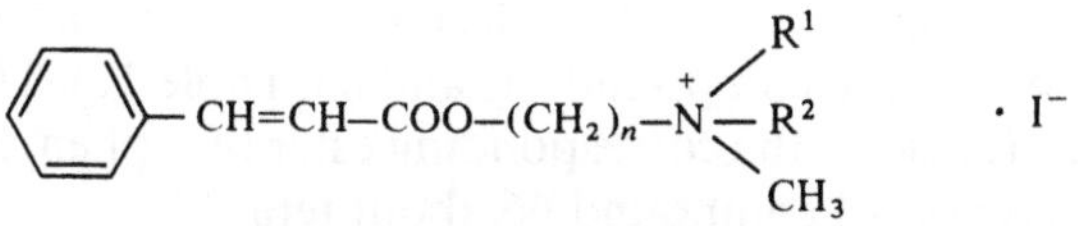

n	Block of transmission from sciatic nerve to gastrocnemius muscle in cats (mg/kg i.v.)[a]		
	NR^1R^2		
	$N(CH_3)_2$	$N(C_2H_5)_2$	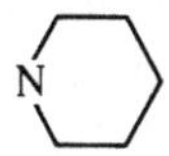
2	6.0–8.0 (54)	14.0–16.0 (59)	
3	0.7–1.2 (55)	5.0–7.0 (60)	6.0–7.0 (63)
4	0.25–0.3 (56)	5.0–6.0 (61)	4.0–6.0 (64)
5	0.5–0.6 (57)	5.0–6.0 (62)	
7	6.0–7.0 (58)		

[a] See Table 1
[b] Numbers of compounds, in parentheses

Table 12. Neuromuscular blocking activity of methiodides of dimethylaminobutyl esters of cinnamic acid derivatives and analogs

$$R{-}COO{-}(CH_2)_4{-}\overset{+}{N}(CH_3)_3 \cdot I^-$$

Compound	R	Block of neuromuscular transmission from sciatic nerve to gastrocnemius muscle in cats (mg/kg i.v.)[a]
65	$C_6H_5{-}CH_2{-}CH_2{-}$	3.0 –6.0
56	$C_6H_5{-}CH{=}CH{-}$	0.25–0.3
66	$C_6H_5{-}CH{=}C(C_6H_5){-}$	2.0 –3.0
67	$C_6H_5{-}CH{=}CH{-}CH{=}CH{-}$	0.5 –0.7
68	$C_6H_5{-}C{\equiv}C{-}$	0.5 –1.0
69	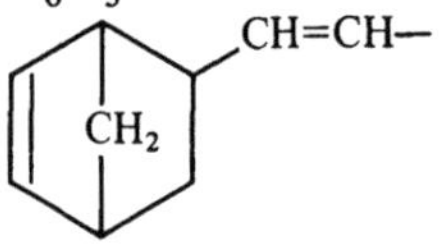	0.35–0.4

C_6H_5 = 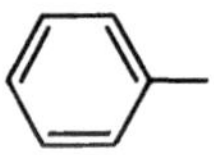

[a] See Table 1

compounds presented in Table 11. On the whole, all the compounds shown in Table 12 appeared to be less potent neuromuscular blocking agents than compound 56, a cinnamic acid derivative.

An increased number of double bonds in the chain of the acidic part of the molecule (compound 67) as well as the replacement of the double by a triple bond (compound 68) had a negative effect on the neuromuscular blocking activity. The presence of a double bond seems to be desirable, since its absence caused a 10- to 20-fold fall in activity (cf. compounds 56 and 65, Table 12). The transition from the ester of cinnamic acid to the corresponding ester of α-phenylcinnamic acid decreased the effectiveness of compound 66 about tenfold.

Quaternary ammonium derivatives of substituted cinnamic acid aminoalkyl esters were especially potent neuromuscular blocking agents (Table 13). Compounds with one or two methoxy groups in the aromatic nucleus (Table 13, compounds 70, 71) and a *p*-nitrocinnamic acid derivative (compound 72) were the most potent. They blocked neuromuscular transmission in cats at doses of 0.03–0.05 mg/kg. Head drop in rabbits occurred on intravenous injection of compound 70 at doses of 0.175–0.200 mg/kg, compound 71 at 0.100–0.125 mg/kg, and compound 72 at 0.125–0.150 mg/kg. Compounds 70–72 caused muscular fasciculations in cats. Neostigmine was their synergist. *p*-Phenylcinnamic acid and *p*-benzyloxycinnamic acid derivatives (Table 13, compounds 73 and 74, respectively) had low neuromuscular blocking activity.

As has been mentioned already, the presence of bulky substituents on the quaternary nitrogen atom results in decreased activity which is confirmed by structure XVII (cf. compound 74, Table 13).

Table 13. Neuromuscular blocking activity of methiodides of substituted cinnamic acid δ-dimethylaminobutyl esters

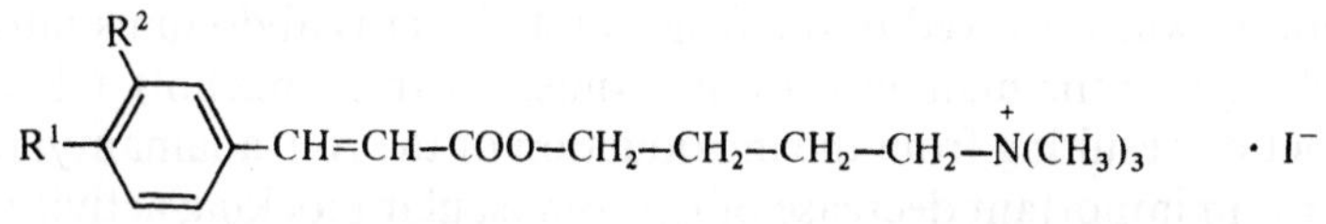

Compounds	R^1	R^2	Block of transmission from sciatic nerve to gastrocnemius muscle in cats (mg/kg i.v.)
56	H	H	0.25–0.3
70	CH_3O	H	0.03–0.05
71	CH_3O	CH_3O	0.03–0.05
72	NO_2	H	0.03–0.05
73	C_6H_5	H	3.0 –4.0
74	$C_6H_5CH_2O$	H	4.0 –5.0

[a] See Table 1

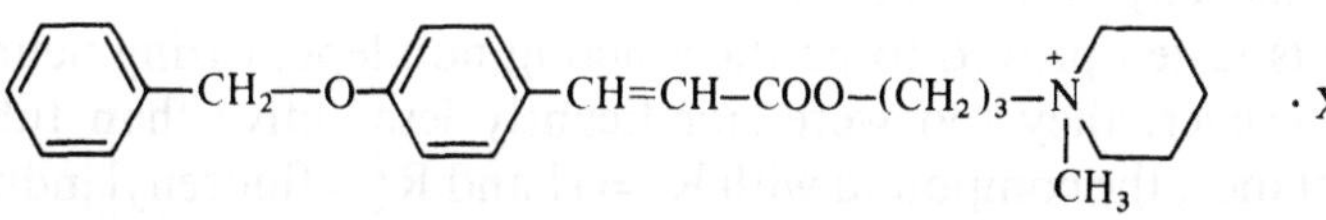

(XVII)

Neuromuscular block was observed in cats only after doses of XVII of 12–15 mg/kg.

In the search for nondepolarizing neuromuscular blocking agents, other compounds with more bulky *N*-substituents were synthetized. Structure XVIII with the *N*-methyl radical replaced by *p*-nitrobenzyl was one of these.

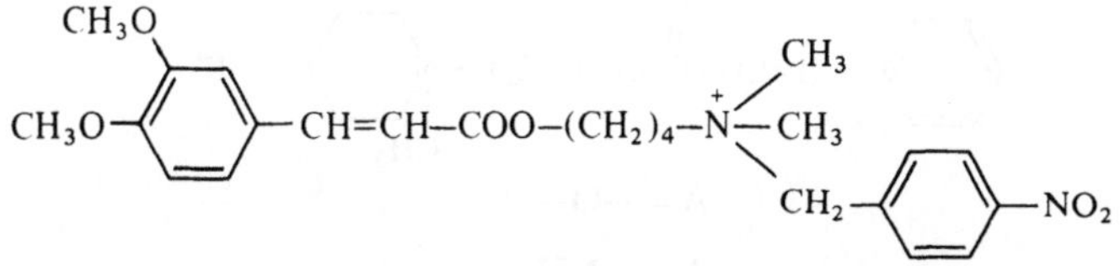

(XVIII)

Structure XVIII differed from the corresponding trimethylammonium analog (compound 71, Table 13) by a significantly lower neuromuscular blocking activity (neuromuscular block in cats occurred only after doses of 7–9 mg/kg). Neostigmine not only failed to antagonize structure XVIII, but quite the opposite, it markedly potentiated its blocking action. The analogous compound with an *N*-1-adamantyl radical (structure XIX) turned out a typical nondepolarizing compound.

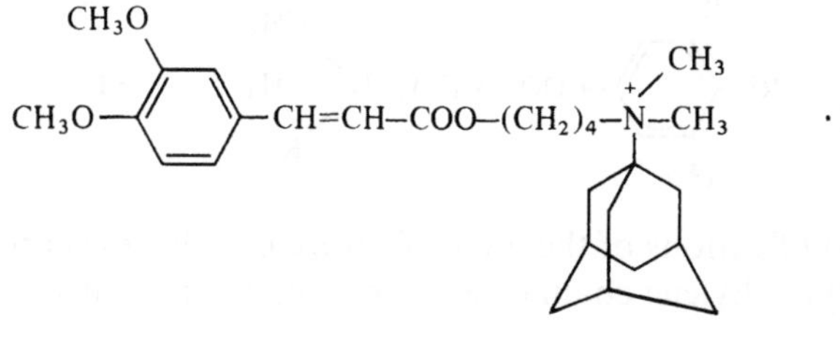

(XIX)

It did not cause fasciculations in cats. Neostigmine antagonized structure XIX. It caused flaccid paralysis in chickens. No stimulating effect on nicotinic receptors was observed in the frog isolated rectus abdominis muscle. Nicotinic blocking was the main effect (antagonism with carbachol). Changes in the mode of action resulting from the introduction of the *N*-1-adamantyl radical was followed by an important decrease of neuromuscular blocking activity. Structure XIX blocked transmission from the sciatic nerve to the gastrocnemius muscle in cats at doses of 7–8 mg/kg (cf. compound 71, Table 13); tubocurarine caused a similar effect at doses of 0.18–0.23 mg/kg.

KITZ et al. (1969), and GINSBURG et al. (1971) carried out a more detailed analysis of the neuromuscular blocking activity of the quaternary ammonium derivatives of cinnamic acids with $n=2$ having the general structure XX.

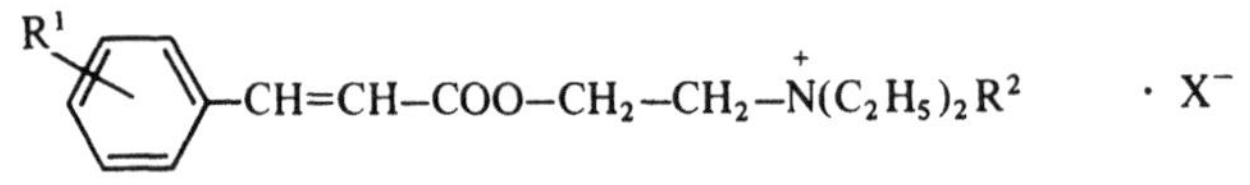

(XX)

where $R^1 = p$-Cl; *o*-, *m*-, *p*-NO_2
R^2 = fluorenyl, *p*-nitrobenzyl.

The agents tested proved to be short-acting nondepolarizing neuromuscular blockers. However, they too were significantly less active than tubocurarine. Thus, for instance, the compound with $R^1 = H$ and R^2 = fluorenyl induced neuromuscular block in cats at 5.0–6.8 mg/kg.

It was shown that the replacement of the ester group by an amide group (structures XXI–XXII) somewhat increased neuromuscular blocking activity. The amide (structure XXII) blocked neuromuscular transmission at doses of 2–3 mg/kg and the ester (structure XXI) at doses of 6–7 mg/kg.

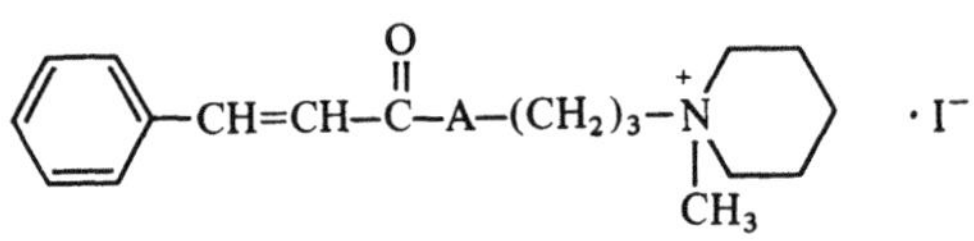

A = –O– (XXI)

A = –NH– (XXII)

II. Benzoic Acid Derivatives

This section presents the data on neuromuscular blocking activity of benzoic acid derivatives designed on the pattern of the cinnamic acid derivatives and having the general structure XXIII (ARENDARUK et al. 1970, 1972a).

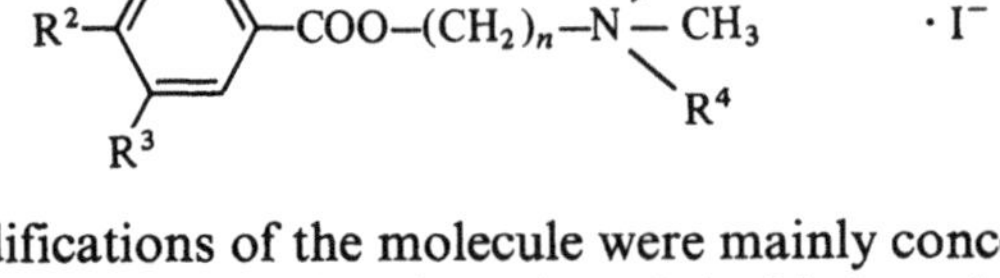

(XXIII)

Structural modifications of the molecule were mainly concerned with: (a) the length of the polymethylene chain (the value of *n*); (b) the substituents in the ar-

Table 14. Neuromuscular blocking activity of quaternary ammonium derivatives of benzoic acid aminoalkyl esters

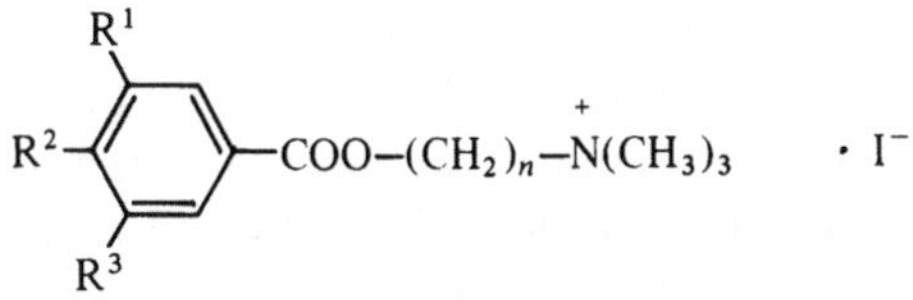

Compound	n	R^1	R^2	R^3	Block of transmission from sciatic nerve to gastrocnemius muscle in cats (mg/kg i.v.)[a]
75	2	H	H	H	8.0–9.0
76	3	H	H	H	2.5–3.0
77	4	H	H	H	0.2–0.25
78	5	H	H	H	1.2–2.0
79	6	H	H	H	0.6–0.8
80	7	H	H	H	4.0–5.0
81	9	H	H	H	4.0–5.0
82	11	H	H	H	10.0–12.0
83	4	H	NO_2	H	0.04–0.06
84	4	H	$SO_2N(CH_3)_2$	H	0.04–0.05
85	4	H	Cl	H	0.3–0.35
86	4	H	CH_3O	H	0.4–0.45
87	4	CH_3O	CH_3O	H	0.4–0.45
88	4	CH_3O	CH_3O	CH_3O	0.45–0.55

[a] See Table 1

omatic ring (R^1, R^2, R^3); and (c) the radicals attached to the onium nitrogen (R^4). As to the length of the polymethylene chain, $n=4$ proved to be optimal, as was the case with cinnamic acid derivatives (Table 14, compound 77). This substance blocked neuromuscular transmission in cats at doses of 0.2–0.25 mg/kg. Gradual shortening of the chain to $n=2$ or lengthening it to $n=11$ resulted in about 50-fold decrease of potency.

Compounds 75–79 (Table 14) cause a short-term (1–5 min) neuromuscular block. The duration of their effect up to complete restoration of the initial amplitude of muscle contractions was from 3–5 to 8–14 min. They are depolarizing agents and cause muscular fasciculations in cats. Neostigmine potentiated their effect. They induced spastic paralysis in pigeons and chickens and contractions of the frog isolated rectus abdominis muscle.

Compounds 80–82 (Table 14) had some peculiarities. On their administration to cats, there first occurred neuromuscular block which then rapidly disappeared (after 0.5–1.5 min) and then after a short-term and usually incomplete restoration, the neuromuscular block developed again, but it was then longer lasting. After neostigmine pretreatment, at first their blocking action persisted and even became more marked, but then decreased. It is likely that the first stage is related to the depolarization of the subsynaptic membrane, and the second one to the development of nondepolarizing block. Two-stage action of the agents described is confirmed by the electromyographic findings. Thus, registration of posttetanic "decurarization" in the gastrocnemius muscle (experiment conducted by V. P. Fisenko) showed that after administration of 2–4 mg/kg compound 81, it was absent in the first stage (which is typical of depolarizing block), and in the second stage it appeared, i.e., neuromuscular block became nondepolarizing.

Quaternary ammonium derivatives of substituted benzoic acid δ-dimethylaminobutyl esters are of interest. The compounds with R^1 and $R^3 = H$ and $R^2 = SO_2N(CH_3)_2$ or NO_2 were the most potent. They caused neuromuscular block at doses of 0.04–0.06 mg/kg (Table 14). The introduction of any number of methoxy groups (compounds 86–88) had an unfavorable effect (cf. compound 77). The introduction of a chlorine atom ($R^2 = Cl$; $R^1 = R^3 = H$; compound 84) had a similar influence, but its action was less marked.

All those compounds (83–88) caused spastic paralysis in pigeons and chickens and contractions of the frog isolated rectus abdominis muscle. In experiments on cats, neostigmine not only failed to antagonize them, but could even somewhat potentiate their effect. The block induced by quaternary nitrogen ammonium salts of substituted benzoic acid aminoalkyl esters was short-acting. Complete block lasted for 1–3 min, and complete recovery of neuromuscular transmission occurred between 4–6 and 12–15 min (from the moment of the injection).

The substitution of an NH group for the oxygen atom in the ester group reduced neuromuscular blocking activity. Thus, structure XXIV blocked neuromuscular transmission in cats at doses of 0.9–1.1 mg/kg and the corresponding ester at a dose of 0.2–0.25 mg/kg (see Table 14, compound 77). The amide (structure XXIV) caused spastic paralysis in pigeons at 0.5–1 mg/kg i.v.

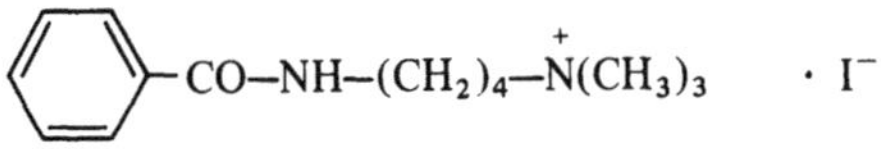

(XXIV)

In the study of the influence of substitution on the onium head, the main attention was given to probable changes in the mode of action. Therefore, the adamantyl radical was used, as it had been previously shown that the attachment of hydrophobic radicals to quaternary nitrogen atoms resulted in the development of nondepolarizing properties (KHARKEVICH and SKOLDINOV 1970). As expected, *N*-1-adamantyl derivatives proved nondepolarizing neuromuscular blocking agents. However, their activity was sharply decreased (Table 15) after the replacement of *N*-methyl by *N*-1-adamantyl radicals as was also the case with the *N*-1-adamantyl derivative of 3,4-dimethoxycinnamic acid (structure XIX).

Thus, the type of action of the majority of the quaternary salts of nonsubstituted benzoic acid aminoalkyl esters is analogous to that of the structurally close cinnamic acid derivatives. Both produced a marked and short-acting neuromuscular block. In both, four methylene groups ($n = 4$) appeared to be the optimal distance for the trimethylammonium salts. The closest acetylcholine analogs ($n = 2$) proved much less active. It was shown in the series of trimethylammonium benzoic acid derivatives that lengthening the polymethylene chain of the amino alcohol part of the molecule up to $n = 7$, 9 or 11 changes the depolarizing action to a mixed action.

Table 15. Effect of adamantyl radicals on the mode of neuromuscular block of δ-dimethylaminobutyl ester methiodides of benzoic acid and its derivatives

$$R^1{-}COO{-}(CH_2)_4{-}\overset{+}{N}(CH_3)(R^2){-}CH_3 \cdot I^-$$

Compound	R^1	R^2	Block of transmission from sciatic nerve to gastrocnemius muscle in cats (mg/kg i.v.)[a]	Type of paralysis in birds (mg/kg i.v.)		Effect on frog isolated rectus abdominis muscle	
				Pigeons	Chickens	Contracture	Carbachol antagonism
76	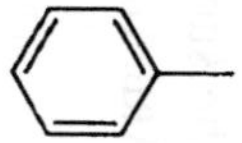	CH_3	0.20–0.25	S(0.4–0.5)	S(0.3–0.4)	+	
89	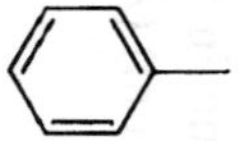	1-Ad	12.0 –14.0	F(8.0–10.0)	F(4.0–5.0)	–	+
83	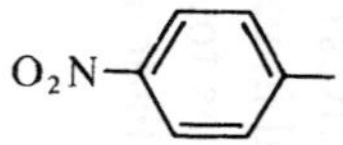	CH_3	0.04–0.06	S(0.1–0.2)	S(0.15–0.20)	+	
90	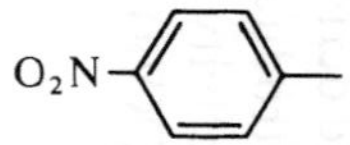	1-Ad	12.0–14.0	F(2.0–3.0)	F(4.0–6.0)	–	+
87	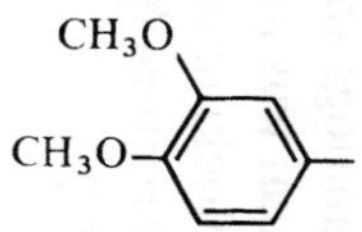	CH_3	0.40–0.45	S(0.3–0.4)		+	
91	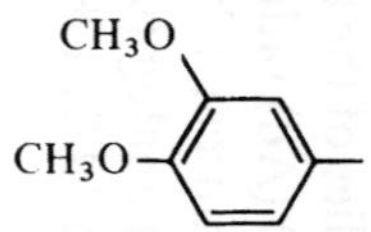	1-Ad	10.0–12.0	F(2.0–4.0)	F(6.0–7.0)	–	+

S spastic paralysis; F flaccid paralysis
[a] See Table 1

The quaternary salts of some substituted benzoic and cinnamic acids were the most potent. In benzoic acid esters, the presence of electron-accepting groups (e.g., NO_2) increased the potency, whereas electron-donating methoxy groups somewhat decreased it. It may be suggested that the presence of an electron-accepting group caused an increase in the partial positive charge on the carbonyl carbon atom, promoting the fixation of the compounds on the subsynaptic membrane, this being manifested by the increase in potency. The partial positive charge on the carbonyl carbon may interact either with nonspecific receptors, or with acetylcholine receptors with a five-atom chain between them. The latter variant might be considered as an alternative one, as there exist bisquaternary ammonium neuromuscular blocking agents in which nitrogen atoms are six methylene groups away from each other (e.g., qualidilum; Chap. 1, Table 1). Partial charges may also appear in the aromatic nucleus. In this case, they will be 6–8 atoms away from the quaternary nitrogen.

Quaternary derivatives of aminoalkyl esters of substituted cinnamic acids are also potent neuromuscular blocking agents. They are comparable to decamethonium, but they are short-acting. Their high activity is probably related either to hydrophobic interactions (MIKHELSON and ZEIMAL 1970) or to the appearance of partial positive charges on the benzene ring (KHROMOV-BORISOV et al. 1968). Additional points of fixation on acetylcholine receptors may also be realized via nucleophilic double bonds or via ester groups by dipole–dipole interaction. The importance of partial charges on the aromatic ring is supported by the lower activity of dihydrocinnamic acid derivatives (Table 12, compound 65) in which the ester group is still present, but the carbonyl group is not conjugated with the phenyl ring which prevents the appearance of partial positive charges on the latter. The substitution of highly hydrophobic radicals (e.g., adamantyl) for one *N*-methyl group in trimethylammonium derivatives of benzoic and cinnamic acids turns them into nondepolarizing, though much less active neuromuscular blocking agents.

D. The Derivatives of Aliphatic Dicarboxylic Acid Esters

I. Neuromuscular Blocking Action

1. Succinylcholine Analogs and Derivatives

In studies of the analogs and derivatives of succinylcholine, the hydrophobic adamantyl (Ad) radical was introduced in various parts of its structure. The highest neuromuscular blocking activity was found in the succinylcholine bis-*N*-1-adamantyl analog (structure XXVa) (diadonium) (ARENDARUK et al. 1972b; KHARKEVICH 1973, 1983).

$$R(CH_3)_2\overset{+}{N}{-}(CH_2)_2{-}OOC{-}(CH_2)_2{-}COO{-}(CH_2)_2{-}\overset{+}{N}(CH_3)_2R \cdot 2p\text{-}CH_3C_6H_4SO_3^-$$

a) R = 1-Ad– (XXV)
b) R = 1-AdCH$_2$–
c) R = 1-AdCH$_2$CH$_2$–

Unlike succinylcholine (structure XXV; $R = CH_3$), diadonium (structure XXV; R = 1-Ad) is a nondepolarizing neuromuscular blocking agent. It has already been mentioned that it did not cause muscular fasciculations. In pigeons or chickens, it caused flaccid paralysis typical of nondepolarizing drugs. It did not stimulate isolated frog rectus abdominis muscle, but it abolished the nicotinomimetic action of carbachol. Neostigmine is a diadonium antagonist. Neuromuscular blocking activity is marked enough in diadonium, but it is lower than that of tubocurarine. Intravenous injection of 0.3–0.35 mg/kg diadonium was followed by neuromuscular block (sciatic nerve–gastrocnemius muscle) in anesthetized cats.

Shifting the 1-adamantyl radical one or two methylene groups further from the quaternary nitrogen atom resulted in a 10- to 20-fold decrease of neuromuscular blocking activity. Thus, intravenous doses at which these compounds caused neuromuscular block in cats for structure XXVb ($R = 1\text{-}AdCH_2$) were 6–8 mg/kg and for structure XXVc ($R = 1\text{-}AdCH_2CH_2$) they were 3–4 mg/kg. Both compounds are nondepolarizing. The introduction of 1-adamantyl radicals in the interonium part of the succinylcholine molecule (structure XXVI) decreased the neuromuscular blocking activity even more.

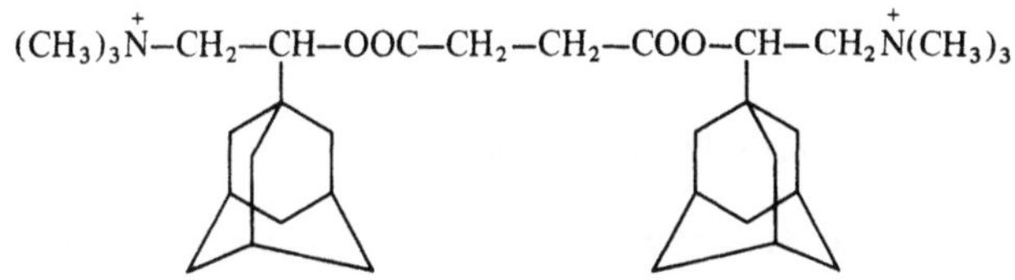

$\cdot\ 2I^-$ (XXVI)

Structure XXVI blocked neuromuscular transmission in cats at doses of 12–15 mg/kg. In spite of the presence of trimethylammonium groups, this compound belongs to the nondepolarizing drugs.

Structure XXVII also appeared to have low activity. It differed from succinylcholine in having a 1,3-adamantyl radical in its molecule instead of two central methylene groups.

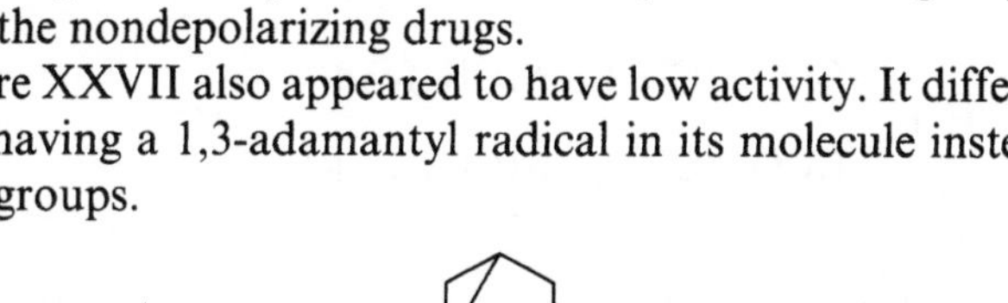

$\cdot\ 2I^-$ (XXVII)

Similar to structure XXVI, it also has two trimethylammonium groups and yet it belongs to the nondepolarizing drugs. It blocked neuromuscular transmission in cats at doses of 20–25 mg/kg. Structure XXVIII was even less active.

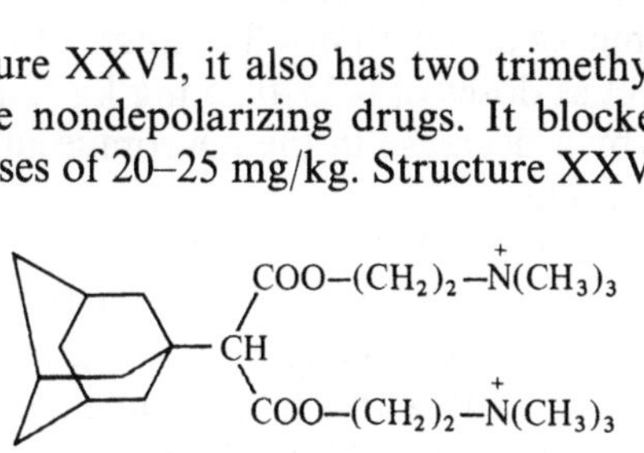

$\cdot\ 2I^-$ (XXVIII)

Neuromuscular block in cats was caused by a dose of 45–50 mg/kg.

Thus, succinylcholine adamantyl analogs and derivatives, irrespective of the position of adamantyl radicals in the molecule, belong to the nondepolarizing neuromuscular blocking agents. This provides further evidence for the importance of the total hydrophobic–hydrophilic balance and not the location of the hydrophobic radical, for the mode of action. However, the location of the 1-adamantyl radical is important for neuromuscular blocking activity. Yet, not only the degree of the hydrophobicity of radicals, but also the structure of the hydrophobic fragment should be taken into account. Thus, for instance, diadonium which contains compact, spherical 1-adamantyl radicals is significantly more potent than structure XXIX with linear aliphatic radicals $n\text{-}C_{10}H_{21}$ having as many carbon atoms as the adamantyl radical, i.e., with formally equal hydrophobicity. Structure XXIX blocked neuromuscular transmission in cats at doses of 18–25 mg/kg, i.e., it was 60–80 times less active than diadonium.

$$(CH_3)_2\overset{+}{\underset{\underset{n\text{-}C_{10}H_{21}}{|}}{N}}-(CH_2)_2-OOC-(CH_2)_2-COO-(CH_2)_2-\overset{+}{\underset{\underset{n\text{-}C_{10}H_{21}}{|}}{N}}(CH_3)_2 \cdot 2I^- \qquad \text{(XXIX)}$$

As has been mentioned already, of the compounds tested, the succinylcholine bis-*N*-1-adamantyl analog, diadonium is of interest for anesthesiologists. It has a combination of rather marked neuromuscular blocking activity and a short duration of the effect.

2. Derivatives of Other Aliphatic Dicarboxylic Acids

Among the derivatives of dicarboxylic acid aminoalkyl esters, two series of compounds have received most attention. The first includes diadonium homologs (structure XXX) – bisquaternary salts which have different interonium distances depending on different dicarboxylic acid (KHARKEVICH 1973, 1983).

$$(CH_2)_m\begin{cases}COO-(CH_2)_2-\overset{+}{N}(CH_3)_2Ad\text{-}1\\ COO-(CH_2)_2-\overset{+}{N}(CH_3)_2Ad\text{-}1\end{cases} \cdot 2I^- \qquad \text{(XXX)}$$

where $m=2, 4, 8$.

The compounds with a longer distance between quaternary nitrogen atoms than diadonium (structure XXX; $m=2$) turned out less active neuromuscular blocking agents. Thus, with $m=4$, the compound blocked neuromuscular transmission at doses of 2.5–3 mg/kg and with $m=8$ at doses of 25–30 mg/kg. Diadonium caused a similar effect at doses of 0.25–0.35 mg/kg. Thus, each twofold increase of m was followed by a decrease in the neuromuscular blocking potency by about an order of magnitude.

The general structure of the second group, dimethiodides of dicarboxylic acids (dimethylaminobutanol esters), is presented in Table 16. It was shown in cats that the compounds were most active at $m=4$ and 6. At $m=2$ or 8 the blocking activity decreased two- to threefold (Table 16). In experiments on rabbits, the compound with $m=4$ caused head drop at 0.115–0.120 mg/kg, and with $m=2$ at 0.55–0.60 mg/kg.

Neostigmine potentiated the neuromuscular block caused in cats by compounds 92, 93, 95, 96. During their administration, muscular fasciculations were

Table 16. Neuromuscular blocking activity of bisquaternary ammonium derivatives of dicarboxylic acids

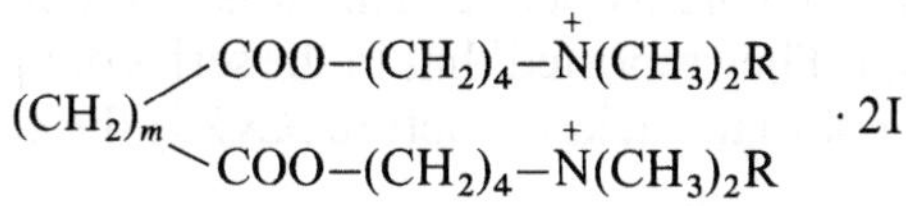

Compound	m	R	Block of transmission from sciatic nerve to gastrocnemius muscle in cats (mg/kg i.v.)	Type of paralysis in pigeons (mg/kg i.v.)	Ability to cause contractions of isolated rectus abdominis muscle in frogs (pD_2 with confidence limits)
92	2	CH_3	0.07–0.08	S(0.35–0.4)	5.0(7.5–4.3)
93	4	CH_3	0.015–0.02	S(0.08–0.1)	5.4(6.4–4.4)
94	4	1-Ad	5.0–6.0	F(0.2–0.25)	[b]
95	6	CH_3	0.04–0.045	S(0.1–0.2)	5.2(6.0–4.4)
96	8	CH_3	0.12–0.14	S(0.12–0.15)	5.0(5.5–4.5)

S spastic paralysis; F flaccid paralysis

1-Ad = 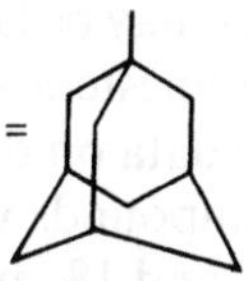

[a] See Table 1
[b] Does not cause contractions; is a carbachol antagonist: $EC_{50} = 1.8 \times 10^{-5}$ $(2.6\text{–}1.0 \times 10^{-5})$ M

observed. Spastic paralysis occurred in pigeons. There was a marked nicotinomimetic effect on the frog isolated rectus abdominis muscle, manifested by muscle contractions. Thus, these bistrimethylammonium derivatives are characterized by a depolarizing mode of action.

Substitution of 1-admantyl for the methyl group at each of the nitrogen atoms (Table 16, compound 94) resulted in the development of nondepolarizing neuromuscular blocking activity. Its potency was significantly reduced as compared with the analogous bistrimethylammonium salt (cf. compounds 93 and 94).

It follows from these data that among the bisquaternary ammonium derivatives of aliphatic dicarboxylic acids and dimethylaminobutanol esters, those with 16 and 18 atoms between cationic centers were the most potent (Table 16). Such an optimum for the interonium distance is also known for some other neuromuscular blocking agents (Imbretil, octadecamethonium, etc.) and it is likely to be related to a corresponding distance between anionic groups of end-plate acetylcholine receptors with which quaternary nitrogen atoms interact.

However, there is every reason to believe that the fixation of the drugs on acetylcholine receptors takes place not only on the anionic centers, but also on the so-called esterophilic areas. Indeed, neuromuscular blocking activity at $m=4$ and $m=6$ (Table 16) was found to be much more marked than in hexadecamethonium (Barlow and Zoller 1964) and octadecamethonium (Paton and Zaimis 1949), respectively, i.e., analogous compounds, but lacking ester groups.

Thus, of the compounds tested, esters of dicarboxylic acids and dimethylaminobutanol were found to be highly active. This was also confirmed in the paper by DANILOV et al. (1968). They reported that in the series of bisquaternary ammonium derivatives of terephthalic acid (structure XXXI) the compound with $n=4$ was the most active.

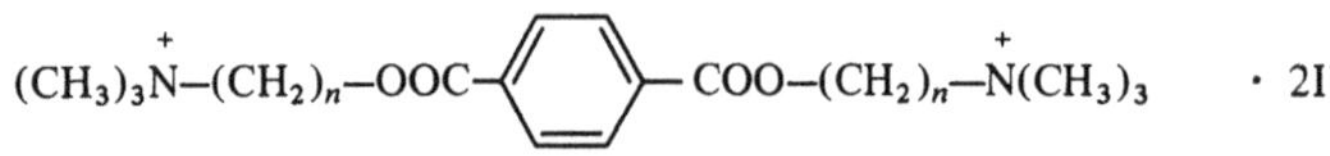

(XXXI)

However, to provide an adequate evaluation of the potency of bisquaternary salts of dicarboxylic acid esters, it is important to know the rate of the hydrolysis of those compounds by cholinesterases as well as their interaction with sites of loss.

II. Hydrolysis by Cholinesterases

It is known that the duration and the intensity of the neuromuscular block induced by quaternary salts of dicarboxylic acid aminoalkanol esters may be largely determined by the rate of their plasma cholinesterase (butyrylcholinesterase) hydrolysis (MIKHELSON and ZEIMAL 1970). This section presents the data on the enzymatic hydrolysis of a number of bisquaternary ammonium compounds which are dicarboxylic acid esters (see structural formulae in Tables 17 and 18; experiments of N. D. Igumnova and E. Y. Lemina; KHARKEVICH 1983). The data on their neuromuscular blocking potency have already been given.

To study the kinetics of enzymatic cholinesterase hydrolysis of substrates, a method based on recording the complete kinetic curve of the reaction with subsequent integration was employed. The reason for this was the low solubility of many drugs which did not allow one to vary the substrate concentration in the reaction mixture significantly nor to use the initial rate technique. However, for some highly soluble substrates (succinylcholine, diadonium), the Michaelis constant and maximal hydrolysis rate were measured by both methods which gave rather close agreement.

In studies of the enzymatic hydrolysis of dicarboxylic acid esters, it was found that under the chosen experimental conditions (exposure time, concentrations of substrate and enzyme, see Table 17), in the majority of the compounds, only one of the ester bonds was hydrolyzed. In several compounds (e.g., Table 17, compounds 103, 104; Table 18, compounds 110, 96) under the same experimental conditions, another ester bond was hydrolyzed, but the hydrolysis rate was significantly lower. The second stage of the reaction developed in other compounds, too, but it took higher plasma cholinesterase concentrations and a longer reaction time. The integration of the curves obtained in these experiments provided a characterization of the reactivity of both the initial substrate (first stage of the reaction) and the monoester (second stage of the reaction). The rate of hydrolysis of the substrates tested (first stage) was unaffected by the products formed during the reaction. Taking into account that the first stage is most important for the design of new compounds, since it leads to an abrupt fall of neuromuscular blocking effect, only these data are presented in Tables 17–19.

Table 17. Kinetic constants of enzymatic hydrolysis of adamantyl derivatives of dicarboxylic acid esters[a]

$$H_3C-\underset{|}{\overset{R}{N}}-(CH_2)_n-OOC-(CH_2)_m-COO-(CH_2)_n-\underset{|}{\overset{R}{N}}-CH_3\cdot 2R^1X$$

Compound	R	m	n	R^1X	Michaelis constant K_m (M)	Maximal hydrolysis rate V_{max} ($M\ min^{-1}\ mg^{-1}$) (first stage)
97	1-Ad	2	2	CH_3I	3.89×10^{-5}	1.06×10^{-8}
98	2-Ad	2	2	CH_3I	9.42×10^{-5}	4.63×10^{-8}
99	$1\text{-}AdCH_2$	2	2	CH_3I	5.2×10^{-5}	2.4×10^{-8}
100	$1\text{-}AdCH_2CH_2$	2	2	$2\cdot p\text{-}CH_3C_6H_4SO_3^-$	4.44×10^{-5}	2.87×10^{-8}
101	1-Ad	4	2	HCl	8×10^{-5}	2.4×10^{-7}
102	1-Ad	4	2	CH_3I	4.82×10^{-5}	3.68×10^{-7}
103	1-Ad	8	2	CH_3I	5.94×10^{-5}	2.18×10^{-6}
94	1-Ad	4	4	CH_3I	8.35×10^{-6}	8.1×10^{-9}
104	$(CH_3)_3\overset{+}{N}-(CH_2)_2-OOC-Ad(I,3)-COO-(CH_2)_2-\overset{+}{N}(CH_3)_3\cdot 2I^-$				1.16×10^{-4}	8.3×10^{-7}
105	$(CH_3)_3\overset{+}{N}-CH_2-\underset{1\text{-}Ad}{CH}-OOC-(CH_2)_2-COO-\underset{1\text{-}Ad}{CH}-CH_2-\overset{+}{N}(CH_3)_3\cdot 2I^-$					[b]

[a] The kinetics of enzymatic hydrolysis was studied using a pH-state with a chart recorder (Radiometer, Denmark) by titration (1×10^{-2} M KOH) of the acid liberated during the hydrolysis. The reaction was conducted at 38 °C in 0.15 M NaCl solution pH = 7.4 in a CO_2-free airflow. A lipophilically dried cholinesterase preparation of horse blood serum with specific activity 98 U/mg was used

[b] The compound was not hydrolyzed under the experimental conditions. No hydrolysis was observed for the monoquaternary salt $H_3C-COO-\underset{1\text{-}Ad}{CH}-CH_2-\overset{+}{N}(CH_3)_3\cdot I^-$, either

Table 18. Kinetic constants of enzymatic hydrolysis of aliphatic dicarboxylic acid ester derivatives[a]

$$(CH_3)_2N{-}(CH_2)_n{-}OOC{-}(CH_2)_m{-}COO{-}(CH_2)_n{-}N(CH_3)_2 \cdot 2RX$$

Compounds	m	n	RX	Michaelis constant K_m (M)	Maximal hydrolysis rate V_{max} (M min^{-1} mg^{-1}) (first stage)
106	1	2	CH_3I	1.36×10^{-4}	2.07×10^{-7}
107	2	2	CH_3I	8.35×10^{-5}	7.75×10^{-8}
108	4	4	HCl	6.94×10^{-5}	3.25×10^{-8}
93	4	4	CH_3I	7.35×10^{-5}	1.57×10^{-8}
95	6	4	CH_3I	5.62×10^{-5}	3.19×10^{-8}
109	7	4	HCl	2.86×10^{-4}	1.25×10^{-8}
110	7	4	CH_3I	3.33×10^{-4}	2.74×10^{-7}
96	8	4	CH_3I	1.13×10^{-4}	2.57×10^{-7}

[a] See Table 17

Table 19. Neuromuscular blocking activity and hydrolysis rate of dimethiodides of dicarboxylic acid aminoesters

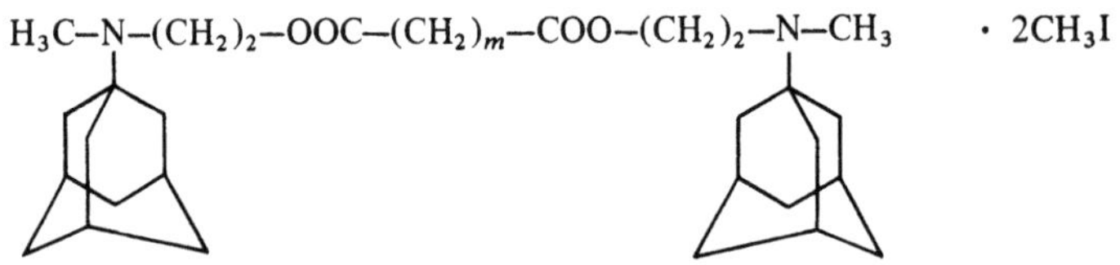

Compound	m	Block of transmission from sciatic nerve to gastrocnemius muscle in cats (mg/kg i.v.)	Maximal hydrolysis rate V_{max} (M min^{-1} mg^{-1}) (first stage)
97	2	0.25–0.35	1.06×10^{-8}
102	4	2.5 –3	3.68×10^{-7}
103	8	25–30	2.18×10^{-6}

It is seen in the tables that for the majority of the compounds tested the values of the Michaelis constant, indirectly reflecting the binding with the active surface of the enzyme were essentially the same. On the other hand, the maximal hydrolysis rates were significantly different.

According to the results of the study of adamantyl derivatives (Table 17) different structures of the hydrophobic adamantyl radicals in the cationic head (1-Ad or 2-Ad, compounds 97 and 98) had some effect on their substrate properties. Maximum hydrolysis rates of 1-adamantyl and 2-adamantyl derivatives differed approximately fourfold. The replacement of adamantyl by the 1-methyleneadamantyl radical (compound 99) or 1-ethyleneadamantyl (compound 100) with similar hydrophobicity failed to produce a noticeable change in the rate of substrate hydrolysis. The hydrolysis rate of these compounds was comparable to that of diadonium.

The structure of the central part of the molecule was a much more important parameter, substantially affecting the hydrolysis rate. Thus, the increase in the distance between cationic centers by lengthening the acidic chain from two methylene groups in diadonium to four in compound 102 was followed by a 35-fold increase in the maximal hydrolysis rate. Further increase in the length of the polymethylene chain up to $m=8$ (compound 103) resulted in an extremely high hydrolysis rate (in the first stage) close to that of acetylcholine (for acetylcholine $V_{max}=3\times10^{-6}$). Thus, increase in the length of the chain in the central part of the molecule from two methylene groups to eight led to an abrupt increase in the hydrolysis rate of the compound: V_{max} differed by two orders of magnitude (cf. compounds 97 and 103).

The increase in the distance between the quaternary nitrogen atoms owing to the lengthening of the amino alcohol part of the molecule led to opposite results. Thus, when the distance between ester and cationic groups corresponded to $n=4$ (compound 94), the maximum hydrolysis rate was 45 times lower than at $n=2$ (compound 102). Compounds 103 and 94 had dissimilar m and n, their interonium distance being equal. V_{max} of those agents differed by two orders of magnitude. Enzymatic hydrolysis (the first stage) of compound 104 (1,3-adamantanedicarboxylic acid derivative) was higher by an order of magnitude than that of succinylcholine (Table 18, compound 107). Compound 104 differed from the latter by the presence of a hydrophobic adamantyl radical in the central part of its molecule. The hydrolysis rate of the first bond in compound 104 was comparable to that of acetylcholine.

In the series of aliphatic dicarboxylic acid esters not containing adamantyl radicals (Table 18), among the dimethylaminobutanol ester derivatives (compounds 93, 95, 110, 96), the hydrolysis rate increased with the lengthening of the central part of the molecule. Increase in the distance between ester groups from $m=4$ (compound 93) to $m=8$ (compound 96) resulted in a 16-fold increase in V_{max}. Thus, in this series as well as among adamantyl derivatives, the structure of the central part was important for the substrate properties of the compound.

The maximum hydrolysis rate of tertiary (compound 108) and quaternary (compound 93) aminoesters with $m=n=4$ differed only twofold. On the lengthening of m to 7 ($n=4$), the maximum hydrolysis rate of the bisquaternary salt (compound 110) was about 22 times higher than that of the tertiary amine (compound 109). Among adamantyl derivatives (Table 17), the tertiary (compound 101) and quaternary (compound 102) ammonium salts with $m=4$ and $n=2$ had rather close V_{max}.

The comparison of the kinetic constants of enzymatic hydrolysis of compounds 93 and 94, 107 and 97 (Tables 17, 18) suggested that the introduction of the hydrophobic adamantyl radical in the cationic head caused a decrease of the hydrolysis rate. The addition of adamantyl radicals to quaternary nitrogen atoms in the succinylcholine molecule was followed by an approximately sevenfold decrease in the maximal hydrolysis rate of diadonium. The observed dissimilarities in the hydrolysis rate of these compounds are most likely responsible for the un-

Table 20. Neuromuscular blocking activity and hydrolysis rate of dimethylaminobutyl ester derivatives

$$(CH_3)_3\overset{+}{N}-(CH_2)_4-OOC-(CH_2)_m-COO-(CH_2)_4\overset{+}{N}(CH_3)_3 \cdot 2I^-$$

Compound	m	Block of transmission from sciatic nerve to gastrocnemius muscle in cats (mg/kg i.v.)	Maximal hydrolysis rate V_{max} (M min^{-1} mg^{-1}) (first stage)
93	4	0.015–0.02	1.57×10^{-8}
95	6	0.04 –0.045	3.19×10^{-8}
96	8	0.12 –0.14	2.57×10^{-7}

equal duration of their action in patients: neuromuscular block on single succinylcholine administration lasted 5–7 min while that induced by diadonium under the same conditions lasted about 12–15 min.

It would be of interest to compare the data on neuromuscular blocking activity and rate of enzymatic hydrolysis of the three compounds tested (Table 19). It can be seen that each twofold increase in the length of the central part of the molecule ($m = 2, 4, 8$) was followed by a tenfold fall in neuromuscular blocking activity. At the same time, the hydrolysis rate was substantially increased and, in compound 103, approached that of acetylcholine. The parallelism between the given parameters suggested that rapid hydrolysis of compounds may play a role in the decrease of their neuromuscular blocking effect. A similar correlation can also be observed among esters which do not contain adamantyl radicals, in particular, among dimethylaminobutanol ester derivatives (Table 20). In this case, changes in effectiveness in vivo also correlate with the ability of agents to be hydrolyzed by plasma cholinesterase.

In laboratory animals, cholinesterase of the neuromuscular junction probably plays an important role in terminating the action of dicarboxylic acid esters (HOBBINGER 1976). Therefore, for analysis of the esters' dissimilar neuromuscular blocking activity in vivo it is also desirable to have information on the comparative activity of agents during the inhibition of skeletal muscle cholinesterase. These data have been obtained by E. Y. Lemina from experiments on isolated frog rectus abdominus muscle in vitro. At frog neuromuscular junctions of tonic fibers, to which rectus abdominis muscle belongs, the concentration of cholinesterase is greater than that of acetylcholinesterase (see HOBBINGER 1976). ED_{50} of some compounds were estimated before and after complete inhibition of cholinesterase by neostigmine. The latter variant of the experiment allows one to characterize more accurately the true potency of the compounds. Without neostigmine, compounds 107 ($m = n = 2$) and 95 ($m = 6$, $n = 4$) with interonium distance 10 and 18 atoms, respectively, turned out most active (Table 21). With inhibited cholinesterase, the activity of esters is increased, but to a different degree, which likely indicates their unequal ability to be hydrolyzed by cholinesterase. Maximal activity was displayed by the agent with 18 atoms separating the cationic heads (compound 95): its ED_{50} was found to be higher by two orders of magnitude than be-

Table 21. Neuromuscular blocking activity, hydrolysis rate, and neostigmine-induced potentiation of nicotinomimetic effect of dicarboxylic acid esters

$$(CH_3)_3\overset{+}{N}-(CH_2)_n-OOC-(CH_2)_m-COO-(CH_2)_n-\overset{+}{N}(CH_3)_3 \cdot 2I^-$$

Compound	*m*	*n*	Neuromuscular block in cats (mg/kg i.v.)	Maximal hydrolysis rate V_{max} (M min^{-1} mg^{-1})	Experiments on frog rectus abdominis muscle[a]		
					ED_{50} before neostigmine administration	ED_{50} after neostigmine administration	Potentiation
107	2	2	0.06 –0.08	7.75×10^{-8}	5.6×10^{-7}	2.3×10^{-7}	×2.4
93	4	4	0.015–0.02	1.57×10^{-8}	3.7×10^{-6}	1.7×10^{-6}	×2.2
95	6	4	0.04 –0.045	3.19×10^{-8}	2.5×10^{-7}	2.1×10^{-9}	×100
96	8	4	0.12 –0.14	2.57×10^{-7}	2×10^{-6}	4.2×10^{-7}	×5

[a] Stimulating effect of the compounds was recorded

fore neostigmine administration. However, these data necessitate additional analysis to rule out the role of direct action of neostigmine on acetylcholine receptors and probable binding with nonspecific receptors (MIKHELSON and ZEIMAL 1970).

The data presented indicate that, among the esters studied, the neuromuscular blocking effect in vivo is correlated more with the plasma cholinesterase hydrolysis rate than with the ability to produce a depolarizing effect on isolated skeletal muscle of the frog. This is true whether cholinesterase activity is maintained or inhibited. Thus, compound 93, most active in vivo and blocking neuromuscular transmission in cats at doses of 0.015–0.02 mg/kg, appears the least active in experiments on isolated muscle. It cannot be excluded that its highest potency in vivo is to a considerable extent due to its lower hydrolysis rate by cholinesterase. On the other hand, the neuromuscular blocking effect of the highly active (in vitro) compound 95 (its ED_{50} after the inhibition of cholinesterase was 2.1×10^{-9} M) was lower in experiments on cats than that of compound 93. Manifestation of the high activity of compound 95 on the whole animal seems to be prevented by its more marked hydrolysis by junctional and plasma cholinesterases. However, the comparisons presented are of a tentative character, since they have been obtained in three animal species. It is known that the hydrolysis rate in different laboratory animals, in particular by plasma cholinesterase, varies over a wide range (FOLDES and FOLDES 1965; HOBBINGER and PECK 1970; HUGHES and CHAPPLE 1976; see also Chaps. 8 and 10). The activity of esters is also different in different animal species (ZAIMIS and HEAD 1976).

Thus, the true potency determined by the direct interaction of the compounds with acetylcholine receptors of muscles, tested in vitro, does not always correlate with the degree of neuromuscular effect in vivo. Apparently, maximal correlation is found in esters with minimal hydrolyzability by cholinesterases and not binding with sites of loss in plasma and tissues.

E. Conclusions

Chemical modifications of the alkaloid thesine resulted in the synthesis of a large number of highly active neuromuscular blocking agents of the nondepolarizing type. The majority of them belong to the bisquaternary ammonium compounds and are derivatives of aminoesters of aminoalkylamides of truxillic acids. For these series of compounds, the relationship between chemical structure and neuromuscular blocking action is established. The derivatives of α-truxillic acid in which the quaternary nitrogen atoms are separated by 13 atoms and the cationic centers are *N*-methyl- (or *N*-ethyl)-piperidinium, *N*-methyl- (or *N*-ethyl)-pyrrolidinium, or triethylammonium groups are the most active. Aminoalkylamide derivatives are characterized by a much higher selectivity of the blocking action on the end-plate acetylcholine receptors than the derivatives of the corresponding aminoesters.

Active nondepolarizing neuromuscular blocking agents were also found among the salts of bistertiary amines belonging to the derivatives of α-truxillic acid amides. Their structure is analogous to that of the corresponding bis quaternary ammonium salts (optimum distance between cationic centers and *N*-substituents). The neuromuscular blocking action of these salts of bistertiary amines is highly selective and neuromuscular transmission is rapidly restored after the block. The derivatives of α-truxillic acid are effectively antagonized by anticholinesterase agents. Of the compounds tested, a number of aminoester derivatives (anatruxonium and cyclobutonium) as well as the aminoalkylamide derivative pyrocurinum are used in patients (see Chaps. 19 and 26).

As far as monoquaternary derivatives of benzoic and cinnamic acids are concerned, some of these have high neuromuscular blocking activity, but they cause a depolarizing block. This is mainly the case with trimethylammonium derivatives of substituted benzoic and cinnamic acids with four methylene groups separating cationic centers and the ester groups.

At the same time, monoquaternary salts are worthy of note as potential short-acting, nondepolarizing neuromuscular blocking agents. The short duration of their effect in general is due to their weaker binding with the end-plate acetylcholine receptors than that of bisquaternary salts. Naturally, monoquaternary salts in the form of rapidly hydrolyzed esters are of great interest. The attempts to obtain sufficiently active neuromuscular blocking agents in this series have so far failed, since the introduction of hydrophobic radicals in the structure of monoquaternary salts was as a rule associated with a prominent fall in neuromuscular blocking potency. However, it should not be considered that this trend has no prospects. The task is to choose a structure and locus for hydrophobic radicals which is optimal for the synthesis of neuromuscular blocking agents, so that their nondepolarizing mode of action may be combined with sufficiently high activity.

It was also observed that the presence of hydrophobic adamantyl radicals in dicarboxylic acid esters affected their pharmacologic activity. The introduction of adamantyl radicals into the structure of dicarboxylic acid derivatives with depolarizing neuromuscular blocking action turned them into nondepolarizing agents, irrespective of the locus of the adamantyl radical in the molecule (see also

Chap. 4). This change in the mode of action is likely to be related to additional hydrophobic interaction with acetylcholine receptors promoting the stabilization of the subsynaptic membrane and probably, also to the interaction with ion channel sites. The change in the mode of action usually results in a decrease of the neuromuscular blocking activity.

As has been pointed out already, the presence of adamantyl radicals also affects the interaction of the compounds with cholinesterases. Thus, cholinesterase hydrolysis of diadonium (an *N*-1-adamantyl succinylcholine analog) developed slower than that of succinylcholine. 1-Adamantyl radicals, when attached to a succinylcholine derivative near the ester groups, hindered the hydrolysis even more. As already mentioned, under the experimental conditions described, plasma cholinesterase was not found to hydrolyze monoquaternary and bisquaternary esters containing an adamantyl radical in the amino alcohol part of the molecule next to the ester group. Of the adamantyl derivatives synthesized, diadonium was proposed for clinical application. It is one of the first nondepolarizing short-term neuromuscular blocking agents. Separate chapters are devoted to its pharmacologic, toxicologic, and clinical characteristics (Chaps. 20 and 26).

Acknowledgment. The authors wish to express their appreciation to Mrs. MARIA LIPMAN for the translation of this chapter.

References

Arendaruk AP (1953) Issledovaniya alkaloidov Thesium Minkwitzianum (Investigations of alkaloids Thesium Minkwitzianum). Thesis, Moscow

Arendaruk AP, Kravchuk LA, Skoldinov AP, Kharkevich DA (1963) Chemical and pharmacological studies in the series of cyclobutanedicarboxylic acids derivatives. In: Kharkevich DA (ed) Sovremennye problemy farmakologii (Recent problems of pharmacology). Medgiz, Moscow, pp 138–157

Arendaruk AP, Gracheva EA, Skoldinov AP, Kharkevich DA (1967) Main esters and amides of substituted cinnamic acids and their analogs (in Russian). Khim-Farm Zh 4:20–25

Arendaruk AP, Gracheva EA, Skoldinov AP, Kharkevich DA (1970) Synthesis and curare-like activity of mono-quaternary ammonium salts of alkaminic esters of cinnamic acid, its derivatives and analogs (in Russian). In: Kharkevich DA (ed) Novye kurarepodobnye i ganglioblokiruyushchie sredstva (New curare-like and ganglion-blocking agents). Meditsina, Moscow, pp 168–179

Arendaruk AP, Skoldinov AP, Smirnova NV, Soloviev VM, Kharkevich DA (1972 a) Quaternary salts of alkaminic esters of benzoic acid (in Russian). Khim-Farm Zh 1:15–18

Arendaruk AP, Skoldinov AP, Smirnova NV, Kharkevich DA (1972 b) Synthesis and pharmacological properties of bis-(*N*-methyl-*N*-adamantyl) aminoethyl ester of succinic acid, diadonium (in Russian). Khim-Farm Zh 3:8–11

Barlow RB, Zoller A (1964) Some effects of long chain polymethylene bis-onium salts on junctional transmission in peripheral nervous system. Br J Pharmacol 23:131–150

Bovet D (1951) Some aspects of the relationship between chemical constitution and curare-like activity. Ann NY Acad Sci 54:407–432

Bovet D (1959) Rapports entre constitution chimique et activité pharmacodynamique dans quelques séries du curarés de synthèse. In: Bovet D, Bovet-Nitti F, Marini-Bettòlo GB (eds) Curare and curare-like agents. Elsevier, Amsterdam, pp 252–287

Bowman WC (1980) Pharmacology of neuromuscular function. Wright, Bristol

Brücke F (1956) Dicholinesters of α,ω-dicarboxylic acids and related substances. Pharmacol Rev 8:265–335

Cheymol J (ed) (1972) Neuromuscular blocking and stimulating agents. International Encyclopaedia of pharmacology and therapy, vol 1–2, Section 14. Pergamon, Oxford
Danilov AF, Eremina ND, Kvitko IYa, Lavrentieva VV, Mikhelson MYa, Porai-Koshitz BA, Rozhkova EK, Shelkovnikov SA (1968) Ability to block neuromuscular transmission in the series of tere- and ortho-phthalic acid derivatives (in Russian). Farmakol Toksikol 4:412–418
Everett AJ, Lowe LA, Wilinson S (1970) Revision of the structures of (+)-tubocurarine chloride and (+)-chondrocurine. J Chem Soc D:1020–1021
Foldes FF, Foldes VM (1965) ω-Aminofatty acid esters of choline. Interaction with cholinesterases and neuromuscular activity in man. J Pharmacol Exp Ther 150:220–230
Ginsburg S, Kitz RJ, Savarese JJ (1971) Neuromuscular blocking activity of a new series of quaternary *N*-substituted choline esters. Br J Pharmacol 43:107–126
Hobbinger F (1976) Pharmacology of anticholinesterase drugs. In: Zaimis E (ed) Neuromuscular junction. Springer, Berlin Heidelberg New York, pp 487–581 (Handbook of experimental pharmacology, vol 42)
Hobbinger F, Peck AW (1970) The relationship between the level of cholinesterase in plasma and the action of suxamethonium in animals. Br J Pharmacol 40:775–789
Hughes R, Chapple DJ (1976) Effects of non-depolarizing neuromuscular blocking agents on peripheral autonomic mechanisms in cats. Br J Anaesth 48:59–68
Kharkevich DA (1965) On pharmacological properties of a new curare-like agent – anatruxonium (in Russian). Farmakol Toksikol 3:305–309 ((1966) Fed Proc 25:T521–T523)
Kharkevich DA (1966a) On the pharmacology of a new nondepolarizing neuromuscular blocking drug, cyclobutonium (in Russian). Farmakol Toksikol 1:47–53
Kharkevich DA (1966b) On the relationship between curare-like activity of the derivatives of cyclobutanedicarboxylic acids and the structure of the central part of their molecule (in Russian). Farmakol Toksikol 6:715–720
Kharkevich DA (1969) Pharmacology of curare-like drugs (in Russian). Meditsina, Moscow
Kharkevich DA (1970) On pharmacological properties of a new antidepolarizing neuromuscular blocking drug diadonium diiodide (in Russian). Farmakol Toksikol 5:531–536
Kharkevich DA (1973) Curare-like agents. In: Kharkevich DA (ed) Uspekhi v sozdanii novykh lekarstvennykh sredstv (Advances in drug design). Meditsina, Moscow, pp 138–187
Kharkevich DA (1974a) New curare-like agents. J Pharm Pharmacol 26:153–165
Kharkevich DA (1974b) On predominant action of a number of pharmacological agents on *m*-cholinoreceptors of the heart (in Russian). Farmakol Toksikol 9:94–102
Kharkevich DA (1975) Hydrophobic properties of the neuromuscular blocking agents. In: Klinge E (ed) Proceedings of the 6th international congress of pharmacology, Helsinki. vol 1, pp 33–47
Kharkevich DA (ed) (1983) Novye miorelaksanty (New neuromuscular blocking agents). Meditsina, Moscow
Kharkevich DA, Kravchuk LA (1961) On the pharmacology of a new curare-like agent truxilonium (in Russian). Farmakol Toksikol 3:318–324
Kharkevich DA, Kravchuk LA (1963) On some relations between chemical structure and curare-like activity in the series of bis-quaternary derivatives of cyclobutanedicarboxylic acids (in Russian). Farmakol Toksikol 6:702–707 (1964) Fed Proc 23:T1327–T1329)
Kharkevich DA, Shorr VA (1980) Cardiotropic antimuscarinic action of some curare-like agents. Arch Int Pharmacodyn Ther 248:238–250
Kharkevich DA, Skoldinov AP (1970) New acetylcholine antagonists (in Russian). Zh Vsesoyuzn khim ob-va im DI Mendeleeva 2:145–156
Kharkevich DA, Skoldinov AP (1971) On the effect of lipophilic radicals in the molecule of curare-like agents on the mechanism of their action (in Russian). Dokl Akad Nauk SSSR 198:985–988

Kharkevich DA, Skoldinov AP (1976) On the significance of hydrophobic radicals for the interaction of curare-like agents with acetylcholine receptors (in Russian). Zh Vsesoyuzn obshchestva im DI Mendeleeva 2:124–129
Kharkevich DA, Skoldinov AP (1980) On some principles of interaction of curare-like agents with acetylcholine receptors of skeletal muscles. J Pharm Pharmacol 32:733–739
Kharkevich DA, Arendaruk AP, Gracheva EA, Skoldinov AP (1967) On curare-like properties of mono-quaternary ammonium derivatives of cinnamic acid (in Russian). Farmakol Toksikol 5:562–567
Khromov-Borisov NV, Indenbom ML, Danilov AF (1968) Tetra-, nona-decamethylene-bis-pyridinium neuromuscular blocking agents. Distribution of π-electron density and relative curare-like activity. Doklady Akademii Nauk SSSR 183:134
Kitz RJ, Karris JH, Ginsburg S (1969) A study in vitro of new short-acting, nondepolarizing neuromuscular blocking agents. Biochem Pharmacol 18:871–881
Kravchuk LA (1966a) Therapeutic range and duration of curare-like effect of bisquaternary salts of dialkylaminoalkyl esters of truxillic acids (in Russian). Farmakol Toksikol 1:53–60
Kravchuk LA (1966b) Relationship between structure and duration of curare-like action in the series of truxillic acids' diesters (in Russian). Farmakol Toksikol 4:420–425
Kravchuk LA (1970) On the duration of curare-like action of bisquaternary ammonium derivatives of truxillic acids. In: Kharkevich DA (ed) Novye kurarepodobnye i ganglioblokiruyushchie sredstva (New curare-like and ganglion-blocking agents). Medgiz, Moscow, pp 35–41
Mashkovsky MD (1955) On curare-like properties of alkaloid thesine and its diiodomethylate (in Russian). Farmakol Toksikol 6:3–9
Mikhelson MYa, Zeimal EV (1970) Atsetilkholin. O molekulyarnom mekhanizme deistviya (Acetylcholine). Nauka, Leningrad
Paton WDM, Zaimis E (1949) The pharmacological action of polymethylene bistrimethylammonium salts. Br J Pharmacol 4:381–400
Samoilov DN (1971a) On anticholinesterase activity of bisquaternary ammonium derivatives of truxillic acids (in Russian). Farmakol Toksikol 2:149–155
Samoilov DN (1971b) Chemical structure and cholinolytic activity of bisquaternary ammonium derivatives of truxillic acids (in Russian). Farmakol Toksikol 4:413–420
Samoilov DN (1971c) O zavisimosti mezhdu khimicheskim stroeniem i kholinergicheskimi svoistvami bis-chetvertichnykh ammonievykh proizvodnykh truksillovykh kislot (On the relationship between chemical structure and cholinergic properties of bisquaternary ammonium derivatives of truxillic acids). Thesis, First Medical Institute, Moscow
Zaimis E (ed) (1976) Neuromuscular junction. Springer, Berlin Heidelberg New York (Handbook of experimental pharmacology, vol 42)
Zaimis E, Head S (1976) Depolarizing neuromuscular blocking drugs. In: Zaimis E (ed) Neuromuscular junction. Springer, Berlin Heidelberg New York, pp 365–419 (Handbook of experimental pharmacology, vol 42)

[illegible] (1970) On the significance of hydrophobic [illegible] for the interaction of [illegible] with acetylcholine receptors. In: [illegible]

[illegible] (1977) On some particular [illegible] agents with [illegible] receptors of skeletal [illegible]

[illegible]

[illegible]

CHAPTER 14

Quinuclidinium Compounds

M. D. MASHKOVSKY, L. N. YAKHONTOV, and V. V. CHURYUKANOV

A. Introduction

Quinuclidine (I) is a bicyclic compound with a tertiary bridgehead nitrogen atom. As regards the chemical structure, it is similar to quinolizidine (norlupinane) (II), pyrrolizidine (III), and to some other 1-azabicyclic compounds which have given rise to numerous pharmacologically active alkaloids and synthetic compounds. Quinuclidine might also be regarded as a closed bicyclic analog of the dialkylaminoalkanes (IV) and *N*-methylpiperidine (V). The dialkylaminoalkane and *N*-methylpiperidine group occur in numerous neurotropic and other pharmacologically active compounds.

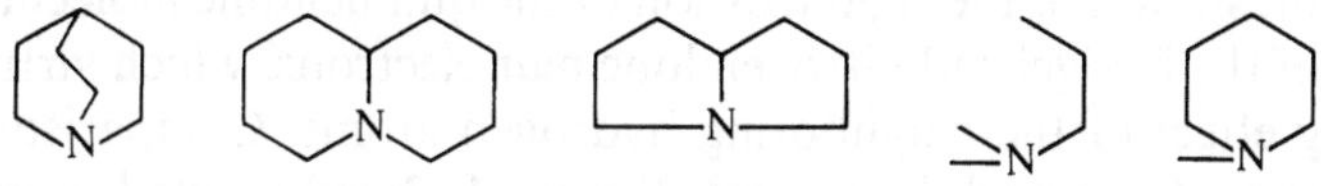

(I) (II) (III) (IV) (V)

The research carried out by Rubtsov, Yakhontov, and co-workers (RUBTSOV et al. 1960; YAKHONTOV 1969, 1977; YAKHONTOV et al. 1980) has made it possible to synthesize a great number of new compounds modeled on quinuclidine. Study of these compounds has led to the design of some pharmacologically active substances. Today, some of these compounds are applied as drugs. For example, a series of 3-hydroxyquinuclidine esters possess cholinergic or anticholinergic activity (MASHKOVSKY 1967). 3-Acetoxyquinuclidine hydrochloride (aceclidinum) selectively stimulates muscarinic receptors and is an effective means of decreasing the intraocular pressure in glaucoma and for the treatment of intestinal atonia (ZAITSEVA and MASHKOVSKY 1961; MASHKOVSKY and ZAITSEVA 1968). Hydrochlorides of diphenylpropionate (aprolidinum) and benzylate 3-hydroxyquinuclidine are powerful anticholinergic substances (MASHKOVSKY and ZAITSEVA 1967 a, b). 2,2,6,6-Tetramethylquinuclidine hydrobromide (temechinum) is a potent ganglion blocker (SHARAPOV 1972). 3-Benzoyloxyquinuclidine hydrochloride (oxylidinum) exhibits sedative, hypotensive, and antiarrhythmic effects (MASHKOVSKY and ZAITSEVA 1962; ZAITSEVA and MASHKOVSKY 1970). Quinuclidyl-3-diphenylcarbinol hydrochloride (phencarolum) is an antihistaminic drug (KAMINKA 1977).

Pharmacologic studies have shown that, as anticholinergic compounds, quinuclidine derivatives are more active than the respective derivatives of 2-diethylaminoethanol or *N*-methylpiperidine (MASHKOVSKY and ZAITSEVA 1967 a, b; ZAITSEVA et al. 1964).

Inasmuch as tubocurarine and many other active neuromuscular blocking drugs, including diplacinum (see Table 3, structure VIIId) (MASHKOVSKY and BRISKIN 1952), contain quaternary nitrogen atoms, it was of interest to examine the neuromuscular blocking activity of bisquaternary compounds containing quinuclidine rings with quaternary nitrogen atoms.

B. Some Chemical Peculiarities of Quinuclidine

Unlike aliphatic amines and *N*-substituted piperidines, quinuclidine is a 1-azabicyclooctane system that has a rigid structure. Each of the two six-membered rings of the quinuclidine molecule has a "boat" configuration, which is unusual for saturated six-membered heterocyclic compounds.

The features of the quinuclidine structure determine the presence in this class of derivatives of a number of characteristic chemical properties (RUBTSOV et al. 1960; YAKHONTOV 1969, 1977; YAKHONTOV et al. 1980). First of all, unlike aliphatic, monocyclic, and other 1-azabicyclic compounds, the conformational changes in quinuclidines induced by free rotation around the bond axes are restricted.

One manifestation of the rigid fixation of the quinuclidine molecule is the specific nature of the bridgehead nitrogen lone pair electrons, which virtually escape the shielding effect of the neighboring hydrogen atoms. Comparison of the infrared spectra of quinuclidines and those of *N*-substituted piperidines or piperazines has demonstrated that the quinuclidine compounds lack absorption at 2,700–2,800 cm^{-1} inherent in monocyclic products, determined by interaction of nitrogen lone pair electrons with neighboring axial C–H bonds (RUBTSOV et al. 1960). A high degree of lone pair electron deshileding at the quinuclidine nitrogen atom provides for a far greater ease of interaction with electrophilic agents as compared with analogous reactions of tertiary aliphatic or monocyclic amines.

The rigidity of the quinuclidine ring system governs not only the deshielding of the bridgehead nitrogen lone pair electrons, but also its definite steric trend along the symmetry axis of the molecule. This in turn leads to fhe fact that free nitrogen p electrons appear to be virtually orthogonal to the π electrons of the multiple bonds, which are semicyclic in position 2 or endocyclic in positions 2, 3, as well as to π electrons of the aromatic systems condensed with the quinuclidine ring in positions 2, 3. This effect excludes the possibility of overlapping p and π orbitals and, therefore, mesomeric interactions between them. As a result, in the quinuclidine derivatives, the effect of nitrogen on the neighboring aromatic or unsaturated groups appears to be purely inductive, without involvement of conjugation, which is not observable in analogous compounds.

The symmetric quinuclidine structure is marked by high stability. The respective quinuclidinium bases are more prone to elimination of the aliphatic alcohol molecule rather than to Hofmann degradation, the energetically favorable bicyclic quinuclidine system being preserved. These features of the quinuclidine molecule offer wide possibilities for a purposeful search for new biologically active compounds.

The rigid structure of the bicyclic system fixes the distances between the reaction centers in the molecule, excludes interaction between the centers to form, for example, intramolecular hydrogen bonds, and suppresses some tautomeric transformation that alter the reactivity toward biochemical receptors. Deshileding and the related high nucleophilicity of the bridgehead nitrogen of the quinuclidine molecules ensure their greater ease of access to the electrophilic centers of biochemical receptors, more stable interaction with them, and a stronger shielding of their reaction center in the complex thus formed.

To illustrate the successful use of the quinuclidine molecule for the synthesis of new biologically active substances, the compounds already mentioned (aceclidinum, aprolidinum, etc.) which are structurally related to acetylcholine may be cited as examples. Unlike acetylcholine and its aliphatic and monocyclic analogs which are capable of different conformational changes, the respective 3-acyloxyquinuclidines have a rigid structure with a fixed distance between the ester and basic centers. At the same time, the basicity and reactivity of the nitrogen atom in these quinuclidine compounds are appreciably higher than those in their aliphatic or monocyclic analogs.

Structural comparison of 3-hydroxyquinuclidine and choline, and higher reactivity of quinuclidine compounds as compared with their aliphatic, monocyclic, and other azabicyclic analogs, including processes at the biochemical level, have formed the basis for the design of highly effective drugs affecting cholinergic transmission in different parts of the nervous system, including neuromuscular blocking drugs.

C. Relationship Between the Structure and Neuromuscular Blocking Activity

In the search for new drugs affecting neuromuscular transmission symmetric bisquaternary quinuclidine compounds (Table 1, structure VI), in which two quinuclidine groups are connected by polymethylene chains, have been synthesized. Oxygen-containing analogs have also been obtained, in which one of the CH_2 groups of the polymethylene chain is replaced by oxygen (Table 2, structure VII), as well as a quinuclidine analog of diplacinum (Table 3, structure VIIIa).

Asymmetric bisquaternary compounds have been synthesized for comparison. *N*-methylpyrrolodinium, *N*-methylpiperidinium, or *N*-methylmorpholinium groups were introduced into these compounds (Table 4, structure IX) instead of one of the quinuclidine groups.

Neuromuscular blocking activity was studied in cats anesthetized with urethane. Contractions of the gastrochemius muscle were recorded in response to electrical stimulation of the sciatic nerve. Relaxation of the nictitating membrane served as an indicator of ganglion-blocking action. The compounds were injected intravenously. In unanesthetized rabbits, the intravenous dose evoking head drop was regarded as an indicator of neuromuscular blocking potency. AD_{50} (the intravenous dose which evokes apnea in 50% of animals) was determined in unanesthetized white mice.

Table 1. Neuromuscular blocking and ganglion-blocking activity of quinuclidinium derivatives

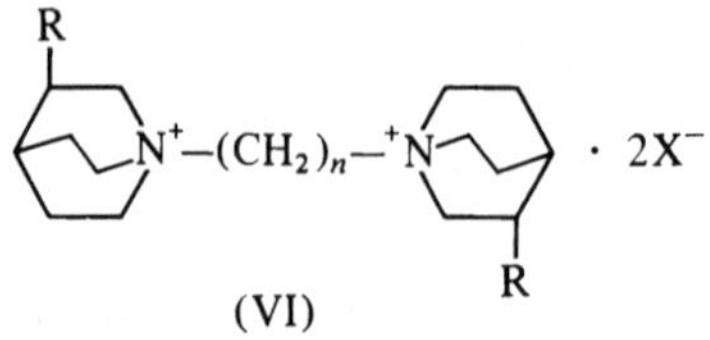

(VI)

Compound	n	R	X	Neuromuscular blocking action[a] (mg/kg i.v.)	Ganglion-blocking action (superior cervical ganglion[b] (mg/kg i.v.)	AD_{50} mice[c] (mg/kg i.v.)
VIa	3	H	I	No effect in doses under 10 mg/kg	1–1.25	28.2
VIb	4	H	I	5	1	16.26
VIc	5	H	I	5	1	16.31
VId	6	H	I	1	1	8.12
VIe	7	H	I	0.5	0.3–0.4	1.5
VIf	9	H	I	0.5	No effect in doses under 0.05 mg/kg	0.92
VIg	10	H	I	0.5	1.5	0.17
VIh	3	CH_3	I	Weak effect in doses under 15 mg/kg	3	32.5
VIi	4	CH_3	I	10	7–10	10.38
VIj	5	CH_3	I	5	3	11.75
VIk	6	CH_3	I	2	5	4.9
VIl	7	CH_3	I	0.75	0.5–0.75	1.65
VIm	9	CH_3	I	0.75	0.5–0.75	0.27
VIn	10	CH_3	I	0.75	0.5	0.21
VIo	4	C_6H_5	Br	2	2	2.1
VIp	5	C_6H_5	Br	2	2	1.6
VIq	6	C_6H_5	Br	2	0.75	1.0
VIr	7	C_6H_5	Br	1	0.5–0.6	0.41
VIs	10	C_6H_5	Br	1.5	0.5	0.38
VIt	4	$CH_2C_6H_5$	Br or Cl	1.5	No effect in doses under 1.5 mg/kg	6.37
VIu	6	$CH_2C_6H_5$	Br or Cl	0.25	Facilitates ganglionic transmission in doses 0.2–0.3	0.7
VIv	7	$CH_2C_6H_5$	Br or Cl	1–1.5	1.5	0.7
VIw	10	$CH_2C_6H_5$	Br or Cl	1	1.5	0.47
VIx	4	$COOCH_3$	I	20	Weak effect in a dose of 15 mg/kg	83.5

Table 1 (continued)

Compound	n	R	X	Neuromuscular blocking action[a] (mg/kg i.v.)	Ganglion-blocking action (superior cervical ganglion[b] (mg/kg i.v.)	AD_{50} mice[c] (mg/kg i.v.)
VIy	6	$COOCH_3$	10		Weak effect in a dose of 10 mg/kg	23.7
VIz	10	$COOCH_3$	I	2	No effect in doses under 20 mg/kg	1.32
Hexamethonium				40	0.2	
Decamethonium				0.05		

C_6H_5 = 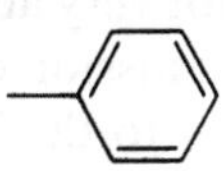

$CH_2C_6H_5$ = $-CH_2-$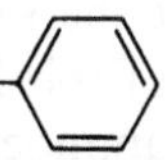

[a] The dose (intravenously) producing block of sciatic nerve–gastrocnemius muscle transmission (ED_{95}) (anesthetized cat)
[b] 50% decrease (ED_{50} i.v.) of nictitating membrane contractions during stimulation of the cervical sympathetic trunk (anesthetized cat)
[c] AD_{50} dose which evokes apnea in 50% of animals

Almost all the compounds of the general structure VI possess, to some degree, both neuromuscular blocking and ganglion-blocking activity (SADRITDINOV 1962a, b) (Table 1). Of the compounds belonging to this group having no substituents in the quinuclidine ring (R=H), only compound VIa (n=3) exhibits no neuromuscular blocking activity.

Neuromuscular blocking activity occurs in compound VIb (n=4). Further elongation (to n=9) of the polymethylene chain increases the blocking activity. In this group, substance VIf (n=9) is the most active. When given in doses of 0.03–0.4 mg/kg, it reduces the amplitude of muscle contraction by 50%. At a dose of 0.05 mg/kg, it completely blocks neuromuscular transmission for 20–30 min. As the number of methylene groups is raised to n=10, the neuromuscular blocking activity diminishes.

Compound VIi with methyl groups in position 3 of the quinuclidine ring and with four methylene groups in the polymethylene chain has a weak neuromuscular blocking action. As the polymethylene chain is made longer, the neuromuscular blocking activity rises, being most pronounced in compounds VI having 7, 9, and 10 methylene groups. At a dose of 0.75 mg/kg, these compounds completely block neuromuscular transmission.

As a result of introducing phenyl or benzyl radicals into position 3 of the quinuclidine rings, and if n is equal to 6–10, the activity of compounds VIq–s,

VIv–w decreases as compared with the analogs containing methyl groups (VIk–n). An exception to the rule is 1,6-bis(3,3′-benzylquinuclidyl-1,1′)-hexane dichloride (compound VIu) which has six carbon atoms in the chain and shows marked neuromuscular blocking activity. In doses of 0.1–0.15 mg/kg, it reduces the amplitude of muscle contractions by 50%. In doses of 0.2–0.3 mg/kg, it blocks the neuromuscular transmission for 5–10 min.

The mode of the neuromuscular block of these compounds was studied by determination of their interaction with neostigmine and diplacinum (nondepolarizing neuromuscular blocking drug). Neostigmine (0.15 mg/kg) and diplacinum were injected intravenously after blocking neuromuscular transmission. It has shown that compounds with substituents in the quinuclidine rings are nondepolarizing neuromuscular blocking agents; compounds having no substituents in the quinuclidine rings are depolarizing ones (SADRITDINOV 1962a, b).

Symmetric bisquaternary derivatives of quinuclidine of the general structure VII containing an oxygen atom in the polymethylene chain are not very active as neuromuscular blocking agents. They block neuromuscular transmission only in relatively high doses (7–50 mg/kg), noticeably yielding in activity to the known neuromuscular blocking drugs (SADRITDINOV 1962b). At the same time, these compounds show pronounced ganglion-blocking activity (Table 2).

As regards neuromuscular blocking activity, 1,3-bis(β-aminoethoxy)benzene-substituted quinuclidinium (VIIIa) was compared with similar compounds containing, instead of quinuclidinium groups, other structurally related quaternary aminoalkane groups (Table 3).

Table 2. Neuromuscular blocking and ganglion-blocking activity of quinuclidinium derivatives

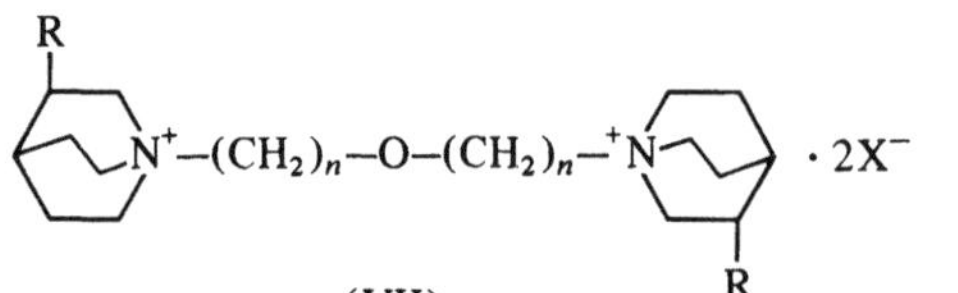

(VII)

Compound	n	R	X	Neuromuscular blocking action (mg/kg i.v.)	Ganglion-blocking action (superior cervical ganglion (mg/kg i.v.)	AD_{50} mice (mg/kg i.v.)
VIIa	1	H	I	20	0.3–0.4	23.75
VIIb	1	CH_3	I	15–20	3	25.25
VIIc	1	C_6H_5	Br	10	3	7.62
VIId	1	$CH_2C_6H_5$	Br	7	2	5.1
VIIe	2	H	I	15	1	16.25
VIIf	2	CH_3	I	6–10	1	11.3
VIIg	2	$COOCH_3$	I	50	3	56.5
Hexamethonium iodide				40	0.2	47

[a] See Table 1

Table 3. Neuromuscular blocking activity of compounds of the general structure

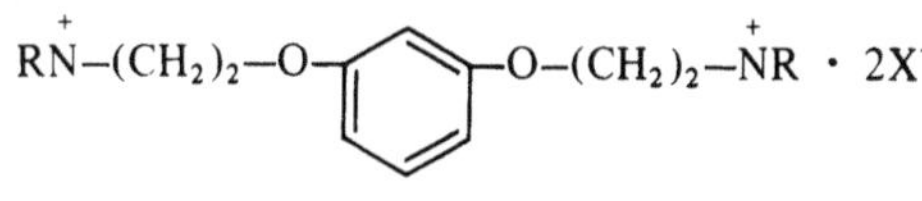

(VIII)

Compound	RN^+	X	Neuromuscular blocking activity (mg/kg i.v.)		AD_{50} mice (mg/kg i.v.)
			Sciatic nerve–gastrocnemius muscle block (ED_{50})	Head drop in rabbits (ED_{50})	
VIIIa	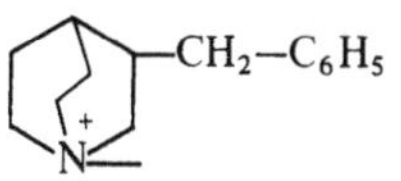	Br	5–6	0.53 (0.51–0.55)	1.55 (1.49–1.61)
VIIIb	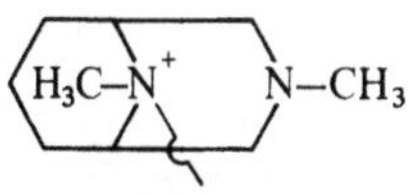	Br	2.5	1.25 (0.06–2.81)	1.4 (1.32–1.48)
VIIIc	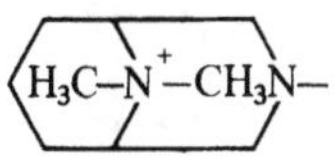	I	3–4	2.59 (2.37–2.81)	7.0 (6.52–7.48)
VIIId	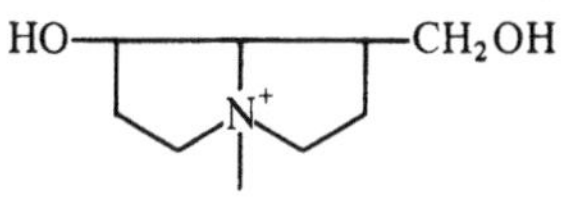	Cl	1.5–1.8	0.52	7.45

In this series, a compound containing platinecinium groups (VIIId), previously studied and recognized under the name of diplacinum (MASHKOVSKY and BRISKIN 1952), appeared to be the most active. The quinuclidine derivative was less active as evaluated from the effect on neuromuscular transmission in cats, but was identical to diplacinum according to head drop in rabbits.

Diazabicycloalkanium derivatives (VIIIb, c) appeared to be more active than quinuclidinium compounds as evaluated from the effect on neuromuscular transmission in cats and less active according to head drop (MASHKOVSKY and MEDVEDEV 1960). Compound VIIIa is a nondepolarizing neuromuscular blocking agent. Its effect is reversed by neostigmine.

Of the asymmetric compounds of the general structure IX (Table 4), compounds IXg with nine methylene groups in the interonium chain and the smallest A group (CH_2) is the most active from the viewpoint of neuromuscular blocking action. Similar compounds containing three methylene groups in the interonium chain have low activity. Thus, it has been shown that a number of bisquinuclidinium derivatives possess neuromuscular blocking activity.

The activity of compounds of the general structure VI (see Table 1) rises with increasing number of CH_2 groups in the interonium chain. The most active in this

Table 4. Neuromuscular blocking and ganglion-blocking activity of asymmetric bisquaternary compounds

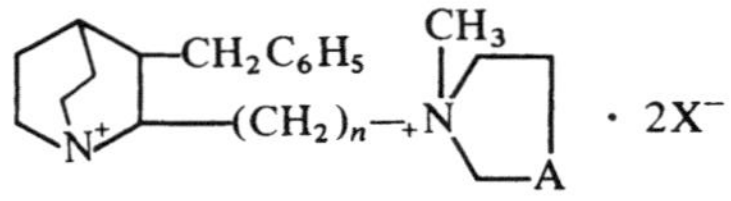

(IX)

Compound	A	n	X	Neuromuscular blocking activity (mg/kg i.v.)		Ganglion.blocking activity (mg/kg i.v.)		AD_{50} mice (mg/kg i.v.)
				Sciatic nerve–gastrocnemius muscle block (anesthetized cats)	Head drop in rabbits (ED_{50})	Doses reducing 50% of the nictitating membrane tonus to electrical stimulation of the cervical sympathetic nerve (anesthetized cats)	Doses inhibiting hypotension in response to vagus nerve stimulation	
IXa	CH_2CH_2	3	Br	25	12.7 ±2.5	0.073±0.008	0.2–0.5	34 ±2.5
IXb	CH_2	3	I	20	10.3 ±0.09	0.085±0.09	0.2–0.5	46.2 ±4.3
IXc	CH_2O	3	Cl	50	15	0.039±0.07	0.2	42 ±4.5
IXd	CH_2CH_2	7	Br	2	0.49±0.037	1	2	5.1 ±0.7
IXe	CH_2	7	Br	0.5	6.62±0.05	0.2 ±0.5	0.5–1	4.8 ±0.5
IXf	CH_2CH_2	9	Br	5	1.0 ±0.1	1.5	2	1.8 ±0.2
IXg	CH_2	9	I	0.08	0.75–0.07			0.94±0.1

series is compound VIf unsubstituted in the quinuclidinium rings and with $n=9$, as is compound VIu having benzyl substituents in the quinuclidinium rings and $n=6$.

Compounds containing nine or ten methylene groups are, as a rule, more active than lower homologs, which is similar to the high activity of decamethonium as compared with that exhibited by its lower homologs (PATON and ZAIMIS 1949). At the same time, in a subgroup where the quinuclidine rings contain benzyl groups, compound VIu, containing six methylene groups is the most active. Compounds that contain phenyl substituents in the quinuclidinium rings are less active than those containing benzyl radicals.

All the compounds of structure VII have low activity, which is likely to be a consequence of a relatively short interonium chain. It should be noted, however, that in this group, the most active is compound VIId which contains benzyl substituents in the quinuclidinium rings.

In a series of asymmetric compounds of structure IX, compounds in which A consists of one CH_2 group and with seven or nine CH_2 groups in the methylene chain appeared to be the most active. In contrast to neuromuscular blocking activity, powerful ganglion-blocking action is shown by compounds with a short polymethylene chain.

It is of interest to note that neuromuscular blocking action exhibited by this group of compounds (VIe–g) is potentiated by succinylcholine and attenuated by the nondepolarizing drug diplacinum. Thus, compounds VIe–g have a depolarizing action. At the same time, the action of compound VId is abolished by neostigmine and suxamethonium and potentiated by diplacinum, thereby attesting to the nondepolarizing blockade induced by this compound.

D. Pharmacology of Qualidilum

Of all quinuclidinium compounds studied, compound VIu (qualidilum) was selected for clinical trials thanks to its relatively high neuromuscular blocking activity, low toxicity, and accessibility to chemical synthesis (MASHKOVSKY and SADRITDINOV 1962).

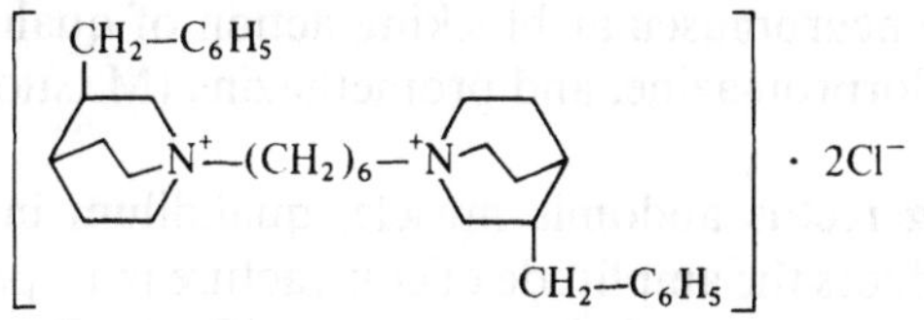

Qualidilum is a nondepolarizing neuromuscular blocking agent. It causes pronounced myorelaxation in different animal species (frogs, mice, rabbits, and cats).

In experiments on mice, qualidilum in a dose of 0.5 mg/kg i.v. produces no marked changes in the general status. Raising the dose to 0.75 mg/kg entails muscle relaxation and ataxia. After 7–10 min, the animals return to the initial state. Administration of the drug in doses of 1–1.25 mg/kg brings about a more prolonged muscle relaxation; some of the mice show clonicotonic convulsions

Table 5. Neuromuscular blocking activity of qualidilum, tubocurarine, and gallamine

Substance	Head drop in rabbits (mg/kg i.v.)	Sciatic nerve–gastrocnemius muscle block in anesthetized cats (mg/kg i.v.)
Qualidilum	0.062	0.2–0.3
Tubocurarine	0.15–0.16	0.3
Gallamine	0.2–0.25	0.85–0.9

ending in death. A dose of 2 mg/kg evokes apnea and is absolutely lethal. AD_{50} for qualidilum is 1.32 mg/kg, that for diplacinum 5.3 mg/kg.

In anesthetized cats, qualidilum reduces or completely blocks gastrocnemius muscle contractions induced by sciatic nerve stimulation. During direct muscle stimulation, administration of qualidilum does not change the contraction amplitude.

The blocking action of qualidilum occurs on intravenous injection of 0.1 mg/kg (MASHKOVSKY and SADRITDINOV 1962). The drug completely blocks neuromuscular transmission in doses of 0.2–0.3 mg/kg. Impairment of the transmission is recorded 20–30 s after the injection and lasts 5–10 min. Under artificial ventilation, the recovery of neuromuscular transmission supervenes 15–30 min after the drug injection.

The lethal dose of qualidilum for cats without artificial ventilation is 0.2–0.4 mg/kg i.v. Under artificial ventilation, the animals tolerate doses of 50–75 mg/kg. In rabbits, the head drop symptom occurs on administration of qualidilum at a dose of 0.05 mg/kg i.v.; $ED_{50}=0.062$ mg/kg (Table 5).

The impairment of neuromuscular transmission induced by qualidilum is reversed by neostigmine. The depolarizing neuromuscular blocking agents succinylcholine and decamethonium reduce the neuromuscular blocking action of qualidilum. On the contrary, nondepolarizing drugs (diplacinum) potentiate the effect of qualidilum. The neuromuscular blocking action of qualidilum is potentiated by barbiturates, chlorpromazine, and promethazine (MASHOVSKY and SADRITDINOV 1962).

In isolated frog rectus abdomis muscle, qualidilum in a concentration of 10^{-5}–10^{-4} g/ml reduces the amplitude of contracture in response to acetylcholine (10^{-6} g/ml) (MASHKOVSKY and SADRITDINOV 1962). Qualidilum also affects ganglionic transmission. In experiments on cats with electrical stimulation of the preganglionic cervical sympathetic trunk and recording of the nictitating membrane contractions, qualidilum exerts a biphasic effect: in doses of 0.3–0.5 mg/kg i.v. it facilitates ganglionic transmission, whereas raising the dose to 0.7–0.8 mg/kg considerably inhibits it. In doses exceeding twice the paralyzing dose (0.5–0.6 mg/kg) qualidilum partially blocks transmission through the parasympathetic ganglia of the heart during stimulation of the vagus nerve. Under artificial ventilation qualidilum in doses appreciably exceeding the paralyzing dose (2 mg/kg i.v.) produces no changes of the cardiovascular system of cats and rabbits, and does not affect the tone of the bronchi.

The drug exerts an antihistaminic action. In cats (0.2–0.3 mg/kg i.v.) it reduces the depressor effect of histamine. It also decreases (at concentrations of 10^{-7}–10^{-4} g/ml) the spasmogenic action of histamine on isolated ileum of guinea pig. Qualidilum (10^{-6}–10^{-5} g/ml) reduces the spasmogenic effect of acetylcholine (10^{-6} g/ml) on isolated ileum of rabbit. Clinical trials have shown qualidilum to be an effective nondepolarizing neuromuscular blocking agent (see Chap. 26).

E. Conclusions

Bisquaternary quinuclidinum compounds posses neuromuscular blocking and ganglion-blocking activity. The high stability of the quinuclidine system, resistance of quinuclidine compounds to conformational changes, and deshielding of the bridgehead nitrogen atom give rise to an easier and more stable interaction of bisquaternary quinuclidinium derivatives with acetylcholine receptors as compared with analogous compounds belonging to aliphatic, monocyclic, and other azabicyclic series.

Apparently, these factors also determine the high neuromuscular blocking activity of symmetric bisquaternary quinuclidinium salts as compared with asymmetric compounds of the same type, as well as a relatively higher activity exhibited by asymmetric substances containing a pyrrolidine group, less in volume and less prone to conformational changes, as compared with analogous piperidinium derivatives. As in other cases, the capacity to bind with acetylcholine receptors is affected by the changes in the magnitude of the valence angles on introduction of oxygen atoms into the chain containing two quaternary nitrogen atoms.

According to the reported data, the neuromuscular blocking activity of bisquaternary ammonium salts is governed not only by the distances between the quaternary nitrogen atoms, but also by the size of the substituents at quaternary atoms. As far as aliphatic or monocyclic compounds are concerned, substances with a decamethylene chain are the most active. Octamethylene compounds appear the most active among the respective tropinium and scopinium derivatives, while tetramethylene ones are the most active among 3,9-diazabicyclo(3,3,1)nonanium derivatives (MASHKOVSKY and MEDVEDEV 1960). Analysis of steric characteristics of the molecules suggests that the optimal length of the polymethylene chain for symmetric bisquaternary quinuclidinium compounds must be between the corresponding parameters for tropinium and 3,9-diazabicyclo(3,3,1)nonanium quaternary salts. Studies of bisquaternary quinuclidinium derivatives have resulted in the introduction into medical practice of qualidilum, 1,6-hexamethylenebis(3-benzylquinuclidine)dichloride.

References

Barlow RB, Ing HR (1948) Curare-like action of polymethylene bis-quaternary ammonium salts. Br J Pharmacol Chemother 3:298–304

Kaminka ME (1977) Phencarol – an antihistaminic agent of the quinuclidyl-carbinols group (in Russian). Farmakol Toksikol 2:158–162

Mashkovsky MD (1967) Synthetic quinuclidine derivatives are new group of drugs (in Russian). Khim-Farm Zh 3:3–8

Mashkovsky MD, Briskin AI (1952) Pharmacological properties of the new curare-like agent diplacinum (in Russian). Farmakol Toksikol 5:24–32

Mashkovsky MD, Medvedev VA (1960) On the pharmacology of the symmetrical bis-quaternary derivatives of 9-methyl-3,9-diazabicyclo-(3,3,1)-nonane (in Russian). Farmakol Toksikol 6:493–499

Mashkovsky MD, Sadritdinov F (1962) Curare-like properties of hexane-1,6-di(3′3′-benzylquinuclidine-1′1′)dichloride ("qualidil") (in Russian). Farmakol Toksikol 6:685–691

Mashkovsky MD, Zaitseva KA (1962) On the pharmacology of 3-hydroxybenzoyl-quinuclidine (in Russian). Farmakol Toksikol 1:32–37

Mashkovsky MD, Zaitseva KA (1967a) The influence of certain cholinolytics on experimental catatonia (in Russian). Biull Eksp Biol Med 8:54–56

Mashkovsky MD, Zaitseva KA (1967b) Comparative cholinolytic activity of amizyl (benactizine), aprophen and of corresponding to them quinuclidine esters (in Russian). Farmakol Toksikol 1:36–41

Mashkovsky MD, Zaitseva KA (1968) The cholinomimetic activity of 3-acetoxyquinuclidine (aceclidine). Arzneimittelforsch 18:320–322

Paton WDM, Zaimis EJ (1949) The pharmacological actions of polymethylene bistrimethylammonium salts. Br J Pharmacol Chemother 4:381–400

Rubtsov MV, Mikhlina EE, Yakhontov LN (1960) Chemistry of quinuclidine derivatives (in Russian). Usp Khim 29:74–105

Sadritdinov F (1962a) Effects of bis-quaternary ammonium quinuclidines on the transmission in sympathetic ganglia and neuromuscular transmission (in Russian). Farmakol Toksikol 3:327–335

Sadritdinov F (1962b) Concerning pharmacology of polymethylene-bis-quinuclidine halogenides (in Russian). Farmakol Toksikol 4:428–433

Sharapov IM (1972) On the pharmacology of temechine – a new ganglion blocking agent (in Russian). Farmakol Toksikol 6:687–690

Yakhontov LN (1969) Chemistry of quinuclidine (in Russian). Usp Khim 38:1038–1071

Yakhontov LN (1977) Recent advances in quinuclidine chemistry. Heterocycles 7:1033–1090

Yakhontov LN, Mashkovsky MD, Mikhlina EE (1980) Research of quinuclidine derivatives for new drugs search (in Russian). Khim-Farm Zh 12:23–32

Zaitseva KA, Mashkovsky MD (1961) Aceclidinum a new drug for treatment of postoperative intestine and bladder atonia (in Russian). Med Tekh 5:42–44

Zaitseva KA, Mashkovsky MD (1970) Antiarrhythmic activity of oxylidine (in Russian). Farmakol Toksikol 3:305–309

Zaitseva KA, Mashkovsky MD, Roshchina LF (1964) Effects produced by some tertiary and quaternary derivatives of quinuclidine and allied to it compounds upon cholinoreactive systems of the body (in Russian). Farmakol Toksikol 6:686–690

CHAPTER 15

Derivatives of Terphenyl

N. V. KHROMOV-BORISOV

A. Introduction

The relationship between neuromuscular blocking activity and the chemical structure of dications is usually investigatedin a single class of compounds exhibiting similar structural parameters. Generally, these parameters characterize the structure of the intercationic part of the molecule or the cationic groups. One can cite as examples compounds containing a polymethylene chain, esters of dicarboxylic (aliphatic and aromatic) acids, derivatives of cyclopentanophenanthrene (steroids), and truxillic acid, adamantyl, and quinuclidine derivatives.

When various substituents are introduced into the cationic groups and the intercationic part of the molecule, it is possible to determine the dependence of neuromuscular blocking activity on the structure of dications within a certain class. In this case, the structural parameters characteristic of a given class are either chosen arbitrarily or imitate the structure of natural compounds.

To make an appropriate choice of structural parameters of the intercationic part, one should study the mechanism of the effect of neuromuscular blocking agents at a molecular level. For this purpose, the type of action and the activity of dications which are very similar to each other in structure (homologous series) are studied. On the basis of this study, hypothetical schemes for nicotinic acetylcholine receptors of skeletal muscles can be developed and suggestions made concerning changes in the mutual disposition of active groups of a nicotinic acetylcholine receptor during its transition into the activated state.

This approach affords in principle the possibility of directed synthesis of neuromuscular blocking agents exhibiting the activity of the desired type without any secondary effects. In this approach, pharmacologic investigation of neuromuscular blocking agents obtained provides an experimental verification of the hypothesis advanced. As will be seen in this chapter, this approach has proved successful in the synthesis of tercuronium, a new neuromuscular blocking agent of high activity and nondepolarizing type of action.

B. The Molecular Complementarity of the Nicotinic Acetylcholine Receptors and Dicationic Neuromuscular Blocking Agents

The formation of the acetylcholine receptor–dication complex is ensured by the following noncovalent bonds: the mutual attraction of ions with opposite charges and dipoles with opposite (antiparallel) directions, and the hydrophobic interaction of nonpolar radicals (van der Waals' forces).

On the basis of the investigation of the activity and selectivity of action of a large number of dicationic neuromuscular blocking agents, a scheme for a tetrameric nicotinic acetylcholine receptor has been suggested (KHROMOV-BORISOV and MICHELSON 1966). According to this scheme, the dications in which the N–N distance (length of the intercationic chain) is 14 or 20 Å exhibit the highest activity. Moreover, for dications with N–N distance of 14 Å, the presence of dipolar groups in the intercationic part of the molecule does not promote the depolarizing activity (MICHELSON and ZEIMAL 1973).

For dications with N–N distance of 20 Å, the dipolar groups in the intercationic part of the molecule can profoundly affect the neuromuscular blocking activity. If the disposition of the dipole with respect to the cationic heads corresponds to that of the ester group in the acetylcholine molecule, then, as the result of complementarity with respect to the esterophilic site of the receptor, the dipole will favor an increase in neuromuscular blocking activity.

This increase is confirmed by comparison of the neuromuscular blocking activity of isomeric dications (I) and (II).

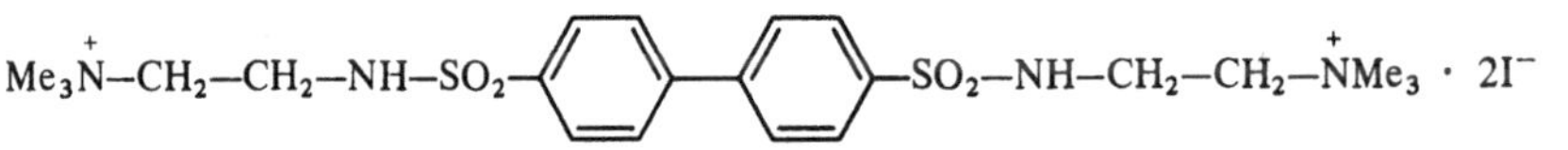

(I)

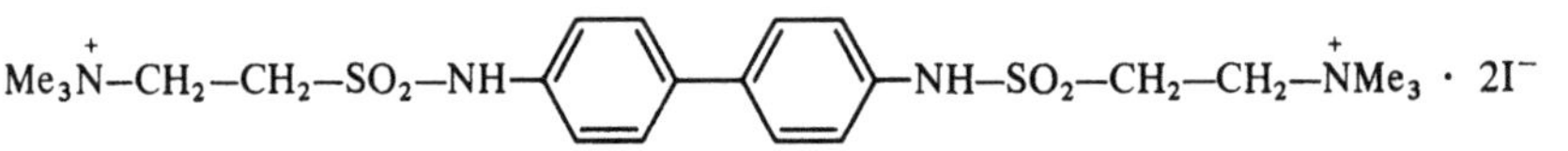

(II)

Dication (I) is a neuromuscular blocking agent of high potency (ED_{50} in cats 0.05 ± 0.01 μM/kg. When the SO_2–NH dipolar group is rotated by 180°, its dipole–dipole attraction (antiparallel dipoles) is replaced by mutual repulsion (parallel dipoles), and dication (II) almost completely loses its neuromuscular blocking activity (KHROMOV-BORISOV and INDENBOM 1966, 1967).

C. Conformational Properties of Dicationic Neuromuscular Blocking Agents

We have suggested a dynamic scheme for the tetrameric acetylcholine receptor of skeletal muscles (KHROMOV-BORISOV 1974, 1975, 1976, 1978) based on the assumption that the acetylcholine agonists cause a change in the conformation of the nicotinic acetylcholine receptor, whereas antagonists hinder this change (TRIGGLE 1965). This scheme takes into account the influence of the conformational flexibility of dications on the type of neuromuscular blocking action. The investigation of dications exhibiting different conformational flexibilities, other conditions being equal, shows that flexibility of dications favors the depolarizing type of action, whereas rigidity favors the nondepolarizing type. This scheme makes it possible to suggest the minimum degree of flexibility required for the dication to exhibit the depolarizing type of action: for a dication with N–N distance of 20 Å, a conformation with N–N distance of 12 Å should be possible, whereas for a dication with N–N distance of 14 Å, a conformation with N–N distance of 8.5 Å should be possible. If a dication is less flexible, it will hinder the conformational change in the acetylcholine receptor and will be a neuromuscular blocking agent of the nondepolarizing type.

In order to verify this concept experimentally, compounds (III) and (IV) ($n=2$ and 4) were synthesized and investigated.

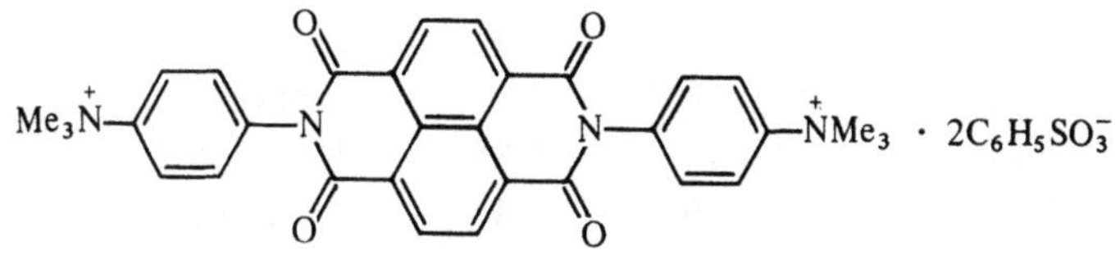

(III)

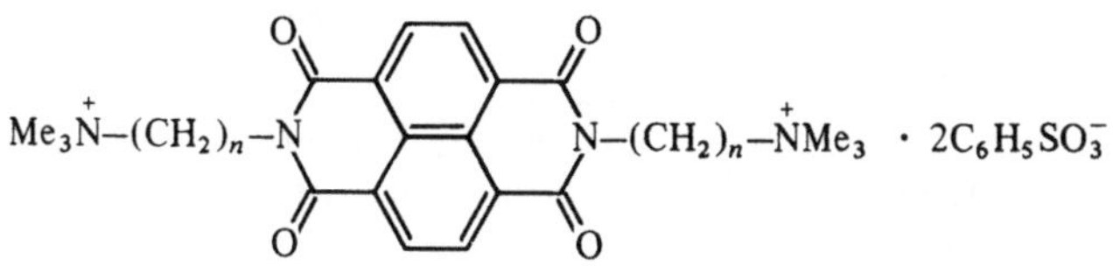

(IV)

Dications (III) and (IV) ($n=4$) are very similar in structure: both compounds are derivatives of the diimide of naphthalene-1,4,5,8-tetracarboxylic acid, they have identical cationic heads (N^+Me_3) and very similar N–N distances (~19 Å). Dication (III), exhibiting a rigid structure, was found to be a highly active neuromuscular blocking agent with the nondepolarizing type of action (Khromov-Borisov et al. 1971). Dication (IV) ($n=4$) exhibits sufficient conformational flexibility: in a contracted conformation its N–N distance is ~10 Å (i.e., less than 12 Å). It was found to be a neuromuscular blocking agent with the depolarizing type of action (Khromov-Borisov et al. 1973).

The most active of the presently known neuromuscular blocking agents with the nondepolarizing type of action exhibit a rigid molecular structure. They include toxiferine and pancuronium (Pavulon). Tubocurarine exhibits a lower activity and specificity. Its molecule is more or less flexible: in a contracted conformation, its N–N distance is 8.5 Å (Khromov-Borisov 1972).

D. The Structure of Cationic Heads

It is known that, when methyl radicals in the trimethylammonium group of acetylcholine (or other agonists) are gradually replaced by ethyl radicals, the depolarizing activity decreases. The data listed in Table 1 (Khromov-Borisov et al. 1975) show that the most pronounced decrease in the depolarizing activity (68-fold) and the change in the type of action take place for the transition from N^+Me_2Et to N^+MeEt_2 (Table 1, compounds 2 and 3). A similar correlation is observed for the analogous derivatives of diphenyl-p,p'-disulfonic acid – dications containing different cationic heads (Danilov et al. 1966).

The data listed in Table 2 confirm the suggestion that, in the series of bis-trialkylammonium dications with the depolarizing type of action, the most pronounced decrease in depolarizing activity (19-fold) and the transition from the de-

Table 1. Effect of the chemical structure of the cationic head on the depolarizing activity of derivatives of acetylcholine with the structure

$$CH_3{-}CO{-}O{-}CH_2{-}CH_2{-}\overset{+}{N}R_3 \cdot I^-$$

Compound	$\overset{+}{N}R_3$	D_2[a] (*M*)	α[b]	Type of action	Increase in D_2
1	$\overset{+}{N}Me_3$	1.3×10^{-7}	1.0	Agonist	
					×9
2	$\overset{+}{N}Me_2Et$	1.2×10^{-6}	1.0	Agonist	
					×68
3	$\overset{+}{N}MeEt_2$	8.2×10^{-5}	0.85	Mixed	
					×2.7
4	$\overset{+}{N}Et_3$	2.8×10^{-4}	0.6	Antagonist	
5	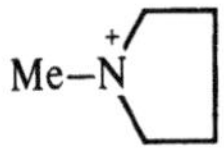	2.2×10^{-6}	1.0	Agonist	

[a] Dose needed to induce 50% of maximal response (a measure of affinity)
[b] Maximal response [a measure of intrinsic activity (ARIËNS, 1969)]; frog rectus abdominis muscle

Table 2. Activity and type of action of dications

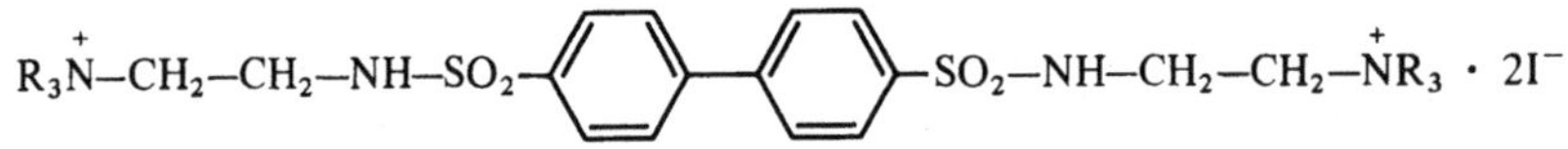

Compound	$\overset{+}{N}R_3$	ED_{50}[a] (m*M*/kg	Type of action	Increase in ED_{50}
1	$\overset{+}{N}Me_3$	0.05	Depolarizing	
				×3.6
2	$\overset{+}{N}Me_2Et$	0.18	Depolarizing	
				×19
3	$\overset{+}{N}MeEt_2$	3.4	Mixed	
				×2.6
4	$\overset{+}{N}Et_3$	8.8	Nondepolarizing	

[a] Cat tibialis anterior muscle

polarizing type of action to the mixed type of action also occurs on passing from N^+Me_2Et to N^+MeEt_2 (Table 2, compounds 2 and 3). This is probably due to the fact that when two ethyl radicals are present in the cationic head, an extended conformation

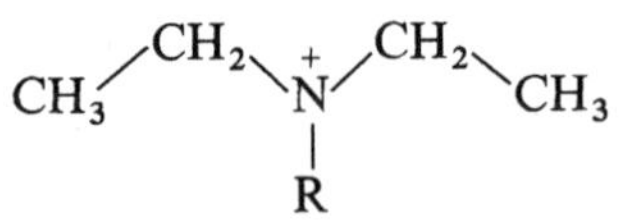

becomes possible. In such a conformation, the cationic head can interact with the hydrophobic radicals of the acetylcholine receptor located outside the active zone of the anionic center and this prevents the transition of the acetylcholine receptor to the activated state.

The mixed (partial depolarizing) type of action of cations containing an N^+MeEt_2 cationic group is probably due to the fact that in

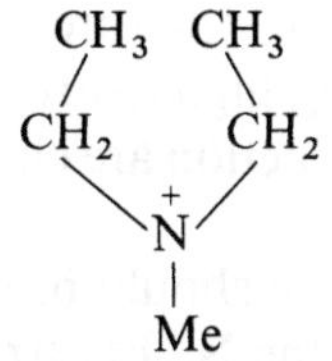

a contracted conformation, this cationic group interacts with the hydrophobic part of the acetylcholine receptor inside the active zone of the anionic center and therefore favors the transition of the acetylcholine receptor to the activated state. Hence, substances with the mixed type of action (partial agonists) containing N^+MeEt_2 cationic groups affect the acetylcholine receptor molecule as a mixture of two conformers with different types of action (agonists and antagonists).

It must be noted that, for the N^+Et_3 cationic group, the probability of existing in a contracted conformation is extremely low and therefore such groups induce nondepolarizing action. This assumption is consistent with the fact that, when the cationic group N^+MeEt_2 is converted to the pyrrolidine ring

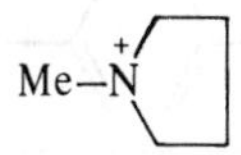

the depolarizing activity increases severalfold and a substance with a mixed type of action acts entirely as an agonist (see Table 1).

Dications with ethylpiperidine cationic heads

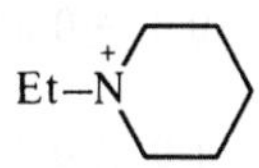

exhibit the nondepolarizing type of action. If the ethyl radical is incorporated into the rigid quinuclidine bicyclic structure

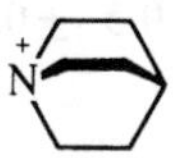

the dication becomes a depolarizing agent (Sadritdinov 1962). The transition from depolarizing to nondepolarizing agents is observed for mono- and dications containing trimethylammonium groups when one methyl radical is replaced by an adamanthyl radical (Kharkevich and Skoldinov 1971).

These relationships show that a decrease in the depolarizing activity of mono- and dications and the transition from depolarizing to nondepolarizing agents depends on the fact that, in some conformations, the molecules acquire the ability to form hydrophobic contacts outside the active zone of the anionic site of the acetylcholine receptor rather than depending on an increase in their molecular weight.

E. Directed Synthesis of Tercuronium

On the basis of the foregoing data, the following conclusions can be made concerning the relationship between the structure of dications and their activity and type of action. Dications exhibiting high neuromuscular blocking activity, selectivity, and nondepolarizing type of action are characterized by the following main structural parameters:

1. The N–N distance of the dication should be close to either 14–20 Å.
2. Cationic heads should contain the N^+Et_3 structure.
3. The intercationic part of the molecule should be rigid.

If the N–N distance for a dication is 14 Å, it is not necessary to incorporate dipolar groups in the intercationic part of the molecule.

On the basis of these requirements, the dibenzenesulfonate of *p′,p″*-bistriethylammonium-*p*-terphenyl, called tercuronium (Table 3, compound 5) was synthesized (KHROMOV-BORISOV et al. 1979). In order to confirm the correctness of these requirements, analogs of tercuronium containing various cationic heads

Table 3. Neuromuscular blocking activity of terphenyl derivatives (nondepolarizing action)

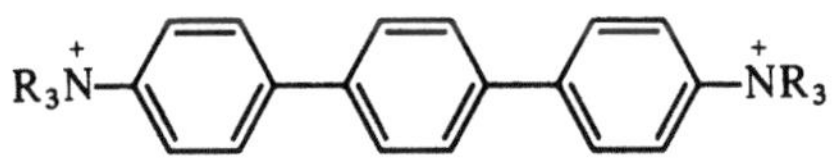

Compound	$\overset{+}{N}R_3$	Neuromuscular blocking doses	
		Cats (m*M*/kg)[a]	Rabbits HDD[b] (m*M*/kg)
1	$\overset{+}{N}Me$	0.3 ±0.01	0.25
2	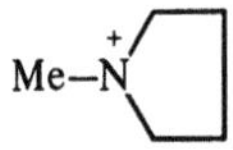	1.0 ±0.2	0.30
3	Me–$\overset{+}{N}Et_2$	0.5 ±0.09	0.20
4	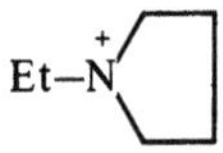	0.5 ±0.09	0.14
5	$Et_3\overset{+}{N}$ (tercuronium)	0.08 ± 0.002	0.022
6	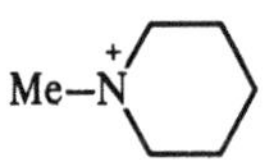	0.5 ±0.08	0.12
7	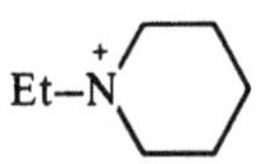	0.3 ±0.04	0.10
8	Me–$\overset{+}{N}Pr_2$	0.5 ±0.10	0.10

[a] Tibialis anterior muscle
[b] Head drop dose after rapid i.v. injection

Table 4. Neuromuscular blocking activity of biphenyl dications and terphenyl dications

Compound	Dication	N–N distance (Å)	Rabbits HDD[a] (mM/kg)
1	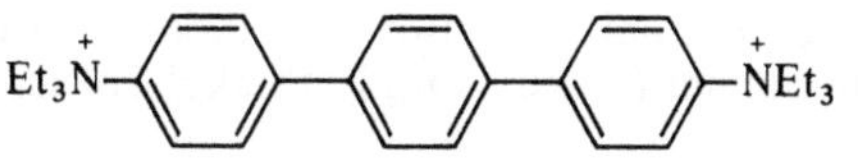	14.2	0.022
2	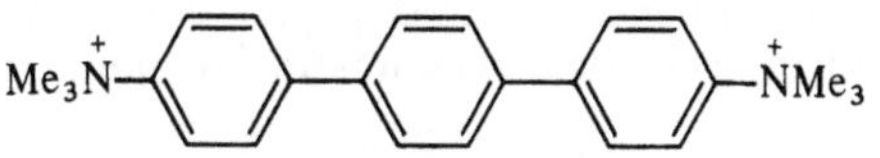	14.2	0.25
3	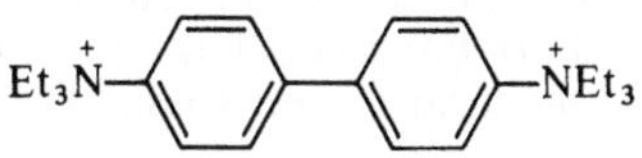	10.0	6.5
4	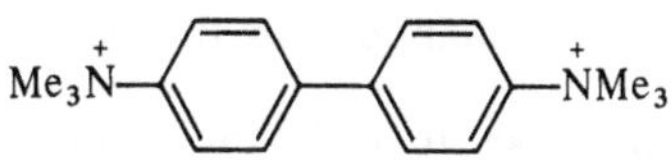	10.0	16.0

[a] See Table 3

have also been synthesized and investigated. The results of investigations are given in Table 3.

Investigations have shown (Table 4), that the decrease in the N–N distance (from 14.2 to 10.0 Å) in rigid dications with identical cationic heads leads to a drastic decrease in their neuromuscular blocking activity (KHROMOV-BORISOV et al. 1979; KHROMOV-BORISOV 1976).

F. Conclusions

The investigation of the effect of structure of dicationic neuromuscular blocking agents on their pharmacologic properties has made it possible to establish three main parameters determining their activity and type of action: the intercationic distance, the conformational flexibility of the molecule, and the dimensions of the hydrophobic environment of the cationic nitrogen atoms.

It should be pointed out that, in different species of animals, the nicotinic acetylcholine receptors of skeletal muscles are different. For example, the hydrophobic zone of the rabbit acetylcholine receptor is much larger than that of the cat acetylcholine receptor (KHROMOV-BORISOV et al. 1980).

Doubtless, apart from these structural parameters, the activity and type of action of dication are also affected by their physicochemical properties and, primarily, by their solubility in water and lipids (hydrophobic–hydrophilic balance). Extensive investigations in this field are in progress (KHARKEVICH and SKOLDINOV 1981; see also Chap. 4).

References

Ariëns EJ (1969) Théories sur les récepteurs et relations entre structure et activité dans le domain des substances cholinergiques et anticholinergiques. Actual Pharmacol 22:261

Danilov AF, Indenbom ML, Michelson MJ, Khromov-Borisov NV (1966) Curare-like activity of some new bis-quaternary compounds (in Russian). Farmakol Toksikol 5:582–587

Kharkevich DA, Skoldinov AP (1971) The influence of lipoidophylic radicals in the molecules of curare-like compounds on their mechanism of action (in Russian). Dokl Akad Nauk (DAN) SSSR 198, 4:985–988

Kharkevich DA, Skoldinov AP (1981) Curare-like activity of quaternary ammonium compounds containing 2-adamantyl radicals (in Russian). Farmakol Toksikol 6:670–672

Khromov-Borisov NV (1972) The structure of d-tubocurarine (in Russian). Farmakol Toksikol 35, 6:583–587

Khromov-Borisov NV (1974) Pharmacochemical investigation on the neuromuscular synapse. Ergeb Exp Med 17:117–125

Khromov-Borisov NV (1975) Acetylcholine and the structure of cholinoreceptors (in Russian). In: Serghejev PV (ed) Kratkii kurs molekularnoi pharmacologii (Short course of molecular pharmacology). Second Medical Institute, Moscow, pp 113–143

Khromov-Borisov NV (1976) The role of conformations in the interaction of biological molecules and receptors (in Russian). Khim-Pharm J 10, 11:5–15

Khromov-Borisov NV (1978) The conformation and flexibility of cholinergic molecules. In: Stoclet JC (ed) Advances in pharmacology and therapeutics, vol 3. Pergamon, Oxford, pp 271–279

Khromov-Borisov NV, Indenbom ML (1966) Derivatives of biphenyl-4,4′-disulfodiamides with two quaternary ammonium groups at a distance of 20 Å from each other (in Russian). J Org Khim 2:125–129

Khromov-Borisov NV, Indenbom ML (1967) 4,4′-bis-(p-dimethylaminoethyl)-biphenyl dimethiodid (in Russian). J Org Khim 3:1684–1686

Khromov-Borisov NV, Michelson MJ (1966) The mutual disposition of cholinoreceptors of locomotor muscles, and the change in their disposition in the course of evolution. Pharmacol Rev 18, 3:1051–1090

Khromov-Borisov NV, Indenbom MJ, Danilov AF (1971) Bis-quaternary ammonium curare-like compounds with rigid structure of the molecule (in Russian). Khim-Pharm J 9:3–7

Khromov-Borisov NV, Indenbom ML, Danilov AF, Jaaro NB (1973) Relationship between the type of action of the bis-quaternary ammonium myorelaxants and their conformational properties (in Russian). Khim-Pharm J 7, 6:5–7

Khromov-Borisov NV, Danilov AF, Indenbom ML, Tikhonova LN, Lavrentjeva VV, Starshinova LA (1975) The cholinomimetic activity of acetylcholine and sebacinyldicholine derivatives with different structure of the cationic groups (in Russian). Farmakol Toksikol 6:683–687

Khromov-Borisov NV, Torf SF, Cherepanova VP, Danilov AF (1979) Synthesis and curare-like activity of p′,p″-bis-quaternary ammonium derivatives of p-terphenyl (in Russian). Khim-Pharm J 7:34–39

Khromov-Borisov NV, Indenbom ML, Danilov AF (1980) Comparative pharmacochemical investigation of the cholinoreceptors of skeletal muscles in cats and rabbits (in Russian). Khim-Pharm J 2:15–20

Michelson MJ, Zeimal EV (1973) Acetylcholine. An approach to the molecular mechanism of action. Pergamon, Oxford, p 167

Sadritdinov F (1962) The influence of bis-quaternary quinuclidine compounds on the transmission of the excitation in the synaptic ganglia and on the neuro-muscular conductance (in Russian). Farmakol Toksikol 3:327–334

Triggle DJ (1965) Chemical aspects of the autonomic nervous system. Academic, New York, pp 70–71

CHAPTER 16

Delphinium Alkaloids

M. D. MASHKOVSKY and V. V. CHURYUKANOV

A. Introduction

Delphinium plants of the family Ranunculaceae are widely spread over the world. Some of the plants belonging to this genus grow in central Asia and other regions of the USSR. These plants are known to be poisonous; animals eating them show the effects of poisoning, the symptoms of muscle relaxation being predominant. As far back as the beginning of the nineteenth century, a crystalline base called delphinine ($C_{34}H_{47}O_9N$) was isolated from *Delphinium staphisagria* L. Later on, a number of alkaloids were obtained from different types of delphinium. All of them are tertiary bases.

It is known that the ability to induce muscle relaxation is characteristic for alkaloids that are quaternary ammonium bases, particularly for tubocurarine. However, there are also alkaloids that are tertiary amines which possess neuromuscular blocking activity. For example, the alkaloid thesine isolated from *Thesium minkwitzianum* L. Fedsch. of the Santalaceae family belongs to a group of symmetric bistertiary compounds displaying a neuromuscular blocking action. This alkaloid is a derivative of *l*-pseudoheliotridane (MASHKOVSKY 1943).

There are also monotertiary alkaloids that exhibit a neuromuscular blocking action. Of these, β-erythroidine (structure I), isolated from the seeds of *Erythrina americana* of the Erythrinae family is the most famous.

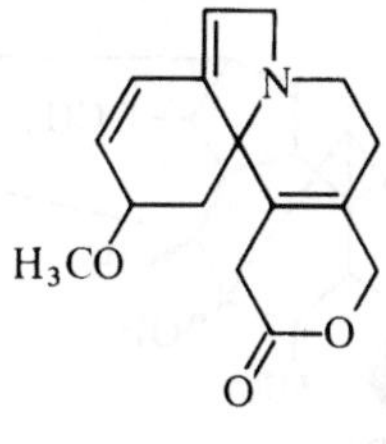

(I)

Quaternary ammonium bases penetrate biologic barriers poorly and are usually ineffective when administered orally. Therefore, tubocurarine and other quaternary neuromuscular blocking agents are applied intravenously, but neuromuscular blocking drugs administered orally are also of interest for medical practice. In this connection, research has been carried out in the USSR since the 1950s to examine the possibility of using alkaloids isolated from delphinium plants as oral neuromuscular blocking drugs.

Presently, 37 alkaloids have been isolated from different types of delphinium plants growing in the USSR (Table 1). Other alkaloids have also been isolated

Table 1. The most important alkaloids isolated from delphinium plants growing in the USSR

Plant	Alkaloid	Chemical composition
D. ajacis L.	Delcosine (delphamine)	$C_{24}H_{39}NO_7$
	Ajacine	$C_{34}H_{48}N_2O_9$
D. biternatum Huth.	Delphatine	$C_{38}H_{55}N_3O_{10}$
	Delbine	$C_{25}H_{41}NO_7$
	Delbiterine	$C_{25}H_{39}NO_7$
D. confusum M. Pop.	Condelphinum	$C_{25}H_{39}NO_6$
D. corumbosum Regel	Methyllicaconitine (delsemidine)	$C_{37}H_{50}N_2O_{10}$
	Delcorine	$C_{26}H_{41}NO_7$
	Deoxydelcorine	$C_{26}H_{41}NO_6$
D. dictyocarpum DC	Methyllicaconitine	$C_{37}H_{50}N_2O_{10}$
	Anthranoyllicoctonine	$C_{32}H_{46}N_2O_8$
	Eldelinum (deltalin)	$C_{27}H_{41}NO_8$
	Dictiocarpine	$C_{26}H_{39}NO_8$
	Delectine	$C_{31}H_{44}N_2O_8$
	Dictionine	$C_{33}H_{37}NO_{10}$
D. elatum L.	Elatinum	$C_{27}H_{41}NO_8$
	Eldelinum	$C_{27}H_{41}NO_8$
D. orientale J. G.	Delsoline	$C_{25}H_{41}NO_7$
D. rotundofolium	Methyllicaconitine	$C_{37}H_{50}N_2O_{10}$
	Brounine	$C_{25}H_{41}NO_7$
	Delseminum	$C_{25}H_{41}NO_7$
	Licoctoninum (delsinum)	
	Delcosine	$C_{24}H_{39}NO_7$

from the plants *D. flexuosum* Biel., *D. freynii* Conr., *D. pyramidatum* Alb., etc. (OREKHOV 1955; YUNUSOV 1981).

In chemical structure, delphinium alkaloids are similar to aconite alkaloids. The basis of their structure is the licoctoninum ring system (structure II).

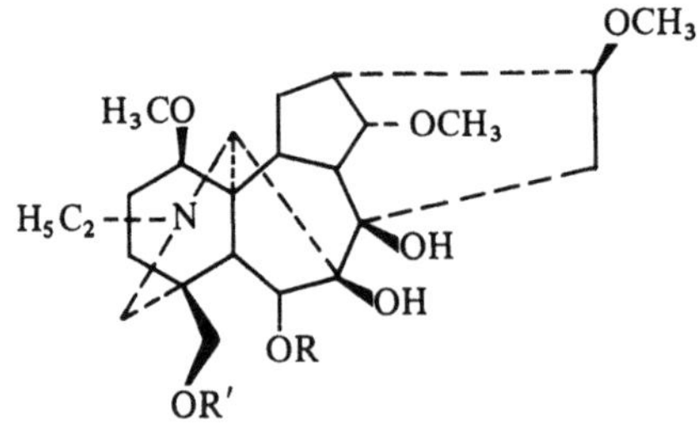

(II)

Different alkaloids belonging to this group differ in the substituents found at varyious positions on the licoctoninum skeleton (KUZOVKOV and BOCHARNIKOVA 1958; KUZOVKOV and PLATONOVA 1959; KUZOVKOV and PLATONOVA 1962; PELLETIER et al. 1967; PELLETIER and BHATTACHARYYA 1977). The following alkaloids have been studied pharmacologically: elatinum (structure III), eldelinum (deltalin, structure IV), methyllicaconitine (delsemidin, structure V), delseminum (structure VI), licoctoninum (delsinum, structure VII), condelphinum (structure VIII).

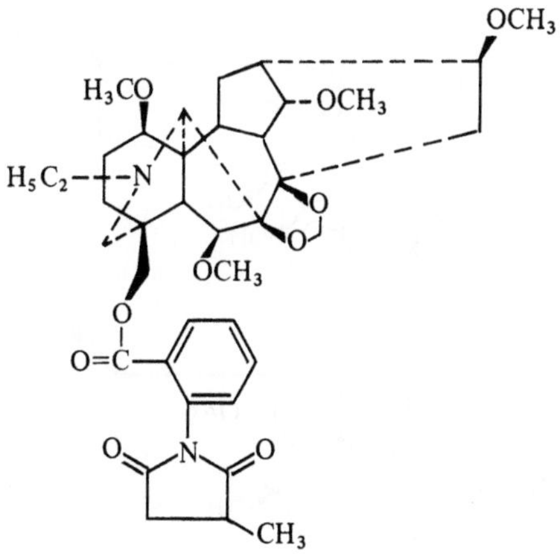
OCH3
H3CO
OCH3
H5C2
N
O
O
OCH3
O
O=C
O
N
O
CH3

(III)

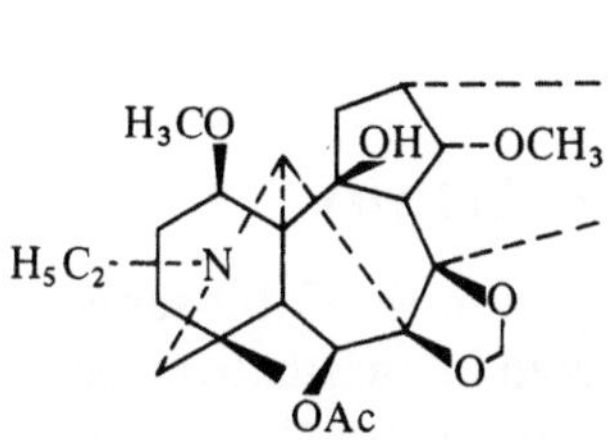
OCH3
H3CO
OH
OCH3
H5C2
N
O
O
OAc

(IV)

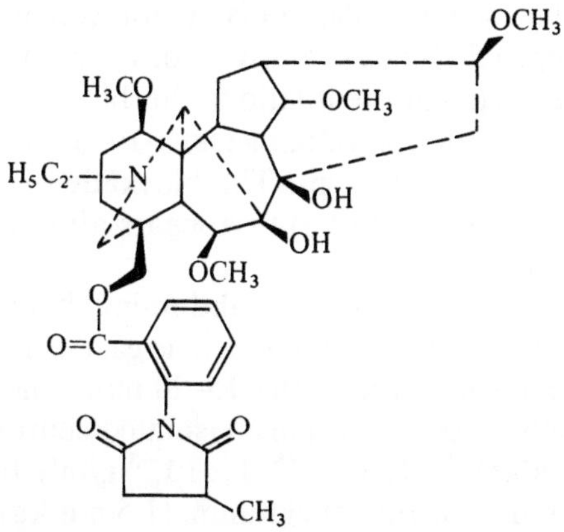
OCH3
H3CO
OCH3
H5C2
N
OH
OH
OCH3
O
O=C
O
N
O
CH3

(V)

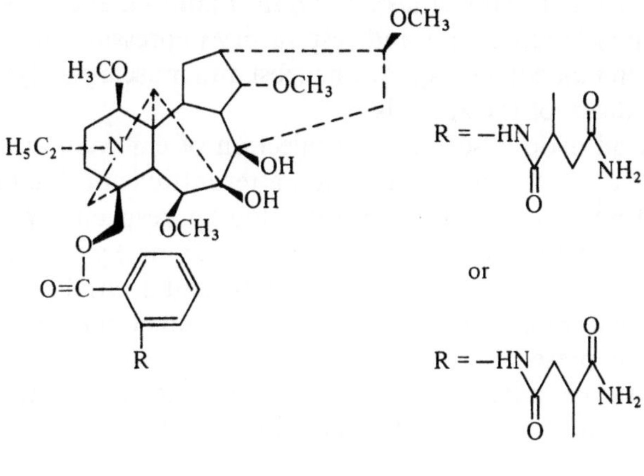
OCH3
H3CO
OCH3
H5C2
N
OH
OH
OCH3
O
O=C
R
R = —HN
O
NH2
O
or
R = —HN
O
NH2
O

(VI)

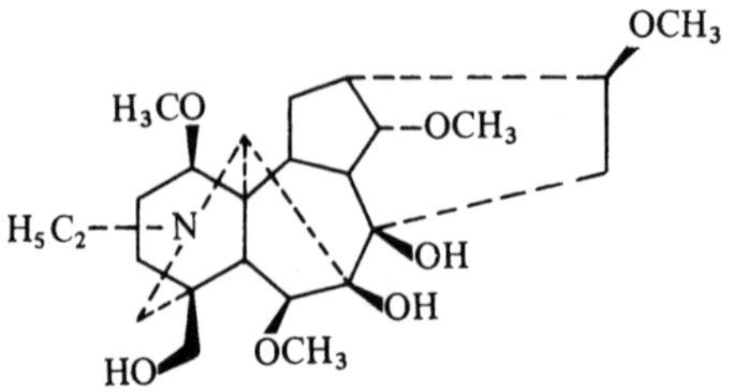

(VII)

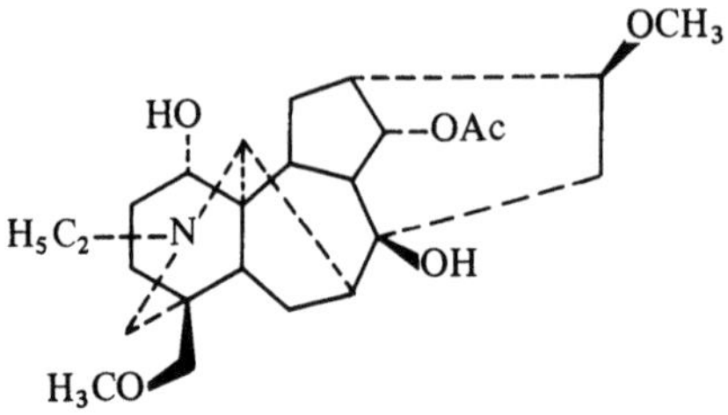

(VIII)

B. Neuromuscular Blocking Activity

The main pharmacologic property of most of this series of alkaloids is a curare-like action. It has been demonstrated in anesthetized cats (by recording gastrocnemicus muscle contractions during electrical stimulation of the sciatic nerve) that elatinum disturbs neuromuscular transmission starting with a dose of 0.3–0.5 mg/kg i.v. It is completely blocked at a dose of 1 mg/kg; respiration is arrested simultaneously. Under artificial ventilation, the blockade of transmission induced by 1 mg/kg lasts 30–40 min (DOZORTSEVA 1956). The drug is a nondepolarizing neuromuscular blocking agent. The blockade of neuromuscular transmission induced by elatinum (2 mg/kg) disappears after administration of neostigmine (0.15 mg/kg i.v.).

In frogs, injection of elatinum into the abdominal lymphatic sac at a dose of 3 mg/kg blocks neuromuscular transmission. The gastrocnemius muscle stops responding to sciatic nerve stimulation after 15–20 min, whereas direct muscle excitability is retained. It has been shown in an isolated neuromuscular preparation of the frog that this alkaloid (2×10^{-5}–1×10^{-5} g/ml) blocks neuromuscular transmission. Intravenous injection of elatinum (1.5 mg/kg) in mice evokes relaxation of limb muscles. The effect lasts 5–7 min. Elatinum at a dose of 2.5 mg/kg produces total muscle relaxation and respiratory depression, while doses of 5–5.5 mg/kg result in immediate respiration arrest and muscle paralysis followed by convulsions and death of the animals.

In rabbits, head drop is seen during injection of elatinum at a dose of 0.6–0.8 mg/kg i.v. Dogs show muscle relaxation after intravenous and subcutaneous injections of elatinum in doses 0.3–0.8 and 2 mg/kg, respectively. Elatinum also exhibits ganglion-blocking activity. At a dose of 1.5 mg/kg in anesthetized cats, it leads to relaxation of the nictitating membrane and arterial pressure decrease. The nictitating membrane contraction and pressor effect in response to epinephrine are completely preserved.

Elatinum is approximately half as active as tubocurarine with reference to the blocking effect on neuromuscular transmission in cats; 0.3–0.4 mg/kg elatinum and 0.15–0.18 mg/kg tubocurarine produced equal effects in cats. At the same time, respiratory arrest in rabbits was induced by 1.5–2 mg/kg elatinum and 0.2–0.25 mg/kg tubocurarine. Neuromuscular blocking effects can be also produced by methyllicaconitine, delseminum, and condelphinum. Methyllicaconitine is less active than elatinum. It impairs neuromuscular transmission in anesthetized cats at a dose of 2 mg/kg i.v. (DOZORTSEVA 1958, 1959).

Under artificial ventilation, the action of the alkaloid lasts 10–15 min. The drug is a nondepolarizing neuromuscular blocking agent. The blockade of neuromuscular transmission induced by methyllicaconitine (2–3.5 mg/kg i.v.) is completely reversed by neostigmine (0.15 mg/kg i.v.). Mice die following intravenous injection of methyllicaconitine at a dose of 3–3.5 mg/kg.

This alkaloid exhibits a neuromuscular blocking effect not only on intravenous injection, but also on oral and rectal administration. In dogs and cats, myorelaxation occurs after oral administration at doses of 40–50 mg/kg and after rectal administration (in the form of solution or suppositories) at doses of 20–25 mg/kg. For these administration routes, the blocking effect develops progressively. The maximum effect is attained after 1–1.5 h and lasts 5–6 h (under artificial ventilation). For these administration routes, neuromuscular block is also eliminated by neostigmine.

In cats, methyllicaconitine at doses of 1–2 mg/kg i.v. reduces arterial pressure. Methyllicaconitine has an inhibitory effect on autonomic ganglia (DOZORTSEVA 1959). Thus, its intravenous injection in cats at doses of 1–2 mg/kg decreases the tone, while doses of 4 mg/kg produce complete relaxation of the nictitating membrane during stimulation of the preganglionic symphathetic fibers. At a dose of 5 mg/kg, methyllicaconitine prevents changes in the heart rate in response to vagus nerve stimulation.

A similar action is also produced by delseminum and condelphinum. Delseminum at a dose of 3 mg/kg slightly inhibits neuromuscular transmission. A pronounced neuromuscular blocking effect is recorded after administration of 4 mg/kg. Condelphinum is less potent. These alkaloids are also nondepolarizing neuromuscular blocking agents. Their effect is abolished by neostigmine. Like methyllicaconitine, they are effective on oral administration, but the doses need to be higher (60–100 mg/kg). They also exert ganglion-blocking and hypotensive effects.

Eldelinum (structure IV) and delsinum (structure VII) differ essentially from elatinum and delseminum. Unlike elatinum, eldelinum, and delsinum produce no neuromuscular blocking action (DOZORTSEVA and MASHKOVSKY 1951).

The similarity in the action of elatinum, methyllicaconitine, and selseminum as well as that in their chemical structure, suggest that the neuromuscular blocking effect of the alkaloids is determined by the presence of the licoctoninum ring system in the molecule. It appears that neuromuscular blocking activity is most characteristic for licoctoninum esters esterified with *N*-acylated anthranilic acid (KUZOVKOV and BOCHARNICOVA 1958). That the neuromuscular blocking activity is a consequence of such an esterification of alkaloids is supported by the data on the lack of neuromuscular blocking activity in eldelinum and delsinum (licoctoninum), which differ from elatinum and delseminum in the absence of the ester group.

C. Conclusion

The neuromuscular blocking effect is the most characteristic pharmacologic feature of most of the delphinium alkaloids studied. The alkaloids are nondepolarizing agents. The alkaloid-induced neuromuscular blockade is abolished by neostigmine. The neuromuscular blocking effect of these alkaloids is determined by the presence in their structure of the licoctoninum ring system esterified with *N*-acylated anthranilic acid.

The pronounced neuromuscular blocking action exhibited by elatinum, methyllicaconitine, delseminum, and condelphinum, form the basis for clinical study of their efficacy. Elatinum, methyllicaconitine, and condelphinum have been applied orally to treat patients with spastic pareses, traumatic injurues of the brain accompanied by increased muscle tone, postencephalic parkinsonism, spinal arachnoiditis, and other pyramidal and extrapyramidal diseases (CHLENOV and SERKOVA 1957; KOBELIANSKAYA 1959; SERCOVA 1956). Application of the drugs reduced muscle spasticity, thereby attenuating pains and improving the general status. Delseminum has been used in anesthesia. Intravenous injection of the drug produced muscle relaxation similar to that induced by tubocurarine.

Today, after the introduction into medical practice of more effective drugs for muscle relaxation, the majority of these alkaloids are not applied. Of all the delphinium alkaloids, methyllocaconitine (mellictinum) is the only one presently in use.

References

Chlenov LC, Serkova MP (1957) Drug treatment of spastic hemipareses (in Russian). Klin Med 35:32–37

Dozortseva PM (1956) On the pharmacology of alkaloids from Delfinium elatum (elatinum and eldelinum) (in Russian). Farmakol Toksikol 3:42–48

Dozortseva PM (1958) Mellictinum is the new drug for treatment of spastic pareses (in Russian). Med Tekh II:54

Dozortseva PM (1959) The pharmacology of alkaloid methyllicaconitine (mellictine) (in Russian). Farmakol Toksikol 1:34–38

Dozortseva PM, Mashkovsky MD (1951) On the comparative pharmacology of alkaloids delsemine and delsin (in Russian). Farmakol Toksikol 4:49–54

Kobelyanskaya LG (1959) Administration of mellictine in a neurological clinic in spastic pareses (in Russian). Farmakol Toksikol 1:38–42

Kuzovkov AD, Bocharnikova AV (1958) Research of aconite alkaloids. X. Elatin. The structure of esterifying acid (in Russian). Zh Obshch Khim 28:556–558

Kuzovkov AD, Platonova TF (1959) The study of aconite alkaloids. XV. On the structure of elatinum, methyllicaconitine, ajacine, delseminum, avadkharidine, licaconitine, and eldelinum (in Russian) Zh Obshch Khim 29:2782–2786

Kuzovkov AD, Platonova TF (1962) Research of aconite alkaloids. XIX. On the structure of eldelin and delfelin (in Russian). Zh Obshch Khim 32:1290–1293

Mashkovsky MD (1943) On the pharmacology of alkaloid thesin (in Russian). Farmakol Toksikol 1:25–32

Orekhov AP (1955) Chimia alkaloidow (Chemistry of alkaloids) Academy of Sciences, Moscow

Pelletier SW, Bhattacharyya G (1977) The nature of "delsemine" from Delphinium semibarbatum (Dwarf Larkspur). Tetrahedron Lett 32:2735–2736

Pelletier SW, Keith IH, Parthasarathy PS (1967) The structures of condelphine, isotalatizidine and talatizidine. J Am Chem Soc 89:4146–4157
Serkova MP (1956) Application of elatin in neurological clinic for spastic pareses treatment (in Russian). Farmakol Toksikol 3:48–51
Yunusov SY (1981) Alkaloidy (Alkaloids) Fun, Tashkent

Preclinical Pharmacology of New Neuromuscular Blocking Drugs

CHAPTER 17

Bisquaternary Steroid Derivatives

K. BIRÓ, E. KÁRPÁTI, and L. SZPORNY

A. Introduction

The aminosteroids have yielded many interesting compounds with potential therapeutic applications (BUCKETT 1972), not least of which have been the steroidal neuromuscular blocking agents reviewed in detail by BUCKETT (1975). From this group of compounds, we were interested in bisquaternary compounds.

A naturally occurring bisquaternary ammonium salt, malouetin (JANOT et al. 1960) was first studied pharmacologically by QUEVAUVILLER and LAINÉ (1960). Compounds derived from the naturally occurring alkaloid paravallarin (LE MEN et al. 1963) have recently been prepared from the natural product, and their activities have been investigated. Synthetic efforts based on the androstane nucleus led to dipyrandium chloride (BIGGS et al. 1964), which was expected to be of short duration. Similar synthetic work on conessine yielded the putative short-acting drug, dimethylconessine from a large series of compounds (BUSFIELD et al. 1968). All these neuromuscular blocking agents were the result of either isolation of natural products and semisynthetic derivatives thereof or the result of using the steroidal nucleus purely as a supporting framework upon which to place active centers such as the quaternary ammonium group. A different concept, the incorporation of acetylcholine fragments directly into the steroidal skeleton, was realized in pancuronium bromide (BUCKETT and BONTA 1966; LEWIS et al. 1967). Further development of neuromuscular blocking drugs, possessing a rapid onset and short duration resulted in the synthesis of dacuronium bromide (BUCKETT and SAXENA 1969; FELDMAN and TYRELL 1970) and Org-6368 (SUGRUE and DUFF 1973).

From these drugs emerged a clinically useful agent, pancuronium bromide, which has superseded the older muscle relaxants. However, pancuronium can produce cardiovascular stimulation in both animals (BUCKETT 1968; SMITH et al. 1970) and humans (KELMAN and KENNEDY 1971; COLEMAN et al. 1972; SPEIGHT and AVERY 1972). According to VERCRUYSSE et al. (1979), the fact that pancuronium acts as a muscarinic antagonist at both pre- and postjunctional sites would explain both the "vagolytic" and "indirect sympathomimetic" properties attributed to this agent.

New attempts to find an ideal neuromuscular blocking agent resulted some novel potent compounds in the azasteroid series. The first of them was HS-342 (SINGH et al. 1972). Chandonium was a newer substance in this series (SINGH and PAUL 1974) and the newest, RGH-4201 (BIRÓ and KÁRPÁTI 1981) is a potent neuromuscular blocking drug in animals and humans. Pipecuronium bromide is another new compound with an androstane neucleus. It differs chemically from pancuronium mainly by the presence of two piperazine groups in place of piperidine groups. Its pharmacologic properties have been described by KÁRPÁTI and BIRÓ (1980).

B. Chandonium and Its Analogs

I. HS-342 and HS-467

SINGH et al. (1972) initiated a program of synthesis of bisonium steroids as potential neuromuscular blocking agents, with one or both cationic heads present as part of the steroid ring skeleton at different interonium distances. They first developed the 4,17*a*-diaza-D-homosteroid system and prepared (SINGH et al. 1972, 1973) 4,17*a*-dimethyl-4,17*a*-diaza-D-homo-5α-androstane dimethiodide (HS-342), which was a further example of the incorporation of quaternary nitrogen atoms within a D-homo-5α-androstane structure.

The compound (Fig. 1) has been the subject of extensive pharmacologic investigation (MARSHALL et al. 1972, 1973a, b). HS-342, with an interonium distance of 8.17 Å falls only just outside the range of values exhibited by tubocurarine (PALMER et al. 1980), and possesses both neuromuscular and ganglion-blocking activities. In anesthetized cats, HS-342 was equiactive with tubocurarine in depressing indirectly elicited twitches of tibialis anterior muscle (290 μg/kg i.v.). The onset and duration of action were considerably shorter than those of tubocurarine. The drug was not investigated clinically.

Continuing the work toward the synthesis of different bisonium azasteroids of potential neuromuscular blocking activity, SINGH and PAUL (1974) prepared two more potent compounds, chandonium iodide (HS-310) and HS-467. HS-467 was only slightly less active than HS-342 as a nondepolarizing neuromuscular blocking drug and vagolytic agent, but possessed considerably less ganglion-blocking activity than HS-342.

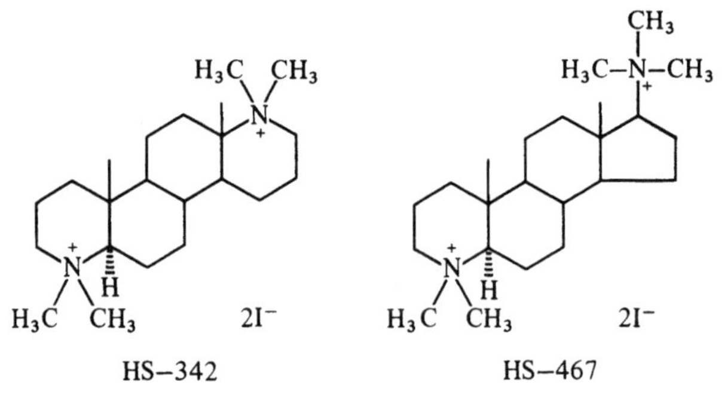

Fig. 1. Structures of HS-342 and HS-467

II. Chandonium Iodide

In animals, chandonium (Fig. 2) is a potent neuromuscular blocking drug with rapid onset and intermediate duration, but it has relatively powerful vagal blocking activity and blocks norepinephrine uptake (MARSHALL and OJEWOLE 1979). In accordance with this, its neuromuscular blocking action in humans is accompanied by pronounced tachycardia (BOWMAN 1980).

According to GANDIHA et al. (1975), the mean dose of chandonium necessary to produce 50% blockade of cat soleus muscle activity was 66 μg/kg. The corresponding dose for tibialis anterior was 70 μg/kg. Chandonium was slightly more potent on the slowly contracting soleus. Chandonium was 4–5 times more potent than the prototype compound of the series HS-342, 14–18 times more potent than AH 8165, and 2.6–4 times more potent than stercuronium. The onset of action of these compounds was closely similar (about 76 s to peak effect at 50% block). The time from the end of injection to full recovery from 50% blockade – the duration of action – was 3.25 min for soleus, and 2.75 min for tibialis. The durations of action of HS-342, AH 8165, and stercuronium were similar to chandonium.

The anticholinesterase activity of chandonium was weak in relation to its neuromuscular blocking potency, especially against mammalian muscle enzyme. The specificity of chandonium was investigated using guinea pig isolated ileum and vas deferens preparations ((HARVEY et al. 1976). The pA_2 values of this compound at muscarinic receptors were 5.7 against acetylcholine and 5.8 against carbachol. The lack of difference in the pA_2 values against acetylcholine and the cholinesterase-stable agonist, carbachol, suggests that inhibition of acetylcholinesterase by chandonium did not influence its blocking action.

In the nonatropinized cat, neuromuscular blocking doses of chandonium produced a slight tachycardia accompanied by a short-lasting rise in blood pressure. At lower doses, chandonium blocked the negative chronotropic effect of vagal stimulation and the bradycardia produced by acetyl-β-methylcholine.

Neuromuscular blocking doses had no ganglion-blocking effect. Increasing the dose to 1 mg/kg, produced only a 6% reduction in the responses of the preganglionically stimulated nictitating membrane. Chandonium, at doses up to 1 mg/kg, had no measurable effect on transpulmonary pressure or airflow rate, which indicates a lack of histamine-releasing effect.

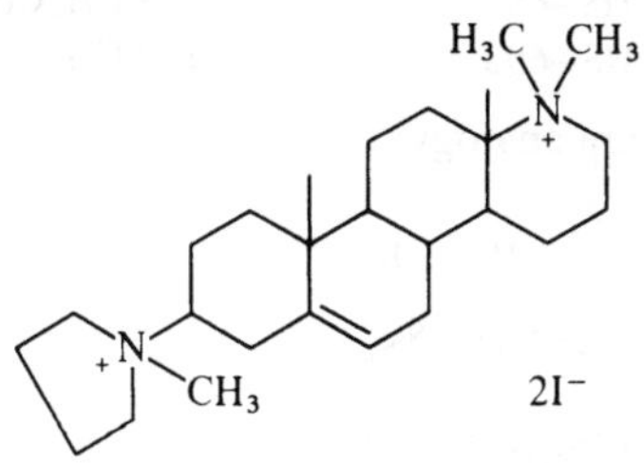

Fig. 2. Structure of chandonium iodide

III. New Derivatives

Four novel analogs (Fig. 3) of chandonium were tested for neuromuscular and autonomic blocking activities in anesthetized cats by TEERAPONG et al. (1977). The order of potencies relative to chandonium (1.0) was HS-692 (0.5) > HS-693 (0.25) > HS-704 (0.1) = HS-705, characteristics such as onset and duration being indistinguishable from those of chandonium. All compounds were antagonized by neostigmine. At neuromuscular blocking doses, none of the compounds exhibited any ganglion-blocking activity and no marked changes in blood pressure or heart rate were observed, but these compounds depressed the response of the heart rate to vagal stimulation: HS-693 (86%) > HS-705 (55%) = HS-704 (55%) > HS-692 (28%) > chandonium (9%). HS-692, the dihydro derivative of HS-310, possessed the most desirable spectrum of activity (SINGH et al. 1979 a; TEERAPONG et al. 1979). Since HS-692, the saturated congener of HS-310 would be a far more rigid molecule than HS-310, this result directly supports an activity–flexibility–unsaturation hypothesis (PALMER et al. 1980).

Using HS-310 as a prototype, new chemical modifications have been made (Fig. 4; SINGH et al. 1979 b). Chandonium iodide analogs containing an acetylcholine-like moiety (HS-627 and HS-626) in anesthetized cats had twice the neuromuscular blocking activity of chandonium and produced less vagolytic action than chandonium at neuromuscular blocking doses.

None of the analogs of chandonium so far synthesized is a sufficient improvement over the original compound to be worthy of further development. However, chandonium has such worthwhile properties that it is to be hoped that the structure–activity knowledge gained from it will shoon lead to a similar compound that is free from cardiovascular side effects (BOWMAN 1981).

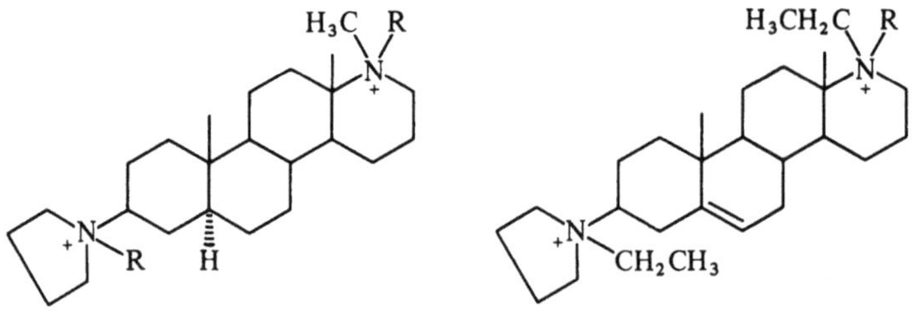

R = CH_3 = HS–692
R = CH_2CH_3 = HS–693

R = CH_2CH_3 = HS–704
R = CH_3 = HS–705

Fig. 3. Structures of chandonium analogs

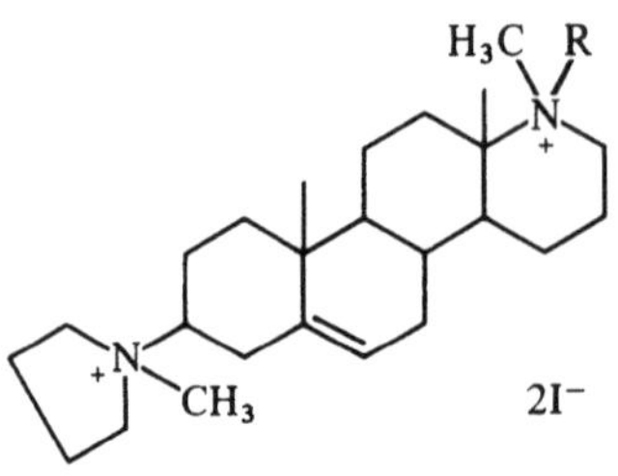

R = CH_2CH_2OH = HS–626
R = $CH_2CH_2OCOCH_3$ = HS–627

Fig. 4. Structures of new chandonium analogs

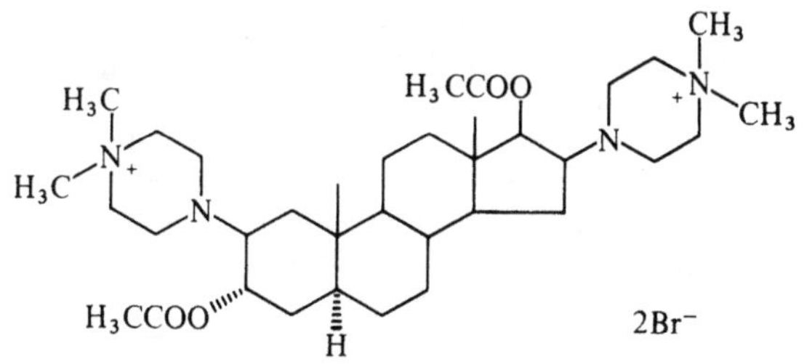

Fig. 5. Structure of pipecuronium bromide

C. Pipecuronium Bromide

I. Introduction

Pipecuronium bromide (Arduan) has been recently introduced into clinical anesthesia and has come to be used as a neuromuscular blocking agent in the Soviet Union and Hungary; it is also being introduced in several other countries.

Pipecuronium bromide (2β,16β-bis(4′-dimethyl-1′-piperazino)-3α,17β-diacetoxy-5α-androstane dibromide; Fig. 5) differs pharmacologically both quantitatively and qualitatively from pancuronium bromide. Kárpáti and Biró (1980) first described the pharmacologic properties of pipecuronium in detail and showed that the agent was about twice as potent as pancuronium. It has no effect on the cardiovascular system and is free from histamine-releasing activity.

II. Action on the Neuromuscular Junction

Neuromuscular blocking action was examined in isolated nerve–muscle preparations and in vivo experiments. Pipecuronium in a concentration of 2.5 μg/ml completely inhibited diaphragm contractions in rats and 0.08 μg/ml caused 80% inhibition of chick biventer cervicis muscle contractions. In both cases, the blockade developed rapidly and, after washout, muscular activity rapidly returned to its original level. On the basis of relative potencies, pipecuronium was 1.7–3 times as effective as pancuronium.

As compared with pancuronium, greater potency of pipecuronium was seen in tibialis and soleus muscles of anesthetized cats, rabbits, and dogs. Pipecuronium was less potent on soleus muscle than on tibialis muscle both in cats and rabbits. With respect to duration of action, pipecuronium acts 1.2–2.1 times longer than pancuronium (Table 1).

In conscious mongrel dogs, the dose causing total relaxation was found to be 40 μg/kg for pipecuronium bromide and 80 μg/kg for pancuronium. Head drop reaction for pipecuronium was observed within the 20–30 s, cessation of respiration at 2 min, its reappearance after 36 min, and standing on four legs (indicating the end of total paralysis) was only observed after more then 60 min. In dogs under pentobarbital anesthesia, 5.6 μg/kg pipecuronium and 16 μg/kg pancuronium produced 90% twitch block on the tibialis muscle. The onset time for pipecuronium was comparable with head drop time in unanesthetized dogs, but the

Table 1. Comparison of the neuromuscular blocking action of pipecuronium and pancuronium in anesthetized cats, rabbits, and dogs

Preparation	Pipecuronium		Pancuronium	
	ED_{50} (µg/kg)	Duration of action (min)	ED_{50} µg/kg)	Duration of action (min)
Cat tibialis	2.0 (1.7–2.4)	24	7.9 (7.1–8.8)	11
Cat soleus	3.1 (2.9–3.4)	23	9.5 (8.8–10.4)	15
Rabbit tibialis	2.7 (2.5–2.8)	16.5	5.8 (5.3–6.4)	14
Dog tibialis	3.7 (3.3–4.2)	16	12 (11–13)	8.5

Table 2. Comparison of the time course of neuromuscular block produced by pipecuronium and pancuronium in dogs

Compound	Preparation	Dose (µg/kg)	Block (%)	Onset (s)	Duration of action (min)
Pipecuronium	Tibialis	5.6	91±5	27±3	26±2.3
	Conscious	40	100	134±21[a]	36±3.7[b]
Pancuronium	Tibialis	16	88±6	19±2.4	11±1.5
	Conscious	80	100	55±4.8	23±1.0

[a] Respiratory failure
[b] Spontaneous respiration

duration of action was considerably shorter in the case of pentobarbital anesthesia (Table 2).

The neuromuscular effects of 20 µg/kg bolus injection and 3.3 µg/kg repetitive doses of pipecuronium are summarized in Table 3. The maintenance dose was one-sixth of the initial dose. Time between injection and return of twitch to 25% of control did not increase after the repetitive doses. Cumulation could not be observed and the slope of recovery after the last dose was fast.

The mechanism of action of pipecuronium was examined in anesthetized cats (Kárpáti and Biró 1980). In experiments where both nerve and muscle action potentials were recorded simultaneously, during complete muscle paralysis, potentials along the nerve were unaffected and direct muscular stimulation caused normal contractions. These results define the action of pipecuronium at the neuromuscular junction. Evidence for a postjunctional action of pipecuronium was consequent on reduction of the sensitivity of the muscle to close-arterial injections of acetylcholine and the fact that blockade due to pipecuronium could be reversed by neostigmine. Suxamethonium and other depolarizing blockers also antagonized the neuromuscular blockade. Under complete blockade, tetanus did not occur after the tetanic stimulation of peroneal nerve or, depending on the frequency of stimulus, occasional contractions occurred. In the case of optimal rhythm of sciatic nerve stimulation (5 Hz), pipecuronium (10–15 µg/kg) elicited the phe-

Table 3. The neuromuscular effect of first and maintenance doses of pipecuronium on anesthetized cat tibialis anterior muscle preparation

First dose (20 μg/kg)			Subsequent maintenance doses (3.3 μg/kg)					Recovery index after last dose (min)
Maximal block (%)	Time to maximal block (min)	Time to 25% recovery[a] (min)	Time to 25% recovery (min) 1st	2nd	3rd	4th	5th	
95±2.4[b]	8.2±1.0	28.1±3.5	11.9+2.3	12.7±2.3	12.5±2.4	12.3±2.1	12.6±2.4	12.1±3.1

[a] Time between administration of dose and recovery of twitch to 25% of control
[b] Mean ± standard error

nomenon of pessimum inhibition. The increase of stimulation frequency to 20 Hz enhanced the pessimal reaction. Under the action of pipecuronium, posttetanic facilitation of synaptic transmission in the gastrocnemius (ALYAUTDIN et al. 1980) and tibialis muscle (KÁRPÁTI and BIRÓ 1980) was observed. All these results proved a greater affinity of pipecuronium for muscle receptors, i.e., postjunctional predominance, a competitive antiacetylcholine effect.

III. Other Pharmacologic Actions

1. Effect on Cardiovascular and Central Nervous Systems

The dose of pipecuronium bromide causing total muscle relaxation did not influence blood pressure, heart rate, on ECG. A dose of 1–2 mg/kg produced a transient biphasic change in blood pressure. On increasing the dose up to 1 g/kg, permanent hypotension and bradycardia appeared in cats. No effect of pipecuronium bromide was noticed upon coronary blood flow and myocardial oxygen consumption at doses of 15–60 μg/kg on anesthetized cats (ALYAUTDIN et al. 1980). Systemic arterial pressure and heart rate showed practically no change either.

The conclusion from work on anesthetized, open-chest dogs (KÁRPÁTI and BIRÓ 1980) is that pipecuronium did not produce significant cardiovascular effects even at 100 times the neuromuscular blocking doses. According to PULAY et al. (1980), studies with pipecuronium were performed using six different widely used methods of anesthesia in dogs. Pipecuronium was administered after the steady state of general anesthesia had been achieved. Heart rate decreased during the induction of anesthesia and it decreased further after the administration of either the first or repeated doses.

The possible effect of pipecuronium (20–100 μg/kg) upon different parts of the CNS was investigated by the method of evoked potentials (ALYAUTDIN 1978; ALYAUTDIN et al. 1979, 1980). On decerebrated cats, pipecuronium at 1–5 times the myoparalytic dose did not affect the monosynaptic and polysynaptic potentials of the spinal cord, or the latency period of evoked potentials. The compound did not affect the jaw-opening reflex and the amplitude of primary responses in the somatosensory zone of the cortex.

Table 4. Inhibitory effects of pipecuronium on the responses of cat tibialis anterior muscle to peroneal nerve stimulation, of the heart rate to vagal stimulation, and of the nictitating membrane to preganglionic nerve stimulation

ED_{50} (mg/kg)			Relative potency	
Neuromuscular block *M*	Vagal block *V*	Ganglion block *G*	*V*/*M*	*G*/*M*
0.006 (0.003–0.013)	1.8 (1.1–3.1)	>10	300	>1,700

2. Ganglion-Blocking and Atropine-Like Activity

In anesthetized cats, pipecuronium did not decrease nictitating membrane contractions evoked by preganglionic stimulation of cervical sympathetic nerve, either in the effective neuromuscular blocking dose or 500 times that dose. A marked ganglionic effect could be observed only at doses greater than 10 mg/kg. Doses of pipecuronium higher than those required to produce neuromuscular block blocked the negative chronotropic effect of stimulation of the vagus nerve (Table 4). A 50% block occurred only after a dose of 1.8 mg/kg, which is 300 times higher than the muscle relaxant dose.

3. Lack of Histamine Release

Large doses of pipecuronium bromide showed no bronchoconstriction in guinea pigs or hypotension in ganglion-blocked cats (Kárpáti and Biró 1980). No increase of histamine concentration in human serum (J. Leibinger et al. 1979, unpublished work) occurred after exposure to pipecuronium.

4. Cholinesterase-Inhibiting Activity

Most curare-like agents belong to the group of onium salts. The presence of several cationic centers provides more stable binding to the nicotine-like acetylcholine (ACh) receptor and a correspondingly higher curare-like activity (Kharkevich and Skoldinov 1980). Not only the receptor, but the inactivating enzyme itself contains ACh binding sites. This fact explains why most agents with nicotinic ACh receptor blocking activity are acetylcholinesterase (AChE) inhibitors as well. The AChE inhibitory effect of pipecuronium was studied (Simon et al. 1980) in human red blood cell AChE and on human plasma cholinesterase (ChE) enzymes. In this system, pipecuronium inhibited the activity of both the human red blood cell AChE, $pI_{50} = 3.99$, and the plasma ChE, $pI_{50} = 4.33$. Pipecuronium showed a reversible mixed inhibitory effect both on AChE and ChE. Simon and co-workers suggested that the duration of the muscle relaxant effect of this agent is not due to its anti-AChE activity. Alyautdin et al. (1980) found that pipecuronium in concentrations of 10^{-7}–10^{-5} mol/l did not affect the activity of rat brain AChE. However, a certain inhibitory action of the agent was observed in respect to pseudocholinesterase of normal horse serum. In concentrations as high as

10^{-5} mol/l, pipecuronium inhibited the activity of the latter by more than 50% which is of no practical consequence.

5. Drug Interactions

The effects of anesthetic agents are complex, involving both stimulant and depressant actions, not only on the transmission process, but also on the contractile mechanism. An additional type of interaction may arise if the anesthetic agent modifies distribution or elimination of the blocking agent (BOWMAN 1980). In general, inhalation anesthetics potentiate the nondepolarizing neuromuscular blockers (MILLER 1975; HUGHES 1970; KATZ and GISSEN 1967). Intravenous anesthetics depress transmission by mechanisms similar to those by which inhalation anesthetics act. Of 18 anesthetics or other drugs used in surgical operations, ketamine and pentazocine considerably increased the toxicity of pipecuronium bromide in mice. It was moderately increased by atropine and propanidid (KÁRPÁTI and BIRÓ 1980).

PULAY et al. (1980) used six different methods of anesthesia, and for the introduction of combined intratracheal anesthesia, thiobutabarbital, methohexital, hydroxydione, and ketamine were given to dogs. In the thiobutabarbital group, the mean duration of apnea was prolonged twofold, which indicates that interaction occurs between the two drugs.

IV. Distribution, Excretion, and Metabolism

The disposition of pipecuronium was studied using ^{14}C-labeled compound in rats (VERECZKEY and SZPORNY 1980). The radioactive substance is rapidly taken up by the liver and to a lesser extent by the kidneys. The compound is retained by the liver for a longer period of time, about 6% of the administered ^{14}C activity can be recovered from the liver 21 days after the i.v. administration of the drug. Only traces of radioactivity could be detected in the brain. Binding to rat plasma protein is about 30%. The compound seems to be eliminated with urine, the amount excreted being 38% and 45% in 24 and 48 h, respectively. No radioactivity was detected in the expired air. Pipecuronium was given to pregnant rats under artificial respiration at a dose causing total paralysis; in fetuses removed at various times (0.5–4.0 h), less than 0.1% of the injected dose of pipecuronium could be detected. The pharmacokinetics of pipecuronium was studied by a two-compartment analysis of plasma disappearance curves following a single i.v. injection of ^{14}C-labeled compound in rats (BODROGI et al. 1980). The biologic half-life of the compound averaged 41 min. the V_1 and V_2 values around 130 ml revealed virtual inner and outer distribution volumes. The small rate constants (mean values of 0.02 for K_1 and 0.05 for K_2) showed a poor binding to plasma proteins and a rapid transfer of the drug between the two compartments. A metabolic clearance pattern of 9.6 ml/min was observed. The rate of hepatic elimination of the compound and/or its metabolites reached 6% after 4 h. Thin layer chromatographic separation and tentative identification of metabolites showed the free dihydroxy metabolite and suggested that a monoacetate was present in plasma and bile. The metabolites have no neuromuscular blocking potency.

D. RGH-4201

I. Introduction

The need for a short-acting nondepolarizing neuromuscular blocking agent to replace suxamethonium is recognized widely. Approaches to this objective have been made using the steroidal neuromuscular blocking agents, but at best the duration of action has been twice that desired and the action has usually been coupled with cardiovascular and other side effects which generally occur on administration of higher i.v. doses of aminosteroid. These derivatives have usually been less potent than the longer-acting agents such as pipecuronium.

The neuromuscular blocking and other pharmacologic properties of a new bisquaternary steroid 3α-pyrrolidino-17β-methyl-17α-aza-D-homo-5α-androstane-dimethobromide, RGH-4201, Duador (Fig. 6) were studied in detail in conscious and anesthetized dogs, anesthetized cats and rats, and in isolated organs, in comparison with some short-acting relaxant agents. RGH-4201 exhibited a nondepolarizing neuromuscular block that was considerably faster in onset and shorter in duration of action than the reference substances suxamethonium, chandonium, diadonium (Biró and Kárpáti 1981), vecuronium, pancuronium (W. C. Bowman and J. Houston 1981, unpublished work), vecuronium, pancuronium, atracurium, and tubocurarine (F. F. Foldes 1981, unpublished work). RGH-4201 blocked cardiac muscarinic receptors like all other short-term nondepolarizing blocking drugs.

II. Comparative Neuromuscular Blocking Effects

Results of the comparative studies on neuromuscular blocking dose in conscious mongrel dogs indicated that complete relaxing doses were equal at 100 μg/kg i.v. for RGH-4201, suxamethonium, and chandonium (Table 5). Note that the duration of action of RGH-4201 was considerably shorter than that of suxamethonium, chandonium, and diadonium.

The ED_{50} values for RGH-4201 were greater than those of suxamethonium and chandonium in anesthetized cats. There was no significant difference between ED_{50} obtained in "fast" and "slow" contracting muscles (Table 6) although the onset of action was relatively more rapid in the tibialis and the duration of action was shorter in the soleus muscle. According to W. C. Bowman and J. Houston (1981, unpublished work), RGH-4201 was more rapid in onset and shorter in du-

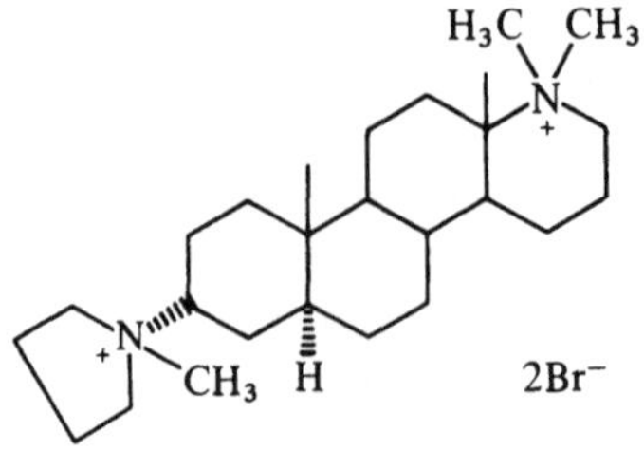

Fig. 6. Structure of RHG-4201

Table 5. Comparison of short-term neuromuscular blocking agents in conscious mongrel dogs. Average ± standard error of mean. Statistical analysis was performed by Student's *t*-test. Each compound was compared with RGH-4201. * = $P<0.05$; ** = $P<0.01$; *** = $P<0.001$

Compound	Complete relaxing dose (μg/kg)	Head drop (s)	Appearance of spontaneous respiration (min)	Head posture regained (min)	End of paralysis (min)
RGH-4201	100	17.5±1.2	5.7±0.4	7.8±0.56	8.8±0.8
Suxamethonium	100	14.3±1.7	9.6±0.9**	11.5±0.8**	11.7±0.7*
Chandonium	100	27.0±4.6	11.3±1.1***	14.6±2.2**	18.4±2.2**
Diadonium	250	22.6±3.1	10.0±2.6	15.5±0.9***	19.6±1.5***

Table 6. Neuromuscular blocking avtivity of RGH-4201, suxamethonium, chandonium, and diadonium in anesthetized cats

Compounds	ED_{50} (μg/kg) fiducial limits		Onset of action[a] (s)		Duration of action (min)	
	Tibialis muscle	Soleus muscle	Tibialis muscle	Soleus muscle	Tibialis muscle	Soleus muscle
RGH-4201	104.8 (86.2–127.4)	113.2 (92.3–139.0)	74.4±6.0	93.0±5.6	12.2	10.9
Suxamethonium	31.8 (19.1–52.9)	51.9 (37.6–71.8)	140.0±12.0	106.0±13.0	13.9	10.1
Chandonium	56.4 (41.2–77.4)	43.3 (32.9–57.3)	94.1±6.0	98.0±8.1	16.1	12.5
Diadonium	190.8 (159.0–230.0)	182.7 (151.0–221.2)	137.0±11.0	152.0±11.0	20.1	15.8

[a] Mean ± standard error

ration of action than vecuronium or pancuronium in anesthetized cats (Table 7). In the case of in vitro and in vivo studies on rats F. F. FOLDES (1981, unpublished work) found similar results when comparing RGH-4201 with vecuronium, atracurium, and tubocurarine. The recovery rate of RGH-4201 was more rapid than those of the other compounds in the rat phrenic nerve–hemidiaphragm preparation.

All of the authors demonstrated that RGH-4201 produced neuromuscular block of rapid onset and disappearance in the species studied. Onset of action was within 30 s in conscious mongrel, beagle, and anesthetized German shepherd dogs, and 1.2–2.2 min in anesthetized cats, depending on the muscle investigated. Spontaneous respiration in dogs appeared at 5–10 min, depending on the method of anesthesia used (ALÁNT and PULAY 1979), and at 5–7 min in the case of conscious, respirated dogs (BIRÓ and KÁRPÁTI 1981).

In the case of repeated doses of 100 μg/kg RGH-4201 on cat tibialis and soleus muscles, after each subsequent dose separated by 10-min intervals, the slope of

Table 7. Effective doses and time courses of effects produced by RGH-4201, vecuronium, and pancuronium in the cat

	RGH-4201		Vecuronium		Pancuronium	
Dose to produce 85%–95% twitch block of tibialis anterior (μg/kg)	160		35		20	
	Tibialis	Soleus	Tibialis	Soleus	Tibialis	Soleus
Onset (min)	1.5±0.3	2.2±0.4	4.8±0.4	4.0±0.5	5.2±0.5	5.1±0.7
Recovery index	1.8±0.2	4.7±1.0	1.96±0.2	3.1±0.4	4.9±0.9	6.3±1.1
Duration	4.0±0.8	6.2±1.2	8.3±1.2	10.2±1.4	14.3±2.0	19.1±2.4

the return of twitches decreased and paralysis was prolonged threefold after the last injection. Cumulation time was 30 min for 80%–90% block. W. C. Bowman and J. Houston (1981, unpublished work) found that RGH-4201 is more cumulative than vecuronium when repeated doses are given (at the time of maximal effect produced by the preceding dose) so that the difference in duration of action tends to disappear by the third or fourth dose, although RGH-4201 remains faster in onset. Neuromuscular block produced by RGH-4201 in the cat tibialis muscle showed the characteristics of a typical nondepolarizing action and it was rapidly antagonized by neostigmine.

III. Cardiovascular and Other Pharmacologic Actions

Two of the main cardiovascular side effects (hypotension and tachycardia) of the neuromuscular blocking agents currently in use are associated with autonomic blocking actions of the compounds (Durant et al. 1979). At low doses, RGH-4201 caused a slight increase of heart rate in anesthetized dogs and cats. The heart rate of conscious beagle dogs was significantly increased during intubation and relaxation: by 69% after higher (500 μg/kg) and by 36% after lower (100 μg/kg) doses. Systolic blood pressure and central venous pressure did not change in dogs anesthetized by different anesthetics, but meaningful tachycardia was measured in the ketamine–fentanyl group (O. Alánt and I. Pulay 1979, unpublished work). Doses of 1 mg/kg and greater produced a transient increase in blood pressure and a total dose of 1 g/kg was well tolerated by anesthetized cats. Ten times the effective blocking dose (10–15 mg/kg i.v.) of RGH-4201 produced tachycardia in rats but, as also found by F. F. Foldes (1981, unpublished work), at neuromuscular blocking doses it did not increase the heart rate.

None of the compounds tested for sympathetic ganglion-blocking activity was active at neuromuscular blocking doses. In doses up to 5 mg/kg i.v. RGH-4201, chandonium (Biró and Kárpáti 1981) and RGH-4201, vecuronium, pancuronium (W. C. Bowman and J. Houston 1981, unpublished work) were without effect on contractions of the nictitating membrane evoked by preganglionic stimulation.

Small doses of RGH-4201, chandonium, diadonium, and pancuronium, unlike suxamethonium and vecuronium, possessed vagolytic action. The vagolytic effect was more than 50% at effective neuromuscular blocking doses and lasted for 35 min. Diminution of acetylcholine-induced bradycardia was even more pronounced and lasted for 30 min, but the depressor action of acetylcholine was unaffected (BIRÓ and KÁRPÁTI 1981).

The histamine content in the serum of German shepherd dogs was not elevated during myorelaxation by RGH-4201 (O. ALÁNT and I. PULAY 1979, unpublished work). The inhibition of acetylcholinesterase (AChE) activity of human red blood cells expressed as I_{50}, was 9.4×10^{-5} mol/l, whereas that of serum cholinesterase (ChE) was 2.3×10^{-5} mol/l. Inhibition was mostly reversible. RGH-4201 showed a mixed inhibitory effect on serum ChE and, by a mixed or noncompetitive type of mechanism, inhibited the AChE of human red blood cells (G. SIMON 1978, unpublished work).

IV. Pharmacokinetics

For studying the fate of RGH-4201 in rats, ^{3}H-labeled compound was used. The volume of distribution was found to be 1.8–0.4 l, suggesting little or no plasma protein binding and this may hint at high tissue binding (RÓNAI-LUKÁCS and VERECZKEY 1982). Thus, the great majority of the compound persists in the liver, 2 weeks after treatment, 27% of the administered dose could be recovered in it. RGH-4201 did not undergo any metabolism in the rat.

V. Clinical Studies

The neuromuscular blocking activity of RGH-4201 was first demonstrated in humans by Foldes, Boros, and Tassonyi. The paper by FOLDES et al. (1983) deals with the results obtained in four studies carried out at several European locations: Hungary (M. BOROS 1980, unpublished work; E. TASSONYI 1981, unpublished work); Austria (FITZAL et al. 1982); and the Netherlands (S. AGOSTON 1981, unpublished work). Premedication and maintenance of anesthesia were uniform in all these studies. The observations made during the clinical use of RGH-4201 for the maintenance of surgical relaxation during balanced anesthesia indicate that

Table 8. The neuromuscular effects of a single 0.4 mg/kg dose of RGH-4201 and of 0.15 mg/kg maintenance doses in patients

Maximal block (%)	Time to maximal block (min)	Clinical duration (min)	Intubation time (min)	Clinical duration of maintenance doses (min)				Recovery rate after last dose (min)
				1st	2nd	3rd	4th	
92.1[b] ±1.2	5.5 ±0.4	26.3 ±1.6	2.3 ±0.1	24.6 ±1.8	22.8 ±1.7	19.4 ±2.7	18.0 ±2.5	20.1 ±2.0

[a] Time between administration of dose recovery of twitch to 25% of control
[b] Mean ± standard error

Table 9. The effect of a 0.4 mg/kg dose of RGH-4201 on heart rate in 89 patients

Before	Heart rate (beats/min)	
	After RGH-4201[a]	after intubation
84.9 ± 2.7[b]	99.2 ± 2.5 (+16.7%)[c]	104.2 ± 2.8 (+22.5%)

[a] Effect of RGH-4201 on heart rate maximal at 2 min
[b] Mean ± standard error
[c] Percentage change compared with pre-RGH-4201 values

RGH-4201 is a neuromuscular blocking agent of relatively short duration of action.

Conditions for endotracheal intubation were satisfactory 2 min after the administration of an initial dose of 0.4 mg/kg RGH-4201 (Table 8). The administration of 0.13–0.15 mg/kg repeat doses whenever the twitch tension returned to 25% of control provided satisfactory surgical relaxation. There was no indication of any cumulative effect after the repeated injection of fractional doses of RGH-4201. Twitch tension recovered following the administration of the last fractional dose of RGH-4201 to 90% of control in about 50 min. At this time, edrophonium reestablished neuromuscular transmission to control levels. Except for the consistent moderate (14%–18%) elevation of heart rate caused by it (Table 9), RGH-4201 compares favorably with other nondepolarizing neuromuscular blocking drugs (vecuronium, atracurium, pancuronium) now under clinical investigation.

E. Conclusions

Bisquaternary steroid compounds have been investigated as neuromuscular blocking agents for more than 20 years. As yet, only pancuronium has been introduced into widespread clinical practice. The other drug, pipecuronium also has a relatively rapid onset, intermediate duration of action, and lack of side effects.

Among the short-acting steroid relaxants, the bisquaternary RGH-4201 and the monoquaternary vecuronium have been extensively studied in clinical anesthesia. The short-acting bisquaternary compounds including RGH-4201, however, have unwanted effects on the cardiovascular system.

It is unlikely that the ideal muscle relaxant drug will ever be synthesized and it is more profitable to consider where the present neuromuscular blocking agents are inadequate, in the expectation that drugs will be synthesized that will avoid some or all of their disadvantages (Feldman 1976). Although the presently available steroid relaxants are in many respects excellent, they are not ideal. Perhaps the most sought after and elusive is the short-acting nondepolarizing agent. Although in the past, several agents have appeared to be satisfactory in animal studies, they have been unsuitable for humans.

Although a short-acting nondepolarizer to replace suxamethonium would be of great value, KATZ and KATZ (1975) suggest that of equal or even greater value would be a moderately long-acting nondepolarizer (similar in duration to pancuronium) which had a rapid onset – equal to or faster than that of suxamethonium – and whose action could be antagonized 30 min after injection. Such an agent might be even more controllable than a short-acting nondepolarizer.

References

Alyautdin RN (1978) On the effect of curare like agents on the synaptic transmission in the spinal cord (in Russian). Farmakol Toksikol 41:397–400

Alyautdin RN, Buyanov VV, Lemina EJU, Muratov VK, Samoilov DN, Fisenko VP, Shorr VA (1979) Appraisal of some pharmacological properties of the new steroid curare like agent RGH-1106 (in Russian). Farmakol Toksikol 42:239–243

Alyautdin RN, Buyanov VV, Fisenko VP, Lemina EJU, Muratov VK, Samoilov DN, Shorr VA (1980) On some properties of a new steroid curare like compound pipecuronium bromide. Arzneimittelforsch 30:355–357

Biggs RS, Davis M, Wien R (1964) Muscle relaxant properties of a steroid bisquaternary ammonium salt. Experientia 20:119–120

Biró K, Kárpáti E (1981) The pharmacology of a new short acting nondepolarising muscle relaxant steroid (RGH-4201). Arzneimittelforsch 31:1918–1924

Bodrogi L, Fehér TI, Váradi A, Vereczkey L (1980) Pharmacokinetics of pipecuronium bromide in the rat. Arzneimittelforsch 30:366–370

Bowman WC (1980) Pharmacology of neuromuscular function. John Wright, Bristol, England

Bowman WC (1981) New neuromuscular blocking drugs and their antagonists. Indian J Pharmacol 13:1–22

Buckett WR (1968) The pharmacology of pancuronium bromide: A new nondepolarising neuromuscular blocking agent. Indian J Med Sci 1:565–568

Buckett WR (1972) Aspects of the pharmacology of aminosteroids. In: Briggs MH, Christie G (eds) Advances in steroid biochemistry and pharmacology, vol 3. Academic, London

Buckett WR (1975) Steroidal neuromuscular blocking agents. In: Harper NJ, Simmonds AB (eds) Advances in drug research, vol 10. Academic, London

Buckett WR, Bonta IL (1966) Pharmacological studies with NA 97 (2β,16β-dipiperidino-5α-androstane-3α,17β-diol diacetate dimethobromide). Fed Proc 25:718

Buckett WR, Saxena PR (1969) The pharmacology of dacuronium bromide – a new short-acting neuromuscular blocking drug of nondepolarising type. Abstract of the 4th international congress on pharmacology, Basel, p 420

Busfield D, Child KJ, Clarke AJ, Davis B, Dodds MG (1968) Neuromuscular blocking activities of some steroidal mono and bisquaternary ammonium compounds with special reference to NN′-dimethyl-conessine. Br J Pharmacol Chemother 32:609–623

Coleman AJ, Downing JW, Leary WP, Moyes DG, Styles M (1972) The intermediate cardiovascular effects of pancuronium, alcuronium and tubocurarine in man. Anaesthesia 27:415–422

Durant NN, Houwertjes MC, Agoston S (1979) Hepatic elimination of Org NC 45 and pancuronium. Anesthesiology 51:S267

Feldman SA (1976) The ideal muscle relaxant. In: Spierdijk J, Feldman SA, Mattie H (eds) Anaesthesia and pharmacology. Williams and Wilkins, Baltimore

Feldman SA, Tyrell MF (1970) A new steroid muscle relaxant Dacuronium-NB 68 (Org). Anaesthesia 25:349–355

Fitzal S, Ilias W, Kalina K, Schwarz S, Foldes FF, Steinbereithner K (1982) Neuromuskulare und kardiovaskulare Effekte von Duador, einem neuen kurz wirksamen nicht depolarisierenden Muskelrelaxans. Anaesthesist 31:674–679

Foldes FF, Nagashima H, Boros M, Tassonyi E, Fitzal S, Agoston S (1983) Muscular relaxation with atracurium, vecuronium and Duador under balanced anaesthesia. Br J Anaesth 55:97S–103S

Gandiha A, Marshall IG, Paul D, Rodger IW, Scott W, Singh H (1975) Some actions of chandonium iodide, a new short-acting muscle relaxant, in anaesthetized cats and on isolated muscle preparations. Clin Exp Pharmacol Physiol 2:159–170

Harvey AL, Paul D, Rodger IW, Singh H (1976) Actions of the muscle relaxant chandonium iodide on guineapig ileum and vas deferens preparations. J Pharm Pharmacol 28:617–619

Hughes R (1970) Effects of anaesthetics and their interactions with neuromuscular blocking agents in cats. Br J Anaesth 42:826–833

Janot M, Lainé F, Goutarel R (1960) Alcaloides steroides. V: Alcaloides du Malouetia bequaertiana E. Woodson (Apocynacées): La funtuphillamine B et la malouétine. Ann Pharm Fr 18:673–677

Katz RL, Gissen AJ (1967) Neuromuscular and electromyographic effects of halothane and its interaction with (+)-tubocurarine in man. Anesthesiology 28:564–567

Katz RL, Katz GJ (1975) Clinical considerations in the use of muscle relaxants. In: Katz RL (ed) Muscle relaxants. Excerpta Medica, Amsterdam (Monographs in anaesthesiology, vol 3)

Kárpáti E, Biró K (1980) Pharmacological study of a new competitive neuromuscular blocking steroid, pipecuronium bromide. Arzneimittelforsch 30:346–454

Kelman GR, Kennedy BR (1971) Cardiovascular effects of pancuronium in man. Br J Anaesth 43:335–338

Kharkevich DA, Skoldinov AP (1980) On some principles of interaction of curare like agents with acetylcholine receptors of skeletal muscles. J Pharm Pharmacol 32:733–739

Le Men J, Kan C, Beugelmans R (1963) Structure de la paravallaridine: alcaloide steroidique du paravallaris microphylla Pitard (Apocynacées). Bull Soc Chim Fr 597–603

Lewis JJ, Martin-Smith M, Muir TC, Ross HH (1967) Steroidal monoquaternary ammonium salts with nondepolarising neuromuscular blocking activity. J Pharm Pharmacol 19:502–508

Marshall IG, Singh H, Paul D (1972) The neuromuscular and other blocking actions of 4,17*a*-dimethyl-4,17*a*-diaza-D-homo-5α-androstane dimethiodide (HS 342). J Pharm Pharmacol 24:146P

Marshall IG, Paul D, Singh G (1973a) Some actions of 4,17*a*-dimethyl-4,17*a*-diaza-D-homo-5α-androstane dimethiodide (HS 342), a new neuromuscular blocking drug. J Pharm Pharmacol 25:441–446

Marshall IG, Paul D, Singh H (1973b) The neuromuscular and other blocking actions of 4,17*a*-dimethyl-4,17*a*-diaza-D-homo-5α-androstane dimethoiodide (HS 342) in the anaesthetized cat. Eur J Pharmacol 22:129–134

Marshall RJ, Ojewole JAO (1979) Comparison of the autonomic effects of some currently-used neuromuscular blocking agents. Br J Pharmacol 66:77–78

Miller RD (1975) Factors affecting the action of muscle relaxants. In: Katz RL (ed) Muscle relaxants. Excerpta Medica, Amsterdam (Monographs in anaesthesiology, vol 3)

Palmer RA, Kalam MA, Singh H, Paul D (1980) Steroids and related studies, part LIII. Structure and function of synthetic bisquaternary aza steroidal neuromuscular blocking agents. J Cryst Mol Struct 10:31–53

Pulay I, Alánt O, Darvas K, Weltner J, Zétény ZS (1980) Respiration paralysing and circulatory effects of a new nondepolarising relaxant, pipecuronium bromide in anaesthetized dogs. Arzneimittelforsch 30:358–360

Quevauviller MA, Lainé MF (1960) Sur la toxité et le pouvoir curarisant du chlorure de malouétine. Ann Pharm Fr 18:678–680

Rónai-Lukács S, Vereczkey L (1982) Identification of RGH 4201, a new short-acting neuromuscular blocking agent in rats. In: Görög S (ed) Advances in steroid analysis. Akademia Elsevier, Budapest, pp 489–491

Simon G, Biró K, Kárpáti E, Tuba Z (1980) The effect of the steroid muscle relaxant pipecuronium bromide on the acetylcholinesterase activity of red blood cells in vitro. Arzneimittelforsch 30:360–363

Singh H, Paul D (1974) Steroids and related studies, part 25. Chandonium Iodide (17*a*-methyl-3*β*-pyrrolidino-17*a*-aza-D-homoandrost-5-ene Dimethiodide) and other quaternary ammonium steroid analogues. J Chem Soc Perkin Trans I:1475–1479

Singh H, Paul D, Parashar VV (1972) Steroid skeleton modification to 4,17*a*-dimethyl-4,17*a*-diaza-D-homo-5α-androstane dimethiodide – a potential neuromuscular blocking agent. Abstracts of the IUPAC symposium on the chemistry of natural products, New Delhi, p 247

Singh H, Paul D, Parashar VV (1973) Steroids and related studies, part 20. 4,17*a*-diaza-D-homo steroids. J Chem Soc Perkin Trans I:1204–1206

Singh H, Bhardwaj TR, Ahuja NK, Paul D (1979a) Steroids and related studies, part 44. 17*a*-methyl-3*β*-(*N*-pyrrolidinyl)-17*a*-aza-D-homo-5α-androstane bis(methiodide) (Dihydrochandonium iodide) and certain other analogues of chandonium iodide. J Chem Soc Perkin Trans I:305–307

Singh H, Bhardwaj TR, Paul D (1979b) Steroids and related studies, part 48. A chandonium iodide analogue possessing an acetylcholine like moiety. J Chem Soc Perkin Trans I:2451–2454

Smith G, Proctor DW, Spence AA (1970) A comparison of some cardiovascular effects of tubocurarine and pancuronium in dogs. Br J Anaesth 42:923–927

Speight TM, Avery GS (1972) Pancuronium bromide: A review of its pharmacological properties and clinical application. Drugs 4:163–226

Sugrue MF, Duff N (1973) Org 6368: A competitive steroidal muscle relaxant with a rapid onset and a short duration of action. Naunyn-Schmiedebergs Arch Pharmacol [Suppl] 279:R30

Teerapong P, Marshall IG, Harvey AL, Singh H, Paul D, Bhardwaj TR, Ahuja NK (1977) Neuromuscular and autonomic actions of four analogues. J Pharm Pharmacol 29 [Suppl]:80P

Teerapong P, Marshall IG, Harvey AL, Singh H, Paul D, Bhardwaj TR, Ahuja NK (1979) The effects of dihydrochandonium and other chandonium analogues on neuromuscular and autonomic transmission. J Pharm Pharmacol 31:521–528

Vercruysse P, Bossuyt P, Hanegreefs G, Verbeuren TJ, Vanhoutte PM (1979) Gallamine and pancuronium inhibit pre- and postjunctional muscarinic receptors in canine saphenous veins. Am Soc Pharmacol Exp Ther 209:225–230

Vereczkey L, Szporny L (1980) Disposition of pipecuronium bromide in rats. Arzneimittelforsch 30:364–366

[illegible]

[illegible]

[illegible]

[illegible]

[illegible]

[illegible]

CHAPTER 18

Vecuronium (ORG-NC-45)

W. C. BOWMAN and G. A. SUTHERLAND

A. Introduction

Vecuronium bromide (ORG-NC-45, Norcuron), henceforth called simply vecuronium, is the monoquaternary homologue of pancuronium. Vecuronium differs from pancuronium in that the methyl group on the 2β-nitrogen atom of the latter is lacking (Fig. 1). Vecuronium is one of a series of amino steroidal derivatives, the synthesis and chemical properties of which have been described by BUCKETT et al. (1973), and more recently by SAVAGE (1980, 1981) and SAVAGE et al. (1980). Vecuronium was selected for further study from a series of closely related analogues on the grounds that, in the cat, it exhibited pronounced neuromuscular blocking activity coupled with only very weak actions at sympathetic ganglia and on the cardiac vagus (DURANT et al. 1979 c). Since the first description of its properties, vecuronium has undergone extensive additional study both in animals (as described in this chapter) and in humans (as described by MILLER in Chap. 28). Much of the work up to the end of 1979 is described in the proceedings of a symposium edited by BOWMAN and NORMAN (1980), and short reviews have also been published by BOOIJ et al. (1981 a) and DURANT (1982). With the knowledge of structure–activity relations available at the time, DURANT et al. (1979 c) concluded that the conformation of the D ring acetylcholine-like fragment of pancuronium, which is known to have a different molecular geometry and electronic structure from that of the A ring (SAVAGE et al. 1971), while intrinsically suited to the neuromuscular cholinoceptor, is relatively incompatible with the cardiac muscarinic receptor. The isomeric acetylcholine fragment in the A ring of pancuronium, on the other hand, seemed to confer affinity for the cardiac muscarinic receptor. Accordingly, when the A ring acetylcholine fragment was modified by making the nitrogen tertiary (as in vecuronium), instead of quaternary, atropine-like action on the heart was virtually lost. Further study of addi-

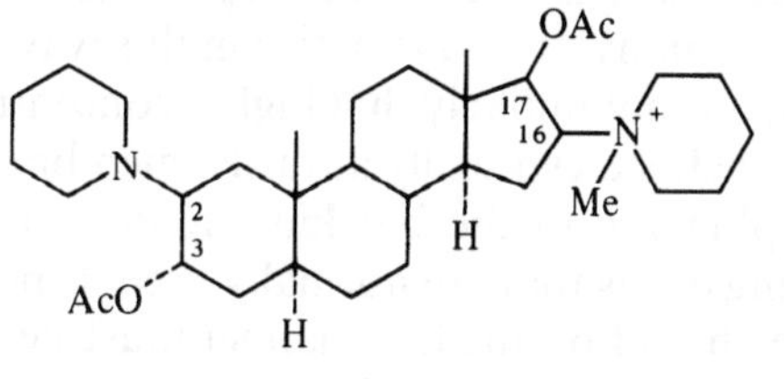

Fig. 1. Vecuronium bromide (ORG-NC-45, 1-[3α,17β-diacetoxy-2β-piperidino-5α-androstan-16β-yl]-1-methylpiperidinium bromide)

tional new compounds in the series by the same authors, and by others indicate, however, that in many structures, abolition of atropine-like action on the heart involves more complex molecular modifications than a simple replacement of a quaternary with a tertiary nitrogen attached to the A ring. Full structure–activity relations have not yet been elucidated.

B. Mechanism of Action

Vecuronium exhibits the characteristics of a nondepolarizing neuromuscular blocking drug of the acetylcholine antagonist type. Figure 2a illustrates an experiment on an anaesthetized cat which is merely a more modern version of the classical type of experiment designed by Claude Bernard during his study of the action of curare in frogs. Thus, during the block of the indirectly evoked twitches of the tibialis anterior muscle, there was no effect on the action potential recorded from the motor nerve, and the muscle responded normally to direct stimulation (Baird et al. 1982). These results locate the site of action of vecuronium at the neuromuscular junction. Figure 2b illustrates another experiment on the tibialis anterior muscle of a cat (Baird et al. 1982), which shows that during the block of the indirectly evoked twitches, contractions of the muscle produced by close-arterial injections of acetylcholine are also blocked. This result indicates that the main action of vecuronium, like that of other drugs in the group, is to block acetylcholine receptors rather than to impair acetylcholine release, although it does not of course exclude the possibility of a minor action on the motor nerve endings. Indeed, electrophysiological evidence indicates that, as well as blocking postjunctional receptors, vecuronium reduces the mean quantal content of the end-plate potential in the sartorius muscle of the toad (Torda and Kiloh 1982a, b). Similar evidence of a prejunctional component of action has been provided for other neuromuscular blocking drugs (tubocurarine Hubbard and Wilson 1973; Galindo 1971; and pancuronium Galindo 1972) but its significance is not yet clear, nor is its existence accepted by all.

Neuromuscular blocking drugs, such as tubocurarine and gallamine, as well as blocking the recognition sites of the cholinoceptors, may also, under appropriate conditions, plug the open forms of the associated ion channels (Katz and Miledi 1978; Colquhoun et al. 1979; Colquhoun and Sheridan 1981; Dreyer 1981; and see Colquhoun in Chap. 3). Vecuronium has only recently been studied by the appropriate electrophysiological techniques. According to Torda and Kiloh (1982b), channel plugging at normal membrane potentials is not produced by vecuronium in concentrations sufficient to depress the amplitudes of miniature end-plate currents. Since many drugs are active in this way providing the concentration is high enough, it is not unlikely that higher concentrations of vecuronium would also produce this effect, especially at raised membrane potentials. The significance of channel plugging in the blocking action, in vivo, of the doses of neuromuscular blocking drugs that are normally used, is not yet known. In order that channel plugging should occur, the channel must be in an open state, and plugging is therefore more pronounced the greater the concentration of agonist (acetylcholine) present. It has therefore been suggested that channel plugging

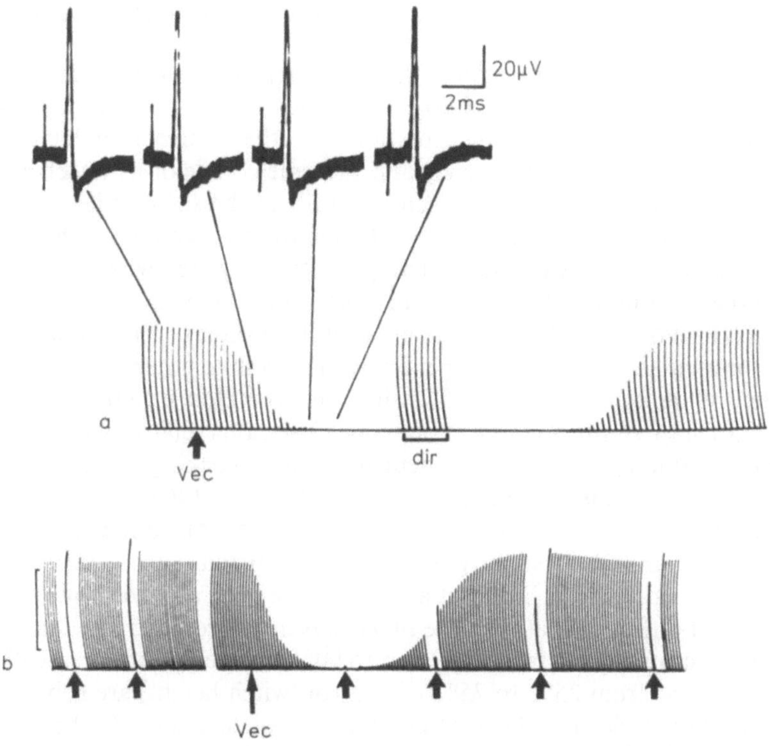

Fig. 2. **a** Cat, chloralose anaesthesia. Maximal twitches of a tibialis anterior muscle and antidromic nerve action potentials recorded from a ventral root were evoked by stimulation of the motor nerve at a frequency of 0.1 Hz. The representative action potentials are associated with the twitches indicated by the *oblique lines*. At *Vec*, vecuronium 100 µg/kg was injected i.v. This dose caused complete abolition of the twitches evoked by nerve stimulation, but was without effect on the nerve action potentials. During the block, the muscle responded normally to direct electrical stimulation (*dir*). **b** Cat, chloralose anaesthesia. Maximal twitches of a tibialis anterior muscle were evoked by stimulation of the motor nerve at a frequency of 0.1 Hz. At intervals, electrical stimulation was temporarily stopped and acetylcholine 8 µg was injected close-arterially into the muscle (*unlabelled arrows*). The contractions above the *unlabelled arrows* are the responses to acetylcholine. At *Vec*, vecuronium 30 µg/kg was injected i.v. During the block of the twitches evoked by nerve stimulation, the contractions produced by close-arterially injected acetylcholine were also blocked. Tension calibrations = 5 N. Both **a** and **b** from BAIRD et al. (1982)

might contribute to the tetanic fade that characteristically occurs during partial block produced by drugs of this class (COLQUHOUN et al. 1979; BOWMAN 1980a; BOWMAN and MARSHALL 1981; DREYER 1981). During a tetanus, the frequency of channel opening is obviously greatly enhanced, and this might allow more molecules of antagonist to enter and plug the open channels, thereby enhancing the block as the tetanus continues (i.e. a so-called use-dependent block). However, there is some controversy as to the importance of channel plugging in the production of tetanic fade. Some workers consider that blockade of prejunctional cholinoceptors may play the more important role, as discussed in Chap. 5.

C. Potency and Time Course of Action: Cumulative Effects

The neuromuscular blocking action of vecuronium has been studied in anaesthetized rats, cats, rhesus monkeys and dogs by various workers, and its actions have been compared with those of pancuronium and other neuromuscular blocking drugs. Table 1 summarizes its potency and the time course of its effects in these four species. The results show that vecuronium is of the same order of potency as pancuronium, being somewhat more potent in the dog, and somewhat less potent in the other three species. Its relative potency compared with pancuronium in the dog (i.e. 1.6 times more potent than pancuronium) resembles that reported for humans (Krieg et al. 1980a; Fahey et al. 1981), whereas, as is not uncommon with neuromuscular blocking drugs generally, its absolute potency in the cat (approximately 40 μg/kg to produce 90% twitch block) is closest to that in humans. (See Miller Chap. 28 for potency in humans.) The rat is relatively insensitive to vecuronium (Table 1), as it is to all neuromuscular blocking drugs other than tubocurarine, to which it is especially sensitive (Bowman 1964).

The onset of action of vecuronium (time from injection to maximum effect) in experimental animals is slightly more rapid than that of pancuronium, although not all workers (e.g. Baird et al. 1982) have been able to show a statistically significant difference between the two drugs in this respect. The duration of action of vecuronium (time from injection to 90% recovery) and its so-called recovery index (time from 25% to 75% of control twitch height) are substantially and statistically significantly shorter than those of pancuronium. In the cat, there is little difference in the sensitivity of the slow-contracting soleus and the fast-contracting tibialis anterior muscles to vecuronium, although with a given dose the duration of the block is longer in the soleus (see Fig. 7).

Foldes et al. (1981) have emphasized that stimulation of nerve–muscle preparations with single shocks at low frequencies to evoke twitches has no physiological counterpart in muscle activity. They adduced evidence that stimulation with brief (0.1 s) trains of 50 Hz every 10 or 20 s is more relevant from a physiological point of view. The effective doses of neuromuscular blocking drugs were smaller, and recovery from a given percentage block was slower, when brief trains of stimuli were applied. For example, in rats the dose of vecuronium necessary to reduce contractions to 50% of control was reduced by about one-fifth when brief tetani were evoked instead of twitches, and recovery time from a steady state block when the infusion was terminated was approximately doubled. The tetanic pattern of stimulation had a substantially greater delaying influence on the rate of recovery from tubocurarine than on that from vecuronium.

A common characteristic of most nondepolarizing neuromuscular blocking drugs is that pronounced cumulation between successive doses occurs when a series of injections is administered. However, in rats, cats, and monkeys, cumulation has been found to be almost negligible with vecuronium (Durant 1978; Marshall et al. 1980a; Durant et al. 1981). For example, when the same dose of vecuronium was repeatedly injected, the second and subsequent doses being administered at the time of full recovery of the twitches after the preceding dose, there was a small degree of cumulation between the first and second doses, but there was little difference between the effects produced by the second and each

Table 1. Potency of vecuronium and time course of action in different animal species

Species, anesthetic and muscle	Dose to produce about 90% twitch block at 0.1 Hz (μg/kg)	Mean block (%)	Time from injection to maximum block; onset (min)	Time from injection to 90% recovery; duration (min)	Recovery time from 25% to 75% control twitch; recovery index (min)	Reference
Rat Sodium pentobarbitone Tibialis anterior	250	88 ± 3.1	1.17 ± 0.15	3.3 ± 0.6	0.9 ± 0.2	MARSHALL et al. (1980a)
Cat α-Chloralose Gastrocnemius	28	85.2 ± 2.6	4.72 ± 0.25	8.18 ± 0.58	1.9 ± 0.24	I. MCINDEWAR quoted in MARSHALL et al. (1980a)
Rhesus monkey Ketamine– sodium pentobarbitone–nitrous oxide, Adductor pollicis	10	82.4 ± 5.5	7.8 ± 0.6	20.4 ± 5.4	6.6 ± 0.6	DURANT et al. (1980)
Dog Halothane Hindlimb extensors	14	~90%	Not reported	42 ± 2[a]	Not reported	BOOIJ et al. (1980)

[a] Time to 50% recovery after three times the dose to produce 90% twitch block. Times are means ± standard error

subsequent dose. The consequence of the minor cumulative effects of vecuronium is that after a succession of doses, duration of action and time of recovery become relatively very much longer for more cumulative drugs, such as pancuronium or tubocurarine, than for vecuronium.

D. Pharmacokinetics

FOLDES et al. (1981) observed that although recovery from vecuronium was considerably faster than that from pancuronium in anaesthetized rats, there was no difference in the rates of recovery from the two drugs following a standardized washing procedure in the isolated phrenic nerve–diaphragm preparation. They suggested, on the basis of these results, that the in vivo difference between the two drugs is a consequence of more rapid removal of vecuronium from the biophase through redistribution, metabolism and excretion, rather than of a weaker binding capacity to the cholinoceptors.

By analogy with pancuronium, it might be expected that vecuronium would be hydrolysed in the plasma (catalysed by plasma cholinesterase) to its 3-desacetyl derivative, or its 17-desacetyl derivative, and finally to its 3,17-bisdesacetyl derivative (SAVAGE et al. 1980). The potencies of these three potential metabolites of vecuronium at the neuromuscular junction and at the cardiac vagal nerve endings in the cat (MARSHALL et al. 1983) are listed in Table 2. In fact, no evidence for any substantial production of these metabolites has yet been obtained and it seems that the elimination of vecuronium from the biophase may be largely the result of redistribution and excretion; metabolism in plasma may occur too slowly to play an important part. At the same time, a suggestion by Savage and his co-workers (SAVAGE et al. 1980; SAVAGE 1981) should be borne in mind for further testing when that becomes possible. They put forward the possibility, on the basis of chemical studies, that the rapid recovery from vecuronium seen in all species, might be partly the result of a chemical inactivation mechanism, which is confined to the junctional cleft because of the high local concentration that presumably exists there as a consequence of the high affinity of vecuronium for the cholinoceptors. In high concentrations, vecuronium undergoes a chemical degradation

Table 2. Potencies of vecuronium and its potential metabolites on the cat soleus muscle and the cardiac vagus (after MARSHALL et al. 1983)

Compound	Dose to produce 50% soleus twitch block (μg/kg) *S*	Dose to produce 50% block of vagal-induced bradycardia (μg/kg) *V*	*V*/*S*
Vecuronium	24 ± 4	1,588 ± 227	80
3-Hydroxyvecuronium	34 ± 11	1,174 ± 235	40
17-Hydroxyvecuronium	725 ± 183	498 ± 50	0.9
3,17-Dihydroxyvecuronium	1,610 ± 138	247 ± 49	0.2

Doses are means ± standard error

arising from the ability of the tertiary basic 2β-nitrogen atom to catalyse the conversion of both the 3- and the 17-acetoxy groups to hydroxyl groups in aqueous solution at pH 7, quite independently of cholinesterase. The tertiary nitrogen atoms of any metabolites formed would also exert this effect, so that the number of catalytic molecules (metabolites plus parent molecules) in the cleft would decrease less rapidly than the number of parent molecules possessing neuromuscular blocking action. An action of this sort might account for the clear increase in the slope of the recovery curve that frequently occurs at the time of about 50% twitch recovery (see Fig. 7 for example), and it might be a factor in the lack of cumulative effect evident with the drug. If such an effect were to occur, the concentrations of the desacetylated compounds, though substantial in the biophase of the junctional clefts, might be negligible and undetectable in the plasma.

At the present time, the most precise pharmacokinetic data relating to vecuronium have been obtained in humans, since it is only in humans that the most specific and sensitive assay method, one based on mass spectrometry, has been used (CRONNELLY et al. 1982). These human data are discussed by MILLER in Chap. 28. BOOIJ et al. (1981 b) made a pharmacokinetic study of vecuronium and pancuronium in anaesthetized dogs using a less sensitive and less specific fluorimetric assay. With this method, it was necessary to administer very large doses of vecuronium and pancuronium, greatly in excess of the neuromuscular blocking doses. The plasma half-life of elimination of vecuronium was short (about 3 min), being about one-third of that of pancuronium. The relatively rapid clearance of vecuronium accounts for the difference in duration of action of the two drugs. The apparent volume of distribution after a 90-min infusion of vecuronium was similar to the extracellular fluid volume, indicating that, despite its marginally greater lipophilicity compared with pancuronium, vecuronium (like pancuronium) does not penetrate cell membranes nor the blood–brain barrier to any appreciable extent. Renal excretion of vecuronium, after the large dose administered, was only in the region of 12% in 6 h. The neuromuscular blocking action of vecuronium in cats, unlike that of pancuronium, was not modified by excluding the kidneys from the circulation (DURANT et al. 1979 a), indicating that renal elimination is not an important mechanism in this species. (The same is true in humans as described by MILLER in Chap. 28.) On the other hand, liver exclusion in the cat significantly increased the depth and prolonged the duration of the blockade produced by vecuronium (DURANT et al. 1979 b), indicating that hepatic elimination of the drug is important. In rats, over 60% of an intravenous dose of unchanged vecuronium may be found in the bile, compared with only 15% of pancuronium. Pancuronium was eliminated primarily in the urine (UPTON et al. 1982). Despite the significant uptake of vecuronium by the cat liver, DURANT et al. (1979 b) did not consider it sufficient to account fully for the short duration of action of the drug, since even when the liver was permanently excluded from the circulation, the time course of action of vecuronium was still shorter than that of pancuronium. Furthermore, administration of vecuronium directly into the liver via the portal vein produced a neuromuscular block that was only slightly smaller than that produced by systemic intravenous injection. These observations indicate that hepatic uptake and elimination in the cat comprise only part of the elimination mechanisms of vecuronium, and lead to further speculation about Savage's suggestion, already mentioned, of an autocatalytic breakdown process in the junctional cleft to account for the rapid termination of the blocking action.

E. Antagonism of Blocking Action

The rapid rate of spontaneous recovery from vecuronium in all animal species (Table 1) suggests that a fast recovery rate might also occur in humans, and allows the speculation that in many cases it might be clinically permissible to allow spontaneous recovery to occur, thereby obviating the need for anticholinesterase drugs and atropine-like drugs. However, it should be borne in mind that, in general (although there are a few exceptions), spontaneous recovery from neuromuscular blocking drugs (and indeed from most drugs) is more rapid the smaller the species, and this is the case with vecuronium as is evident from Table 1. Thus, amongst the species studied, rate of recovery decreases in the order: rat > cat > dog > monkey. Consequently, on the basis of animal tests it might be supposed that it will not always be possible to wait for spontaneous recovery in humans, and in fact a reversal agent has usually been used in clinical trials, although were there to be any contraindication to an anticholinesterase drug or to atropine-like drugs a reversal agent could probably be avoided without problems (BAIRD et al. 1982).

The antagonistic actions of neostigmine, pyridostigmine, and 4-aminopyridine against vecuronium have been studied in rats, cats, and rhesus monkeys (DURANT et al. 1980; BOOIJ et al. 1980; MARSHALL et al. 1980a). 4-Aminopyridine acts by increasing the evoked release of acetylcholine (see Chap. 33) rather than by prolonging its life. The most detailed comparative study of antagonists to vecuronium was carried out by BOOIJ et al. (1980) who administered vecuronium to rats by constant intravenous infusion, and calculated the cumulative dose of antagonist (given by incremental bolus injections) necessary to restore the partially blocked twitches of the tibialis anterior muscle to 50% of control amplitude. When given by infusion, vecuronium and pancuronium were antagonized to essentially the same extent by the same doses of the antagonists. The infusion method would of course remove any contribution of spontaneous recovery from the reversal process.

It is well known that a tetanus, interposed during a partial block of twitches produced by a nondepolarizing drug, gives rise to a temporary posttetanic antagonism of the twitch block arising from posttetanic facilitation of neuromuscular transmission (anticurare effect of a tetanus). BAIRD et al. (1982) observed in anaesthetized cats that the antagonistic action of a tetanus was generally more pronounced during block produced by vecuronium than during a similar depth of block produced by tubocurarine or pancuronium (Fig. 3). This difference is presumably a manifestation of the more rapid recovery rate from vecuronium that occurs in any case, and merely reflects the fact that vecuronium, for whatever reason, is more readily removed from the biophase. However, the difference suggested the possibility that edrophonium might be a more effective antagonist of vecuronium than of longer-acting drugs, such as tubocurarine and pancuronium. This in fact proved to be the case, as was shown by the finding that a submaximal

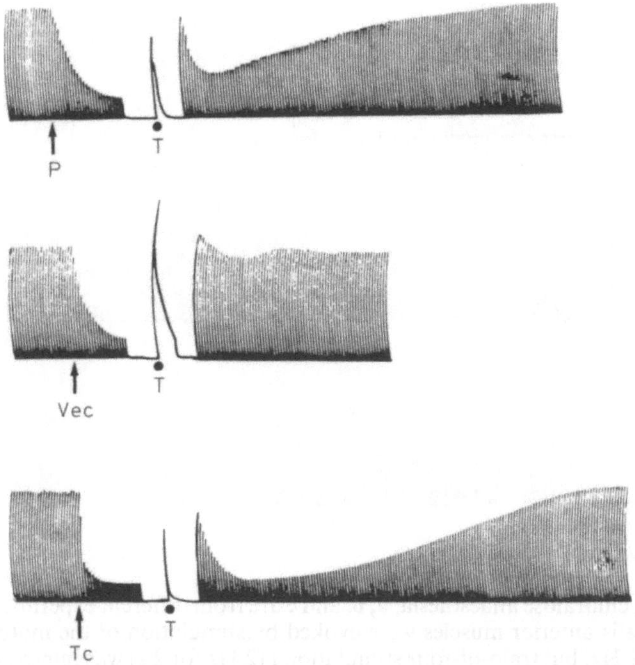

Fig. 3. Cat, chloralose anaesthesia. Maximal twitches of a tibialis anterior muscle were evoked by stimulation of the motor nerve at a frequency of 0.1 Hz. At *P*, pancuronium 20 µg/kg, at *Vec*, vecuronium 35 µg/kg, and at *Tc*, tubocurarine 300 µg/kg were injected. About 4.5 min after injection, during the partial blocks produced, a tetanus *T* (100 Hz for 10 s) was interposed. The paper speed was increased during the tetani. The duration of tetanic stimulation in each case is indicated by the tetanus evoked during vecuronium block, since tetanic fade in this instance was less than complete. Tetanic fade was least and post-tetanic antagonism was greatest during vecuronium block. In these respects, pancuronium came next and tubocurarine last. At least 3 h elapsed between the effects of the injections illustrated. Two smaller doses of vecuronium and three smaller doses of tubocurarine (effects not shown) had been injected before each of the illustrated blocks produced by these drugs. These omitted doses were administered in unsuccessful attempts to match the pancuronium block. Tension calibration = 5 N. (BAIRD et al. 1982)

dose of edrophonium (0.25 mg/kg) produced a greater antagonism of vecuronium than of the other blocking drugs (Fig. 4). With a fully effective dose (0.5–0.75 mg/kg) of edrophonium, any difference between blocking drugs is less obvious. Nevertheless, the results pointed to the possibility that edrophonium might be particularly valuable as an antagonist of relatively short-acting drugs such as vecuronium. Apart from its more rapid antagonistic action, edrophonium, even in the large doses necessary, may have weaker actions on autonomic cholinergic junctions than has neostigmine, and, if this is so, less atropine will be required to prevent its muscarinic effects.

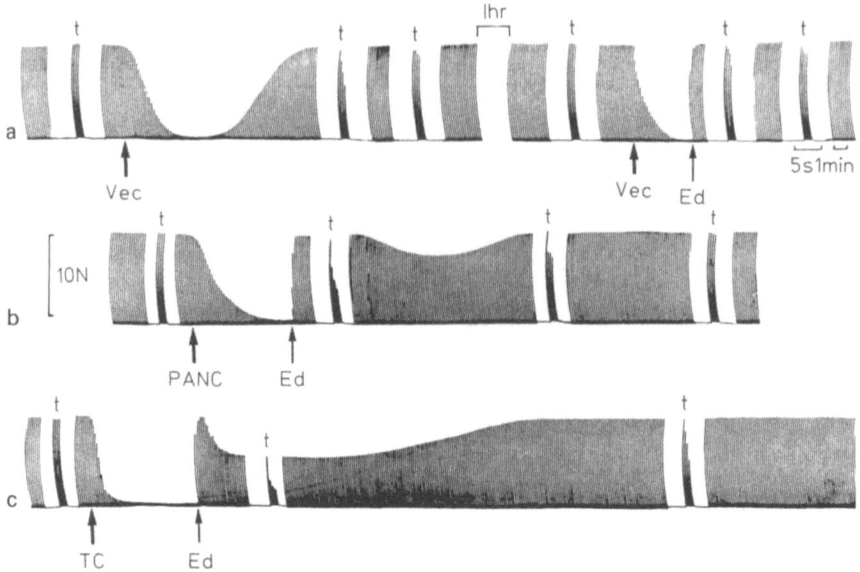

Fig. 4 a–c. Cats, chloralose anaesthesia; **a, b,** and **c** are from different experiments. Maximal twitches of tibialis anterior muscles were evoked by stimulation of the motor nerves at a frequency of 0.1 Hz, but train-of-four stimulation *t* (2 Hz for 2 s) was interposed where appropriate. At *Vec*, vecuronium 40 μg/kg, at *PANC*, pancuronium 22 μg/kg and at *TC*, tubocurarine 400 μg/kg were injected. At *Ed*, edrophonium 0.25 mg/kg was injected. This dose of edrophonium completely restored the twitches blocked by vecuronium to control, but the responses to train-of-four stimulation were 3–4 min slower to recover to control amplitude than were the responses to single-shock (0.1 Hz) stimulation. The antagonistic action of this dose of edrophonium was clearly less against pancuronium and tubocurarine than against vecuronium. Tension calibration = 10 N. (BAIRD et al. (1982)

F. Unwanted Effects

Vecuronium is relatively free from the unwanted effects commonly associated with drugs of this class. Such effects can therefore be considered only in comparison with other drugs. The main unwanted effects of nondepolarizing neuromuscular blocking drugs that experience has taught might be expected (see, for example, BOWMAN 1980b, c, 1981, 1982) are: (a) histamine release; (b) anticholinesterase activity; (c) ganglion block; (d) atropine-like block of cardiac muscarinic receptors and of similar muscarinic receptors on dopaminergic interneuron-like cells in sympathetic ganglia (SIF cells), and on noradrenergic nerve endings; (e) impairment of cardiac vagal activity by a mechanism independent of ganglion block or muscarinic receptor block; (f) block of noradrenaline reuptake; and possibly (g) stimulation of noradrenaline release.

I. Histamine Release

Many organic bases can release histamine from mast cells and basophil leucocytes, and most neuromuscular blocking drugs have been shown to produce this effect when sufficiently large doses or concentrations have been used. Amongst the commonly used neuromuscular blocking drugs, tubocurarine is the most potent histamine liberator, and the consequence of such liberation (urticaria, itching, erythema at the injection site, hypotension and bronchconstriction) are well documented in both the basic pharmacological literature and in the clinical literature. A sensitive radioenzymatic assay for histamine has recently been developed, and its use has enabled a clear correlation to be demonstrated between the hypotension and the elevation of plasma histamine produced in humans by normal doses of tubocurarine (MOSS et al. 1981). At the time of writing, most other drugs, including vecuronium, have not been tested by such a sensitive technique. However, indirect evidence both from animal experiments, and from tests in humans (BOOIJ et al. 1980), indicate that histamine release is not a problem associated with the use of vecuronium. This does not of course obviate the possibility that, as with all drugs, an occasional idiosyncratic reaction to vecuronium leading to histamine release might occur in a susceptible individual.

Figure 5 illustrates a previously unpublished experiment by Dr. I. W. Rodger in which arterial pressure, heart rate, transpulmonary pressure, rate of airflow and tidal volume were recorded simultaneously with maximal twitches of a tibialis anterior muscle in an artificially ventilated cat under chloralose anaesthesia. Tubocurarine, in a dose just sufficient to produce complete block of the twitches, produced a decrease followed by an increase in arterial pressure, an increase in

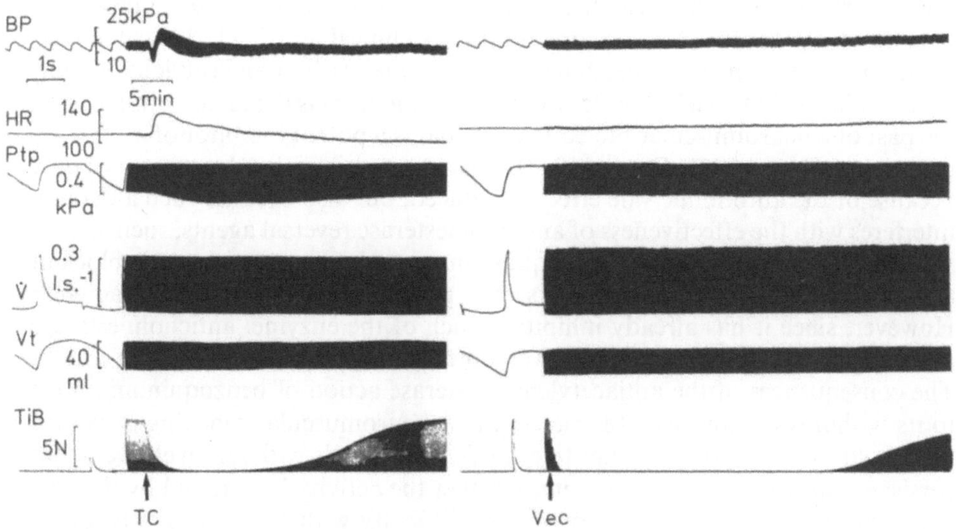

Fig. 5. Cat, chloralose anaesthesia, artificially ventilated. Records from above downwards: *BP* arterial blood pressure; *HR* heart rate; *Ptp* transpulmonary pressure; $\dot{V}$ rate of airflow; *Vt* tidal volume; *TIB* maximal twitches of a tibialis anterior muscle (0.1 Hz). At *TC* tubocurarine 0.25 mg/kg, and at *Vec* vecuronium 0.15 mg/kg injected intravenously 2 h later. Note that tubocurarine produced a biphasic change in blood pressure, a rise followed by a fall in heart rate and an increase in transpulmonary pressure. Vecuronium, despite the large dose, produced none of these effects. After tubocurarine administration, the recovery from vecuronium was slower than normal. Figure supplied by Dr. I. W. Rodger

heart rate, and a slowly developing increase in transpulmonary pressure without change in rate of airflow or tidal volume (see also GANDIHA et al. 1975). A similar effect on transpulmonary pressure is produced by a small dose of histamine and may be attributed to constriction of alveolar ducts and respiratory bronchioles. The effect of histamine and the effect of tubocurarine on the cardiovascular system and on transpulmonary pressure were reduced by the antihistamine drug mepyramine, and were abolished by a combination of mepyramine and cimetidine. These effects of tubocurarine on arterial pressure and transpulmonary pressure may therefore be attributed to histamine release. Not all cats appear to react to tubocurarine, in doses of this order of magnitude, by releasing sufficient histamine to affect blood pressure and respiration; in some, the cardiovascular and other changes produced appear to be mainly the result of ganglion block. However, in the same cat as that in which tubocurarine was effective (Fig. 5), and in all other cats, vecuronium, in a dose about three times greater than that necessary to produce complete twitch block, was without effect on blood pressure, heart rate or respiratory variables. Even in considerably larger doses, vecuronium has not been found to produce any histamine-like effects. It may therefore be concluded that histamine release is unlikely to be a problem in the use of vecuronium.

II. Anticholinesterase Activity

All neuromuscular blocking drugs are capable of inhibiting cholinesterases, but the important question is whether they, or their metabolites, produce this effect in concentrations close to those present during clinical use. Table 3 lists the IC_{50} values of some neuromuscular blocking drugs against human acetylcholinesterase and plasma butyrylcholinesterase. Benzoquinonium is the main example from the past of a neuromuscular blocking drug that is a potent inhibitor of acetylcholinesterase (HOPPE 1951; FOLDES 1978). This was a serious disadvantage, not only because of the autonomic side effects produced, but also because such an action interferes with the effectiveness of anticholinesterase reversal agents, such as neostigmine and pyridostigmine. Benzoquinonium produces neuromuscular block of a nondepolarizing nature despite its own powerful anticholinesterase activity. However, since it has already inhibited much of the enzyme, anticholinesterase agents have little or no ability to overcome its neuromuscular blocking action. The consequences of the antiacetylcholinesterase action of benzoquinonium led to its withdrawal from use. In selecting new neuromuscular blocking drugs for further study it is also important to exclude compounds with relatively weak but persistent anticholinesterase activity, whether the activity be extered by the parent drug or by its metabolites, otherwise difficulty with reversal may be unexpectedly experienced after a long-lasting operation in which repeated maintenance doses of the neuromuscular blocking drug have been injected.

Pancuronium is the main example of a neuromuscular blocking drug that inhibits plasma butyrylcholinesterase. Fortunately, inhibition of this enzyme is of little practical importance except possibly in so far as it may impair the metabolism of other drugs, such as procaine and succinylcholine, that are inactivated by this enzyme.

Table 3. Anticholinesterase activities of some neuromuscular blocking drugs (listed in increasing order of potency against acetylcholinesterase)[a]

Drug	Concentration to produce 50% inhibition (*M*)	
	Human acetylcholinesterase	Human plasma butyrylcholinesterase
Suxamethonium	1.3×10^{-3}	6.4×10^{-4}
Metocurine	3.2×10^{-3}	4.8×10^{-4}
Pancuronium	3.0×10^{-4}	5.6×10^{-8} [b]
Gallamine	4.5×10^{-4}	2.4×10^{-4}
Alcuronium	6.7×10^{-4}	5.4×10^{-5}
Tubocurarine	7.2×10^{-4}	3.0×10^{-4}
Decamethonium	3.9×10^{-5}	2.4×10^{-5}
Atracurium	5.6×10^{-5}	Not known
Vecuronium	6.6×10^{-5}	6.2×10^{-7}
Fazadinium	2.0×10^{-6}	Not known
Benzoquinonium	2.2×10^{-7} [b]	2.1×10^{-5}

[a] With the exception of those relating to atracurium and fazadinium, the data are from FOLDES and co-workers (reviewed in FOLDES 1978 and 1980, personal communication) who used erythrocytes as the source of acetylcholinesterase. The data for atracurium and fazadinium are from HUGHES and CHAPPLE (1981) who used skeletal muscle as the source of acetylcholinesterase. Since the data for these two drugs were obtained by different workers with a different tissue source of enzyme, they can serve only as a rough guide for comparison in this context

[b] Clinically important

Vecuronium is about ten times less potent than pancuronium in inhibiting plasma butyrylcholinesterase, and, because vecuronium, in humans, is slightly more potent than pancuronium in its neuromuscular blocking action, its antibutyrylcholinesterase activity is even less important than that of pancuronium. Vecuronium is about 4.5 times more potent than pancuronium as an inhibitor of acetylcholinesterase, but the effective inhibitory concentration is still many times greater than that necessary to produce neuromuscular block, so that this action has no practical importance. In the anaesthetized cat, doses of vecuronium at least five times greater than that just necessary to produce complete twitch block, do not potentiate the depressor responses to small intravenous doses of acetylcholine, confirming that any anticholinesterase activity is unimportant at neuromuscular blocking doses.

It is well known that plasma cholinesterase (acetylcholine acyl hydrolase) exists in ten variants (including one inactive one) controlled by four allelic genes. The genotypes of individuals can be determined by measurement of the so-called dibucaine numbers and fluoride numbers (see, for example, FOLDES 1978). WHITTAKER and BRITTEN (1980) have shown that pancuronium can be used to differentiate the usual enzyme from the atypical dibucaine-resistant enzyme, since the drug was found to be about 150 times more effective in inhibiting the normal than the atypical enzyme. In contrast, vecuronium had less affinity than pancuronium for the normal enzyme (Table 3), but greater affinity than pancuronium for the atypical enzyme, with the result that there was relatively little difference in the affinities of vecuronium for the two types of enzyme.

III. Ganglion Block

Sympathetic ganglion block produced by tubocurarine and fazadinium may contribute to the hypotension seen in humans with these agents (McDOWELL and CLARKE 1969; BLOGG et al. 1973; UNGERER and ERASMUS 1974). Tubocurarine may have a somewhat more powerful action on parasympathetic than on sympathetic ganglia (GUYTON and REEDER 1950), but the difference is slight. Ganglion-blocking action is readily detected in the anaesthetized cat in which the nictitating membrane is made to contract by preganglionic stimulation. Tubocurarine produces a degree of blockade of transmission through the superior cervical ganglion of the cat in doses just sufficient to block twitches of the tibialis anterior muscle (BOWMAN and WEBB 1972), and fazadinium impairs ganglionic transmission in doses that are relatively only slightly larger (MARSHALL 1973). However, many other neuromuscular blocking drugs have only very weak ganglion-blocking potency (MARSHALL 1980), and vecuronium is one of these. Thus, doses of vecuronium more than 500 times greater than the neuromuscular blocking dose were required to depress contractions of the cat nictitating membrane evoked by preganglionic stimulation (DURANT et al. 1979c).

IV. Block of Muscarinic Receptors

In general, neuromuscular blocking drugs have only very weak actions in blocking muscarinic receptors in glands, and in the smooth muscles of the eye, the bladder and the gastrointestinal tract, and vecuronium is no exception. For example, vecuronium, in concentrations (10^{-2} *M*) nearing the limits of its solubility, was found to produce only a small depression of responses of the guinea-pig ileum to acetylcholine. Thus, vecuronium blocked the muscarinic receptors of the guinea-pig ileum only in concentrations several thousand times higher than those necessary to produce neuromuscular block (R. J. MARSHALL 1981, personal communication).

Although gallamine has very little action in blocking muscarinic receptors in the gut, it was found, shortly after its introduction into clinical use, to be quite potent in blocking muscarinic receptors in the heart (RIKER and WESCOE 1951). Some years later, pancuronium was found to exert a similar effect on the heart (SAXENA and BONTA 1970), and so it began to appear that the cardiac muscarinic receptors differ in their characteristics from those in smooth muscles and glands, even though both types are about equally sensitive to atropine. LEUNG and MITCHELSON (1982a, b) made a detailed study of the relative selectivity of pancuronium for cardiac muscarinic receptors compared with those in the gut. A convenient method for estimating the selectivity of a drug for the neuromuscular junction compared with the heart, is to record, concurrently, twitches of a skeletal muscle, and bradycardia responses to vagal stimulation and to a muscarinic agonist (such as metacholine), in, for example, the anaesthetized cat. The method is illustrated by reference to Fig. 6. Cumulative log dose–response curves for the cardiac and the skeletal muscle effects can be constructed, and a measure of selectivity is then given by comparing the dose to produce 50% inhibition of maximal skeletal muscle twitches with the dose to produce 50% inhibition of the bra-

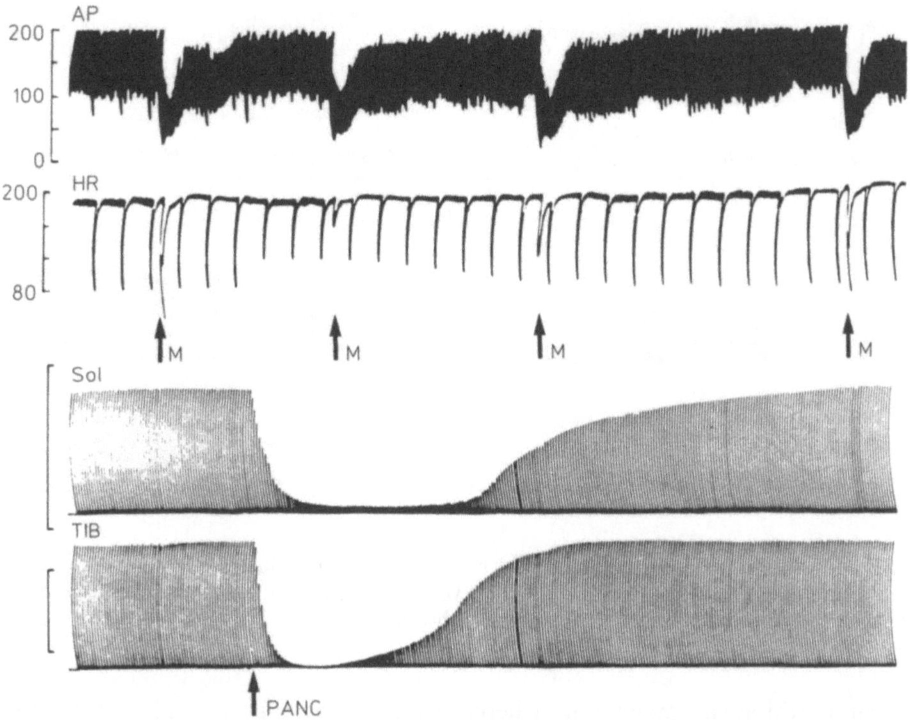

Fig. 6. Cat, chloralose anaesthesia. Records from above downwards: *AP* arterial pressure (mmHg); *HR* heart rate (beats/min); *Sol* maximal twitches of a soleus muscle; *Tib* maximal twitches of a tibialis anterior muscle (evoked by stimulation of their motor nerves at a frequency of 0.1 Hz). Muscle twitch tension calibration = 5 N. The right vagus nerve was stimulated every 100 s for 5 s at 8 Hz to evoke decreases in heart rate. *M* i.v. bolus injections of metacholine, each of 10 μg/kg. At *PANC* pancuronium 30 μg/kg was injected. Note that, as well as producing neuromuscular block, pancuronium simultaneously produced a decrease in the bradycardia responses to both vagal stimulation and to metacholine, but it did not affect the depressor response to metacholine. From an experiment by J. HOUSTON published by BOWMAN (1982)

dycardia responses to vagal stimulation. Some workers prefer to compare the 90% twitch blocking dose with the 10% vagal blocking dose. Whatever method is used, the comparison in terms of receptor sensitivity is not accurate, since the dose–response curves for the effects are not parallel, and the ratio varies to some extent according to the frequency of stimulation used. Nevertheless, providing that all the experiments are carried out in the same way, comparison of the selectivity ratios for different drugs is meaningful. MARSHALL (1980) has reviewed and summarized the data obtained from a large number of experiments of this type. It is evident from Fig. 6 that pancuronium produces some vagal block in doses just big enough to produce complete twitch block. In fact, if the two effects are compared at the 50% level, a ratio of around 4:1 is obtained (MARSHALL 1980). That is, about four times as much pancuronium is required to produce 50% block of the response to vagal stimulation as to produce 50% muscle twitch block. Al-

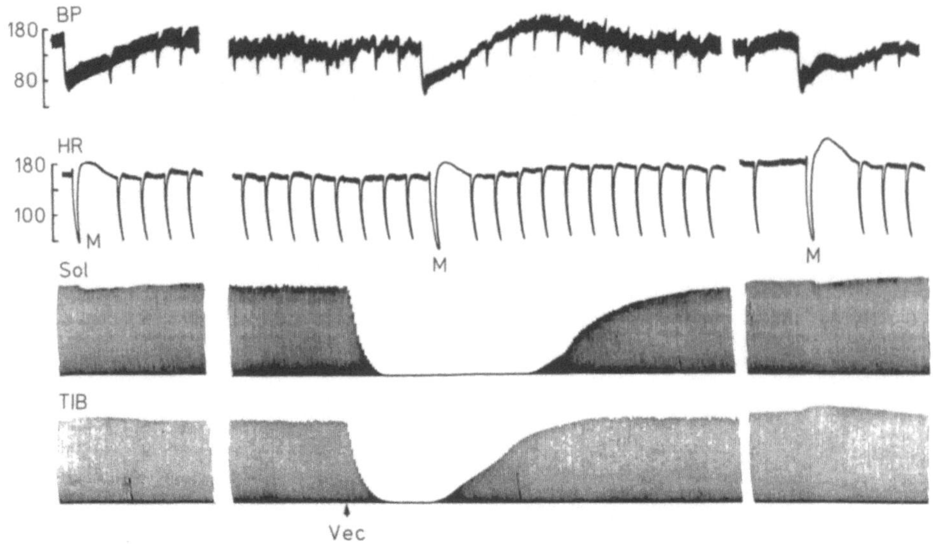

Fig. 7. Cat, chloralose anaesthesia. Similar experiment to that illustrated in Fig. 6. At *Vec*, vecuronium 40 μg/kg was injected. Note that vecuronium was without effect on bradycardia produced by vagal stimulation or by metacholine

curonium exhibits a similar ratio, whereas with gallamine and fazadinium the ratio is around 1 : 1 or even less. Vecuronium is outstanding in this respect in that Durant et al. (1979 c) obtained a ratio for this drug of around 60 : 1, Marshall et al. (1983) around 80 : 1, and, in unpublished experiments, other workers (R. J. Marshall 1982, J. Houston 1982, personal communications) have independently found similar high degrees of selectivity for the neuromuscular junction. The absence of effect of a neuromuscular blocking dose of vecuronium on bradycardia responses is illustrated in Fig. 7. The strikingly weak effect of vecuronium compared with pancuronium on cardiac muscarinic receptors has also been demonstrated in the pithed rat and in isolated guinea-pig atria (Marshall et al. 1980 b).

In the cat, at least, the block of the bradycardia responses to vagal stimulation by gallamine, pancuronium and some other drugs is the result of blockade of the muscarinic receptors in the sinoatrial node, since the bradycardia responses to metacholine (a directly acting muscarinic agonist) are blocked simultaneously. Figure 6 illustrates this effect of pancuronium and Fig. 7 the absence of this effect of vecuronium. Nevertheless, the vagal blocking action may not be entirely a consequence of muscarinic receptor block in all species, under all conditions, as explained later. However, when it is due to muscarinic receptor block, there is evidence that the antagonism is competitive in nature, at least with low concentrations of antagonist (Rathbun and Hamilton 1970; Clark and Mitchelson 1976; Marshall and Ojewole 1979; Leung and Mitchelson 1982 a). With high concentrations of antagonist (e.g. of gallamine and stercuronium), so-called metaffinoid antagonism has been proposed to contribute to the effect (Clark and Michelson 1976; Li and Mitchelson 1978).

Figure 6 also illustrates that there is a difference between the muscarinic receptors in blood vessels that mediate vasodilation and the consequent depressor response to metacholine, and those in the heart that mediate bradycardia, for while pancuronium blocked the latter, it did not block the former, even though both are sensitive to atropine. Although arteriolar muscarinic receptors may differ from others in that they appear to require the presence of an as yet unidentified endothelial factor to mediate vasodilatation (see FURCHGOTT 1981 for a concise review), they resemble most of the other noncardiac muscarinic receptors (e.g. those in the gut) in their agonist and antagonist selectivity.

The anaesthetized cat lacks spontaneous cardiac vagal tone, and hence it is necessary to stimulate the vagus electrically in order to demonstrate interference with its action. This has the advantage that vagal activity is then controllable at a constant level. Obviously, in a species, and under conditions, in which vagal tone is pronounced (e.g. anaesthetized humans), an action to impair vagal activity such as that illustrated for pancuronium in Fig. 6, would result in a degree of tachycardia with a ceiling limited by the amount of background vagal tone. Vecuronium, because it is virtually free from vagal blocking action, would be expected not to produce a tachycardia of this type.

Cardiac muscarinic receptors are not the only muscarinic receptors that are blocked by certain neuromuscular blocking drugs. Stimulation of the vagus has been shown to reduce the release of noradrenaline from concomitantly stimulated sympathetic nerves to the heart (LÖFFELHOLZ and MUSCHOLL 1970; LEVY and BLATBERG 1976; VANHOUTTE and LEVY 1980). It is therefore proposed that vagal nerve terminals impinge not only upon the nodal and atrial cells of the heart, but also on the sympathetic nerve endings where they act to inhibit the release of noradrenaline. The cholinoceptors on the sympathetic nerve endings are of the muscarinic type (for a brief review see SHEPHERD et al. 1978). Indeed, muscarinic receptors seem to be present on a variety of different types of nerve endings where they mediate inhibition of transmitter release (e.g. FOSBRAEY and JOHNSON 1980; ABBS and JOSEPH 1981; COX and ENNIS 1982), and this may be so whether or not a cholinergic innervation is present. The muscarinic receptors present on sympathetic nerve terminals are blocked by gallamine and pancuronium (VERCRUYSSE et al. 1979).

Where noradrenaline release from sympathetic nerve endings is negatively modulated by activity in the parasympathetic nerves, as is presumably the case in the heart, the actions of gallamine and pancuronium would be expected to cause an increase in noradrenaline release and a consequent tachycardia that would add to that arising from blockade of the action of the vagus on the sinoatrial node.

Most blood vessels are generally believed not to possess a parasympathetic innervation, but in those that do, activity in parasympathetic nerves presumably acts to inhibit noradrenaline release from the sympathetics. Pancuronium and gallamine, by blocking this inhibitory effect, would then be expected to give rise to vasoconstriction. Whether or not particular blood vessels receive a parasympathetic innervation, their sympathetic nerve terminals may possess inhibitory muscarinic receptors, and therefore they may serve as models for testing neuromuscular blocking drugs for this particular kind of atropine-like activity. The iso-

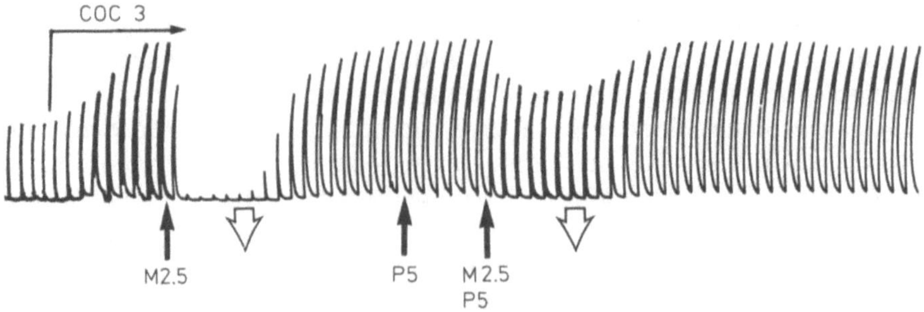

Fig. 8. Isolated perfused rabbit ear artery stimulated transmurally (8 Hz for 10 s every 2 min) to excite the intramural adrenergic nerve endings. From *COC* 3, cocaine (3 μg/ml) was present in the perfusion fluid for the remainder of the experiment. At *M* 2.5, 2.5 μg/ml metacholine was present in the perfusion fluid and was removed at the *inverted open arrows*. At *P* 5, 5 μg/ml pancuronium was present in the perfusion fluid

lated perfused dog saphenous vein (VANHOUTTE 1977) or ear artery of the rabbit (DE LA LANDE and RAND 1965) are examples of this kind. Figure 8 illustrates an experiment on the isolated perfused ear artery of the rabbit stimulated to constrict through its intramural sympathetic nerves. Cocaine was added to the perfusion fluid to block neuronal uptake of noradrenaline because, in the absence of cocaine, certain neuromuscular blocking drugs (including pancuronium) may themselves produce this effect, which would complicate the interpretation of any interaction with muscarinic receptors. Muscarinic agonists are without effect on vasoconstriction produced by noradrenaline, but they impair the responses to sympathetic nerve stimulation be reducing the evoked release of noradrenaline from the nerve endings (RAND and VARMA 1970). This effect of metacholine is illustrated in Fig. 8. Pancuronium, subsequently added to the perfusion fluid, in the presence of cocaine, was without effect on the vasoconstrictions evoked by nerve stimulation, but it greatly impaired the ability of metacholine to inhibit them. In further experiments, gallamine was shown similarly to block the action of metacholine, but vecuronium in concentrations up to four times those of pancuronium (the largest tested) was without effect on the metacholine-induced inhibition (W. C. BOWMANN and J. HOUSTON 1980, unpublished work).

Another location of muscarinic cholinoceptors that are blocked by pancuronium and gallamine is on the small dopaminergic interneurones (so-called small intensely fluorescent or SIF cells) in sympathetic ganglia. These SIF cells are activated through muscarinic receptors stimulated by acetylcholine that is released from collaterals of the preganglionic cholinergic fibres. Dopamine, released from the SIF cells onto the postganglionic neurones, hyperpolarizes them and therefore suppresses ganglionic transmission (GREENGARD and KEBABIAN 1974). Block, produced by gallamine or pancuronium, of activation of the SIF cells may, therefore, at appropriate stimulation frequencies, facilitate transmission through the ganglion by inactivating the inhibitory modulating influence of the dopaminergic SIF cell loop (GARDIER et al. 1978). Vecuronium has not yet been tested for a possible action on the SIF cell loop, but, since the muscarinic receptors involved seem to resemble those in the heart and on sympathetic nerve endings (BOWMAN 1982), it seems likely that vecuronium will be without influence on this mechanism in clinically used concentrations.

V. Inhibitory Action on the Postganglionic Cardiac Vagus

It is clear from the foregoing description that, in the cat at least, the main mechanism through which certain neuromuscular blocking drugs depress cardiac vagal activity is by blocking the postjunctional muscarinic cholinoceptors. However, this may not be the only mechanism involved in all species. Thus, from a series of elegant experiments on the isolated guinea-pig heart, Waud and co-workers (LEE SON and B.E. WAUD 1977; LEE SON and D.R. WAUD 1978, 1980; WAUD et al. 1978) have calculated that the concentrations of pancuronium and gallamine that block the responses to postganglionic vagal stimulation are too low to exert significant blocking action on the postjunctional muscarinic receptors, and they conclude (LEE SON and WAUD 1980) that the main site of action of these two drugs in blocking the cardiac vagus is on the postganglionic nerve terminals. Vecuronium is again very much less potent than pancuronium in producing this type of vagal inhibition (LEE SON et al. 1981).

VI. Noradrenaline Release and Reuptake Block

Pancuronium and gallamine have been shown to exert indirect sympathomimetic effects (BROWN and CROUT 1970; DOMENACH et al. 1976; MARSHALL and OJEWOLE 1979) and both pancuronium and fazadinium inhibit the neuronal reuptake of noradrenaline (IVANKOVICH et al. 1975; QUINTANA 1977; DOCHERTY and MCGRATH 1978; TOMLINSON 1979; MARSHALL and OJEWOLE 1979; SALT et al. 1980; MARSHALL et al. 1980b). Pancuronium has been shown to increase plasma catecholamine concentration (NANA et al. 1973), presumably as a consequence of its uptake blocking action, and it is likely that it is this same action that may lead to serious drug interactions with other drugs with Uptake 1 blocking action, such as imipramine (EDWARDS et al. 1979).

Vecuronium is considerably less potent than pancuronium in blocking the neuronal uptake of noradrenaline (MARSHALL and OJEWOLE 1979; DOCHERTY and MCGRATH 1980; SALT et al. 1980), the relative potency differing according to the species and the technique used. Thus, vecuronium was about 30 times less potent than pancuronium in potentiating the effects of sympathetic cardioaccelerator stimulation in pithed rats (DOCHERTY and MCGRATH 1980), although it was only 4 times less potent than pancuronium in inhibiting the neuronal uptake of tritiated noradrenaline in perfused rat hearts (SALT et al. 1980).

VII. Cardiovascular Effects

Pancuronium, gallamine and fazadinium are examples of neuromuscular blocking drugs that tend to produce tachycardia and hypertension (KELMAN and KENNEDY 1971; WALTS 1975; HUGHES et al. 1976; FRAHLEY et al. 1978; ROIZEN and

Freeley 1978; Pauca and Skovsted 1981). Peripheral resistance is usually not increased, although Lee et al. (1980) have described a constrictor effect of pancuronium on veins under some circumstances.

A combination of the unusual type of atropine-like action, the Uptake 1 blocking action, and the postganglionic cardiac vagal inhibitory action of gallamine, pancuronium and fazadinium amply accounts for the cardiovascular effects of these drugs in humans. There is considerable species difference with regard to the relative importance of these effects. For example, block of cardiac muscarinic receptors is readily observed in the cat, whereas Uptake 1 block seems to be the predominant effect in the rat, and block of cardiac vagal nerve endings predominates in the guinea-pig. It is not known which effect is the most predominant in humans, but it is quite likely that the relative importance varies from patient to patient, according to such factors as the preexisting autonomic balance, the type of premedication given, the anaesthetic used and any concurrent therapy with other drugs. In so far as screening methods are concerned, freedom from cardiac vagal blocking action in the cat, and from other effects on the heart rate and blood pressure as illustrated in Fig. 7, seems to be a useful indication that a new drug will be free from unwanted cardiovascular effects in humans.

In an elegantly designed series of experiments in which equiactive doses of neuromuscular blocking drugs (pancuronium, tubocurarine, metocurine, and vecuronium) were administered repeatedly to anaesthetized dogs in such a way that on different occasions each dog received each of the drugs, Booij et al. (1980) confirmed that the haemodynamic effects of vecuronium were trivial compared with those of the other drugs studied. A similar lack of cardiovascular effects of vecuronium in anaesthetized cats, was described by Marshall et al. (1980 b).

G. Influence of Acid–base Balance

The influence of changes in acid–base balance on the depth of neuromuscular block produced by vecuronium has been studied in the isolated phrenic nerve–diaphragm preparation of the rat, and in anaesthetized cats (Funk et al. 1980). Both in vitro and in vivo, a fall in pH enhanced the blocking action of vecuronium, possibly because of increased protonation of the tertiary nitrogen, although a pharmacokinetic basis for the potentiation cannot be ruled out. A rise in pH, especially a metabolic alkalosis, had the opposite effect although the change was less striking than the effect of acidosis.

H. Interactions with Other Drugs

Potential interactions of vecuronium with other drugs at the neuromuscular junction have been studied on the isolated diaphragm muscle of the rat, and in anaesthetized rats and cats (Krieg et al. 1980 b; McIndewar and Marshall 1981). In general, vecuronium behaved like other nondepolarizing neuromuscular blocking drugs, being potentiated by halogenated anaesthetics (halothane, enflurane, methoxyflurane) and by certain antibiotics (streptomycin, clindamycin, neomycin, gentamicin) and to a lesser extent by some intravenous anaesthetics (Althesin, ketamine, propanidid). In one respect, however, vecuronium appeared to exhibit an unusual interaction, in that it (but not pancuronium) was potentiated by metronidazole (McIndewar and Marshall 1981). The interaction was not evident immediately after metronidazole was administered, and it did not occur in vitro. A striking potentiation of vecuronium in vivo became apparent, however, an hour or so after the administration of metronidazole had terminated. The authors concluded, in view of the absence of the interaction in vitro and the slow onset in vivo, that the effect may be caused by a metabolite of metronidazole rather than by the parent compound, and since the interaction was not evident with pancuronium they proposed that its basis lay in some pharmacokinetic characteristic of vecuronium, rather than in an action of a metronidazole metabolite on neuromuscular transmission. No similar interaction between vecuronium and metronidazole has been recorded in humans.

J. Conclusions

The results of animal experiments suggested that vecuronium would be a useful, short to medium duration neuromuscular blocking drug in humans. Generally, the cat is useful for forecasting the human dose (approximately 40 μg/kg to produce 90% twitch block in the cat), but no one animal species can reliably allow a forecast of the duration of action in humans. Potency of highly ionized molecules is usually determined mainly by pharmacodynamic variables, and it appears that the motor end-plate cholinoceptors in the cat must closely resemble those in humans. Duration of action, on the other hand, is mainly determined by pharmacokinetic and metabolic characteristics, and only chance could allow any one animal species to resemble another with respect to a particular drug molecule. However, the fact that vecuronium exhibits a relatively rapid recovery rate in all the animal species studied suggested that this would also be so in humans.

Animal experiments also suggested that vecuronium would have a number of additional advantages in that it is easily reversible, relatively noncumulative and virtually free from histamine-releasing action, ganglion-blocking action, atropine-like action, and from cardiovascular effects in general. The extent to which the forecasts from animal experiments are borne out in humans may be judged from the description by R. D. Miller in Chap. 28.

References

Abbs ET, Joseph DN (1981) The effects of atropine and oxotremorine on acetylcholine release in rat phrenic nerve-diaphragm preparations. Br J Pharmacol 73:481–482

Baird WLM, Bowman WC, Kerr WJ (1982) Some actions of Org NC45 and of edrophonium in the anaesthetized cat and in man. Br J Anaesth 54:375–385

Blogg CE, Savege TM, Simpson JC, Ross LA, Simpson BR (1973) A new muscle relaxant – AH 8165. Proc R Soc Med 66:1023–1027

Booij LHDJ, Edwards RP, Sohn YJ, Miller RD (1980) Cardiovascular and neuromuscular effects of Org NC45, pancuronium, metocurine and d-tubocurarine in dogs. Anesth Analg (Cleve) 59:26–30

Booij LHDJ, Vree TB, Crul JF (1981 a) Org NC45: A new nondepolarizing muscle-relaxant with a steroid structure (in Dutch). TGO Tijdschr Geneesmidelenonderzoek 6:1055–1059
Booij LHDJ, Vree TB, Hurkmans F, Reekers-Ketting JJ, Crul JF (1981 b) Pharmacokinetics and pharmacodynamics of the muscle relaxant drug Org NC45 and each of its hydroxy metabolites in dogs. Anaesthesist 30:329–333
Bowman WC (1964) Neuromuscular blocking agents. In: Laurence DR, Bacharach AL (eds) Evaluation of drug activities. Academic, London, pp 325–351 (Pharmacometrics, vol 1, chap 16)
Bowman WC (1980 a) Prejunctional and postjunctional cholinoceptors at the neuromuscular junction. Anesth Analg (Cleve) 59:935–943
Bowman WC (1980 b) A new nondepolarizing neuromuscular blocking drug. Trends Pharmacol Sci 1:263–266
Bowman WC (1980 c) Pharmacology of neuromuscular function. John Wright, Bristol
Bowman WC (1981) New neuromuscular blocking drugs and their antagonists. Indian J Pharmacol 13:1–22
Bowman WC (1982) Non-relaxant properties of neuromuscular blocking drugs. Br J Anaesth 54:147–160
Bowman WC, Marshall IG (1981) Die Rolle prä- und postsynaptischer cholinergischer Rezeptoren bei der neuromuskulären Übertragung und deren Beeinflußbarkeit durch Muskelrelaxantien. In: Buzello W (ed) Muskelrelaxantien. Georg Thieme, Stuttgart, pp 34–48
Bowman WC, Norman J (eds) (1980) Symposium on Org NC45. Br J Anaesth 52 [Suppl 1]:1–72S
Bowman WC, Webb SN (1972) Neuromuscular blocking and ganglion blocking activities of some acetylcholine antagonists in the cat. J Pharm Pharmacol 24:762–772
Brown BR, Crout JR (1970) The sympathomimetic effect of gallamine on the heart. J Pharmacol Exp Ther 172:266–273
Buckett WR, Hewett CL, Savage DS (1973) Pancuronium bromide and other steroidal neuromuscular blocking agents containing acetylcholine fragments. J Med Chem 16:1116–1124
Clark AL, Mitchelson F (1976) The inhibitory effect of gallamine on muscarinic receptors. Br J Pharmacol 58:323–331
Colquhoun D, Sheridan RE (1981) Modes of action of gallam ine. Proc R Soc Lond [Biol] 211:181–203
Colquhoun D, Dreyer F, Sheridan RE (1979) The actions of tubocurarine at the frog neuromuscular junction. J Physiol (Lond) 293:247–284
Cox B, Ennis C (1982) Tryptamine-induced inhibition of serotonin release from rat hypothalamic slices is mediated via a cholinergic interneurone. Br J Pharmacol 76:242P
Cronnelly R, Gencarelli P, Miller RD, Fisher DM, Nguyen L (1982) Pharmacokinetics of pancuronium and Org NC45 (Norcuron). Anesth Analg (Cleve) 61:176–177
De La Lande IS, Rand MJ (1965) A simple isolated nerve-blood vessel preparation. Aust J Exp Biol Med Sci 43:639–656
Docherty J, McGrath JC (1978) Sympathomimetic effects of pancuronium bromide on the cardiovascular system of the pithed rat: a comparison with the effects of drugs blocking the neuronal uptake of noradrenaline. Br J Pharmacol 64:589–599
Docherty JR, McGrath JC (1980) A comparison of the effects of pancuronium bromide and its monoquaternary analogue Org NC45 on autonomic and somatic neurotransmission in the rat. Br J Pharmacol 71:225–233
Domenach JS, Garcia RC, Sasiain JMR, Loyala A, Oroz JS (1976) Pancuronium bromide: an indirect sympathomimetic agent. Br J Anaesth 48:1143–1148
Dreyer F (1981) Molekulare Grundlagen der neuromuskulären Blockade. In: Buzello W (ed) Muskelrelaxantien. Georg Thieme, Stuttgart, pp 15–34
Durant NN (1978) Studies on neuromuscular blocking and facilitatory agents. PhD Thesis, University of Strathclyde
Durant NN (1982) Norcuron®, a new nondepolarizing neuromuscular blocking agent. Semin Anesth 1:47–56

Durant NN, Houwertjes MC, Agoston S (1979a) Renal elimination of Org NC45 and pancuronium. Anesthesiology 51:S266
Durant NN, Houwertjes MC, Agoston S (1979b) Hepatic elimination of Org NC45 and pancuronium. Anesthesiology 51:S267
Durant NN, Marshall IG, Savage DS, Nelson DJ, Sleigh T, Carlyle IC (1979c) The neuromuscular and autonomic blocking activities of pancuronium analogues in the cat. J Pharm Pharmacol 31:831–836
Durant NN, Houwertjes MC, Crul JF (1980) Comparison of the neuromuscular blocking properties of Org NC45 and pancuronium in the rat, cat, and rhesus monkey. Br J Anaesth 52:723–730
Durant NN, Nguyen N, Lee C, Katz RL (1981) The cumulative effects of norcuron and atracurium. Anesthesiology 55:A209
Edwards RP, Miller RD, Roizen MF, Ham J, Way WL, Lake CR, Roderick L (1979) Cardiac responses to imipramine and pancuronium during anesthesia with halothane or enflurane. Anesthesiology 50:421–425
Fahey MR, Morris RB, Miller RD, Sohn YJ, Cronnelly R, Gencareli P (1981) Clinical pharmacology of Org NC45 (Norcuron™). Anesthesiology 55:6–11
Foldes FF (1978) Enzymes of acetylcholine metabolism. In: Foldes FF (ed) Enzymes in anesthesiology. Springer, Berlin Heidelberg New York, pp 91–168
Foldes FF, Chaudry I, Ohta Y, Amaki Y, Nagashima H, Duncalf D (1981) The influence of stimulation parameters on the potency and reversibility of neuromuscular blocking agents. J Neural Transm 52:227–249
Fosbraey P, Johnson ES (1980) Release-modulating acetylcholine receptors on cholinergic neurones of the guinea-pig ileum. Br J Pharmacol 68:289–300
Frahley DS, Lemoncelli GL, Coleman A (1978) Severe hypertension associated with pancuronium bromide. Anesth Analg (Cleve) 57:265–267
Funk DI, Crul JF, Van Der Pol FM (1980) Effects of changes in acid-base balance on neuromuscular blockade produced by Org NC45. Acta Anaesth Scand 24:119–124
Furchgott RF (1981) The requirement of endothelial cells in the relaxation of arteries by acetylcholine and some other vasodilators. Trends Pharmacol Sci 2:173
Galindo A (1971) Prejunctional effect of curare: its relative importance. J Neurophysiol 34:289
Galindo A (1972) Curare and pancuronium compared: effects on previously undepressed myoneural junction. Science 178:753
Gandiha A, Marshall IG, Paul D, Rodger IW, Scott W, Singh H (1975) Some actions of chadonium iodide, a new short-acting muscle relaxant in anaesthetised cats and isolated preparations. Clin Exp Pharmacol Physiol 2:159–170
Gardier RW, Tsevdos EJ, Jackson DB, Delaunois AL (1978) Distinct muscarinic mediation of suspected dopaminergic activity in sympathetic ganglia. Fed Proc 37:2422–2428
Greengard P, Kebabian JW (1974) Role of cyclic AMP in synaptic transmission in the mammalian peripheral nervous system. Fed Proc 33:1059–1067
Guyton AC, Reeder RC (1950) Quantitative studies on the autonomic effects of curare. J Pharmacol Exp Ther 98:188–193
Hoppe JO (1951) A new series of synthetic curare-like compounds. Ann NY Acad Sci 54:395–406
Hubbard JI, Wilson DF (1973) Neuromuscular transmission in a mammalian preparation in the absence of blocking drugs, and the effect of d-tubocurarine. J Physiol (Lond) 228:307–326
Hughes R, Chapple DJ (1981) The pharmacology of atracurium: a new competitive neuromuscular blocking agent. Br J Anaesth 53:31–44
Hughes R, Payne JP, Sugai N (1976) Studies on fazodinium bromide (AH8165): a new nondepolarizing neuromuscular blocking agent. Can Anaesth Soc J 23:36–47
Ivankovich AD, Miletich DJ, Albrecht RF, Zahed B (1975) The effect of pancuronium on myocardial contraction and catecholamine metabolism. J Pharm Pharmacol 27:837–841

Katz B, Miledi R (1978) A re-examination of curare action at the motor end-plate. Proc R Soc Lond [Biol] 203:119–133
Kelman GR, Kennedy BR (1971) Cardiovascular effects of pancuronium in man. Br J Anaesth 43:335–338
Krieg N, Crul JF, Booij LHDJ (1980a) Relative potency of Org NC45, pancuronium, metocurine and tubocurarine in anaesthetized man. Br J Anaesth 52:783–788
Krieg N, Rutten JMJ, Crul JF, Booij LHDJ (1980b) Preliminary review of the interactions of Org NC45 with anaesthetics and antibiotics in animals. Br J Anaesth 52 [Suppl 1]:33–36S
Lee C, Yang E, Lippmann M (1980) Constrictive effect of pancuronium on capacitance vessels. Br J Anaesth 52:261–263
Lee Son L, Waud BE, Waud DR (1981) A Comparison of the neuromuscular blocking and vagolytic effects or Org NC45 and pancuronium. Anesthesiology 55:12–18
Lee Son S, Waud BE (1977) Potencies of neuromuscular blocking agents at the receptors of the atrial pacemaker and motor end plates of the guinea-pig. Anesthesiology 47:34–36
Lee Son S, Waud DR (1978) A vagolytic action of neuromuscular blocking agents on the cardiac vagus nerve in the guinea-pig atrium. Anesthesiology 48:191–194
Lee Son S, Waud DR (1980) Effects of nondepolarizing neuromuscular blocking agents on the cardiac vagus nerve in the guinea-pig. Br J Anaesth 52:981–987
Leung E, Mitchelson F (1982a) The interaction of pancuronium with cardiac and ileal muscarinic receptors. Eur J Pharmacol 81:1–9
Leung E, Mitchelson F (1982b) Modification by hexamethonium of the muscarinic receptor blocking activity of pancuronium and homatropine in isolated tissues of the guinea-pig. Eur J Pharmacol 80:11–17
Levy MN, Blatberg B (1976) Effects of vagal stimulation on the overflow of norepinephrine into the coronary sinus during cardiac sympathetic nerve stimulation in the dog. Circ Res 38:81–85
Li CK, Mitchelson F (1978) The effect of stercuronium on cardiac muscarinic receptors. Eur J Pharmacol 51:251–259
Löffelholz K, Muscholl E (1970) Inhibition by parasympathetic nerve stimulation of the release of the adrenergic transmitter. Naunyn Schmiedeberg's Arch Pharmacol 267:181–184
Marshall IG (1973) The ganglion blocking and vagolytic actions of three short-acting neuromuscular blocking drugs in the cat. J Pharm Pharmacol 25:530–536
Marshall IG (1980) Actions of non-depolarizing neuromuscular blocking agents at cholinoceptors other than at the motor endplate. In: Conseiller C et al. (eds) Curares and curarisation. Elsevier/North Holland, New York, pp 257–274
Marshall IG, Agoston S, Booij LDHJ, Durant NN, Foldes FF (1980a) Pharmacology of Org NC45 compared with other non-depolarizing neuromuscular blocking drugs. Br J Anaesth 52:11–19S
Marshall IG, Gibb AJ, Durant NN (1983) The neuromuscular and vagal blocking actions of pancuronium bromide, its metabolites, and vecuronium bromide and its potential metabolites in the anaesthetized cat. Br J Anaesth 55:703–714
Marshall RJ, McGrath JC, Miller RD, Docherty JR, Lamar JC (1980b) Comparison of the cardiovascular actions of Org NC45 with those produced by other non-depolarizing neuromuscular blocking agents in experimental animals. Br J Anaesth 52:21S
Marshall RJ, Ojewole JAO (1979) Comparison of the autonomic effects of some currently used neuromuscular blocking agents. Br J Pharmacol 66:77–78p
McDowell SA, Clarke RSJ (1969) A clinical comparison of pancuronium with d-tubocurarine. Anaesthesia 109:190–219
McIndewar IC, Marshall RJ (1981) Interactions between the neuromuscular blocking drug Org NC45 and some anaesthetic, analgesic and antimicrobial agents. Br J Anaesth 53:785–792
Moss J, Rosow CE, Savarese JJ, Philbin DM, Kniffen KJ (1981) Role of histamine in the hypotensive action of tubocurarine in humans, Anesthesiology 55:19–25
Nana A, Cardan E, Domokos M (1973) Blood catecholamine chanbges after pancuronium. Acta Anaesth Scand 17:83–87

Pauca AL, Skovsted P (1981) Cardiovascular effects of pancuronium in patients anaesthetized with enflurane and fluroxene. Can Anaesth Soc J 28:39–45
Quintana A (1977) Effect of pancuronium bromide on the adrenergic reactivity of the isolated rat vas deferens. Eur J Pharmacol 46:275–277
Rand MJ, Varma B (1970) The effects of cholinomimetic drugs on responses to sympathetic nerve stimulation and noradrenaline in the rabbit ear artery. Br J Pharmacol 38:758–770
Rathbun FJ, Hamilton JT (1970) Effect of gallamine on cholinergic receptors. Can Anaesth Soc J 17:754–790
Riker WF, Wescoe WC (1951) The pharmacology of Flaxedil with observations on some of its analogs. Ann NY Acad Sci 54:373–391
Roizen MF, Freeley T (1978) Diagnosis and treatment. Drugs five years later. Pancuronium bromide. Ann Intern Med 88:64–68
Salt PJ, Barnes PK, Conway CM (1980) Inhibition of neuronal uptake of noradrenaline in the isolated perfused rat heart by pancuronium and its homologues, Org 6368, Org 7268 and Org NC 45. Br J Anaesth 52:313–317
Savage DS (1980) Mechanisms of action of muscle relaxants and relationships between structure and activity. In: Conseiller C et al. (eds) Curares and curarisation. Elsevier/North Holland, New York, pp 21–31
Savage DS (1981) Prinzipien der chemischen Struktur des idealen Muskelrelaxans. In: Buzello E (ed) Muskelrelaxantien. Georg Thieme, Stuttgart, pp 190–195
Savage DS, Cameron AF, Ferguson G, Hannaway C, Mackay IR (1971) Molecular structure of pancuronium bromide (3α,17β-diacetoxy-2β-dipiperidino-5α-androstane dimethobromide), a neuromuscular blocking agent. Crystal and molecular structure of the water: methylene chloride solvate. J Chem Soc B 410–415
Savage DS, Sleigh T, Carlyle I (1980) The emergence of Org NC 45, 1-[(2β,3α,5α,16β,17β)-3,17-bis(acetyloxy)-2-(1-piperidinyl)-androstan-16-yl]-1-methylpiperidinium bromide, from the pancuronium series. Br J Anaesth 52:3–9S
Saxena PR, Bonta IL (1970) Mechanism of selective cardiac vagolytic action of pancuronium bromide. Specific blockade of certain cardiac muscarinic receptors. Eur J Pharmacol 11:332–341
Shepherd JT, Lorenz RR, Tyce GM, Vanhoutte PM (1978) Acetylcholine-inhibition of transmitter release from adrenergic nerve terminals mediated by muscarinic receptors. Fed Proc 37:191–194
Tomlinson DR (1979) On the mechanism of pancuronium-induced super sensitivity to noradrenaline in rat smooth muscle. Br J Pharmacol 65:473–478
Torda TA, Kiloh N (1982a) Org NC 45 reduces quantal release of acetylcholine; a preliminary communication. Anaesth Intensive Care 10:127–129
Torda TA, Kiloh N (1982b) Myoneural actions of Org NC 45. Br J Anaesth 54:1217–1222
Ungerer MJ, Erasmus FR (1974) Clinical evaluation of a new non-depolarizing relaxant. S Afr Med J 48:2561–2564
Upton RA, Nguyen TL, Miller RD, Castagnoli N (1982) Renal and bilary elimination of vecuronium (Org NC 45) and pancuronium in rats. Anesth Analg (Cleve) 61:313–316
Vanhoutte PM (1977) Cholinergic inhibition of adrenergic transmission. Fed Proc 36:2444–2447
Vanhoutte PM, Levy MN (1980) Prejunctional cholinergic modulation of adrenergic neurotransmission in the cardiovarscular system. Am J Physiol 238:H275
Vercruysse P, Bossuyt P, Hanegreefs G, Verbeuren TJ, Vanhoutte PM (1979) Gallamine and pancuronium inhibit prejunctional and postjunctional muscarinic receptors in canine saphenous veins. J Pharmacol Exp Ther 209:225–230
Walts LF (1975) Complications of muscle relaxants. In Katz RL (ed) Muscle relaxants. Elsevier/North Holland, New York, p 209 (Monographs in anaesthesiology, vol 3)
Waud DR, Lee Son S, Waud BE (1978) Kinetic and empirical analysis of dose-response curves illustrated with a cardiac example. Life Sci 22:1275–1286
Whittaker M, Britten JJ (1980) Inhibition of the plasma cholinesterase variants by pancuronium bromide and some of its analogues. Clin Chim Acta 108:89–94

CHAPTER 19

The Derivatives of α-Truxillic Acid

D. A. KHARKEVICH

A. Introduction

The study of structure-activity relationships in the series of truxillic acid (1,3-diphenylcyclobutane-2,4-dicarboxylic acid) derivatives prompted the evaluation of the optimal structures of the cationic centers and of the interonium part of the molecules. The derivatives of α-truxillic acid basic esters and amides with $n=3$ and 4 and with bis-*N*-methyl (or ethyl)piperidinium, bis-*N*-methyl (or ethyl)pyrrolidinium, bis-*N*-methyldiethylammonium or bistriethylammonium groups seemed most interesting for neuromuscular blocking drug design (see Tables 2, 3, 7, 8 in Chap. 13).

Not only bisquaternary ammonium salts, but also bistertiary salts appeared to be worthy of attention for the derivatives of α-truxillic acid aminoalkylamides. Therefore, this section deals with drugs of three groups: (a) bisquaternary ammonium salts of α-truxillic acid aminoalkylesters; (b) bisquaternary ammonium salts of α-truxillic acid aminoalkylamides; and (c) bistertiary salts of α-truxillic acid aminoalkylamides. The compounds were synthesized at the Institute of Pharmacology, Academy of Medical Sciences of the USSR, Moscow.

B. Bisquaternary Salts of Aminoalkylester Derivatives: Anatruxonium, Cyclobutonium, Truxilonium, and Pyrocyclonium

It was mentioned in Sect. A that many α-truxillic acid ester derivatives had high neuromuscular blocking activity. Of the compounds tested, those most promising for clinical application were selected for more detailed examination (Table 1). The comparative pharmacologic evaluation of these four compounds is presented in Table 1 (KHARKEVICH and KRAVCHUK 1961; KHARKEVICH 1965, 1966, 1969, 1970, 1973, 1983; KHARKEVICH et al. 1974, 1977).

I. Neuromuscular Blocking Action

Neuromuscular blocking compounds were studied in experiments on rabbits, cats, rats, chickens, pigeons, and frogs. Anatruxonium and cyclobutonium were studied most comprehensively. In experiments on rabbits, head drop after intravenous bolus injection was determined. All four compounds turned out active neuromuscular blocking agents. They caused head drop in doses of about 20–

Table 1. Neuromuscular blocking activity in the series of bisquaternary ammonium salts of α-truxillic acid derivatives

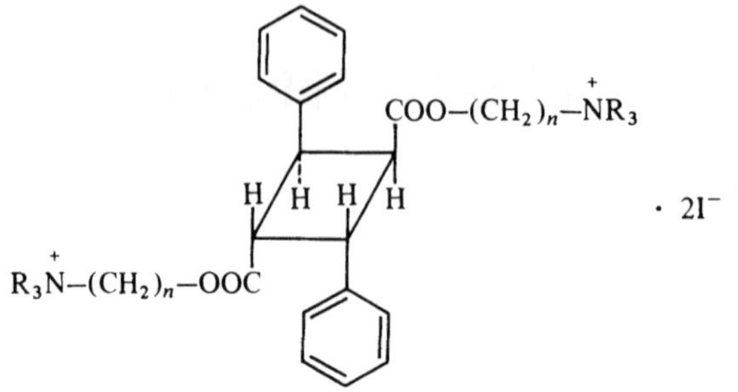

Compound	n	$\overset{+}{N}R_3$	Neuromuscular blocking activity (μg/kg i.v.)		
			Head drop in rabbits (ED_{50} with confidence limits $P=0.05$)	Neuromuscular block in cats[a]	Myorelaxation in patients
Anatruxonium	3	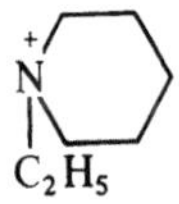	21.7 (17.3–27.1)	100–130	100–150
Cyclobutonium	3	$\overset{+}{N}(C_2H_5)_2CH_3$	37.5 (32.4–43.3)	130–180	120–150
Pyrocyclonium	3	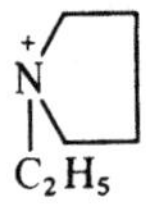	24.2 (21.8–26.9)	120–140	[b]
Truxilonium	4	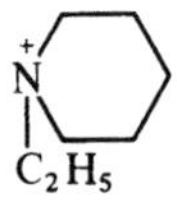	41 (39–43)	150–180	120–150
Tubocurarine			137 (105–178)	180–230	400–500

[a] Transmission from the sciatic nerve to the gastrocnemius muscle in anesthetized cats (chloralose 60 mg/kg plus urethane 300–400 mg/kg i.v.) was recorded. The nerve was stimulated by supramaximal rectangular pulses (0.5 s) at stimulus per second. The contractions were recorded mechanographically

[b] Clinical trial was not carried out

40 μg/kg, i.e., they were 3.5–7 times more active than tubocurarine (Table 1). The effects occurred after 30–60 s without muscular fasciculations and lasted 1–3 min.

Cats were either decerebrated or anesthetized (urethane, 400 mg/kg plus chloralose, 60 mg/kg i.v.). The contractions of the gastrocnemius muscle were recorded mechanographically. The peripheral section of the sciatic nerve was stimulated by rectangular impulses (0.5 ms; 1 stimulus per second). The compounds tested blocked neuromuscular transmission within the dose range of 100–180 μg/kg i.v. No muscular fasciculations or facilitation of neuromuscular transmission prior to the blockade were observed. The neuromuscular blockade was maintained for 3–8 min and it was associated with more or less marked suppression of respiration and even apnea. Thus, in experiments on cats, α-truxillic acid ester derivatives proved more active than tubocurarine (Table 1).

Repeated administration of the drugs to cats in equal doses at 1.5- to 2-h intervals was followed by a marked potentiation of the effect. Each administration made the blockade of neuromuscular transmission more pronounced and prolonged. Under the same experimental conditions, the effect of anatruxonium, truxilonium, and cyclobutonium on neuromuscular transmission was studied after administration of various general anesthetics. It was shown that diethyl ether potentiated the neuromuscular blocking action of those drugs. Thiopental sodium (30 mg/kg i.v.) and hexobarbital sodium (30 mg/kg i.v.) have essentially no influence on the neuromuscular blocking action of the drugs mentioned.

Special attention was devoted to the comparative sensitivity of muscles of various groups to anatruxonium, truxilonium, and cyclobutonium, and also to the influence of diethyl ether and halothane on the order of muscle relaxation after their administration (LEPAKHIN 1970). Comparative sensitivity was studied for the following muscles: m. masseter, m. gastrocnemius, m. triceps brachii, m. obliquus abdominis, m. transversus abdominis, mm. intercostales, and m. phrenicus. Tubocurarine was employed in similar experiments for comparison. The experiments were conducted on cats anesthetized by chloralose plus urethane as they are known to have no effect on the sensitivity of the end-plate acetylcholine receptors to neuromuscular blocking agents. Moreover, it was shown that the results of the experiments on cats anesthetized by chloralose plus urethane did not differ from those on decerebrated animals. The state of neuromuscular transmission was estimated on the basis of the amplitude of evoked potentials in the muscles studied in response to stimulation of the corresponding motor nerves. Peripheral sections of the nerves were stimulated by supramaximal rectangular impulses of 0.1 ms duration at a frequency of 5 stimuli per second. The animals were artificially ventilated before the administration of neuromuscular blocking agents. The degree of blockade of neuromuscular transmission was estimated by the mean values of the amplitude of the sixth, seventh, and eighth potentials in each series which corresponded to their maximal decrease (BLACKMAN 1963). The values obtained were expressed as percentages of the initial amplitude of the potentials. The duration of muscle relaxation was also recorded. To estimate the latter, the index t_{75} was used, i.e., the time from administration to restoration of the amplitude of potentials to 75% of the initial value. Neuromuscular blocking agents were used in doses causing a 80%–100% decrease of the potential amplitude in m.masseter which is the most sensitive.

A comparative study of the sensitivity of muscles to tubocurarine showed that it blocked most the transmission in the neuromuscular synapses of the masseter muscle and limb muscles and it blocked it less in the abdominal muscles. Respiratory muscles, especially m. phrenicus, turned out the least sensitive to tubocurarine which is consistent with the known experimental and clinical data.

The comparative sensitivity of the end-plate acetylcholine receptors to neuromuscular blocking agents of the α-truxillic acid derivatives appeared to be different from their sensitivity to tubocurarine. They blocked most the neuromuscular transmission in the masseter muscle and limb muscles and, to a lower degree, in those of the respiratory muscles. The abdominal muscles turned out to be the least sensitive.

To exclude the individual variations of the animals' sensitivity to neuromuscular blocking agents, experiments were carried out in which 1.5–2 h after the injection of an α-truxillic acid derivative, the same animals were given tubocurarine. In this case, the order of muscle relaxation was of a tubocurarine type. Thus, it was found that the comparative sensitivity of acetylcholine receptors of various groups of muscles to the derivatives of α-truxillic acid and to tubocurarine was different. The dissimilarity of the sensitivity in respiratory and abdominal muscles appeared to be the most characteristic. In contrast to tubocurarine, α-truxillic acid derivatives blocked neuromuscular transmission in the diaphragm and intercostal muscles more than in the abdominal muscles. This specific feature was most marked in anatruxonium and least marked in cyclobutonium (see also Chap. 6).

In addition, the effect of inhalation anesthesia, diethyl ether and halothane, on the comparative sensitivity of acetylcholine receptors of different muscles to anatruxonium and tubocurarine was studied. Prior to the administration of the drug to be tested, inhalation of diethyl ether or halothane was started in concentrations of 4–6 and 1.5–2.5 vol.%, respectively. Neuromuscular blocking agents were administered at doses in which, after the effect of the general anesthetics, they suppressed the amplitude of the evoked potentials of the masseter muscle (the most sensitive) by 85%–100%.

Diethyl ether was found to produce unequal effects on the sensitivity of acetylcholine receptors of various muscles to tubocurarine. After anesthesia, it was the sensitivity of the respiratory muscles which increased most. At the moment of maximal effect of tubocurarine, the degree of inhibition of neuromuscular transmission in the diaphragm and the abdominal muscles became almost equal. Repeated tubocurarine treatment under diethyl ether anesthesia was followed by an even greater increase of the sensitivity of acetylcholine receptors of respiratory muscles. However, the restoration of neuromuscular transmission in the respiratory muscles was faster than in other muscles.

The effect of halothane on the sensitivity of various muscles to tubocurarine was equal to that of diethyl ether. The difference between halothane and diethyl ether was that, under halothane anesthesia, tubocurarine neuromuscular block in the respiratory muscles was even greater than in the abdominal ones.

To provide control, after the amplitude of potentials in all the muscles studied was restored and reached the initial level, the inhalation was stopped, the animals were given urethane instead, and the experiment was repeated 2 h later. It was thus revealed that, in the same animal, the degree and duration of the neuromuscular block became characteristic of tubocurarine as it was previously shown under urethane plus chloralose.

The effect of these general anesthetics on the sensitivity of muscles to anatruxonium was different from that of tubocurarine. Diethyl ether caused a greater increase in sensitivity to anatruxonium in the abdominal muscles, while the difference in the degree of neuromuscular block, induced by anatruxonium after diethyl ether anesthesia, between respiratory and abdominal muscles became less marked and the duration of the abdominal muscle relaxation was increased more than that of the respiratory muscles.

Under halothane anesthesia, the sensitivity to anatruxonium in the abdominal muscles was also increased more than that of the respiratory ones. In this case, the duration of abdominal muscle relaxation was significantly longer than that of respiratory muscle.

In control experiments, 2 h after the inhalation of diethyl ether or halothane was stopped and the animal was given urethane instead, anatruxonium blocked the neuromuscular transmission in the muscles studied in the order earlier observed for that drug under urethane plus chloralose anesthesia.

The comparison of the data obtained in this series with the results in decerebrated or chloralose plus urethane anesthetized cats suggested that diethyl ether and halothane had different influences on the sensitivity of acetylcholine receptors of different groups of muscles to tubocurarine and anatruxonium. These anesthetics increased the sensitivity to tubocurarine most of all in the respiratory muscles and that to anatruxonium in the abdominal muscles. However, in both cases, the duration of abdominal muscle relaxation was increased much more than that of the respiratory muscle.

The mechanism and the locus of the drugs' neuromuscular blocking action were analyzed using the following data. Thus, for instance, in experiments on isolated frog rectus abdominis muscle, none of the derivatives of α-truxillic acid esters caused muscle contractions. On the other hand, they decreased or completely abolished acetylcholine or carbachol stimulation of the muscle. Thus, anatruxonium in concentrations of 10^{-7}–3×10^{-7} g/ml decreased the stimulating effect of acetylcholine by 50%–60% and at 10^{-6} g/ml it completely blocked the muscle response to acetylcholine.

In experiments on chickens and pigeons, the compounds caused flaccid paralysis typical of nondepolarizing neuromuscular blocking agents. For example, intravenous cyclobutonium (15–20 μg/kg) led to the development of flaccid paralysis in chickens which persisted for 3–5 min. A similar effect was observed after administration of 10–15 μg/kg anatruxonium.

The mechanism of action of the compounds was also studied in anesthetized cats (urethane plus chloralose) while recording the evoked potentials of the gastrocnemius muscle. The peripheral section of the sciatic nerve was stimulated by supramaximal impulses of 0.1 ms duration. During the stimulation of the nerve at a frequency of 20–50 stimuli per second, anatruxonium and cyclobutonium promoted the development of pessimal inhibition (tetanic fade), the amplitude of potentials gradually decreasing, starting from the second impulse. Besides, both drugs caused posttetanic "decurarization" (Fig. 1). All this is typical of nondepolarizing neuromuscular blocking agents.

It was shown in experiments on anesthetized cats that neostigmine (30–50 μg/kg i.v.) pretreatment 5 min prior to the neuromuscular blocking agents to be studied or its injection after they induced complete block resulted in a substantial decrease of the degree and duration of the neuromuscular block.

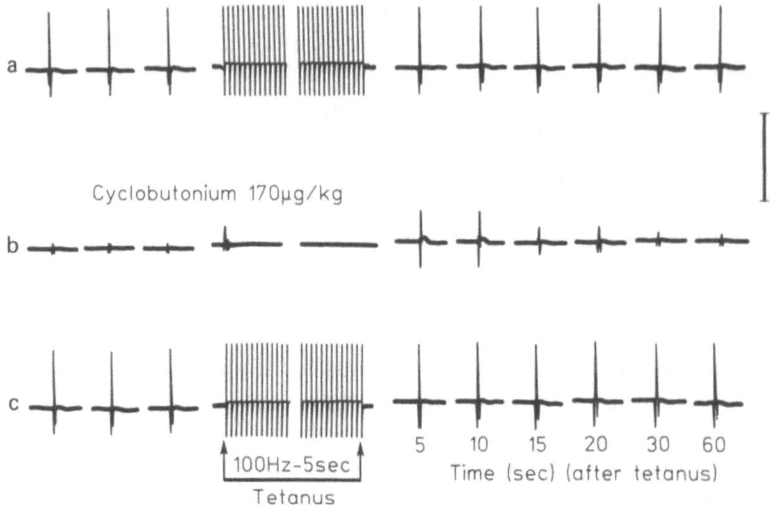

Fig. 1 a–c. Effect of cyclobutonium on posttetanic responses in gastrocnemius muscle. Evoked potentials before **a** and 2 min after i.v. cyclobutonium injection (170 μg/kg) **b**; **c** restoration. Peripheral end of sciatic nerve was stimulated by single rectangular stimuli (0.1 ms). Tetanic stimulation was applied within 5 s (rectangular supramaximal stimuli, 100 Hz, 0.1 ms). Stages of tetanus are shown. *Vertical line* amplitude scale 1 mV. Experiment on cat anesthetized with urethane (600 mg/kg) and chloralose (70 mg/kg i.v.)

Thus, the data presented indicate that antruxonium, pyrocyclonium, truxilonium, and cyclobutonium are nondepolarizing neuromuscular blocking agents. Comparison of the cumulative curves from the experiments on frog isolated rectus abdominis muscle provided evidence of the character of their interaction with acethylcholine receptors. Carbachol served as agonist. Anatruxonium and cyclobutonium (10^{-7}–5×10^{-6} *M*) were shown to shift the cumulative curves to the right, the maximum amplitude and the parallelism of the curves remaining unchanged. Hence, within this range of concentrations, anatruxonium and cyclobutonium act competitively. To specify the locus of their action, the experiments on rat isolated phrenic nerve–diaphragm preparations were carried out. Anatruxonium (6×10^{-8} *M*) and cyclobutonium (1.25×10^{-8} *M*) appeared to decrease the amplitude of miniature potentials without affecting their frequency. Membrane potential remained unchanged. The results suggest that anatruxonium and cyclobutonium act subsynaptically, blocking end-plate acetylcholine receptors.

II. Assessment of Side Effects

In addition to neuromuscular blocking activity, anatruxonium, cyclobutonium, and truxilonium are characterized by a marked cardiotropic antimuscarinic action. They block the transmission from the vagus nerve to the heart and eliminate the cardiotropic effect of acetylcholine on the heart without affecting its hypotensive action (Fig. 2). The initial inhibition of muscarinic acetylcholine receptors of the heart was observed after intravenous injection of 5 μg/kg anatruxonium and

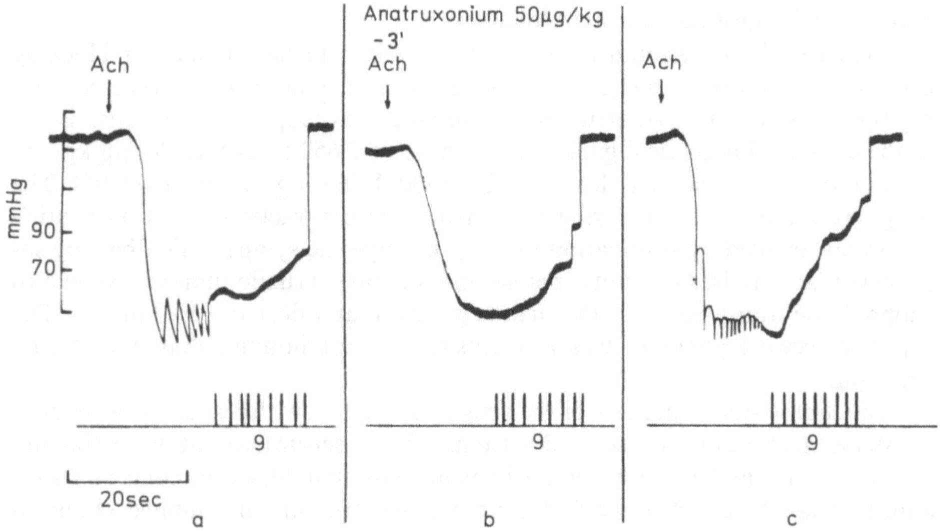

Fig. 2 a–c. Effect of anatruxonium on acetylcholine-induced bradycardia, hypotension, and salivation. Downwards: arterial pressure; salivation from the parotid in drops (the number of drops is indicated by *vertical lines* and their sum). **a** before injection of anatruxonium, **b** 3 min after injection of anatruxonium (50 μg/kg i.v.), **c** 33 min after injection of anatruxonium. *Ach*=acetylcholine (20 μg/kg i.v.). Experiment on cat (3.2 kg) anesthetized with urethane (500 mg/kg) and chloralose (50 mg/kg i.v.)

15 μg/kg cyclobutonium and complete block resulted from 40–50 μg/kg and 100–120 μg/kg, respectively. The effect of the drugs in blocking doses was maintained for 30–40 min. ED_{50} for the elimination of the negative chronotropic effect of acetylcholine was 19.8 (12.4–27.2) μg/kg i.v. for anatruxonium and 43 (26.5–59.5) μg/kg i.v. for cyclobutonium ($P = 0.05$). Sinus tachycardia observed after those drugs seems likely to be related to the cardiotropic antimuscarinic effect.

Anatruxonium and cyclobutonium were studied with respect to their influence on the effects of acetylcholine on the bronchi, ileum, urinary bladder, and salivary glands. The experiments were carried out on urethane plus chloralose anesthetized cats. To measure the muscle tone of the bronchi, the technique of Konzett and Rössler was used. The contractions of the small intestine in situ were recorded according to Nikolayev-Subbotin. The contractions of the urinary bladder were recorded via water–air transmission through a catheter introduced into the urinary bladder. The intensity of salivation was estimated according to the amount of saliva flowing from both parotid glands (the number of drops per minute was registered).

Anatruxonium in neuromuscular blocking doses (150 μg/kg i.v.) and cyclobutonium (200 μg/kg i.v.) were found to decrease the bronchial response to acetylcholine by 49.5% and 46.8%, on average, respectively. However, even in very high doses (1–2 mg/kg), they did not completely suppress the bronchial response to acetylcholine. On the whole, the sensitivity of the bronchial muscarinic receptors is substantially lower than that of the cardiac receptors. Cyclobutonium and anatruxonium had essentially no effect on muscarinic receptors of the salivary glands, small intestine, or urinary bladder.

Anatruxonium, cyclobutonium, or truxilonium in neuromuscular blocking doses did not induce a ganglion-blocking effect. The latter only occurred after high doses of the drugs. Anatruxonium caused a 50% suppression of transmission in the cat superior cervical ganglion at a dose of 1,650 (1,210–2,090) μg/kg and cyclobutonium did so at a dose of 1,230 (900–1,560) μg/kg i.v. ($P=0.05$). The drugs were administered during sustained tonic contraction of the nictitating membrane evoked by stimulation of the preganglionic sympathetic fibers by supramaximal impulses (6 stimuli per second; 0.5 ms). Truxilonium was shown to suppress the transmission in the stellate ganglion at a dose of 0.5–2 mg/kg. The superior cervical ganglion was less sensitive to truxilonium than the stellate ganglion.

Anticholinesterase activity of the drugs was studied with respect to acetylcholinesterase (rabbit brain) and cholinesterase (horse serum) according to Hestrin. Anatruxonium and cyclobutonium in neuromuscular blocking concentrations appeared not to affect acetylcholinesterase. Anatruxonium inhibited cholinesterase by 50% at a concentration of 1.2×10^{-6} *M* and cyclobutonium did so at 1.2×10^{-5} *M*.

Anatruxonium and cyclobutonium in neuromuscular blocking doses caused an insignificant decrease of the arterial pressure. Increased dose resulted in the enhancement of the hypotensive action. This was parallel to the ganglion-blocking action. Sometimes slight tachycardia was observed. No other alterations of the cardiac functions induced by neuromuscular blocking agents in doses of 100–1,000 μg/kg i.v. were observed on the ECG (experiments on anesthetized cats and intact rabbits). Truxilonium at a dose of 500 μg/kg i.v. was shown not to affect the coronary blood flow in cats. This is confirmed by the resistance of the coronary vessels remaining unaffected by truxilonium. Cardiac oxygen consumption was also unchanged.

Anatruxonium and cyclobutonium are unlikely to have myotropic spasmolytic activity, as in concentrations of 2×10^{-6}–2×10^{-3} g/ml they failed to decrease the amplitude of the contractions of the isolated rat ileum induced by barium chloride (4×10^{-4} g/ml). The drugs have no α-adrenoceptor-blocking activity. Moreover, after high doses of both drugs, the vasopressor effect of epinephrine may somewhat increase (experiments on anesthetized cats).

It was shown in the isolated guinea pig ileum that anatruxonium and cyclobutonium in concentrations of 2×10^{-6}–10^{-3} g/ml failed to eliminate histamine-induced (2×10^{-7} g/ml) contractions of the ileum. In addition, it was observed that both drugs at neuromuscular blocking doses did not alter the hypotensive effect of histamine (5 μg/kg i.v.) in anesthetized cats.

Anatruxonium and cyclobutonium showed no local anesthetic or irritating effect after the application of 0.5%–1% solutions to the rabbit eye. Furthermore, the influence of anatruxonium and cyclobutonium on the circulating blood was analyzed in artificially ventilated rabbits. No changes were observed in hemoglobin, red cells, reticulocytes, leukocytes (white cell count), platelets, or clotting time of blood after high doses of the drugs (0.5–2 mg/kg i.v.).

III. Toxicologic Study

The toxicity of the neuromuscular blocking agents was studied on artificially ventilated nonanesthetized rabbits and anesthetized cats (urethane plus chloralose). The arterial pressure and ECG were recorded. The animals were found to tolerate intravenous injection of 200–300 mg/kg anatruxonium. Increasing the dose to as high as 400 mg/kg resulted in the death of one cat of three. Two cats were administered 500 mg/kg anatruxonium and one of them died. No deaths were observed in intact rabbits after administration of 200–400 mg/kg intravenous cyclobutonium, while anesthetized cats could tolerate 600 mg/kg of the drug. Higher doses were not administered.

The toxicity of truxilonium was studied after single and repeated administration over 27 days. Single doses of truxilonium were increased until cardiac arrest occurred (under artificial ventilation). In anesthetized cats, cardiac arrest occurred after more than 1,000 times the dose blocking transmission from the sciatic nerve to the gastrocnemius muscle. To cause cardiac arrest in nonanesthetized rabbits, it was necessary to apply $3–5 \times 10^3$ times the dose required to evoke head drop.

During the 27-day treatment, truxilonium was injected intravenously to rabbits daily at the dose causing head drop. Long-term truxilonium administration did not result in marked changes in the general state of animals. Body weight was essentially unchanged. ECG remained unchanged. No negative influence on the composition of the circulating blood – hemoglobin content, amount of red cells, leukocytes, white blood cell count, clotting time of the blood – was observed.

After truxilonium administration was stopped, postmortem examination was performed. Some regions of the cerebral cortex, subcortical ganglia, hypothalamic area, heart, lungs, liver, kidneys, spleen, small intestine, and adrenals were subjected to microscopic examination. In addition to hematoxylin–eosin staining and Nissle brain staining, histologic treatment of the slices involved several histochemical tests to detect fats, lipids, glycogen, and ferrous oxide salts.

The comparative macroscopic and microscopic examination of the state of internal organs and nervous system of control and test rabbits revealed essentially no pathologic changes in the latter. Thus, the study of α-truxillic acid aminoalkylester derivatives showed that they have low toxicity.

C. Bisquaternary Salts of Aminoalkylamides: Dipyronium and Amidonium

Research in the series of diphenylcyclobutane dicarboxylic acid derivatives (see Chap. 13) showed that the substitution of amide for ester groups in α-truxillic acid quaternary salts resulted in an important increase in neuromuscular blocking potency and a decrease of side effects. Therefore, it was of interest to study amide analogs of the most active esters of this series in terms of the design of highly selective neuromuscular blocking agents. We synthesized the α-truxillic acid quaternary amides: dipyronium (α-truxillic acid bis[1,3-(*N*-pyrrolidinyl)propylamide]dimethiodide) and amidonium (α-truxillic acid bis[1,3-(*N*-piperidinyl)propyl-

Table 2. Neuromuscular blocking activity of bisquaternary ammonium salts of α-truxillic acid aminoalkylamides

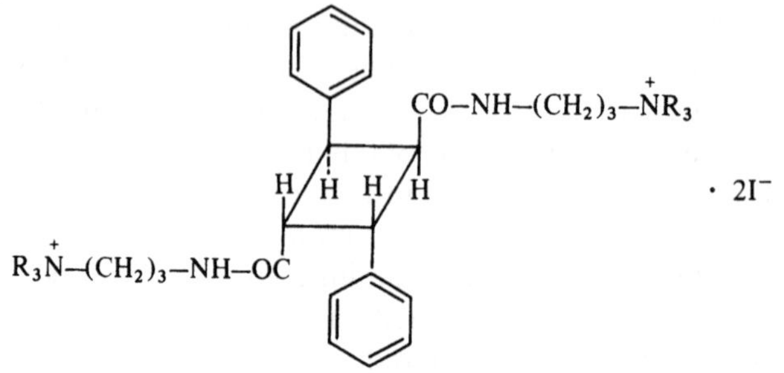

Compound	$\overset{+}{N}R_3$	Head drop in rabbits (ED_{50} μg/kg i.v. with confidence limits, $P=0.05$)	Neuromuscular block in cats (μg/kg i.v.)[a]	Mechanism of action
Amidonium	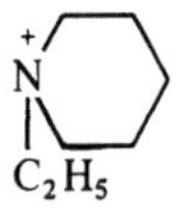	10.8 (10.7–10.9)	18–22	Nondepolarizing
Dipyronium		10 (7.2–12.8)	18–22	Nondepolarizing
Tubocurarine		137 (105–178)	180–230	Nondepolarizing

[a] See Table 1

amide]diethiodide). Pharmacologic study of dipyronium and amidonium comprised the evaluation of their neuromuscular blocking action, their probable side effects, and toxicity (LEMINA 1978; KHARKEVICH 1983).

I. Neuromuscular Blocking Action [1]

Neuromuscular blocking activity of the compounds was studied in rabbits and cats. The ability of dipyronium and amidonium to cause head drop on their intravenous injection was studied on nonanesthetized rabbits. Head drop developed 1.5–2 min after the i.v. injection and lasted 2–4 min. No muscular fasciculations were observed. The head drop test indicated that dipyronium and amidonium were about 14 times more active than tubocurarine (Table 2).

The high neuromuscular blocking potency of the compounds was confirmed in experiments on anesthetized cats in which the ability of dipyronium and amidonium to interfere with the neuromuscular transmission from the sciatic nerve to the gastrocnemius muscle was studied. Intravenous injection of the compounds was followed by a complete neuromuscular block after doses of 18–22 μg/kg.

1 Sections C.I and C.II were written in collaboration with E. Y. U. LEMINA

The action started 1.5–2 min after the injection of dipyronium or amidonium and reached its maximum at 3–6 min. The duration of complete block was 4–8 min, recovery of the initial amplitude of muscle contraction was reached 20–27 min after the injection. The neuromuscular block developed without a preceding facilitation of neuromuscular transmission. Repeated injection of equal doses of dipyronium or amidonium at 2 h intervals resulted in the potentiation of their neuromuscular blocking action.

The mode of neuromuscular blocking action of both agents was studied in pigeons, anesthetized cats (interaction with neostigmine and electromyographic analysis), and in experiments on frog isolated rectus abdominis muscle. Dipyronium (0.02–0.025 μg/kg) and amidonium (0.012–0.025 μg/kg) caused flaccid paralysis in pigeons; the effect developed after 30–90 s and lasted 3–4 min.

The interaction of the compounds tested with neostigmine was determined in anesthetized cats. Neostigmine (30–50 μg/kg) was injected intravenously either 3 min prior to the injection of neuromuscular blocking agents, or after they induced a complete block. Neostigmine was found to antagonize dipyronium and amidonium: its pretreatment substantially reduced their blocking action and its injection during the block was rapidly followed by recovery of neuromuscular transmission.

In experiments on anesthetized cats, the neuromuscular blocking action of the agents was also examined electromyographically. The potentials of the gastrocnemius muscle evoked by electrical stimulation of the peripheral section of the sciatic nerve were recorded. The changes of the evoked potentials in response to stimulation at a frequency of 5 and 20 stimuli per second and the influence of the agents on posttetanic responses were estimated. It follows from these experiments that neuromuscular block induced by the drugs tested was associated with the phenomena typical of nondepolarizing neuromuscular blockers – tetanic fade (Fig. 3) and posttetanic decurarization. These findings suggested dipyronium and amidonium to be nondepolarizing neuromuscular blocking agents.

To specify the mechanism of neuromuscular blocking action of the agents, the antagonism of dipyronium and amidonium with carbachol (method of cumulative curves) on frog isolated rectus abdominis muscle was investigated. Dipyronium and amidonium appeared to cause a parallel rightward shift of cumulative concentration–effect curves of carbachol, the slope and the initial maximum of the contractile reaction remaining unchanged, which is typical of competitive antagonists. Affinity values (pA_2) to acetylcholine receptors of the skeletal muscles were: dipyronium 6.95 (6.7–7.2); amidonium 7.0 (6.65–7.35). Thus, dipyronium and amidonium are competitive neuromuscular blocking agents.

The safety margin of dipyronium and amidonium was estimated in experiments on albino mice by intravenous injection. Mean effective doses (with confidence limits at $P=0.05$) in which the agents inhibited the righting reflex (paralyzing dose PD_{50}) and mean effective doses in which they caused respiratory arrest (apneic dose AD_{50}) were determined. PD_{50}/AD_{50} was used as a criterion of safety. Tubocurarine and pancuronium served as controls. The safety margin was found to be 2.25 for dipyronium and amidonium, 1.7 for tubocurarine, and 1.5 for pancuronium. Thus, dipyronium and amidonium have a wider safety margin than tubocurarine or pancuronium.

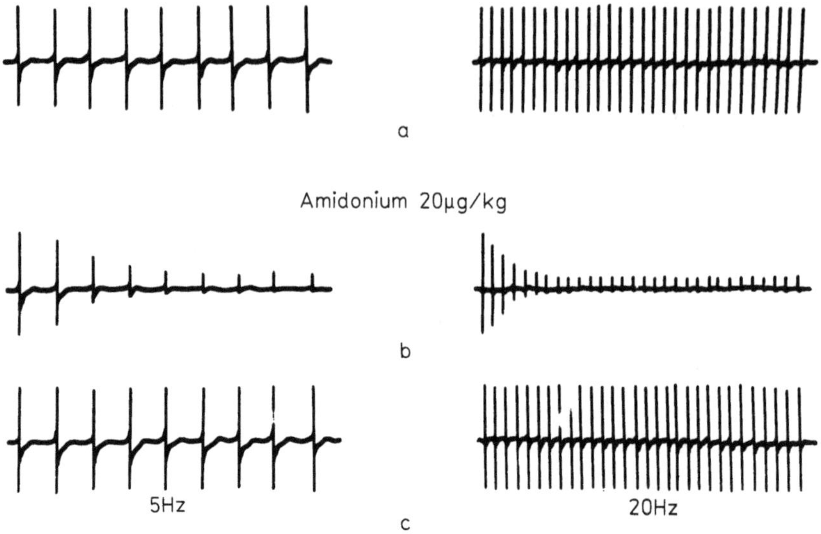

Fig. 3 a–c. Effect of amidonium on gastrocnemius muscle evoked potentials. Peripheral end of sciatic nerve was stimulated by supramaximal rectangular stimuli (5 and 20 Hz, 0.5 ms). **a** before amidonium injection, **b** 2 min after amidonium injection (20 μg/kg i.v.), **c** restoration after 27 min. Experiment on cat anesthetized with urethane (500 mg/kg) and chloralose (50 mg/kg i.v.)

The influence of general anesthetics (diethyl ether, thiopental sodium, and hexobarbital sodium) on the neuromuscular blocking action of dipyronium and amidonium was studied on decerebrated cats. The contractions of the gastrocnemius muscle evoked by the stimulation of a peripheral section of the sciatic nerve were mechanographically recorded. The effect of dipyronium and amidonium at a dose causing an incomplete neuromuscular block (10–15 μg/kg) was registered prior to, and after administration of diethyl ether (inhalation for 20 min), thiopental sodium (30 mg/kg i.v.), or hexobarbital sodium (30 mg/kg i.v.). It follows from the results obtained that diethyl ether potentiated the blocking action of dipyronium and amidonium, while thiopental sodium and hexobarbital sodium failed to affect it.

II. Assessment of Side Effects

To estimate the selectivity of the neuromuscular blocking action of dipyronium and amidonium, their probable side effects were studied. First, it was tested whether or not they have antimuscarinic, ganglion-blocking, anticholinesterase, and histamine-releasing activity, since such activities can be found in many non-depolarizing neuromuscular blocking agents. In addition, the influence of the agents on histamine receptors, adrenoceptors, and smooth muscles was investigated. Some experiments dealt with the anesthetic and irritant activity of the drugs.

The antimuscarinic activity of dipyronium and amidonium was tested on anesthetized cats and in experiments on rat isolated ileum.

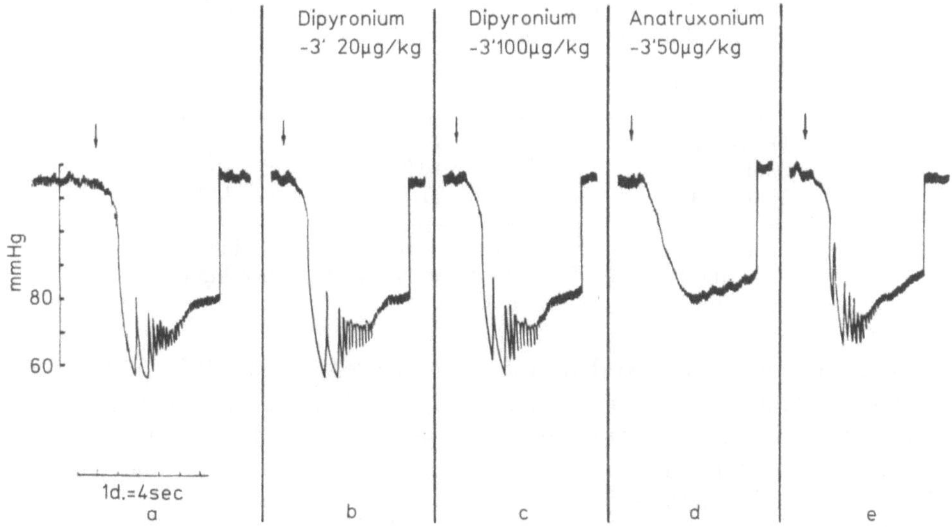

Fig. 4a–e. Effect of dipyronium and anatruxonium on acetylcholine-induced bradycardia and hypotension. Changes of arterial pressure: **a** before dipyronium administration, **b** 3 min after dipyronium administration (20 μg/kg i.v.), **c** 3 min after dipyronium administration (100 μg/kg i.v.), **d** 3 min after anatruxonium administration (50 μg/kg i.v.), **e** 48 min after anatruxonium administration. Acetylcholine (*arrows*) was injected (20 μg/kg i.v.). Experiment on cat anesthetized with urethane (500 mg/kg) and chloralose (50 mg/kg i.v.)

The ability to prevent acetylcholine-induced (10–20 μg/kg i.v.) bradycardia and hypotension was used as a criterion for antimuscarinic activity. The drugs tested in neuromuscular blocking doses failed to decrease bradycardia and depressor response to acetylcholine, i.e., they had no effect on muscarinic receptors of the heart or vessels (Fig. 4). Only an additional injection of five times the neuromuscular blocking dose (100 μg/kg) resulted in a slight inhibition of the cardiac muscarinic receptors, causing an initial decrease of acetylcholine-induced bradycardia (less than 20%). It should be noted that even when used at doses of 0.5–1 mg/kg (25–30 times as high as those causing neuromuscular block) dipyronium and amidonium failed to block the effect of acetylcholine on the cardiac muscarinic receptors completely. Thus, dipyronium and amidonium have a rather low cardiotropic antimuscarinic activity and it appears only in doses substantially higher than those blocking neuromuscular transmission.

It was shown in experiments on the isolated rat ileum that these agents in concentrations up to 10^{-4} g/ml did not antagonize the effect of acetylcholine on muscarinic receptors of the intestine. The ganglion-blocking activity of the agents was studied in anesthetized cats. Their effect on the transmission in the superior cervical sympathetic ganglion was studied using the method of KHARKEVICH (1967). Arterial pressure was also used as an indirect criterion of the action of the agents on the sympathetic ganglia. Their influence on transmission in the parasympathetic ganglia was estimated according to their ability to prevent bradycardia and depressor response to electrical stimulation of peripheral section of the vagus nerve. The agents were injected intravenously.

Table 3. The effect of dipyronium and amidonium on the activity of horse serum cholinesterase

Compound	I_{50} [a]
Dipyronium	1.12×10^{-4} (0.96×10^{-4}–1.28×10^{-4})
Amidonium	1.03×10^{-5} (0.91×10^{-5}–1.15×10^{-5})
Neostigmine	1.4×10^{-7} (1.32×10^{-7}–1.54×10^{-7})

[a] Concentrations (*M*) in which the compounds inhibit cholinesterase by 50%. In parentheses confidence limits at $P=0.05$

The experiments showed that even when injected in doses 100–200 times as high as the neuromuscular blocking dose (2–4 mg/kg), dipyronium and amidonium did not decrease the amplitude of contractions of the nictitating membrane induced by the stimulation of preganglionic sympathetic fibers and did not change the arterial pressure.

Dipyronium and amidonium in neuromuscular blocking doses (0.02 mg/kg) and in doses 4–5 times as high as the latter (0.08–0.1 mg/kg) failed to decrease bradycardia induced by electrical stimulation of the vagus nerve. On the other hand, at doses of 0.5–1 mg/kg (25–50 times as high as the neuromuscular blocking dose) they prevented bradycardia. This short-term (5–10 min) vagolytic action is likely to be due to their blockade of the intramural parasympathetic cardiac ganglia, although it might also be related to a slight cardiotropic antimuscarinic activity appearing at high doses. Thus, neither of the agents possesses ganglion-blocking action at neuromuscular blocking doses.

The anticholinesterase activity of dipyronium and amidonium was studied in vitro using Hestrin's colorimetric method. Rat brain homogenate (1:10) served as a source of acetylcholinesterase, normal horse serum (1:2) as a source of cholinesterase. Dipyronium and amidonium were shown not to inhibit the activity of brain acetylcholinesterase in the concentrations tested (10^{-6}–10^{-3} *M*). At the same time, in high concentrations they had a certain blocking effect on horse serum cholinesterase (Table 3). Evidently, when used in neuromuscular blocking doses, dipyronium and amidonium will have no anticholinesterase activity.

The histamine-releasing activity of both neuromuscular blocking agents was tested on anesthetized cats. The histamine content was determined in the venous blood (from the right atrium) prior to, and after intravenous injection of 80 μg/kg dipyronium and amidonium. The histamine concentration in the blood samples was measured fluorimetrically. In addition, arterial pressure served as an indirect criterion of the histamine-releasing action of the agents.

The results of the experiments are presented in Table 4. It can be seen that after dipyronium or amidonium was injected at four times the neuromuscular blocking dose, the histamine content in the blood did not increase, the arterial pressure remained stable, thus indicating that dipyronium and amidonium had no histamine-releasing activity.

The antihistaminic, adrenoceptor-blocking, and myotropic spasmolytic activity of dipyronium and amidonium were tested on isolated organs. The histamine-induced (2×10^{-7} g/ml) contractions of the guinea pig ileum, epinephrine-in-

Table 4. The effect of neuromuscular blocking agents on histamine content in blood and on arterial pressure in anesthetized cats

Compound	Cat	Histamine content in blood from right atrium (μg/ml)		Arterial pressure (mm Hg)	
		Before injection	1 min after injection	Before injection	1 min after injection
Dipyronium 80 μg/kg	1	0.072	0.072	100	100
	2	0.082	0.086	150	150
	3	0.030	0.028	170	170
	4	0.022	0.019	170	170
	5	0.035	0.035	160	160
Amidonium 80 μg/kg	1	0.018	0.017	150	150
	2	0.046	0.038	130	130
	3	0.012	0.010	140	140
	4	0.044	0.037	130	130
	5	0.02	0.017	140	140
	6	0.012	0.015	170	170
Tubocurarine chloride 1 mg/kg	1	0.009	0.06	100	50
	2	0.020	0.039	130	110
	3	0.024	0.059	180	50
	4	0.035	0.061	130	60
	5	0.042	0.093	180	60
	6	0.11	0.58	180	80

duced (2×10^{-7} g/ml) contraction of rabbit spleen strips, and barium chloride-induced (2×10^{-4} g/ml) contractions of the rat ileum were recorded. Dipyronium and amidonium were injected in two or three increasing concentrations (10^{-8}–10^{-4} g/ml). At the end of each experiment, antagonists (mebhydrolin hydrochloride 10^{-8} g/ml; phentolamine 5×10^{-7} g/ml; papaverine hydrochloride 10^{-6} g/ml) were used for controls. It follows from these experiments that even in high concentrations dipyronium and amidonium failed to decrease the contractions of isolated organs in response to histamine, epinephrine, and to barium chloride. Thus, dipyronium and amidonium are devoid of H_1-blocking, α-adrenoceptor-blocking and myotropic activity.

The local anesthetic and irritant activity of the agents was studied in rabbits and guinea pigs. Solutions (1%–2%) of dipyronium and amidonium were applied to the eye and their effects on the corneal reflex and conjunctiva were tested within 3 h of the application. The state of the conjunctiva was checked again after 24 h. It was thus revealed that dipyronium and amidonium had neither local anesthetic, nor irritant activity.

III. Toxicologic Study

Acute toxicity of dipyronium and amidonium was tested on artificially ventilated anesthetized rabbits and cats, arterial pressure and ECG being recorded. The drugs tested were injected intravenously in increasing doses (up to 2–4 mg/kg). It was found that the animals could tolerate 100–200 times the neuromuscular blocking doses. Within 4–6 h of observation, the arterial pressure remained stable, and the ECG showed no changes of heart activity. After that period, spontaneous respiration was restored. These findings indicate low toxicity of the agents.

Since the preclinical study of dipyronium suggested that it could be of clinical interest, its chronic toxicity was tested (KOROBOV 1978). Tests were carried out on albino rats, rabbits, and mongrel dogs of either sex. In the course of the experiment, all the animals were kept in a nursery and given water and food according to a standard diet. A sterile solution of dipyronium was administered once daily in the morning, to rats in the lateral tail vein, to rabbits in the marginal auricular vein, and to dogs in the muscle of the external femoral surface. The rats were given the drugs for 10 days, the rabbits for 14 days, and the dogs for 10–12 days. Control animals were given saline in equal volumes. Rabbits and dogs were injected with dipyronium in doses causing neuromuscular block without respiration suppression, 15 and 40 μg/kg, respectively; the rats received doses of 400 μg/kg which required artificial ventilation, the latter being performed using a box respiratory system (see Chap. 11).

The toxicity of dipyronium was estimated according to its influence on the general state and weight gain of animals, the composition and clotting time of the circulating blood, biochemical values of the blood serum, macrostructure and microstructure of the visceral organs. The ECG was recorded in rabbits. The probable hemolytic action of the drug was tested. Reticulocyte content in the circulating blood, total bilirubin in the blood serum, and spleen weight were used as criteria of hemolytic action of the drug.

The following data were gained after 10–14 days dipyronium treatment of rabbits, rats, and dogs. The general state and weight of dipyronium-treated animals were essentially equal to controls (control animals were injected with saline). The ECG of rabbits, recorded on the last day of the test, seemed to indicate that dipyronium had no unfavorable effect on the myocardium. The heart rate of the test rabbits was 238 ± 7 beats/min, the duration of the PQ interval was 0.08 s, QRS was 0.02 s. In control animals, these values were 225 ± 11 beats/min, 0.08 s, and 0.02 s, respectively.

Dipyronium did not cause any statistically significant changes of the morphological composition or clotting time of blood of cats or rabbits as compared with controls, neither did it cause any regular shifts of the blood composition in dogs. Blood serum tests of dipyronium-treated rats, rabbits, and dogs failed to show pathologic changes of any values of phosphorus–calcium metabolism, liver or kidney functions. Only in rabbits, dipyronium caused a certain decrease of the cholesterol level in the blood serum (Table 5).

Repeated intravenous injections of dipyronium to rats and rabbits did not result in hyperemia or infiltrates in the injection sites. Microscopic examination of the vessels in which the drug was introduced revealed no thrombophlebitis or other pathologic changes of the vascular wall. Macroscopic or microscopic examination of tissues after a single administration of 0.5 ml 0.5% dipyronium solution in the marginal auricular vein of anesthetized rabbits revealed no alteration or thrombogenesis.

Table 5. Biochemical values of the blood serum of rats and rabbits after 10–14 days i.v. dipyronium

Values	Rats ($N=9$)		Rabbits ($N=6$)	
	0.9% Saline	Dipyronium 400 µg/kg	0.9% Saline	Dipyronium 15 µg/kg
Ca_2 (mg%)	10.1±0.1	9.8±0.1	12.7±0.4	13.0±0.3
Mineral phosphorus (mg%)	7.5±0.1	7.5±0.1	4.5±0.3	4.5±0.2
Glucose (mg%)	139 ±5	121 ±4	145 ±5	146 ±5
Urea nitrogen (mg%)	9 ±1	11 ±1	11 ±2	9 ±1
Uric acid (mg%)	1.8±0.1	1.9±0.1	0.7±0.1	0.8±0.1
Cholesterol (mg%)	70 ±3	74 ±3	51 ±5	34 ±6
Total protein (g%)	6.9±0.1	6.9±0.1	6.4±0.2	6.4±0.1
Albumins (g%)	3.6±0.1	3.5±0.1	3.6±0.1	3.8±0.1
Albumins/globulins	1.1	1.0	1.3	1.5
Total bilirubin (mg%)	0.1	0.1	0.1	0.1
Alanine aminotransferase (u/ml)	47 ±7	56 ±5	103 ±11	105 ±10
Aspartate aminotransferase (u/ml)	93 ±10	91 ±14	64 ±6	46 ±3

Blood samples taken from rats and rabbits after repeated injections showed essentially no difference in reticulocyte content between the treated group and controls. The quantity of total bilirubin in the blood serum was under 0.1 mg%. Moreover, no difference in spleen weight between test and control animals was observed.

The color of the blood serum samples taken from anesthetized artificially ventilated rabbits 5 min after the injection of 0.5 ml 0.5% dipyronium solution did not differ from control samples taken from the same animal prior to the injection.

Visceral organs of rats treated daily with intravenous dipyronium at doses of 400 µg/kg over 10 days were exposed to morphological examination. The drug did not affect the absolute or relative weight of the adrenals, spleen or thymus, or the structure of the two latter organs. The lipid content in the zona fasciculata or zona glomerulosa of the adrenal cortex was somewhat increased (3.7 arbitrary units; control 3.5 arbitrary units).

The morphology of the heart did not differ from controls except for one rat in which plethora of capillaries of the myocardium was observed. In the kidneys of only two rats, increases were observed in the number of swollen glomerules (15–19; control 4–5), and of granular ones (10–12, control 5–6); a moderate granular degeneration of epithelial cytoplasm of several convoluted tubules was also noted, sometimes with inclusions of the smallest fatty drops. In the liver of all the dipyronium-treated rats, there were observed disseminated large-focal small-drop fatty infiltrations of hepatocyte cytoplasm, predominantly in the center of the lobules, associated with its granular degeneration. In three rats, there were dilated and plethoric capillaries in the periphery of the liver lobules. Microscopic examination of other organs (lungs, esophagus, stomach, small and large intestine, pancreas, thyroid gland, testes, and urinary bladder) showed essentially the same structure of these organs in control and test rats.

Thus, dipyronium had essentially no effect on the structure of the majority of the organs tested in rats. However, it should be noted that in all the rats tested, dipyronium caused signs of moderate fatty infiltration and albuminoid degeneration of the liver cell cytoplasm, and in some animals early signs of fatty and albuminoid degeneration in the epithelial cells of renal convoluted tubules.

Taking into consideration the changes of the liver structure in the rats, a special test was devoted to the rate and reversibility of these alterations. With this goal, morphological examination of the liver structure was performed 10 and 30 days after the 10-day administration of 400 μg/kg (5–6 rats in a series). It was found that 10 days after withdrawal, only in one rat of six were there small- or medium-drop fatty inclusions in the hepatocyte cytoplasm of some liver lobules; 10 days after withdrawal, liver structure of the test animals was the same as in controls.

Morphological examination of the visceral organs of six rabbits treated daily with intravenous dipyronium at a dose of 15 μg/kg for 2 weeks gave the following results. No noticeable drug-related changes of absolute or relative weight of adrenals, spleen, or thymus were observed. Microscopic examination revealed in the adrenal cortex plethora of the zona reticularis and zona fasciculata and also an increased number of lipids in all the zonae, especially in the zona fasciculata (2.8 arbitrary units; control 2.0 arbitrary units).

There was no difference in the heart between test and control animals. The kidneys of several rabbits had a moderate plethora, of medulla or cortex, or total plethora with signs of disseminated granular degeneration of epithelial cytoplasm of straight and convoluted tubules. In the same animals, there was observed an increased number of granular renal glomerules (10–15; control 4–5) and especially in swollen ones (20–30; control 4–6). In the liver of all test rabbits there was a moderate plethora of capillaries along the periphery of the liver lobules, which in some animals was associated with slight diffuse-focal fatty infiltration of hepatocyte cytoplasm, predominantly in the center of the lobules. Microscopy of other rabbit organs tested (lungs, esophagus, stomach, small and large intestine, pancreas and thyroid gland, testes, and urinary bladder) did not show structural difference from controls.

The data obtained indicate that dipyronium did not significantly affect the structure of the majority of rabbit organs tested. The action is limited to slight vascular disorders in the adrenals, liver, and kidneys, and in some animals it is manifested by insignificant fatty infiltration in the liver and initial metabolic disturbance in the cytoplasm of epithelial cells of renal tubules.

Morphological examination of three dogs given 40 μg/kg dipyronium intramuscularly for 2 weeks did not reveal any significant changes of macrostructure or microstructure of the majority of the organs tested (heart, lungs, esophagus, stomach, small or large intestine, pancreas, thyroid and thymus glands, gonads, urinary bladder, and prostate). The observed deviations from the normal structure of other organs were not regular and are unlikely to be drug related. Thus, two dogs showed small- and medium-drop lipid inclusions in the epithelial cells of some convoluted, and even more in the straight renal tubules, and in one dog there was observed a moderate decrease of the number of lipids in the adrenals (2.8 arbitrary units; control 3.6 arbitrary units), predominantly in the internal layer of the cortical zona fasciculata, and an increased number of splenic reticulocytes repleted with hemosiderin.

Thus, the neuromuscular blocking agent dipyronium, when repeatedly injected intravenously in rats and rabbits and intramuscularly in dogs, did not essentially affect the general state or weight gain of animals, the clotting time or morphological composition of circulating blood, the main biochemical values of the blood serum, or histologic structure of the majority of visceral organs, and had no hemolytic effect. The observed small-drop fatty infiltration of hepatocytes in the liver of rats and rabbits was not associated with disturbed main liver functions and was easily reversible. Dipyronium had no local irritant action. Thus, experimental study of the nondepolarizing neuromuscular blocking agents dipyronium and amidonium indicates that they are highly potent and highly selective. In doses causing neuromuscular block, they had no side effects. They have low toxicity.

D. Bistertiary Salt of Aminoalkylamide: Pyrocurinum

The search for active neuromuscular blocking agents among the bistertiary amines was based on the assumption that the interaction of such compounds with acetylcholine receptors of the skeletal muscles must be less stable than that of bisquaternary salts or even bisonium compounds with one tertiary and one quaternary nitrogen atom. It is obvious that, after block of neuromuscular transmission induced by these agents, it should be rapidly restored which is of great practical importance. Furthermore, it does not seem unreasonable to suppose that antagonists of neuromuscular blocking agents (anticholinesterase agents, Pymadin) must be more effective in eliminating the action of bistertiary amine salts than that of bisquaternary salts, if both groups of neuromuscular blocking agents interact with the antagonists competitively. Pyrocurinum is one such compound belonging to the group of bistertiary amine salts and having a marked neuromuscular blocking activity (Kharkevich 1983).

Pyrocurinum is α-truxillic acid bis[1,3-(*N*-pyrrolidinyl)propylamide]dihydrochloride

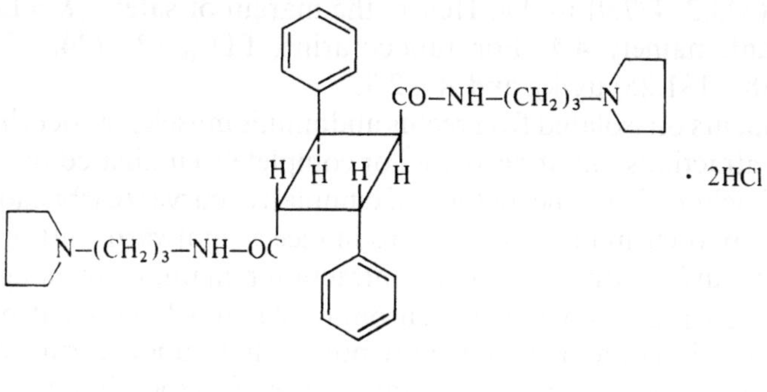

I. Neuromuscular Blocking Action

The neuromuscular blocking activity of pyrocurinum was studied in cats, rabbits, mice, and frogs. In experiments on rabbits, intravenous pyrocurinum caused head drop starting from a dose of 20 µg/kg. $ED_{50}=28.5$ (26.1–30.9) µg/kg ($P=0.05$). The effect developed 20–50 s after the injection. No muscular fasciculations were observed. Head drop was usually associated with muscle relaxation of the limbs. Within the dose range 20–40 µg/kg, a short respiration arrest occurred in only 1 rabbit of 36 (40 mg/kg i.v.). Under the same conditions, tubocurarine caused head drop in much higher doses: $ED_{50}=137$ (105–178) µg/kg ($P=0.05$).

In experiments on cats (anesthetized by chloralose 60 mg/kg plus urethane 400 mg/kg i.v.) the transmission from the sciatic nerve to the gastrocnemius muscle was recorded mechanographically. The peripheral section of the nerve was stimulated by supramaximal impulses of 0.5 ms duration at a frequency of 1 stimulus per second. Pyrocurinum induced neuromuscular block at doses of 120–130 µg/kg i.v. The block developed after 30–50 s. No facilitation of the transmission or muscular fasciculations were observed. The duration of the complete block was 2–4 min. The blocking effect was potentiated on repeated injection of equal doses (100–130 µg/kg) at 2-h intervals. Neuromuscular block was usually associated with more or less marked depression of respiration. Tubocurarine in these experiments blocked the neuromuscular transmission at doses of 180–230 µg/kg.

Neostigmine pretreatment (30 µg/kg) 5 min prior to pyrocurinum administration significantly decreased the blocking effect of the latter. After ether inhalation, the effect of pyrocurinum was potentiated. Hexobarbital sodium (25–30 mg/kg i.v.) or thiopental sodium (25–30 mg/kg i.v.) has essentially no effect on pyrocurinum-induced neuromuscular block.

In experiments on intact pigeons, the minimal doses at which pyrocurinum caused neuromuscular block of 20 s and more were determined; the type of paralysis was also identified. It was shown that, starting from intravenous doses of 25–30 µg/kg, pyrocurinum caused flaccid paralysis of the duration specified.

In experiments on intact mice, the effective doses inhibiting the righting reflex in half of the animals (PD_{50}) and the mean apneic dose (AD_{50}) with confidence limits at $P=0.05$ were determined. Pyrocurinum was injected intravenously. Under such conditions PD_{50} for pyrocurinum was 90.5 (75.4–108.6) µg/kg and $AD_{50}=405$ (343.2–477.9) µg/kg. Hence, the margin of safety ($K=LD_{50}/ED_{50}$) was significant, namely 4.4. For tubocurarine, $PD_{50}=36$ (24.5–52.9) µg/kg, $AD_{50}=105$ (84–131.25) µg/kg and $K=2.9$.

In experiments on isolated frog rectus abdominis muscle, pyrocurinum caused no muscle contractions, but it decreased or completely eliminated the stimulating action of carbachol. Using the method of cumulative curves (carbachol was taken as an agonist) pyrocurinum was shown to induce a rightward shift of the curves, the parallelism and maximum muscle contraction remaining unchanged.

Thus, pyrocurinum is a potent neuromuscular blocking agent of comparatively short duration of action. It is more potent than tubocurarine. A comparatively wide safety margin is an important feature of pyrocurinum. It is a nondepolarizing (competitive) neuromuscular blocking agent. The following indicate its nondepolarizing mode of action: (a) the absence of muscular fasciculations and initial facilitation of neuromuscular transmission in rabbits and cats; (b) antagonism with neostigmine; (c) flaccid paralysis in pigeons; and (d) its failure to stimulate frog rectus abdominis muscle and antagonism with carbachol.

The data from experiments on frog rectus abdominis muscle using the method of cumulative curves indicated the competitive action of pyrocurinum.

II. Assessment of Side Effects

Pyrocurinum (100–500 μg/kg i.v.) failed to affect the arterial pressure in cats (anesthetized by chloralose 60 mg/kg plus urethane 400 mg/kg i.v.). ECG recordings showed essentially no alterations of the heart activity (120–250 μg/kg i.v.). Coronary blood flow and myocardial oxygen consumption were also unchanged. Thus, pyrocurinum did not affect systemic or coronary circulation which is its important advantage over tubocurarine.

In contrast to tubocurarine, pyrocurinum in neuromuscular blocking doses had no effect on transmission in the superior cervical ganglion of anesthetized cats. The interaction of pyrocurinum and acetylcholine was studied in different experimental conditions. Thus, in anesthetized cats pyrocurinum (150–200 μg/kg) caused only partial and short-term decrease of acetylcholine-induced (10 μg/kg) bradycardia, the hypotensive effect of acetylcholine being unchanged. Acetylcholine-increased salivation after pyrocurinum administration was equal to controls.

Pyrocurinum ($10^{-6}-10^{-4}$ g/ml) did not affect acetylcholine-induced (2×10^{-6} g/ml) contractions of the isolated rat ileum. It had no effect on acetylcholinesterase (from rat brain homogenates). At 10^{-4} *M*, it caused a 50% inhibition of cholinesterase activity (horse serum). I_{50} of neostigmine with respect to acetylcholinesterase and cholinesterase is 1.5×10^{-7} *M*. Thus, pyrocurinum in neuromuscular blocking doses had essentially no ganglion-blocking, antimuscarinic, or anticholinesterase activity.

No influence on the effects of epinephrine were observed in anesthetized cats, in which pressor responses to epinephrine (10–15 μg/kg) were registered, or on isolated trabeculae of rabbit spleen (epinephrine was used in concentrations of 2×10^{-7} g/ml and pyrocurinum 10^{-6}–10^{-4} g/ml). The influence of this agent on the central nervous system is described in Chap. 9.

The ability of pyrocurinum to release histamine was given special attention. Tubocurarine was used for comparison. Both drugs were used in doses more than four times as high as those blocking the transmission from the sciatic nerve to the gastrocnemius muscle. It was demonstrated in anesthetized cats that pyrocurinum, in contrast to tubocurarine, increased the histamine content in plasma in only some of the animals, which is another advantage of this agent over tubocurarine.

Pyrocurinum had no antihistaminic activity; this was shown in experiments on isolated guinea pig ileum. In concentrations of 2×10^{-6}–2×10^{-4} g/ml, it was found to have no effect on histamine-induced (2×10^{-7} g/ml) ileum contractions. Pyrocurinum seems to have no spasmolytic activity, either; its introduction into the bath medium (2×10^{-6}–2×10^{-4} g/ml) failed to cause any changes in barium chloride-induced (2×10^{-4} g/ml) contractions of the isolated rat ileum. The application of 1%–5% pyrocurinum solution to the guinea pig or rabbit eye did not cause irritation. Thus, pyrocurinum has a highly selective neuromuscular blocking activity. In doses required for neuromuscular block, it has essentially no side effects.

III. Toxicologic Study

Neuromuscular blocking agents belonging to the salts of tertiary amines are more lipophilic as compared with bisquaternary ammonium salts and easily penetrate tissue barriers. For example, erythrine alkaloids β-erythroidine and dihydro-β-erythroidine are easily absorbed from the gastrointestinal tract, penetrate the blood–brain barrier, and produce an inhibitory effect on the central nervous system (SALEMA and WRIGHT 1951; PATON 1959; NANNA et al. 1960). Delphinium alkaloids belonging to the tertiary amines also easily pentrate tissue barriers (delseminum, elatinum, condelphinum, mellictinum) (DOZORTSEVA and MASHKOVSKY 1951; DOZORTSEVA 1956). Pyrrolizidine alkaloids (heliotrine, laziocarpine) have a neuromuscular blocking activity similar to that of tubocurarine. They are hepatotoxic (GALLAGER 1960). Several hours after oral administration of the alkaloids to rats, necrotic foci appeared in the liver, and hepatocyte division was slowed down (ROGERS and NEWBORNE 1970). The ability of pyrrolizidine alkaloids to penetrate into the intracellular medium of hepatocytes is due to the presence of tertiary nitrogen atoms in their molecules and their correspondingly high lipophilicity.

Pyrocurinum also belongs to the tertiary amines. As already mentioned, in acute experiments, it demonstrated a highly selective neuromuscular blocking effect and had essentially no side effects. For a more adequate evaluation of the safety of pyrocurinum, its effect on animals on repeated long-term administration was studied (KOROBOV 1978). To detect probable toxic effects, it was studied in rats, rabbits, and dogs. Detailed description of the methods is given in Chaps. 11 and 20). In chronic experiments, pyrocurinum was administered to rats intravenously over 10 days under artificial ventilation at a dose of 3 mg/kg causing a respiratory arrest and at three times the apneic dose (9 mg/kg). It was given in subapneic doses to rabbits (40 μg/kg i.v.) and to dogs (100 μg/kg i.v.) over 12–14 days.

The general state and weight of pyrocurinum-treated animals were essentially the same as in controls. The ECG of rabbits on day 14 of the experiment indicate the absence of any unfavorable influence on the heart. The heart rate of the test rabbits was 235 ± 6 beats/min, the duration of PQ interval was 0.08 s, QRS 0.02 s. In controls, these values were 225 ± 11 beats/min, 0.08 s, and 0.02 s, respectively.

Pyrocurinum did not cause statistically significant changes of the clotting time or composition of the circulating blood of rats and rabbits as compared with controls. No regular changes of blood composition in dogs were found either. Studies of blood serum of pyrocurinum-treated rats, rabbits, and dogs, failed to show changes of some biochemical constants (see Table 5) characterizing phosphorus–calcium metabolism or liver or kidney function. It should be noted that in the blood serum of rats given a high dose of the drug (9 mg/kg), the levels of glucose and uric acid were somewhat decreased. However, the latter was within the physiologic range for this species.

After repeated injection of pyrocurinum solution to rats and rabbits, no hyperemia or infiltrates could be seen at the injection site. Microscopic examination of the vessels in which the drug was injected showed no thrombophlebitis or irritation of the vascular wall. After a single injection of 1% solution of pyrocurinum in the marginal auricular vein of the rabbit, microscopic examination of the vein wall and the adjacent tissues did not detect any signs of irritation, alteration, or thrombogenesis.

In the blood samples taken from rats and rabbits after chronic pyrocurinum administration, the content of reticulocytes was equal to control values. Total bilirubin was within the normal range. No difference in the spleen weight of test and control animals was found. It was found that 5 min after a single administration (i.v.) of 0.5 ml 1% solution of pyrocurinum to artificially ventilated anesthetized rabbits, the color of the blood serum samples was the same as in control samples taken from the same animals before the injection. These findings indicate the absence of hemolytic activity in pyrocurinum.

Morphological investigation of pyrocurinum-treated animals was carried out on the day after the administration was finished. Visceral organs of rats after 10-day injections of 3 and 9 mg/kg were subjected to microscopic and macroscopic examinations. No noticeable effect on absolute or relative weight of adrenals, spleen, or thymus was observed. The structure of the spleen and thymus in test and control animals was the same. The quantity of lipids in the zona reticularis or zona fasciculata of the adrenal cortex of rats given 3 mg/kg pyrocurinum was somewhat decreased (2.4 arbitrary units; control 3.5 arbitrary units). In one of the rats given 9 mg/kg, the quantity of lipids in these areas of the adrenal cortex was 2.7 arbitrary units and in the rest of rats of that group it was essentially equal to the control value.

The majority of rats did not have any drug-related pathologic changes in parenchymatous organs (heart, liver, kidneys). A slightly marked small-drop fatty infiltration of hepatocyte cytoplasm in the center of some lobules was detected in only two rats. In two rats, there was a focal plethora of hepatic capillaries, predominantly in the peripheral part of the lobules. Similar changes were observed in several control animals.

The examination of other organs (lungs, esophagus, stomach, small and large intestine, pancreas, thyroid gland, prostate, testes, and urinary bladder) showed no noticeable deviations from the normal structure. These findings indicate an insignificant influence of pyrocurinum on the morphology of the main organs of rats.

The examination of the visceral organs of six rabbits revealed no important effect on the absolute or relative weight of the adrenals, spleen, or thymus. Microscopic examination of the adrenals showed a moderate decrease of the lipid content in the zona reticularis and zona fasciculata of the adrenal cortex in all the animals (1.8 arbitrary units; control 2.0 arbitrary units). In one rabbit, a moderate hyperplasia of the lymphoid tissue of the spleen was observed, in another rabbit, the number of hemosiderin-overloaded reticulocytes was increased in the spleen.

Alterations of parenchymatous organs were observed in only a few animals. In one rabbit, a diffuse small-drop fatty infiltration of the myocardium developed; in its liver there appeared foci of round cell infiltration surrounded by an area of hepatocytes undergoing granular or fatty degeneration. In its kidneys, the number of swollen and granular glomerules was increased. In two control rabbits out of six, similar changes were detected. In one rabbit, there was a plethora of the renal capillaries of the cortical and medullary substance, the number of swollen glomerules was increased (15–20 per 100 glomerules; control 4–5). In its lungs there were signs of plethora or round cell infiltration of the interstitial tissue. No pathologic changes were observed in gastrointestinal tract, pancreas, thyroid gland, testes, or urinary bladder.

In such experimental conditions pyrocurinum was shown to have no pathologic influence on the structure of the visceral organs of the rabbit. Deviations in the structure of the heart, liver, and kidneys of several animals were not drug related, as they were equally observed in control rabbits.

Morphological examination of the visceral organs of three dogs failed to detect any significant alterations of macrostructure or microstructure in the majority of the organs tested (heart, lungs, esophagus, stomach, large and small intestine, spleen, pancreas, thyroid and thymus glands, gonads, urinary bladder, and prostate). In one dong, there was plethora of the vessels of adrenal medulla, partially affecting the zona reticularis of the adrenal cortex, and a moderate decrease of the lipid content (2.6 arbitrary units; control 3.4 arbitrary units) in the cells of the zona fasciculata of the adrenal cortex. In the same dog, small-drop fatty infiltrations were detected in the cytoplasm of the epithelium of some straight tubules of the kidneys and small foci of round cell infiltration in the medulla were observed. In the liver of another dog, focal large-drop fatty infiltration of hepatocytes of several lobules was observed.

The data obtained indicate that chronic injection of pyrocurinum does not cause regular changes in the structure of visceral organs in dogs and the observed deviations do not seem to be drug related. In acute experiments, intravenous injections of pyrocurinum to rats and rabbits at 50 or 100 times the apneic dose under artificial ventilation did not result in the animals' death. Thus, pyrocurinum is an active nondepolarizing neuromuscular blocking agent with low toxicity.

E. Conclusions

Hence, in the series of bisquaternary salts of aminoalkylesters a number of compounds (anatruxonium, cyclobutonium, truxilonium, pyrocyclonium) have a high neuromuscular blocking activity. According to the type of action, they are typical nondepolarizing agents. They have a marked cardiotropic antimuscarinic activity which is responsible for a certain amount of tachycardia and they also induce moderate hypotension. They are antagonized by anticholinesterase agents and Pymadin. Anatruxonium, cyclobutonium, and truxilonium have successfully passed clinical trials (Chap. 26). The two former drugs have been recommended for anesthesiologic practice.

As already mentioned, the data indicating that the transition from the ester derivatives of α-truxillic acid to the corresponding amides resulted in increased neuromuscular blocking potency and decreased side effects, laid the basis for the design of dipyronium and amidonium. The study of dipyronium and amidonium (aminoalkylamide analogs of the α-truxillic acid aminoesters anatruxonium and pyrocyclonium) confirmed this regularity. Indeed, corresponding esters were less potent neuromuscular blocking agents than the amides tested (5–6 times in experiments on cats). Amidonium and dipyronium, as well as their bisester analogs, behaved as competitive neuromuscular blocking agents.

The majority of α-truxillic acid aminoester derivatives are known to have an unfavorable proportion of cardiotropic antimuscarinic and neuromuscular blocking activity: they develop the cardiotropic antimuscarinic effect before the neuromuscular blocking one. Thus, for instance, anatruxonium, even at 5 μg/kg (neuromuscular blocking dose 130 μg/kg), somewhat decreased the sensitivity of cardiac muscarinic receptors to acetylcholine and at 50 μg/kg it blocked those receptors completely. In contrast to the esters, aminoalkylamides dipyronium and amidonium, injected in doses causing neuromuscular block, had no antimuscarinic action on the heart, and only at doses significantly higher than neuromuscular blocking doses could they slightly inhibit cardiac muscarinic receptors.

The study of ganglion-blocking activity of dipyronium and amidonium showed that their ability to interact with nicotinic receptors of autonomic ganglia was significantly less than that of the analogous esters. It was found that even at 100–200 times the neuromuscular blocking doses, they did not have this activity while the corresponding esters blocked sympathetic ganglia in doses only 3–4 times higher than neuromuscular blocking doses. Hence, the substitution of amide for ester groups resulted in a significant decrease of the affinity to ganglionic nicotinic receptors.

The comparison of anticholinesterase properties of the agents revealed that dipyronium and amidonium did not inhibit acetylcholinesterase and only in high concentrations did they have an inhibitory influence on butyrylcholinesterase. However, this effect was much less marked than it was in esters. Another advantage of the amides over esters is their lack of histamine-releasing activity. α-Truxillic acid ester derivatives have a moderately marked ability to cause histamine release (but less than tubocurarine) while dipyronium and amidonium have no such activity.

Thus, amidonium and dipyronium, in contrast to the derivatives of α-truxillic acid aminoesters, do not essentially block cardiac or other muscarinic receptors, nicotinic receptors of autonomic ganglia, or cholinesterases; they do not cause histamine release i.e., they are much more selective neuromuscular blocking agents.

One of the drugs tested, dipyronium, underwent preliminary clinical trials. It was found that, being a sufficiently active and selective agent dipyronium was long-acting and neuromuscular transmission was restored relatively slowly after its administration. Therefore, it seems promising in providing a long-term neuromuscular block (for example, for long-lasting surgery or in the treatment of tetanus).

Marked nondepolarizing neuromuscular blocking activity was also found in a bistertiary salt of aminoalkylamide pyrocurinum. On its intravenous injection, neuromuscular block develops rapidly and it persists for a comparatively short time. The recovery of neuromuscular transmission occurs rather rapidly. The neuromuscular blocking action of pyrocurinum is highly selective. It has essentially no side effects. It is antagonized by anticholinesterase agents. Assessment of acute and chronic toxicity indicated low toxicity of pyrocurinum. Pyrocurinum has successfully passed clinical trials and has been recommended for use in anesthesiology as a neuromuscular blocking agent of medium duration of action (Chap. 26).

Acknowledgments. The author wishes to express his appreciation to N. V. KOROBOV for his collaboration in writing the sections on the chronic toxicity of agents and to Mrs. MARIA LIPMAN for the translation of this chapter.

References

Blackman JG (1963) Stimulus frequency and neuromuscular block. Br J Pharmacol 20:5–16
Gallager CH (1960) The effect of pyrrolizidine alkaloids on liver enzyme system. Biochem Pharmacol 3:220–231
Dozortseva PM (1956) On the pharmacology of alkaloids from Delphinium elatum (elatinum and condelphinum) (in Russian). Farmakol Toksikol 3:42–48
Dozortseva PM, Mashkovsky MD (1951) On comparative pharmacology of alkaloids delseminum and delsinum (in Russian). Farmakol Toksikol 4:49–54
Kharkevich DA (1965) On pharmacological properties of a new curare-like drug anatruxonium (in Russian). Farmakol Toksikol 3:305–309 or (1966) Fed Proc 25:T 521–T 523
Kharkevich DA (1966) On the pharmacology of a new nondepolarizing neuromuscular blocking agent cyclobutonium (in Russian). Farmakol Toksikol 1:47–53
Kharkevich DA (1969) Farmakologiya kurarepodobnykh sredstv (Pharmacology of curare-like agents). Meditsina, Moscow
Kharkevich DA (ed) (1970) Novye kurarepodobnye i ganglioblokiruyushchie sredstva (New curare-like and ganglion-blocking agents). Meditsina, Moscow
Kharkevich DA (ed) (1973) Uspekhi v sozdanii novykh lekarstvennykh sredstv (Advances in drug research). Meditsina, Moscow, pp 138–187
Kharkevich DA (ed) (1983) Novye miorelaksanty (New muscle relaxants). Meditsina, Moscow
Kharkevich DA, Kravchuk LA (1961) On the pharmacology of a new curare-like agent truxilonium (in Russian). Farmakol Toksikol 3:318–324
Kharkevich DA, Skoldinov AP, Arendaruk AP, Kozakova TP, Muratov VK (1974) Anatruxonium – a new curare-like drug of nondepolarizing action (in Russian). Khim-Farm Zh 4:56–62
Kharkevich DA, Skoldinov AP, Arendaruk AP, Kozakova TP, Egorov NV, Astashina IA, Gurevich GI, Shmaryan MI, Maisky VV (1977) Cyclobutonium – a new domestic curare-like drug (in Russian). Khim-Farm Zh 2:145–150
Korobov NV (1978) Issledovanie khronicheskoi toksichnosti novykh kurarepodobnykh sredstv (Chronic toxicological study of new curare-like agents). Cand Med Sci Thesis, First Medical Institute, Moscow
Lemina, EYU (1978) Farmakologiya novykh kurarepodobnykh sredtsv dipironiya i amidoniya (Pharmacology of new curare-like drugs). Cand Med Sci Thesis, First Medical Institute, Moscow
Lepakhin VP (1970) Sensitivity of different muscles to alpha-truxillic acid derivatives (in Russian). In: Kharkevich DA (ed) Novye kurarepodobnye i ganglioblokiruyushchie sredstva (New curariform and ganglion-blocking agents). Meditsina, Moscow, pp 48–63

Nanna C, Macmillan H (1960) Neuromuscular blocking action of dihydro-β-erythroidine. Arch Int Pharmacodyn Ther 124:445–454
Paton WDM (1959) The effects of muscle relaxants other than muscle relaxation. Anesthesiology 20:454–463
Rogers AE, Newborne PM (1971) Lasiocarpine: factors influencing its toxicity and effects on liver cell division. Toxicol Appl Pharmacol 18:356–366
Salama S, Wright S (1951) Action of calabash curare and related curariform substances on the central nervous system on the cat. Br J Pharmacol 6:459–463

CHAPTER 20

Adamantyl Compounds

D. A. KHARKEVICH

A. Introduction

It was mentioned earlier that the introduction of hydrophobic adamantyl radicals into the structure of depolarizing agents turns them into nondepolarizing ones (Chap. 4). This principle laid the basis for the synthesis of *N*-adamantyl analogs of succinylcholine and decamethonium called respectively, diadonium and decadonium (KHARKEVICH 1970a, b). These agents were synthesized at the Institute of Pharmacology, of the USSR Academy of Medical Sciences, Moscow.

B. Diadonium

The physiologic activity of diadonium is caused by the biscation of succinic acid bis[*N*-dimethyl-(1-adamantyl)-aminoethyl ester]. Experiments have been mainly devoted to diadonium diiodide (structure I; X = I).

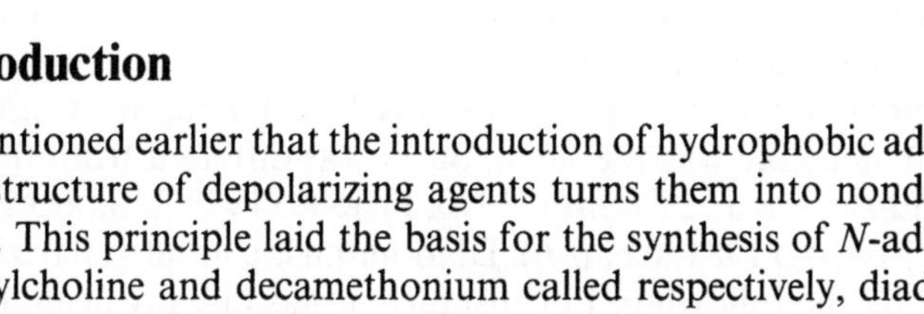

(I)

(I)

In patients diadonium is used as a highly soluble salt di-*p*-toluolsulfomethylate (its pharmacologic properties are identical to those of diadonium diiodide).

I. Neuromuscular Blocking Action

Diadonium has marked neuromuscular blocking activity. Thus, when administered to intact rabbits by intravenous bolus, it caused head drop at doses of 130–180 μg/kg. No muscular fasciculations were observed. Intravenous doses in which diadonium inhibited the righting reflex (paralyzing dose PD_{50}) and apnea and death (apneic dose AD_{50}) were estimated in intact mice. PD_{50} appeared to be 1,000 (934–1,070) μg/kg and AD_{50} = 3,700 (2,913–4,699) μg/kg (P = 0.05). Hence, the safety margin $K = AD_{50}/PD_{50} = 3.7$. For tubocurarine, PD_{50} = 36 (24.5–52.9) μg/kg, AD_{50} = 105 (84–131.25) μg/kg (P = 0.05) and K = 2.9.

Neuromuscular block was quite distinct in anesthetized cats (chloralose + urethane). Square wave supramaximal stimuli (0.5 ms) at 1 stimulus per second were applied on the peripheral end of the sciatic nerve. The contractions of the gastrocnemius muscle were recorded mechanographically. Complete neuromuscular block was observed on i.v. administration of diadonium at doses of 250–350 μg/kg. There was observed a parallel respiratory inhibition up to apnea. No drug-related muscular fasciculations were observed. The block persisted for 3–10 min. After the repeated administration of equal doses of diadonium at 1.5–2 h intervals, the blocking effect became longer and more marked with every injection. After ether anesthesia, the effect of diadonium on neuromuscular transmission was enhanced and prolonged.

The order of block of different groups of muscles after administration of diadonium and succinylcholine was the same, but it was different from that caused by tubocurarine (LEPAKHIN and FISENKO 1970; FISENKO 1972; KHARKEVICH 1973; KHARKEVICH and FISENKO 1981, Chap. 6). Diadonium is a nondepolarizing agent. This is confirmed by the following. For instance, it has already been mentioned that no muscular fasciculations were observed after diadonium was administered to cats or rabbits. The registration of neuromuscular transmission in cats demonstrated that neostigmine (50–100 μg/kg) was a diadonium antagonist. Electromyographic recordings showed that, after diadonium administration, pessimal inhibition (tetanic fade) and posttetanic "decurarization" occurred. In intact pigeons and chickens, diadonium caused flaccid paralysis. It failed to stimulate the rectus abdominis muscle of the frog, but it prevented the development of ace-

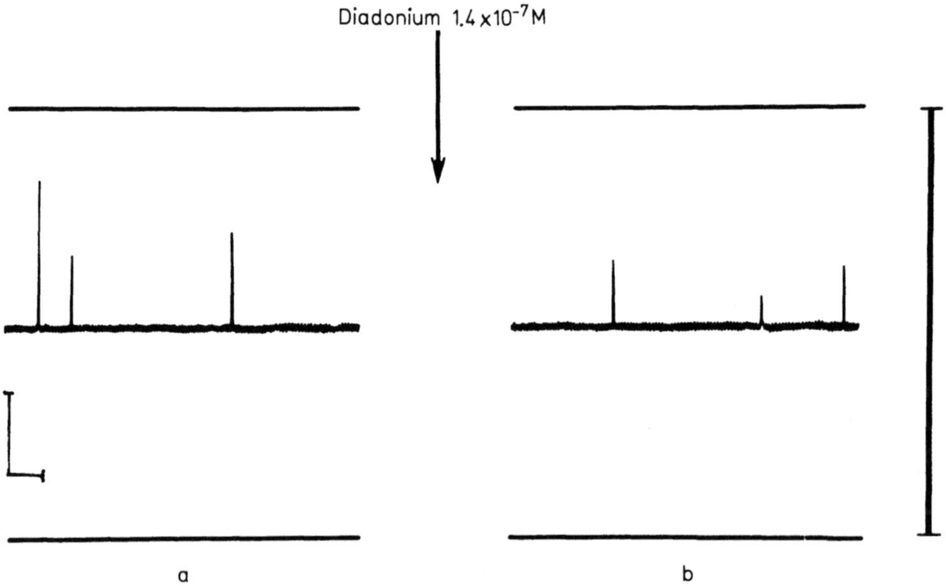

Fig. 1 a, b. Miniature end-plate potentials **a** before and **b** after diadonium administration. *Vertical line on right* membrane potential amplitude scale (60 mV); *on left: vertical line* miniature potential amplitude scale (1 mV), *horizontal line* time scale (500 ms). Rat isolated phrenic nerve–diaphragm preparations

tylcholine-induced muscle contractions. It failed to affect the membrane potential of the rat isolated diaphragm.

It was shown that cumulative curves of the contraction of frog isolated rectus abdominis muscle were shifted rightward after diadonium administration (10^{-7}–5×10^{-6} *M*), but their slope and the amplitude of contractions remained unchanged. This indicates the competitive character of the neuromuscular block caused by diadonium in this concentration range.

The localization of diadonium action may be determined by its influence on miniature potentials. It turned out that, on rat isolated phrenic nerve–diaphragm preparations, diadonium decreased the amplitude of miniature potentials without affecting their frequency (Fig. 1). It is evident that diadonium fails to affect the presynaptic endings and that the neuromuscular block it induces may be related to the block of acetylcholine receptors of the end-plate. Thus, diadonium is a nondepolarizing agent with a subsynaptic site of action (KHARKEVICH 1970b; FISENKO 1971; FISENKO et al. 1973).

Hydrolysis of diadonium in the body seems to be followed mainly by the formation of compounds II and III. The maximum hydrolysis rate of diadonium by horse blood serum cholinesterase (first stage) in vitro is about seven times lower than that of succinylcholine.

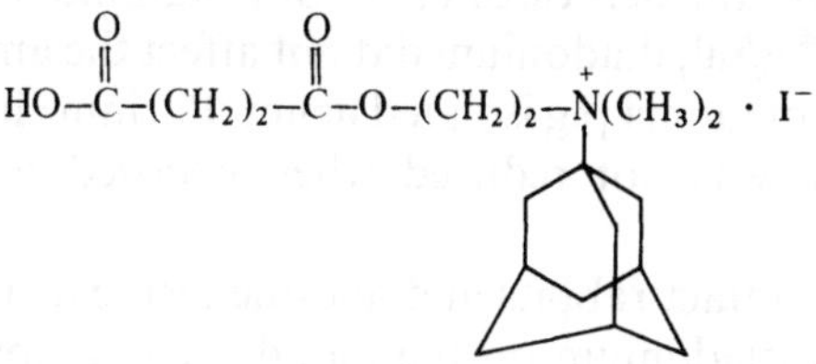

(II)

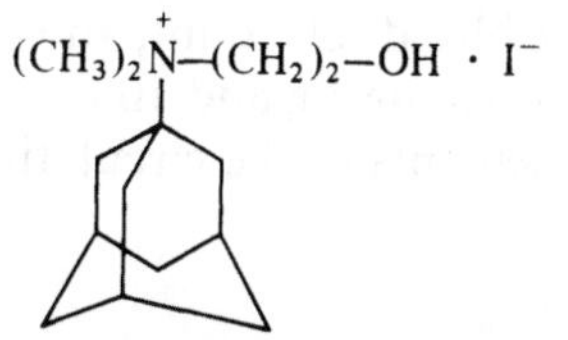

(III)

Both metabolites were synthesized and pharmacologically tested. It appeared that their neuromuscular blocking activity was very low. Neuromuscular block in cats appeared after compound II was administered at doses of 25–30 mg/kg and compound III at 30–40 mg/kg i.v. No muscular fasciculations were observed after their administration. They caused flaccid paralysis in chickens. Hence, the most probable products of diadonium hydrolysis also belong to the nondepolarizing neuromuscular blocking agents, but of very low activity.

II. Assessment of Side Effects

Diadonium has essentially no effect on circulation. For instance, after neuromuscular blocking doses of diadonium, a very slight and short-lasting elevation of arterial pressure might occur (for 10–30 s) which may be followed be slight hypotension. However, even after high doses of diadonium (1–3 mg/kg), changes of the arterial pressure (predominantly hypotension) are negligible.

The ECG in rabbits and cats showed that diadonium did not cause any disorders of heart activity. Sometimes, a short-term sinus tachycardia was observed. The latter is likely to be related to the block of cardiac muscarinic receptors caused by neuromuscular blocking doses of diadonium. Cardiotropic effects induced by stimulation of the peripheral end of the vagus nerve and by acetylcholine are not manifested in this case, and yet, the acetylcholine hypotensive effect persisted (see Chap. 7, Sect. A).

Close to the neuromuscular blocking dose (250–500 μg/kg) diadonium did not affect interneuronal transmission in the superior cervical ganglion of the cat (during electrical stimulation of preganglionic fibers). Even after 3 mg/kg, the tone of nictitating membrane was decreased by no more than 20%–25%.

Diadonium has a certain anticholinesterase activity. It was shown in experiments in vitro with human blood cell acetylcholinesterase, that diadonium caused inhibition of a mixed type. The inhibition constant $K_i^{comp} = 1.17 \times 10^{-5}$ *M*. Neostigmine under prolonged incubation blocked acetylcholinesterase by 50% (I_{50}) at 5×10^{-8} *M*. Diadonium competitively inhibited horse blood serum cholinesterase ($K_i^{comp} = 3 \times 10^{-5}$ *M*). In doses of 250–500 μg/kg i.v., it failed to change the vasopressor action of epinephrine (anesthetized cats).

Diadonium is devoid of antihistamine activity. In neuromuscular blocking doses, it failed to affect the depressor effect of histamine on the arterial pressure in cats. At 10^{-7}–10^{-4} g/ml, diadonium did not affect the amplitude of histamine-induced contractions of guinia pig isolated ileum. Barium chloride-induced spasm of rat isolated ileum was not reduced when exposed to diadonium (10^{-6}–10^{-4} g/ml).

In experiments on intact rabbits and anesthetized cats (under artificial ventilation), the action of diadonium on the blood (hemoglobin, red cells, reticulocytes, white blood cells, white blood cell count, and clotting time) were measured. It was shown that in doses of 500 μg/kg and 10 mg/kg, diadonium did not affect the composition of cellular elements of the circulating blood or its clotting time.

III. Toxicologic Study

Taking into account species-related variations of the sensitivity to neuromuscular blocking agents, diadonium was tested on rats, rabbits, and dogs (KOROBOV 1978; KOROBOV and YAVORSKY 1981). Daily repeated administration lasted for 10–14 days which is enough to detect the toxicity of agents intended for single application. The chronic toxicity of diadonium was tested on male mongrel rats, male rabbits, and mongrel dogs of either sex. Throughout the experiments, the animals were kept in a nursery on a standard water–food schedule. Each group consisted of ten rats, six rabbits, and three dogs.

Sterile solutions of diadonium, in isotonic sodium chloride, were administered once daily in the morning to rats in the lateral tail vein, to rabbits in the marginal auricular vein, and to dogs intramuscularly. Daily administration lasted 10 days for rats, 14 days for rabbits, and 12–14 days for dogs. Control animals were injected with isotonic solutions of sodium chloride.

Rats were treated with diadonium in apneic doses (LD_{95}) of 3 mg/kg under artificial ventilation via a box respirator device of the "iron lung" type (see

Chap. 11, Sect. C) and also in subapneic doses of 1.5 mg/kg under spontaneous respiration. Rabbits and dogs were given the agent in the maximum doses not causing respiratory arrest (rabbits 200 μg/kg equal to ED_{95} for head drop: dogs 500 μg/kg which caused a 5–15 min inhibition of the righting reflex).

Immediately after the administration, artificial ventilation of rats was started. To prevent cooling, the animals were warmed during the experiments. They were weighed the day before the experiment and on the last day of the experiment. Circulating blood was tested the day before the first injection and 1 day after the last injection. After another day, blood was taken from rats and rabbits for biochemical examination, autopsies were performed, and samples for histologic examination were taken. For dogs, the blood and histologic samples were taken 1 day after the last injection. Control animals were exposed to the same examinations on the same days as the test animals.

Using conventional methods, the following values of circulating blood were estimated: hemoglobin, red blood cells, reticulocytes, platelets, white blood cells, white blood cell count, and clotting time.

Using automatic analyzers the following were determined in the blood serum: content of calcium ions, mineral phosphorus, glucose, urea nitrogen, uric acid, cholesterol, total protein, albumins, and total bilirubin. In addition, in dogs, the activity of alkaline phosphatase and lactate dehydrogenase was estimated. The value of albumins/globulins was calculated. The activity of serum aminotransferases (glutamate–oxaloacetic and glutamate–pyruvic) was estimated according to the method of REITMAN and FRANKEL (1957) and modified by KAPETANAKI (1962).

Morphological study consisted of macroscopic examination of visceral organs, and measuring the weight of adrenals, spleen, and thymus. Histologic sections of visceral organs were stained with hematoxylin–eosin and scarlet red. The content of fat in slices of the organs tested was measured.

At day 10–14 of diadonium treatment, no noticeable changes of general state or body weight of test rats, rabbits, or dogs were observed, as compared with control animals. ECG of rabbit recorded at day 14 of the experiment indicate the absence of cardiotoxic effect of diadonium. Heart rate of the test animals was 245 ± 7 beats/min, the PQ interval was 0.08 s, QRS 0.02 s. In controls, these values were 225 ± 11 beats/min, 0.08 s, and 0.02 s. Repeated diadonium administration did not cause statistically significant changes of clotting time or blood composition of rats or rabbits, as compared with controls. In dogs, the changes in blood composition were not drug related, either.

Biochemical examination of blood serum showed that diadonium did not affect the functional state of liver. Protein synthesis (general protein and albumins/globulins) was unchanged. Lipid and carbohydrate metabolism, characterized by serum cholesterol and glucose content, was also unaffected. Liver function of transformation and excretion of some pigments was unaffected (indicated by normal content of serum bilirubin). Normal levels of serum aminotransferase in rats and rabbits as well as alkaline phosphatase and lactate dehydrogenase in dogs indicated that diadonium did not produce a damaging action on the liver or other parenchymatous organs. The functional state of the kidneys which is important for the elimination of diadonium and its metabolites was apparently unaffected, too. This is confirmed by the normal content of urea and uric acid in the animals' blood.

No drug-related changes of the content of calcium or inorganic phosphorus in the blood serum of rabbits or dogs were observed after repeated diadonium administration. Increased content of calcium ions in rats (by 1 mg%), inorganic phosphorus content being normal, cannot be regarded as related to the toxic action of the drug.

After repeated administration of diadonium, the content of reticulocytes in the samples of circulating blood of rats and rabbits was the same as in controls. Total bilirubin in the serum was no higher than 0.1 mg%. Moreover, the spleen weight was the same in test and control animals. It was found that 5 min after i.v. bolus injection of 0.5 ml 5% diadonium solution to artificially ventilated anesthetized rabbits, the color of the serum samples was the same as in control samples taken from the same animals before diadonium administration. Thus, the absence of spleen reaction, and normal content of serum bilirubin and reticulocytes in the blood indicate that diadonium does not cause hemolysis.

The data presented, indicating low toxicity of diadonium, are confirmed by those of macroscopic and microscopic studies of visceral organs of diadonium-treated animals. For instance, the values of absolute and relative weight of adrenals, spleen, and thymus were the same in test and control animals. A certain decrease of lipid content observed in the cortical substance of adrenals of rats and rabbits is likely to be related to a nonspecific adaptation response to diadonium administration. The changed amount of lipids in the adrenal cortex is observed after administration of numerous chemical agents and within certain limits does not indicate toxic action.

The structure of parenchymatous organs of the majority of animals was unaffected after repeated diadonium administration. Structural changes observed in the liver or kidneys of some animals are unlikely to be drug related, since they occurred as often in control animals. Diadonium did not affect the histologic structure of esophagus, small or large intestine, pancreas or thyroid gland, gonads, prostate, or urinary bladder.

Microscopic examination of the vessels to which the agent was administered showed no thrombophlebitis manifestations or other pathologic alterations of the vascular wall. After single i.v. injection of 0.5 ml 5% solution of a highly soluble salt of diadonium, the examination of the marginal auricular vein walls and the tissues adjacent to the injection site detected no irritation, alteration, or thrombogenesis. No irritation of rabbit eye mucous was observed after administration of 0.5% diadonium.

An acute toxicity study of diadonium was carried out on anesthetized artificially ventilated rats, rabbits, and dogs to which intravenous 50- to 100-fold apneic doses (up to 300, 20, and 30 mg/kg i.v., respectively) were injected. Within the 2 h observation period, no cardiac arrest or pathologic ECG changes which could be considered as drug related were observed, which indicates good tolerance of the drug. Artificially ventilated rabbits and cats could tolerate 100–200 mg/kg i.v.

C. Decadonium

Decadonium is 1,10-[methyl-*N*(1′-adamantyl)-amino]-decane diiodomethylate. It is a bis-*N*-adamantyl analog of decamethonium.

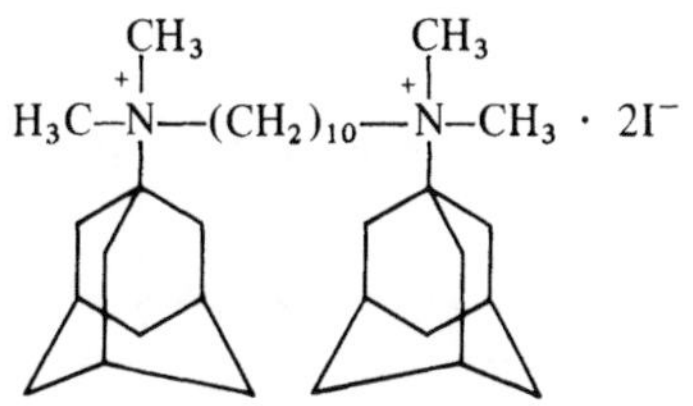

I. Neuromuscular Blocking Action

In experiments on different species, decadonium was shown to have marked neuromuscular blocking potency. It evoked head drop in rabbits in doses of 100–110 μg/kg i.v. Its safety margin *K* is comparatively narrow. According to the inhibition of righting reflex in mice, its $PD_{50} = 385$ (320–462) μg/kg, and $AD_{50} = 780$ (693–877) μg/kg i.v. ($P = 0.05$). Hence, $K = AD_{50}/PD_{50} = 2$. In the same experiments for tubocurarine $K = 2.9$.

Decadonium blocks the transmission from the sciatic nerve to the gastrocnemius muscle in anesthetized cats at doses of 250–300 μg/kg i.v. No muscular fasciculations were observed. In the same doses, decadonium depressed respiration up to apnea. On repeated administration (every 1.5–2 h) of the same doses of decadonium, neuromuscular block became more marked. Under rhythmic stimulation of the sciatic nerve, decadonium promotes the development of tetanic fade (Fig. 2). In addition, a posttetanic "decurarization", typical of nondepolarizing agents was observed (FISENKO 1971).

Diethyl ether potentiated the neuromuscular blocking action of decadonium. Hexobarbital sodium (30 mg/kg i.v.) fails to affect the neuromuscular blocking activity of decadonium. Neostigmine is its antagonist. Premedication of the former (50–100 μg/kg) reduces and shortens the blocking action of decadonium on acetylcholine receptors of the gastrocnemius muscle of anesthetized cats.

Decadonium causes flaccid paralysis in pigeons and chickens. It was shown in rat isolated phrenic nerve–diaphragm preparations that decadonium did not affect membrane potential or the frequency of miniature potentials, but reduced the amplitude of the latter. Decadonium fails to stimulate frog isolated rectus abdominis muscle, but decreases the stimulating action of acetylcholine and carbachol. The comparison of cumulative curves showed that, in the presence of decadonium (10^{-7}–5×10^{-6} *M*), they were shifted rightward along the abscissa, their parallelism and the maximum amplitude being unchanged (carbachol was used as anagonist). The tests carried out indicate that decadonium is an active competitive neuromuscular blocking agent with a subsynaptic site of action (KHARKEVICH 1970a; FISENKO 1971).

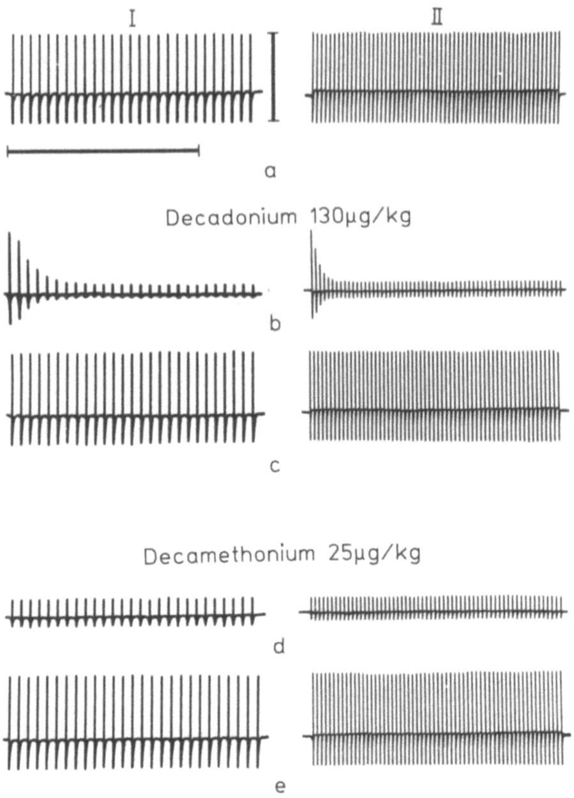

Fig. 2 a–e. Changes of evoked potentials of the gastrocnemius muscle in response to decadonium, and decamethonium injected 2 h later. **a** before drug administration, **b** 3 min after decadonium administration (130 µg/kg), **c** restoration, **d** 7 min after decamethonium administration (25 µg/kg), **e** restoration. Peripheral end of the sciatic nerve was stimulated at 20 Hz (I) and 50 Hz (II). *Vertical line* amplitude scale (1 mV). *Horizontal line* time scale (1 s). Experiment on cat anesthetized with urethane (600 mg/kg) and chloralose (70 mg/kg i.v.)

II. Assessment of Side Effects and Toxicity

Decadonium reduces the arterial pressure to a small extent in anesthetized cats at doses of 100–500 µg/kg i.v. Usually, arterial pressure is decreased by 10–30 mmHg within a short time. Hypotension is sometimes followed by a slight and very short increase in arterial pressure. Decadonium has essentially no effect on transmission in the superior cervical ganglia in cats (on electrical stimulation of preganglionic sympathetic fibers) in neuromuscular blocking doses (250–300 µg/kg i.v.). After administration of 1 mg/kg decadonium, the amplitude of contractions of the nictitating membrane is reduced by 30%–35%. Decadonium (100–500 µg/kg) slightly intensifies the vasopressor effect of epinephrine in anesthetized cats.

Decadonium prevents the negative chronotropic effect of acetylcholine on the heart (Fig. 3) without decreasing its hypotensive effect (experiments on anesthe-

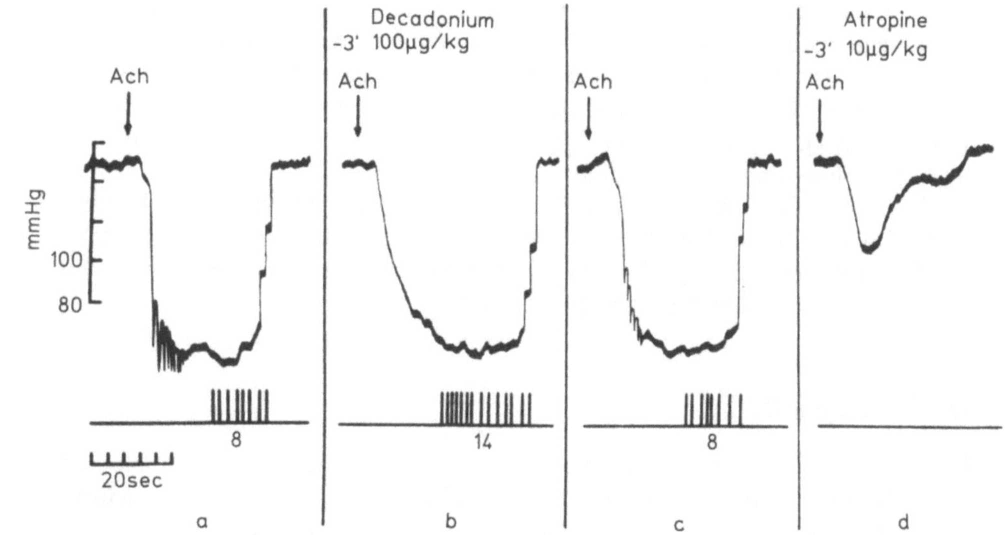

Fig. 3 a–d. Effect of decadonium and atropine on acetylcholine-induced bradycardia, hypotension, and salivation. Downward: arterial pressure, salivation from the parotid in drops (the number of drops is indicated by *vertical lines* and their sum). **a** before injection of decadonium, **b** 3 min after injection of decadonium (100 μg/kg i.v.), **c** 18 min after injection of decadonium, **d** 3 min after injection of atropine (10 μg/kg i.v.). *Ach* acetylcholine (20 μg/kg i.v.). Experiment on cat (4 kg) anesthetized with urethane (500 mg/kg) and chloralose (50 mg/kg i.v.)

tized cats) in doses lower than the neuromuscular blocking dose (see Chap. 7, Sect. A). Cardiotropic antimuscarinic action is likely to cause sinus tachycardia observed in a number of animals.

Decadonium has a marked anticholinesterase activity. It inhibited acetylcholinesterase of rabbit brain by 50% (I_{50}) at a concentration of 2.2×10^{-6} *M* and cholinesterase from horse blood serum at of 7.9×10^{-7} *M*. I_{50} for neostigmine was 1.5×10^{-7} *M* (for both cholinesterases). If we take the activity of neostigmine as 100, then the antiacetylcholinesterase activity of diadonium is 6.8, and anticholinesterase activity is 19.0 (SAMOILOV 1971). This is manifested by increased salivation, prolonged hypotensive action of acetylcholine, and enhanced tone of isolated ileum (KHARKEVICH 1970a). Decadonium fails to affect the influence of histamine on arterial pressure in cats and guinea pig isolated ileum (10^{-6}–10^{-4} g/ml). Decadonium (10^{-7}–10^{-5} g/ml) did not decrease barium chloride-induced (4×10^{-4} g/ml) spasm of rat isolated ileum. At a concentration of 0.5%, decadonium did not irritate the mucous membrane of rabbit eye.

The effect of decadonium on the circulating blood of cats and rabbits was also studied. Hemoglobin, erythrocytes, reticulocytes, leukocytes, platelets, white blood cell count, and clotting time were estimated. Injections of decadonium in doses of 0.5 and 15 mg/kg did not change the parameters studied.

The acute toxicity of decadonium was tested on artificially ventilated rabbits and cats. Intravenous administration of 15 mg/kg decadonium induced apnea of

30–90 min duration, the neuromuscular block lasted for 45–115 min, followed by a complete restoration of the initial neuromuscular transmission and the state of animals. After administration of 50–150 mg/kg, the majority of animals survived more than 4–5 h. Cats could tolerate 15–30 mg/kg, but they died after receiving 100–150 mg/kg i.v. Hence, toxic doses of decadonium are significantly different from neuromuscular blocking doses.

Thus, decadonium is an active nondepolarizing neuromuscular blocking agent. It causes slight hypotension and tachycardia in animals. It has a marked anticholinesterase activity. Toxicity of decadonium is comparatively low, and its therapeutic index wide. Neostigmine is its antagonist.

D. Conclusions

Diadonium is a potent nondepolarizing short-acting neuromuscular blocking drug. Its safety margin is larger than that of tubocurarine. The order of relaxation of different groups of muscles is identical to that of succinylcholine. On its repeated administration, the action is enhanced and prolonged. Diethyl ether potentiated its neuromuscular blocking action. In animals, diadonium has a weak, short-term action on the arterial pressure (predominantly hypotensive). In neuromuscular blocking doses, it blocked the transmission from the vagus nerve to the heart and eliminated acetylcholine bradycardia without affecting the hypotensive action of the latter. Therefore, diadonium may induce tachycardia. Neostigmine antagonizes its neuromuscular blocking action. The results of toxicologic studies of diadonium suggest that it is a neuromuscular blocking agent of low toxicity, adequate for clinical use. Diadonium has successfully passed clinical trials and has been recommended for practical use as a short-term neuromuscular blocking agent (Chap. 7, Sect. A).

Decadonium is also a nondepolarizing neuromuscular blocking agent. Its activity is insignificantly higher than that of diadonium, but its action is longer. Decadonium may be classified as a neuromuscular blocking agent of medium duration of action. It has a rather marked anticholinesterase activity. It blocks cardiac muscarinic receptors which results in a moderate tachycardia. Its administration to animals in neuromuscular blocking doses is followed by slight hypotension. However, preliminary clinical trials of decadonium have shown that it causes a marked fall of arterial pressure.

Acknowledgments. The author wishes to express his appreciation to N. V. KOROBOV for his collaboration in writing the section on the chronic toxicity of agents and to Mrs. MARIA LIPMAN for the translation of this chapter.

References

Fisenko VP (1971) O mekhanizme deistviya novykh kurarepodobnykh sredstv (On the mechanism of action of new curare-like agents). Cand Med Sci Thesis, First Medical Institute, Moscow

Fisenko VP (1972) Electromyographic analysis of the mechanism of action of new curare-like agents (in Russian). Farmakol Toksikol 3:284–287

Fisenko VP, Polgar AA, Smirnov VS (1973) Microelectrophysiological studies of the mechanism and localization of action of a number of new curare-like drugs (in Russian). Farmakol Toksikol 2:206–209

Kapetanaki KG (1962) On the method of evaluation of the activity of transaminases (aminotransferase) in the serum (in Russian). Lab Delo 1:19–23

Kharkevich DA (1970a) On neuromuscular blocking activity of decadonium diiodide (in Russian). Farmakol Toksikol 4:395–399

Kharkevich DA (1970b) On pharmacological properties of a new antidepolarizing curare-like drug diadonium diiodide (in Russian). Farmakol Toksikol 5:531–563

Kharkevich DA (1973) Curare-like drugs (in Russian). In: Kharkevich DA (ed) Uspekhi v sozdanii novykh lekarstvennykh sredstv (Advances in drug research). Meditsina, Moscow, pp 138–187

Korobov NV, Yavorsky AN (1981) On the toxicity of a new antidepolarizing curare-like agent diadonium (in Russian). Farmakol Toksikol 3:338–342

Lepakhin VK, Fisenko VP (1970) On comparative sensitivity of neuromuscular junctions of various muscles to diadonium and decadonium (in Russian). Farmakol Toksikol 3:288–292

Reitman S, Frankel S (1957) A colorimetric method for the determination of serum glutamic oxalacetic and glutamic pyruvic transaminases. Am J Clin Pathol 28:56–63

Samoilov DN (1971) O zavisimosti mezhdu khimicheskim stroeniem i kholinergicheskimi svoistvami bis-chetvertichnykh ammonievykh proizvodnykh truksillovykh kislot (On the relationship between chemical structure and cholinergic properties of bisquaternary ammonium derivatives of truxillic acids). Cand Med Sci Thesis, First Medical Institute, Moscow

CHAPTER 21

Tercuronium

A. F. DANILOV

A. Introduction

In search of specific neuromuscular blocking agents, we have paid attention to the fact that C-toxiferine and its derivative diallylnortoxiferine are the most specific neuromuscular blocking agents (WASER 1972). They are bisquaternary ammonium substances with completely rigid molecules. The highly specific neuromuscular blocking agents pipecuronium (Arduan) and vecuronium (ORG-NC-45) have rigid molecules. On the contrary, the comparison of antimuscarinic, ganglion-blocking, and curare-like potency of several series of ammonium compounds with flexible molecules, including polymethylene bistrialkylammonium compounds, has revealed no substances with selective action on neuromuscular transmission (DANILOV et al. 1981). It has therefore been suggested that only compounds which are incapable of changing their conformation may act specifically on nicotinic acetylcholine receptors of skeletal muscle. According to this suggestion, bisquaternary *p*-terphenyl derivatives have been synthesized by KHOROMOV-BORISOV and his collaborators (see Chap. 15). The highest neuromuscular blocking potency and specificity has been found in *p,p''*-bistriethylammonium-*p*-terphenyl dibenzenesulfonate, referred to as tercuronium (DANILOV et al. 1978).

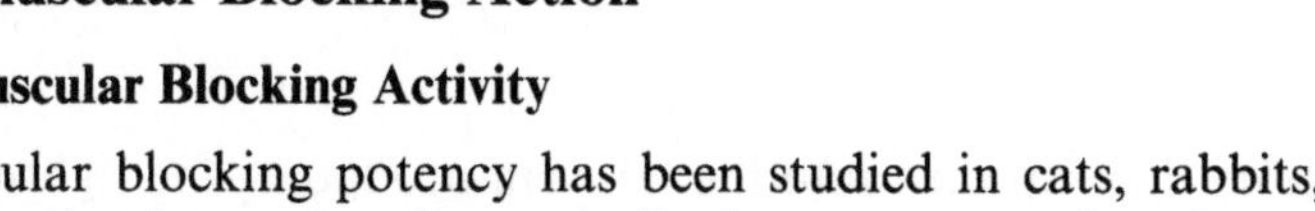

B. Pharmacology of Tercuronium

I. Neuromuscular Blocking Action

1. Neuromuscular Blocking Activity

Neuromuscular blocking potency has been studied in cats, rabbits, mice, and pigeons as well as in rat phrenic nerve–diaphragm preparation, frog rectus abdominis muscles, and frog cutaneous pectoris muscles.

a) Cats

The neuromuscular blocking activity of tercuronium as compared with tubocurarine and pancuronium is shown in Table 1. In cats under chloralose (50 mg/kg) and urethane (500 mg/kg) anesthesia, the mean dose of tercuronium producing block (95%) of tibialis anterior muscle twitches (0.2 Hz) is 0.08 mg/kg intravenously (ED_{95}). Thus, the neuromuscular blocking potency of tercuronium in cats is five times the potency of tubocurarine and about one-half that of pancuronium. The onset time and duration are 40–60 s and 2.0–2.5 min, respectively. The full block lasts 2–4 min. Muscle twitches are completely recovered in 10 ± 1.5 min. The time of action of tercuronium is a little less than that of tubocurarine and similar to pancuronium. The same results have been obtained in decerebrated cats (CHEREPANOVA et al. 1982).

Tercuronium possesses a cumulative property similar to that of tubocurarine and pancuronium. If tercuronium is repeatedly injected every 40 min, the maximum decrease in neuromuscular blocking dose is observed following the third or fourth injection, but does not change thereafter. The equieffective tercuronium dose decreases by 60%–80%. This is a characteristic feature of nondepolarizing neuromuscular blocking agents which may be accounted for by two main factors. First, a significant part of this kind of agent is lost on the way from the blood to acetylcholine receptors, owing to its binding to plasma proteins (COHEN et al. 1965; SKIVINGTON 1972; COHEN 1974), to capillary endothelium, and other "sites of loss" (CAVALLITO and SANDY 1959). The most considerable loss of substance occurs during the first application when all sites of loss are free. Second, the recovery of muscle twitches inhibited by neuromuscular blocking agent does not mean that the whole receptor pool is free of the substance; at this moment, only about 20% of the receptors are free (see PATON and WAUD 1967; BARNARD et al. 1971). Here, a lesser quantity of the substance is necessary to occupy the rest of the receptors compared with the first application of the drug. Therefore, nondepolarizing agents with stable molecule must possess cumulative properties. One can imagine that only unstable compounds may be devoid of cumulative properties.

b) Rabbits

The neuromuscular blocking potency of tercuronium in rabbits is 7.5 times the potency of tubocurarine and approaches that of pancuronium (Table 1). The mean tercuronium dose which produces head drop (HDD_{50}) in 50% of the rabbits used in experiments is 0.016 mg/kg. The maximum effect develops 1.5–2.0 min after injection and lasts 3–5 min. The onset and termination times of action for the three agents do not differ significantly. The value of LD_{50}/HDD_{50} is 1.4 (KHARKEVICH 1969).

c) Mice

The tercuronium dose injected into the tail vein producing the righting reflex in 50% of the animals used (ED_{50}) is equal to 0.044 mg/kg and for pancuronium it is 0.028 mg/kg. The LD_{50}/ED_{50} values for both agents are approximately equal to 1.5.

d) Rat Phrenic Nerve–Diaphragm Preparations

In these preparations (BÜLBRING 1946), tercuronium ($1.1 \pm 0.2\ \mu M$) inhibits the diaphragm twitches evoked by indirect stimulation (0.1 Hz) by 50% 3 min after application. The same effect is obtained with $4.4 \pm 1.0\ \mu M$ tubocurarine or

Table 1. Neuromuscular blocking potency of tercuronium, tubocurarine, and pancuronium[a]

Agents	Cats[b]	Rabbits[c]		Mice[d]		Rats[e]
	Block of indirectly elicited tibialis anterior twitches ED_{95} (μg/kg)	HDD_{50} (μg/kg i.v.)	LD_{50} (μg/kg i.v.)	EC_{50} (μg/kg i.v.)	LD_{50} (μg/kg i.v.)	Phrenic nerve–diaphragm preparation EC_{50} (*M*)
Tercuronium	80 ± 10 (12)	16 ± 5 (20)	22 ± 4 (5)	44 ± 5 (12)	67 ± 0 (12)	$1.1 \pm 0.2 \times 10^{-6}$ (12)
Tubocurarine	300 ± 40 (6)	120 ± 20 (6)	175 ± 3 (5)			$4.4 \pm 1.0 \times 10^{-6}$ (10)
Pancuronium	40 ± 6 (6)	12 ± 4 (6)	18 ± 6 (5)	28 ± 4 (12)	42 ± 5 (12)	$4.0 \pm 0.2 \times 10^{-6}$ (10)

[a] Doses are given in μg/kg; numbers of experiments are given in parentheses.

[b] Cats of either sex were anesthetized with chloralose (50 mg/kg) and urethane (500 mg/kg) mixture intraperitoneally. Tibialis anterior muscle twitches were elicited by single supramaximal pulses of 0.2 ms duration applied to n. peroneus every 5 s. Mean (± standard error) doses of agents injected into v. jugularis externa producing 95% block of twitches were determined from dose–response curves.

[c] Unesthetized rabbits of either sex, the mean (± standard error) dose of agents intravenously injected producing head drop in 50% of animals (HDD_{50}) and death (LD_{50}) of 50% of animals is shown.

[d] In mice, the dose preventing the righting reflex in 50% of animals (ED_{50}) and producing death (LD_{50}) of 50% of animals was determined (mean ± standard error).

[e] Rat phrenic nerve–diaphragm preparations were mounted in a bath (100 ml) with aerated solution at 20°–22 °C. The phrenic nerve was stimulated with single supramaximal pulses of 0.1 ms duration every 10 s. The agents were introduced into the bath with a syringe, the amount never exceeding 1 ml. The mean (± standard error) molar blocking concentration of the agent resulting in 50% decline of diaphragm twitches (EC_{50}) was obtained from the plot.

4.0 ± 0.2 μ*M* pancuronium (Table 1). Thus, among these agents, tercuronium proved to be the most potent when applied to rat phrenic nerve–diaphragm preparations.

2. Site and Mode of Action

The neuromuscular blocking effect of tercuronium is due to its action on the postjunctional membrane. In cats, tercuronium simultaneously blocks the tibialis anterior muscle twitches evoked both by close-arterial injection of acetylcholine (BROWN et al. 1936) and by nerve stimulation (Fig. 1). Direct stimulation of the muscle produced normal contractions in this case.

Tercuronium is a nondepolarizing agent. The neuromuscular blockade due to tercuronium is never preceded by an augmentation of muscle twitches and no muscle fasciculation occurs either. In cats, tubocurarine potentiated the action of tercuronium whereas both succinylcholine and decamethonium antagonized it. The effect of tercuronium was also increased by a higher frequency of nerve stimulation. When the muscle twitches are partially blocked, motor nerve tetanization results in the failure of muscle contraction (Vedensky inhibition). The muscle twitches are restored following nerve tetanization (posttetanic decurarization). In pigeons, tercuronium 0.2 mg/kg i.v. causes flaccid paralysis. Neostigmine and other anticholinesterase substances antagonize the effect of tercuronium. Its action is also reversed by 4-aminopyridine (PASKOV et al. 1973; HORN et al. 1979).

In frog rectus abdominis muscle, tercuronium does not possess nicotinomimetic properties. When applied at doses of 0.1–0.3 μ*M*, it produces a dose-dependent shift of the acetylcholine dose–response curves without significantly affecting their slope or maxima (Fig. 2). The slope and 95% confidence limits for the

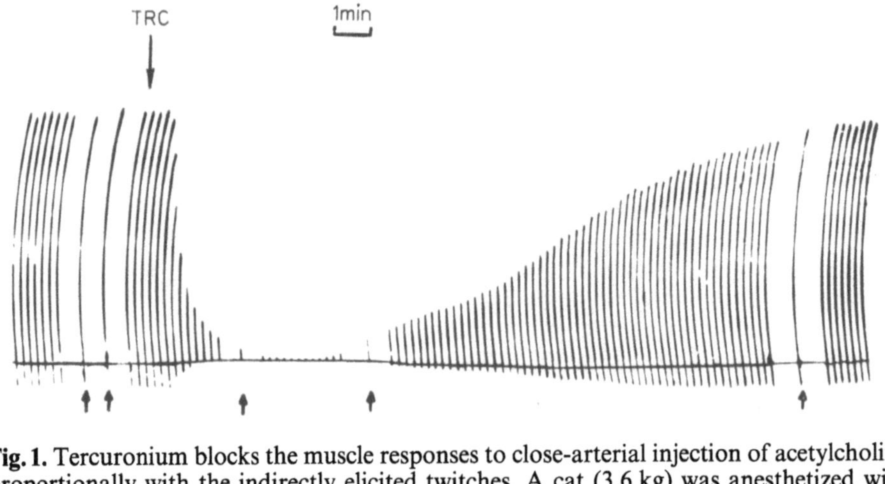

Fig. 1. Tercuronium blocks the muscle responses to close-arterial injection of acetylcholine proportionally with the indirectly elicited twitches. A cat (3.6 kg) was anesthetized with chloralose (50 mg/kg) and urethane (500 mg/kg) intraperitoneally; atropine 1 mg/kg was applied intraperitoneally. The maximal twitches of the tibialis anterior muscle were evoked by stimulation of nervus peroneus communis every 7 s. During the arterial injection of acetylcholine (3 μg/kg *upward arrows*) electrical stimulation of the nerve was stopped. Tercuronium (*downward arrows*) was administered at 0.08 mg/kg

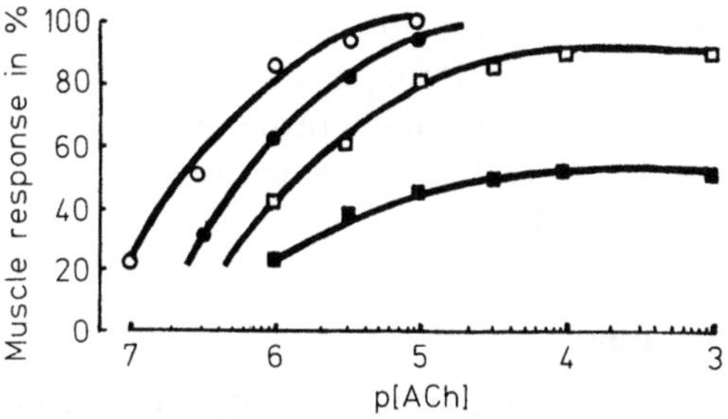

Fig. 2. Cumulative dose–response curves for acetylcholine with respect to increased concentrations of tercuronium in frog rectus abdominis muscle. Each point represents the mean value of ten experiments. The parallel shift in the curves following treatment by tercuronium 10^{-7} *M* (*full circles*); 3×10^{-7} *M* (*open squares*) relative to controls (*open circles*) shows a competitive antagonism. The decrease of the maximum response shows a noncompetitive antagonism of tercuronium at 10^{-6} *M* (*full squares*)

regression of log(DR-1) on log*B* are 0.97 (0.91–1.05) which is not significantly different from unity. These results show that antagonism of tercuronium in these concentrations is due to its competition with acetylcholine. The apparent dissociation constant K_B found from the regression equation is 0.15 μ*M*. However, while tercuronium concentrations are increased severalfold, the shift of the dose–response curves is accompanied by a decline in their maxima. Tercuronium (1.0 μ*M*) accounts for a twofold decrease of the curve maximum (A'_2 after VAN ROSSUM 1966). Thus, tercuronium in high concentrations acts as a noncompetitive antagonist. This distinguishes it from tubocurarine which acts as a competitive antagonist, even in high concentrations (JENKINSON 1960).

In frog cutaneous pectoris muscle (*Rana temporaria*), tercuronium did not affect the resting membrane potential, abolished the muscle action potentials, and did not affect the nerve action potentials. It decreased the end-plate potentials

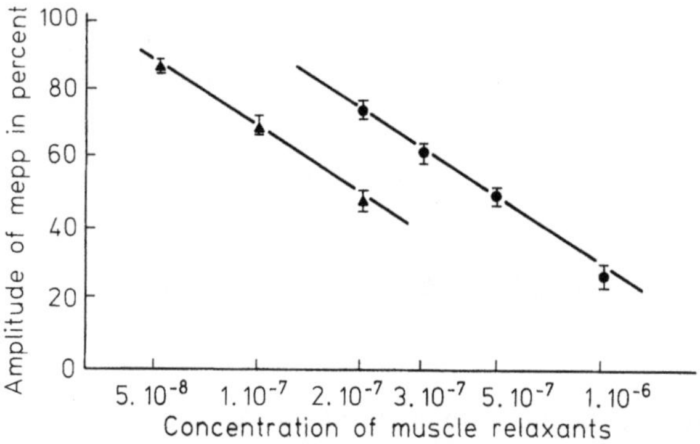

Fig. 3. Effect on MEPP of increased concentration of tercuronium (*triangles*) and tubocurarine (*circles*). Frog cutaneous pectoris muscle, intracellular electrodes. Abscissa: molar concentration of agents; ordinate: amplitude of MEPP as a percentage of control. Each point represents the mean ± standard error of 3–10 experiments. The data obtained confirm that tercuronium, like tubocurarine, affects postjunctional sites

(EPP) as well as miniature end-plate potentials (MEPP) by 50% in 0.1 μ*M*. Tercuronium did not change the MEPP frequency nor the shape of MEPP and EPP. The equieffective concentration of tubocurarine was 0.5 μ*M* (Fig. 3). Thus, tercuronium is a highly active nondepolarizing neuromuscular blocking agent without apparent presynaptic action.

3. Neostigmine Antagonism

The degree of neostigmine antagonism is one of the determining factors in estimating the usefulness of a neuromuscular blocking agent for clinical practice. To estimate the fitness of tercuronium for anesthesia the neostigmine antagonism to it was compared with that of tubocurarine, gallamine, and pancuronium (MALYGIN 1977; CHEREPANOVA et al. 1982). Experiments were carried out on cats under anesthesia. Taking into consideration the cumulative properties of nondepolarizing muscle relaxants, ED_{95} of the agent was established in each experiment following the third or fourth administrations, the drug being applied every 40 min. The last administered dose of the agent, referred to as $ED_{95}R$, was usually 60%–80% less than the first administration (see Sect. B.I.1). Then 2–6 $ED_{95}R$ muscle relaxant was administered intravenously, followed by 0.1 mg/kg neostigmine in 2 min with the view to determining the block duration. Neostigmine antagonism appears to be the most pronounced in the case of tercuronium (Fig. 4). Even when 5 $ED_{95}R$ tercuronium was applied, the recovery of tibialis anterior muscle twitches started within 2–5 min after neostigmine administration and complete restoration of twitches followed 15–20 min later. Neostigmine antagonism to other agents was less pronounced. When administering 5 $ED_{95}R$ pancuronium, muscle twitches were not observed until 9–10 min after neostigmine application, while the full amplitude twitches reappeared in about 30 min only. In the case of 5 $ED_{95}R$ tubocurarine application, the onset and the complete recovery of muscle twitches did not occur until 20 and 48 min, respectively, after

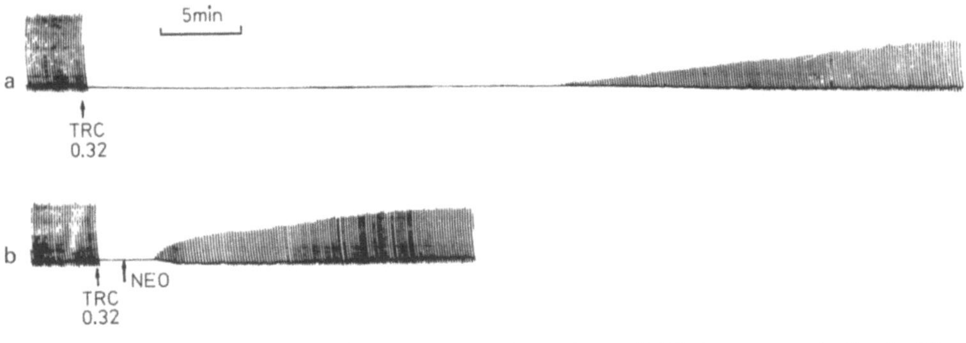

Fig. 4 a, b. Neostigmine antagonism to tercuronium. A cat (3.2 kg) was anesthetized with chloralose (50 mg/kg) and urethane (500 mg/kg) intraperitoneally; atropine sulfate (1 mg/kg) was applied intravenously. the indirectly elicited maximal twitches of the tibialis anterior muscle occurred every 10 s. **a** effect of 5 ED_{95} tercuronium intravenously (*TRC*), **b** repetitive administration of 5 ED_{95} tercuronium (*TRC*) and neostigmine (*NEO*) intravenously. Time gap between **a** and **b** is 1 h. Before the experiment, tercuronium was administered three times at the neuromuscular blocking dose

Table 2. Neostigmine antagonism to tercuronium, tubocurarine, pancuronium, and gallamine

Agents	Dose 4 ED_{95} (mg/kg)	Mean (± standard error) time of full blockade (min)		A/B
		Agents alone A	Agents + neostigmine B	
Tercuronium	0.32	30.6±6.6	3.3±0.8	9.3±2.3
Tubocurarine	1.2	27.1±10.5	6.3±2.9	4.3±2.0
Pancuronium	0.16	24.1±3.9	8.5±1.0	2.9±0.6
Gallamine	3.9	28.7±7.2	11.7±2.8	2.4±1.3

[a] Cats under chloralose plus urethane anesthesia. Tibialis anterior muscle twitches were elicited by single supramaximal pulses every 5 s; 4 ED_{95} neuromuscular blocking agent were injected into v. jugularis externa. Neostigmine 0.1 mg/kg was applied intravenously within 2 min following the neuromuscular blocking agent injection. The mean (± standard error) time of onset of the muscle twitsch recovery (10% of the control amplitude) is shown.

neostigmine administration. In some experiments, even repeated neostigmine injection could not evoke immediate complete restoration of muscle twitches blocked by tubocurarine.

Other experiments on cats enabled one to estimate; first, the mean time of onset of muscle twitch recovery (10% of control amplitude) after the administration of 4 ED_{95} tercuronium, tubocurarine, pancuronium, or gallamine; second, the decrease of time course of action of these agents owing to neostigmine 0.1 mg/kg intravenously applied within 2 min after injection of the agent. Following the administration of 4 ED_{95} tercuronium, tubocurarine, and gallamine, the block was of similar duration, whereas the pancuronium action was shorter (Table 2). It was tercuronium whose duration of action was most shortened by neostigmine (9.3 times), the action of 4 ED_{95} tubocurarine, pancuronium, and gallamine being shortened by 4.3, 2.9, 2.4 times, respectively. Thus, these experiments reveal the strongest neostigmine antagonism to tercuronium. It seems therefore that tercuronium can provide more controllable and, hence, safer muscle relaxation than other agents. As far as differences in the antagonism of neostigmine to these neuromuscular blocking agents is concerned, it is probably due to their different antiacetylcholinesterase action at neuromuscular junctions. This suggestion is confirmed by the data presented in Sect. B.II.4.

II. Side Effects

1. Effect on Blood Pressure

In cats and in rabbits under anesthesia, blood pressure was recorded from the carotid artery which was cannulated and connected to a transducer. Tercuronium at neuromuscular blocking doses did not affect blood pressure. Even a tenfold increase of the blocking dose resulted in only a moderate (10–20 mmHg) and temporary arterial hypotension. A further 15- to 20-fold dose increase resulted in a drop of blood pressure by 40–60 mmHg. Atropine (1 mg/kg) premedication failed to prevent the fall in blood pressure. Tubocurarine always produced a temporary arterial hypotension. Pancuronium in the blocking dose did not affect blood pressure whereas increasing the dose three- to fivefold resulted in a considerable and stable pressor effect.

2. Ganglion-Blocking Action

Ganglion-blocking activity of tercuronium and other muscle relaxants was compared in cats under ether and chloralose (80 mg/kg i.v.) anesthesia. The contraction of the nictitating membrane was recorded with a force displacement transducer. The cervical sympathetic trunk was stimulated with restangular pulses of supramaximal voltage, 1 ms duration at a frequency of 5 Hz. The dose of the agent which caused a 50% decline of the membrane contraction (ED_{50}) was obtained from the plot. For tercuronium ED_{50} was 0.8±0.07 mg/kg (Table 3) which showed a tenfold increase in its neuromuscular blocking dose and approached the ganglion-blocking dose of hexamethonium. The maximum effect lasted for 2–3 min. The nictitating membrane contraction reached the initial level 8–10 min after the tercuronium injection. The blood pressure had been restored earlier; 0.4±0.02 mg/kg tubocurarine (almost equal to its neuromuscular blocking dose) produced a similar decrease in the nictitating membrane contraction (Table 3). The ganglion-blocking dose of pancuronium was 30 times more than its myoparalytic dose. Thus, significant differences in ganglion-blocking potency

Table 3. Ganglion-blocking and vagolytic potency of neuromuscular blocking agents in cats[a]

Agents	Neuromuscular blocking dose (mg/kg) *A*	Ganglion-blocking dose ED_{50} (mg/kg) *B*	*B/A*	Vagolytic dose ED_{50} (mg/kg) *C*	*C/A*
Tercuronium	0.08±0.01 (12)	0.8±0.007 (6)	10	0.8 ±0.1 (5)	10
Tubocurarine	0.3 ±0.04 (6)	0.4±0.02 (4)	1	0.4 ±0.1 (3)	1
Pancuronium	0.04±0.006 (6)	1.2±0.1 (4)	30	0.02±0.005 (3)	0.5

[a] Cats under chloralose plus urethane anesthesia; the tibialis anterior muscle twitches were elicited by single supramaximal pulses (0.2 Hz). The mean (± standard error) dose of the neuromuscular blocking agent producing 95% block of twitches (ED_{95}) is shown. The heart rate was derived from the electrocardiogram (lead II). The peripheral segment of the vagus nerve was stimulated periodically for 7 s with pulses of 1 ms duration at a frequency of 20 Hz. The mean (± standard error) dose of the agent which decreased the bradycardia in response stimulation of the vagus nerve by 50% (ED_{50}) is shown. In cats under either and chloralose (80 mg/kg i.v.) anesthesia, the nictitating membrane contraction was recorded with a force displacement transducer. The cervical sympathetic trunk was stimulated with supramaximal pulses of 1 ms duration at a frequency of 5 Hz. The mean (± standard error) dose of the agent which caused 50% decline of the membrane contraction (ED_{50}) was obtained from the plot. The agents were injected into the external jugular vein through a permanent cannula. Numbers of experiments are given in parentheses

of the muscle relaxants tested could be found when comparing their ganglion-blocking and neuromuscular blocking doses. The ratio of ganglion-blocking to neuromuscular blocking dose is 30 : 1 in the case of pancuronium, 10 : 1 in the case of tercuronium, and is approximately 1 : 1 with tubocurarine.

3. Vagolytic Action

Tercuronium failed to inhibit bradycardia and hypotension in cats in response to stimulation of the vagus nerve with doses similar to those producing neuromuscular blockade (Fig. 5). Bradycardia and hypotension were reduced only by doses

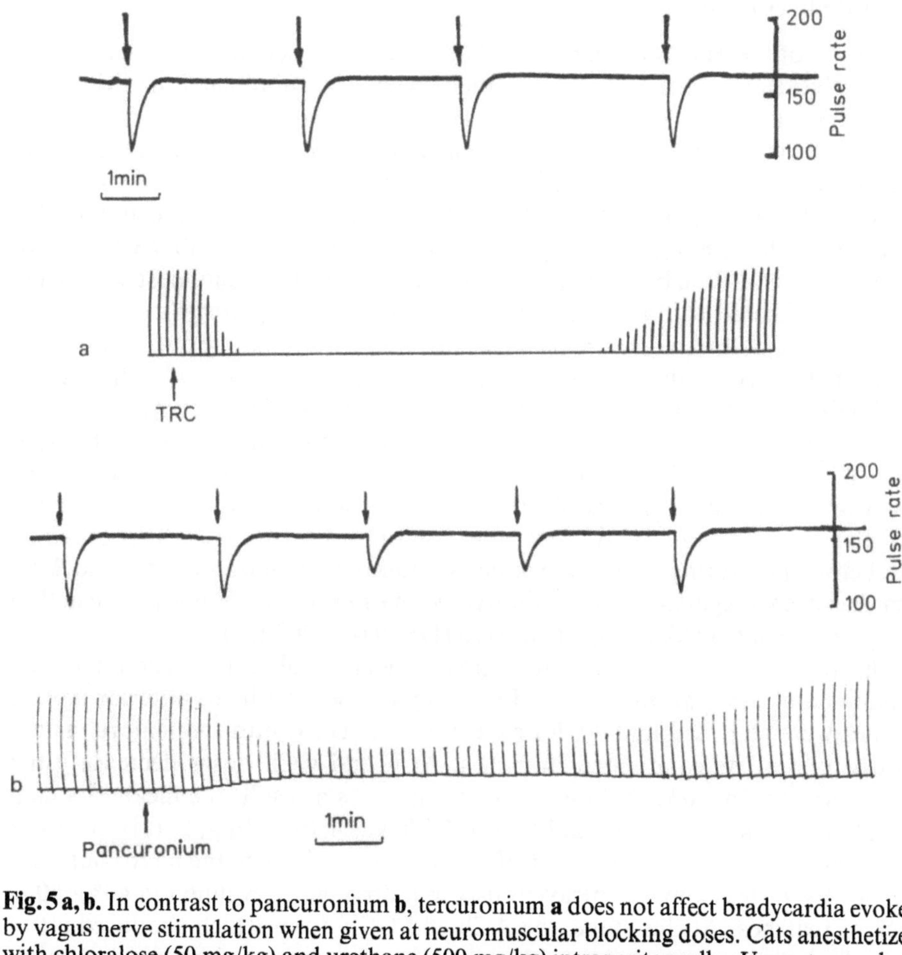

Fig. 5 a, b. In contrast to pancuronium **b**, tercuronium **a** does not affect bradycardia evoked by vagus nerve stimulation when given at neuromuscular blocking doses. Cats anesthetized with chloralose (50 mg/kg) and urethane (500 mg/kg) intraperitoneally. *Upper traces* show the heart rate recorded electrocardiographically with superficial needle electrodes, using a tachometer. Stimulation of the distal segment of the right vagus nerve with supramaximal pulses of 1 ms duration at 20 Hz for 7 s (*arrows*). *Lower traces* show the twitches of the tibialis anterior muscle evoked by indirect stimulation. **a** tercuronium 0.1 mg/kg i.v., **b** pancuronium 0.02 mg/kg i.v.

10–20 times higher than the neuromuscular blocking dose (ED_{95}). The same effect was evoked by tubocurarine at doses of 0.4 mg/kg, that is, near to those producing neuromuscular block. Pancuronium has the strongest vagolytic effect. Even below the blocking dose (at 0.02 mg/kg), pancuronium antagonized the bradycardia elicited by vagus nerve stimulation.

Thus, comparison of the neuromuscular blocking and vagolytic potency of these agents showed that tercuronium was a most specific compound. In other experiments with anesthetized cats, tercuronium at doses of 0.1–1.0 mg/kg did not affect the hypotension produced by acetylcholine (3 μg/kg) injected intravenously. To sum up, tercuronium has no atropine-like action when given in doses 10–15 times its neuromuscular blocking dose.

4. Anticholinesterase Action

Estimation of the anticholinesterase action of neuromuscular blocking agents in vivo involves serious difficulties. First, their enzyme inhibitory action in vivo does not always completely correlate with antiacetylcholinesterase potency obtained in purified enzyme in vitro. Second, biochemical methods with tissue homogenization are hardly applicable in defining the anticholinesterase action of neuromuscular blocking agents in vivo because ammonium compounds form unstable complexes with the enzyme which breaks down owing to the solubility of the homogenate. In addition, biochemical methods enable the total cholinesterase in tissue to be estimated with respect to both extracellular and intracellular enzymes. Effects of anticholinesterase agents have been supposed to be due to the inhibition of external acetylcholinesterase (AChE;EC 3.1.1.7) referred to as functional AChE (KOELLE 1957, 1963; McISAAK and KOELLE 1959). The external AChE accounts for only about 20%–30% of the total cholinesterase activity of a homogenate (BARRNETT and PALADE 1959; NAMBA and GROB 1968) and is easily accessible to ammonium compounds. Therefore, the method of comparison of a neuromuscular blocking dose of acetylcholine (ACh) and carbachol (CCh) obtained both before and during partial blockade produced by a neuromuscular blocking agent has been suggested for quantitative estimation of anticholinesterase action of neuromuscular blocking agents in vivo (DANILOV 1967 a, b).

This technique has been used to estimate the anticholinesterase action of tercuronium and other agents in cats. For this purpose the tibialis anterior muscle twitches were recorded and ACh or CCh were injected into the femoral artery through one of its cannulated branches. ACh and CCh doses decreasing the muscle twitches by 50%–60% were determined. As a result, the mean blocking dose of ACh was 15.2 μ*M*/kg and that of CCh was 0.16 μ*M*/kg, that is, the ratio of their blocking doses in controls (DR_C) was about 94. Then the neuromuscular blocking agent was applied intravenously in a dose which resulted in 40%–60% block of twitches. At this stage, the ACh and CCh doses which produced a full blockade were determined. Their blocking doses are increased by nondepolarizing agents. However, the ACh dose augmentation was less than that of CCh, owing to the anticholinesterase action of neuromuscular blocking agents and, as a result, to a decrease of the ACh hydrolysis rate. Then the dose ratio of ACh to CCh in this case (DR_A) would be less than that in controls (DR_C), corresponding to the degree of enzyme inhibition. Thus, the value of DR_C divided by DR_A can be considered as the index I of anticholinesterase inhibition by a neuromuscular blocking agent at the neuromuscular junction.

$$I = \frac{DR_C}{DR_A}.$$

The index of inhibition of each agent has been determined in five experiments (MALYGIN 1977). Tercuronium has a lower I (1.6) than tubocurarine (1.7), pancuronium (2.2), and gallamine (7.2). Compared with other agents, tercuronium gives rise to the weakest inhibition of AChE at neuromuscular junctions.

The antiacetylcholinesterase action of tercuronium and pancuronium has also been compared in rat phrenic nerve–diaphragm preparations (MALYGIN 1977) using a modified (see BALASHOVA et al. 1976) Ellman's technique (ELLMAN et al. 1961). Acetylcholinesterase inhibition by tercuronium at neuromuscular blocking doses has been shown to be weaker than that in the case of pancuronium.

C. Toxicity of Tercuronium

Cats and rabbits under chloralose plus urethane anesthesia and artificial ventilation tolerated intravenous injection of tercuronium at 200 times its neuromuscular blocking dose (ED_{95}). In these cases, the blood pressure decreased to 60–80 mmHg. Then the blood pressure gradually recovered, approximating or reaching the initial level by the 3–4 h.

Evaluation of the chronic toxicity has been carried out in six nonanesthetized rabbits. Four animals were given daily injections of 16 µg/kg tercuronium in the marginal ear vein over a period of 12 days. This dose produced head drop. Two control rabbits were injected with 0.9% sodium chloride solution. The animals were killed by exsanguination: the two control rabbits and two of the experimental series were killed the day following the last tercuronium injection, and the rest 2 weeks later. The sensitivity of the animals to tercuronium remained constant in the course of its injection. No changes in their behavior were noticed. The body weight did not change significantly. The rectal temperature remained constant during the experiment. No hematologic, blood biochemical, or pathologic changes in the urine were found. Repeated injections of 10^{-4} μM tercuronium at the same vein site caused neither thrombosis, local inflammatory reaction, nor any edema of the perivascular tissue. Histologic analysis of the vein revealed no signs of irritation.

D. Conclusions

The experimental results show tercuronium to be a potent, highly specific nondepolarizing neuromuscular blocking agent with a duration of blocking action comparable to that of tubocurarine and pancuronium. The neuromuscular blocking potency of tercuronium on different species is 5–8 times that of tubocurarine. Tercuronium does not reveal any presynaptic action. In contrast to tubocurarine, pancuronium, and gallamine, tercuronium does not affect autonomic ganglionic transmission and does not possess atropine-like action, even after several neuromuscular blocking doses have been given. It can block ganglionic transmission and the transmission from the vagus nerve to the heart only when its dose is 10–20 times the myoparalytic dose. Therefore, tercuronium does not cause undesirable side effects. Tercuronium does not produce a significant acetylcholinesterase inhibition in the neuromuscular junction. The anticholinesterase action of tercuronium in vivo is less than that of tubocurarine, pancuronium, and gallamine. This agrees with the quantitative estimation of neostigmine antagonism to these agents. The most pronounced antagonism has been revealed with respect to tercuronium. As far as the specificity of tercuronium is concerned, it is probably due to its high affinity to acetylcholine receptors of the skeletal muscle and its rigid molecular structure which is incapable of conformational changes. It is not unlikely that a flexible molecule can change its conformation to fit different types of acetylcholine receptors and therefore may possess a high affinity to them. That is probably, why among flexible ammonium compounds, specific neuromuscular blocking agents have not been found (DANILOV et al. 1981). The probability of possessing an equal affinity to different types of acetylcholine receptors is much less for rigid compounds.

Thus, the application of tercuronium in anesthesia should allow one to obtain safe and well-controlled muscle relaxation. This has been confirmed by clinical observations (see Chap. 26).

References

Balashova EK, Danilov AF, Rosengart VI, Sherstobitov OE (1976) Functional acetylcholinesterase of rat diaphragm neuromuscular junction (in Russian). In: Turpaev TM (ed) Physiologiy i biochimiy mediatornih processov (Physiology and biochemistry of transmission). Nauka, Moscow, pp 13–14

Barnard EA, Wieckovsky J, Chiu TH (1971) Cholinergic receptor molecules and cholinesterase molecules at mouse skeletal muscle junctions. Nature 234:207–209

Barrnett RJ, Palade GE (1959) Enzymatic activity in the M Bond. J Biophys Biochem Cytol 6:163–169

Brown GL, Dale HH, Feldberg W (1936) Reactions of the normal mammalian muscle to acetylcholine and to eserine. J Physiol (Lond) 87:394–424

Bülbring E (1946) Observation on the isolated phrenic nerve diaphragm preparation of the rat. Br J Pharmacol 1:38–61

Cavallito CJ, Sandy P (1959) Anticholinesterase activities of some bis-quaternary ammonium salts. Biochem Pharmacol 2:233–342

Cherepanova VP, Danilov AF, Khromov-Borisov NV, Malygin VV, Starshinova LA, Torf SP (1982) A new nondepolarizing myorelaxant with high activity and specificity. Eur J Pharmacol 81:245–255

Cohen EN (1974) Uptake, distribution and elimination of the muscle relaxants. In: Scientific foundations of anaesthesia. Heineman, London, pp 350–356

Cohen EN, Carbascio A, Fleishli G (1965) The distribution and fate of d-tubocurarine. J Pharmacol Exp Ther 147:120–129

Danilov AF (1967a) the role of the anticholinesterase action of myorelaxants in their blocking action (in Russian). Pharmakol Toksikol 30:124

Danilov AF (1967b) Inhibition of the cholinesterase of the neuromuscular synapses by decamethonium and succinylcholine (in Russian). Pharmakol Toksikol 30:664–669

Danilov AF, Malygin VV, Starshinova LA (1978) Tercuronium a new nondepolarizing myorelaxant. Abstr 7th Int Congr Pharmacol Paris. Pergamon, New York, p 153
Danilov AF, Starshinova LA, Fedorov DJ, Shelkovnikov SA (1981) The rigidity of ammonium compounds as obliged condition for selective action on acetylcholine receptors of skeletal muscle (in Russian). Dokl Akad Nauk SSSR 258:1234–1242
Ellman GL, Courtney KD, Andres V, Featherstone RM (1961) A new and rapid colorimetric determination of acetylcholinesterase activity. Biochem Pharmacol 7:88–95
Horn AS, Lambert JJ, Marshall JG (1979) A comparison of the facilitatory action of 4-aminopyridine methiodide and 4-aminopyridine on neuromuscular transmission. Br J Pharmacol 65:53–59
Jenkinson DH (1960) the antagonism between tubocurarine and substances which depolarize the motor end-plate. J Physiol (Lond) 152:309–324
Kharkevich DA (1969) Pharmakologya kurarepodolnik szedstv (Pharmacology of curare-like substances). Meditsina, Moscow
Koelle GB (1957) Histochemical demonstration of reversible anticholinesterase action at selective cellular sites in vivo. J Pharmacol Exp Ther 120:488–503
Koelle GB (1963) Cytological distributions and physiological functions of cholinesterases. In: Koelle GB (ed) Cholinesterases and anticholinesterases agents. Springer, Berlin, pp 187–299 (Handbuch der experimentellen Pharmakologie, vol 15)
Malygin VV (1977) Anticholinesteraznoe deistvie postsinaptisheskih myorelaxantov (The anticholinesterase action of postsynaptic muscle relaxants). Thesis, Pavlov's Medical Institute, Leningrad
Mc Isaak R, Koelle GB (1959) Comparison of the effects of inhibition of external, internal, and total acetylcholinesterases upon ganglionic transmission. J Pharmacol Exp Ther 126:9–20
Namba T, Grob D (1968) Cholinesterase activity of the motor end-plate in isolated muscle membrane. J Neurochem 15:1445–1454
Paskov DS, Staenov EA, Mitsov VV (1973) The new anticurare and analeptic drug, pimadin (in Russian). Eksp Khir Anesteziol 4:48–52
Paton WDM, Waud DR (1967) The margin of safety of neuromuscular transmission. J Physiol (Lond) 191:59–90
Skivington MA (1972) Protein binding of three tritiated muscle relaxants. Br J Anaesth 44:1030–1034
Van Rossum JM (1966) Limitations of molecular pharmacology. Adv Drud Res 3:189–234
Waser PG (1972) Chemistry and pharmacology of natural curare compounds. In: Cheymol J (ed) Neuromuscular blocking and stimulating agents. Pergamon, Oxford, vol 1, pp 205–239 (International encyclopedia of pharmacology and therapeutics, sect 14)

CHAPTER 22

Dioxonium

A. A. ĶIMENIS

A. Introduction

The neuromuscular blocking agent, dioxonium, was synthesized by dimerizing the structure of cholinomimetic compounds. This procedure is known to lead to the loss of muscarinomimetic properties and the appearance of a pronounced blocking activity with respect to cholinoreceptors of skeletal muscles (BOVET et al. 1951). Imbretil, containing two carbachol molecules connected by six methylene groups, also fails to exert cholinomimetic effects characteristic of carbachol, but shows a strong and protracted neuromuscular blocking effect (KLUPP et al. 1953). This approach has proved to be useful in the case of acetylcholine derivatives containing a dioxolan moiety, e.g., F-2268 (compound 1), showing strong muscarinomimetic activity (FOURNEAU et al. 1944).

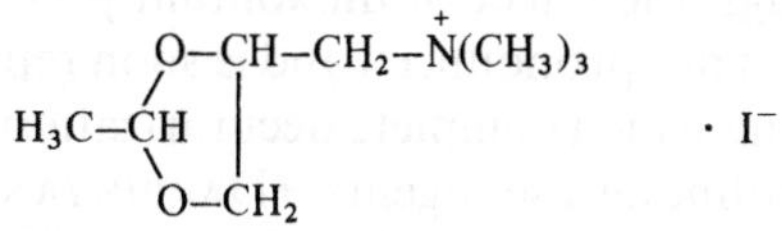
(1)

The compound formed by dimerizing molecules of F-2268 and its analogs was obtained during the synthesis of cyclic diacetals of succinic, maleic, and other dialdehydes derived from 2,5-dihydrofuran (SOKOLOV et al. 1964; SOKOLOV and HILLER 1967). Among other compounds obtained in this synthesis were the bisammonium derivatives of succinic aldehyde cyclic acetal (alkyl halide salts of bisaminomethyldioxolanylethane) having the general formula (2)

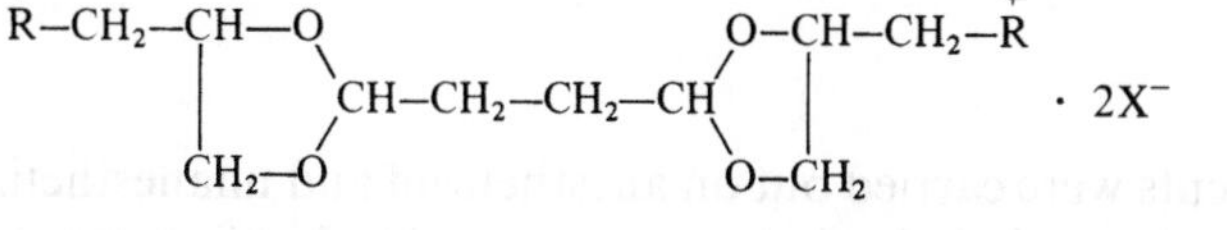
(2)

where R^+ is trialkylammonium, N-methylpiperidinium, N-methylpyrrolidinium, N-methylmorpholinium, and $X^- = I^-$ or Br^- (SOKOLOV et al. 1968).

Compound (3), which may be regarded as a dimerized molecule of compound F-2268 shows a high neuromuscular blocking activity, its ED_{50}, as tested on

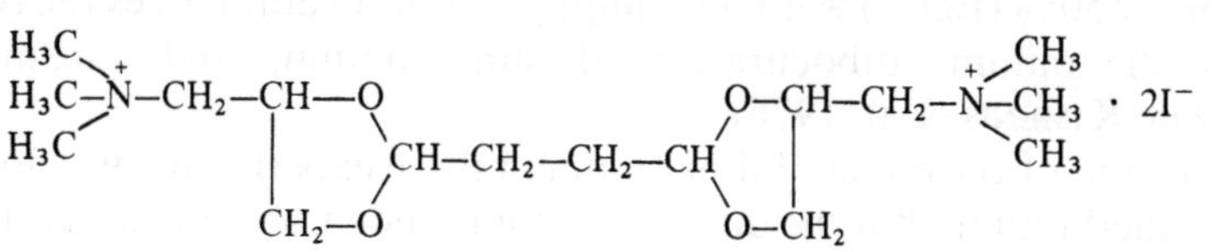
(3)

anesthetized cats by measuring contractions of gastrocnemius muscle, being 0.012 (0.010–0.014) mg/kg. Under similar conditions, the ED_{50} of suxamethonium (Ditilinum) amounts to 0.019 (0.014–0.024) mg/kg. Thus, the compound obtained by dimerizing the molecule of F-2268 is nearly twice as active a neuromuscular blocking agent as suxamethonium, the molecule of which is made up of two connected acetylcholine molecules. The compound where the trimethylammonium groups are replaced by methylpyrrolidinium (compound 4, dioxonium) has been found to be twice as active as the F-2268 dimer (SOKOLOV and HILLER 1970).

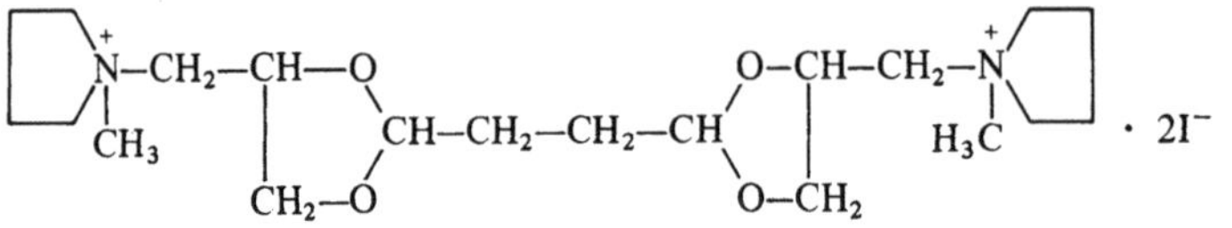

(4)

It should be noted that the substitution of trimethylammonium for methylpyrrolidinium in the suxamethonium series and related compounds fails to bring about a significant increase in neuromuscular blocking activity (ĶIMENIS 1960a). The effects of compounds (3) and (4) on neuromuscular junctions are characterized by a relatively high selectivity. They are practically inactive to muscarinoreceptors and do not affect hemodynamics. Compound (4), known as dioxonium, has been recommended for clinical trial and is described as a muscle relaxant of medium duration of action, exceeding 13-fold the activity of tubocurarine (GINTERS 1970). Although the effect of dioxonium proceeds in two phases, the first depolarizing phase is not pronounced and is soon replaced by the stable nondepolarizing phase during which complete decurarization can be achieved by administration of anticholinesterase agents (DARBINYAN and VAISBERG 1970; VANEVSKY and KARGAPOLOV 1970; VANEVSKY et al. 1972; ARTEMOV and YERIVANTSEV 1972). Owing to these properties, dioxonium was recommended for clinical use.

B. Experimental

I. Neuromuscular Blocking Activity

1. Cats

The experiments were carried out on anesthetized and unanesthetized animals. In urethane-anesthetized (1–2 g/kg) cats, the amplitude of gastrocnemius contractions was recorded during stimulation of the sciatic nerve with single square wave supramaximal stimuli (0.1 Hz, 0.5 ms). Intravenous administration of dioxonium elicited a temporary suppression of muscle contraction in the 0.006–0.01 mg/kg dose range and at higher doses. The dose which reduced the amplitude of muscle contraction by 50% (ED_{50}) was 0.005 mg/kg. Table 1 summarizes the relative activities of dioxonium, tubocurarine, decamethonium, and suxamethonium (KLUŠA 1968; ĶIMENIS et al. 1976).

The duration of the effect of dioxonium in these experiments was longer than that of decamethonium, but the difference from tubocurarine was not statistically

Table 1. Relative activities of dioxonium and some other neuromuscular blocking agents in cats under anesthesia

Compound	ED_{50} (mg/kg) confidence limits $P=0.05$	Relative activity of dioxonium		
Dioxonium	0.005 (0.003–0.007)	22	2	3.8
Tubocurarine	0.113 (0.077–0.149	1		
Decamethonium	0.011 (0.008–0.014)		1	
Suxamethonium	0.019 (0.014–0.024)			1

significant. When applied to unanesthetized cats (ĶIMENIS and VĒVERIS 1973), dioxonium injected into the femoral vein was more active than other muscle relaxants used in the head drop test (Table 2).

2. Rabbits

The comparison of head drop doses (HDD) measured in rabbits during slow injection of dioxonium into the marginal auricular vein demonstrated a higher neuromuscular blocking activity than for tubocurarine and decamethonium, although the difference in activity was not so marked as in the case of cats (Table 2).

Relaxation of cervical muscles following dioxonium administration is accompanied by a slight relaxation of muscles in the extremities. At doses eliciting head drop, contractions of respiratory muscles are also somewhat inhibited. The duration of head drop in response to dioxonium injection was 2–6 min, and was practically the same as in the case of tubocurarine and decamethonium.

3. Dogs

The HDD determinations in unanesthetized male dogs were performed by injecting the test agents into a subcutaneous vein in the hindlimbs. As can be seen from Table 2, dioxonium in these animals is twice as effective as tubocurarine, but does not differ in activity from decamethonium (VĒVERIS 1974).

Table 2. Head drop doses (HDD) of dioxonium and other neuromuscular blocking agents (mg/kg)

Compound	HDD (mg/kg) confidence limits $P=0.05$		
	Rabbit	Cat	Dog
Dioxonium	0.096 (0.082–0.110)	0.035 (0.033–0.037)	0.059 (0.052–0.066)
Tubocurarine	0.232 (0.221–0.243)	0.40 (0.37–0.43)	0.125 (0.085–0.165)
Decamethonium	0.140 (0.124–0.156)	0.05 (0.04–0.06)	0.090 (0.055–0.125)
Suxamethonium	0.194 (0.143–0.245)	0.120 (0.108–0.132)	0.130 (0.105–0.155)

Table 3. Relative neuromuscular blocking activity of dioxonium in unanesthetized mice and rats

Compound	ED_{50} (mg/kg) confidence limits $P = 0.05$	
	Mice	Rats
Dioxonium	0.300 (0.250–0.360)	0.220 (0.177–0.273)
Tubocurarine	0.095 (0.083–0.108)	0.062 (0.050–0.076)
Decamethonium	0.630 (0.525–0.756)	2.050 (1.628–2.589)
Suxamethonium	0.210 (0.182–0.240)	0.290 (0.220–0.380)

4. Mice and Rats

The curare-like activity of dioxonium was assayed in mice by the rotarod test, and in rats by the impaired ability of the animals to hold onto a sloping screen (a metal net placed at an angle of 60°; KHARKEVICH 1969). The mean effective doses for i.v. administration of the agents in the study are given in Table 3. It is obvious that rats and mice are significantly less sensitive to dioxonium than other animals.

5. Pigeons

The working solutions were injected into the wing vein of unanesthetized pigeons to attain relaxation of cervical muscles (flaccid paralysis) or spastic contractions of muscles in the lower extremities (spastic paralysis). A comparison of effective doses (Table 4) demonstrates that pigeons are more sensitive to dioxonium than to decamethonium, suxamethonium and, in particular, to tubocurarine (VĒVERIS 1974; ĶIMENIS et al. 1976).

6. Rat Isolated Phrenic Nerve–Diaphragm Preparation

Stimulation of n. phrenicus by square wave supramaximal stimuli at 0.1 Hz for 0.5 ms was used to determine the mean effective concentrations (EC_{50}) producing a 50% reduction in the height of diaphragm contractions in response to nerve stimulation (Table 5). The effect of dioxonium in these experiments was comparable with that of suxamethonium, but was somewhat inferior to that of tubocurarine.

Table 4. Experimental data obtained in experiments with pigeons

Compound	Myoparalytic dose (mg/kg) confidence limits $P = 0.05$
Dioxonium	0.046 (0.037–0.099)
Tubocurarine	0.260 (0.198–0.322)
Decamethonium	0.130 (0.102–0.158)
Suxamethonium	0.170 (0.140–0.200)

Table 5. The results of experiments on isolated rat phrenic nerve–diaphragm preparation

Compound	EC_{50} (mg/ml) confidence limits $P=0.05$
Dioxonium	2.75×10^{-6} ($1.61–3.83 \times 10^{-6}$)
Tubocurarine	1.34×10^{-6} ($1.15–1.58 \times 10^{-6}$)
Suxamethonium	2.36×10^{-6} ($1.82–2.91 \times 10^{-6}$)

Consequently, species differences in the sensitivity to dioxonium appear significant. According to our data obtained with unanesthetized animals, the curarizing dose found in the least sensitive animals (mice) was 8.6 times as high as in the most sensitive species (cats). According to their sensitivity to dioxonium, the animals of different species can be ranged in the following sequence: cat > pigeon > dog > rabbit > rat > mouse. This order is a little different in the case of decamethonium: cat > dog > pigeon = rabbit > mouse > rat; and tubocurarine: rat > mouse = dog > rabbit = pigeon > cat.

Thus, the order of animals belonging to different species ranged according to their sensitivity to dioxonium resembles the appropriate sequence characteristic of depolarizing compounds – decamethonium and suxamethonium. Unlike decamethonium, dioxonium is more active in rats than in mice, which is typical of tubocurarine chloride; on the other hand, dioxonium is different both from decamethonium and tubocurarine owing to its higher activity in pigeons which are inferior in sensitivity only to cats (ĶIMENIS and VĒVERIS 1973).

II. Mechanism of Action

The depolarizing properties of dioxonium are demonstrated by its ability to induce contractions of frog rectus abdominis muscle; however, the cholinomimetic activity of dioxonium is 250 times lower than that of acetylcholine (ĶIMENIS et al. 1976). To elucidate the character of interaction between dioxonium and cholinoreceptors of skeletal muscles we used isolated preparations of frog rectus abdominis muscle by the VAN ROSSUM (1963) method in order to obtain cumulative dose–response curves (KLUŠA and ĶIMENIS 1968a). These results (Fig. 1, Table 6) show that, like suxamethonium and decamethonium, dioxonium exhibits significant affinity for specific nicotinic receptors. However, in terms of intrinsic activity, dioxonium is inferior to suxamethonium and decamethonium.

Prior exposure of a muscle to dioxonium solution alters the character of acetylcholine curves. Dioxonium used at low concentrations (10^{-7}–10^{-6} M) increases the maximum effect of acetylcholine, whereas at higher concentrations (10^{-5}–10^{-3} M), the maximum effect of acetylcholine is diminished and the angle between the curve and the abscissa is altered (Fig. 2).

Besides increasing maximal values, dioxonium used at low concentrations is capable of shifting the acetylcholine curves leftward, suggesting its involvement in "allosteric facilitation" (KOMISSAROV 1969), based on noncompetitive interaction (sensitization) between dioxonium and acetylcholine. The observed drop in

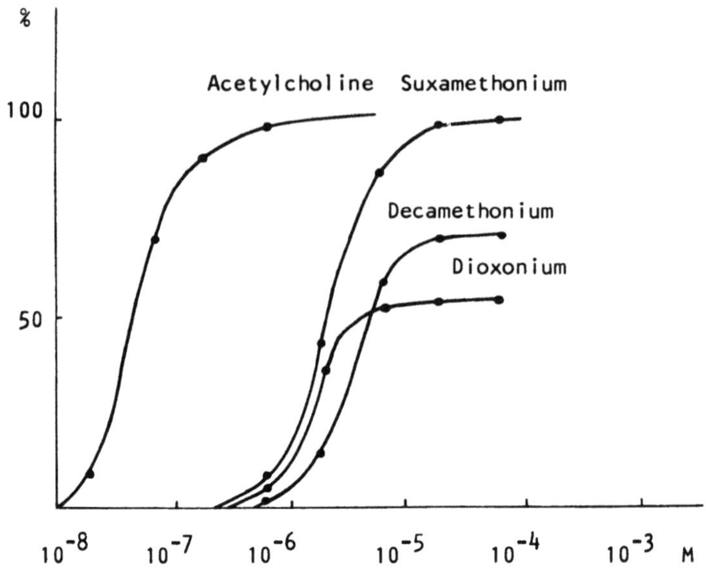

Fig. 1. Cumulative dose–response curves plotted using experimental data obtained with isolated preparations of frog rectus abdominis muscle. Abscissa: drug concentration, logarithmic scale; ordinate: relative activity, maximal effect taken as 100%

the peak values indicates noncompetitive antagonism or "allosteric inhibition." Thus, the experimental evidence obtained demonstrates a dual action of dioxonium: a combination of agonism and antagonism, in other words, a noncompetitive and depolarizing mode of action. A mixed or biphasic mechanism of action is also proposed for dioxonium on the basis of in vivo evidence. Following i.v. administration of dioxonium, pigeons develop spastic paralysis replaced by features of flaccid paralysis. Immediately after its application to unanesthetized cats, dioxonium elicits in most cases a slight increase in the muscle twitch height and induces muscle fasciculations which are, however, less manifest than after decamethonium administration. These manifestations are commonly absent on repeated administration. On a second application of dioxonium, posttetanic facilitation of soleus muscle was observed, the same effect was registered on tibialis muscle after a third or fourth injection (Vēveris 1974). At the onset of continuous

Table 6. Parameters characterizing the agonistic effect of dioxonium in comparison with suxamethonium and decamethonium

Compound	Affinity pD_2	Intrinsic activity α, confidence limits $P=0.05$
Acetylcholine	7.20	1.00
Suxamethonium	5.64	1.00
Decamethonium	5.40	0.70 (0.67–0.73)
Dioxonium	5.85	0.53 (0.49–0.55)

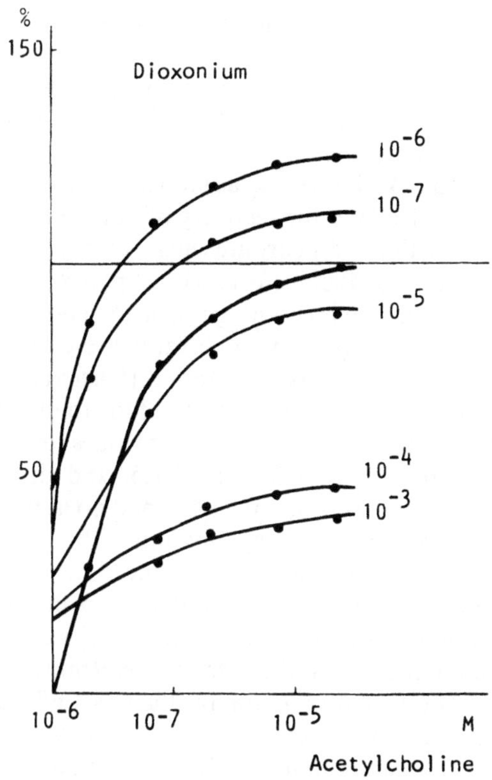

Fig. 2. Cumulative dose–response curves of acetylcholine in the absence (o) or presence of various concentrations of dioxonium

i.v. infusion of dioxonium to cats under anesthesia, we observed deepening of the neuromuscular block with neostigmine and choline and some reversal of the block with tubocurarine. On continued infusion, neostigmine exhibited antagonism, whereas tubocurarine was synergistic with dioxonium. At the onset of infusion, dioxonium, like decamethonium, blocked neuromuscular transmission in tibialis (white) muscle to a greater extent than in soleus (red) muscle. After 60–90 min, the two muscles were equally sensitive to dioxonium, but subsequently the pattern of events resembled that observed with muscle relaxants of the nondepolarizing type when soleus is more readily affected. On continuous infusion of dioxonium to anesthetized rabbits and dogs, alterations in the character of neuromuscular block set in earlier than in the case of cats.

The biphasic action of dioxonium is also supported by the findings in intact rabbits where the head drop dose of dioxonium is decreased to some extent by prior administration of neostigmine, indicating the occurrence of a depolarizing phase at the onset of drug action. However, the duration of head drop in the presence of neostigmine is considerably diminished from 7.2 (4.9–9.4) min to 2.9 (1.8–4.0) min. This decrease in the duration of effect of dioxonium is apparently due to the rapid development of the second, nondepolarizing phase of neuromuscular block in the course of which neostigmine is capable of exhibiting its antagonism (Ķimenis 1972).

Registration of gastrocnemius contractions in anesthetized rabbits during stimulation of the sciatic nerve has revealed that injection of dioxonium immediately induces some increase in twitch height of the muscle prior to the onset of partial neuromuscular block. However, administration of neostigmine 10 min after the drug application leads to the recovery of muscle contractions.

The mechanism of action of neuromuscular blocking agents can also be judged by the modification of their effects with dipyroxim (TMB-4) (Danilov and Lavrentieva 1967). Dipyroxim is antagonistic to depolarizing myorelaxants, while the effects of nondepolarizing myorelaxants are either unaffected or potentiated. With intact pigeons, it has been found (Ķimenis and Vēveris 1974a) that a prior i.v. injection of dipyroxim (5 mg/kg) fails to affect the myoparalytic dose of dioxonium significantly, whereas the same values for suxamethonium and decamethonium are increased by a factor of 1.5, and decreased 2.3-fold for tubocurarine. In anesthetized rabbits, dipyroxim had no appreciable effect on the partial neuromuscular block produced by dioxonium, but weakened the curarizing effect of decamethonium and somewhat enhanced the effect of tubocurarine. In cats under anesthesia, dipyroxim (2.5–5 mg/kg i.v.) produced some weakening of the effect of dioxonium on tibialis muscle following its first administration (in the first phase), but did not do so in the case of soleus muscle.

On subsequent injections of dioxonium (at the onset of the second phase) dipyroxim exerted no effect on tibialis muscle. Dipyroxim was clearly antagonistic to the blocking action of decamethonium. Thus, dioxonium, in terms of the effect of dipyroxim on its curarizing activity, is intermediate between depolarizing myorelaxants and tubocurarine. This is indicative of the mixed mechanism of action for this drug.

III. Peculiarities of Effect on Repetitive Administration

To study changes in the magnitude of effect and mode of action on repeated applications, dioxonium was administered to rabbits (Ķimenis and Vēveris 1974b). Similar experiments were carried out with repeated doses of tubocurarine and suxamethonium. The test solutions were injected i.v. at 5–10 times greater intervals than the duration of cervical relaxation with the respective myorelaxant. These intervals for dioxonium and tubocurarine chloride amounted to 30 and 60 min, 10 and 20 min, 35 and 70 min, respectively. The head drop dose and duration of the effect were determined. It has been found that the effective doses of dioxonium and tubocurarine tend to decline on repeated administration, suggesting a cumulative effct. On the seventh application of dioxonium, its dose is diminished approximately 3.4-fold as compared with the first injection (intervals between injections were 30 min). No further decrease in the dose was observed. A similar pattern of changes was observed on repeated administration of tubocurarine.

When injected at 60-min intervals, the cumulative effect of dioxonium is significantly less manifest, the drop in effective doses being equal to only one-third of the initial values. It should be pointed out that on repeated administration of dioxonium, a slight increase (no more than 1.5-fold) in the duration of its action is also observed, although it is not statistically significant in most cases. As compared with the first application of dioxonium and tubocurarine, their subsequent administration leads to a stronger effect on respiratory muscles; in some cases cervical relaxation was accompanied by impaired respiration. In contrast to dioxonium and tubocurarine chloride, the effects of suxamethonium subsided on subsequent applications, the effective doses having been increased 2.5-fold.

IV. Possible Side Effects

1. Effects on Arterial Pressure and Autonomic Nervous System

In anesthetized cats, dioxonium used in doses blocking neuromuscular transmission in gastrocnemius muscle had practically no effect on arterial pressure under spontaneous respiration (Ķimenis et al. 1972). Under artificial respiration, we did not observe any alterations in arterial pressure and heart rate, even at doses as high as 10 mg/kg, i.e., 2,000 times the ED_{50} of its neuromuscular blocking action.

Injected in cats at a dose (0.5 mg/kg) 100 times its neuromuscular blocking dose, dioxonium elicited a partial block of transmission from vagus to heart, but failed to affect the depressor effect of acetylcholine. Administration of dioxonium to cats in doses up to 1 mg/kg had no effect on the tone of the nictitating membrane in experiments with stimulation of preganglionic fibers of the sympathetic trunk (Kluša 1968).

2. Cardiac Effects

A comparative study has been performed of the effects exerted by dioxonium and suxamethonium on cardiac activity (Ķimenis and Vītoliņa 1973), since the latter compound is known to provoke cardiac irregularities. Unlike suxamethonium, dioxonium in rabbits under artificial respiration in curarizing or even 10–15 times higher doses elicits no impairment of heart activity. Rapid administration (2 s) of high doses of dioxonium (2–10 mg/kg) sometimes produces bradycardia and extrasystole. When injected for 30 s at a dose as high as 20 mg/kg, dioxonium has no effect on heart rate.

In anesthetized cats, dioxonium used in the curarizing dose (0.01 mg/kg) failed to affect the coronary blood flow. At higher doses (up to 0.5 mg/kg) dioxonium produced a slight (10%–20%) increase in coronary circulation. The effect had its peak 6–10 min after administration and lasted 5–10 min. Hence, it can be concluded that dioxonium does not act adversely on coronary circulation.

3. Anticholinesterase Activity

Anticholinesterase activity of dioxonium and its derivatives was assayed in vitro (Ķimenis et al. 1974). Cholinesterase activity was measured colorimetrically as in Augustinsson (1957) with some modifications to provide quantitative assessment of the inhibitory effect of compounds on cholinesterase activity (Ķimenis 1961). It has been found that dioxonium at relatively high concentrations exhibits anticholinesterase activity. The pI_{50} value for human blood serum cholinesterase was 4.67 (4.48–5.00), for acetylcholinesterase from human erythrocytes its value was 4.36 (4.23–4.56), and for cat brain acetylcholinesterase it was equal to 3.09 (3.02–3.16). Consequently, anticholinesterase activity of dioxonium is more readily demonstrable with respect to blood serum cholinesterase than to acetylcholinesterase. It has been reported earlier (ĶIMENIS 1960b) that the corresponding methylpyrrolidine derivative of suxamethonium is 27 times as active in inhibiting acetylcholinesterase in the brain ($pI_{50}=4.52$). The dioxolan structures in the dioxonium molecule, in contrast to ester groups in the appropriate suxamethonium derivative, apparently fail to provide favorable conditions for interaction with acetylcholinesterase.

Compared with neostigmine (ĶIMENIS 1960b), dioxonium is 80 times less inhibitory to human blood serum cholinesterase and 160 times less inhibitory to human erythrocyte acetylcholinesterase. With respect to acetylcholinesterase from cat brain, neostigmine is 3,000 times as active as dioxonium.

It must be noted that in relation to its anticholinesterase activity, dioxonium has been found to possess no side effects, as revealed by experiments in animals (KLUŠA 1968). Because dioxonium has a high curare-like activity, the concentrations capable of blocking neuromuscular transmission are much below the level required for inhibition of cholinesterase activity. However, intraperitoneal administration of relatively high doses of dioxonium (0.6 mg/kg, i.e., 1.5 times its LD_{50}) to poorly responding animals results in acetylcholinesterase inhibition in skeletal muscles by 30%; hepatic cholinesterase is inhibited by 27% (MOZGOVAYA et al. 1984).

4. Effect on Potassium Ion Concentration in Blood Plasma

It is known that depolarizing muscle relaxants stimulate efflux of potassium ions from skeletal muscles, leading to an increase in their concentration in blood plasma. Nondepolarizing myorelaxants do not possess this ability. Furthermore, they even tend to counteract this effect of depolarizing agents (KLUPP et al. 1954; PATON 1956).

The effct of dioxonium on the blood plasma potassium level has been studied in rabbits by comparing its activity with that of a typical depolarizing agent, suxamethonium (ĶIMENIS and VĒVERIS 1974b). Both muscle relaxants were administered at 60-min intervals in doses capable of eliciting head drop. Dioxonium injected intraperitoneally was found (Fig. 3) to increase the potassium concentration in blood plasma, the concentration peak was observed 4–5 min after its administration. In the case of suxamethonium, the maximum concentration of potassium was registered after 2.5 min. A second injection of dioxonium has a less pronounced effect, while the next injection was even less effective. By contrast, suxamethonium led to a considerable increase in potassium level both on the first and on subsequent applications. The lack of response to repetitive injections of dioxonium indicates failure of the drug to induce depolarization in these animals on repeated dosage. On the other hand, suxamethonium in the same experiments, besides a rapid rise in potassium concentration observed immediately after its application, also elicited a slow secondary increase in its level, especially following

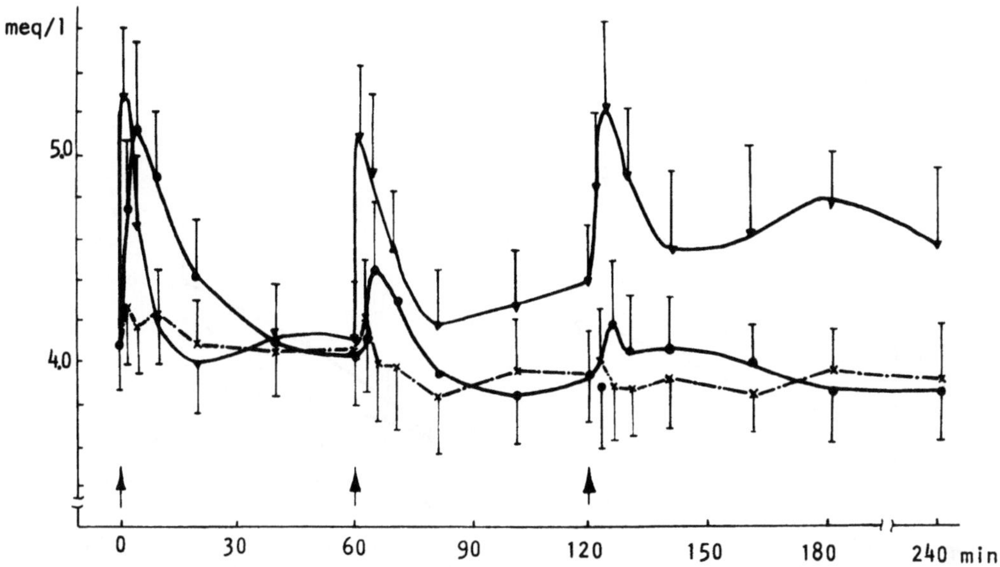

Fig. 3. Potassium concentration in blood plasma of rabbits in response to repeated administration of dioxonium (*circles*) and suxamethonium (*triangles*) as compared with control animals (*crosses*). Confidence limits $P = 0.05$. *Arrows* indicate injection of drug

the third injection, when the potassium concentration in blood plasma remained at a much higher level than in the control animals. This was still observed as late as 2 h after the application (Fig. 3). This sustained elevation of blood plasma potassium may be due to microlesions in muscles caused by uncoordinated activity in response to a depolarizing neuromuscular blocking agent (RACK and WESTBURY 1966). The absence of secondary increase in potassium concentration following dioxonium administration indicates that, unlike suxamethonium, dioxonium fails to produce a protracted increase in permeability or damage to muscle membrane.

5. Effect on Creatine Phosphokinase Activity in Blood Plasma

The use of muscle relaxants of the depolarizing type, especially suxamethonium, is known to increse the activity of blood creatine phosphokinase. This phenomenon is due to the action of these agents on muscle membranes or microinjury caused in muscles by fasciculation (TAMMISTO et al. 1967; VENTAFRIDDA et al. 1970). As dioxonium demonstrates depolarizing properties in some animals, especially cats, a comparative study was undertaken to investigate the effects of dioxonium and suxamethonium on creatine phosphokinase activity in blood plasms using cats and rabbits (VĒVERIS and ĶIMENIS 1974). Anesthetized cats (urethane 500 mg/kg plus chloralose 50 mg/kg or phenobarbital sodium 40 mg/kg intraperitoneally) were given three i.v. injections of dioxonium (0.01 mg/kg) or suxamethonium (0.05 mg/kg) at 15-min intervals. Rabbits were used without anesthesia or were given small doses of phenobarbital sodium (25 mg/kg) prior to experiments. Dioxonium and suxamethonium were administered into the auricular vein

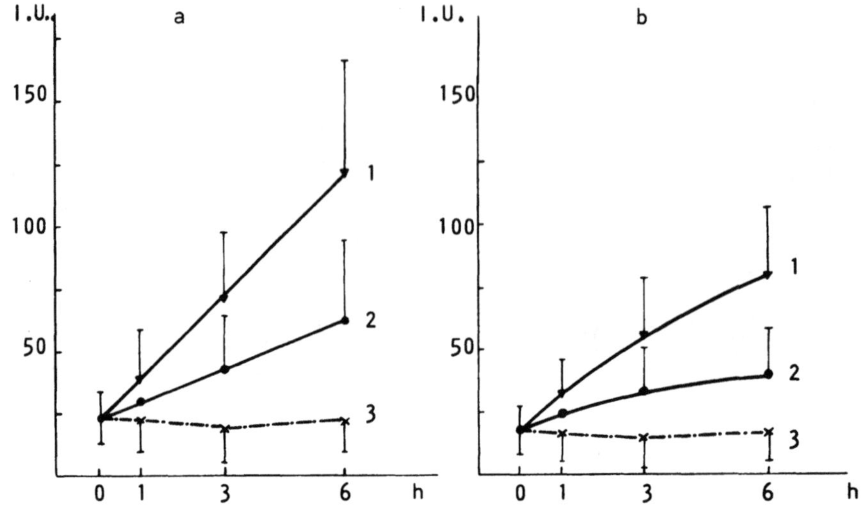

Fig. 4 a, b. Effects of dioxonium and suxamethonium on creatine phosphokinase activity in cats under **a** urethane plus chloralose and **b** phenobarbital anesthesia. *1* suxamethonium; *2* dioxonium; *3* control

three times at 25-min intervals in doses that caused relaxation of cervical muscles. It was found that following suxamethonium administration, creatine phosphokinase activity in the blood plasma of cats increased much more appreciably than after administration of dioxonium (Fig. 4). A less pronounced increase in creatine phosphokinase activity was noted under phenobarbital than under urethane plus chloralose anesthesia. Similar results were obtained in rabbits. In animals which received a prior injection of a small dose of phenobarbital sodium, the increase in creatine phosphokinase activity in response to the studied neuromuscular blocking agents, especially suxamethonium, was reduced as compared with intact animals.

A certain correlation between the intensity of muscle fasciculation and increase in creatine phosphokinase activity was observed. For instance, in phenobarbital-anesthetized cats, muscle fasciculation caused by suxamethonium was less pronounced than in animals under urethane plus chloralose anesthesia and the enzyme activity was not increased so dramatically. Dioxonium, which was less effective in increasing creatine phosphokinase activity in the blood plasma of cats than suxamethonium, was found to provoke only slightly noticeable muscle fasciculations (urethane plus chloralose anesthesia) or did not do so at all (phenobarbital anesthesia). In rabbits, dioxonium was also found to increase to some extent creatine phsophokinase activity in the absence of appreciable fasciculations. It has been reported that elevated activity of creatine phosphokinase may also be due to hypoxia (LOEGERING and CRITZ 1971). In our experiments with rabbits, the effect of hypoxia cannot be excluded because on repetitive administration of dioxonium, besides relaxation of cervical muscles, it produces in some cases a partial relaxation of the respiratory musculature.

6. Effect on Ca^{2+}-Dependent ATPase Activity in the Sarcolemma of Skeletal Muscles

Dioxonium (3×10^{-6} *M*) in vitro does not appear to have a significant effect on the activity of Ca^{2+}-dependent ATPase in the sarcolemma of skeletal muscles in rabbits. In analogous conditions, tubocurarine activates this enzyme to some extent, but suxamethonium (Myo-Relaxin) supresses its activity. Dioxonium used in vivo at a dose producing complete relaxation of skeletal muscles had no effect on ATPase activity in rats, while tubocurarine and suxamethonium inhibited it (MOZGOVAYA and BRAZALUK 1978).

7. Histamine Release

The ability of dioxonium to release bound histamine has been investigated in rabbits (VĒVERIS 1974). For the sake of comparison, tubocurarine was applied in parallel experiments. The drugs were administered i.v. in head drop doses. Quantitative measurements of histamine in blood plasma were made according to the procedure described by CODE and MCINTIRE (1956). The data obtained (Table 7) show that dioxonium applied at curarizing doses does not affect the histamine concentration in blood plasma, but, at the same time, the level of histamine rises approximately tenfold in response to tubocurarine.

A suspension of peritoneal mast cells was also used in experiments in vitro. The percentage of released histamine was determined in relation to the total histamine contained in the corresponding volume of mast cells. Dioxonium used in concentrations of 10^{-4} and 5×10^{-4} *M* was practically unable to liberate histamine from mast cells, whereas tubocurarine at 10^{-4} *M* released almost one-third of the total histamine. Only at a higher concentration (10^{-3} *M*) dioxonium produced a considerable release of histamine (12%–16%). Tubocurarine used in the same concentration released 77%–82% of histamine (Table 8). Thus, the release of histamine in vitro proceeds more readily in response to tubocurarine than to dioxonium, the concentration of the drugs being equal.

By comparing the histamine-releasing properties of several neuromuscular blocking agents, it has been concluded by some authors (FRISK-HOLMBERG and STANDBERG 1971) that the presence of a free OH group is crucial for the manifestation of these properties. The molecule of dioxonium does not carry a free OH

Table 7. Effect of dioxonium and tubocurarine on blood plasma histamine in rabbits

Compound	Histamine level (mkg/ml) confidence limits $P=0.05$	
	Control	After drug injection
Dioxonium	0.005 (0.001–0.009)	0.008 (0.002–0.014)
Tubocurarine	0.005 (0.001–0.009)	0.054 (0.024–0.084)
Isotonic NaCl solution	0.005 (0.001–0.009)	0.006 (0.001–0.011)

Table 8. Release of histamine from the peritoneal mast cells of rats by dioxonium and tubocurarine

Concentration of compounds (M)	Amount of histamine (%) confidence limits $P=0.05$	
	Isolated intestine of guinea pig	Arterial pressure in cats
Control	1.5 (0.6–2.4)	1.0 (0.0–2.0)
Dioxonium		
10^{-4}	2.0 (0.9–3.1)	1.0 (0.0–2.0)
5×10^{-4}	4.1 (1.5–6.7)	2.0 (0.1–3.9)
10^{-3}	12.0 (8.4–15.6)[a]	16.0 (9.9–22.1)[a]
Tubocurarine		
10^{-4}	28.0 (24.0–32.0)[a]	34.0 (24.5–43.5)[a]
5×10^{-4}	42.0 (35.5–48.5)[a]	45.0 (34.0–56.0)[a]
10^{-3}	77.0 (68.0–86.0)[a]	82.0 (72.0–92.0)[a]

[a] $P=0.05$ compared with spontaneous release of histamine

group. This may be regarded as one of the factors responsible for its low capacity to liberate histamine. The study of the effects produced on the mast cells by chemical substances with different chemical structures suggests a strong relationship between the lipophilic nature of these substances and their ability to release histamine (FRISK-HOLMBERG 1972). The mechanism of neuromuscular blocking action of bisammonium derivatives is also determined by their lipophilic properties. More lipophilic compounds act as nondepolarizing agents, while those with less lipophilicity are depolarizers (PATON 1956; KHARKEVICH and SKOLDINOV 1971). The lipophilic : hydrophilic atom ratio is higher in the compounds of the nondepolarizing type as compared with depolarizing compounds; e.g., for tubocurarine the ratio is 20 : 1, and it is 5 : 1 for suxamethonium and acetylcholine. It has been found, too, that depolarizers, in contrast to tubocurarine, fail to release histamine from a suspension of mast cells (FRISK-HOLMBERG and STANDBERG 1971). The molecule of dioxonium contains six more carbons than the molecule of suxamethonium, the number of nitrogen atoms being equal. This implies stronger lipophilic properties for dioxonium which are apparently responsible not only for differences in the mechanism of action of dioxonium and suxamethonium, but also for the ability of high concentrations of dioxonium to release histamine in vitro.

It goes without saying that whether a neuromuscular blocking agent will liberate histamine or not in clinical usage depends on differences in the concentration levels capable of releasing histamine and those eliciting block of neuromuscular transmission. As dioxonium is a much more active neuromuscular blocking agent than tubocurarine, but is far less active in releasing histamine, it is believed that some histamine release observed at high concentrations of dioxonium in vitro is negligible from a practical viewpoint.

V. Pharmacokinetics

Pharmacokinetic studies were performed (VĒVERIS et al. 1975) with unlabeled dioxonium and dioxonium carrying a ^{14}C-labeled *N*-methyl group (specific radioactivity 0.5 mCi/mmol). Experiments with radioactive dioxonium were carried out on female white rats under artificial respiration. Dioxonium (0.3, 1.0, and 5 mg/kg) was administered i.v. It has been established that the labeled drug in blood plasma undergoes rapid elimination for 10–20 min after its administration, followed by a decline in the elimination rate lasting several hours, the process being exponential with time (Fig. 5). The half-life of radioactivity in blood plasma ($t_{1/2}$) is 42–45 min. By reaching a certain level corresponding, on average, to 0.39 (0.28–0.50) μg of dioxonium ^{14}C in 1 ml blood plasma, neuromuscular transmission is restored in all cases, regardless of the doses used and the initial level of the drug in blood. The uptake of radioactive products by organs and tissues following dioxonium ^{14}C administration is shown in Table 9. One can see that the concentration of radioactive products in the lever and lungs following dioxonium ^{14}C injection increases rapidly and after 5–10 min reaches its level in the blood plasma. In skeletal muscles, peak radioactivity is several times lower than in the lungs and liver, but in the muscles the decrease of radioactivity occurs much more slowly. However, since the mass of muscle tissue in rats constitutes up to 45% of the total organism mass, muscles take the lead in the distribution of the drug: 20–30 min after administration about 40% of the total radioactivity is found in skeletal muscles (Fig. 6).

The highest radioactivity was detected in kidneys where 20–30 min after administration its level was 30 times as high as that in the muscles. But, as kidneys account for only 1% of the total mass of the animal, they contain at this time about 20% of the applied radioactive dose. Kidneys provide the main route for

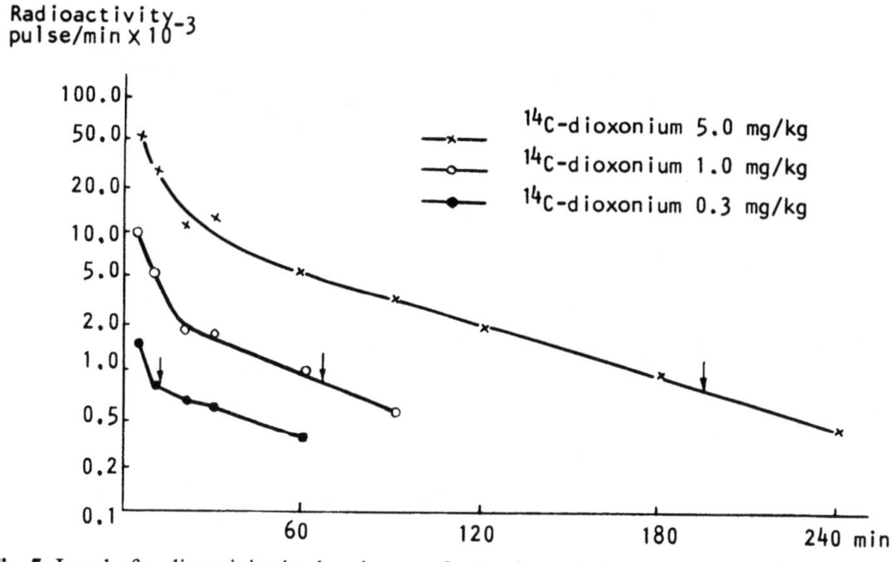

Fig. 5. Level of radioactivity in the plasma of rats after administration of various doses of dioxonium ^{14}C. Reversal of neuromuscular block is indicated by *arrows*

Table 9. Level of radioactive products (10^3 counts min^{-1} g^{-1}) in rat tissues after injection of dioxonium ^{14}C (5 mg/kg)

Tissue	Time after injection (min)								
	5	10	20	30	60	90	120	180	240
Liver	14	15	15.8	15	7.5	5.4	4.3	1.8	1.2
Lung	20	25	17	9	5	3.8	2.9	2.0	1.8
Kidney	75	190	173	170	172	190	173	138	112
Spleen	8	9	10	10.5	6	4.5	4	1.5	2.3
Muscle	4	5	5.35	5.3	4.2	3.6	2.9	2.1	1.75
Adipose tissue		0.6	0.8	1	1.2	1	0.8		0.5
Plasma	45	24	10.5	12	5.5	3.5	2.2	1	0.4

the removal of dioxonium ^{14}C. In 48 h, 80%–90% of the radioactive label is excreted in the urine. The clearance of the labeled drug from blood plasma by excretion in the urine proceeds at a high rate. At a dioxonium ^{14}C dose of 0.3 mg/kg, the renal clearance is 1.2 ml/min, which practically does not differ from the total clearance value (1.29 ml/min). A comparison of the dynamics of labeled dioxonium in blood plasma with the uptake of radioactivity by different organs and excretion of the labeled drug in the urine suggests a biphasic model for the kinetics of radioactivity. During the first phase, the decline in radioactivity in

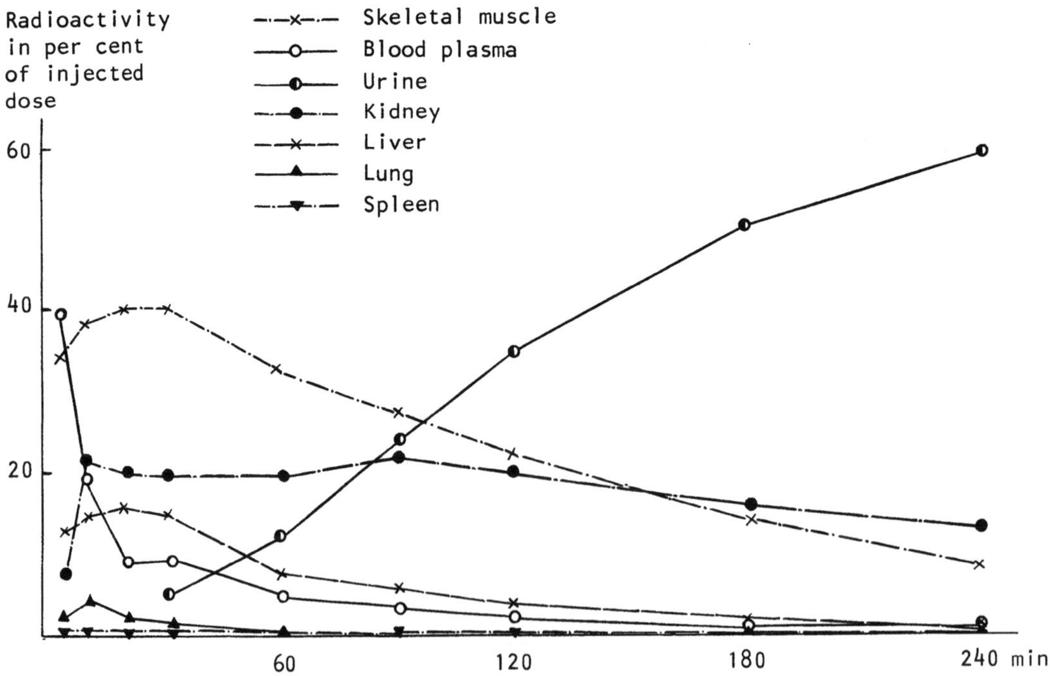

Fig. 6. Distribution of radioactivity in rats after administration of dioxonium ^{14}C (5 mg/kg)

blood plasma is accompanied by its rapid growth in various organs. During this period, 10–20 min after administration, insignificant quantities of radioactive products are found in the urine. During the second phase, a fall in radioactivity in blood plasma and in the organs is observed, whereas the level of radioactivity in the urine is increased.

The elimination of dioxonium in the urine is not responsible for the cessation of its neuromuscular blocking effects. For instance, administration of 1 mg/kg dioxonium ^{14}C which is three times higher than the average curarizing dose, leads to the reversal of neuromuscular block after about 75 min. During this period, excretion in the urine accounts for no more than 30% of the drug. Therefore, it can be expected that the abolition of neuromuscular blocking effect is mainly due to the redistribution of the drug in the organism.

We are still unaware of the possible biotransformations of dioxonium. Metabolic conversion of dioxonium is not likely to play an important role because, as evidenced by experimental findings in cats, in the course of 4 h, approximately 40% of the administered dose is excreted in the urine in a pharmacologically active form. The insignificant extrarenal clearance (0.04 ml/min per 100 g rat body weight for a 0.3 mg/kg dose) suggests stability of dioxonium in the organism.

VI. Toxicology

1. Acute Toxicity

Acute toxicity of dioxonium is due to its specific neuromuscular blocking activity. the LD_{50} values were measured in white rats and mice. Following intraperitoneal administration of nonlethal doses, the animals developed muscular relaxation lasting 15–20 min. Administration of higher doses caused dyspnea, cyanosis, and clonicotonic seizures, that usually led to death. The LD_{50} of dioxonium was found to be 1.0 (0.70–1.47) mg/kg in white mice and 0.4 (0.32–0.49) mg/kg in white rats.

2. Subchronic Toxicity

Young rats were given, every other day for 2 weeks, intraperitoneal injections of dioxonium solution at 50% of LD_{50} (0.2 mg/kg). Dioxonium had no effect on the growth of animals, as compared with animals in the control group. Repeated administration of dioxonium does not change the cell morphology of peripheral blood. The activity of serum and liver aminotransferase was also unaffected. The drug did not influence the protein composition of blood serum. Postmortem investigation of liver, kidneys, spleen, lungs, adrenals, and myocardium from experimental and control rats did not reveal any pathologic changes in these organs.

C. Conclusion

The results of investigations in the bisammonium series of succinic aldehyde cyclic acetals have revealed that the dimerization of strong muscarinomimetic structures may be used in the synthesis of active curare-like agents. Apparently, the same structural features responsible for high muscarinomimetic activity are also responsible for the observed selective neuromuscular blocking effect of the dimer molecule. In the molecule of dioxonium, the dioxalan rings probably represent the structural elements corresponding to the structure of cholinoceptors of skeletal muscles, the degree of complementarity being sufficiently high to provide strong affinity and high blocking activity. At the same time, these rings are responsible for a certain rigidity of the molecule which makes impossible its interaction with other cholinoceptors. Thus, the structural properties of dioxonium provide both high activity and selectivity of action. Furthermore, experimental findings with dioxonium indicate the existence of species differences both in regard to neuromuscular blocking activity and the mechanism of action.

References

Artemov IA, Yerivantsev NA (1972) Application of a novel muscle relaxant dioxonium during surgery in elderly patients (in Russian). In: Eksp Klin Farmakoter 3:75–79

Augustinsson KB (1957) Assay methods for cholinesterases. In: Glick D (ed) Methods of biochemical analysis, vol 5. Interscience, New York, pp 1–63

Bovet D, Bovet-Nitti F, Guarino S, Longo V, Fusco R (1951) Recherches sur les poisons curarisants de synthése. III-e Partie: Succinylcholine et dérivés aliphatiques. Arch Int Pharmacodyn Thér 88:1–50

Code CF, McIntire FC (1956) Quantitative determination of histamine. In: Glick D (ed) Methods of biochemical analysis, vol 3. Interscience, New York, pp 49–95

Danilov AF, Lavrentieva VV (1967) Influence of cholinesterase reactivators TMB-4 and 2-PAM on the effectivity of muscle relaxants (in Russian). Farmakol Toksikol 5:567–572

Darbinyan TM, Vaisberg LA (1970) Dioxonium – a novel muscle relaxant (in Russian). Vestn Khir 10:84–88

Fourneau E, Bovet D, Bovet F, Montezin G (1944) Activité acetylcholinique particulierment intense dans une nouvelle série d'aminoacetals de polyols. Bull Soc Chim Biol 26:516–528

Frisk-Holmberg M (1972) Drug-induced changes in the release and uptake of biogenic amines. A study of mast cells. Acta Physiol Scand [Suppl] 84:376

Frisk-Holmberg M, Standberg K (1971) Histamine release from rat peritoneal mast cells and cat paws induced by some neuromuscular blocking agents. Acta Physiol Scand 81:367–375

Ginters YY (1970) Experience in the use of a novel Soviet muscle relaxant dioxonium (in Russian). In: Special methods of investigation in clinical practice. XI scientific-practical conference, Riga

Kharkevich DA (1969) Farmakologiya kurarepodobnikh sredstv (Pharmacology of curare-like compounds). Meditsina, Moscow

Kharkevich DA, Skoldinov AP (1971) Effect of lypophilic radicals in the molecule of curare-like compounds on the mechanism of their action (in Russian). Dokl Akad Nauk SSSR 4:985–988

Ķimenis AA (1960a) Pharmacological characteristics of bis-piperidine and bis-pyrrolidine analogue of dithyline (in Russian). Izv Akad Nauk Latv SSR 3:169–176

Ķimenis AA (1960b) Anticholinesterase properties of dithyline and some of its heterocyclic derivatives (in Russian). Izv Akad Nauk Latv SSR 10:135–141

Ķimenis AA (1961) Curare-like and anticholinesterase activity of some heterocyclic analogues of dithyline (dicholine ester of succinic acid) (in Russian). Cand med Sci thesis, Riga

Ķimenis AA (1972) Effect of neostigmine on the action of dioxonium and its analogues in rabbits (in Russian). Izv Akad Nauk Latv SSR 7:113–116

Ķimenis AA, Vēveris MM (1973) Species sensitivity to dioxonium (in Russian). Farmakol Toksikol 6:690–692
Ķimenis AA, Vēveris MM (1974a) Effect of dipiroxim on dioxonium action (in Russian). Izv Akad Nauk Latv SSR 11:66–70
Ķimenis AA, Vēveris MM (1974b) Action of dioxonium, D-tubocurarine and succinylcholine of their repeated administration in rabbits (in Russian). Izv Akad Nauk Latv SSR 3:110–115
Ķimenis AA, Vitoliņa RO (1973) Influence of dithyline and dioxonium on cardiac activity in rabbits (in Russian). Izv Akad Nauk Latv SSR 11:107–110
Ķimenis AA, Kluša VE, Ginters JJ (1972) Pharmacology of a novel myorelaxant – dioxonium (in Russian). Farmakol Toksikol 2:172–175
Ķimenis AA, Viluma LT, Sokolov GP (1974) Anticholinesterase activity of bis-ammonium derivatives of some cyclic acetals (in Russian). Izv Akad Nauk Latv SSR 3:77–82
Ķimenis AA, Kluša VE, Ginters JJ, Veveris MM (1976) Dioksoniy. Farmakologiya i anesteziologiceskoye primeneniye (Dioxonium. Pharmacology and anaesthesiologic applicā tion). Zinatne, Riga
Klupp H, Kraupp O, Stormann H, Stumpf C, Honetz N, Kobinger W (1953) Über die pharmakologischen Eigenschaften einiger Polymethylen-di-carbaminsäure-bischolinester. Arch Int Pharmacodyn Thér 96:161–182
Klupp H, Kraupp O, Honetz N, Kobinger W, Loudon M (1954) Über die Freisetzung von Kalium aus der Muskulatur unter der Einwirkung einiger Muskelrelaxantien. Arch Int Pharmacodyn Thér 98:340–354
Kluša VE (1968) Influence of some quaternary salts of bis-aminomethyldioxolanylethane and -ethylene on N-cholinoreceptors (in Russian). Thesis, Tartu State University
Kluša VE, Ķimenis AA (1968a) Comparative pharmacology of some quaternary salts of bis-aminomethyldioxolanylethane and -ethylene (in Russian). Izv Akad Nauk Latv SSR 1:118–123
Kluša VE, Ķimenis AA (1968b) A relationship between chemical structure and character of action on the N-cholinoreactive systems in some quaternary salts of bis-aminomethyldioxolanylethane and -ethylene (in Russian). Izv Akad Nauk Latv SSR 2:138–146
Komissarov IV (1969) Elementi teorii receptorov v molekuljarnoy farmakologii (Elements of receptor theory in molecular pharmacology). Meditsina, Moscow
Loegering DG, Critz IB (1971) Effect of hypoxia and muscular activity on plasma enzyme levels in dogs. Am J Physiol 220:100–104
Mozgovaya EN, Brazaluk AZ (1978) Effect of halothane, muscle relaxants and anaesthetics on Ca^{2+} dependent ATPase activity in the sarcolemma of skeletal muscles (in Russian). Ukr Biokhim Zh 50:230–233
Mozgovaya EN, Brazaluk AZ, Ershov VV, Muratova IA, Shpilevskaya LT, Yarovenko LT (1984) Muscle relaxant dioxonium and metabolic reaction of organism (in Russian). In: Eksp Klin Farmakoter 11:105–110
Paton WDM (1956) Mode of action of neuromuscular agents. Br J Anaesth 28:470–480
Rack PM, Westbury DR (1966) The effects of suxamethonium and acetylcholine on the behaviour of cat muscle spindles during dynamics, stretching and during fusimotor stimulation. J Physiol (Lond) 196:698–713
Rossum JM (1963) Cumulation dose-response curves. II. Technique for the making of dose-response curves in isolated organs and the evaluation of drug parameters. Arch Int Pharmacodyn Thér 143:299–330
Sokolov GP, Hiller SA (1967) Author's certificate N 196883. Izobreteniya 12:42
Sokolov GP, Hiller SA (1970) Author's certificate N 267634
Sokolov GP, Voronkov MG, Hiller SA (1964) Author's certificate N 165468. Izobreteniya N 19:17
Sokolov GP, Kluša VE, Ķimenis AA, Hiller SA (1968) Synthesis and curare-like action of quaternary salts of bis-aminomethyldioxolanylethane and -ethylene (in Russian). Khim-Farm Zh 3:3–7
Tammisto T, Leikkonen P, Airaksinen M (1967) The inhibitory effect of d-tubocurarine on the increase of serum-creatine-kinase activity produced by intermittent suxamethonium administration during halothane anaesthesia. Acta Anaesth Scand 11:333–340

Vanevskii VL, Kargapolov LN (1970) Application of a novel myorelaxant dioxonium in anaesthesiology (in Russian). Vestn Khir 6:74–77
Vanevskii VK, Kotomina GL, Kargapolov LN (1972) Clinical application of a novel myorelaxant dioxonium (in Russian). Eksp Klin Farmakoter 3:67–74
Ventafridda V, Terno G, Sciancalepore G (1970) Comportamento della creatinofosfochinasi a del postassio serico in rapporto alla mialgia da succinilcolina. Minerva Anesth 36:200–205
Vēveris MM (1974) Experimental investigation of the peripheral myorelaxant dioxonium (in Russian). Thesis, Kaunas
Vēveris MM, Ķimenis AA (1974) Effect of dioxonium and succinylcholine on the creatinephosphokinase activity in the blood plasma of cats and rabbits (in Russian). Izv Akad Nauk Latv SSR 5:104–108
Vēveris MM, Ķimenis AA, Gilev AP, Sokolov GP, Hiller SA (1975) Pharmacokinetics of dioxonium (in Russian). Farmakol Toksikol 6:669–672

CHAPTER 23

Fazadinium Dibromide

M. B. Tyers

A. Introduction

The major problems associated with the use of currently available competitive neuromuscular blocking drugs are that they are slow to produce muscle paralysis and their actions are prolonged to the extent that the depth of paralysis is not easily controlled. Reversal of paralysis may be achieved with neostigmine, but the paralysis may return later as the effect of this anticholinesterase agent declines. Other problems are reflected in cardiovascular changes resulting from blockade of nicotinic receptors in autonomic ganglia, muscarinic receptors at the sinoatrial node and through the release of histamine from mast cells. Newer agents differ only slightly from tubocurarine in most of these effects. The depolarising agent succinylcholine, despite a substantial number of side effects is still very widely used in anaesthetic practice because the muscle paralysis is rapid in onset and of short duration. The need for a rapid onset, short-acting, competitive neuromuscular blocking drug to replace succinylcholine has been recognised for many years. Many of the newer drugs have been heralded from animal studies to have achieved this objective, but in humans they have not been convincing. Nevertheless, each of these drugs has contributed to our knowledge of the pharmacology of these agents and has led to a clearer definition of the expectations that anaesthetists have for the ideal muscle relaxant. This chapter describes some of the results of investigations into the pharmacology of fazadinium dibromide in animals and humans.

B. Azobisarylimidazo(1,2-*a*)pyridinium Dihalides

Structure–Activity Relationships

Investigations into the pharmacological properties of a short chemical series of azobisarylimidazo(1,2-*a*)pyridinium dihalides led to the identification of several compounds which possess competive neuromuscular blocking activity in the sciatic nerve–tibialis anterior muscle preparation of the anaesthetised cat (Table 1). Of particular interest was that for several of these compounds muscle paralysis was rapid in onset and of short duration and was not accompanied by adverse changes in blood pressure or heart rate. More detailed examination of the structure–activity relationships of this series shows that 2-phenyl (AH 8165) is preferred to 2-methyl (AH 11245) for neuromuscular blocking activity. Substitutions with phenyl in the 3 position (AH 8607 and AH 8608) reduced potency and

Table 1. Neuromuscular blocking and cardiovascular effects of some azobisarylimidazo [1,2-*a*] pyridinium dihalides in the peroneal nerve–tibialis anterior muscle preparation of the chloralose-anesthetized cat

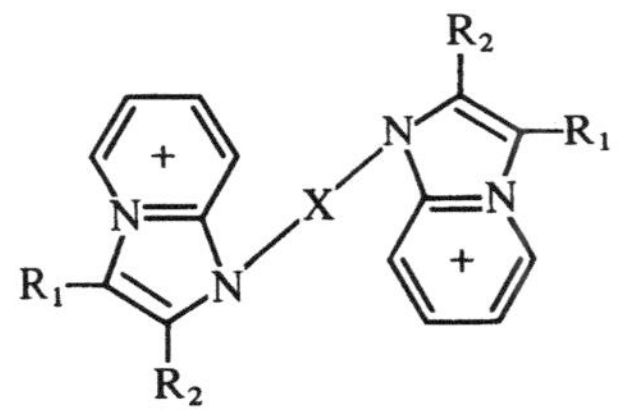

Compound	Structural formula			Neuromuscular blocking properties			Cardiovascular changes[d]	
	R_1	R_2	X	Inhibition of muscle twitch[a] ED_{90} (mg/kg i.v.)	Onset time to maximum block (s)[b]	Duration of block at ED_{90}[c] (min:s)	Blood pressure	Heart rate
AH 7394	H	H	–N=N–	Inactive at 10				
AH 11245	CH_3	CH_3	–N=N–	2.0	10	7:39	+	+
AH 7060	H	Ph	–N=N–	0.6	12	2:30	–	0
AH 8165	CH_3	Ph	–N=N–	0.35	12	2:30	0	0
AH 8608	Ph	CH_3	–N=N–	1.4	24	6:40	+	0
AH 8607	Ph	Ph	–N=N–	2.0	29	13:20	–	–
AH 8117	Br	Ph	–N=N–	0.6	18	2:30	–	–
AH 7486	H	(195)	–N=N–	2.5	18	3:12	+	++
AH 10409	CH_3	(196)	–N=N–	0.8	23	3:46	–	–
AH 8627	CH_3	Ph	$–CH_2–$	0.8	20	5:10	––	––
AH 8628	CH_3	Ph	$–CH_2CH_2–$	0.7	24	14:30	––	––
AH 11172	CH_3	Ph	$–CH_2C{\equiv}CCH_2–$	0.25	16	2:47	––	+
AH 10133	CH_3	Ph	$–(CH_2CO_2CH_2)_2–$	1.5	15	2:30	––	+
Tubocurarine				0.1	72	14:36	–	0/–
Pancuronium				0.01	126	8:24	0	0

[a] Indirect stimulation at 1 Hz
[b] From i.v. injection to peak effect
[c] From first indication of block to 90% recovery of twitch height
[d] –/–– slight/moderate fall; +/++ slight/moderate rise; 0 no effect

significantly prolonged the duration of neuromuscular blockade. Bromo (AH 7486) or aceto (AH 10409) substitution in the *para* position of the 2-phenyl ring in AH 7486 reduced potency and cardiovascular effects were more marked. Replacement of the azide bond (AH 11172, AH 8627, AH 8628) affected potency and duration of neuromuscular blockade and caused marked falls in blood pressure and increased heart rate. The succinate derivate (AH 10133) is potentially degradable by pseudocholinesterase, but this compound was less potent than the corresponding azide and was longer lasting. Incubation with plasma showed that AH 10133 was degraded, but at a considerably slower rate than succinylcholine. All of these compounds were competitive neuromuscular blocking drugs since their actions on the tibialis muscle could be reversed with neostigmine.

C. Animal Pharmacology of Fazadinium Dibromide

I. Neuromuscular Blocking Properties

The neuromuscular blocking properties of fazadinium on indirectly evoked contractions of skeletal muscle in the anaesthetised rat, cat, dog, cynomolgus monkey, and cotton-eared marmoset (*Callithrix jacchus*) have been described by several authors (BRITTAIN and TYERS 1972; BOLGER et al. 1972; BOUYARD et al. 1973; MARSHALL 1973 a, b; TYERS 1975; HUGHES et al. 1976). The neuromuscular blocking potencies of competitive neuromuscular blocking drugs in anaesthetised animals do not vary to any great extent between species (see MACLAGEN 1976). The same is true for fazadinium which was most potent in the anaesthetised cotton-eared marmoset and least potent in the cat. But the potency ratio for fazadinium between these two species was only about 3 : 1 (Table 2). In the anaesthetised cat, fazadinium is more potent than gallamine, but is less potent than succinylcholine, tubocurarine, and pancuronium (Table 3).

Table 2. Neuromuscular blocking effects of fazadinium dibromide in the anesthetized cat, dog, cynomolgus monkey and cotton-eared marmoset

Species (*N*)	Neuromuscular blockade[a]		
	Inhibition of tibialis anterior muscle, 1 Hz, ED_{90} (mg/kg i.v.)	Onset time; injection to peak effect (s)	Duration; Injection to 90% recovery (min : s)
Cat (10)	0.4 (0.3–0.47	16.4 (13.0–21.3)	2:27 (2:01–2:47)
Dog (3)	0.18 (0.11–0.27)	28.7 (18.7–36.2)	3:47 (3:12–4:16)
Cynomolgus monkey (3)	0.13 (0.09–0.17)	27.1 (22.3–32.5)	9:12 (7:42–12:03)
Cotton-eared marmoset (2)	0.10 (0.06–0.14)	53.0 (38.1–60.1)	20:55 (17:47–24:0)

[a] Values are means (range)

Table 3. Neuromuscular blocking properties of fazadinium, tubocurarine, gallamine, pancuronium, and succinylcholine in the anesthetized cat

Neuromuscular blocking drug (N)	Neuromuscular blockade[a]		
	Inhibition of tibialis anterior muscle, 1 Hz, ED_{90} (mg/kg i.v.)	Onset time; injection to peak effect (s)	Duration; Injection to 90% recovery (min:s)
Fazadinium (10)	0.4 (0.3–0.47)	16.4 (13.0–21.3)	2:27 (2:01–2:47)
Tubocurarine (10)	0.08 (0.06–0.1)	80.0 (64.0–96.1)	14:16 (12:33–17:47)
Gallamine (5)	0.9 (0.75–1.2)	45.6 (33.6–75.0)	8:34 (7:49–10:16)
Pancuronium (9)	0.01 (0.007–0.015)	125 (102–141)	8:18 (7:49–10:05)
Succinylcholine (6)	0.062 (0.045–0.09)	37.0 (29–43)	3:36 (2:57–3:52)

[a] Values are means (range)

1. Onset of Action

In contrast to other competitive neuromuscular blocking drugs, the rate of onset of paralysis with fazadinium is particularly rapid, even when compared with succinylcholine (Table 3). It is most likely that fazadinium achieves maximum neuromuscular blockade in its first circulation through the muscle. A study carried out in the anaesthetised cat (TYERS 1978) to compare the time course of the circulation of an intravenous bolus of fazadinium with the onset of inhibition of twitches of the tibialis anterior muscle clearly shows that this is the case.

2. Duration of Action

Species variation in the duration of action of competitive neuromuscular blocking drugs is well known, particularly for the newer synthetic molecules. Fazadinium is no exception and the data given in Table 2 show clearly that the duration of paralysis in the cotton-eared marmoset is almost ten times greater than in the anaesthetised cat. Similar data have been obtained for other neuromuscular blocking drugs in cats and in larger primates (MUSCHIN and MAPLESON 1964).

3. Site and Mechanism of Neuromuscular Block

Fazadinium causes muscle paralysis through an antagonist action at postjunctional nicotinic receptors (BRITTAIN and TYERS 1973). This was demonstrated in the anaesthetised cat by showing that tibialis anterior muscle twitches evoked by direct electrical stimulation of the skeletal muscle were not affected by fazadinium while responses induced by close-arterial injections of acetylcholine were abolished by doses as low as 0.05 mg/kg i.v.

Experiments to demonstrate that the neuromuscular blocking action of fazadinium is "competitive" in nature (BRITTAIN and TYERS 1973; MARSHALL 1973a; BOUYARD et al. 1973) are largely qualitative rather than quantitative in that pharmacodynamic data for direct interactions of fazadinium with an agonist on nicotinic receptors have not been determined. In the isolated chick biventer cervicis muscle preparation, fazadinium ($EC_{50}=0.41$ mg/ml) inhibited indirectly evoked muscle twitches without any preceding potentiation or contractures. In the conscious, 7-day-old chick, fazadinium causes a short-lasting flaccid paralysis ($ED_{50}=0.06$ mg/kg i.v.) contrasting with the spastic paralysis obtained with depolarising agents. In the anaesthetised cat, there are no muscle fasciculations prior to blockade. Low doses of neostigmine rapidly restore neuromuscular transmission. These studies demonstrate that the mechanism of the neuromuscular blocking action of fazadinium is competitive in nature.

II. Termination of Action

Knowledge of the reasons for the termination of action of neuromuscular blocking drugs has important clinical implications. If renal clearance is the rate-limiting process, as for gallamine (FELDMAN and LEVI 1963; CHURCHILL-DAVIDSON 1967), tubocurarine (LOGAN et al. 1974) and pancuronium (SOMOGYI et al. 1977) then problems may be encountered in patients with renal insufficiency. Similarly, succinylcholine causes a prolonged apnoea in patients deficient in pseudocholinesterase.

Renal and hepatic clearance of fazadinium appear not to be important, at least for single doses, since in the anaesthetised cat with circulation occluded to either the liver or kidneys several single doses of the drug can be given without any indication of accumulation or prolonged paralysis (TYERS 1978). The initial rapid plasma clearance phase ($t_{1/2}=0.83$–2.12 min in humans, TYERS 1978; HULL et al. 1980) is probably a result of the rapid distribution of fazadinium throughout the extracellular fluid volume. The second slower redistribution ($t_{1/2}=42$ min; HULL et al. 1980) occurs at a later stage with clearance ultimately via both liver and kidneys (see Sect. D). Thus, the rate-limiting step for the termination of drug action cannot be the rate at which fazadinium is cleared from the plasma, but may be dependent upon the rate at which fazadinium dissociates from the nicotinic receptors (HASHIMOTO et al. 1979; TYERS 1978). Extensive metabolism of fazadinium does not occur (see Sect. D) and is certainly insufficient to account for the duration of action.

III. Selectivity of Action

In artificially ventilated anaesthetised cats, dogs, and monkeys, neuromuscular blocking doses of fazadinium (0.1–0.8 mg/kg i.v.) cause only slight falls in blood pressure and negligible effects on heart rate (BRITTAIN and TYERS 1973; MARSHALL 1973a; HUGHES and CHAPPLE 1976). Higher doses (1.0–5.0 mg/kg i.v.) cause dose-dependent falls in blood pressure (5–50 mm Hg) and small but variable effects on heart rate. The effects last for about the same duration as the blockade of tibialis anterior muscle twitches and are unaffected by pretreatment with atropine, mepyramine or bilateral vagotomy (TYERS 1975).

1. Ganglion Blockade

The vasodepressor responses caused by high doses of fazadinium are probably due to blockade of transmission in sympathetic ganglia (BRITTAIN and TYERS 1973; HUGHES and CHAPPLE 1976; MARSHALL 1973a; TYERS 1975). In these studies in vivo the potency ratio for neuromuscular block (inhibition of 0.1 Hz twitches) to sympathetic ganglion block for fazadinium was approximately 10:1 compared with a value of 2:1 for tubocurarine (HUGHES and CHAPPLE 1976).

2. Histamine Release

Several neuromuscular blocking drugs in high doses can release histamine from mast cells and basophil leucocytes. Tubocurarine is particularly effective and releases histamine with doses used clinically (COMROE and DRIPPS 1946). In conscious guinea-pigs, fazadinium, 20–100 mg intradermally, does not cause the development of weals and in anaesthetised guinea-pigs, 0.1–10 mg/kg i.v. has no effect on resting bronchial resistance while tubocurarine 4 mg/kg i.v. causes a maximal increase in resistance (BRITTAIN and TYERS 1973). In addition the vasodepressor responses induced by large doses of fazadinium in the anaesthetised cat and dog are not reduced by mepyramine and are not accompanied by increases in plasma histamine (MARTIN and HARRISON 1973). Thus, there is no evidence in animals that even high doses of fazadinium cause a release of histamine.

3. Actions on the Heart

The vagal inhibitory effects of competitive neuromuscular blocking drugs cause tachycardias in humans as a result of reduced parasympathetic tone (KELMAN and KENNEDY 1971). In anaesthetised animals, sympathetic and parasympathetic tone are low and these drugs rarely produce tachycardias. However, bradycardias induced by electrical stimulation of the vagus nerve or by intravenous injections of acetyl-β-methylcholine provide useful models for determining vagolytic activity (SAXENA and BONTA 1970).

In the anaesthetised cat and monkey fazadinium 0.06–2.0 mg/kg i.v. reduces the bradycardias induced by vagal nerve stimulation (BRITTAIN and TYERS 1973; MARSHALL 1973a; HUGHES et al. 1976) through a direct antagonist action on muscarinic receptors (MARSHALL and OJEWOLE 1979; DALTON and TYERS 1981).

D. Pharmacokinetics and Metabolic Fate of Fazadinium

The metabolism of fazadinium by laboratory animals in vivo has been studied using radiolabelled drug (BELL 1981). The excretion of radioactivity was found to be species dependent. In the rat and rabbit, the principal routes for the excretion of drug-related material are bile and urine respectively whereas in the dog, both routes are important. In the rat, although the urinary radioactivity (< 10% dose) was present largely as fazadinium, no unchanged drug was detected in bile. Although the biliary metabolite has not been identified, differential pulse polarography indicated that it lacks the tetrazene (N−N=N−N) linkage of fazadinium and may be hydrolysed by acid to produce 3-methyl-2-phenylimidazo[1,2-*a*]pyridine.

After bolus doses given to rabbits, less than 5% of the dose of fazadinium was excreted unchanged in urine. Treatment with *β*-glucuronidase has indicated that the principal urinary metabolite of fazadinium is a glucuronide conjugate of 3-methyl-2-*p*-hydroxyphenylimidazo[1,2-*a*]pyridine. In dogs, fazadinium is excreted in urine and bile largely as unchanged drug. In each of these species, 3-methyl-2-phenylimidazo[1,2-*a*]pyridine was only a *minor* metabolite in vivo.

Radiolabelled studies in humans (BLOGG et al. 1973a) have shown that the principal route for the excretion of drug-related material is urine where up to 76% of a dose of ^{3}H-fazadinium was excreted within 96 h. In urine samples collected up to 4 h after administration of ^{3}H-fazadinium, >90% of the radioactivity was present as unchanged drug. After this time increasing amounts of a metabolite were excreted so that at 12 h it represented the majority of urinary radiolabelled material. This unidentified metabolite, accounting for approximately 20% of the dose, is highly polar. DUVALDESTIN et al. (1978) reported that, in humans, metabolites of fazadinium accounted for no more than 3% of the dose. However, the fluorimetric technique (PASTORINO 1978) used by these workers would not have detected the polar metabolite shown by the radiolabelled studies.

E. Conclusions

Fazadinium dibromide (AH 8165) is the preferred compound of a series of azobisarylimidazo[1,2-*a*]pyridinium dihalides. From studies in several species it is clear that fazadinium produces a competitive neuromuscular blockade which is very rapid in onset and achieves maximum effect in its first circulation. As for most other drugs of this type, the duration of blockade varies between species, being longest in primates, particularly the marmoset. Fazadinium is not extensively metabolised and is excreted by both renal and hepatic routes. The rate-limiting step for the termination of drug action is most likely to be the rate at which fazadinium dissociates from the nicotinic receptors. Blood pressure, heart rate and the electrocardiogram are generally unaffected in anaesthetised animals. However, bradycardias induced by vagal stimulation are antagonised through an action on cardiac muscarinic receptors. Autonomic ganglionic transmission is blocked only by high doses. Fazadinium does not release histamine. Chapter 29 describes the clinical profile of fazadinium which has now been in use for several years.

Acknowledgement. I am very grateful to Dr. J. A. BELL, Miss S. T. O'DRISCOLL, and Miss A. WARD for their assistance in the preparation of this manuscript.

References

Bell A (1981) Metabolic studies with fazadinium dibromide. Ph.D. thesis (Council for National Academic Award)
Blogg CE, Simpson BR, Martin LE, Bell JA (1973) Metabolism of ^{3}H-AH 8165 in man. Br J Anaesth 45:1233–1234
Bolger L, Brittain RT, Jack D, Jackson MR, Martin LE, Mills J, Poynter D, Tyers MB (1972) Short lasting competitive neuromuscular blocking activity in a series of azobis-arylimidazo-(1,2-a)-pyridinium dihalides. Nature 238:354–355
Bouyard P, Mesdjian E, Balansard P, Gastant JA, Jadot G (1973) Pharmacologie du dibromure de 1-l'azo'bis-3-methyl-2-phenyl-1H-imidazo-(1,2-a)pyridinium, AH 8165, nouveau curarimimetic. Ann Anaesthiol Fr XIV:397–403
Brittain RT, Tyers MB (1972) AH 8165: a new short-acting competitive neuromuscular blocking drug. Br J Pharmacol 45:158–159P
Brittain RT, Tyers MB (1973) The pharmacology of AH 8165: a rapid-acting, short-lasting, competitive neuromuscular blocking drug. Br J Anaesth 45:837–843
Churchill-Davidson HC (1967) Muscle relaxant and renal failure. Anesthesiology 28:540–546
Comroe JH, Dripps RD (1946) The histamine-like action of curare and tubocurarine injected intracutaneously and intra-arterially in man. Anesthesiology 7:260–262
Dalton D, Tyers MB (1981) The selective muscarinic antagonist actions of pancuronium and alcuronium. Br J Pharmacol 74:784P
Duvaldestin P, Hanzel D, Demetriou M, Desmonte JM (1978) Pharmacokinetics of fazidinium in man. Br J Anaesth 50:773–779
Feldman SA, Levi JA (1963) Prolonged paresis following gallamine. Br J Anaesth 35:804–806
Hashimoto Y, Shima T, Matsukawa S, Satou M (1979) Neuromuscular blocking potency of fazadinium in man. Anaesthesia 34:10–13
Hughes R, Chapple DJ (1976) Effects of non-depolarizing neuromuscular blocking agents on peripheral autonomic mechanisms in cats. Br J Anaesth 48:59–68
Hughes R, Payne JP, Sugai N (1976) Studies on fazadinium bromide (AH 8165): a new non-depolarizing neuromuscular blocking agent. Can Anaesth Soc J 23:36–47
Hull CJ, English MJM, Sibbald A (1980) Fazadinium and pancuronium: a pharmacodynamic study. Br J Anaesth 52:1209–1213
Kelman GR, Kennedy BR (1971) Cardiovascular effects of pancuronium in man. Br J Anaesth 43:355–342
Logan DA, Howie HB, Crawford J (1974) Anaesthesia and renal transplantation: an analysis of fifty-six cases. Br J Anaesth 46:69–72
Maclagen J (1976) Competitive neuromuscular blocking drugs. In: Zaimis E (ed) Neuromuscular junction. Springer, Berlin Heidelberg New York (Handbook of experimental pharmakology, vol 42)
Marshall IG (1973a) The effects of three short-acting neuromuscular blocking agents on fast and slow-contracting muscles of the cat. Eur J Pharmacol 21:299–304
Marshall IG (1973b) The ganglion blocking and vagolytic actions of three short-acting neuromuscular blocking drugs in the cat. J Pharm Pharmacol 25:530–536
Marshall RJ, Ojewole JAO (1979) Comparison of the autonomic effects of some currently used neuromuscular blocking agents. Br J Pharmacol 66:77P
Martin LE, Harrison C (1973) An automated fluorometric method for the determination of histamine in biological samples. Biochem Med 8:292–308
Muschin WW, Mapleson WW (1964) Relaxant action in man of dipyrandium chloride ((M&B 9105A). Br J Anaesth 36:761–768
Pastorino AM (1978) Fluorimetric determination and pharmacokinetic studies of fazadinium bromide in dogs. Arzneimittelforsch 28:1728
Saxena PR, Bonta IL (1970) Mechanism of selective cardiac vagolytic action of pancuronium bromide. Specific blockade of cardiac muscarinic receptors. Eur J Pharmacol 11:332–340

Somogyi AA, Shanks CA, Triggs EJ (1977) Effect of renal failure on the disposition and neuromuscular blocking action of pancuronium bromide. Eur J Clin Pharmacol 12:23–29
Tyers MB (1975) Pharmacological studies on new, short-lasting competitive neuromuscular blocking drugs. Ph.D (CNAA) thesis
Tyers MB (1978) Factors limiting the rate of termination of the neuromuscular blocking action of fazadinium dibromide. Br J Pharmacol 63:287–293

CHAPTER 24

Atracurium

R. HUGHES

A. Introduction

Neuromuscular blocking agents are widely used to produce relaxation of skeletal muscles during surgical operations. Since the introduction of curare into anaesthesia by Griffith and Johnson in 1942, anaesthetic practice has been completely revolutionised because there is now no longer the need to produce skeletal muscular relaxation by deep anaesthesia with the inherent risks involved.

All competitive neuromuscular blocking agents in current use, in addition to their actions at the neuromuscular junction, inhibit cholinergic transmission at autonomic sites and may produce undesirable cardiovascular side effects (HUGHES and CHAPPLE 1976a). Furthermore, all show a significant increase in the duration of neuromuscular blockade when excretion is impaired by renal insufficiency (WINGARD and COOK 1977).

Atracurium is one of a new series of competitive neuromuscular blocking agents which has been developed to overcome the disadvantages of existing drugs (STENLAKE 1979). It is degraded mainly by Hofmann elimination at physiological pH and temperature which proceeds independently of hepatic and renal function (HUGHES and CHAPPLE 1981).

B. Neuromuscular Blocking Activity

I. Chick Isolated Biventer Cervicis Preparation

In the chick preparation of GINSBORG and WARRINER (1960), atracurium (0.5–1.0 μg/ml), similar to the nondepolarising agents tubocurarine (0.5–1.0 μg/ml) and gallamine (10 μg/ml), inhibited the responses to indirect stimulation without producing a contracture. In contrast, the blocking action of the depolarising drugs suxamethonium (0.1 μg/ml) and decamethonium (0.5 μg/ml) characteristically was preceded by an initial contracture.

II. Anaesthetised Cats, Dogs, and Rhesus Monkeys

In nerve–muscle preparations of anaesthetised cats, dogs, and rhesus monkeys, described by HUGHES and CHAPPLE (1976a, b), intravenous doses of atracurium 0.25–0.5 mg/kg were sufficient to cause complete neuromuscular blockade of the single twitch and tetanic responses of the gastrocnemius muscles and to arrest

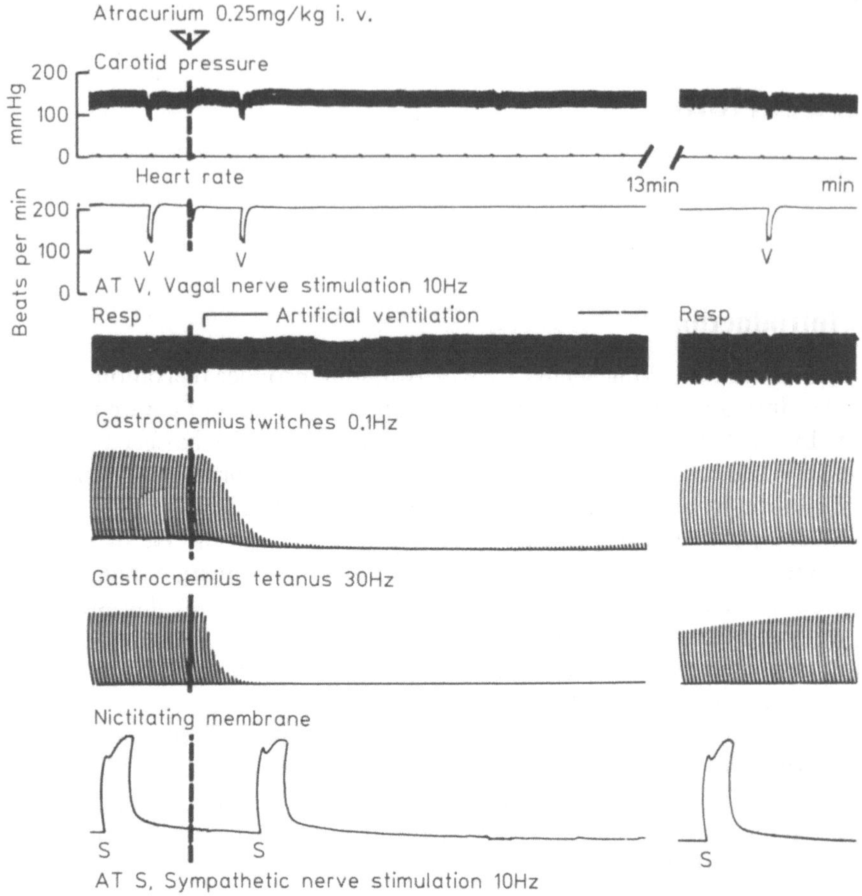

Fig. 1. Tracing from a cat anaesthetised with chloralose. A dose of atracurium 0.25 mg/kg i.v. arrested breathing and abolished the single twitch and tetanic response (30 Hz) of the gastrocnemius muscle to indirect stimulation. Effects on carotid pressure, heart rate, vagal induced bradycardia and contractions of the nictitating membrane caused by sympathetic nerve stimulation were minimal. (HUGHES and CHAPPLE 1981)

breathing (Fig. 1; Table 1; HUGHES and CHAPPLE 1981). Although atracurium was significantly more potent in dogs than in either cats or rhesus monkeys, the time course of neuromuscular blockade was similar throughout these species. This observation may be of importance since it has been reported that the duration of some competitive neuromuscular blocking agents was longer in monkeys, as in humans, than in other species (BAMFORD et al. 1967; BIGGS et al. 1964; BUSFIELD et al. 1968; HUGHES 1972; HUGHES and CHAPPLE 1976b; HUGHES et al. 1976a, b; MUSHIN and MAPLESON 1964). In anaesthetised monkeys, the rate of recovery from neuromuscular blockade by atracurium was significantly faster than that from tubocurarine and dimethyltubocurarine, but was comparable with that from pancuronium, fazadinium, gallamine, and alcuronium (HUGHES and CHAPPLE 1981). In each species studied, neuromuscular blockade by atracurium

Table 1. Neuromuscular blocking effects of atracurium in groups of anesthetized cats, dogs and rhesus monkeys. Maximal block of the single twitches and the tetanic concentrations (30 Hz for 1 s) of the gastrocnemius muscles is expressed as a percentage of initial value. Mean values are quoted (± standard error where relevant). *Double asterisks* denote that potency in dogs was significantly different from that in cats and rhesus monkeys as the 1% level. (HUGHES and CHAPPLE 1981)

Species	Dose (mg/kg i.v.)	Single twitch			Tetanic response (30 Hz)			50% Paralysing dose (mg/kg i.v.)	
		Block (%)	Onset to maximal effect (min)	Full recovery (min)	Block (%)	Onset to maximal effect (min)	Full recovery (min)	Single twitch	Tetanic response
Cats	0.0625	0	0	0	3[a]	2[a]	5[a]		
($N=6$)	0.125	54±12.9	5.0±0.7	11±2.2	93±3.2	5.9±1.0	21±1.4		
	0.25	99±1.43	3.0±0.8	29±2.5	100	2.1±0.5	37±2.0		
	0.5	100	0.9±0.05	45±4.1	100	0.5±0.04	51±3.5	0.129	0.095
	1.0	100	0.5±0.04	61±6.2	100	0.4±0.06	65±4.1	±0.008	±0.005
	2.0	100	2.2±1.8	80±5.5	100	0.4±0.06	85±5.5		$P<0.01$
	4.0	100	0.8±0.4	>80	100	0.3±0.04	> 84		
Dogs	0.0625	16±11.5	4.0±0.3	6.5±1.1	80±5.1	4.3±0.9	14±1.5		
($N=4$)	0.125	97±2.0	4.9±0.8	22 ±2.8	99±0.3	3.4±0.9	34±2.4		
	0.25	100	2.3±0.3	43 ±5.5	100	1.0±0.2	56±5.0	0.080**	0.050**
	0.5	100	1.8±0.3	60 ±4.0	100	1.1±0.2	70±3.6	±0.007	±0.002
	1.0	100	1.8±0.2	80 ±5.7	100	1.1±0.2	93±8.3		$P<0.05$
	2.0	100	1.4±0.4	88 ±11.4	100	1.1±0.3	>100		
	4.0	100	1.1±0.2	>41	100	0.9±0.2	>42		
Rhesus	0.0625	11[a]	5.5[a]	23[a]	20±9.0	2.1±0.2	8±4.2		
monkeys	0.125	51±3.5	4.6±0.3	17±3.0	90±1.8	3.7±0.3	25±2.0		
($N=7$)	0.25	99±0.6	4.0±0.5	32±2.9	100	1.8±0.4	39±1.8	0.124	0.083
	0.5	100	1.0±0.04	53±5.7	100	0.6±0.07	56±4.0	±0.002	±0.008
	1.0	100	0.7±0.09	80±8.0	100	0.4±0.05	79±6.7		$P<0.01$
	2.0	100	0.5±0.04	>96	100	0.4±0.12	> 96		
	4.0	100	0.5±0.2	>68	100	0.3±0.08	> 68		

[a] One result only

was readily antagonized by intravenous neostigmine 0.05 mg/kg and by edrophonium 0.2 mg/kg.

C. Changes in Acid–Base Balance

Effects of respiratory and metabolic acidosis and alkalosis on neuromuscular blockade by atracurium have been investigated in anaesthetised cats by the techniques described by HUGHES (1970). During respiratory and metabolic acidosis, when the mean blood pH was reduced from 7.35 to 6.82, neuromuscular paralysis was increased and prolonged; some changes achieved significance (HUGHES and CHAPPLE 1981). Conversely, when the mean blood pH was increased from 7.29 to 7.66 during respiratory and metabolic alkalosis, neuromuscular paralysis by atracurium was significantly reduced and recovery from block was significantly shortened (Fig. 2). Such changes were not unexpected since the inactivation of atracurium by Hofmann elimination is enhanced by raising the pH. However, within the physiological range of pH encountered during clinical anaesthesia, it is unlikely that the neuromuscular blocking action of atracurium will be significantly changed. Furthermore, patients are often hyperventilated during surgical anaesthesia which would tend to shorten the action of atracurium.

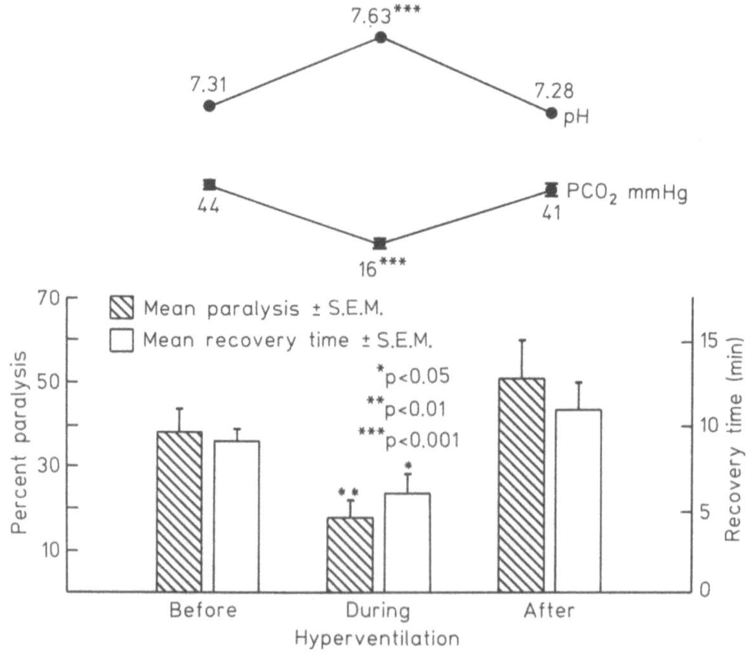

Fig. 2. Effects of respiratory alkalosis induced by hyperventilation on paralysis by atracurium 0.125 mg/kg i.v. of the single twitch response of the gastrocnemius muscle in a group of eight anaesthetised cats

D. Other Actions of Atracurium

I. Effects on Autonomic Mechanisms

Recordings from an anaesthetised cat in Fig. 1 show that the bradycardia induced by vagal nerve stimulation and the contraction of the nictitating membrane in response to sympathetic nerve stimulation were unimpaired after a dose of atracurium 0.25 mg/kg i.v. Dose–response curves for atracurium obtained from the results in anaesthetised cats, dogs, and rhesus monkeys (Fig. 3) show that there is a wide separation between the doses required for neuromuscular paralysis and those that inhibit autonomic mechanisms (HUGHES and CHAPPLE 1981). The inhibition of catecholamine uptake may contribute in part to the cardiovascular effects of pancuronium (DOCHERTY and MCGRATH 1977; IVANKOVICH et al. 1975). A study of catecholamine uptake by perfused rat hearts (IVERSEN 1963; SALT et al. 1980) showed that atracurium partially blocked both uptake processes, but only at concentrations in excess of those likely to occur after neuromuscular paralysing doses. $Uptake_2$ (extraneuronal disposition) was inhibited more strongly than $Uptake_1$ (neuronal disposition); the corresponding IC_{50} values were 20 and 100 μg/ml, respectively. No doubt the absence of effects on autonomic mechanisms has contributed to the cardiovascular stability associated with atracurium.

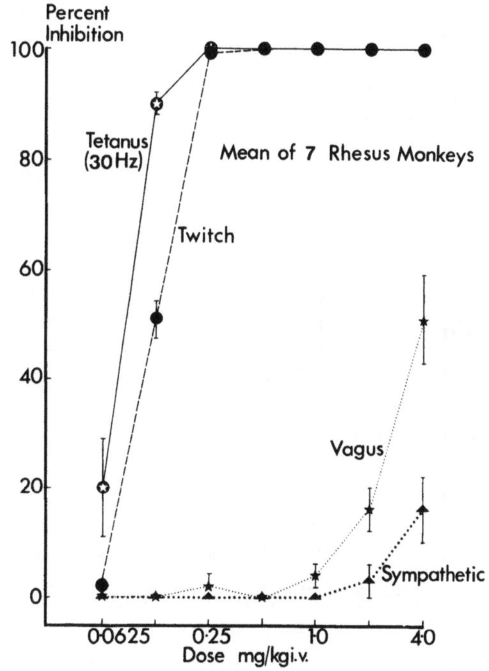

Fig. 3. Dose–response curves for atracurium showing the wide separation between doses causing neuromuscular blockade of the tetanic (30 Hz) and single twitch (0.1 Hz) responses of the gastrocnemius muscles and those causing vagal and sympathetic blockade in anaesthetised rhesus monkeys. Mean values were measured at time of maximum effect and *vertical lines* indicate standard error. (HUGHES and CHAPPLE 1981)

II. Cardiovascular Effects

The lack of cardiovascular effects is illustrated in Fig. 1 which shows a tracing from an anaesthetised cat given atracurium 0.25 mg/kg i.v. The wide separation between the neuromuscular and cardiovascular effects of atracurium is shown in Fig. 4. Small insignificant increases in mean arterial blood pressure occurred after i.v. doses of 0.125–1 mg/kg; mean heart rate was unchanged. Significant hypotension and slight bradycardia were evident after a dose of 4 mg/kg i.v., but this dose was 16 times greater than that required for full neuromuscular paralysis. Similar results were obtained in rhesus monkeys, but in anaesthetised dogs smaller doses of atracurium (2 mg/kg i.v.) reduced mean arterial pressure to 53% ± 20.2% of the control value, nevertheless this dose was 8 times greater than that required for full neuromuscular paralysis (HUGHES and CHAPPLE 1981). Haemodynamic studies in open-chest anaesthetised dogs showed that circulatory depression was slight with doses of atracurium 2–4 times that needed for full neuromuscular blockade, but was severe after 8–16 times this dose. In other experiments high concentra-

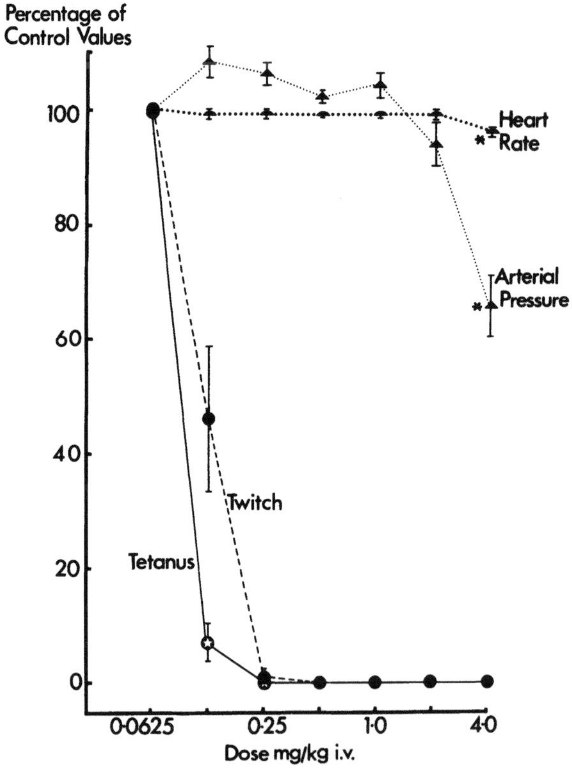

Fig. 4. Effects of atracurium on the tetanic (30 Hz) and single twitch (0.1 Hz) responses of the gastrocnemius muscle and on mean arterial blood pressure and heart rate in a group of six anaesthetised cats. Mean values were measured at time of maximum effect and *vertical lines* indicate standard error. *Asterisks* denote values of mean arterial blood pressure and heart rate significantly different from the control levels at the 5% level. (HUGHES and CHAPPLE 1981)

tions of atracurium (100 μg/ml) had no inotropic or chronotropic effects on isolated spontaneously beating guinea-pig atria.

III. Histamine Release

In anaesthetised dogs, intravenous administration of the H_1-receptor antagonist mepyramine 10 mg/kg and the H_2-receptor antagonists burimamide 15 mg/kg or cimetidine 15 mg/kg reduced the moderate hypotension caused by supramaximal doses of atracurium 2 mg/kg i.v. (Fig. 5) and the severe hypotension induced by neuromuscular blocking doses of tubocurarine 0.4 mg/kg i.v. Thus, this indirect evidence of histamine release by atracurium in dogs occurred only after administration of eight times the neuromuscular blocking dose and after the full neuromuscular blocking dose of tubocurarine.

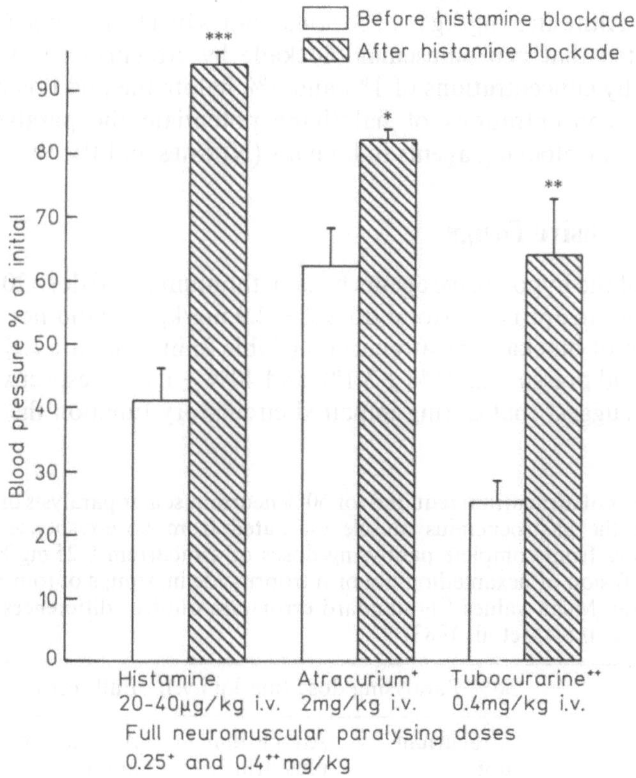

Fig. 5. Effects of histamine, atracurium and tubocurarine on mean arterial blood pressure before and after blockade of histamine receptors in four anaesthetised dogs. Mean values are shown and *vertical lines* indicate standard error. *Asterisks* denote significant differences after histamine blockade at the 5% level, *double asterisks* 1%; and *treble asterisks* 0.1%, respectively. (HUGHES and CHAPPLE 1981)

E. Drug Interactions

It is known that various drugs used during clinical anaesthesia interact with competitive neuromuscular agents (MACLAGAN 1976; BOWMAN 1980) and such studies with atracurium have been carried out in anaesthetised cats (CHAPPLE et al.1983).

I. Premedicants

Pretreatment intravenously with diazepam 0.5 mg/kg, morphine 0.5 mg/kg, pentazocine 2 mg/kg, or pethidine 4 mg/kg did not significantly alter the dose of atracurium required to produce 50% neuromuscular paralysis as estimated from dose–response curves.

II. Anaesthetics

The neuromuscular paralysing potency of atracurium was not significantly altered by the intravenous anaesthetics ketamine 2 mg/kg i.v., methohexitone 0.2% infusion or Althesin 3 mg/kg i.v. Previous studies by HUGHES and CHAPPLE (1981) demonstrated that neuromuscular blockade by atracurium was enhanced and prolonged by concentrations of 1% and 2% halothane and it is recognised that anaesthetic concentrations of halothane potentiate the paralysing action of neuromuscular blocking agents in humans (HUGHES and PAYNE 1979).

III. Hypotensive Drugs

Infusions of the hypotensive agents hexamethonium (total dose 20–26 mg/kg i.v.) and sodium nitroprusside (total dose 2.3–2.9 mg/kg i.v.) did not alter the depth or duration of blockade by atracurium (Table 2) in spite of a reduction in mean arterial blood pressure af 21% ±3.1% and 40% ±1.7%, respectively). These observations suggest that during impaired circulatory function the inactivation of

Table 2. Doses of atracurium required for 50% neuromuscular paralysis of the single twitch response of the gastrocnemius muscle estimated from dose-response curves and full recovery times from complete paralysing doses of atracurium 0.25 mg/kg i.v. alone and during an infusion of hexamethonium or nitroprusside in groups of four anesthetized cats for each drug. Mean values (± standard error) are quoted; differences did not achieve significance (CHAPPLE et al. 1983)

	50% Paralysing dose (mg/kg i.v.)		Full recovery (min)	
	Atracurium alone	Atracurium plus drug	Atracurium alone	Atracurium plus drug
Hexamethonium (20–26 mg/kg i.v.)	0.116±0.013	0.136±0.006	28±0.8	32±3.2
Nitroprusside (2.24–2.88 mg/kg i.v.)	0.122±0.016	0.106±0.011	20±2.6	24±4.7

atracurium was unaffected, since it is not dependent on hepatic or renal function or on a fast circulation to transport the drug to these organs from its site of action in the muscles. Thus, it can be inferred that atracurium may be used in severely ill patients with slow circulations with little risk of a prolonged block or recurarisation.

IV. Drugs Used for Resuscitation

The neuromuscular blocking action of atracurium was slightly increased by lignocaine 1–4 mg/kg i.v. and reduced by adrenaline 1–2 µg/kg i.v. (Table 3). Adrenaline is known to antagonise the action of competive agents by a direct stimulant effect on the muscle (BOWMAN and NOTT 1969). Neither i.v. propranolol 0.07 mg/kg nor $CaCl_2$ 0.5 mequiv. affected the neuromuscular block produced by atracurium.

Table 3. Effects of lignocaine, adrenaline, propranolol, and calcium chloride on partial block of the single twitch response of the gastrocnemius muscle by atracurium 0.1–0.2 mg/kg i.v. in anesthetized cats. Mean values (± standard error where relevant) are quoted; differences did not achieve significance (CHAPPLE et al. 1983)

Drug	Dose of drug[a] (mg/kg i.v.)	Block of twitch (% of control)	Change in block of twitch (%)
Lignocaine	1 (N=4)	27±2.6	+12±7.3
	2 (N=4)	40±8.0	+14±5.5
	4 (N=1)	20	+15
Adrenaline	1 µg/kg (N=4)	70±18.1	−40±15.9
	2 µg/kg (N=4)	69±16.5	−45±11.2
Propranolol	0.07 mg/kg (N=4)	44±6.8	0
Calcium chloride	0.5 meq i.v. (N=3)	35±30.2	0

[a] Drug given at time of maximum block by atracurium

V. Antibiotics

Of clinical importance is the interaction of the aminoglycoside antibiotics with neuromuscular blocking agents (PITTINGER and ADAMSON 1972). The mechanism of action has not been clearly defined, but a mixture of competitive and depolarising block is often seen and these drugs may also have presynaptic effects. It is not surprising therefore that the response to neostigmine in humans is unpredictable with instances of little or no reversal of the combined block (FOGDALL and MILLER 1974; SOKOLL and GERGIS 1981).

In experiments in anaesthetised cats gentamicin 1 and 3 mg/kg, neomycin 3 mg/kg, and polymyxin B 10,000 u/kg caused a slight increase in the neuromuscular blocking action of atracurium when given intravenously at the time of maximum block (Table 4); the effect of polymyxin B 20,000 u/kg was significant

Table 4. Effects of gentamicin, neomycin, polymyxin B and co-trimoxazole on partial block of the single twitch response of the gastrocnemius muscle by atracurium 0.1–0.175 mg/kg i.v. in anesthetized cats. Mean values (± standard error where relevant) are quoted (CHAPPLE et al. 1983)

Drug	Dose[a] (mg/kg i.v.)	Block of twitch (% control)	Increase in block of twitch (%)	Significance
Gentamicin ($N=4$)	1	52 ± 10.8	6 ± 2.9	None
Gentamicin ($N=2$)	3	45	17	None
Neomycin ($N=5$)	3	34 ± 11.2	21 ± 5.9	None
Polymyxin ($N=3$)	1 (10,000 u/kg)	45 ± 14.7	11 ± 4.4	None
Polymyxin ($N=4$)	1 (20,000 u/kg)	34 ± 20.1	51 ± 16.9	$P<0.05$
Co-trimoxazole ($N=4$)	12	36 ± 5.4	0	None

[a] Drug given at time of maximum block by atracurium

($P<0.05$). Neomycin 3 mg/kg when given i.v. 5 min prior to atracurium significantly ($P<0.01$) enhanced the block whereas co-trimoxazole 12 mg/kg when given i.v. either 5 min before or during the block by atracurium had no effect. However, the combined neuromuscular block produced by atracurium and neomycin 3 mg/kg and polymyxin B 20,000 u/kg was reversed by intravenous neostigmine 250 µg, preceded by atropine 300 µg, but it has yet to be shown whether this reversal of the combined block occurs in humans.

VI. Neuromuscular Blocking Agents

It is well established that the depolarising and competitive neuromuscular blocking agents are mutually antagonistic. Hence, the action of atracurium 0.1–0.2 mg/kg i.v., like that of other competitive neuromuscular blocking agents (LEBOWITZ et al. 1980) was significantly potentiated ($P<0.01$) by tubocurarine 0.2 mg/kg i.v. when given at the time of maximum block whereas suxamethonium 0.2 mg/kg i.v. caused a transient reduction in neuromuscular block by atracurium. However, when suxamethonium 0.2 mg/kg i.v. was given 10–15 min before atracurium 0.2 mg/kg the degree of neuromuscular block was not significantly changed. This lack of interaction should permit the use of suxamethonium when very rapid endotracheal intubation is required without impairing the use of atracurium for producing subsequent muscular relaxation.

F. Breakdown Products and Related Substances

The breakdown products from Hofmann elimination are laudanosine and the quaternary monoacrylate, and from the ester hydrolysis, the quaternary acid and quaternary alcohol (see Chap. 10). The breakdown products together with two related substances metholaudanosine and the monoquaternary analogue have been tested for neuromuscular and cardiovascular activity in anaesthetised cats (CHAPPLE and CLARK 1983).

The results showed that doses of 4 mg/kg i.v. of either laudanosine, the quaternary acid or metholaudanosine had little or no neuromuscular blocking activity (Table 5). Three compounds, the quaternary monoacrylate, the quaternary alcohol and the monoquaternary analogue produced neuromuscular block at this dose. The neuromuscular blocking effects of the quaternary monoacrylate and the quaternary alcohol could be due in part to the presence of atracurium as an impurity (6.7% and 1.7% respectively).

Doses of 4 mg/kg i.v. of either the quaternary monoacrylate. laudanosine, the quaternary alcohol, metholaudanosine or the monoquaternary analogue caused distinct hypotension, but only the quaternary monoacrylate caused significant sympathetic blockade. Significant vagal blockade was evident after administration of the quaternary monoacrylate, the quaternary alcohol and the monoquaternary analogue; changes in heart rate were minimal (Table 5).

It is important that the breakdown products of atracurium are without neuromuscular and cardiovascular effects at doses which are clinically effective. PAYNE and HUGHES (1981) have shown that intravenous doses of 0.3–0.6 mg/kg facilitated endotracheal intubation and provided adequate surgical relaxation in anaesthetised humans. The doses of the breakdown products and related substances studied (0.5–4 mg/kg i.v.) were up to ten times the equivalent molarity for therapeutic doses of atracurium. Even if atracurium was broken down to only the most potent of these substances, then the results indicate that it would have no important neuromuscular or cardiovascular effects even after doses of 1 mg/kg atracurium. In practice it is unlikely that a mixture of these compounds will be present since it has been demonstrated that both Hofmann elimination and ester hydrolysis occur in anaesthetised cats and all the products of these reactions were present in urine and bile (NEILL and CHAPPLE 1982). Furthermore, the product of each route may not be present in equal amounts since each product is capable of further breakdown or may differ in its respective rate of elimination. It has been shown that all the metabolites were cleared rapidly with 70% of drug-related material being eliminated in 5 h (NEILL and CHAPPLE 1982) and up to 90% in 7 h (NEILL et al. 1983).

Since doses of 2 or 4 mg/kg i.v. of each of the breakdown products or related substances were required to produce distinct neuromuscular or cardiovascular effects, it is unlikely that the quantities present either as an impurity or formed by metabolism after clinical blocking doses of atracurium are of pharmacological importance.

Table 5. Effects of breakdown products of atracurium and related substances on the single twitch (0.1 Hz) and tetanic (30 Hz) responses of the gastrocnemius muscle, mean arterial blood pressure, heart rate, the vagal induced bradycardia and contractions of the nictitating membrane in anesthetized cats. Inhibition of the neuromuscular, vagal and sympathetic responses are expressed as percentages and effects on arterial blood pressure and heart rate as percentage of initial value. Mean values (± standard error where relevant) are quoted. The monoquaternary analogue was tested in six cats and the other compounds in four cats (CHAPPLE and CLARK 1983)

Compound (maximum in atracurium)	Dose (mg/kg i.v.)	Single twitch		Tetanic response		Mean arterial pressure (initial=100)	Heart rate (initial =100)	Vagal response (% inhibition)	Nictitating membrane (% inhibition)
		Block (%)	Full recovery (min)	Block (%)	Full recovery (min)				
Quaternary monoacrylate (2%)	0.5	0	0	0	0	100	100±0.8	12± 5.6	5± 2.9
	1.0	1± 1.0	2±1.5	58±15.0	12±1.5	93± 3.2	100±0.9	28± 9.5	11± 4.6
	2.0	58±21.1	9±5.9	100	21±3.6	90± 5.9	102±2.8	48±12.2	16± 5.1
	4.0	100	31±8.4	100	35±9.7	76± 8.3	98±2.1	88± 5.7	34±14.6
Laudanosine (2% with quaternary acid)	0.5			0		109± 6.8	103±1.5	10±10	1± 1
	1.0	None		0		108± 4.5	101±1.3	13±13	1± 0.8
	2.0			0		74± 3.9	105±5.1	15±14	5± 3.8
	4.0			18± 8.6	3±2.1	66± 2.3	104±5.1	26±17.1	8± 6.6
Quaternary acid (2% with laudanosine)	0.5	None		None		101± 2.4	100±0.5	0	0
	1.0					108± 7.2	100±1.4	0	0
	2.0					110±10	99±1.8	1± 1.5	0
	4.0					106±14.4	99±2.2	13± 2.7	1± 1.3
Quaternary alcohol (1%)	0.5	0		0		93+ 3.1	98±1.0	1± 1.5	1± 0.5
	1.0	0		10± 9.7	2±2.0	90± 4.9	98±2.7	4± 2.8	4± 2.7
	2.0	10±10	2±1.5	81±10.2	13±1.8	77± 6.8	93±3.8	11± 6.3	13± 7.1
	4.0	50±22	9.8±1.9	100± 0.3	25±2.6	63± 9.4	86±6.2	47±17.0	22±12.6
Metholaudanosine (1%)	0.5	None		0		96± 4.3	99±0.5	1± 0.8	2± 2.5
	1.0			0		89± 5.5	97±1.4	5± 2.3	3± 3.5
	2.0			0		71±10.8	101±2.9	8± 6.2	3± 3.3
	4.0			8± 7.8	0.3±0.3	56± 6.6	117±7.4	27±12.6	8± 8.3
Monoquaternary analogue (3%)	0.5	0		0	0	99± 1.5	100±0.3	3± 4	1± 1.2
	1.0	0		6± 4.6	5±3.5	97± 1.5	100	7± 3.4	4± 2.3
	2.0	6±4.4	1±0.5	48±14.4	7±1.6	95± 1.7	102±1.4	21± 5.6	10± 2.5
	4.0	72±4.5	4±0.6	97± 2.8	12±0.9	73± 7.8	96±4.1	49±10.4	16± 1.7

G. Cholinesterase Inhibition

The enzyme was extracted from human skeletal muscle and its activity measured as described by ELLMAN et al. (1961). At a concentration of atracurium dibesylate of 70 μg/ml the enzymic activity was inhibited by 50% (HUGHES and CHAPPLE 1981). If the normal therapeutic blood level is taken as 5 μg/ml, extrapolation from these results shows less than 5% inhibition at this concentration.

H. Conclusions

Atracurium is a novel neuromuscular blocking agent which was designed to undergo degradation at physiological temperature and pH by a self-destruction mechanism called Hofmann elimination which proceeds independently of hepatic and renal function. The drug may also undergo an enzymic ester hydrolysis. The decomposition products from both pathways are unlikely to cause neuromuscular and cardiovascular effects in the amounts that are likely to be present after clinical dosage of atracurium.

Atracurium is a highly selective nondepolarising neuromuscular blocking agent of intermediate duration (about 35 min) in anaesthetised cats, dogs, and rhesus monkeys. Intravenous doses of 0.25 or 0.5 mg/kg were sufficient to cause complete neuromuscular blockade in 1–3 min; paralysis was effectively antagonised by neostigmine and edrophonium. Neuromuscular blockade by atracurium was reduced by alkalosis and enhanced by acidosis, but, within the physiological range of pH, it is unlikely that the neuromuscular blocking action of atracurium will be significantly altered in clinical practice.

Vagal blockade became appreciable only at doses 8–16 times the full neuromuscular blocking dose and in this respect atracurium possesses a distinct advantage over available neuromuscular blocking agents, such doses of atracurium had a minimal effect on sympathetic function. Arterial blood pressure was reduced only by doses of atracurium 16 times the full neuromuscular blocking dose in anaesthetised cats and rhesus monkeys; effects on the heart rate were minimal. In anaesthetised dogs hypotension and bradycardia became of importance at doses 4 times the full neuromuscular blocking dose and circulatory depression was severe at 8 times this dose. Such effects caused by excessive doses of atracurium could be attributable mainly to histamine release which is prominent in dogs.

Drug interaction studies were carried out in anaesthetised cats with those drugs likely to be used during anaesthetic practice. Neuromuscular block by atracurium was not significantly changed by clinical doses of diazepam, morphine, pentazocine, pethidine, ketamine, Althesin, methohexitone, co-trimoxazole, sodium nitroprusside, hexamethonium, lignocaine, propranolol, or calcium chloride. However, the neuromuscular blocking action of atracurium was enhanced by tubocurarine, halothane, gentamicin, neomycin, and polymyxin B and was anatagonised by adrenaline and transiently antagonised by suxamethonium.

References

Bamford DG, Biggs DF, Davis M, Parness EW (1967) Neuromuscular blocking properties of stereoisomeric and androstane-3,17-bisquaternary ammonium salts. Br J Pharmacol 30:194–202

Biggs RS, Davis M, Wien R (1964) Muscle relaxant properties of a steroid bis-quaternary ammonium salt. Experientia 20:119–120

Bowman WC (1980) Pharmacology of neuromuscular function. Wright, Bristol, pp 109–115

Bowman WC, Nott MW (1969) Actions of sympathomimetic amines and their antagonists on skeletal muscle. Pharmacol Rev 21:27–72

Busfield D, Child KJ, Clark AJ, Davis B, Dodds MG (1968) Neuromuscular blocking activities of some steroidal mono and bis-quaternary ammonium compounds with special reference to N,N′-dimethyl conessine. Br J Pharmacol 32:609–623

Chapple DJ, Clark JS (1983) Pharmacological action of breakdown products of atracurium and related substances. Br J Anaesth 55:11–15S

Chapple DJ, Clark JS, Hughes R (1983) Interaction between atracurium and drugs used in anaesthesia. Br J Anaesth 55:17–22S

Docherty JR, McGrath JC (1977) Potentiation of cardiac sympathetic nerve response in vivo by pancuronium bromide. Br J Pharmacol 61:472P

Ellman GL, Courtney KD, Andres V, Featherstone RM (1961) A new and rapid colorimetric determination of acetylcholinesterase activity. Biochem Pharmacol 7:88–95

Fogdall RP, Miller RD (1974) Prolongation of a pancuronium-induced neuromuscular blockade by polymyxin B. Anaesthesiology 40:84–87

Ginsborg BL, Warriner J (1960) The isolated chick biventer-cervicis nerve-muscle preparation. Br J Pharmacol 15:410–411

Griffith HR, Johnson GE (1942) the use of curare in general anaesthesia. Anesthesiology 3:418–420

Hughes R (1970) the influence of changes in acid base balance on neuromuscular blockade in cats. Br J Anaesth 42:658–668

Hughes R (1972) Evaluation of the neuromuscular blocking properties and side-effects of the two new isoquinolinium bisquaternary compounds (BW252C64 and BW403C65). Br J Anaesth 44:27–42

Hughes R, Chapple DJ (1976a) Effects of non-depolarizing neuromuscular blocking agents on peripheral autonomic mechanisms in cats. Br J Anaesth 48:59–68

Hughes R, Chapple DJ (1976b) Cardiovascular and neuromuscular effects of dimethyl tubocurarine in anaesthetised cats and rhesus monkeys. Br J Anaesth 48:847–852

Hughes R, Chapple DJ (1981) the pharmacology of atracurium: a new competitive neuromuscular blocking agent. Br J Anaesth 53:31–44

Hughes R, Payne JP (1979) Interactions of halothane with non-depolarizing neuromuscular blocking drugs in man. Br J Clin Pharmacol 7:485–490

Hughes R, Ingram GS, Payne JP (1976a) Studies on dimethyl tubocurarine in anaesthetised man. Br J Anaesth 48:969–974

Hughes R, Payne JP, Sugai N (1976b) Studies on fazadinium bromide (AH 8165), a new non-depolarizing neuromuscular blocking agent. Can Anaesth Soc J 23:36–47

Ivankovich AD, Miletich, Albrecht RF, Zahed B (1975) The effect of pancuronium on myocardial contraction and catecholamine metabolism. J Pharm Pharmacol 27:837–841

Iversen LL (1963) The uptake of noradrenaline by the isolated perfused heart. Br J Pharmacol 21:523–537

Lebowitz PW, Ramsey FM, Savarese JJ, Alli HH (1980) Potentiation of neuromuscular blockade in man produced by combination of pancuronium and metocurarine and pancuronium and d-tubocurarine. Anaesth Analg 59:604–609

Maclagan J (1976) Competitive blocking drugs. In: Zaimis E (ed) Neuromuscular junction. Springer, Berlin, p 421

Mushin WW, Mapleson WW (1964) Relaxant action in man of dipyrandium chloride (M&B 9105A). Br J Anaesth 36:761–768

Neill EAM, Chapple DJ (1982) Metabolic studies in the cat with atracurium: a neuromuscular blocking agent designed for non-enzymic inactivation at physiological pH. Xenobiotica 12:203–210
Neill EAM, Chapple DJ, Thompson CW (1983) the metabolism and kinetics of atracurium: an overview. Br J Anaesth 55:23–25S
Payne JP, Hughes R (1981) Evaluation of atracurium in anaesthetised man. Br J Anaesth 53:45–54
Pittinger C, Adamson R (1972) Antibiotic blockade of neuromuscular function. Ann Rev Pharmacol 12:169–184
Salt P, Barnes PK, Conway CM (1980) Inhibition of neuronal uptake of noradrenaline and its homologues, Org 6368, Org 7688 and NC 45. Br J Anaesth 52 Suppl:313–317
Sokoll MD, Gergis SD (1981) Antibiotics and neuromuscular function. Anesthesiology 55:148–159
Stenlake JB (1979) Ions-cyclic nucleotids-cholinergy. In: Stoclet JC (ed) Advances in pharmacology and therapeutics. Pergamon, Oxford, p 303
Wingard LB, Cook DR (1977) Clinical pharmacokinetics of muscle relaxants. Clin Pharmacokin 2:330–343

Clinical Pharmacology of New Neuromuscular Blocking Drugs

CHAPTER 25

General Principles and Methods of Evaluation of Neuromuscular Blocking Agents in Anesthesiology

F. F. FOLDES, H. NAGASHIMA, and D. DUNCALF

A. Introduction

There is considerable species variation in the pharmacodynamic effects (PATON and ZAIMIS 1949; FOLDES 1954) and pharmacokinetics (AGOSTIN 1976) of neuromuscular (NM) blocking agents (NMBA). Because of this, the information obtained in animal experiments on the potency, onset, and duration of the NM effect of NMBA and on their side effect liability and disposition has only limited relevance to the effects of these compounds in humans. Thus, for example, pancuronium (Pavulon) is about five and ten times more potent than tubocurarine (Tc) in humans (BAIRD and REID 1967) and cats (BUCKETT et al. 1968), respectively, but about five times less potent than Tc in rats (CHAUDHRY and FOLDES 1981). Vecuronium (Org-NC-45, Norcuron) is more potent than pancuronium in humans (CRUL and BOOIJ 1980) and dogs (MARSHALL et al. 1980), but less potent than pancuronium in rats, cats, and rhesus monkeys (MARSHALL et al. 1980). To predict the effects of NMBA in the clinical setting, it is necessary to investigate their pharmacodynamics and pharmacokinetics, with accepted pharmacologic methods, in humans. Preliminary observations should be carried out in conscious human subjects. These should be followed by more comprehensive studies in anesthetized volunteers and/or patients. The clinical pharmacologic screening of NMBA should also extend to the observation of the influence of pathophysiologic (e.g. acidosis, alkalosis, hypo- and hyperthermia) and inherited and acquired pathologic conditions and drugs on the desired and unwanted effects and on the disposition of these compounds.

B. Screening of Neuromuscular Blocking Agents in Conscious Subjects

Before undertaking the evaluation of a new NMBA in human subjects, the investigation has to be approved by the Institutional Review Board for Human Studies and the subjects must sign informed consent forms. Adequate animal pharmacology and short-term toxicity studies are prerequisites of the screening of NMBA in human subjects. The testing of new NMBA should be performed by, or in the presence of, physicians who are experienced in the clinical use of such compounds. Equipment and drugs necessary for the management of respiratory, circulatory, and allergic emergencies must be immediately available.

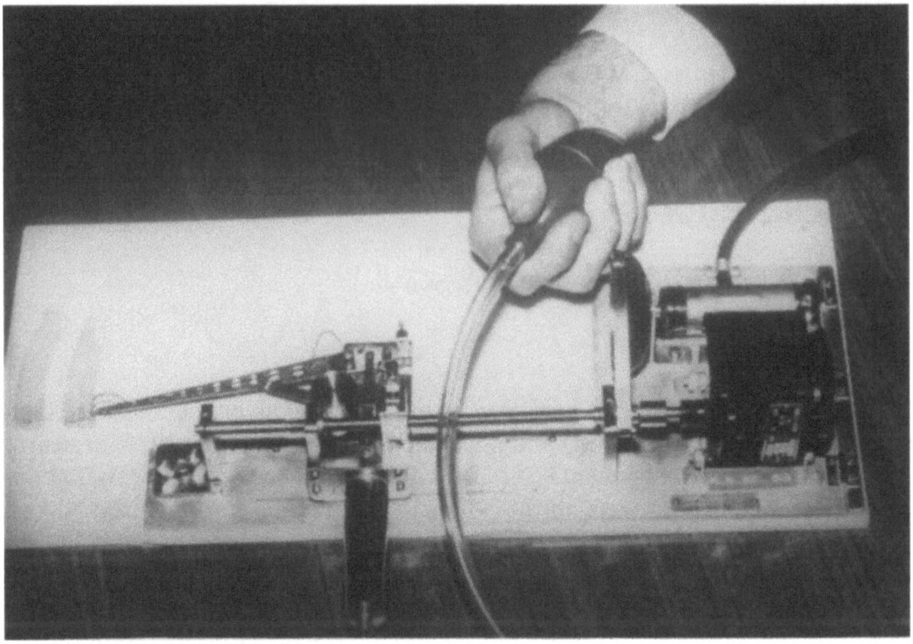

Fig. 1. Ergograph

Before the administration of the NMBA to be tested, the subject's grip strength should be determined with a dynamometer and the fatigability of the flexor muscles of the hand with an ergograph (Fig. 1; OSSERMAN 1958; FOLDES et al. 1961) or a similar device. At the same time, the subject's vital capacity, heart rate, and blood pressure should be recorded. After completion of these baseline measurements, the NMBA to be tested should be administered intravenously (i.v.). In the first subject, starting with a dose that is expected to cause little or no NM block, gradually increasing fractional doses should be injected 5 min apart. The baseline measurements should be repeated 4 min after the injection of each fractional dose. The administration of fractional doses should be continued until the grip strength decreases to 10%–30% of control, vital capacity decreases to 60% of control, there is significant change in heart rate or blood pressure, or allergic manifestations develop, whichever occurs first. With nondepolarizing NMBA, it is usually possible to decrease grip strength to 10%–30% of control without decreasing vital capacity to the point of respiratory embarrassment (Fig. 2; UNNA and PELIKAN 1951; FOLDES et al. 1961). With depolarizing NMBA, the relative sparing effect on the respiratory muscles is less and the grip strength can only be decreased to 40%–60% of control before subjective ventilatory difficulties develop (Fig. 3; FOLDES et al. 1961).

A single dose, that is about 20%–30% less than the sum of the fractional doses administered to the first subject, can be expected to decrease the grip strength to 10%–30% of control. It may be necessary to adjust the mg/kg dose until this effect is obtained in most subjects. In anesthetized subjects, twice this dose will usually cause an approximately 95% NM block.

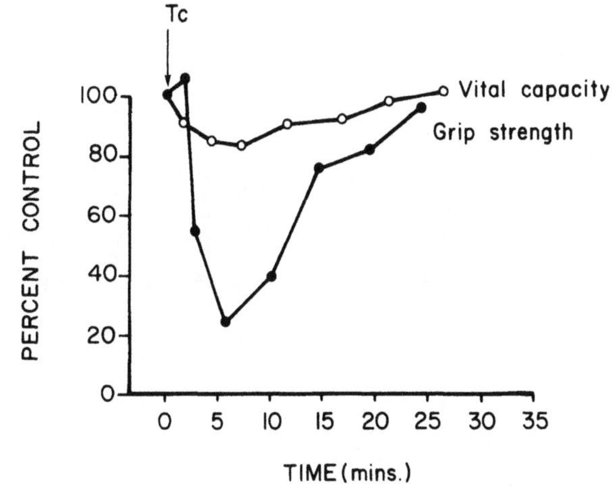

Fig. 2. The effect of tubocurarine (Tc) on vital capacity and grip strength. At the *arrow*, 0.1 mg/kg Tc was injected i.v. Note that Tc decreased grip strength by about 80%, but caused less than 20% decrease of vital capacity

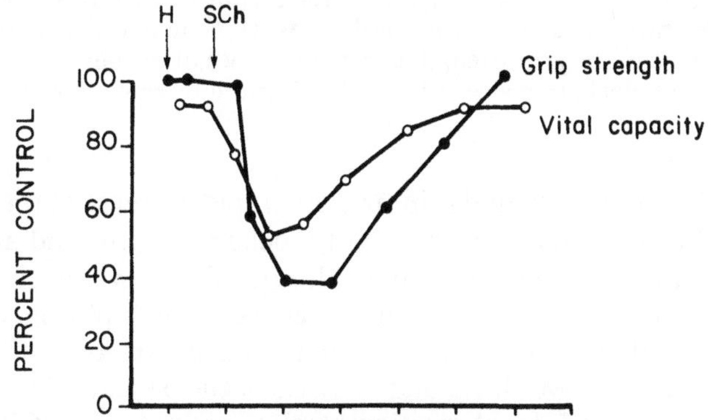

Fig. 3. The effect of succinylcholine (SCh) on vital capacity and grip strength. At the *first arrow*, 0.3 mg/kg hexafluorenium bromide (Mylaxen) was injected i.v. to prevent hydrolysis of SCh by plasma butyrylcholinesterase. At the *second arrow*, 0.1 mg/kg SCh was injected i.v. Note that the decrease of grip strength and vital capacity is similar

A more reliable, but also more time-consuming approach to the assessment of NM potency in conscious subjects, is the determination of the cumulative or individual dose–response regression line. Because of the considerable individual variation to muscle relaxants (PELIKAN et al. 1953), especially in the lower dose range, the cumulative dose–response should be determined in at least eight subjects. In these experiments, from the return of the grip strength to its control value, the duration of action of the NMBA can be assessed.

The type of NM block produced by the NMBA investigated, can be readily determined from experiments on conscious subjects. Large differences between

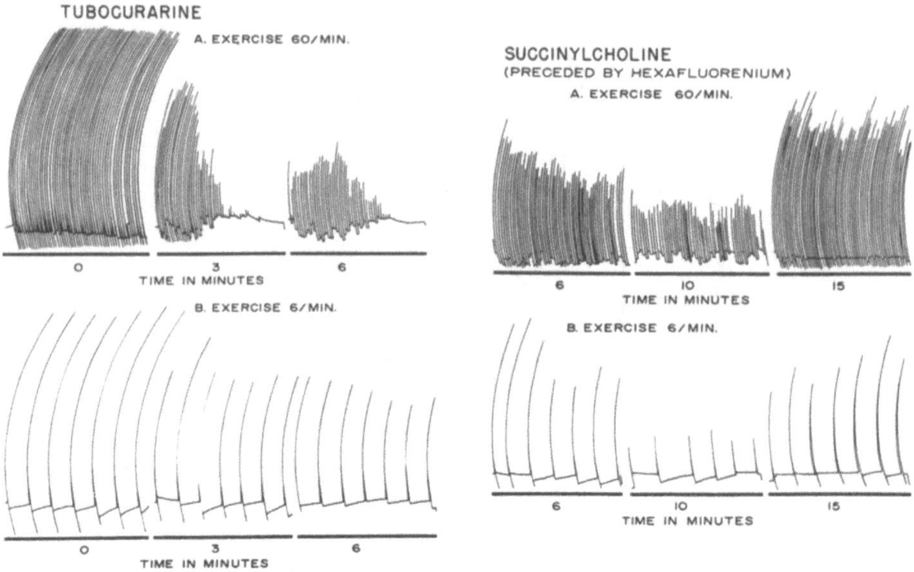

Fig. 4. Influence of rate of exercise on the grip strength during partial neuromuscular block induced by tubocurarine (Tc) or succinylcholine (SCh). Note that during partial Tc block there is rapid fatigue of the grip strength during fast, but not during slow exercise. During partial SCh-induced block, grip strength is not influenced by rate of exercise

the decrease of vital capacity and grip strength, rapid decrease of the grip strength when exercised rapidly during partial NM blockade (Fig. 4) and antagonism of the NM block (Fig. 5) by edrophonium chloride (Tensilon) are characteristic of nondepolarizing block. Little difference between the blocking effect on the muscles of the hand and respiration, well-maintained strength with rapid rate of exercise during partial NM block, and increase of the partial NM block by edrophonium indicate phase I depolarization block (FOLDES et al. 1957).

Significant increase of the heart rate in the course of the experiment is suggestive of an atropine-like effect on the muscarinic receptors of the heart (RIKER and WESCOE 1951), or inhibition of the reuptake of endogenous catecholamines (BOWMAN 1980). Decrease of the heart rate and excessive salivation and bronchial secretion is the result of vagal stimulation (FOLDES 1951). Fall of blood pressure, especially if accompanied by whealing and reddening along the injected vein and/or bronchiolar constriction is indicative of histamine release (ALAM et al. 1939; BROB et al. 1947; LANDMESSER 1947; MACLAGAN 1976), or to ganglionic blockade of the sympathetic innervation of the arterioles induced by the NMBA (THOMAS 1957).

In evaluating the findings on the NM and other effects of NMBA in conscious subjects, it should be taken into account that they are influenced by psychologic factors such as motivation, anxiety, or discomfort. The changes in the grip strength are a manifestation of not only the effect of NMBA, but also that of the subjects voluntary efforts. Similarly, changes in heart rate in conscious subjects may be caused not only by the NMBA, but also by psychologic factors.

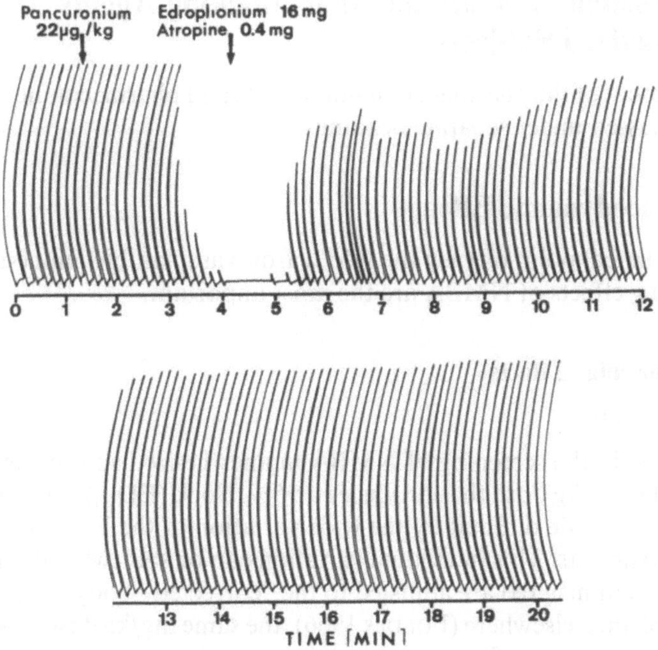

Fig. 5. Ergographic tracing of grip strength, demonstrating the prompt reversal of pancuronium-induced neuromuscular block by edrophonium

By using electrical stimulation of the ulnar nerve at the wrist and recording the changes of the isometric twitch tension of the adductor pollicis muscle, measured by a force displacement transducer on a polygraph, the NM effects of the NMBA can be determined without interference by uncontrollable psychologic factors. Electrical stimulation, however, causes considerable discomfort that is not tolerated by most conscious subjects. The discomfort can be eliminated by blocking conduction of the ulnar nerve at the elbow.

The effect of NMBA on NM transmission can also be assessed by electromyography during voluntary activity (LIPPOLD 1952) and indirect electrical stimulation (CHURCHILL-DAVIDSON and RICHARDSON 1952). The amplitude of the integrated electromyogram, however, is only proportional to the isometric twitch tension during voluntary, but not during electrically induced contractions (LIPPOLD 1952).

The main advantage of the screening of NMBA in conscious subjects is the possibility of observing their effects in the absence of any other drug. Another advantage is that, despite the modifying effect of psychologic factors, the results obtained in conscious subjects furnish a more useful background for the testing of NMBA in anesthetized subjects than the findings of animal experiments. It should be remembered, however, that from the point of view of their clinical use, more relevant information can be obtained on the pharmacodynamic effects of NMBA in anesthetized subjects.

C. Assessment of Neuromuscular Blocking Agents in Anesthetized Subjects

The assessment of the pharmacodynamic effects and pharmacokinetics of NMBA will be discussed under separate headings.

I. Pharmacodynamic Effects

From the clinical point of view the NM, cardiovascular, histamine-releasing and K^+-releasing effects of NMBA are the most important.

1. Neuromuscular Effects

a) In Volunteers

The first step in the screening of NMBA in anesthetized subjects is the determination of the mg/kg dose that produces a 95% block (ED_{95}). The determination of the ED_{95} is made difficult by the wide variation of the NM effects of NMBA in different subjects. This individual variation can be somewhat diminished if the doses to be administered are adjusted to the "corrected" body weight (BW_c). For reasons explained elsewhere (FOLDES 1966), the same mg/kg dose of a NMBA will have relatively greater effect in subjects who are heavier than in those who are lighter then 70 kg. From the actual body weight (BW) BW_c can be calculated by the empirically derived formula

$$BW_c = \frac{BW}{2} + 35 .$$

The dose of NMBA may also be determined on the basis of "lean" body weight (GUBNER 1961; WULFSOHN 1972) or body surface area (DAVIS and KARP 1951).

The ED_{95} is usually determined from the cumulative or individual dose–response regression line. The dose–response is determined by measuring, with a force displacement transducer, and continuously recording, the isometric twitch tension of the adductor pollicis muscle elicited by indirect stimulation of the ulnar nerve at the wrist with supramaximal impulses of 0.2 ms duration administered at 0.1 Hz.

The cumulative dose–response is determined as follows. Based on the information obtained on conscious subjects, a mg/kg dose is selected that is expected to cause about 15%–25% decrease of the isometric twitch tension of the adductor pollicis muscle. About one-third of this initial dose is injected at the time of development of the maximal effect of the first dose. The same dose is repeated, at the time of development of the maximal effect of every preceding dose, until 80%–95% NM block develops. From the data so obtained, the slope of the log dose–response regression line and the ED_{50} and ED_{95} of the NMBA can be calculated. In evaluating the dose–response data, doses having no effect or causing complete block should not be used for the calculation of the regression lines. Depending on the variation in the individual ED_{50} and ED_{95} values, cumulative dose–response regression lines should be determined on 6–10 subjects.

When determining the dose–response with the single-dose method, each subject only receives one dose of NMBA. Usually, four groups of subjects are used. Members of the first group receive a mg/kg dose of NMBA that is expected to cause about 15%–25% block. The mg/kg doses in the second, third, and fourth groups are about 33%, 66%, and 100% greater than those in the first group. From the mean effect of the doses in each group, the slope of the log dose–response regression lines and the ED_{50} and ED_{95} values can be calculated. Depending on the variation in the effect of the same dose in different subjects, each dose has to be administered to 5–10 subjects. Those subjects in whom a dose caused no or complete NM block are excluded from the calculations.

Determining ED_{95} by the cumulative or individual dose method both have their advantages and disadvantages. The main advantage of the cumulative dose method is that ED_{95} can be determined on relatively fewer (6–10) subjects than the number of subjects (20–40) required for the single-dose method. Theoretically, however, the cumulative method is less reliable. The reason for this is that, with the cumulative dose method, at the time of administration of the second dose, the plasma concentration of the NMBA is less than the highest level obtained with the first dose, primarily because of distribution to tissues and excretion and with some agents metabolism. The second dose will increase the plasma concentration, but only to a lower level than that which would have been obtained if the sum of the two doses had been administered together. The same considerations also apply to the plasma concentrations obtained after the administration of subsequent fractional doses. Because of this, the sum of fractional doses required to obtain the same plasma concentration will be greater than a single dose that produces the same plasma level.

In agreement with these considerations, the cumulative dose ED_{95} (55 μg/kg) of the relatively short-acting vecuronium was found to be higher (Fig. 6; NAGASHIMA et al. 1981) than the single-dose ED_{95} (44 μg/kg). With the relatively

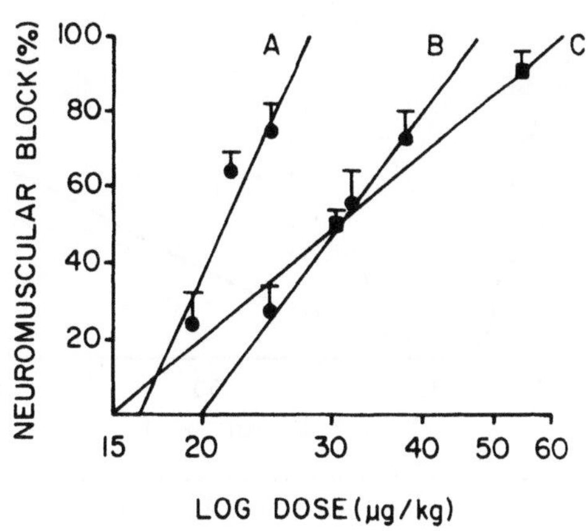

Fig. 6. The log dose–response regression lines of vecuronium: *A* individual dose–response during enflurane anesthesia; *B* individual dose–response under neuroleptanesthesia; *C* cumulative dose–response under neuroleptanesthesia. Note the increase of the neuromuscular potency of vecuronium under enflurane anesthesia and the difference in the slopes of the individual dose and cumulative dose regression lines

long-lasting tubocurarine (Tc) and pancuronium, however, ED_{50} and ED_{95} determined with the cumulative and single-dose method were about the same (DONLON 1980).

Many anesthetic agents influence the NM actions of NMBA (AUER and MELTZER 1914; FOLDES 1960; MILLER 1975). The potentiating effect of different anesthetic agents on NMBA is variable and with the same agent concentration dependent. It is therefore important to determine NM potency during steady state anesthesia and to indicate the type of anesthesia employed when reporting the potency data.

b) In Patients not Requiring Muscular Relaxation

After ED_{95} has been determined, this dose should be administered i.v. at constant speed (e.g., in 5 s) to 6–10 patients and its maximal NM effect; the time required from the start of injection to the development of the maximal effect (onset time); the time required from the start of injection to the return of the twitch tension to 25% of control (clinical duration); the time required for the increase of the twitch tension from 25% to 75% of control (recovery rate); and the time required from the start of injection to the return of the twitch to 95% of control (recovery time) should be determined. The T_4/T_1 ratio (Fig. 7; ALI et al. 1973, 1975) should be determined when the twitch tension has returned to 95% of control, and again after the i.v. injection of 0.01 mg/kg atropine sulfate in 15 s, followed immediately by the injection of 0.5 mg/kg edrophronium in 30 s. With this sequence of administration of the two compounds, changes in heart rate in either direction can be minimized (FOLDES et al. 1981).

When subjects employed for the determination of the dose–response regression line of NMBA are surgical patients, the operation should be delayed until the necessary data have been collected. The determination of the clinical duration, recovery rate, and recovery time after a single dose of NMBA can only be determined in surgical patients who do not require muscular relaxation during the

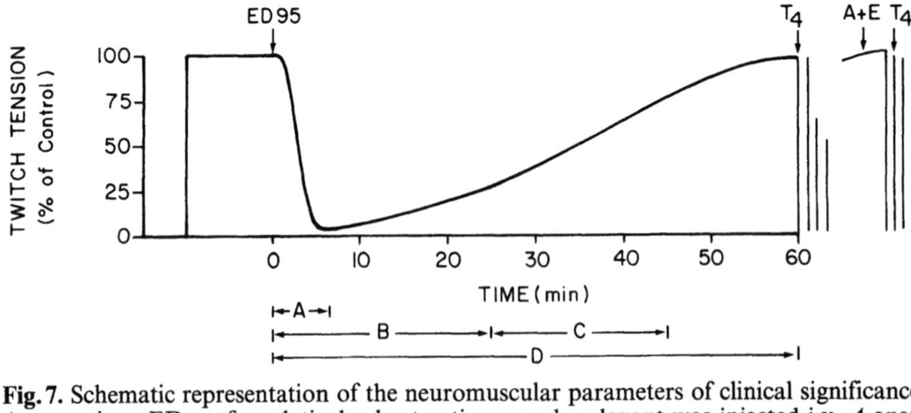

Fig. 7. Schematic representation of the neuromuscular parameters of clinical significance. At zero time, ED_{95} of a relatively short-acting muscle relaxant was injected i.v. *A* onset time; *B* clinical duration; *C* recovery rate; *D* recovery time. *A* + *E* indicates atropine + edrophonium and T_4 denotes the application of four supramaximal impulses at 2 Hz. For further explanation see text

operative procedure. These patients should be intubated only after the development of the maximal effect of the NMBA with the help of topical anesthesia applied to the larynx. During long-lasting operations, not requiring muscular relaxation, the time interval between the return of the twitch tension to 95% of control and the time required for the increase of the T_4/T_1 ratio close to unity, without the use of an antagonist, can also be determined.

c) In Patients Requiring Muscular Relaxation

One of the objectives of the search for new NMBA has been the synthesis of nondepolarizing compounds which would produce adequate conditions for endotracheal intubation as rapidly as succinylcholine chloride (Anectine, SCh). From the point of view of muscular relaxation, conditions for endotracheal intubation are ideal when not only the muscles of the jaw, tongue, pharynx, and larynx, but also the diaphragm is completely paralyzed. Under these circumstances, the cords will be immobile and there will be no "bucking" even when the depth of anesthesia is inadequate to suppress vagal reflexes, which initiate uncoordinated diaphragmatic movements. Because of the resistance of the diaphragm to nondepolarizing NMBA (FOLDES et al. 1961), this can only be achieved with large doses of NMBA. With large doses, even of the relatively short-acting NMBA, the clinical duration may be unduly prolonged. If, however, the level of general anesthesia is deep enough to obtund sensory reflexes originating from the larynx, satisfactory conditions for endotracheal intubation can be achieved with smaller doses of NMBA. Bucking, when the cords are wide open and immobile, is an indication for deepening of the level of anesthesia, not for more NMBA. Those who prefer lighter planes of general anesthesia may obtund bucking by the use of topical anesthesia. It is of interest that muscular relaxation with the short-acting NMBA vecuronium (AGOSTON et al. 1980), atracurium (NGUYEN et al. 1982), and Duador (RGH-4201) (FOLDES et al. 1983) was found to be adequate when the twitch tension of the adductor pollicis muscle was only depressed to 30%–40% of control. The situation is the same with SCh (F. F. FOLDES 1981, unpublished work).

When comparing the conditions produced by different NMBA for endotracheal intubation, it is important to attempt to use the same depth of general anesthesia. Conditions for endotracheal intubation should either be checked at a fixed time, e.g., 2 min after the start of injection of the NMBA, or at a time when the twitch tension of the adductor pollicis is decreased to 20%–30% of control. Conditions for endotracheal intubation can be scored as: 3 = no movement of cords, no bucking; 2 = no movement of cords, slight bucking; 1 = cords moving but can be easily separated, bucking; and 0 = cords adducted.

In general, twice ED_{95} is used for the facilitation of endotracheal intubation. Further increase of the dose of NMBA does not make earlier intubation possible, but increases the clinical duration of short-acting NMBA (e.g., vecuronium) (AGOSTON et al. 1980).

In patients requiring prolonged muscular relaxation, it is also possible to determine the size of fractional doses required for the maintenance of relaxation (Fig. 8). For good surgical relaxation, the twitch tension of the adductor pollicis muscle should be 25% of control or less. The fractional dose, administered when-

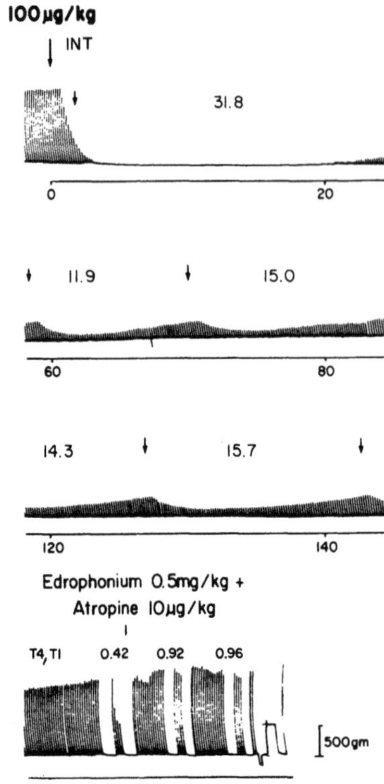

Fig. 8. Experimental design of the clinical investigation of the effect of muscle relaxants in surgical patients. Continuous recording of the indirectly elicited isometric twitch tension of the adductor pollicis muscle. At zero time, 100 µg/kg, and when ever the twitch tension recovered to 25% of control, 15 µg/kg vecuronium was injected i.v. during neuroleptanesthesia. After the last fractional dose, twitch tension was allowed to recover spontaneously to 95% of control. The T_4/T_1 ratio was measured at this time and again 2 and 5 min after the i.v. injection of atropine and edrophonium. Note that the clinical duration of action of subsequent fractional doses was similar

ever the twitch tension returns to 25% of control, should decrease twitch tension to less than 10% of control. Similar duration of action of a series of fractional doses indicates the absence of cumulation.

d) Spontaneous Recovery and Reversibility by Anticholinesterases

After the administration of the last fractional dose, the recovery parameters, i.e., recovery rate and recovery time can be observed (see Fig. 7). The ease of reversibility of the NM block by anticholinesterases should also be determined. This can be achieved by administering 0.5 mg/kg edrophonium, 0.02 mg/kg neostigmine methylsulfate (Prostigmin), or 0.1 mg/kg pyridostigmine bromide (Mestinon) at various stages of spontaneous recovery. These doses of anticholinesterases, if administered when the twitch tension has recovered to at least 50% of control, should increase twitch tension to above 90% of control and the T_4/T_1 ratio to above 0.75 in 2 min after edrophonium and 5–10 min after neostigmine or pyridostigmine administration (Table 1; FOLDES et al. 1981).

Table 1. Effect of antagonists administered at various stages of spontaneous recovery from vecuronium-induced neuromuscular block[a]

Antagonist	Dose (mg/kg)	Before antagonist		After antagonist			
				2 min		5 min	
		Twitch tension[b] (% of control)	T_4/T_1	Twitch tension (% of control)	T_4/T_1	Twitch tension (% of control)	T_4/T_1
Edrophonium	0.5	26.8±4.4(5)[c]	0.03±0.003	89.4±5.0	0.58±0.06	103.9±7.8	0.67±0.06
		62.7±3.0(10)	0.33±0.05	90.9±4.8	0.83±0.02	92.2±4.3	0.87±0.02
		97.2±1.8(28)	0.55±0.03	110.1±2.1	0.93±0.01	111.9±2.2	0.95±0.01
Pyridostigmine	0.1	82.4±7.9(9)	0.36±0.04	100.5±6.4	0.63±0.04	108.5±6.2	0.75±0.04

[a] Note that, during early stages of spontaneous recovery, reversal of the NM effect of anticholinesterases is incomplete, even with the relatively short-acting vecuronium
[b] Isometric twitch tension of indirectly stimulated adductor pollicis muscle; supramaximal stimuli of 0.2 ms duration at 0.1 Hz
[c] Means ± standard error of number of observations indicated in parentheses

e) Influence of Various Factors on Neuromuscular Activity

Systematic investigation of some of the factors that may influence NM activity or side effect liability of NMBA in human volunteers or patients is often difficult. For example, subjecting volunteers to extremes of respiratory acidosis or alkalosis, wide variations of body temperature, or large doses of antibiotics can not be justified. The influence of these factors must be observed on patients in whom these conditions occur spontaneously, inadvertently, or are produced deliberately for medical reasons. Gathering the necessary data for the evaluation of the influence of these factors in a single institution may take a very long time. Pooling the data collected in several clinics would facilitate earlier evaluation of the influence of such factors.

In contrast, the influence of other factors, such as the preliminary administration of SCh on the effect of nondepolarizing NMBA, the inhalation anesthetic used, or the effect of conventional doses of antibiotics administered orally, parenterally, or via peritoneal lavage can be easily studied during anesthetic management of surgical patients.

α) Effect of Succinylcholine on Nondepolarizing Muscle Relaxants. Preliminary administration of SCh usually increases the intensity of the NM effects of nondepolarizing NMBA (FOLDES et al. 1957). This effect can be investigated in patients intubated after administration of 1 mg/kg SCh; 2 min after the effects of SCh have dissipated and the twitch tension becomes stable (not infrequently the post-SCh twitch tension is higher than control), the previously determined ED_{70} is injected i.v. Depending on the effect of this dose on the twitch tension, the dose is adjusted to obtain a greater than 90% NM block in most patients. The onset time, maximal effect, and clinical duration of the first dose, the effect of repeat doses, and recovery parameters can then be investigated as described. A more time-consuming approach to the investigation of the effect of SCh on nondepolarizing NMBA is the determination of their dose–response after a single dose of SCh.

β) Respiratory Acidosis and Alkalosis. The influence of respiratory acidosis and alkalosis on the potency of different NMBA is variable. In the cat, elevation of the arterial PCO_2 increases the potency of Tc, but decreases that of metocurine, gallamine, and SCh (PAYNE 1958). In human subjects, Tc (BARAKA 1964) and pancuronium (NORMAN et al. 1970) are potentiated by respiratory acidosis and antagonized by respiratory alkalosis. In humans, for obvious reasons, the possibilities for the investigation of the effects of respiratory acidosis and alkalosis on the potency of NMBA are limited. The effects of changes in PCO_2 on the potency of NMBA may be observed on healthy young volunteers. At first, 40%–60% steady state NM block is produced by the continuous i.v. infusion of the NMBA tested during controlled ventilation, adjusted to keep the arterial PCO_2 at 38–42 Torr. Subsequently, the PCO_2 is increased to 78–82 Torr by excluding the CO_2 absorber or by adding CO_2 to the inhaled gas mixture. The elevated PCO_2 is maintained for as long as necessary for the NM block to become stabilized. After this, the PCO_2 is gradually decreased and stabilized at 38–42 Torr for 10 min. At this time, the PCO_2 is gradually decreased to 18–22 Torr by hyperventilation and kept at this level until the NM block becomes stable.

γ) Temperature. The influence of temperature on the effect of NMBA is controversial. In vitro, decreasing the temperature from 37° to 17 °C, despite a 60% decrease of presynaptic stimulated acetylcholine release, increases twitch tension of the indirectly stimulated muscle by about 100% (FOLDES et al. 1978). Since the effect of temperature on the twitch tension of the completely curarized, directly stimulated muscle is about the same, the site of this effect must be the muscle fiber (BOWMAN 1980; FOLDES et al. 1978). In vitro, lowering the temperature from 37° to 27 °C increased the NM effect of Tc, pancuronium, toxiferine, SCh, and decamethonium, but slightly antagonized that of alcuronium and gallamine. Further decrease of the temperature to 17 °C significantly increased the potency of all NMBA. In humans, reduction of temperature to 26 °C appeared to antagonize Tc, but increased the effect of decamethonium (CANNARD and ZAIMIS 1959). It is impossible to determine, however, whether changes in pharmacodynamics or pharmacokinetics are responsible for the influence of temperature on the NM effects of NMBA. While the influence of temperature on the potency of NMBA seems to require further study, there can be no doubt that the duration of action of Tc (HAM et al. 1978) and pancuronium (MILLER et al. 1978), probably because of lower excretion and metabolism, is prolonged. The influence of temperature on the potency of NMBA should be investigated in patients who require hypothermia during surgery. Steady state NM block should be established by the continuous i.v. infusion of the NMBA before cooling. Changes in the intensity of the NM blockade, caused by the NMBA infused at the unchanged rate can then be observed during both cooling and rewarming.

δ) Halogenated Inhalation Anesthetics. Inhalation anesthetics like ether (AUER and MELTZER 1914) and other anesthetic vapors and gases (FOLDES 1959), including the halogenated agents (MILLER 1971; FOLDES et al. 1980) increase significantly the NM effect of NMBA (MILLER et al. 1971; FOLDES et al. 1980). It is conceivable that the influence of the same inhalation anesthetic of different NMBA is variable. The influence of inhalation anesthetics on the NM effect of NMBA is concentration dependent both in vitro (CHAUDHRY et al. 1980) and in vivo (OHTA et al. 1980). Consequently, equipotent steady state concentrations of the various anesthetic agents should be employed for the determination of the dose–response of NMBA.

After establishment of steady state inhalation anesthesia, the dose–response should be determined with the single-dose method. From the ratios of the ED_{95} values determined under balanced and inhalation anesthesia, the approximate initial and repeat doses of the NMBA to be used for its clinical investigation under inhalation anesthesia can be selected. Once these doses have been selected, the maximal effect, time of development of the maximal effect and clinical duration of the initial dose, the variation if any, in the duration of repeat doses, recovery rate, and recovery time can be determined. The effect of antagonists administered at various stages of spontaneous recovery of NM function can also be assessed (FOLDES et al. 1981).

ξ) Antibiotics. Certain antibiotics (e.g., aminoglycosides, polymyxins, tetracyclines, lincomycin, clindamycin) in large doses can produce NM block (for references see BOWMAN 1980) in much smaller doses, they potentiate the effect of

NMBA (BURKETT et al. 1979). From the clinical point of view, it is important to determine whether or not patients on conventional doses of antibiotics before, during, or after surgery exhibit increased sensitivity to NMBA. This can be determined by comparing the effects of the NMBA on two groups of identically managed patients, only one of which receives antibiotics. Observation of the effect of peritoneal lavage with antibiotic solutions on the intensity and duration of the NM block should be part of this study.

2. Side Effects

From the clinical point of view, the circulatory side effects, histamine release, and K^+ release are the most important.

a) Cardiovascular Effects

The circulatory side effects of NMBA may be manifested by variations in heart rate and rhythm and blood pressure. Various NMBA may cause one or more of these circulatory side effects by blocking or stimulating nicotinic or muscarinic cholinergic receptors at various sites in the autonomic nervous system or by inhibiting the reuptake of noradrenaline (BOWMAN 1980; FOLDES 1981). Blood pressure may also be affected by histamine release (ALAM et al. 1939).

Since heart rate and rhythm and blood pressure are affected by peripheral stimuli such as endotracheal intubation, operative pain or traction on the eyeball, the circulatory effects of NMBA should be assessed before endotracheal intubation and the start of surgery.

To observe the circulatory effects, intra-arterial blood pressure, electrocardiogram, and heart rate should be monitored continuously. When after induction of anesthesia, these parameters have become stabilized, $2 \times ED_{95}$ of the NMBA should be injected i.v. The circulatory effects of NMBA are usually fully developed before the development of their maximal NM effect. If $2 \times ED_{95}$ causes no significant circulatory changes, from the clinical viewpoint, the NMBA may be considered free of circulatory side effects.

b) Histamine Release

The histamine-releasing effect of NMBA may be determined by bioassay (COLLIER and MACAULEY 1952) or fluorometry (SHORE et al. 1959). Reddening and whealing along the injected vein, flaring around the site of intracutaneous injection, generalized urticaria, edema of face and neck, epiglottic edema, bronchospasm, and fall in blood pressure are suggestive of histamine release. Most of these signs of histamine release after clinical doses of NMBA will only be encountered in patients with allergic diathesis. To avoid the frequently severe histamine reactions, before the use of new NMBA, intracutaneous skin wheal tests must be performed on such patients. It is also advisable to test plasma histamine levels before and at 1, 2, 5, 10, and 20 min after i.v. injection of $2 \times ED_{95}$ of the NMBA.

c) Potassium Release

Potassium release only occurs with depolarizing NMBA, especially if large doses are injected rapidly. The potassium-releasing effect should be determined by measuring the plasma K^+ concentration before and at 1, 2, 4, 6, 8, and 10 min after the i.v. injection of $2 \times ED_{95}$ of the NMBA.

II. Pharmacokinetics

1. Volumes of Distribution, Half-lives, Clearance

After i.v. administration, 40%–90% of various NMBA become bound to plasma proteins (FOLDES et al. 1982). The unbound (free) NMBA is at first distributed to those tissues with good blood supply which have capillaries that allow the passage of the highly charged NMBA from the plasma into the extracellular fluid and which have high affinity for NMBA. In the pharmacokinetic nomenclature, the plasma, together with those tissues which take up drugs (e.g., NMBA) rapidly, are referred to as the central compartment and the volume of this compartment V_1 is expressed in mg per kilogram body weight (VAN ROSSUM 1971; GREENBLATT and KOCH-WESER 1975). The rapid distribution of NMBA into the central compartment changes the initial equilibrium between bound and free NMBA and, as the plasma level of free NMBA decreases, more and more of the protein-bound NMBA is released and is available for distribution into the central compartment.

Because of distribution from plasma into other parts of the central compartment, the concentration of NMBA in plasma drops rapidly and the slope of the log concentration–time line is steep. This first, rapid phase of distribution is called the α phase and the time required for the concentration of the NMBA to decrease to one-half of its original value is called the α half-life ($t_{1/2\alpha}$). After the equilibration of the concentration of NMBA in the various parts of the central compartment, the plasma level of the NMBA will continue to fall at a much slower rate, partly because of distribution to a peripheral compartment, consisting of relatively poorly perfused tissues, and partly because of elimination by excretion through the kidney and liver and by metabolism. This second phase of slow distribution is called the β phase. During this phase, the slope of the log plasma concentration–time regression line is less. The time required for the concentration of NMBA to decrease to one-half of its concentration present at the beginning of this second, or elimination phase is called the β half-life ($t_{1/2\beta}$). The hypothetical volume of the compartment into which the NMBA is distributed during this second phase is designated as V_2. It should be realized, however, that in reality no such compartment exists and that V_2 is composed of the volume of poorly perfused tissues and a hypothetical volume of tissues. Distribution of NMBA into this hypothetical volume would cause the same decrease of the plasma concentration of the NMBA as that caused by metabolism and/or excretion.

In addition to V_1 and $t_{1/2\alpha}$ and V_2 and $t_{1/2\beta}$, the clearance of the NMBA is also of great importance from the point of view of expected duration of action of NMBA. Clearance (C) represents the volume of plasma per unit body weight from which NMBA is eliminated per unit time and is expressed as ml kg^{-1} min^{-1}.

For the determination of the various pharmacokinetic parameters of NMBA, heparinized venous blood samples are usually withdrawn at 1, 2, 4, 7, 10, 15, 20, 30, 45, and 60 min and every 30 min thereafter for 2–3 h more. The concentration of NMBA in the plasma can be determined by a suitable method such as radioimmunoassay (MATTEO et al. 1974), fluorometry (KERSTEN et al. 1973), high performance liquid chromatography, or mass spectrometry. The various pharmacokinetic parameters may be calculated from these data (VAN ROSSUM 1971; GREENBLATT and KOCH-WESER 1975).

The length of $t_{1/2\alpha}$ and $t_{1/2\beta}$ respectively supply information on the rate of onset and the duration of action of NMBA to be expected. Short or long $t_{1/2\alpha}$ are indicative of fast or slow onset of action, respectively. Short or long $t_{1/2\beta}$ are associated with brief or long duration of action, respectively. Since the degree of NM block depends on the plasma concentration of the NMBA (MATTEO et al. 1974; WAUD 1975; AGOSTON et al. 1977), knowledge of this parameter is important for the evaluation of the pharmacologic profile of NMBA.

2. Protein Binding and Interaction with Cholinesterases

The binding of NMBA to plasma proteins and their interaction with cholinesterases has significant influence on the pharmacodynamic and pharmacokinetic characteristics of these drugs. As already mentioned, only the free drug is available for immediate distribution to its sites of action. As long as the binding sites on plasma proteins are not fully saturated by NMBA, the ratios of the concentrations of bound and free NMBA will remain constant. After the binding sites have become saturated, even a relatively small increase in the total drug concentration may lead to a relatively large increase in the concentration of free drug immediately available for distribution. This will result in a more rapid onset and increased intensity of action and may lead to overdosage. With some of the NMBA, e.g., with metocurine, this can occur in the clinical range. Decrease of the concentration of plasma proteins as in liver disease, nephrosis, or malnutrition, or occupation of the binding sites by other substances may have similar effects (GREENBLATT and KOCH-WESER 1975).

NMBA, e.g., SCh or suxethonium, may be substrates of plasma butyrylcholinesterase (EC 3.1.1.8) or may inhibit this enzyme or red cell and muscle acetylcholinesterase (EC 3.1.1.7). There is a wide species variation in the ability of plasma butyrylcholinesterase to hydrolyze SCh and other ester-type NMBA (FOLDES and FOLDES 1965). This results in significant variation in the potency and duration of action of ester-type compounds between humans and most laboratory animals (HOPPE 1955). It is therefore essential to determine the hydrolysis rate of any new ester type NMBA in human plasma.

Although most NMBA inhibit to a variable extent both butyrylcholinesterase and acetylcholinesterase (FOLDES 1978; DEERY et al. 1982), from the clinical point of view of the available NMBA, only the inhibitory effect of pancuronium on plasma butyrylcholinesterase or that of benzoquinonium chloride (Mytolon) on muscle acetylcholinesterase, may have clinical significance. NMBA which inhibit butyrylcholinesterase intensify and prolong the action of hydrolyzable NMBA and those that inhibit acetylcholinesterase may interfere with the reversal of the NM blocking effect of these compounds (BOWMAN 1980). The methods used for the determination of the interaction of NMBA with cholinesterases and that of their binding to plasma proteins are described in Chap. 8.

3. Placental Transfer

After clinical doses administered to the mother, none of the presently used NMBA penetrate into the fetal circulation in high enough concentration to effect NM transmission in the newborn (POPPERS and FINSTER 1975). The placental transfer of NMBA can be tested by obtaining blood samples at the time of delivery from the umbilical vein by radioimmunoassay (MATTEO et al. 1974), fluorometry (KERSTEN et al. 1973), or some other sufficiently sensitive method.

4. Influence of Pathologic Conditions on Pharmacokinetics

In the presence of inherited abnormalities of plasma butyrylcholinesterase (FOLDES et al. 1963), the metabolic transformation of hydrolyzable NMBA is very slow. This will cause marked increase in the potency and excessive prolongation of action of these compounds. The duration of action of nondepolarizing relaxants, especially after administration of several doses, may be prolonged in patients with kidney disease (MILLER 1975). Many nondepolarizing NMBA are metabolized by the microsomal enzymes of the liver. This metabolism and the uptake of NMBA by the liver may be decreased by liver disease (AGOSTON 1976). Liver disease may also adversely affect the biliary excretion of NMBA. Since the primary route of excretion of NMBA and that of their metabolites is through the kidney, diseases of this organ will also prolong the $t_{1/2\beta}$ and prolong the duration of action of NMBA.

D. Summary and Conclusions

Because of the wide species variation in the NM effects, side effect liability, and disposition of NMBA, it is essential that their pharmacodynamics and pharmacokinetics should be investigated with classical pharmacologic techniques before attempting their introduction into clinical practice. Useful information can be obtained, with no danger and little discomfort, on many facets of the pharmacologic profile of new NMBA in conscious subjects. On the basis of such information, more detailed investigation of the NMBA can be planned on anesthetized subjects.

The information available on all new NMBA before introduction into clinical practice should include: type of block produced by the compound; ED_{95}; onset time (from start of injection to development of maximal effect); clinical duration (from start of injection to return of the twitch tension to 25% of control); size and duration of action of repeat doses that decrease twitch tension from 25% to less than 10% of control; variation in the effect of serially administered repeat doses; recovery rate (time for the increase of twitch tension from 25% to 75% of control after the last dose), recovery time (time between the injection of the last dose and recovery of the twitch to 95% of control); T_4/T_1 ratio at the time of maximal spontaneous recovery. The effects of the preliminary administration of an intubating dose of SCh on ED_{95} of new nondepolarizing NMBA, the influence of respiratory acidosis or alkalosis and temperature should also be investigated. Measurement of the relationship between plasma concentration and the intensity of NM blockade and calculation of some of the characteristic pharmacokinetic parameters, such as distribution half-life $t_{1/2\alpha}$, volume of central compartment V_1, elimination half-life $t_{1/2\beta}$, volume of peripheral compartment V_2, and plasma clearance (c) give useful information. Knowledge of the metabolism, pharmacologic effects of the metabolites, and routes of excretion of the unchanged compounds and their metabolites is especially important for the safe use of NMBA in patients in whom the metabolism and/or excretion of drugs is influenced by pathologic processes (e.g., liver or kidney disease). Finally, the cardiovascular side effect liability, histamine-releasing and K^+-releasing effect, and placental transfer of any new NMBA must also be assessed before it can be safely used under all circumstances by clinicians.

References

Agoston S (1976) Disposition and effects of some non-depolarizing neuromuscular blocking agents in animal and man. Doctoral thesis, University of Groningen, VRB Groningen

Agoston S, Crul JF, Kersten UW, Scaf AHJ (1977) Relationship of the serum concentration of pancuronium to its neuromuscular activity in man. Anesthesiology 47:509–512

Agoston S, Salt P, Newton D, Bencini A, Boomsma P, Erdmann W (1980) The neuromuscular blocking action of ORG-NC45, a new pancuronium derivative in anaesthetized patients. Br J Anaesth 52:53S–59S

Alam M, Anrep CV, Barsoum GS, Taloat M, Weininger E (1939) Liberation of histamine from the skeletal muscles by curare. J Physiol (Lond) 95:148–158

Ali HH, Utting JE, Gray TC (1973) Quantitative assessment of residual antidepolarizing block (Part 1). Br J Anaesth 43:473–485

Ali HH, Wilson RS, Savarese JJ, Kitz RJ (1975) The effect of tubocurarine on indirectly elicited train-of-four muscle response and respiratory measurements in humans. Br J Anaesth 47:570–574

Auer J, Meltzer SJ (1914) The effect of ether inhalation upon the skeletal motor mechanism. J Pharmacol Exp Ther 5:521–522

Baird WLM, Reid AM (1967) The neuromuscular properties of a new steroid compound, pancuronium bromide. Br J Anaesth 39:775–780

Baraka A (1964) The influence of carbon dioxide on the neuromuscular block caused by tubocurarine chloride in the human subject. Br J Anaesth 36:272–278

Bowman WC (1980) Pharmacology of neuromuscular function. University Park, Baltimore, pp 99–115

Buckett WR, Narjoribanks CEB, Marwick FA, Morton MB (1968) The pharmacology of pancuronium bromide (Org. NA97), a new potent steroidal neuromuscular blocking agent. Br J Pharmacol 32:671–682

Burkett L, Bikhazi GB, Thomas KC Jr, Rosenthal DA, Wirta MG, Foldes FF (1979) Mutual potentiation of the neuromuscular effects of antibiotics and relaxants. Anesth Analg 58:107–115

Cannard TH, Zaimis E (1959) The effect of lowered muscle temperature on the action of neuromuscular blocking drugs in man. J Physiol 149:112–119

Chaudhry I, Foldes FF (1981) In vitro effect of three new neuromuscular blocking agents. Anesthesiology 55:A220

Chaudhry I, Ohta Y, Nagashima H, Deery A, Foldes FF (1980) Inhalation anesthetic relaxant interactions in vitro. Anesthesiology 53:S266

Churchill-Davidson HC, Richardson AT (1952) Decamethonium iodide (C_{10}): Some observations on its action using electromyography. Proc R Soc Med 45:179–185

Collier HOJ, Macauley B (1952) The pharmacological properties of "Laudolissin" – A long-acting curarizing agent. Br J Pharmacol 7:398–408

Crul JF, Booij LHD (1980) First clinical experience with ORG-NC45. Br J Anaesth 52:49S–52S
Davis N, Karp M (1951) Dosage guide for administration of halogen salts of d-tubocurarine dimethyl ether. Anesth Analg 30:47–51
Deery A, Foldes FF, Benad G, McCloskey MA (1982) Interaction of neuromuscular blocking agents with human cholinesterase. Anesthesiology 56:A275.
Donlon JV Jr, Savarese JJ, Ali HH, Teplik RS (1980) Human dose-response curves for neuromuscular blocking drugs. Anesthesiology 53:161–166
Foldes FF (1951) The use of mytolon chloride in anesthesiology. Ann NY Acad Sci 54:503–511
Foldes FF (1954) The mode of action of quaternary ammonium type neuromuscular blocking agents. Br J Anaesth 26:394–398
Foldes FF (1959) Factors which alter the effects of muscle relaxants. Anesthesiology 20:464–504
Foldes FF (1960) The pharmacology of neuromuscular blocking agents in man. Clin Pharmacol Ther 1:345–395
Foldes FF (1966) The choice and mode of administration of relaxants. In: Foldes FF (ed) Muscle relaxants. Davis, Philadephia, pp 4–5
Foldes FF (1978) Enzymes of acetylcholine metabolism. In: Foldes FF (ed) Enzymes in anesthesiology. Springer, New York Heidelberg Berlin, p 124
Foldes FF (1981) Circulatory effects of neuromuscular blocking agents. Proceedings of the 10th International Anaesthesia Postgraduate Course, June 29 to July 3, Vienna, Austria. Egermann, Vienna, pp 71–78
Foldes FF, Foldes VM (1965) ω-amino fatty acid esters of choline. Interaction with cholinesterases and neuromuscular activity in man. J Pharmacol Exp Ther 150:220–230
Foldes FF, Wnuck AL, Hodges RJ, Thesleff S, deBeer EJ (1957) The mode of action of depolarizing relaxants. Anesth Analg 36:23–37
Foldes FF, Monte AP, Brunn HM Jr, Wolfson B (1961) Studies with muscle relaxants in unanesthetized subjects. Anesthesiology 22:230–236
Foldes FF, Foldes VM, Smith JC, Zsigmond EK (1963) The relation between plasma cholinesterase and prolonged apnea caused by succinylcholine. Anesthesiology 24:208–216
Foldes FF, Kuze S, Vizi ES, Deery A (1978) The influence of temperature on neuromuscular performance. J Neural Transm 43:27–45
Foldes FF, Bencini A, Newton D (1980) Influence of halothane and enflurane on the neuromuscular effects of ORG-NC45 in man. Br J Anaesth 52:64S–65S
Foldes FF, Yun H, Radnay PA, Badola RP, Kaplan R, Nagashima H (1981) Antagonism of the NM effect of ORG-NC45 by edrophonium. Anesthesiology 55:A201
Foldes FF, Deery A, Benad G, McCloskey MA, Strauch R (1982) The binding of neuromuscular blocking agents to plasma proteins. Anesthesiology 56:A274
Foldes FF, Nagashima H, Boros M, Tassonyi E, Fitzal S, Agoston S (1983) Muscular relaxation with atracurium, vecuronium and Duador under balanced anaesthesia. Br J Anaesth 55:97S–103S
Greenblatt DJ, Koch-Weser J (1975) Clinical pharmacokinetics. N Engl J Med 293:702–705
Grob D, Lilienthal JL, Harvey AM (1947) On certain vascular effects of curare in man: the "histamine" reaction. Bull Johns Hopkins Hosp 80:299–322
Gubner RS (1961) Simple arthropometric indices of body fatness and heart size. Clin Res 9:15
Ham J, Miller RD, Benet LZ, Matteo RS, Roderick LL (1978) The effect of temperature on the pharmacokinetics and pharmacodynamics of d-tubocurarine. Anesthesiology 49:324–328
Hoppe JO (1955) Observations on the potency of neuromuscular blocking agents with particular reference to succinylcholine. Anesthesiology 16:91–124
Kersten UW, Meijer DKF, Agoston S (1973) Fluorimetric and chromatographic determination of pancuronium bromide and its metabolites in biological materials. Clin Chim Acta 44:59–66

Landmesser CN (1947) A study of the bronchoconstrictor and hypotensive actions of curarizing drugs. Anesthesiology 8:506–523
Lippold OCJ (1952) Relation between integrated action potentials in human muscle and its isometric tension. J Physiol 117:492–499
Maclagen J (1976) Competitive neuromuscular blocking drugs. In: Zaimis E (ed) Neuromuscular junction. Handbook of experimental pharmacology, vol 42. Springer, Berlin Heidelberg New York, pp 421–486
Marshall IG, Agoston S, Booij LHDJ, Durant NN, Foldes FF (1980) Pharmacology of ORG-NC45 compared with other non-depolarizing neuromuscular blocking drugs. Br J Anaesth 52:11S–19S
Matteo RS, Spector S, Horowitz P (1974) Relation of serum d-tubocurarine concentration to neuromuscular blockade in man. Anesthesiology 41:440–443
Miller RD (1975) Factors affecting the action of muscle relaxants. In: Katz RL (ed) Muscle relaxants, Monographs in anaesthesiology, no 3. Excerpta Medica, Amsterdam, pp 163–192
Miller RD, Eger EI II, Way WL (1971) Comparative neuromuscular effects of forane and halothane alone and in combination with d-tubocurarine in man. Anesthesiology 35:38–42
Miller RD, Agoston S, van der Pol F, Booij LHDJ, Crul JF, Ham J (1978) Hypothermia and the pharmacokinetics and pharmacodynamics of pancuronium in the cat. J Pharmacol Exp Ther 207:532–538
Nagashima H, Yun H, Radnay PA, Duncalf D, Kaplan R, Foldes FF (1981) Influence of anesthesia on human dose-response of ORG-NC45. Anesthesiology 55:A202
Nguyen HD, Nagashima H, Kaplan R, Lauber R, Yun H, Foldes FF (1982) Relaxation with BW33A under neurolept and enflurane anesthesia. Anesthesiology 57:A277
Norman J, Katz RL, Seed RF (1970) The neuromuscular blocking action of pancuronium in man during anesthesia. Br J Anaesth 42:702–710
Ohta Y, Nagashima H, Lofrumento R, Foldes FF (1980) Halothane-isoflurane relaxant interactions in vivo. Anesthesiology 53:S265
Osserman KE (1958) Myasthemia Gravis. Grune and Stratton, New York, pp 100–101
Paton WDM, Zaimis EJ (1949) The pharmacological actions of polymethylene bistrimethylammonium salts. Br J Pharmacol 4:381–400
Payne JP (1958) The influence of carbon dioxide on the neuromuscular blocking activity of relaxant drugs in the cat. Br J Anaesth 30:206–216
Pelikan EW, Tether JE, Unna KR (1953) Sensitivity of myasthenia gravis patients to d-tubocurarine and decamethonium. Neurology 3:284–296
Poppers PJ, Finster M (1975) The use of muscle relaxants in obstetrics. In: Katz RL (ed) Muscle relaxants, Monographs in anesthesiology no 3. Excerptia Medica, Amsterdam, pp 205–208
Riker WF, Wescoe WC (1951) The pharmacology of flaxedil with observations on certain analogs. Ann NY Acad Sci 54:373–392
Shore PA, Burkhalter A, Cohn VH (1959) A method for the fluorometric assay of histamine in tissues. J Pharmacol Exp Ther 127:182–186
Thomas ET (1957) The effect of tubocurarine chloride on the blood pressure of anaesthetized patients. Lancet 2:772–773
Unna KR, Pelikan EW (1951) Evaluation of curarizing drugs in man. VI. Critique of experiments on unanesthetized subjects. Ann NY Acad Sci 54:480–492
Van Rossum JN (1971) Significance of pharmacokinetics for drug design. In: Ariens EJ (ed) Drug design, vol 1. Academic, New York, pp 495–503
Waud BE (1975) Serum d-tubocurarine concentration and twitch height. Anesthesiology 43:381–382
Wulfsohn NL (1972) Ketamine dosage for induction based on lean body mass. Anesth Analg 51:299–305

CHAPTER 26

Neuromuscular Blocking Agents of Different Chemical Structure

A. A. BUNATIAN

A. Introduction

At present, a depolarizing neuromuscular blocking agent (NMBA), succinylcholine, mainly applied for intubation, and two nondepolarizing NMBA, tubocurarine and pancuronium, are widely used in anesthesiologic practice. Interest has been shown in the new steroid NMBA pipecuronium (Arduan) and vecuronium. However, intensive search for new NMBA of selective action upon acetylcholine receptors of skeletal muscles is still in progress. This is not only due to imperfections of the NMBA available. The high standard of modern anesthesiology requires the availability of a great number of different NMBA for anesthesiologists to choose NMBA not only for different patients, but at different stages of the operation in the same patient.

The problem of combined use of depolarizing and nondepolarizing NMBA, though unwanted has become pressing. A nondepolarizing NMBA of brief duration of action, diadonium, seems to solve it. Diadonium and other NMBA synthesized and introduced into clinical use in the USSR are described in this chapter.

B. Bisquaternary Adamantyl-Containing Ester

Diadonium

Diadonium is a nondepolarizing NMBA of short duration of action. It was pharmacologically studied and introduced into clinical use by KHARKEVICH 1970b). Diadonium was synthesized at the Pharmacological Institute, USSR Academy of Medical Science, Moscow.

1. Neuromuscular Blocking Action

A number of anesthesiologists, have investigated clinical and electromyographic (EMG) characteristics of the development, maintenance, and elimination of the neuromuscular blocking effect of diadonium (SVADZHAN and SHITOV 1975; MICHELSON et al. 1977; YEFREMOV 1979; OSTROVSKY and KISELEV 1980).

According to clinical and EMG research, diadonium belongs to the nondepolarizing agents. Muscle relaxation under diadonium possesses all the features of a nondepolarizing blockade (Fig. 1). Similar to tubocurarine, it causes a drop in the amplitude of the single potential of the muscle and progessive decrease at low

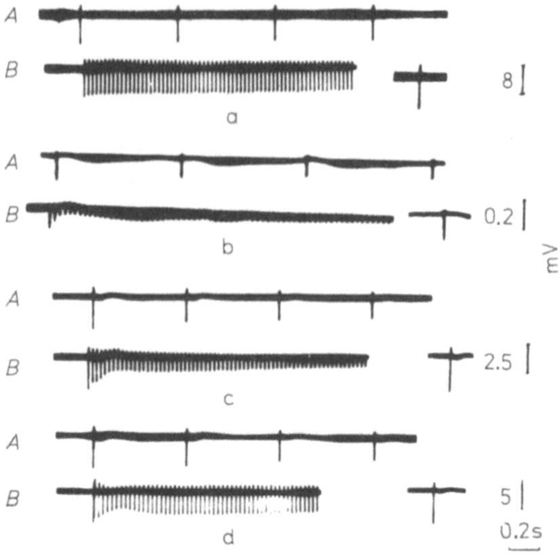

Fig. 1 a–d. Development and course of neuromuscular blockade after administration of diadonium (2 mg/kg) **a** initial EMG pattern, **b–d** EMG patterns 3, 12, 20 min after administration of diadonium. *A*, 1 Hz stimulation; *B*, 30 Hz stimulation

and high frequency electrical stimulation of the motor nerve (pessimal response); following tetanic stimulation, the amplitude of the single potential rises (posttetanic facilitation phenomenon); anticholinesterase is used to treat the paralysis caused by diadonium. In the case of normal muscular tone, i.v. administration of diadonium leads to relaxation of masticatory, thoracic, and abdominal muscles, muscles of the extremities, and diaphragm. Within 1 min, 3–5 mg/kg diadonium causes relaxation of skeletal muscles, hyporeflexia, and in half the cases, 20–30 s apnea. In 5–6 min, recovery of the tone of various muscular groups occurs in reverse order. According to the EMG data, 10–12 min after single administration of the drug, there is no sign of neuro-muscular blockade.

At doses exceeding 6 mg/kg, diadonium decreases and eliminates pharyngeal, laryngeal, and corneal reflexes in nearly all patients within 90–120 s. Besides, as a rule, 30–35 s after administration of 6–8 mg/kg diadonium and even some 20–30 s later (dose = 9–11 mg/kg), respiratory depression and apnea ensue. Toward the end of the second minute, intubation can be safely performed. According to YEFREMOV (1979), 10 mg/kg diadonium provides "smooth intubation" in practically all cases, but in 12.3%, there is a need for additional administration of the drug. Usually, to the end of the second minute EMG demonstrates maximum fall in the amplitude of the single potential (by 90%–95% of the initial level) under low and medium frequences of stimulation (2–10 Hz) and fails to register at high rates of stimulation (20–50 Hz). Distinct posttetanic facilitation is apparent.

Thus, the period of neuromuscular blockade is relatively short and resembles that under succinylcholine. Judging by the EMG data. 6–8 mg/kg diadonium achieve peak effect in 120–150 s and 9–11 mg/kg in 90–120 s, resulting in profound relaxation, areflexia, and moderate dilatation of the pupils (86.5% of cases). The clinical picture remains stable for some 6–16 min, depending on the initial dose of the drug and the agents used to maintain anesthesia (SVADZHAN and SHITOV 1975). The state of neuromuscular blockade permits one to perform various traumatic intracavitary interventions and a number of diagnostic procedures. Electromyographically, this period is characterized by a fall in the amplitude of the single potential, ranging from 90% to 60% of the initial level and poor response (10%–30%) to low and high frequency stimulation. The period of recovery of neuromuscular transmission is short and is characterized by a brief period of residual muscle relaxation. Clinically, profound muscle relaxation is preserved for some 2–6 min, permitting one to perform cavitary manipulations and artificial pulmonary ventilation.

At the same time, restoration of the corneal, pharyngeal, and laryngeal reflexes and of muscular tone is going on, mydriasis and anisocoria decrease and disappear. Spontaneous though inadequate breathing occurs simultaneously with the rise in the amplitude of the potential by 40%–60% of the initial level. Residual disturbances of neuromuscular transmission disappear some 5–6 min after complete restoration of spontaneous respiration. The amplitude of the potential and muscular response to stimulation become normal, there is no sign of posttetanic facilitation. Repeated administration of diadonium (half of the first dose) provides the same intensity of muscle relaxation, but of somewhat shorter period (10–12 min), owing to its poor cumulative effect. (BUNATIAN et al. 1979). Repeated administration of doses equal to the initial one results in prolongation of the maximum blockade.

The period of restoration of muscular tone and spontaneous breathing varies between 5 and 10 min after termination of the operation when the total dose of the drug does not exceed 3 g and duration of the procedure ranges from approximately 60 to 90 min. There is usually no sign of residual muscle relaxation or of inadequate artificial ventilation, and therefore no antagonists are used. Multiple administration of diadonium for a period of 3–4 h can result in prolonged blockade of neuromuscular transmission and a slowing down of the process of recovery of respiratory muscles. Thus, in the case of prolonged procedures it is preferable to use nondepolarizing NMBA of medium or prolonged action.

On the other hand, at the final stage of anesthesia, along with residual muscle relaxation caused by other nondepolarizing NMBA (tubocurarine, pancuronium, tercuronium), administration of 5 mg/kg diadonium provides controlled muscle relaxation of brief, profound action. Residual muscle relaxation is effectively treated with antagonists. At the same time, metabolic acidosis and hypokalemia weaken the effect of neostigmine (YEFREMOV 1979). After restoration of spontaneous respiration, no recurarization is observed (KHARKEVICH and SHITOV 1983; DARBINYAN and SAAKYAN 1979). Thus, it follows that the clinical investigation of the paralyzing properties of diadonium place the drug in the group of nondepolarizing NMBA. However, diadonium, unlike tubocurarine, causes more rapid develoment of neuromuscular blockade (within 90–120 s).

Period of optimal muscle relaxation is relatively short (10–12 min). Neuromuscular transmission is soon normalized. Similar to tubocurarine, ether and halothane potentiate the effect of diadonium, but clinically, 2–4 min prolongation of apnea does not signifficantly influence the most important properties of the drug.

Drugs used for neuroleptanalgesia and ataralgesia do not seem markedly to influence the process of muscle relaxation with diadonium. The common feature of all nondepolarizing NMBA is intensification and prolongation of the paralyzing effect under the influence of metabolic acidosis. Anticholinestrase agents are not very effective in the treatment of this complication. YEFREMOV 1979; MICHELSON et al. 1977).

2. Effect on Cardiovascular System

Investigation of the clinical pharmacology of diadonium and its influence on circulation in particular demonstrates its moderate hypotensive action of short duration. It is often observed in patients with vascular disease and also in patients with hypovolemia.

However, in essentially healthy individuals, diadonium causes a fall in the arterial pressure reaching its maximum level some 2–3 min later and remains at the same level for several minutes. Figure 2 illustrates the comparative study of the effect of diadonium, tercuronium, and tubocurarine on hemodynamics. Duration and intensity of hypotension are determined by the speed of administration, dose of diadonium, and accompanying hemodynamic effects of the drugs used for induction and maintenance of the anesthesia. Thus, halothane, owing to its depressive effect on circulation, in combination with diadonium can lead to marked hypotension. On the contrary, the sympathicomimetic effect of ether prevents hypotension. A fall in the arterial pressure can be accompanied by moderate tachycardia. In the case of additional administration of diadonium, hemodynamic effects are not so distinct. Clinical and experimental studies of the effect of diadonium on the central hemodynamics demonstrate that a moderate decrease in the arterial pressure (by 10–15 mmHg) is accompanied by a certain reduction of the maximum value of the first derivative of the left ventricular pressure (dp/dt_{max}) and also of the left ventricular function. Minute volume remains practically stable, though the stroke volume changes by 14%–18%.

Thus, diadonium has minimal effect on the central hemodynamic indices and myocardium contractility. The state of the latter was assessed by the dynamics of such indices as the period of ejection, diastole, or systole. The ganglion-blocking effect of diadonium resulting in the decrease of diastolic pressure in the aorta can be responsible for the changes mentioned. Mydriasis and stable histamine level in the blood after administration of diadonium confirm the suggestion. Cardiac rhythm disturbances which could be attributed to the effect of diadonium were not observed.

Fig. 2 a–m. Changes in some indices of central hemodynamics in patients with intravenously injected diadonium (10–12 mg/kg), tercuronium (0.15 mg/kg), and tubocurarine (0,25 mg/kg). *Dashed lines,* diadonium; *dashed–dotted lines,* tercuronium; *full lines,* tubocurarine; *asterirsks* $P<0.05$ **a** systolic AP, **b** diastolic AP, **c** mean AP, **d** heart rate, **e** stroke volume (% of initial), **f** cardiac output (% of initial), **g** maximum value of the first derivative of left ventricular pressure, **h** left ventricular work, **i** general peripheral resistance (% of initial); **k** duration of ejection period, **l** duration of diastole, **m** duration of systole. (YEFREMOV 1979)

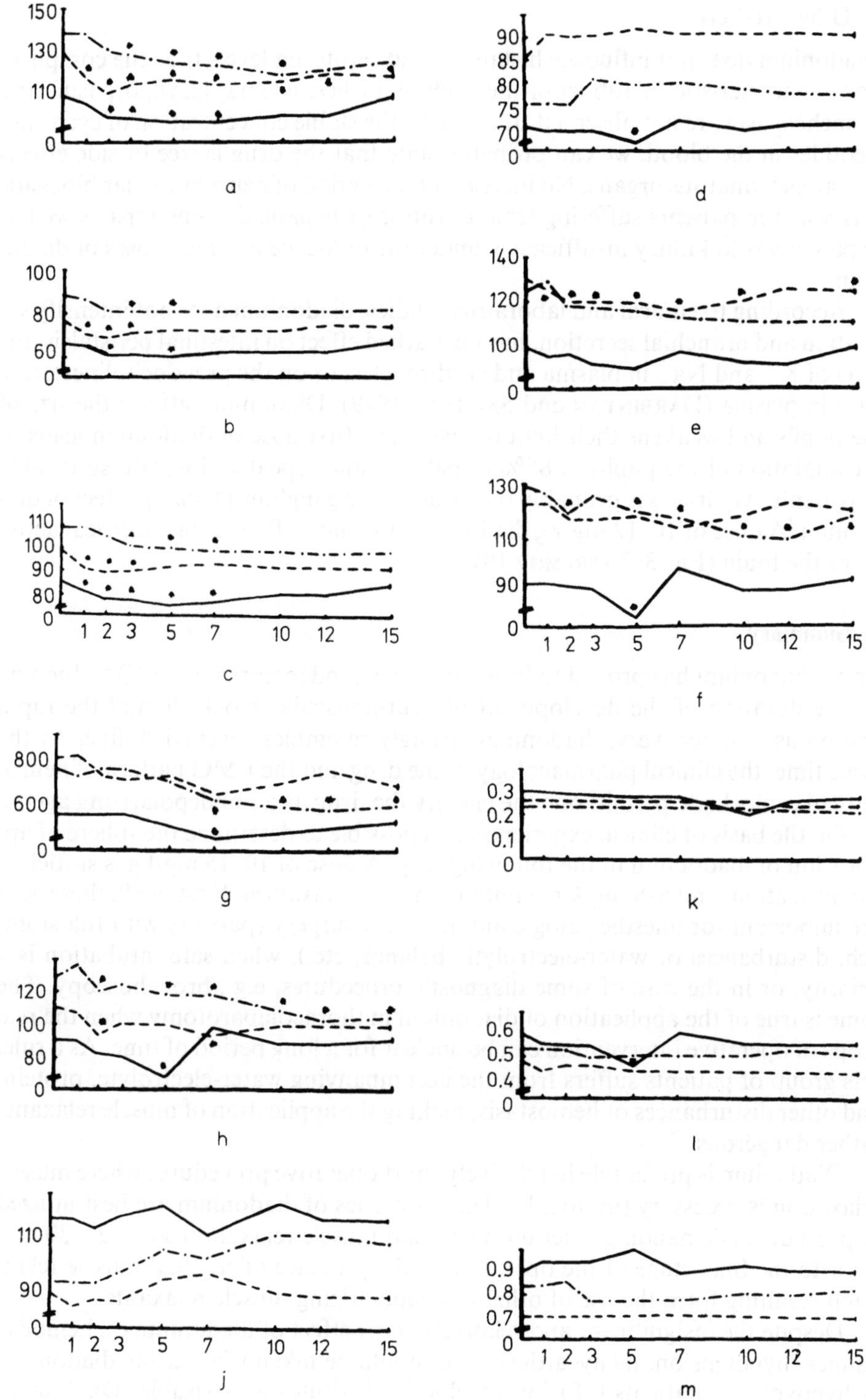
150
130
110
0
a
90
85
80
75
70
0
d
100
80
60
0
b
140
120
100
0
e
110
100
90
80
0
1 2 3 5 7 10 12 15
c
130
110
90
0
1 2 3 5 7 10 12 15
f
800
600
0
g
0.3
0.2
0.1
0
k
120
100
80
0
h
0.6
0.5
0.4
0
l
110
90
0
1 2 3 5 7 10 12 15
j
0.9
0.8
0.7
0
1 2 3 5 7 10 12 15
m

3. Other Actions

Diadonium does not influence histamine and serotonin levels thus the complications characteristic of tubocurarine such as rashes, edema, laryngospasm, and bronchospasm are not observed. Judging by the stable concentration of cyclic nucleotides in the blood, we can probably state that the drug is free of side effects on parenchymatous organs. No increase of the period of neuromuscular blockade was noted in patients suffering from cirrhosis or hepatic dysfunctions as well as in patients with kidney insufficiency under the influence of single doses of diadonium.

According to clinical and laboratory studies, diadonium does not intensify salivation and bronchial secretion, has no marked effect on intestinal peristalsis, the level of K^+ and Na^+ in plasma and erythrocytes, or on the pseudocholinesterase level in plasma (DARBINYAN and SAAKYAN 1979). Diadonium affects the size of the pupils and weakens their light reflexes. The first dose of diadonium leads to the dilatation of the pupils in 85% of patients and repeated doses do so in 94% of patients. Mydriasis is probably the result of the ganglion-blocking effect of diadonium. A dose of 10–12 mg/kg diadonium does not influence bioelectrical activity of the brain (Fig. 3; YEFREMOV 1979).

4. Summary

Thus, diadonium has proved to be an effective nondepolarizing NMBA. Judging by the duration of the development of neuromuscular blockade and the rapid neuromuscular recovery, diadonium strongly resembles succinylcholine. At the same time, the clinical pharmacology of the drug and the EMG pattern of neuromuscular blockade permit, one the classify the drug as a nondepolarizing agent.

On the basis of clinical experience, it is possible to determine the sphere of application of diadonium in the following way. A dose of 10–15 mg/kg is sufficient for intubation, and 6–8 mg/kg maintains muscle relaxation. First of all, these data are important for anesthesiologic aid in urgent surgery (patients with full stomach, disturbances of water-electrolytic balance, etc.), when safe intubation is a priority, or in the case of some diagnostic procedures, e.g., bronchoscopy. The same is true of the application of diadonium in thoracolaparotomy when the real scope of operative intervention can be unclear for a long period of time. As a rule, this group of patients suffers from the accompanying water-electrolyte, protein, and other disturbances of hemostasis, making the application of muscle relaxants rather dangerous.

Diadonium is preferable in relatively short operative procedures where muscle relaxation is necessary (up to 2 h). The properties of diadonium are best utilized in procedures demanding brief but profound muscle relaxation some 20–30 min prior to the final stage of the operation in the presence of residual muscle relaxation resulting form the use of other nondepolarizing muscle relaxants.

Despite the insignificant and relatively short effect of diadonium on hemodynamics (hypotension, tachycardia), care should be taken when using diadonium in hypovolemic patients.Infusion of blood substitutes is advisable. Diadonium does not release histamine, which makes it the drug of choice in allergic patients.

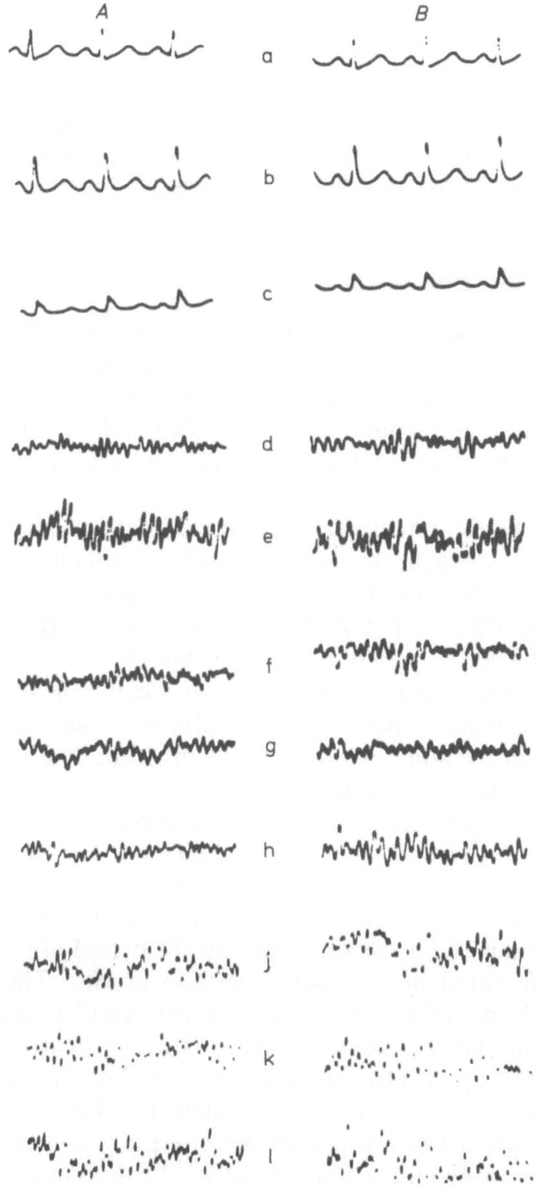

Fig. 3 a–l. ECG and EEG dynamics after intravenous administration of diadonium (10–12 mg/kg) **a–c** ECG in standard leads I, II, III, **d–l** EEG (**d** frontal, **e** temporal, **f** sincipital, **g** occipital, **h** frontotemporal left, **i** sincipitoccipital left, **k** frontotemporal right, **l** sincipitoccipital right). *A* initial indices; *B* 9 min after administration of diadonium (YEFREMOV 1979)

C. α-Truxillic Acid Derivatives

Pyrocurinum

Pyrocurinum belongs to the nondepolarizing NMBA of medium duration of action. It was studied pharmacologically and introduced into clinical practice by KHARKEVICH (1983). Pyrocurinum was synthesized at the Pharmacological Institute, USSR Academy of Medical Science, Moscow.

1. Neuromuscular Blocking Action

To achive total muscle relaxation, i.v. administration of at least 2 mg/kg pyrocurinum is required. Relaxation of skeletal muscles has the classical features of nondepolarizing blockade. Clinical signs of relaxation of the skeletal muscles are observed some 3–5 min after drug administration. Pyrocurinum may take up to 9.2 ± 3.1 min to reach its peak effect (most profound neuromuscular blockade).

Relaxation of the respiratory muscles, muscles of the anterior abdominal wall, and extremities occurs practically simultaneously. Based on the clinical signs, the duration of muscle relaxation with 2–3 mg/kg pyrocurinum varies between 40 and 70 min. Besides the fact that the duration of muscle relaxation is dose dependent, it is also influenced by the effect of the drugs chosen for general anesthesia (anesthetics, analgesics), depth of anesthesia, artificial lung ventilation regimen, and other factors responsible for peripheral and central mechanisms of apnea. Beginning 30 min after the administration of the drug, gradual recovery of normal neuromuscular transmission develops.

The following 15 min demonstrate a two to three fold rise in the amplitude of the single potential (by 40%–45% of the initial level). Breathing becomes spontaneous when the level of the single potential reaches $44\% \pm 8\%$ of the initial value. Judging by the EMG data, total neuromuscular blockade develops earlier with pyrocurinum than with tubocurarine, but the period of stable neuromuscular blockade is somewhat shorter than with tubocurarine. These facts are most easily revealed in the period of restoration of neuromuscular transmission. Rapid, "bursting" neuromuscular recovery is typical of pyrocurinum. Recovery of muscular tone occurs practically simultaneously in all groups of skeletal muscles. Restoration of spontaneous breathing is uneventful following a single dose of 2–3 mg/kg pyrocurinum if the duration of the procedure does not exceed 60 min. Anticholinesterase agents are seldom required. At the same time, insignificant respiratory depression is observed, but it can be easily compensated. In the case of repeated use of pyrocurinum (first half, then one-third of the first dose to provide 70–80 min relaxation), administration of anticholinesterase agents is necessary (neostigmine, galanthamine, etc.). Even under adequate respiration, with fractional use of pyrocurinium, in more than 50% of patients the EMG demonstrates marked depression of neuromuscular transmission in the immediate postanesthesia period. Thus, treatment of the residual myorelaxation is necessary to prevent any possible complication. Indices of respiratory function in operated patients showed decrease of a number of parameters of ventilation some 30–40 min after the final administration of the drug. Tone recovers simultaneously in all groups of muscles. Single administration of pyrocurinum seldom demands anticholinesterase therapy. Decurarization results in rapid recovery of spontaneous breathing. Hemodynamic data studied during the period of restoration of spontaneous ventilation without application of anticholinesterase agents are of great interest. Extubation is performed some 76±15 min after the final administration of the drug.

2. Effect on Cardiovascular System

According to WEISBERG and SHULUTKO (1983), pyrocurinum has no significant effect on circulation. In a number of patients only, during induction a moderate decrease in arterial pressure, probably owing to the effect of the drugs used for induction, and increased cardiac rhythm were observed. Cardiac rhythm kept within 10%–20% of the initial level in 46% of patients.

Following administration of 2–3 mg/kg pyrocurinum, systolic arterial pressure remained stable in 16% of patients, increased by 25% in 8% of patients, and decreased by 5%–20% in 76% of patients.

For more precise study of the effect of pyrocurinum on central hemodynamics, an analog–digital computerized system was used, permitting real-time evaluation of 22 cardiovascular parameters for each cardiac cycle (BUNATIAN et al. 1983). Patients who underwent aortocoronary bypass by heart–lung machine, cold cardioplegia, reconstructive operations on vessels of the lower extremities, and interventions for neoplasms in the lungs were observed. Induction was performed by means of intravenous administration of 1% solution of 2 mg/kg hexobarbital sodium and 0.04 mg/kg fentanyl along with inhalation of nitrous oxide plus oxygen in the ratio 1 : 1.

Anesthesia was maintained with nitrous oxide plus oxygen and 0.5–0.7 vol % halothane with fractional administration of fentanyl and droperidol. Moderate hyperventilation was performed. Pyrocurinum was injected at a dose of 2 mg/kg after intubation under artificial lung ventilation and against the background of stable hemodynamics. In order to reveal any possible effect of pyrocurinum on central hemodynamics, no other drug was injected, no manipulations were performed during the study. Arterial pressure (AP) was measured directly. EEG and central venous pressure were also studied. Analog signals of AP, central venous pressure, and EEG were recorded using a bioamplifier and, after operation, input into the fast store of the computer via amplifiers and digital–analog converters. The data were processed according to special programs (BUNATIAN et al. 1977). The results of the processing were printed out in the form of tables and diagrams, reflecting changes taking place during the study at intervals of 1 min. Stroke and minute volume, general peripheral resistance, and left ventricular work were calculated according to Warner's method (initial values taken as 100%). For estimation of the myocardial contractility, the method of automated measurement of systolic intervals was applied (BUNATIAN et al. 1981).

Hemodynamic data were processes in the fast store of the computer and parameters were collected into a file and output ion the form of a punched tape. The file was used for computerized statistical processing of the findings. Statistical sig-

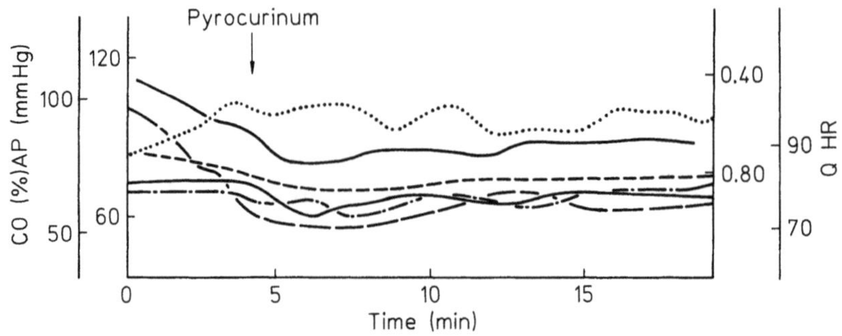

Fig. 4. Dynamics of the parameters of central hemodynamics after administration of 2 mg/kg pyrocurinum. *Full lines,* AP systolic and AP diastolic; *dashed line,* mean AP; *dashed–dotted line,* heart rate (HR); *dotted line,* coefficient of myocardial contractility (Q); *circles* cardiac output (CO) (% of initial)

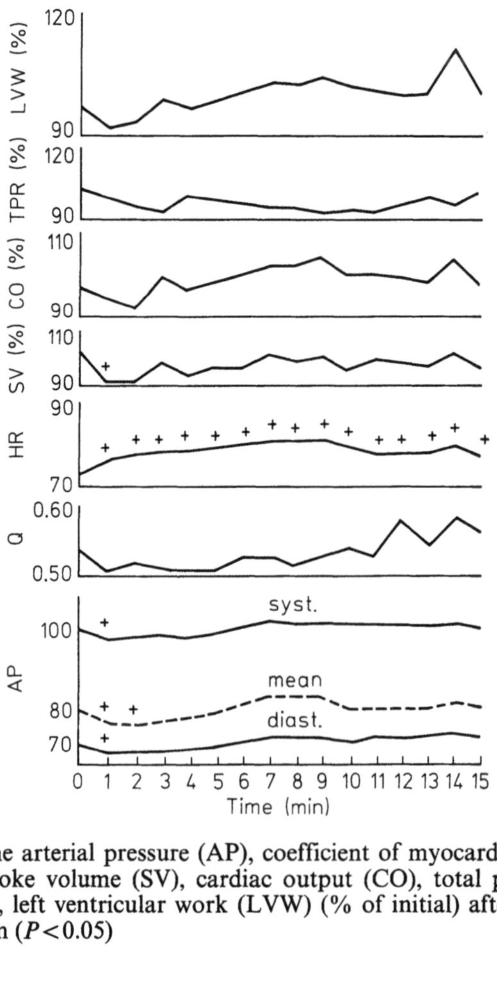

Fig. 5. Changes of the arterial pressure (AP), coefficient of myocardial contractility (Q), heart rate (HR), stroke volume (SV), cardiac output (CO), total peripheral resistance (TPR) (% of initial), left ventricular work (LVW) (% of initial) after administration of 2 mg/kg pyrocurinum ($P<0.05$)

nificance of fluctuations compared with the initial values of parameters was performed by means of a paired test.

The first minute following administration of pyrocurinum showed a statistically significant drop in AP by 5% of the initial level. Later, AP remained practically stable. Phase analysis of the cardiac cycle did not reveal statistically significant changes in preejection and ejection periods. Heart rate increased by 8% of the initial level and remained such during the period of the study (Fig. 4). Stroke volume also decreased during the first minute by 11% ($p < 0.05$). Later, the values of the stroke volume remained unchanged and close to the initial ones. At the same time, cardiac output remained stable during the whole period of investigation. The same was true of the general peripheral resistance and left ventricular work (Figs. 4, 5). Thus, only during the first minute of pyrocurinum injection was a significant decrease in the AP and stroke volume observed. A slight but more persistent increase in the heart rate occured. Relative hypotension registered prior to the injection of pyrocurinum can be related to the second phase of hemodynamic depression which occurred following the induction period. Decreased AP and stroke volume, and increased heart rate during the first minute of drug administration may be signs of reduced venous return to the heart resulting from venous dystonia. This dystonia is due to the effect of pyrocurinum. Decreased stroke volume does not lead to reduction of the cardiac output owing to a small compensating increase of the heart rate in response to the reduced venous return. It is of interest that administration of pyrocurinum to patients during aortocoronary bypass, i.e., in patients with labile hemodynamics, did not result in cardiodepression. This fact proves the minimal influence of the drug on central hemodynamics.

3. Other Actions

Pyrocurinum does not affect the size of the pupils. It causes no allergic response. The histamine concentration in blood remained stable under pyrocurinum (Weisberg and Shulutko 1983).

4. Summary

Thus, pyrocurinum has proved to be a nondepolarizing NMBA of medium duration of action meeting the requirements of modern clinical anesthesiology. In combined anesthesia intravenous administration of pyrocurinum at doses of 2–2.5 mg/kg provides adequate relaxation for approximately 70.68 ± 10.2 min. For more prolonged operations, administration of repeated doses of pyrocurinum does not require antagonists, as the muscle tone recovers simultaneously in all groups of muscles; breathing remains adequate after termination of the effect of the drug. Use of anticholinesterase agents permits one to perform extubation some 2–3 min after administration in patients with adequate spontaneous ventilation. Pyrocurinum at a dose of 2 mg/kg has minimal effects on central hemodynamics and myocardium contractility.

II. Anatruxonium

Anatruxonium is an active nondepolarizing NMBA. It was pharmacologically studied and introduced into clinical use by KHARKEVICH (1965). The drug was synthesized at the Pharmacological Institute, USSR Academy of Medical Science, Moscow.

1. Neuromuscular Blocking Action

Clinically used doses of anatruxonium vary between 0.07 and 0.5 mg/kg. Most authors note that 0.07–0.12 mg/kg anatruxonium is sufficient for profound relaxation of skeletal muscles and breathing. Total relaxation of skeletal muscles and apnea occur with 0.15–0.3 mg/kg anatruxonium. In children, the same effect is achieved with larger doses: 0.3–0.5 mg/kg (MANEVICH et al. 1970). Intravenous injection of 0.3% solution of 0.15–0.2 mg/kg anatruxonium leads to relaxation of facial muscles, muscles of the abdominal wall, extremities, and diaphragm. Relaxation occurs in 40–60 s. Anatruxonium takes up to 3–5 min to reach its peak effect.

After 10–15 min, the blocking effect weakens, resulting in spontaneous movements of the diaphragm. Adequate respiration is restored in 30–60 min. It must be stressed that the duration of neuromuscular blockade depends on a number of factors; dose of anatruxonium, type of basic anesthetic drug, depth of anesthesia, physical state of the patient, duration of the operative prodcedure, and its severity. These facts are responsible for the marked descrepancy in the duration of action of anatruxonium. Thus, according to KOTOMINA (1970) the duration of the minimum relaxing dose of anatruxonium (0.07 mg/kg) is 20–25 min. Doses of 0.1–0.12 mg/kg prolong the activity of the drug to 65 ± 4 min (in some patients to 90 ± 12 min). Results of investigations by KUZIN et al. (1970) are somewhat different. Doses of 0.08–0.1 mg/kg anatruxonium, in combination with superficial ether plus oxygen anesthesia (III_1), bring about profound muscle relaxation for about 97 min. The period of total muscle relaxation and apnea is approximately 120–130 min when the dose of the relaxant ranges from 0.18 to 0.3 mg/kg. In physically strong patients, adequate respiration can be restored 20–25 min after administration of 0.25 mg/kg anatruxonium and in weak patients the same dose can result in 60–90 min apnea with subsequent gradual restoration of adequate spontaneous ventilation. The duration of relaxation of the muscles of the abdominal wall under ether plus oxygen anesthesia is 1–1.5 h and under nitrous oxide plus oxygen it is 30–40 min (DOLYNA and PTUSHKINA 1970).

Owing to closure of the glottis and insufficient relaxation of the masticatory muscles, intubation of the trachea under anatruxonium is rather complicated (FIRSOV and STAHZADZE 1970). Repeat doses of anatruxonium are between one-quarter and one-half of the initial dose. The duration of apnea increases by a factor of 1.5–2, demonstrating cumulative properties of the drug. A characteristic feature of anatruxonium and of other derivatives of α-truxillic acid (cyclobutonium, truxilonium) is satisfactory relaxation of the muscles of the anterior abdominal wall under spontaneous though often inadequate breathing (KHARKEVICH 1974). This is usually observed when minimum doses of the NMBA are

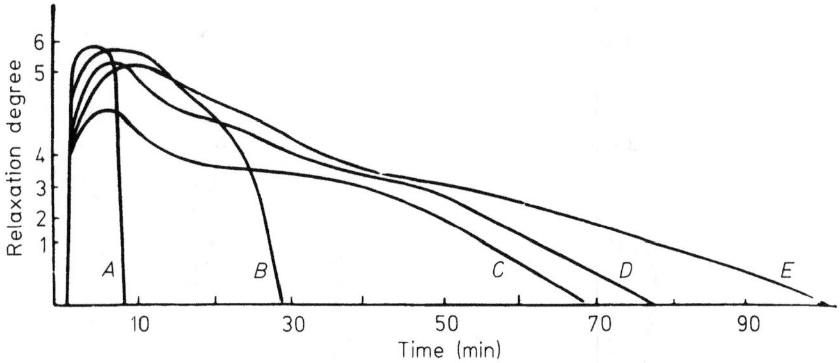

Fig. 6. General characteristics of relaxation with various neuromuscular blocking agents. Ordinate: consecutive involvement of muscles: 1 of the face; 2 extremities; 3 lower part of abdominal wall; 4 upper part of abdominal wall; 5 diaphragm; 6 apnea. Abscissa: time after drug administration (min). *A* succinylcholine; *B* tubocurarine; *C* cyclobutonium; *D* anatruxonium; *E* truxilonium. (MICHELSON 1970)

used (Fig. 6). Unlike tubocurarine, all of the three derivatives of α-truxillic acid cause prolonged relaxation of abdominal muscles with restoration of spontaneous respiration. The latter demands vigilance on the part of the anesthesiologist because spontaneous respiration is often inadequate and assisted ventilation is required (Fig. 7).

Spirogram analysis of 22 patients showed that 0.07–0.1 mg/kg anatruxonium led to reduction of inspiration volume by 50% and of minute lung ventilation by 44% with tachypnea (13%) within the first 15 min of drug action (KOTOMINA 1970). Restoration of adequate spontaneous ventilation is slow. Up to the 60 min, inspiration volume and minute lung ventilation constitute correspondingly 64.6% and 65.6% of the initial values, and are normalized in 120 min (FIRSOV and

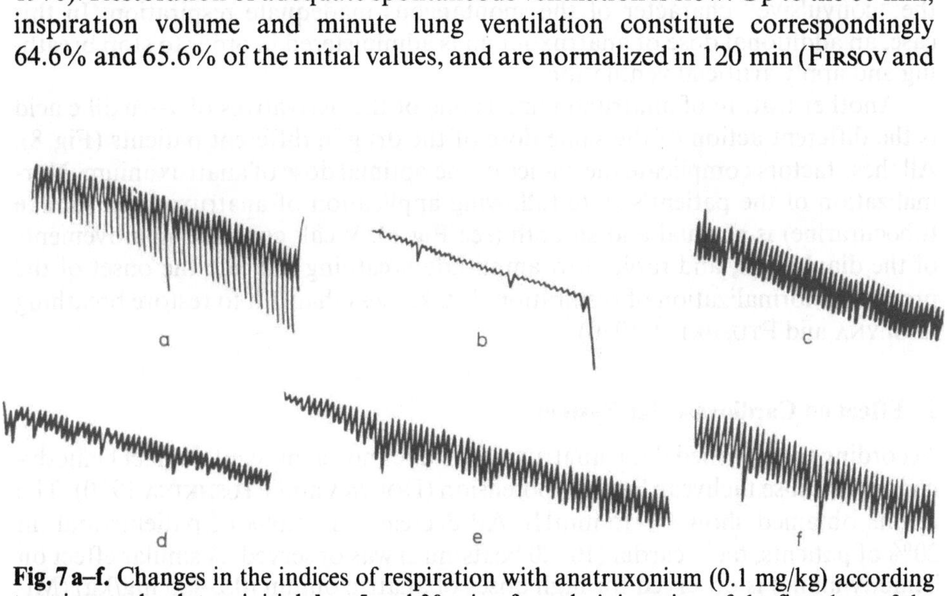

Fig. 7 a–f. Changes in the indices of respiration with anatruxonium (0.1 mg/kg) according to spirography data **a** initial, **b, c** 5 and 30 min after administration of the first dose of the drug, **d, e** 15–30 min after repeated administration of anatruxonium (0.05 mg/kg), **f** spirogram prior to extubation. (KOTOMINA 1970)

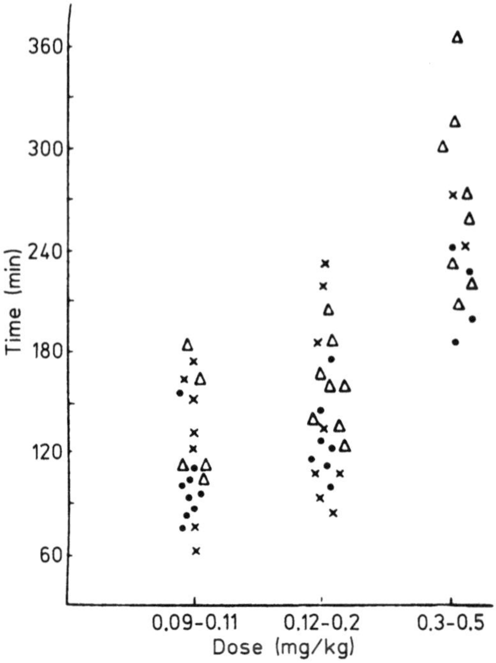

Fig. 8. Duration of the action of bisquaternary ammonium α-truxillic acid derivatives depending on the dose of truxilonium (*traingles*); cyclobutonium (*circles*); and anatruxonium (*crosses*). (MICHELSON 1970)

STAHZADZE 1970). Sometimes it is not easy to apply assisted ventilation owing to the "convulsive" character of the spontaneous inadequate respiration. In this case, an additional dose of anatruxonium is administered in order to stop breathing and apply artificial ventilation.

Another feature of anatruxonium as one of the derivatives of α-truxillic acid is the different action of the same dose of the drug in different patients (Fig. 8). All these factors complicate the choice of the optimal dose of anatruxonium. Normalization of the patient's state following application of anatruxonium (unlike tubocurarine) is gradual and smooth (see Fig. 6). Weak contracting movements of the diaphragm, and rapid, low amplitude breathing indicate the onset of the process of normalization of respiration. It takes less than 2 h to restore breathing (DOLYNA and PTUSHKINA 1970).

2. Effect on Cardiovascular System

According to published data, anatruxonium does not significantly affect hemodynamics or cause tachycardia or hypotension (DOLYNA and PTUSHKINA 1970). The results obtained show 10–15 mmHg AP decrease in 25.6% of patients and, in 20% of patients, tachycardia (10–20 beats/min) was observed. A similar effect on hemodynamics is observed for high doses of anatruxonium (0.2–0.5 mg/kg) (MICHELSON 1970). At the same time DOLYNA and PTUSHKINA (1970) point to more definite changes (10–30 beats/min increase in pulse rate and 30 mmHg AP decrease) resulting from injection of a moderate dose of the drug. Somewhat different data are presented by FIRSOV and STAHZADZE (1970). They point to a fall in the pulse rate in patients with initial tachycardia. Thus, it follows that the effect of anatruxonium on hemodynamics is insignificant. At the same time, taking into consideration the fact that the skin remains dry and warm with certain clinical signs of adequate circulation, one can assume that the hypotensive effect and ganglion-blocking properties of anatruxonium are closely connected (KOTOMINA and KHARKEVICH 1968).

3. Other Actions

The experience gained demonstrates that anatruxonium does not affect the EEG (KOTOMINA 1970; KUZIN et al. 1970; MANEVICH et al. 1970). Nor does it significantly influence the size and form of the pupils and eye reflexes (MANEVICH et al. 1970; DOLINA and PTUSHKINA 1970). At the same time, some authors note moderate dilatation of the pupils and weak reaction to light during operation and 3–4 h postoperatively. The onset of pupil dilatation is observed 5–7 min after administration of the drug. Some 10–15 min later, the pupils aquire submaximum size. Mydriasis is observed for 60–90 min with subsequent gradual contraction of the pupils and normalization of their form. The corneal reflex is poor; it recovers simultaneously with restoration of breathing. Secretion of tears is not disturbed. Intraocular pressure remains unchanged (MICHELSON 1970). Anatruxonium has practically no influence on electrolyte balance (MICHELSON 1970; YERIVANTZEV et al. 1970; NADHZARYAN and YERIVANTZEV 1970). It supresses secretion of salivary, tracheobronchial, and sweat glands, does not change peripheral circulation and biochemical indices of blood, nor does it cause oliguria (MANEVICH et al. 1970; KOTOMINA 1970, DOLINA and PTUSHKINA 1970; KUZIN et al. 1970). The vagolytic effect of anatruxonium permits one to avoid the use of atropine prior to administration of neostigmine in cases of high pulse rate (DOLINA and PTUSHKINA 1970).

4. Summary

Judging by the evidence quoted, anatruxonium belongs to the nondepolarizing NMBA. Preserved spontaneous ventilation, along with relaxation of the anterior abdominal wall is characteristic of the drug (0.07–0.1 mg/kg). The optimal dose for total muscle relaxation is 0.12–0.2 mg/kg. Duration of action is aproximately 60–80 min, depending on the type of basic anesthetic drug, depth of the anesthesia, physical state of the patient, etc. Anatruxonium causes moderate tachycardia and hypotension, dilatation and deformation of the pupils. Different action of the same dose of the drug in different patients can be regarded as a drawback of the drug. Anticholinesterase agents (neostigmine, galanthamine) are effective antidotes of anatruxonium.

III. Cyclobutonium

Cyclobutonium like anatruxonium, belongs to the nondepolarizing NMBA. It was synthesized at the Pharmacological Institute, USSR Academy of Medical

Science, Moscow, and was pharmacologically studied and introduced into clinical use by KHARKEVICH (1966).

1. Neuromuscular Blocking Action

Administration of 0.08–0.1 mg/kg cyclobutonium does not result in total myorelaxation. Superficial diaphragmatic breathing is observed; thus, artificial lung ventilation (ALV) is required. Increased doses (0.22–0.3 mg/kg) provide total muscle relaxation and apnea of average duration 94 min, and relaxation of the anterior abdominal wall for approximately 100–180 min. FIRSOV et al. (1970) state that doses exceeding 14 mg (i.e., higher than 0.2 mg/kg) lead to prolonged muscle relaxation (up to 180 min). Children are especially sensitive to cyclobutonium; this should be taken into consideration in practical application of the drug. Intravenous administration of the drug results in relaxation of masticatory muscles, abdominal wall, extremities, and diaphragm. Recovery of the muscles occurs in reverse order. Muscletone recovery occurs in 5–15 min, depending on the dose (FIRSOV et al. 1970); TER-KHASPAROVA 1970; ORLOV 1970). The period immediately following administration of cyclobutonium is unfavorable for intubation. Tension of masticatory muscles, glottis spasm, coughing, and movements of the extremities are observed. Intubation is possible for 10–15 min during the peak effect of the drug. Inhalational anesthetic drugs (except nitrous oxide) prolong the effect of cyclobutonium. Owing to the cumulative properties of cyclobutonium repeated doses should comprise between one-third and one-half of the first one. In exhausted, dehydrated patients, especially in those suffering from cancerous diseases, the duration of neuromuscular blockade is significantly increased. A case of 2 h apnea in an exhausted patient with carcinoma of the stomach was observed following administration of the minimum dose of cyclobutonium (0.08 mg/kg). (TER-KHASPAROVA 1970).

Restoration of adequate spontaneous ventilation occurs not earlier than 1–2 h after termination of the narcosis, that is why early extubation should be avoided (FIRSOV et al. 1970). Cyclobutonium provides profound muscle relaxation of the anterior abdominal wall and quick resumption of ventilation. Doses of 0.06–0.1 mg/kg cyclobutonium lead to an increase in respiration rate of 10–12 beats/min along with a respiration volume increase of 25%–50%. Spontaneous ventilation remains inadequate for 15–45 min; thus, assisted lung ventilation is required (BARSUKOV 1970). In repeated administration of the drug and in weak patients, ventilation remains inadequate for several hours. Therefore, extubation should be performed judging by the clinical signs of complete resumption of adequate spontaneous ventilation. Neostigmine is used to antagonize the effect of cyclobutonium. Anticholinesterase agents are more effective, providing neuromuscular transmission has partially recovered.

2. Effect on Cardiovascular System

Judging by the clinical investigations, cyclobutonium causes moderate hypotension. A decrease of 5–10 mmHg in systolic arterial pressure in normotensive and hypotensive patients is observed, and 20–30 mmHg in hypertensive patients. Di-

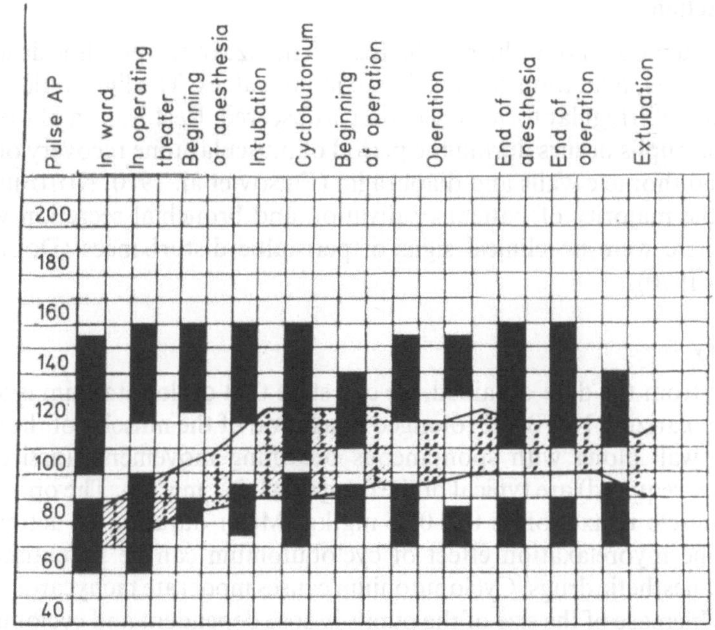

Fig. 9. Changes in the hemodynamic indices upon injection of cyclobutonium (average data). *Full columns* AP (upper row: fluctuations of systolic pressure; lower row: fluctuations of diastolic pressure mmHg). *Hatched columns* pulse fluctuation limits (beats/min). (BARSUKOV 1970)

astolic pressure remains stable or decreases by 10–16 mmHg, correspondingly (FIRSOV et al. 1970). The maximum decrease in pressure occurs some 15–25 min after drug administration with subsequent restoration up to initial values within 20–30 min (TER-KHASPAROVA 1970). Figure 9 illustrates the hemodynamic effect of 0.1–0.11 mg/kg cyclobutonium in 60 patients operated on for various gynecologic diseases.

These hemodynamic fluctuations can be attributed to cardiotropic antimuscarinic and poor ganglion-blocking effects of cyclobutonium. Tachycardia (average 20–30 beats/min) was less marked compared with that under the effect of anatruxonium (KUZIN et al. 1970). Prior to drug administration, marked tachycardia (up to 120–140 beats/min) was observed. Under the influence of cyclobutonium, tachycardia remained at the same level. In 30% of patients, heart rate did not change (FIRSOV et al. 1970). There were no special fluctuations in ECG. A favorable effect of cyclobutonium on the vascular wall, causing decrease of its tone and leading to restoration of amplitude of oscillations, manifests itself 10–20 min after drug administration. Oscillations are clearly outlined (in the area of the shoulder 15 mm; at the femoral level 20 mm), compact, and have irregular, rhombic shapes (FIRSOV et al. 1970).

3. Other Actions

Cyclobutonium does not influence the EEG. The size of the pupils is dose dependent. Sometimes, dilatation can reach maximum values. The light reflex is weak, the pupils are of irregular form, anisocoria is observed. Restoration of the regular shape of the pupils occurs during the period of muscular tone recovery of the abdominal and thoracic walls and diaphragm (FIRSOV et al. 1970; KOTOMINA et al. 1967). In the majority of patients, salivation and bronchial secretion were depressed. There were no clinical signs of peristaltic disturbances (DOLYNA and PTUSHKINA 1970).

4. Summary

Proceeding from the data obtained, we can state that cyclobutonium is an effective nondepolarizing NMBA. Prolonged relaxation of the muscles of the anterior abdominal wall along with spontaneous breathing movements (assisted lung ventilation is required) are typical of the drug (0.08–0.1 mg/kg). The optimal dose for total muscle relaxation is 0.2–0.25 mg/kg. Mean duration of action is 90–100 min. The myorelaxation effect of cyclobutonium can be potentiated by a number of anesthetic drugs. Cyclobutonium causes moderate tachycardia and hypotension. Increase of the size of the pupils is dose dependent and cyclobutonium affects their shape. Anisocoria is often seen. To antagonize the effect of cyclobutonium, anticholinesterase agents are used. Repeated administration of cyclobutonium leads to persistent apnea, demanding prolonged artificial lung ventilation.

IV. Truxilonium

Truxilonium, as well as other derivatives of α-truxillic acid (anatruxonium, cyclobutonium) is a nondepolarizing NMBA. After pharmacologic studies, it was introduced into clinical practice by KHARKEVICH and KRAVCHUK (1961). The drug was synthesized at the Pharmacological Institute, USSR Academy of Medical Science, Moscow.

1. Neuromuscular Blocking Action

The effect of truxilonium is dose dependent. According to MICHELSON (1970), 0.08–0.11 mg/kg truxilonium causes muscle relaxation in the following order: facial, chest, abdominal muscles, and diaphragm (Fig. 6). In the majority of patients, along with marked relaxation of the muscles of the abdominal wall, inadequate spontaneous ventilation occurs, demanding assisted lung ventilation. Tidal volume decreases by 40.3% of the initial value, frequency of breathing increases by 41%, minute ventilation volume decreases by 22.5%. Breathing is completely stopped for 30–40 min in 16.7% of patients (NADZHARYAN 1965). Doses of 0.15–0.3 mg/kg truxilonium provide, as a rule, stable relaxation of the muscles of the abdominal wall for 1.5–2 h. In this case, spontaneous breathing, mainly due to movements of diaphragm, is observed in 50% of patients. Some 1–1.5 min following administration of 0.4–0.6 mg/kg truxilonium 2–4 h apnea and total muscle relaxation ensue.

It was noticed that 0.2–0.3 mg/kg truxilonium led to apnea in 79% of patients (average duration 65 min) and 0.32–0.4 mg/kg truxilonium caused 100 min apnea in 91% of patients (Fig. 8). Only maximum doses of the drug provide optimal conditions for intubation. However, this is not reasonable because significant doses of truxilonium lead to more than 3 h apnea. Considerable differences between the dose leading to profound muscle relaxation of the abdominal wall and that resulting in apnea are a characteristic feature of truxilonium as well as of other derivatives of α-truxillic acid (KOTOMINA 1965; MICHELSON 1970; KUZIN and NADHZARYAN 1970). Inhalational anesthetic drugs prolong the myoplegic effect of truxilonium. That is why it is reasonable to reduce the dose of the relaxant by one-third in this case. Repeat doses of truxilonium comprise between one-third and one-half of the first dose, owing to the cumulative effect of the drug. To antagonize the effect of truxilonium, anticholinesterase agents are used. Some authors (NADHZARYAN 1965) avoid administration of atropine prior to injection of anticholinesterase agents, owing to the cardiotropic antimuscarinic effect of truxilonium. Following decurarization, patients should be kept under observation for not less than 40–60 min as some cases of recurarization 10–20 min after complete resumption of spontaneous ventilation have been described.

2. Effect on Cardiovascular System

According to clinical observations, 0.2–0.5 mg/kg truxilonium causes moderate tachycardia and hypotension. This can be related to the ganglion-blocking effect of the drug. The average pulse rate increase is 10–15 beats/min. Arterial pressure is elevated by 10–20 mmHg (KUSIN and NADHZARYAN 1970). Smaller doses correspond to less marked changes in hemodynamic indices. The latter gradually stabilize in 20–30 min. ECG studies revealed only tachycardia (NADHZARYAN 1965).

3. Other Actions

In 97.6% of patients, truxilonium caused dilatation of the pupils, depression of salivation and of tracheobronchial tree secretion resulting from the vagolytic effect of the drug. The former effect, however, can be also attributed to the ganglion-blocking action of truxilonium (KUZIN and NADHZARYAN 1970). EEG studies showed that truxilonium exerts no significant influence on the bioelectrical activity of the brain (NADHZARYAN 1965).

4. Summary

Thus, truxilonium has proved to be a nondepolarizing drug. There is a significant difference between the dose required for relaxation of the anterior abdominal wall and that causing apnea, which is typical of other derivatives of α-truxillic acid. This fact permits one to perform some operative procedures under spontaneous breathing with assisted lung ventilation. It is not reasonable to use truxilonium for intubation, as the doses required may contribute to persistent apnea (up to 3 h or more). Doses leading to relaxation of the anterior abdominal wall vary between 0.08 and 0.12 mg/kg; spontaneous breathing is not affected. Optimal doses under which artificial lung ventilation can be successfully conducted are 0.2–0.3 mg/kg. Inhalation anesthetic drugs prolong the neuromuscular blocking effect of truxilonium. Owing to the cumulative effect of the drug, repeat doses of truxilonium should not exceed one-third or one-half of the first dose. Application of truxilonium leads to moderate tachycardia and hypotension. Dilatation of the pupils is observed. Under the effect of truxilonium, salivation and bronchial secretion are usually depressed. Truxilonium does not influence ECG and EEG patterns.

D. Bisquaternary Derivative of Terphenyl

Tercuronium

Tercuronium is an active nondepolarizing muscle relaxant. It was pharmacologically studied and introduced into clinical use by Danilov et al. (1979). The drug was synthesized at the Institute of Experimental Medicine, USSR Academy of Medical Science, Leningrad.

1. Neuromuscular Blocking Action

Adequate relaxation is achieved with 0.1–0.25 mg/kg. Clinical myorelaxation with tercuronium occurs in 1–1.5 min, no preceding fasciculation is observed. First, relaxation involves facial muscles, muscles of the neck, and extremities, followed by abdominal and intercostal muscles; finally, relaxation of the diaphragm occurs. Recovery of muscular tone occurs in reverse order. By the third minute following intravenous administration of 0.15 mg/kg tercuronium, apnea is observed, total myorelaxation develops in 4–5 min (Yefremov 1979). The results of EMG study of tercuronium are shown in Fig. 10. The EMG demonstrates the low amplitude of potentials in response to the first stimulus and no response to repeated stimulation (pessimal inhibition). Following 50–100 Hz tetanic stimulation of the nerve, "posttetanic decurarization" is observed: 10- to 15-fold rise in the amplitude of the first response on single stimulus.

Corneal, pharyngeal, and laryngeal reflexes disappear in 3–4 min. By the 5–7 min, the pupils are moderately dilated and remain such during the whole period of action of the drug. The pupils narrow some 15–20 min prior to the onset of diaphragmatic breathing. More than 0.16 mg/kg tercuronium provides excellent conditions for intubation in 54.8% of patients; good in 25.8%; satisfactory in 19.4% of patients. The duration of muscle relaxation was determined on the basis of clinical signs (0.15 mg/kg gives 30–70 min, 0.2 mg/kg gives 50–140 min). The first clinical signs of insufficient muscle relaxation (irregular movements of the diaphragm) appeard during the recovery of neuromuscular transmission; the amplitude of the first response was elevated by 40%–50% of the initial level. The amplitude of the repeated response increased as high as 5%–10% of the initial level. Administration of neostigmine accelerates the restoration of neuromuscular transmission by a factor, of 3–4 (Osypova et al. 1977; Yefremov 1979). Spontaneous ventilation becomes adequate within 15–20 min (Vanevsky et al. 1978).

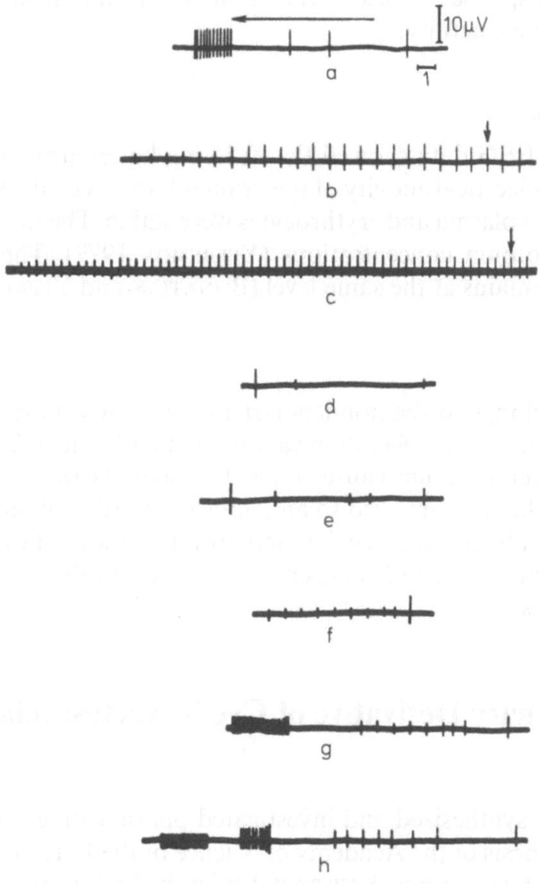

Fig. 10 a–h. EMG pattern with tercuronium (0.15 mg/kg) **a** initial EMG pattern, **b** upon injection of 100 mg succinylcholine, **c** 10 mg tercuronium (0.15 mg/kg), **d** after 20 min, **e** after 40 min, **f** after 60 min, **g** 5 min after administration of neostigmine, **h** in 8 min after administration of neostigmine. (VANEVSKY et al. 1978)

2. Effect on Cardiovascular System

According to the data reported, tercuronium does not produce any marked effect on the cardiovascular system (OSYPOVA et al. 1977; BUNATIAN and YEFREMOV 1979; SAAKJAN 1977). Aiming at more precise evaluation of the effect of tercuronium on hemodynamics, we performed some special investigations by means of continuous control of indices of central hemodynamics using a real-time computerized system in the operating room. A dose of 0.15 mg/kg i.v. tercuronium was used. Anesthesia was maintained with nitrous oxide plus oxygen under neuroleptanalgesia. Intubation was performed following administration of 10–12 mg/kg diadonium. The influence of the given NMBA of hemodynamics was followed up for 15 min. Figure 2 shows the results of the investigations. According to the data obtained, tercuronium causes a brief insignificant hypotensive effect which may be attributed to a vasoplegic (ganglion-blocking) effect of the drug or reduced venous return in response to muscle relaxation. Tercuronium has no negative effect on myocardial contractility.

3. Other Actions

Analysis of the EEG data showed that 0.15 mg/kg tercuronium has no marked effect on the bioelectrical activity of the brain (OSYPOVA et al. 1977). K^+ and Na^+ concentrations in plasma and erythrocytes were stable. The same is true of the histamine and serotonin concentrations (YEFREMOV 1979). The cyclic nucleotide concentration remains at the same level (BUNATIAN and YEFREMOV 1979).

4. Summary

Tercuronium belongs to the nondepolarizing NMBA with medium duration of action. Optimal doses for relaxation vary between 0.15 and 0.2 mg/kg. According to EMG data, tercuronium causes typical nondepolarizing blockade. Neostigmine and galanthamine are used to antagonize the effect of tercuronium. Except for a brief, insignificant hypotension, tercuronium does not cause changes in the cardiovascular system, EEG, concentration of electrolytes, acetylcholine, and cyclic nucleotides.

E. Bisquaternary Derivative of Cyclic Acetosuccinylaldehyde

Dioxonium

Dioxonium was synthesized and investigated pharmacologically at the Institute of Organic Synthesis of the Academy of Science of the Latvian SSR, Riga (SOKOLOV et al. 1968; KLUSHA and KHIMENIS 1967a, b; 1968a, b). According to pharmacologic studies, dioxonium proved to be a highly active NMBA of mixed type of action. It was introduced into clinical practice by KLUSHA et al. (1970) and GHINTERS (1971).

1. Neuromuscular Blocking Action

Clinically, dioxonium reveals mainly its nondepolarizing character (VANEVSKY et al. 1970, 1972; DARBINYAN and WEISBERG 1970). Muscle relaxation, in particular, occurs without preceding fasciculation, though some findings demonstrate the presence of certain signs of depolarization. The EMG shows poor spontaneous muscle activity. However depolarization under dioxonium is less marked than under succinylcholine. Complete muscle relaxation occurs 2–3 min after administration of dioxonium. Profound neuromuscular blockade is preserved for 5–7 min. Tetanic and single stimulations do not evoke potentials. Gradual elevation of the amplitude of potentials occurs some times later. The phenomenon of posttetanic decurarization indicates the onset of the nondepolarizing phase of the action of dioxonium. Later, along with restoration of neuromuscular transmission and weakening of the blockade, posttetanic decurarization becomes more prominent and the EMG pattern preserves only the features of nondepolarizing block-

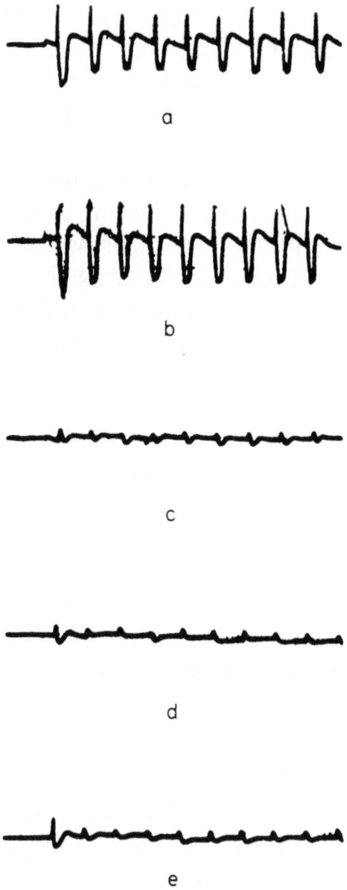

Fig. 11 a–e. Neuromuscular blockade with dioxonium (0.04 mg/kg) **a** initial EMG pattern, **b** 30 s after drug administration: fall in the amplitude of potential and onset of the development of depolarizing phase of blockade, **c** 3 min after dioxonium administration: depolarizing neuromuscular blockade (first phase), **d, e** EMG pattern 40–50 min after dioxonium administration: pessimum phenomenon, nondepolarizing blockade, second phase, 10 Hz stimulation

ade (Fig. 11). Clinical and EMG findings revealed purely antagonistic interaction of dioxonium and anticholinesterase agents. Minimal neuromuscular blocking doses of dioxonium are 0.04–0.06 mg/kg, while for total paralysis of skeletal muscles 0.07–0.09 mg/kg is required. The time of onset of the paralyzing effect fo dioxonium is dose dependent. Muscle relaxation occurs some 1.5–3 min after the administration of the drug. The duration of the effect varies between 20 and 40 min. Dioxonium provides satisfactory conditions for intubation. Analysis of the rate of consumption of dioxonium during anesthesia indicated that the dose of the drug depends on the depth of anesthesia. Comparative study of this index was performed in patients operated under similar or identical anesthesia. It turned out that, at the stage of analgesia, the average specific consumption of dioxonium was 0.06 $mg/kg^{-1}/h^{-1}$ under halothane plus ether anesthesia as well as under nitrous oxide plus oxygen in combination with neuroleptanalgesia. (KIMENIS et al. 1976). Dioxonium consumption changes markedly under the influence of various pathologic states such as anemia, intraoperative hemorrhage, and hypoproteinemia. The duration of clinically adequate muscle relaxation strongly resembles that under tubocurarine. Thus according to KIMENIS et al. (1976) the duration for equivalent doses of dioxonium and tubocurarine is 25.25 ± 1.70 and 28.25 ± 2.19 min ($P < 0.02$) respectively. During abdominal and thoracic operations, repeat doses of dioxonium were used, provided diaphragmatic contraction was observed. Usually, repeat doses of the drug should not exceed one-quarter or one third of the initial dose. Dioxonium causes optimal and stable relaxation of skeletal muscles. Anticholinesterase agents are used to remove neuromuscular blockade, providing inadequate restoration of muscular tone is observed when the patient wakes up.

Application of dioxonium may lead to a number of complications which are mainly characterized by certain disturbances in the process of recovery of skeletal muscle tone and in restoration of breathing (KIMENIS et al. 1976). In most cases, complications result from overdosage of the drug (more than 0.2 mg/kg). Thus, according to the data obtained, in 49 cases (2.86%) of 1683 cases of general anesthesia with dioxonium, prolonged depression of skeletal muscle tone and absence of adequate spontaneous respiration were observed mainly owing to overdosage of the drug. Decurarization is reuqired to prevent recurarization (KIMENIS et al. 1976).

2. Effect on Cardiovascular System

Clinical investigations have proved that dioxonium exerts no influence on the cardiovascular system. This fact permits one to use it in patients with various circulatory disorders and decreases the danger of possible cardiovascular complications during anesthesia because dioxonium does not potentiate negative hemodynamic effects of other components of anesthesia. (DARBINYAN et al. 1970; VANEVSKY and KHARGAPOLOV 1970). Dioxonium also has no ganglion-blocking effect. Compared with succinylcholine, dioxonium does not lead to significant changes in cardiac rhythm during manipulations in reflexogenic zones (DARBINYAN and WEISBERG 1970).

3. Other Actions

Application of dioxonium does not lead to any marked increase in intrapectoral pressure, unlike tubocurarine (GHINTERS 1971). Pneumotachometric studies of dioxonium revealed no significant influence of the drug on bronchial tone (DARBINYAN and WEISBERG 1970). Administration of the drug is not accompanied by rise in histamine level in blood. On intracutaneous administration of dioxonium, no histamine release was observed, either (KIMENIS et al. 1976). Dioxonium caused no elevation of K^+ and Na^+ levels in plasma (KLUSHA et al. 1970; OGHAY et al. 1974). Indices of cholinesterase activity in blood remained stable prior to and after administration of dioxonium. No influence on the urea and protein fraction was registered. Dioxonium does not increase salivation, nor does it affect the EEG (GHINTERS 1971).

4. Summary

Thus, dioxonium is a highly active NMBA of medium duration of action. According to its mechanism of neuromuscular blockade, dioxonium is a biphasic drug. However, its first phase of action (depolarization) is very short and can be determined only by means of the EMG. The optimal dose for total myorelaxation is 0.07–0.09 mg/kg. Selective action of dioxonium on neuromuscular transmission permits one to use the drug in patients with various cardiovascular disorders as well as in patients susceptible to allergic, spastic, and autonomic reactions. The speed of development of the myoparalytic effect of the clinically optimal dose of dioxonium permits one to use it, if necessary, as the only NMBA in various surgical interventions.

F. Quinuclidine Derivative

Qualidilum

Qualidilum is a nondepolarizing neuromuscular blocking drug. It was studied pharmacologically and introduced into clinical practice by MASHKOVSKY and SADRITDINOV (1962). It was synthesized at the All-Union Chemical Pharmaceutical Institute, Moscow.

1. Neuromuscular Blocking Action

Clinically used doses of qualidilum vary between 1 and 3 mg/kg. Administration of 1 mg/kg results in moderate muscle relaxation and respiratory depression. Use of 1.5 mg/kg was accompanied by 2–4 min apnea in 50% of patients. Duration of muscle relaxation is about 10–15 min. Clinically optimal doses, sufficient for 20–35 min apnea and muscle relaxation are 1.8–2.0 mg/kg. Duration of relaxation is determined by the properties of the inhalational anesthetic drug and its concentration. The shortest period of relaxation is observed under nitrous oxide (21.6 ± 1.2 min).

The longest and most profound muscle relaxation is observed under ether: 33.6 ± 1.2 min. (DEGTYAREVA 1967; SADRITDINOV 1966). The onset of neuromuscular blockade is registered in 30–40 s. Muscle relaxation under qualidilum is gradual, without fasciculation of muscles, involving facial muscles, muscles of the trunk, abdominal wall, and diaphragm. Breathing gradually becomes superficial because 1–2 min after drug administration only the diaphragm remains active. In 2.5–4 min, apnea ensues.

Not less than 3 mg/kg is required for adequate intubation. A dose of 2 mg/kg, although enough for muscle relaxation and to stop breathing, provides relatively satisfactory conditions for intubation in only 50% of patients. Intubation can be accompanied by fibrillation of the diaphragm and closure of the glottis. However, even more significant doses of qualidilum do not provide optimal conditions for intubation. Repeat doses of qualidilum, if indicated, comprise one-quarter or one-third of the initial dose.

Sensitivity to qualidilum can be higher in exhausted patients, especially against the background of blood loss. Duration of relaxation and apnea in elderly and weak patients can exceed 1 h. Therefore, the dose in weak patients with signs of dehydratation can be as low as 1.0 mg/kg. Muscle tone recovery occurs gradually. Resumption of adequate spontaneous ventilation is registered in 35–40 min. Assisted lung ventilation is required. Neostigmine and galanthamine proved to be effective antagonists of qualidilum, provided the following conditions of decurarization are observed: partial restoration of neuromuscular transmission (diaphragmatic breathing); application of adequate doses of anticholinesterase agents and absence of acid–base and electrolyte imbalance.

2. Effect on Cardiovascular System

Qualidilum at doses of 0.5–1.0 mg/kg does not affect hemodynamics. Large doses (2.0–3 mg/kg) cause tachycardia. In 70% of patients, heart rate increases by 15–20 beats/min (DEGTYAREVA 1967). Arterial pressure is less affected. In 46 of 59 patients, it changed by less than ± 10 mmHg; in 10 patients it was elevated by 10 mmHg; and in 3 patients it decreased by 30 mmHg. Qualidilum does not affect the EEG. Hemodynamic changes following repeated administration of the drug are insignificant.

3. Other Actions

Bioelectrical activity of the brain and intraocular pressure under qualidilum were stable. The size of the pupils is not practically influenced by the drug. The light reflex is marked. Plasma tolerance to heparin decreases owing to the effect of qualidilum. Increases in fibrinolytic activity and the concentration of heparin in the blood are registered. Under the effect of qualidilum, the concentration of K^+ in plasma is usually reduced by 2.7 mequiv./l (DEGTYAREVA and MICHELSON 1964), but there is practically no change in the levels of Na^+ and Ca^{2+}.

This fact should be taken into consideration during anesthesia and the period of muscle relaxation in patients with water-electrolyte disturbances. Qualidilum sometimes causes urticarial rashes; 15 min later, rashes spontaneously disappear. It is worth mentioning that antihistamine drugs (N-blockers) fail to prevent urticarial rashes. Repeated administration of the drug does not lead to this effect. Other side effects of this drug are: increased salivation and tracheobronchial secretion, sweating and lacrimation (RUMIANTZEV and FEIN 1965; MENIAJLOV et al. 1966; LOGHINOV and TRUBNIKOV 1973).

4. Summary

Judging by the these results, qualidilum belongs to the nondepolarizing neuromuscular blocking drugs of medium duration of action. The optimal dose is 1.8–2 mg/kg. It provides total relaxation of the muscles for 20–35 min. Owing to the cumulative effect of qualidilum, repeat doses of the drug should not exceed one-third or one-half of the initial dose. It is not advisable to use the drug for intubation. Qualidilum does not cause marked changes in the EEG and arterial pressure. Increased salivation, tracheobronchial secretion, urticarial rashes, and tachycardia are the main side effects of the drug.

G. Combined Use of Neuromuscular Blocking Agents with Identical Modes of Action

At present, i.e., almost half a century since the first clinical utilazation of curare, anesthesiologists have formed certain opinions concerning the rules of application of muscle relaxants. They admit that nondepolarizing NMBA are more acceptable for relaxation. They also realize that drugs with different modes of action should not be mixed up. Nevertheless, they continue to use the desirable combinations of depolarizing and nondepolarizing agents. This paradox can be attributed to the rapid, brief, and profound effect of depolarizing muscle relaxants, e.g., succinylcholine, which is so important for intubation of the trachea. Unfortunately, modern nondepolarizing NMBA of medium and prolonged duration of action are lacking this property. The duration of the myoplegic effect is rather long, and optimal conditons for intubation last only 3–4 min.

It is quite obvious that in order to solve the problem it was necessary to synthesize a nondepolarizing NMBA of rapid and short action. For many years, all the attempts of chemists and pharmacologists ended in failure. Then diadonium appeared – a nondepolarizing muscle relaxant of short duration of action (KHARKEVICH 1970b).

It was great advance in the development of controlled myorelaxation. Now during anesthesia the anesthesiologist can manage nondepolarizing drugs of various duration of action, leaving the character of neuromuscular blockade unchanged. Advantages of the combination are obvious (BUNATIAN and YEFREMOV 1979).

I. Diadonium plus Tercuronium

Doses of 10–12 mg/kg diadonium were used for intubation. Subsequently, tercuronium was administered (average dose 0.11–0.12 mg/kg) to maintain muscle relaxation. The first dose of tercuronium was usually administered immediately after intubation. In some cases, tercuronium was injected when the first signs of termination of the action of diadonium were observed. The course of muscle relaxation was uneventful. Moderate dilatation of the pupils was observed. Diadonium-induced anisocoria was preserved.

To study perculiarities of neuromuscular blockade under combined myorelaxation (diadonium plus tercuronium), the EMG data in 28 patients operated on under ataralgesia and neuroleptanalgesia were analyzed. The general tendency of the EMG changes is identical to that under 0.15 mg/kg tercuronium. However there were several peculiarities. The peak effect came earlier, demonstrating that the action of tercuronium had been developing along with partial neuromuscular blockade caused by diadonium. The period of total blockade and the time necessary for 50% recovery of neuromuscular transmission in response to nerve stimulation was about 15–20 min, shorter than in the case of isolated administration of 0.15 mg/kg tercuronium. At the moment of 40%–50% recovery of neuromuscular transmission following the first stimulus, the amplitude of the repeated response reached 10%–15%; contraction of the diaphragm was observed. However, spontaneous restoration of neuromuscular transmission was slow. Five- to sixfold facilitation of neuromuscular transmission occurred under the influence of anticholinesterase agents on single stimulus, and eight- to tenfold facilitation on repeated stimulation (YEFREMOV 1979).

Repeat doses of tercuronium comprising one-third or one-half of the initial dose were used to maintain muscle relaxation. The period of clinically adequate relaxation after repeated administration of 0.05–0.06 mg/kg tercuronium proved to be equal to that following the first dose (0.11–0.12 mg/kg). Some cases of deformation of the pupils were observed (YEFREMOV 1979).

II. Diadonium plus Tercuronium plus Diadonium

In the case of prolonged anesthesia the anesthesiologist faces problems connected with restoration of neuromuscular transmission at the final stage of the operation – putting stitches in a wound. If the operation involves the thoracic cavity or limbs, analgesics (fentanyl, piritramide) are usually preferred rather than NMBA, and hyperventilation is required. Abdominal interventions demand application of certain doses of NMBA. Use of nondepolarizing drugs of prolonged or medium duration of action does not seem reasonable as it will demand long-term artifical lung ventilation and administration of anticholinesterase agents. When there were no nondepolarizing drugs of brief action available, anesthesiologists used depolarizing relaxants thus changing the character of the neuromuscular blockade. This was an erroneous technique which resulted in residual muscle relaxation with poor response to the anticholinesterase drug treatment, demanding prolonged artificial lung ventilation. Besides the situations mentioned, the need for more profound muscle relaxation occurs intraoperatively.

All these problems have been solved with the introduction of diadonium into clinical practice. Some 20–30 min prior to the final stage of the operation, 63 patients received 5–6 mg/kg diadonium after tercuronium for short-term muscle relaxation. Usually in 40–60 s apnea and adequate myoplegia ensued and were observed for 18.4 ± 1.9 min under neuroleptanalgesia, and 16.3 ± 2.3 min ($P > 0.05$) under ataralgesia. If necessary, diadonium was readministered (30 patients) in the same dose. The general tendency of the EMG changes under diadonium at the end of the operation appeared to be similar to those observed during the first administration of the drug. Within 1–2 min, complete blockade developed and lasted for 5–6 min. The time during which neuromuscular transmission recovered to 50% in response to the first stimulation remained the same (14–18 min), despite the fact that the dose of the drug was twice as low. The mean speed of spontaneous recovery of neuromuscular transmission was also twice as low. The process was accelerated some five- or sixfold under the influence of antagonists.

III. Summary

Thus, it follows that the combined use of nondepolarizing NMBA of short and medium duration of action promotes adequate controlled relaxation during the whole period of general anesthesia. The nondepolarizing character of the blockade is also preserved during the whole operating procedure. If necessary during the anesthesia or at the terminating stage of the procedure, diadonium can be safely introduced along with nondepolarizing drugs of medium duration of action (tercuronium, tubocurarine, pancuronium, etc.). In such cases, that required intensity and duration of neuromuscular blockade is achieved, and no unwanted effects on the vital organs are observed.

H. Conclusions

Clinical and electromyographic characteristics of eight NMBA synthesized during the last two decades in the Soviet Union have been presented. Judging by the mechanism of action of these drugs (except dioxonium) they belong to the nondepolarizing NMBA. Dioxonium is a drug of mixed type of action: a brief depolarizing blockade is followed by a stable nondepolarizing one.

The NMBA are of different duration of action. A special place is occupied by diadonium, a drug of short action. Rapid development of neuromuscular nondepolarizing blockade solves the problem of intubation of the trachea without application of depolarizing drugs. The nondepolarizing character of the neuromuscular blockade permits, the combined use of diadonium with other NMBA of more prolonged action.

NMBA of medium duration of action, pyrocurnium and tercuronium, are very promising. Profound relaxation producing no significant influence on hemodynamics and function of the vital organs place these drugs together with other modern safe anesthetic agents.

Good muscle relaxation of the abdominal muscles under continued spontaneous ventilation is characteristic of the derivatives of α-truxillic acid (truxilonium, cyclobutonium, anatruxonium). It permits one to perform a number of abdominal interventions without intubation of the trachea, using a mask for artificial ventilation. Dioxonium and qualidilum are also used in clinical practice as nondepolarizing drugs of medium duration of action.

References

Barsukov PJ (1970) Employment of cyclobutonium during endotracheal ether-oxygen anesthesia in gynecological practice (in Russian). In: Kharkevich DA (ed) Novye kurarepodobnye i ganglioblokiruyushchie sredstva (New curarelike and ganglioblocking agents). Meditzina, Moscow, pp 167–179

Bunatian AA, Ruzaykina TI (1983) Myoparalytic and hemodynamic effect of pyrocurinum. (in Russian). In: Kharkevich DA (ed) Novye miorelaksanty (New muscle relaxants). Meditzina, Moscow, pp 212–219

Bunatian AA, Yefremov AV (1979) Combined use of non-depolarising muscle relaxants (diadonium, tercuronium, tubocurarine, pavulon) (in Russian). Abstracts of the 3rd Congress of anesthesiologists and reanimatologists of the Ukrainian Soviet Socialist Republic, Chernovtzi, Medical Institute, pp 404–405

Bunatian AA, Kosenko RP, Sablin IN, Flerov EV (1977) Computers in anesthesiology (in Russian). Anest Reanim 1:41–44

Bunatian AA, Yefremov AV, Vinnitzky LI, Flerov EV, Vinnitzkaja KB, Meshcheriakova SA, Vorobyova NG (1978) Application of a new non-depolarising muscle relaxant of brief action diadonium in practical anesthesiology (in Russian). Anest Reanim 6:79–80

Bunatian AA, Sablin IN, Flerov EV, Rostunova NV, Friman MY, Shmirin MM (1981) Application of the technique of noninvasive evaluation of myocardial contractility perioperatively using computer (in Russian). Anest Reanim 6:4–10

Danilov AF, Malygin VV, Starshinova LA, Khromov-Borisov NV, Torf SF, Tcherepanova VP (1979) Tercuronium – a new non-depolarising muscle relaxant of highly active and elective type of action (in Russian). Farmakol Toksikol 5:478–481

Darbinyan TM, Saakyan ES (1979) Interaction of diadonium, tercuronium and succinylcholine in succesive application to maintain prolonged myoplegia under general anesthesia (in Russian). Anest Reanim 6:18–22

Darbinyan TM, Weisberg LA (1970) Dioxonium – a new muscle relaxant (in Russian). Vestn Chir 10:84–88

Degtyareva LG (1967) Clinical evaluation of qualidilum – a new non-depolarising muscle relaxant (in Russian). Thesis, First Medical Institute, Moscow

Degtyareva LG, Michelson VA (1964) Experience of clinical application of qualidilum (in Russian). Eksp Khir Anesteziol 6:88–91

Dolyna OA, Ptushkina SG (1970) Application of anatruxonium and cyclobutonium in abdominal operations (in Russian). In: Kharkevich DA (ed) Novye kurarepodobnye i ganglioblokiruyushchie sredstva. (New curare-like and ganglioblocking agents). Meditzina, Moscow, pp 148–152

Firsov AA, Stazhadze LL (1970) Data of clinical study of anatruxonium – a new curare-like agent (in Russian). In: Kharkevich DA (ed) Novye kurarepodobnye i ganglioblokiruyushchie sredstva. (New curare-like and ganglioblocking agents). Ed: Meditzina, Moscow, pp 127–133

Firsov AA, Zhilis BG, Stazhadze LL (1970) The experience of application of a muscle relaxant cyclobutonium in anesthesiology (in Russian). In: Kharkevich DA (ed) Novye kurarepodobnye i ganglioblokiruyushchie sredstva. (New curare-like and ganglioblocking agents). Meditzina, Moscow, pp 152–159

Ghinters JJ (1971) Pharmacology of dioxonium – a new muscle relaxant of the selective type of action (in Russian). Farmakol Toksiol 9:9–13

Kharkevich DA (1965) On pharmacological properties of anatruxonium – a new curare-like agent (in Russian). Farmakol Toksikol 3:305–309

Kharkevich DA (1966) On pharmacology of a new non-depolarising muscle relaxant cyclobutonium (in Russian). Farmakol Toksikol 1:47–53

Kharkevich DA (1970a) Pharmacology of the new non-depolarising agents – anatruxonium, cyclobutonium, truxilonium and pyrocyclonium (in Russian). In: Kharkevich DA (ed) Novye kurarepodobnye i ganglioblokiruyushchie sredstva (New curare-like and ganglioblocking agents). Meditzina, Moscow, pp 41–47

Kharkevich DA (1970b) On pharmacological properties of a new non-depolarising curare-like agent – diadonium iodide (in Russian). Farmacol Toksikol 5:531–536

Kharkevich DA (1974) Anatruxonium – a new curare-like agent of non-depolarising type of action (in Russian) Khem Farm 4:51–68

Kharkevich DA (1983) Pharmacology of pyrocurinum (in Russian). In: Kharkevich DA (ed) Novye miorelaksanty (New muscle relaxants). Meditzina, Moscow, pp 187–191

Kharkevich DA, Kravchuk LA (1961) On pharmacology of a new curare-like agent truxilonium (in Russian). Farmakol Toksikol 3:318–324

Kharkevich DA, Shitov VV (1983) Clinical study of diadonium (in Russian). In: Kharkevich DA (ed) Novye miorelaksanty (New muscle relaxants). Meditzina, Moscow, pp 137–149

Ķimenis AA, Klusha VE, Ghinters JJ, Veveris MM (1976) Dioksony. Farmakologiya i anesteziologicheskoe primenenie (Dioxonium. Pharmacology and anesthesiological use). Riga, Znanije

Kluša VE, Ķimenis AA (1967a) Neuromuscular blocking activity of some quaternary salts of bis-aminomethyldioxoalanylethane (in Russian). Izv AN Latv SSR 7:137–143

Kluša VE, Ķimenis AA (1967b) On the mechanism of neuromuscular blocking action of some quaternary salts of bis-aminomethyldioxoalanylethane (in Russian). Izv AN Latv SSR 8:104–110

Kluša VE, Ķimenis AA (1968 a) Comparative pharmacology of some quaternary salts of bis-aminomethyldioxoalanylethane and -ethylene (in Russian). Izv AN Latv SSR 1:118–123
Kluša VE, Ķimenis AA (1968 b) A relationship between chemical structure and character of action of the N-cholinesterase systems in some quaternary salts of bis-aminomethyldioxoalanylethane and -ethylene (in Russian). Izv AN Latv SSR 2:138–146
Kluša VE, Ķimenis AA, Ghinters JJ (1970) Clinical use of dioxonium (in Russian). Eksp Klin Farm 2:55–74
Kotomina GL (1965) Clinical characteristics of truxilonium (in Russian). Vestn Chir 4:65–68
Kotomina GL (1970) Comparative clinical evaluation of the new muscle relaxants and derivatives of diphenylcylobutadicarbonic acids (in Russian). In: Kharkevich DA (ed) Novye kurarepodobnye i ganglioblokiruyushchie sredstva (New curare-like and ganglioblocking agents). Meditzina, Moscow, pp 73–82
Kotomina GL, Kharkevich DA (1968) Clinical and pharmacological characteristics of anatruxonium (in Russian). Khirurgiya 4:23–26
Kotomina GL, Leosko VA, Korjukin VM (1967) Application of the new muscle relaxants – derivatives of α-truxillic acid in anesthesiology (in Russian). Vestn Chir 1:153–157
Kuzin MI, Nadzharyan TL (1970) Clinical pharmacology of truxilonium (in Russian). In: Kharkevich DA (ed) Novye kurarepodobnye i ganglioblokiruyushchie sredstva (New curare-like and ganglioblocking agents). Meditzina, Moscow, pp 91–104
Kuzin MI, Litkin YI, Sachkov TL (1970) Comparative characteristics of the myorelaxants of α-truxillic acid (in Russian). In: Kharkevich DA (ed) Novye kurarepodobnye i ganglioblokiruyushchie sredstva (New curare-like and ganglioblocking agents). Meditzina, Moscow, pp 112–121
Loghinov EF, Trubnikov VV (1973) Application of qualidile in traumatology (in Russian). In: Anesthesiology and reanimation in some special branches of surgery. Medical Institute, Kalinin
Manevich AZ, Fominikh VP, Petrov MN, Tjukov VL (1970) Anatruxonium in children's anesthesiology (in Russian). In: Kharkevich DA (ed) Novye kurarepodobnye i ganglioblokiruyushchie sredstva (New curare-like and ganglioblocking agents). Meditzina, Moscow, pp 133–138
Mashkovsky MD, Sadritdinov F (1962) Curarelike properties of dichloridum 1.6-di-(3.3-benzil-quinuclidinum-1.1)-hexane (qualidilum) (in Russian). Farmakol Toksikol 6:685–691
Meniajlov NV, Koroleva El, Nevrajev OG, Volodko NA (1966) Application of qualidilum (in Russian). Eksp Khir Anesteziol 4:93–94
Michelson VA (1970) Clinical evaluation of muscle relaxants of the α-truxillic acid group (in Russian). In: Kharkevich DA (ed) Novye kurarepodobnye i ganglioblokiruyushchie sredstva (New curare-like and ganglioblocking agents). Meditzina, Moscow, pp 83–91
Michelson VA, Tzipin LE, Makarov LE, Vorontzov JP, Yusipova TA (1977) Experience of the first application of diadonium in children (in Russian). Vestn AMN USSR 2:77–81
Nadzharyan TL (1965) Clinical pharmacology of truxilonium (in Russian). Thesis, Moscow
Nadzharyan TL, Yerivantzev EA (1970) Application of truxilonium and anatruxonium in geriatrics (in Russian). In: Kharkevich DA (ed) Novye kurarepodobnye i ganglioblokiruyushchie sredstva (New curare-like and ganglioblocking agents). Meditzina, Moscow, pp 138–148
Oghay SV, Ivashchenko EA, Ivlev AM, Synitza YF (1974) Application of dioxonium (in Russian). Sov Med 3:20–22
Orlov VJ (1970) Experience of the clinical use of cyclobutonium in abdominal operations (in Russian). Medical Institute Scientific Works 31:204–208, Ryazan
Osypova NA, Svetlov VA, Kolesnikov AI (1977) Prospective use of ritebronium and tercuronium in anesthesiology (in Russian). In: Ed Darbinyan TM. 2nd All-Union Congress of anesthesiologists and reanimatologists, p 476, Tashkent

Ostrovsky VY, Kiselev SO (1980) Clinical application of diadonium (in Russian). Khirurgiya 12:35–39
Rumiantzev ES, Fein OD (1965) Qualidile – muscle relaxant (in Russian). Khirurgiya 12:35–39
Saakyan ES (1977) Application of a new non-depolarising muscle relaxant tercuronium (in Russian). In: 2nd All-Union Congress of anesthesiologists and reanimatologists, pp 494–496
Sadritdinov F (1966) Effect of the narcotic agents and drugs used for premedication upon curare-like action of qualidilum (in Russian). Eksp Khir Anesteziol 4:87
Sokolov GP, Kluša VE, Ķimenis AA, Hiller SA (1968) Synthesis and curare-like action of quaternary salts of bis-aminomethyldioxoalanylethane and ethylene (in Russian) Khim-Farm Zh 3:3–7
Svadzhan EP, Shitov VV (1977) Clinical use of a new curarelike agent diadonium (in Russian). Eksp Khir Anesteziol 6:62–66
Ter-Khasparova NJ (1970) Curarisation with cyclobutonium (in Russian) In: Kharkevich DA (ed) Novye kurarepodobnye i ganglioblokiruyushchie sredstva (New curare-like and ganglioblocking agents). Meditzina, Moscow, pp 164–167
Vanevsky VL, Khargapolov LN (1970) The experience of application of a new muscle relaxant dioxonium in anesthesiological practice (in Russian). Vestn Chir 6:74–77
Vanevsky VL, Kotomina GL, Kahrgapolov LN (1972) Clinical use of a new muscle relaxant dioxonium (in Russian). Eksp Klin Farmacol 3:67–74
Vanevsky VL, Danilov AF, Kotomina GL, Malighin VV, Starshinova LA (1978) Development of the new antidepolarising muscle relaxants and experience of their first application (in Russian). Anest Reanim 4:36–38
Weisberg LA, Shulutko EM (1983) Electromyographic study of pyrocurinum in clinic (in Russian). In: Kharkevich DA (ed) Novye miorelaksanty (New muscle Relaxants). Meditzina, Moscow, pp 204–212
Yefremov AV (1979) Combined use of non-depolarizing muscle relaxants of different duration of action (diadonium, tercuronium) (in Russian). Thesis, Moscow
Yerivantzev NA, Yelshansky VI, Brusenko EY (1970) Clinical application of anatruxonium in surgery (in Russian). In: Kharkevich DA (ed) Novye kurarepodobnye i ganglioblokiruyushchie sredstva (New curare-like and ganglioblocking agents). Meditzina, Moscow, pp 121–127

CHAPTER 27

Pipecuronium Bromide (Arduan)

E. TASSONYI, G. SZABÓ, and L. VIMLÁTI

A. Introduction

Muscle relaxants belong to the basic anesthetic drugs and it is quite understandable that newer compounds have come into being. Among these drugs, steroid derivatives have an important place and, besides pancuronium, many similarly structured muscle relaxants have been tried in the past few years (SUGRUE et al. 1975; BOWMAN 1980; TUBA 1980).

The aim with the newer drugs is to develop muscle relaxants which have no side effects. Elimination of cardiovascular effects is the main point; after all, muscle relaxants in use at present all have this type of side effects (BRAND 1978). The leading role belongs to the nondepolarizing agents, and the character of their neuromuscular effects shows great variation. Safe anesthesia demands a fast and reliable neuromuscular effect, good elimination, the possibility of antagonization, and lack of cumulation.

A new member of the steroid muscle relaxants, is pipecuronium bromide (Arduan, Gedeon Richter, Budapest, Hungary). During preclinical trials, it has proven to be a fast, potent, nondepolarizing neuromuscular blocking agent, which did not produce cardiovascular effects, even with many times the therapeutic dose (ALYANTDIN et al. 1980; KÁRPÁTI and BIRÓ 1980; PULAY et al. 1980). Pipecuronium does not paralyze the sympathetic ganglia and does not release histamine (KÁRPÁTI and BIRÓ 1980). These are the properties which, at the start of the clinical application of pipecuronium, have drawn attention to the fact that this drug has significant advantages over the previous ones.

B. Clinical Pharmacodynamics

I. The Neuromuscular Blocking Effect

Neuromuscular effects of pipecuronium were studied in patients of ASA status I or II scheduled for elective abdominal surgery, and its effects were compared with pancuronium (Pavulon, Organon). Average age of patients was 49 ± 5 years (mean $\pm$ standard error of the mean), their body weight was 63 ± 3 kg (corrected body weight = 35 kg + half of body weight).

Premedication consisted of atropine 0.5 mg, droperidol 0.05–0.1 mg/kg, or diazepam 0.15 mg/kg, and fentanyl 1 µg/kg intramuscularly 45 min before induction of anesthesia. Induction of anesthesia was with diazepam 0.2 mg/kg intravenously, fentanyl 1–2 µg/kg in fractional doses and thiobutabarbital (Inactin) 1–

3 mg/kg, or methohexital (Brietal) 0.6–1 mg/kg until the eyelid reflex disappeared. Ventilation was assured by face mask, and intubation carried out when neuromuscular block was 90%–95%. Intubation was facilitated either by succinylcholine or by pipecuronium. In a control group of patients, pancuronium was used. Recovery from succinylcholine was achieved and pipecuronium given when twitch response was stable over 5 min (KRIEG et al. 1980). Neuromuscular ED_{95} of pipecuronium was determined after cumulative doses and bolus injections. When twitch response returned to 25% of the control, repetitive doses of pipecuronium were given to maintain surgical relaxation.

Spontaneous recovery of twitch after the last dose of pipecuronium and that of pancuronium was achieved by at least 75% of controls, and 0.02 mg/kg neostigmine with 0.01 mg/kg atropine was given to reverse neuromuscular block. Anesthesia was maintained with 67% nitrous oxide in oxygen. Small, incremental doses of 0.05–0.1 mg fentanyl were added as needed to maintain analgesia.

1. Measurement of Neuromuscular Block

The left ulnar nerve was stimulated via transcutaneous electrodes near the wrist by square wave pulses of 0.2 ms duration with supramaximal amplitude. Three kinds of stimulus parameter were used.

1. Short trains of 100 ms duration at 50 Hz (five stimuli in one train) were given every 10 s. This kind of stimulation was proposed by FOLDES et al. (1981).
2. Single twitches of 0.1 Hz.
3. Trains of 2 s duration at 2 Hz (train-of-four) for measurement of reversal of the block.

Contraction force of the corresponding adductor pollicis muscle was measured using a force displacement transducer and recorded on a polygraph.

The following parameters were determined (Fig. 1) according to WESTRA et al. (1980).

1. Onset time: time between injection and maximal effect.
2. Clinical duration (D_{25}). Time between injection and return of twitch to 25% of control.

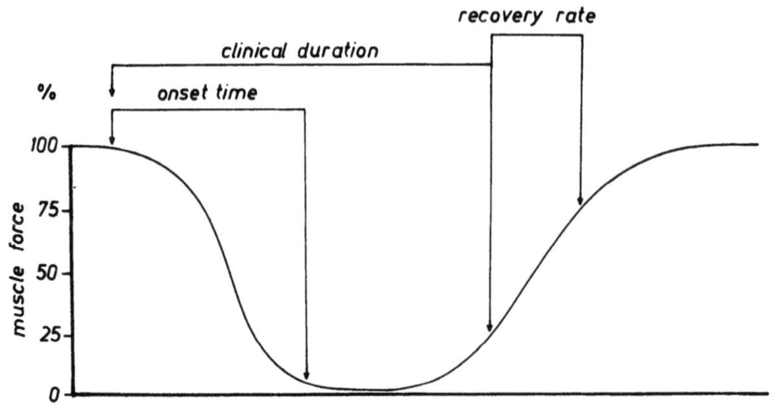

Fig. 1. Characterization of the neuromuscular block

3. Recovery rate. Time for the return of twitch from 25% to 75% of control.
4. T_4/T_1 ratio. With train-of-four stimulation, the ratio of the fourth and first twitch was calculated. This type of parameter was determined before, at 2, 5, and 8 min after antagonist administration.

2. Results

Cumulative dose–response curves are shown in Fig. 2. Two kinds of stimulation were used, and the block occurred with or without succinylcholine (Fig. 2). With short-train stimulation, ED_{95} was 20 ± 0.9 µg/kg with succinylcholine. The corresponding value without succinylcholine, was 32 ± 2 µg/kg. With single twitches, ED_{95} was 30 ± 1.1 µg/kg with succinylcholine and 55 ± 5.1 µg/kg without succinylcholine. For comparison, the dose–response curve of pancuronium is also shown in Fig. 2. ED_{95} for pancuronium was found to be 75 ± 4.9 µg/kg with single stimuli of 0.1 Hz.

To achieve surgical muscle relaxation, pipecuronium was injected, generally as a bolus. The clinical duration of 30 µg/kg bolus injection given after succinylcholine was 25 ± 3.5 min ($N = 5$), onset time was 2.8 ± 0.8 min single twitches). The neuromuscular effects of 50 µg/kg pipecuronium injected as a bolus without previous succinylcholine and stimulated by single twitches are summarized in Table 1. Table 2 shows the effects of repetitive doses of pipecuronium. The maintenance doses of pipecuronium were 12.5 µg/kg. This value is one-quarter of the initial dose. The clinical duration of consecutive doses did not increase; cumulation could not be observed (Table 2).

Neostigmine was used for the reversal of neuromuscular block induced by pipecuronium (Table 3). At the moment of its administration, spontaneous recovery of twitch height was $80\% \pm 3\%$. Twitch height returned to control values within 2 min after neostigmine administration, and a further 6 min were neces-

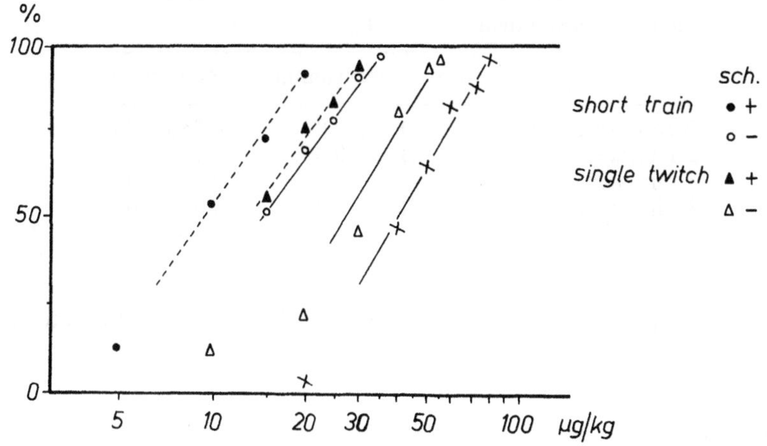

Fig. 2. Cumulative dose–response curves of pipecuronium. Stimulation parameters are indicated in the figure and detailed in text. Pipecuronium was administered with or without previous succinylcholine (*sch.*) as indicated by the *plus* and *minus signs*. *Crosses* indicate cumulative dose–response curve of pancuronium

Table 1. Comparison of the effects of intravenous bolus of pipecuronium and pancuronium without succinylcholine

	Pipecuronium	Pancuronium
Dose (μg/kg)	50	75
Onset time (min)	5.1 ± 0.8[a]	6.8 ± 1.0
Block (%)	95.4 ± 1.1	96.0 ± 1.1
Duration 25[b] (min)	42.5 ± 3.7	59 ± 2.5[c]
Recovery rate (min)	23 ± 4	32 ± 5
Number of patients	14	15

a Mean ± standard error of the mean
b Clinical relaxation time
c $P < 0.001$ (Student's test)

Table 2. Time course of maintenance of neuromuscular blockade with pipecuronium ($N = 6$). Repetitive doses are 12.5 μg/kg which is one-quarter of the initial 95% blocking dose

	Onset time (min)	Block (%)	D_{25} (min)
1st repeat	2.4 ± 0.5	94.3 ± 1.2	34.3 ± 2.0
2nd repeat	2.8 ± 0.4	93.0 ± 1.1	36.0 ± 2.7
3rd repeat	2.7 ± 0.8	93.1 ± 5.0	38.0 ± 3.8

Table 3. Effect of 0.02 mg/kg neostigmine on the T_4/T_1 ratio after administration of pipecuronium ($N = 8$) and pancuronium bromide ($N = 8$)

Time of measurement	T_4/T_1	
	Pipecuronium	Pancuronium
Before neostigmine	0.25 ± 0.04	0.25 ± 0.04
2 min later	0.50 ± 0.04	0.49 ± 0.04
5 min later	0.72 ± 0.05	0.57 ± 0.06
8 min later	0.78 ± 0.05	0.67 ± 0.05

sary to reach a T_4/T_1 ratio of 0.75. The same results could be observed with pancuronium (Table 3).

3. Discussion

The numerical values of the neuromuscular effects of pipecuronium, expressed by ED_{95} and duration time, depend on the kinds of stimulation used. It is also influenced by succinylcholine administered before pipecuronium. Succinylcholine increases the potency of this drug by 30%–40%, while according to KRIEG et al. (1980) it slightly reduces the block induced by pancuronium. When muscles were stimulated with short tetanic trains, ED_{95} was 32 µg/kg for pipecuronium, and at 0.1 Hz stimulus frequency this value was found to be 50 µg/kg. This latter dose correlated so much better to clinical symptoms and to surgical relaxation that we have decided to do all comparative tests with 0.1 Hz stimuli, and also to conform to standards set down in the literature (BOWMAN 1980; KRIEG et al. 1980; WESTRA et al. 1980).

Equipotent doses (ED_{95}) of pipecuronium and pancuronium were 50 µg/kg and 75 µg/kg, respectively. These values deviate from earlier ones. According to BOROS et al. (1980) and TASSONYI and SZABÓ (1981), pipecuronium has proven to be only 20% more potent than pancuronium. This deviation comes probably from the semiquantitative character of the methods used in these studies.

With ED_{95} doses, the clinical relaxation time (D_{25}) for pipecuronium was 42.5 ± 3.7 min, and for pancuronium, 59 ± 2.5 min. This is in disagreement with the statement that the duration of effect of the two drugs are nearly the same (BOROS et al. 1980; TASSONYI and SZABÓ 1981). This deviation may come from the fact that no equipotent doses were compared, and the 20% dose difference means a relative overdosing of pipecuronium. After administration of succinylcholine, ED_{95} of pipecuronium is smaller and the duration of action is shorter: 30 µg/kg and 23 min, respectively.

The maintenance dose of pipecuronium is one-quarter of the initial dose. This was administered at 25% twitch recovery and was proved to be satisfactory in reestablishing a 90%–95% block and maintaining it. The duration and intensity of effect of the consecutive doses did not increase, cumulation did not happen, in contrast to pancuronium, which cumulated slightly (KRIEG et al. 1980). The effect of pipecuronium can be reliably antagonized with 0.02 mg/kg neostigmine; with train-of-four stimuli, the T_4/T_1 ratio reaches 75% in 5–8 min. Respiration returns to normal in the same time.

In conclusion, it can be stated that the neuromuscular blocking effect of pipecuronium is reliable in humans; its duration is of medium length, and it can be repeated without danger of cumulation. Its elimination is relatively fast, the residual effect can easily be put an end to, and therefore, it is perfectly suitable for anesthesiologic use.

II. Effects on Heart Rate

Effects of pipecuronium on the heart rate were checked in 13 patients. The general state of patients was ASA I; their average age 40.7 ± 4.4 years. Some 45 min before induction of anesthesia, 0.5 mg atropine and 0.15 mg/kg diazepam was given intramuscularly as premedication. At induction, 1.5–3 µg/kg fentanyl was followed by 1 mg/kg methohexital. During induction, oxygen was given through a face mask. After the onset of sleep, 67% nitrous oxide in oxygen was given. When a circulatory steady state was established, 50 µg/kg pipecuronium bromide was administered as intravenous bolus.

Table 4. Effect of pipecuronium on the heart rate ($N = 13$)

	Heart rate (beats/min)
Before anesthesia	78.5 ± 4.4
Before pipecuronium	72.9 ± 4.4
2 min later	72.5 ± 3.3
5 min later	71.3 ± 4.2
15 min later	74.2 ± 4.4
30 min later	70.8 ± 4.7

1. Measurement of Heart Rate

Before induction of anesthesia, precardial ECG electrodes were placed on the patients and the second lead monitored continuously on an ECG scope. Heart rate was registered on a polygraph as an analog signal obtained from the R wave of the ECG, and was continuously registered alongside the muscle strength. The sensitivity of the polygraph was 2 beats/min.

2. Results

Pipecuronium did not alter the heart rate (Table 4). Neither 2 min after administration of pipecuronium nor after 5 min, i.e., after the total neuromuscular effect had been reached, did the heart rate change. The pulse remained steady as well.

3. Discussion

During preclinical investigation, pipecuronium did not influence the heart rate (KÁRPÁTI and BIRÓ 1980). No such effect was observed in the first human tests (ALÁNT et al. 1980). Later clinical trials reported bradycardia (WITTEK et al. 1980). Our measurements proved that pipecuronium brings about neither tachycardia, nor bradycardia. These results are justified by the observations of BUNJATJAN and MIHEEV (1980), ALÁNT et al. (1980), and BARANKAY (1980). The fact that pipecuronium does not change the heart rate is a major advantage over currently available nondepolarizing muscle relaxants.

III. Hemodynamic Effects

Hemodynamic effects of pipecuronium have been tested on 12 patients, who were to have major abdominal surgery and whose general state was ASA III or ASA IV. This gave cause for full hemodynamic monitoring. Their average age was 50.6 ± 6.4 years, body weight 65.4 ± 9.1 kg, and average body surface area 1.69 ± 0.20 m^2.

The same anesthetic was used except that the induction was started fractionally with 2.5–5 mg diazepam. Incremental doses of fentanyl were given, then 100–

150 mg Inactin induced sleep, meanwhile giving oxygen. When the circulatory steady state was established, control measurements were done. After administration of 50 μg/kg pipecuronium, the measurements were repeated at 3 and 10 min. No surgical intervention occurred during this time.

1. Methods of Hemodynamic Measurement

Before anesthesia, a three-way thermistor balloon catheter was directed into the pulmonary artery via the left subclavian vein. The position of the catheter was ascertained through pressure curves and X-ray. Bloodless arterial pressure was conscientiously checked. Cardiac output was measured by thermal dilution; central venous pressure, pulmonary artery pressure, and pulmonary capillary wedge pressure were measured via the catheter. Total peripheral resistance, pulmonary vascular resistance, cardiac index, and right and left cardiac work index were calculated from these data (Table 5).

2. Results

Results of hemodynamic measurements are contained in Table 6. Results were analyzed by one-dimensional analysis of variance. Following administration of pipecuronium, no, significant changes were found in the measured parameters.

Table 5. Measured and calculated hemodynamic parameters

HR	Heart rate (beats/min)
SBP	Systolic blood pressure (mm Hg)
DBP	Diastolic blood pressure (mm Hg)
MAP	Mean arterial pressure (mm Hg)
PAP	Pulmonary artery pressure (mm Hg)
PAEDP	Pulmonary artery end-diastolic pressure (mm Hg)
PAMP	Pulmonary artery mean pressure (mm Hg)
CVP	Central venous pressure (mm Hg)
PCWP	Pulmonary capillary wedge pressure (mm Hg)
CO	Cardiac output (l/min)
CI	Cardiac index (l/min)
SV	Stroke volume (ml)
SVI	Stroke volume index (ml/m^2)
TPR	Total peripheral resistance (dyn s cm^{-5})
PVR	Pulmonary vascular resistance (dyn s cm^{-5})
RCWI	Right cardiac work index (kg m m^{-2})
LCWI	Left cardiac work index (kg m m^{-2})

$$\text{TRP} = \frac{\text{MAP} - \text{CVP}}{\text{CO}} \times 80$$

$$\text{PVR} = \frac{\text{PAMP} - \text{PCWP}}{\text{CO}} \times 80$$

$$\text{RCWI} = \text{CI} \times \text{PAMP} \times 0.0136$$

$$\text{LCWI} = \text{CI} \times \text{MAP} \times 0.0136$$

Table 6. Hemodynamic effects of pipecuronium ($N=12$)

Parameters	Pipecuronium		
	Before	After 3 min	After 10 min
HR	92.3±3.1	87.5±3.8	85.7±3.8
SBP	119 ±7.2	115 ±6.5	128 ±9.2
DBP	88 ±4.9	81 ±3.7	87 ±4.5
MAP	99 ±5.6	94 ±4.7	102 ±5.2
PAP	28.1±3.4	25.6±2.9	25.1±3.4
PAEDP	7.2±1.3	9.7±1.6	10.3±3.4
PAMP	15.9±2.1	14.8±2.2	15.7±2.3
CVP	11.0±1.2	9.6±1.5	11.3±1.7
PCWP	5.0±0.9	6.2±1.1	6.8±1.4
CO	4.6±0.3	4.2±0.2	4.1±0.2
CI	2.7±0.2	2.5±0.1	2.4±0.1
SV	50.8±4.2	50.3±4.2	49.1±3.8
SVI	30.0±2.4	29.7±2.3	28.7±2.5
TPR	1,687±114	1,666±137	1,850±169
PVR	198±17	163±21	175±25
RCWI	0.58±0.04	0.51±0.04	0.52±0.07
LCWI	3.7 ±0.2	3.2 ±0.3	3.4 ±0.3

3. Discussion

Circulatory side effects of nondepolarizing muscle relaxants are well known (BRAND 1978). KELMAN and KENNEDY (1971), COLEMAN et al. (1972), and STOELTING (1972) carried out detailed cardiovascular studies, and reported increased mean arterial pressure, heart rate, and cardiac output following on injection of pancuronium. The drug caused no change in peripheral vascular resistance, unlike tubocurarine. We found that, in contrast to pancuronium, pipecuronium is free from cardiovascular side effects. Its use during anesthesia of patients with poor cardiovascular function will be discussed later.

C. Clinical Pharmacokinetics

Inactivation and pharmacokinetic tests of pipecuronium have been done on rats, but the data are not suitable for comparison with other steroidal muscle relaxants (BODROGI et al. 1980). The kinetics of pipecuronium have been measured in clinical conditions (TASSONYI et al. 1981). Data obtained make pharmacokinetic comparison between pipecuronium and pancuronium possible, therefore we can compare pipecuronium with other nondepolarizing muscle relaxants (e.g., FAHEY et al. 1981).

I. Pharmacokinetics in Normal Patients

1. Methods

Pharmacokinetic tests of pipecuronium have been effected on eight patients undergoing abdominal surgery. Average age of patients was 54 ± 9 years, average body weight 59 ± 5 kg. Anesthesia and pipecuronium administration were carried out as described. The serum concentration of pipecuronium was measured by a colorimetric method (SZABÓ and TASSONYI 1981). Drugs employed in premedication did not influence determination of pipecuronium. This method does not measure the possible desacetylated metabolites of pipecuronium. Data were fitted to a biexponential equation (Eq. 1) and constants calculated.

$$c = A\,e^{-\alpha t} + B\,e^{-\beta t}\,. \tag{1}$$

After determining the constants, pharmacokinetic parameters were calculated as in Table 7, such as: distribution and elimination half-lives $t^{\alpha}_{1/2}$, $t^{\beta}_{1/2}$; area under curve (AUC); apparent distribution volume V_β; volume of the central compartment V_c; volume of distribution at steady state V_{dss}; and plasma clearance *Cl*.

2. Results

Results are shown in Table 7. For reasons of comparison, we quote a few data for pancuronium according to SOMOGYI et al. (1976). Quoted data were also calculated by the biexponential model. To calculate the pharmacokinetic parameters by a three-compartment open model would not give any pharmacologic advantage, and it would not be right mathematically either, because of the comparatively short measuring time.

3. Discussion

Kinetic data of pancuronium were calculated by mono- and biexponential equations (BUZELLO and AGOSTON 1978). It seems that for clinical purposes the two-compartment model is quite sufficient (SOMOGYI et al. 1976). Comparing the data

Table 7. Pharmacokinetic parameters of pipecuronium ($N = 8$) and pancuronium ($N = 7$)

Parameters	Pipecuronium	Pancuronium[a]
$t_{1/2} = \ln 2/\alpha$ (min)	4.1 ± 1.4[b]	12.5 ± 1.6
$t_{1/2} = \ln 2/\beta$ (min)	44 ± 7[b]	132 ± 9
$V_\beta = D/\beta$ AUC[c] (ml/kg)	261 ± 28	338 ± 28
$V_{dss} = V_c/k_{12} + k_{21}/k_{21}^{-1}$[d] (ml/kg)	387 ± 44	261 ± 27
$V_c = D/)A + B)$ (ml/kg)	111 ± 22	100 ± 10
$Cl = D/\text{AUC}$ (ml/min)	320 ± 55[b]	123 ± 16

[a] According to SOMOGYI et al. (1976)
[b] $P < 0.01$ pipecuronium vs pancuronium
[c] $\text{AUC} = A/\alpha + B/\beta$
[d] Microconstants of the two-compartment open model

of pipecuronium and pancuronium, we can see that there is substantial difference between the half-lives of the two drugs. Accordingly, pipecuronium disappears from the serum considerably faster than does pancuronium.

We have already obtained similar results for the elimination of pancuronium (TASSONYI and SZABÓ 1977), but at the same time BUZELLO and AGOSTON (1978) have calculated considerably shorter half-lives of pancuronium. The shorter half-life of pipecuronium corresponds to the faster onset of its effect and shorter clinical duration. Comparison of the distribution volumes point to the fact that the two nondepolarizing relaxants are distributed in similar volumes.

The plasma clearance value of pipecuronium is surprisingly high, in fact more than twice the clearance value of pancuronium. The reason for the high plasma clearance is not clear, but comparing the chemical structures of the two relaxants shows that pipecuronium is much more hydrophilic than pancuronium and this could result in faster renal elimination (see Sect. C.II). The other reason for the high clearance values could be the fast enzymatic and chemical desacetylation which can occur in the whole body, not just in the liver.

In conclusion: the pharmacokinetics of pipecuronium are similar in part to pancuronium. The greatest difference is in the distribution and elimination rates of the two drugs. It is worth mentioning that the relationship between concentration in plasma and effect of nondepolarizing muscle relaxants has not yet been made clear (STANSKI and SNEIDER 1979).

II. Pharmacokinetics in Patients with Impaired Renal Function

The elimination pathways of nondepolarizing steroidal muscle relaxants are the following: renal elimination, excretion with bile, and transformation into ineffective derivatives. If elimination of pipecuronium occurs mainly through the kidneys, then muscle relaxation must be done with great care in the case of patients with impaired renal function. Examining the pharmacokinetics of pipecuronium in this situation is of basic importance for reliable clinical application. Results published here have been given in part by TASSONYI et al. (1980).

1. Methods

We have examined the clinical pharmacokinetics of pipecuronium in eight surgical patients with renal failure. The impaired renal function was established with technetium-labeled DTPA. Anesthesia and relaxation were effected as described earlier, with the slight difference that 20% less pipecuronium was used. Results were evaluated by the method given.

2. Results

In view of the fact that the degree of renal impairment was different, data is given individually in Table 8. Figures 3 and 4 are intended as aids to clearer understanding. In Fig. 3, we have marked the half-lives in relation to DTPA clearance. The hatched columns refer to normal behavior. The full columns are for $t^{\alpha}_{1/2}$ and the open columns for $t^{\beta}_{1/2}$. The inverse relation between elimination and plasma

Table 8. Pharmacokinetic parameters of pipecuronium in patients with impaired renal function

Code	$t_{1/2}^{\alpha}$ (min)	$t_{1/2}^{\beta}$ (min)	V_{β} (ml/kg)	V_{dss} (ml/kg)	V_c (ml/kg)	*Cl* (ml/min)	DTPA *Cl* (ml/min)
1	1.3	24	359	396	79	397	90
2	11.3	91	191	272	132	97	55
3	1.6	45	170	195	63	123	50
4	0.9	45	213	239	48	227	50
5	6.4	68	346	455	219	233	48
6	5.2	104	148	171	74	50	24
7	5.9	110	190	210	52	85	19
8	6.1	139	122	143	49	57	17

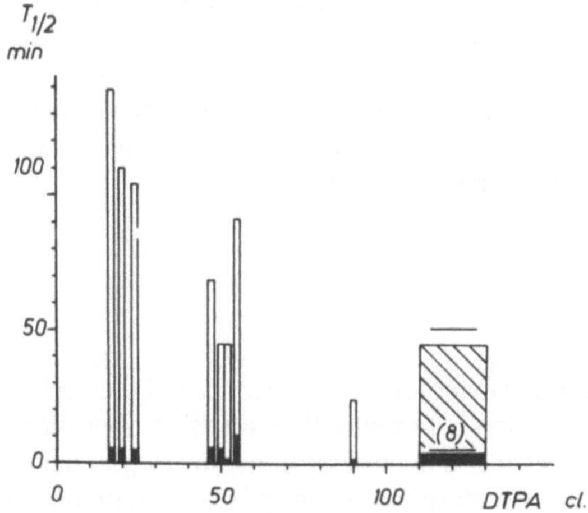

Fig. 3. Relation between the distribution and elimination half-lives of pipecuronium and DTPA clearance. Details are in the text

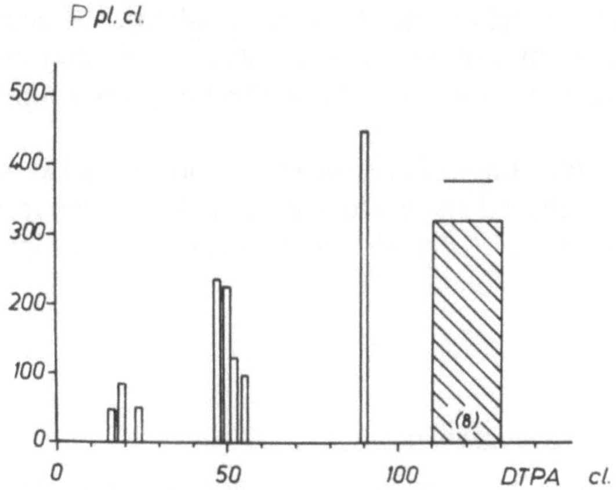

Fig. 4. Relation between the plasma clearance of pipecuronium *P*, and DTPA clearance

clearance is very striking. In Fig. 4, values of pipecuronium plasma clearance are shown in relation to DTPA clearance. Markings are the same as in Fig. 3. Here too, the correlation is pronounced.

3. Discussion

Pharmacokinetic parameters of pancuronium change considerably in the case of impaired renal function (McLOAD et al. 1976), and this points to the fact that elimination by the kidneys plays an important role. Metabolic inactivation and excretion through the bile is not too significant in the case of pancuronium, though biliary obstruction lessens the plasma clearance value considerably (SOMOGYI et al. 1977).

The pharmacokinetic parameters of pipecuronium deviate greatly in the case of normal patients and those with impaired renal function, and they depend on the degree of renal dysfunction. The fact that pipecuronium is more hydrophilic than pancuronium heightens the importance of elimination through the kidneys, but at the same time, the relative chemical and enzymatic instability of pipecuronium stresses the importance of other elimination pathways. Practical usage of the pharmacokinetic data shown and usage of pipecuronium in the case of impaired renal function will be discussed in Sect. D.

D. Clinical Use

The results of clinicopharmacologic tests encouraged the application of pipecuronium, and after employing it in many thousands of operations, we can say that its favorable properties make it successful in clinical practice. In view of the fact that the age of patients undergoing surgery is getting higher, the number of operations carried out on people suffering from cardiovascular diseases is considerable and since with today's level of anesthesia, even the very severely ill reach the operating table, safe muscle relaxants with no side effects are essential. It is also important that anesthesiologists should know all the properties of the available muscle relaxants thoroughly, and the wide possibilities of dosage. Schematic usage narrows the efficiency of one or other relaxants. By chosing the right dosage, the duration, onset time, and intensity of effect of pipecuronium can be controlled.

Our data for applications of pipecuronium come from two sources. Data from 725 patients were collected and evaluated from 11 different institutions. We have applied pipecuronium in 1,020 abdominal operations, of which 210 were acute interventions.

I. Intubation

1. Intubation with Pipecuronium

It is a great advantage in clinical anesthesiology if intubation is facilitated by a nondepolarizing muscle relaxant instead of succinylcholine. On the one hand, this

Table 9. Classification of intubation

Score	
3	Ideal conditions, no bucking
2	Satisfactory conditions, open glottis, bucking
1	Gurgling sound formation, coughing, intubation possible
0	Closed glottis, intubation impossible

eliminates the side effects of succinylcholine, and on the other, it is not necessary to wait for the effect of succinylcholine to cease.

Against these advantage stands the fact that, compared with succinylcholine, the nondepolarizing muscle relaxants act much slower, oxygenation through a face mask can be uncertain, and in the case of any difficulty of intubation, prolonged respiratory paralysis caused by a longer-lasting muscle relaxant could be a catastrophe.

It reflects today's clinical practice, that in 86.6% of cases succinylcholine and in 13.4% of cases, pipecuronium were used to facilitate intubation at 11 institutions. Of our 1,020 patients, 255 (25%) were given pipecuronium for intubation. Induction of anesthesia was by thiobutabarbital, or methohexital–fentanyl combination and no local anesthetic spray was used routinely. Intubation was classified according to Table 9. A dose of 50 μg/kg pipecuronium was administered to 125 patients who were then intubated in 5–6 min. Intubation score was 2.7 ± 0.3.

We judge the 5–6 min waiting time too long, so, by increasing the dosage of pipecuronium, we aimed at shortening this period. We have found that, after administration of 80 μg/kg, intubation could be effected within 2 min. Intubation score was 2.8 ± 0.3. But duration of pipecuronium after such a dose increases to 80–90 min so this amount could only be used for induction in long operations.

2. Pipecuronium Given After Succinylcholine

It can be seen in the previous section that most frequently pipecuronium is given following succinylcholine, after intubation. The smallest effective dose in these cases is 30 μg/kg and with this amount we gain approximately 20 min surgical relaxation. This low 30–40 μg/kg dose is advantageous, especially for shorter, lower abdominal operations. I.e., appendectomies, gynecologic interventions, some urologic surgery, or for anesthesia in diagnostic procedures. If the operation is prolonged, and it is necessary to administer pipecuronium again, it is advisable to give one-third or one-half of the initial dose, as the effect of succinylcholine will have abated by the time the repeated dose is given. If the maintenance dose is too small, relaxation will be unsatisfactory.

For anesthesia in upper abdominal operations, 50–60 μg/kg pipecuronium is the right dose after succinylcholine administration. With this, a 40–60 min duration of muscle paralysis can be assured. The maintenance dose after this amount is one-quarter of the inital dose.

We have administered pipecuronium 5–10 min after succinylcholine. By this time, clinical signs of succinylcholine's lessening effect were showing in a great number of the patients. The effect of pipecuronium was established rapidly, very often within 1 min. This fast effect after succinylcholine administration is a decidedly advantageous manifestation.

II. The Reversal of Pipecuronium Block with Anticholinesterases

The effect of pipecuronium at the end of operation has been discontinued with anticholinesterases in 83% of cases. For antagonization we used 1–2.5 mg neostigmine with 0.5–1.0 mg atropine in two-thirds of the cases, and for the further one-third, galanthamine (Nivalin) was given in 10–20 mg doses.

Antagonization was dispensed with only if all clinical symptoms pointed to the passing of the muscle relaxing effect (normal respiratory tidal volume, satisfactory forced respiration, patient is able to lift own head). It is not advisable to antagonize the block without monitoring up to 30 min after administration of the last dose of pipecuronium, because the muscle relaxant effect is still too strong, and inadequate antagonization could easily occur. In these cases, antagonization must be done by using a portable peripheral nerve stimulator as a semiquantitative way of monitoring neuromuscular block. Antagonization was classed as satisfactory only if the previously mentioned clinical criteria were achieved. In 10% of cases, one dose of anticholinesterase did not give the required result, so antidote was repeatedly given and a satisfactory effect reached. With such antagonization, neither muscle weakness nor respiratory deficiency caused by the relaxant occurred after operations (recurarization).

In our opinion, the effect of pipecuronium can satisfactorily be discontinued with neostigmine or galanthamine. We suggest that galanthamine in the dose mentioned is indicated if spontaneous recovery is well on the way, for instance, if tidal volume is already satisfactory, but peripheral muscle weakness persists. Galanthamine was also administered to patients, where we judged neostigmine to be dangerous: cardiac arrhythmia, conduction defect, asthma, pancreatitis, etc.

III. The Use of Pipecuronium in Patients with Impaired Cardiovascular Function

The circulatory effect of pipecuronium during anesthesia of patients with severe cardiovascular and ischemic heart disease was first studied by BARANKAY (1980). BUNJATJAN and MIHEEV (1980) found a remarkable stability of hemodynamic parameters. In addition, no changes in ECG were observed. WITTEK et al. (1980) observed a considerable bradycardia in 40.4% of the cases, and BOROS et al. (1980) found that pipecuronium reduced heart rate. These authors stress that the mitigation of tachycardia occurring during induction of anesthesia can be advantageous.

The fact that the drug does not cause tachycardia is of great importance, especially in the anesthesia of patients suffering from ischemic heart diseases, since the

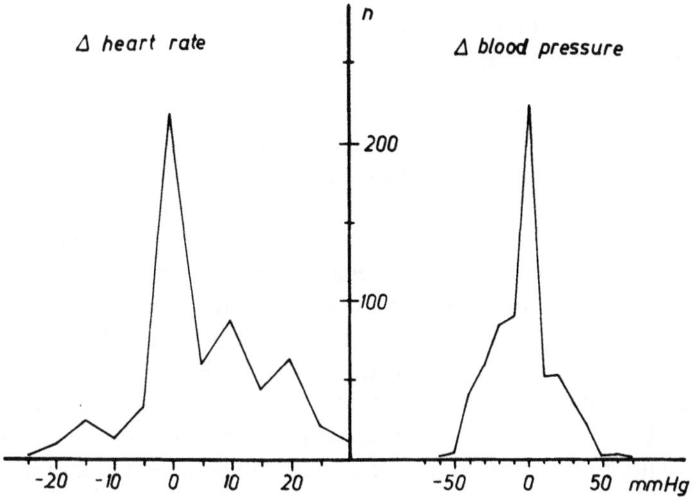

Fig. 5. Distribution pattern of heart rate and blood pressure, respectively, in the 10-min period after administration of pipecuronium

increase of myocardial oxygen requirement due to tachycardia can be ruled out (BARANKAY 1980).

The stability of peripheral resistance (see Sect. B.III) indicates that the effect of pipecuronium on peripheral resistance need not be taken into consideration. This fact makes the drug applicable in those cases in which the drop of peripheral resistance at the beginning, and in the course of anesthesia, should be avoided (i.e., hypovolemia, compensated shock, cardiogenic shock).

ALÁNT et al. (1980) and BUNJATJAN and MIHEEV (1980) investigated the changes of serum potassium level under anesthesia and did not find any significant change when pipecuronium was used. Therefore, cardiac arrhythmia due to serum potassium disturbances elicited by pipecuronium should not be taken into account.

KELMAN and KENNEDY (1971), STOELTING (1972), and BOROS et al. (1980) found hypertensive and tachycardiac effects of pancuronium bromide, therefore the use of this drug is likely to be disadvantageous in patients with arterial hypertension. On the contrary, pipecuronium did not cause changes in blood pressure (BOROS et al. 1980), consequently it can safely be given to patients with hypertension. These data correlate with our own hemodynamic measurements reviewed earlier in this chapter.

The use of pipecuronium during anesthesia of patients with disturbed cardiovascular function has become everyday practice. Pipercuronium was used during anesthesia of 200 patients with different cardiovascular and circulatory disturbances. Changes of blood pressure and heart rate in the 10-min period following pipecuronium administration are illustrated in Fig. 5. No severe blood pressure or heart rate changes attributable to pipecuronium could be observed. It should be emphasized that pipecuronium may be safely used in cases of the most severe ischemic heart disease or arterial hypertension, as well as in patients with aortic stenosis.

IV. The use of Pipecuronium in Patients with Impaired Renal Function

It was first published by BUNJATJAN and MIHEEV (1980) that pipecuronium had been successfully applied in the case of two patients who were in the terminal phase of renal insufficiency. With these patients, the first dose of pipecuronium was 50–60 μg/kg, but the duration of action has not been published.

In the early phase of our clinical investigations, 60 μg/kg pipecuronium was injected into a patient with impaired renal function (creatinine clearance 21 ml/min), and muscle paralysis lasted for about 140 min. This observation led us to seek a correlation between renal function and the pharmacokinetics of pipecuronium. The results are presented in the first part of this chapter. It was found that increased duration of elimination was remarkably consistent when renal function (measured by DTPA clearance) decreased below 15%–20% of normal. It was also found that 40 μg/kg pipecuronium as first dose can be given in these cases without danger of prolonging muscle paralysis.

On the basis of these findings, the following practice is to be followed in patients with severely decreased renal function. When the serum potassium level is not elevated, succinylcholine is used to facilitate intubation. As stated earlier in this study, 30–40 μg/kg pipecuronium is the paralyzing dose after succinylcholine administration. In patients with an elevated serum potassium level, low doses (i.e., 40 μg/kg) of pipecuronium combined with 10% lidocaine spray locally can be used to facilitate intubation. A dose of 40 μg/kg pipecuronium results in approximately 85% block, and further increase of muscle paralysis can be assured by deepening of anesthesia. Repetitive doses of pipecuronium can be decided individually, and it is mandatory to check neuromuscular function with a nerve stimulator. If a repeat dose is necessary, one-quarter of the initial dose is indicated.

V. Interaction Between Pipecuronium and Other Drugs Used is Anesthesiology and Surgery

ALÁNT et al. (1980) investigated the interaction of pipecuronium with different anesthetic agents, i.e., fentanyl, droperidol, halothane, and hydroxydione. No significant changes occurred during this study. However, WITTEK et al. (1980) found a slight prolongation of muscle relaxation when halothane was used as compared with fentanyl or methoxyflurane.

PULAY et al. (1980) studied the duration of action of pipecuronium in mongrel dogs. They found a singificant prolongation of apnae from 38 to 74 min when thiobarbital was used, as compared with methohexital, hydroxydione, and ketamine. Prolongation of duration was attributed to drug interaction.

In humans, we did not find any prolongation of the neuromuscular blocking effect of pipecuronium when anesthesia was induced with thiobutabarbital as compared with methohexital (see Sect. B.I). On the basis of the study of PULAY et al. (1980) it can only be stated that the respiratory paralyzing effect of thiobarbital was significantly longer than that of other drugs. Neuromuscular block cannot be judged from the duration of apnae.

We studied the effect of antibiotics on the intensity and duration of pipecuronium-induced blockade in patients during abdominal surgery when antibiotic administration was indicated by absominal infection (E. TASSONYI 1979, unpublished work). Tetracycline, clindamycin, ampicillin, and gentamicin were given as intravenous infusions during the operation. Neuromuscular block was measured by indirect stimulation (tetanic trains) and registration was carried out as well. No prolonged neuromuscular block, muscle paralysis, or apnea could be observed. It is concluded that pipecuronium is safe in septic patients when antibiotics are necessary intraperitoneally or intravenously in the generally used clinical doses.

E. Conclusions

According to clinical pharmacologic studies, pipecuronium bromide (Arduan) is a medium to long-acting nondepolarizing neuromuscular blocking agent. The 95% blocking dose of pipecuronium bromide is 50 μg/kg; it is more potent than pancuronium bromide. This dose elicits surgical relaxation for 40–45 min. Larger doses produce prolonged neuromuscular blockade up to 80–100 min. Intubation can be carried out within 3 min when 70–80 μg/kg is given.

For maintenance of muscle relaxation, 12–14 μg/kg can be used without danger of cumulation. When larger doses are given, i.e., 20 μg/kg, cumulation may occur. The neuromuscular effect of pipecuronium can be effectively reversed by 20–40 μg/kg neostigmine.

The main advantage of pipecuronium over pancuronium bromide is that it has no cardiovascular side effects, as has also been proven with invasive methods. Its elimination is primarily through the kidneys. With normal renal function, pipecuronium disappears from the serum rapidly. Impaired renal function slows its elimination and lengthens its duration. Pipecuronium bromide is a safe muscle relaxant for patients of poor cardiovascular status and other high risk patients.

References

Alánt O, Darvas K, Pulay I (1980) First clinical experience with a new neuromuscular blocker pipecuronium bromide. Arzneimittelforsch 20:374–379

Alyantdin RN, Buyanon VV, Fisenco VP, Lemina E, Muratov VK, Samoilov DN, Shorr VA (1980) On some properties of a new steroidal curare-like compound pipecuronium bromide. Arzneimittelforsch 30:355–357

Barankay A (1980) Circulatory effects of pipecuronium bromide during anesthesia of patients with severe valvular and ischemic heart diseases. Arzneimittelforsch 30:386–389

Bodrogi L, Fehér T, Váradi A, Vereckey L (1980) Pharmacokinetics of pipecuronium bromide in the rat. Arzneimittelforsch 30:366–370

Boros M, Szenohradszky J, Marosi GY, Tóth J (1980) Comparative clinical study of pipecuronium bromide and pancuronium bromide. Arzneimittelforsch 30:389–393

Bowman WC (1980) A new non-depolarizing neuromuscular blocking drug. Trends Pharm Sci 3:263–266

Brand L (1978) Hemodynamic changes due to anesthesia using curare and curarizing agent. In: Hemodynamic changes in anesthesia, vol 3. Academic Européene d'Anesthesiologie, Paris, pp 1123–1135

Bunjatjan AA, Miheev VI (1980) Clinical experience with a new steroid muscle relaxant: pipecuronium bromide. Arzneimittelforsch 30:383–385

Buzello W, Agoston S (1978) Pharmacokinetics of pancuronium in patients with normal and impaired renal function. Anaesthesist 27:291–297

Coleman AJ, Downing JW, Leary WP, Moyes DG, Styles M (1972) The immediate cardiovascular effects of pancuronium alcuronium and tubocurarine in man. Anesthesia 27:415–421

Fahey MR, Morris RB, Miller RD, Nguyen TL, Upton RA (1981) Pharmacokinetics of Org NC 45 (Norcuron) in patients with and without renal failure. Br J Anaesth 53:1049–1054

Foldes FF, Chaudry J, Ohta Y et al. (1981) The influence of stimulation parameters on the potency and reversibility of neuromuscular blocking agents. J Neural Transm 52:227–249

Kárpáti E, Biró K (1980) Pharmacological study of a new competitive neuromuscular blocking steroid, pipecuronium bromide. Arzneimittelforsch 30:346–354

Kelman GR, Kennedy BR (1971) Cardiovascular effects of pancuronium in man. Br J Anaesth 43:335–356

Krieg N, Crull JF, Booij LMD (1980) Relative potency of Org NC 45, pancuronium, alcuronium and tubocurarine in anesthetized man. Br J Anaesth 52:783–794

McLoad K, Watson MJ, Rawlins MD (1976) Pharmacokinetics of pancuronium in patients with normal and impaired renal function. Br J Anaesth 48:341–349

Pulay I, Alánt O, Darvas K, Weltner J, Zétény ZS (1980) Respiration paralysing and circulatory effects of a new non-depolarizing relaxant, pipecuronium bromide, in anesthetized dogs. Arzneimittelforsch 30:358–360

Somogyi AA, Shanks CA, Triggs EJ (1976) Clinical pharmacokinetics of pancuronium bromide. Eur J Clin Pharmacol 10:367–372

Somogyi AA, Shanks CA, Triggs EJ (1977) Disposition kinetics of pancuronium bromide in patients with total biliary obstruction. Br J Anaesth 49:1103–1107

Stanski DR, Sneider LB (1979) Pharmacokinetics and dynamics of muscle relaxants. Anesthesiology 51:103–105

Stoelting RK (1972) The hemodynamic effects of pancuronium and d-tubocurarine in anesthetized patients. Anesthesiology 36:612–618

Sugrue MF, Duff N, McInderwar J (1975) On the pharmacology of Org 6368 [2β,16β-dipiperidino-5α-androstan-3α-ol acetate dimethobromide], a new steroidal neuromuscular blocking agent. J Pharm Pharmacol 27:721–727

Szabó G, Tassonyi E (1981) Determination of pipecuronium bromide, a new non-depolarizing neuromuscular blocking agent, in human serum. Arzneimittelforsch 31:1013–1015

Tassonyi E, Szabó G (1977) Correlation entre le bloc n-m et taux serique du pancuronium. Central European Anaesthesia Congress, Genova, p 3351

Tassonyi E, Szabó G (1981) Pharmakokinetische und pharmakodynamische Untersuchungen mit einem neuen steroidartigen Muskelrelaxans. In: Haid HB, Mitterschiffthaler G (eds) Zentraleuropäischer Anaesthesiekongreß 2. Springer, Berlin Heidelberg New York (Anaesthesiology and intensive care medicine, vol 140)

Tassonyi E, Szabó G, Vereckey L (1980) Pharmacokinetics of pipecuronium bromide in patients with renal failure. In: Rügheimer E, Wawersik J, Zindler M (eds) 7th World congress of anaesthesiologists. Excerpta Medica, Amsterdam, p 168 (International congress series, no 533)

Tassonyi E, Szabó G, Vereckey L (1981) Pharmacokinetics of pipecuronium bromide, a new non-depolarizing neuromuscular blocking agent, in humans. Arzneimittelforsch 31:1754–1756

Tuba Z (1980) Synthesis of 2β,16β-bis[4′-dimethyl-1-piperazino]-3α,17β-diacetoxy-5α-androstane di-bromide and related compounds. Arzneimittelforsch 30:342–345

Westra P, Houwertjes MC, DeLange AR, Scaf AH, Hindriks FR, Agoston S (1980) Effect of experimental cholestasis on neuromuscular blocking drugs in cats. Br J Anaesth 52:747–756

Wittek L, Gecsényi M, Barna B, Hargitay Z, Adorján K (1980) Report on clinical test of pipecuronium bromide. Arzneimittelforsch 30:379–383

CHAPTER 28

Vecuronium (ORG-NC-45)

R. D. Miller

A. Introduction

For several years, anesthesiologists have stated that two new types of neuromuscular blocking drugs (curare-like drugs, muscle relaxants) are needed in clinical anesthesia (Savarese and Kitz 1975). One type of neuromuscular blocking drug would be one that is nondepolarizing with a very rapid onset time (time from muscle relaxant administration until peak effect) and short duration of action, similar to that of succinylcholine, but without the latter drug's well-known side effects and complications. BW 785U seemed to meet these qualifications (Savarese et al. 1980). However, because of significant histamine release (Rosow et al. 1980), clinical trials with BW 785U were terminated. Thus, the goal of developing this type of neuromuscular blocking drug for clinical anesthesiology has yet to be realized.

A neuromuscular blocking drug with an intermediate duration of action (i.e., between succinylcholine and pancuronium) would also be useful in adding flexibility to the muscle relaxant administration practices of clinical anesthesiologists. Desirable characteristics would include lack of cumulative and cardiovascular effects (Savarese and Kitz 1975). In addition, a neuromuscular blocking drug that is metabolized and/or is not dependent on the kidney for its elimination would be helpful in patients with impaired renal function, since all the currently available nondepolarizing neuromuscular blocking drugs primarily depend on the kidney for their elimination. Initial studies in animals indicated that vecuronium (ORG-NC-45) had most of the qualifications indicated for a neuromuscular blocking drug with an intermediate duration of action (Marshall et al. 1980; Durant et al. 1979). In Chapt. 18, the basic pharmacology of vecuronium is described.

Vecuronium has undergone extensive clinical trials, and has been approved for routine clinical use in the United States and several European countries. Several investigators have described the clinical pharmacology of vecuronium. These clinical data represent the foundation upon which the clinical use of vecuronium in anesthesiology is based. Pancuronium will be used as a basic comparison because it probably is the most commonly used neuromuscular blocking drug by clinical anesthesiologists.

B. Neuromuscular Blocking Characteristics

I. Potency

Vecuronium is slightly more potent than or equipotent to pancuronium. Using nitrous oxide–narcotic anesthesia, KRIEG et al. (1980) found the relative potency of vecuronium to that of pancuronium to be 1.74; that is, vecuronium was 1.74 times more potent than pancuronium. However, they used the incremental or cumulative method for producing dose–response curves. DONLON et al. (1980) found that the incremental dose method and the conventional single bolus injection method produced nearly identical dose–response curves with both pancuronium and tubocurarine. However, FISHER et al. (1982) found that the incremental dose method produces a blockade of lesser magnitude with vecuronium, which is probably due to its rapid clearance from plasma. Using a conventional single bolus injection method under halothane anesthesia, FAHEY et al. (1981 a) found that the ED_{50} (dose of neuromuscular blocking drug causing a 50% depression of twitch tension) for pancuronium and vecuronium was 22 and 15 μg/kg, respectively, for a potency ratio of 1.5 (22/15). CRUL and BOOIJ (1980), using nitrous oxide–narcotic anesthesia, found that the ED_{50} of vecuronium was 28 μg/kg, and pancuronium 42 μg/kg. However, in this study, the incremental or cumulative dose method for determining dose–response curves was used. In summary, despite variations in experimental technique and absolute ED_{50} values, vecuronium has been shown to be either slightly more potent than or equipotent to pancuronium.

II. Duration of Action

Several investigators have studied the duration of action of vecuronium in anesthetized patients. In Table 1, the results of several of these studies are summarized. Obviously, duration of neuromuscular blockade will depend on several factors, including the anesthetic being utilized. However, the clinical data uniformly support the conclusion that vecuronium has a duration of action that is one-third to one-half that of pancuronium. Vecuronium, therefore, fulfills the requirement of a neuromuscular blocking drug with an intermediate duration of action (SAVARESE and KITZ 1975).

III. Cumulative Effects

Clinical anesthesiologists define a cumulative effect of a neuromuscular blocking drug to be an increase in duration of neuromuscular blockade from repetitive doses. Gallamine, pancuronium, and tubocurarine all have cumulative effects; that is, if the same dose of neuromuscular blocking drug is administered at the same point in recovery from the previous neuromuscular blockade, a longer duration of neuromuscular blockade will occur. Preliminary studies indicated that vecuronium did not have a cumulative effect (CRUL and BOOIJ 1980; KRIEG et al. 1980; AGOSTON et al. 1980). Cumulative effects of pancuronium and vecuronium were compared in anesthetized patients under halothane anesthesia (FAHEY et al.

Table 1. Onset time and duration of action of several doses of vecuronium in anesthetized patients

Dose (mg/kg)	Anesthetic	Onset time[a] (min)	Duration of action[b] (min)	Reference
0.010	Halothane	6.7	14	Fahey et al. (1981a)
0.0175	Halothane	6.3	11	Walts et al. (1981)
0.020	Halothane	6.0	27	Fahey et al. (1981a)
0.020	Halothane	6.2	8	Walts et al. (1981)
0.036	Nitrous oxide–narcotic	4.5	25	Krieg et al. (1980)
0.043	Nitrous oxide–narcotic	2–3	28	Crul and Booij (1980)
0.070	Halothane	3.8	34	Fahey et al. (1981a)
0.070	Halothane	4.7	18	Agoston et al. (1980)
0.080	Nitrous oxide–narcotic	3.9	28	Agoston et al. (1980)
0.080	Enflurane	4.8	41	Duncalf et al. (1981)
0.080	Nitrous oxide–narcotic	5.9	36	Duncalf et al. (1981)
0.100	Nitrous oxide–narcotic	2.2	27	Agoston et al. (1980)
0.100	Nitrous oxide–narcotic	5.3	35	Nagashima et al. (1981)
0.140	Halothane	2.8	104	Fahey et al. (1981a)
0.280	Halothane	2.1	174	Fahey et al. (1981a)

[a] Time from vecuronium administration until peak effect
[b] Time from vecuronium administration until 50% or 90% recovery of control twitch tension

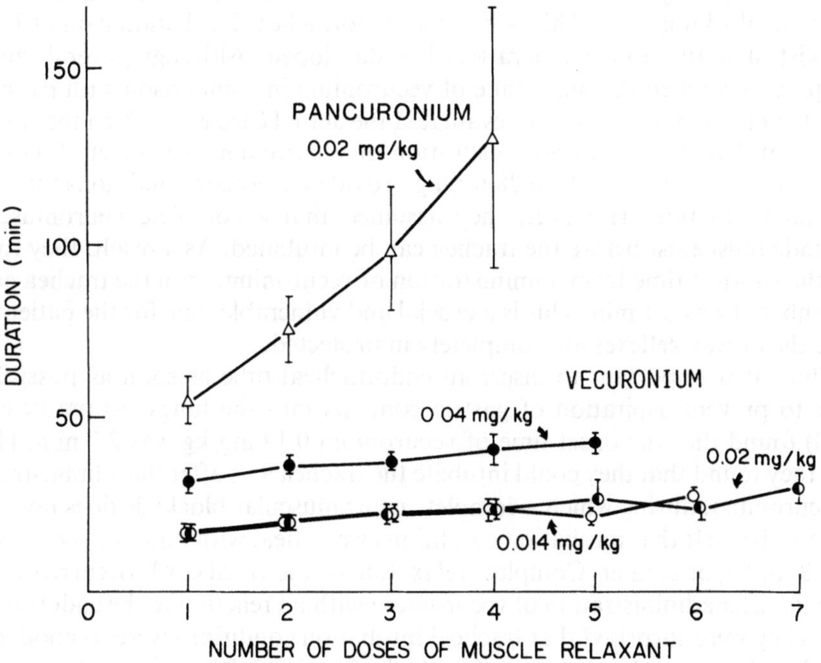

Fig. 1. Effect of repetitive doses on duration of neuromuscular blockade seen with vecuronium and pancuronium. Each symbol and *vertical bar* represents the mean ± the standard error, respectively, for that muscle relaxant dose number. *Lines* have been drawn to enhance the visual interpretation. (Fahey et al. 1981a)

1981 a). Specifically, three doses of vecuronium and one dose of pancuronium were studied. When an initial dose of either drug had been given, and when twitch tension had recovered to 25% of the control tension, the same dose of vecuronium or pancuronium was then given. To be included in the study, patients had to receive vecuronium or pancuronium at least three times in succession. The duration of neuromuscular blockade with each subsequent dose of pancuronium increased (Fig. 1). Although repetitive doses of vecuronium caused a slight increase in duration of neuromuscular blockade, these increases were significantly less than those from pancuronium. Also, these data substantiate the shorter duration of neuromuscular blockade from vecuronium as compared with pancuronium (Fig. 1).

The reasons for this relative lack of cumulative effect are related to the short elimination half-life (see Sect. C.I) and rapid clearance from plasma. The lack of cumulative effect makes it less likely that a neuromuscular blockade will be too intense to antagonize at the end of surgery with vecuronium as compared with pancuronium. This lack of cumulative effect provides an element of safety with vecuronium as compared with pancuronium in that the incidence of prolonged paralysis is attenuated when vecuronium is used.

IV. Endotracheal Intubation

An important question with which anesthesiologists are concerned is related to how rapidly paralysis will ensue after the intravenous administration of a neuromuscular blocking drug. This is because endotracheal intubation cannot be accomplished until sufficient paralysis has developed. Although several clinical groups have studied the onset time of vecuronium in comparison with pancuronium (Table 1), only a few have evaluated the actual ease with which the trachea can be intubated after the administration of vecuronium. FAHEY et al. (1981 a) found that vecuronium 0.07–0.28 mg/kg provided adequate conditions for endotracheal intubation. However, they assumed that a complete neuromuscular blockade must exist before the trachea can be intubated. As a result, they found that the shortest time from administration of vecuronium until the trachea could be intubated was 2.1 min. This is a crucial and vulnerable time for the patient because the airway reflexes are completely unprotected.

Thus, it is important to insert an endotracheal tube as soon as possible in order to prevent aspiration of gastric contents into the lungs. AGOSTON et al. (1980) found that the onset time of vecuronium 0.12 mg/kg was 2.7 min. However, they found that they could intubate the trachea 95 s after the administration of vecuronium, during which a complete neuromuscular blockade does not exist. Overall, they felt that intubation conditions were ideal with doses of vecuronium of 0.08 mg/kg or greater. Complete relaxation of the vocal cords occurred within 90–100 s after administration of vecuronium with no reaction to the endotracheal tube. They were surprised that tracheal intubation conditions were so good, even though only a partial neuromuscular blockade existed. They postulated that the dose of neuromuscular blocking drug required to paralyze laryngeal muscles for satisfactory endotracheal intubation is less than that needed for limb muscle paralysis, which is where monitoring of neuromuscular blockade occurs. Of prime importance is whether vecuronium offers any advantages over pancuronium in this regard. In my experience, the trachea can be intubated in about half the time with vecuronium as required with pancuronium. However, HARRISON and FELDMAN (1981) compared the ease of endotracheal intubation using two doses of vecuronium (0.1 and 0.15 mg/kg) with that of pancuronium (0.1 mg/kg) at 60, 90, and 120 s after neuromuscular blocking drug administration. They found that vecuronium did not offer any significant benefit within 2 min of the administration of the neuromuscular blocking drugs as compared with pancuronium. However, even if there is no advantage to vecuronium as compared with pancuronium with regard to onset time, the shorter duration of action of vecuronium will lessen the chance of a prolonged paralysis occurring. In summary, the consensus of opinion appears to be that the trachea can be intubated more quickly when vecuronium is given in contrast to pancuronium. The ultimate test of vecuronium in regard to time from administration until adequate endotracheal intubation conditions exist can only be determined by many clinicians under a variety of clinical settings.

C. Pharmacokinetics and Pharmacodynamics

I. Comparison with Pancuronium

The first pharmacokinetic study to be performed in anesthetized patients was done by VAN DER VEEN and BENCINI (1980), who found the elimination half-life of vecuronium to be less than that published for pancuronium (31 versus 133 min) by SOMOGYI et al. (1976, 1977). However, VAN DER VEEN and BENCINI (1980) determined the elimination half-life from plasma concentration values apparently collected over a 1-h period following administration of vecuronium, and compared those to at least 6 h of sampling time with pancuronium (SOMOGYI et al. 1976, 1977). VAN DER VEEN and BENCINI (1980) were limited by their relatively insensitive high pressure liquid chromatography method of assay (lower limit of detection 50 ng/ml). SOHN et al. (1982) compared the pharmacokinetics of vecuronium and pancuronium, utilizing a fluorimetric method of analysis, which is more sensitive than the high pressure liquid chromatography method, but is nonspecific; that is, it does not distinguish between the unchanged drug and its metabolites. They found that the total plasma clearance of vecuronium (4.4 ml kg^{-1} min^{-1}) was approximately three times more rapid than that of pancuronium, while the elimination half-lives of vecuronium (103 min) and pancuronium (110 min) were similar. More recently, utilizing a highly sensitive and specific mass spectrometric method for analysis (lower limit less than 1 ng/ml), CRONNELLY et al. (1983) found the elimination half-life to be much shorter than that of pancuronium. Furthermore, the rate of total plasma clearance was markedly increased with vecuronium as compared with pancuronium (Table 2). Thus, the shorter duration of action of vecuronium as compared with pancuronium appears to be primarily on a pharmacokinetic basis.

In regard to the pharmacodynamics, SOHN et al. (1982) found the plasma concentration of vecuronium at 50% paralysis was only 0.6 times the value of pan-

Table 2. Comparative pharmacokinetics of vecuronium and pancuronium in patients with and without renal failure (mean ± standard deviation)

Study	Group	$t_{1/2\beta}$ [a] (min)	V_{Dss} [b] (ml/kg)	Clearance		C_{Pss50} [c] (μg/ml)
				(ml/min)	(ml kg^{-1} min^{-1})	
FAHEY et al. (1981b)	Normal (vecuronium)	80 ± 14	194 ± 41	228 ± 37	3.0 ± 0.3	N.D. [d]
FAHEY et al. (1981b)	Renal failure (vecuronium)	97 ± 38	238 ± 63	176 ± 34	2.5 ± 0.6	N.D.
CRONNELLY et al. (1983)	Normal (vecuronium)	63 ± 15	250 ± 40	N.D.	8.5 ± 0.02	0.1 ± 0.03
CRONNELLY et al. (1983)	Normal (pancuronium)	123 ± 30	250 ± 60	N.D.	1.9 ± 0.06	0.9 ± 0.03
SOMOGYI et al. (1977)	Normal (pancuronium)	133 ± 25	261 ± 46	123 ± 41	N.D.	N.D.
SOMOGYI et al. (1977)	Renal failure (pancuronium)	257 ± 128	296 ± 158	53 ± 16	N.D.	N.D.

[a] Elimination half-life
[b] Volume of distribution at steady state
[c] Plasma concentration of muscle relaxant at which 50% depression of twitch tension occurs
[d] N.D. not done

curonium, indicating that vecuronium has a greater potency, which is consistent with the dose–response data reported here. However, CRONNELLY et al. (1983) found the C_{Pss50} (that plasma concentration of neuromuscular blocking drug required to produce a 50% paralysis) of pancuronium and vecuronium to be virtually identical.

II. Renal Failure

Most of the currently available nondepolarizing neuromuscular blocking drugs are primarily dependent on the kidney for their elimination. Preliminary studies in rats indicated that vecuronium is primarily eliminated unchanged in bile. Specifically, about 50% and 15% of an injected dose of vecuronium appeared unchanged in the bile and urine, respectively (UPTON et al. 1982). The remaining 35% was unaccounted for. FAHEY et al. (1981 b) then compared the pharmacokinetics and neuromuscular blockade of vecuronium in patients with and without renal failure. The elimination half-life tended (statistically insignificant) to be slightly prolonged in patients without renal function (Table 2). Furthermore, the onset, duration, and recovery times of neuromuscular blockade by vecuronium 0.14 mg/kg were no different in patients with normal renal function from those with no renal function. Despite using these large doses of vecuronium, the absence of an increase in neuromuscular blockade and little change in the pharmacokinetics when comparing patients with normal renal function and those with absence of renal function, led FAHEY et al. (1981 b) to conclude that renal excretion probably is not the main route of excretion of vecuronium in humans. Therefore, vecuronium may be a suitable nondepolarizing neuromuscular blocking drug for patients with impaired renal function.

D. Cardiovascular Effects

I. Patients Without Cardiovascular Disease

In animal studies, vecuronium appears to be devoid of cardiovascular effects (BOOIJ et al. 1980; MARSHALL et al. 1980). During all the neuromuscular studies in humans, with routine clinical monitoring of blood pressure and heart rate, no cardiovascular effects have been noted from vecuronium. Recently, BARNES et al. (1982) compared the effect of vecuronium and pancuronium on heart rate and arterial blood pressure in anesthetized humans. In contrast to routine clinical monitoring, in which blood pressure is monitored intermittently, BARNES et al. (1982) monitored blood pressure continuously via a catheter inserted into the radial artery, which was then quantitated by a pressure transducer. They found that vecuronium 0.12 mg/kg produced a slight decrease in heart rate with no change in blood pressure. Conversely, pancuronium 0.1 mg/kg caused a predictable increase in heart rate with a lesser increase in blood pressure. These results are entirely consistent with previous clinical observations and confirm the lack of vagolytic effct of vecuronium as compared with pancuronium.

Endotracheal intubation is a potent stimulus and often increases heart rate and blood pressure. BARNES et al. (1982) found that when the trachea was intubated following pancuronium administration, the increases in blood pressure and heart rate were greater than when paralysis had been induced by vecuronium. They postulated that these results may be explained by the differential effect on the sympathetic nervous system of vecuronium and pancuronium. In addition to its vagolytic effect, pancuronium "sensitizes" the sympathetic nervous system by inhibiting the reuptake of norepinephrine into sympathetic nerves. In the presence of a sympathetic discharge from pancuronium, an augmented cardiovascular response to endotracheal intubation would not be surprising. However, because vecuronium does not "sensitize" the sympathetic nervous system, an exaggerated response in mean arterial pressure and heart rate would not occur during endotracheal intubation. Clinicians fear that the increase in heart rate and blood pressure during endotracheal intubation may be harmful to a patient with a compromised myocardial blood flow, such as a patient with ischemic heart disease. If this is a valid fear, then vecuronium would be preferred to pancuronium for endotracheal intubation.

In a patient with a normal cardiovascular status, the relative cardiovascular effects of the various neuromuscular blocking drugs are relatively unimportant. The real test of the cardiovascular effects of any drug would be when it is given to patients who have compromised cardiovascular status, as will be described in the following sections.

II. Patients Undergoing Coronary Artery Bypass Surgery

MORRIS et al. (1981 a) studied the cardiovascular effects of vecuronium in patients with ischemic heart disease sufficient to warrant coronary artery bypass grafting. All patients had good left ventricular function (ejection fraction greater than 50%) as determined by cardiac catheterization. Following a very large dose of vecuronium (0.28 mg/kg), only a small increase in cardiac output (9%) and a slight decrease in systemic vascular resistance (12%) were found. This dose of vecuronium (0.28 mg/kg) is equivalent to three times that dose necessary to intubate the trachea and nine times the ED_{90} (dose of drug required to depress twitch tension by 90%). Despite administration of this very large dose, no change in heart rate, cardiac rhythm, pulmonary capillary wedge pressure, or blood pressure occurred. MORRIS et al. (1981 a) concluded that large doses of vecuronium had minimal cardiovascular effects in patients with ischemic heart disease. Because all other nondepolarizing neuromuscular blocking drugs produce significant cardiovascular effects, vecuronium may offer an advantage when anesthetizing patients with ischemic heart disease.

III. Patients Undergoing Resection of a Pheochromocytoma

The administration of vecuronium to patients with a pheochromocytoma represents another test of the cardiovascular effects of this drug when given to patients with an unusually sensitive cardiovascular system. Pheochromocytoma is a tumor found in the adrenal medullary tissue, and is characterized by large secretions of catecholamines. GENCARELLI et al. (1981) administered vecuronium to three patients undergoing removal of a pheochromocytoma. These are patients who had a history of hypertension and documented biochemical evidence of elevated urinary and plasma catecholamine secretion. In summary, they found that vecuronium, 0.01–0.14 mg/kg, given to these patients, caused a small increase in plasma catecholamine concentration, but no change in arterial blood pressure or heart rate. They concluded that these observations, in addition to the theoretical advantage of vecuronium having little or no cardiovascular effects, indicated that it may be an appropriate neuromuscular blocking drug for patients with a pheochromocytoma. These observations, along with those observed in patients undergoing coronary artery bypass surgery (MORRIS et al. 1981 a), indicate that vecuronium produces few, if any, cardiovascular changes in patients with a compromised cardiovascular status.

E. Antagonism

Initial clinical studies indicated that the neuromuscular blockade from vecuronium was easily antagonized by the anticholinesterase drugs, neostigmine and pyridostigmine. In fact, FAHEY et al. (1981 a) found that less neostigmine was required for antagonism of a vecuronium neuromuscular blockade as compared with pancuronium. However, they administered intermittent intravenous boluses of neostigmine without compensation of the different pharmacokinetics and durations of action of vecuronium and pancuronium. Thus, it was impossible to ascertain whether the decreasing neuromuscular blockade was due to neostigmine and/or a decreasing plasma concentration of neuromuscular blocking drugs; in other words, the neuromuscular blockade from vecuronium would terminate more rapidly than the blockade from pancuronium; this factor may contribute to the apparent decreased neostigmine requirement with vecuronium. To eliminate the factor of a varying neuromuscular blockade, GENCARELLI and MILLER (1982) determined neostigmine dose–response curves during a constant infusion of vecuronium or pancuronium. They found that there was no difference in the dose of neostigmine required to antagonize a vecuronium or pancuronium neuromuscular blockade. They concluded that vecuronium and pancuronium are effectively and equally (independent of their different pharmacokinetics and durations of effect) antagonized by neostigmine in humans. From a clinical point of view, however, I believe less neostigmine will be required to antagonize a vecuronium neuromuscular blockade than that required for pancuronium, because the former drug leaves the plasma more readily, and therefore, less antagonist will be required. As already indicated, the lesser requirement for neostigmine or pyridostigmine should add an element of safety in attenuating the chance for a prolonged paralysis in patients recovering from surgery and anesthesia.

Because of a more rapid onset and fewer muscarinic effects, edrophonium has been proposed to be possibly preferred to neostigmine or pyridostigmine when antagonizing a nondepolarizing neuromuscular blockade (MORRIS et al. 1981 b). BAIRD et al. (1982) and FOLDES et al. (1981) have confirmed that edrophonium, 0.5–1.0 mg/kg, is a reliable antagonist of vecuronium neuromuscular blockade.

F. Conclusions

Vecuronium is a nondepolarizing neuromuscular blocking drug which is slightly more potent than or equipotent to pancuronium. Its onset time is either similar to or shorter than that of pancuronium, although its duration of action is between one-half and one-third that of pancuronium. The prolongation of neuromuscular blockade after repeated doses of vecuronium is less than with pancuronium. Anticholinesterase drugs easily antagonize a vecuronium neuromuscular blockade. The endotracheal intubation dose of vecuronium is between 0.07 and 0.14 mg/kg. No significant cardiovascular effects have been noted, even with very large doses of vecuronium (0.28 mg/kg). Vecuronium appears to fulfill the requirements proposed by SAVARESE and KITZ (1975) for a nondepolarizing neuromuscular blocking drug with an intermediate duration of action. Vecuronium has significantly added to the flexibility of neuromuscular blocking drug administration and thereby has improved patient care.

References

Agoston S, Salt P, Newton D, Bencini A, Boomsma P, Erdmann P (1980) The neuromuscular blocking action of ORG NC45, a new pancuronium derivative, in anaesthetized patients. Br J Anaesth 52:53S–60S

Baird WLM, Bowman WC, Kerr WJ (1982) Some actions of ORG NC45 and of edrophonium in the anaesthetized cat and in man. Br J Anaesth 54:375–386
Barnes PK, Smith GB, White WD, Tennant R (1982) Comparison of the effects of ORG NC45 and pancuronium on heart rate and arterial pressure in anaesthetized man. Br J Anaesth 54:435–440
Booij LHDJ, Edwards RP, Sohn YJ, Miller RD (1980) Cardiovascular and neuromuscular effects of ORG NC45, pancuronium, metocurine and d-tubocurarine in dogs. Anesth Analg 59:26–31
Cronnelly R, Fisher DM, Miller RD, Gencarelli PJ, Nguyen-Gruenki L, Castagnoli N Jr (1983) Pharmacokinetics and pharmacodynamics of vecuronium (ORG NC45) and pancuronium in anesthetized humans. Anesthesiology 58:405–408
Crul JF, Booij LHDJ (1980) First clinical experience of ORG NC45. Br J Anaesth 52:49S–52S
Donlon JV Jr, Savarese JJ, Ali HH, Teplik RS (1980) Human dose-response curves for neuromuscular blocking drugs. Anesthesiology 53:161–166
Duncalf D, Nagashima H, Hollinger I, Badola RP, Kaplan R, Foldes FF (1981) Relaxation with ORG NC45 during enflurane anesthesia. Anesthesiology 55:A203 (abstract)
Durant NN, Marshall IG, Savage DS, Nelson DJ, Sleigh T, Carlyle IC (1979) The neuromuscular and autonomic blocking activities of pancuronium, ORG NC45, and other pancuronium analogues, in the cat. J Pharm Pharmacol 31:831–836
Fahey MR, Morris RB, Miller RD, Sohn YJ, Cronnelly R, Gencarelli P (1981 a) Clinical pharmacology of ORG NC45 (Nocuron™): A new nondepolarizing muscle relaxant. Anesthesiology 55:6–11
Fahey MR, Morris RB, Miller RD, Nguyen T-L, Upton RA (1981 b) Pharmacokinetics of ORG NC45 (Nocuron) in patients with an without renal failure. Br J Anaesth 53:1049–1053
Fisher DM, Fahey MR, Cronnelly R, Miller RD (1982) Potency determination for vecuronium (ORG NC45): Comparison of cummulative and single-dose techniques. Anesthesiology 51:309–310
Foldes FF, Yun H, Radnay PH, Badola RP, Kaplan R, Nagashima H (1981) Antagonism of the neuromuscular effect of ORG-NC45 by edrophonium. Anesthesiology 55:A201 (abstract)
Gencarelli PJ, Miller RD (1982) Antagonism of ORG NC45 (vecuronium) and pancuronium neuromuscular blockade by neostigmine. Br J Anaesth 54:53–56
Gencarelli PJ, Roizen MF, Miller RD, Hoyce J, Hunt TK, Tyrrell JB (1981) ORG NC45 (Nocuron™) and pheochromocytoma: A report of three cases. Anesthesiology 55:690–693
Harrison P, Feldman SA (1981) Intubating conditions with ORG NC45. Anaesthesia 36:874–877
Krieg N, Crul JF, Booij LHDJ (1980) Relative potency of ORG NC45, pancuronium, alcuronium and tubocurarine in anaesthetized man. Br J Anaesth 52:783–787
Marshall RJ, McGrath JC, Miller RD, Docherty JR, Lamar J-C (1980) Comparison of the cardiovascular actions of ORG NC45 with those produced by other non-depolarizing neuromuscular blocking agents in experimental animals. Br J Anaesth 52:21S–32S
Morris RB, Wilkinson PL, Miller RD, Cahalan M, Quasha A, Robinson SL (1981 a) Cardiovascular effects of ORG NC45 (Nocuron™) in patients undergoing coronary artery bypass grafting. Anesthesiology 55:A205 (abstract)
Morris RM, Cronnelly R, Miller RD, Stanski DR, Fahey MR (1981 b) Pharmacokinetics of edrophonium and neostigmine when antagonizing a d-tubocurarine neuromuscular blockade in man. Anesthesiology 54:399–403
Nagashima H, Kaplan R, Radnay P, Yun H, Duncalf D, Foldes FF (1981) Relaxation with ORG NC45 during neurolept anesthesia. Anesthesiology 54:A200 (abstract)
Rosow CE, Basta SJ, Savarese JJ, Ali HH, Kniffer KJ, Moss J (1980) BW785U – correlations of cardiovascular effects with increases in plasma histamine. Anesthesiology 53:S270 (abstract)
Savarese JJ, Kitz RJ (1975) Does clinical anesthesia need new neuromuscular blocking agents? Anesthesiology 42:236–238

Savarese JJ, Ali HH, Basta SJ, Ramsey FM, Rosow CE, Lebowitz PW, Lineberry CG, Cloutier G (1980) Clinical neuromuscular pharmacology of BW785, an ultra-short-acting nondepolarizing ester neuromuscular blocking agent. Anesthesiology 53:S274 (abstract)

Sohn YJ, Scaf AHJ, Bencini A, Gregoretti S, Agoston S (1982) The comparative pharmacokinetics of vecuronium (ORG NC45) and pancuronium in man. Fed Proc 41:1336 (abstract)

Somogyi AA, Shanks A, Triggs EJ (1976) Clinical pharmacokinetics of pancuronium bromide. Eur J Clin Pharmacol 10:367–372

Somogyi AA, Shanks A, Triggs EJ (1977) The effect of renal failure on the disposition and neuromuscular blocking action of pancuronium bromide. Eur J Clin Pharmacol 12:23–29

Upton RA, Nguyen T-L, Miller RD, Castagnoli N Jr (1982) Renal and biliary elimination of vecuronium (ORG NC45) and pancuronium in rats. Anesth Analg 61:313–316

Van der Veen F, Bencini A (1980) Pharmacokinetics and pharmacodynamics of ORG NC45 in man. Br J Anaesth 52:37S–42S

Walts LF, Stirt JA, Katz RL (1981) A comparison of neuromuscular blocking effects of norcuron and pancuronium. Anesthesiology 55:A210 (abstract)

CHAPTER 29

Fazadinium Dibromide

M. B. TYERS

A. Introduction

Pharmacological studies in animals show that fazadinium produces muscle paralysis through a competitive mechanism of action on nicotinic receptors at the motor end-plate. The duration of action is considerably shorter in lower species than in primates. Direct cardiovascular effects are minimal and extensive toxicological testing in the rat, rabbit, and dog (TYERS 1975; BRITTAIN and TYERS 1973) shows fazadinium to be a very safe drug. Disappointingly, the longer action in primates indicated that fazadinium would probably be of a similar duration in humans. However, the onset time was similar to that of succinylcholine and considerably faster than all other competitive neuromuscular blocking drugs. The possibility that fazadinium may be used for intubation, particularly for "crash induction" in high risk patients encouraged progression into clinical trials.

B. Clinical Pharmacology of Fazadinium

I. Neuromuscular Blocking Properties

The first clinical studies of fazadinium were carried out by SIMPSON et al. (1972). Using the isolated forearm technique (FELDMAN and TYRELL 1970) fazadinium, is 5 mg injected into a vein in the dorsum of the hand, inhibited twitches of the adductor pollicis muscle evoked by stimulation (0.5 Hz) of the ulnar nerve (Fig. 1). The onset of action was particularly rapid (25 s) and no muscle fasciculations were observed. When the tourniquet was released 3 min after the injection of the drug the neuromuscular block persisted for a further 12 min before the muscle twitches returned to preinjection values. CHURCHILL-DAVIDSON (1973)

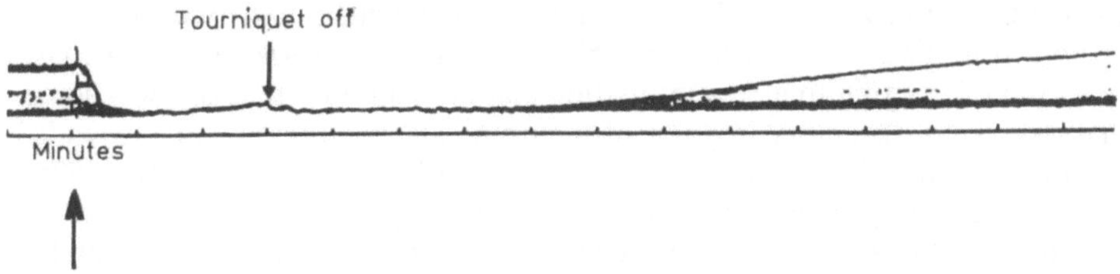

Fig. 1. Effects of fazadinium (5 mg) on indirectly evoked twitches of the isolated forearm: ulnar nerve–adductor pollicis muscle (2 Hz) preparation in humans. (SIMPSON et al. 1972)

found in conscious volunteers that fazadinium, 0.13–0.19 mg/kg i.v., reduced grip strength (50%–70%), but in anaesthetised subjects doses of 0.49–1.0 mg/kg were necessary for intubation. The duration of action was similar to that observed in subhuman primates.

In anaesthetised patients BLOGG et al. (1973) and HUGHES et al. (1976) showed that fazadinium was about half the potency of tubocurarine on muscle twitch responses while SCHUH (1975) found fazadinium to be slightly more potent with a relative potency to tubocurarine of 1.3. Many studies have since been carried out to determine the neuromuscular blocking properties of fazadinium in humans (BRACALI 1980; CECONELLO 1980; UMEH 1980; VENDRAMIN 1980; FAMEWO 1981 b; KIENLEN and DUCAILAR 1981).

1. Onset of Action and Intubation

The rate of onset for muscle paralysis can be assessed by twitch response, time to apnoea, jaw relaxation, paralysis of the vocal cords or time to intubation. All of these, except for muscle twitch responses, are dependent upon subjective assessments and thus there is considerable variation between observers. Despite these variations authors are universally agreed that fazadinium has a very rapid onset of action which is similar to that of succinylcholine. Using muscle twitch responses to assess onset time it is clear that fazadinium is more rapid than tubocurarine, alcuronium and gallamine and is only slightly slower than succinylcholine (BLOGG et al. 1973; BLACKBURN and MORGAN 1978; WILLIAMS et al. 1980; MINSAAS and STROVNER 1980). COLEMAN et al. (1973) measured onset time to intubation using a high dose of fazadinium (1.5 mg/kg) and reported that the glottis was easily and atraumatically seen 25 s after rapid central venous administration and that crash induction could be carried out as readily with fazadinium (31 s) as with succinylcholine (30 s). In contrast, the speed of onset for tubocurarine ranges from 3 to 5 min (SCHUH 1975) and for pancuronium 2 min (UNGERAR and ERASMUS 1974). The speed of onset of fazadinium-induced paralysis increases with increasing dose (COLEMAN et al. 1973; UNGERAR and ERASMUS 1974) and reaches a plateau at about 1 mg/kg (BLOGG et al. 1973).

The assessment of onset times to reach conditions satisfactory for intubation are influenced not only by dose, but also by the anatomy of the patient and the ability and determination of the anaesthetist. BLOGG et al. (1973) and INOUE et al. (1974) reported no great difficulty in rapid intubation with the dose range of 0.5–1.25 mg/kg, but considered a dose of 0.75 mg/kg to produce optimal conditions for easy intubation. SCHUH (1975) was content with 0.4–0.6 mg/kg, and ROWLANDS and FIDDLER (1978) in a series of 500 patients found excellent or satisfactory conditions with a mean dose of 0.65 mg/kg. But ARORA et al. (1973) using a dose of 1 mg/kg to assess intubation conditions found that this was somewhat less than satisfactory in a few of their patients and subsequently extended their investigations in a larger series using 1.25 mg/kg (YOUNG et al. 1974, 1975). In this latter study in which fazadinium was compared with succinylcholine 1.0 mg/kg in 240 patients they reported that there was no difference in the ability of fazadinium and succinylcholine to produce good jaw relaxation, but succinylcholine produced a higher incidence of good relaxation of the vocal cords and significantly less reaction to intubation. DELIGNE et al. (1973) reported that intubation improved with increasing doses of fazadinium and was "usually easy" after a dose of 0.75 mg/kg and "very easy" following administration of 1 mg/kg. MEHTA et al. (1977) found fazadinium 1 and 1.5 mg/kg to be similar to succinylcholine 1.25 mg/kg for crash induction both in terms of onset time and ease of intubation.

2. Prevention of Succinylcholine Muscle Pains

The postoperative muscle pains experienced in about half the patients receiving succinylcholine may be prevented by the prior administration of small doses of a competitive neuromuscular blocking drug (FOLDES 1960). The rapidity of onset of the paralysing action suggests that fazadinium would be ideally suited to this application. Indeed, three studies have shown that small doses of fazadinium, as low as 3 mg i.v. given immediately prior to succinylcholine prevented muscle fasciculations and postoperative muscle pains without prolonging unduly the time to intubation (BENNETTS and KHALIL 1981; FAMEWO 1981a; C.E. BLOGG 1982, personal communication).

3. Duration of Action

There have been several studies to determine the duration of action of fazadinium on twitch response (Table 1). There are some variations in the data reported from these studies largely because different rates of stimulation and end points were used.

Clinical methods to determine the duration of action of neuromuscular blockade include return of spontaneous respiratory or limb movements, decreased compliance, hiccoughing or a comment by the surgeon that the patient feels "tight". Numerous other factors during surgery such as respiratory depression from analgesics, the anaesthetic agent and the depth of anaesthesia, pH changes, etc. will also influence the potency and duration of neuromuscular blocking drugs. Several authors have investigated the duration of action of fazadinium during clinical anaesthesia. A summary of some of the results obtained is given in Table 2. The duration of action of fazadinium is clearly dose dependent and is comparable to that produced by pancuronium. The effects of repeated doses of fazadinium have been determined by ARORA et al. (1973), DELIGNE et al. (1973), INOUE et al. (1974), and ROWLANDS and FIDDLER (1978). Incremental doses ranging from 5 to 20 mg appear to be satisfactory for prolonging paralysis for the required duration. A dose of one-half the initial dose appears to have the same duration of action (DELIGNE et al. 1974).

4. Reversal of Neuromuscular Block

Anticholinesterase reversal of fazadinium-induced neuromuscular block has been demonstrated by many investigators. F.P. BUCKLEY (1974, personal communication) compared the reversibility of fazadinium and tubocurarine with neostigmine using the technique described by MONKS (1972). In 19 patients small incremental

Table 1. The duration of the neuromuscular blocking action of fazadinium in humans as measured by return of muscle twitch

End point	Dose of fazadinium (mg/kg i.v.)	Mean duration (min)	Reference
Time to first sign of recovery of twitch (0.5 Hz)	0.25	8.7	UNGERER and ERASMUS (1974)
	0.50	13.0	
	1.00	30.0	
Time to 10% recovery of twitch (0.5 Hz)	0.25	10.7	KEAN (1975)
	0.50	41.6	
	0.75	49.3	
	1.00	79.4	
Time to 10% recovery of twitch (0.5 Hz)	0.1	4.0	SCHUH (1975)
	0.2	16.0	
	0.4	32.0	
	0.8	113.0	
Time to 100% recovery of twitch (2 Hz)	0.50	25.0	BLOGG et al. (1973)
	0.75	41.7	
	1.00	49.0	
Time to 100% recovery of twitch (0.2 Hz)	0.67	50	DELIGNE et al. (1973)
	0.75	60	
	1.0	70	
Time for 25%–75% recovery (0.2 Hz)	0.2	9.9	HASHIMOTO et al. (1979)

Table 2. Duration of muscle relaxation with fazadinium in humans

End point	Dose (mg/kg i.v.)	Duration (min)	Reference
Return of diaphragm movement	1.0	58	ARORA et al. (1973)
Return to tidal volume of 50 ml (normal)	0.5	8	INOUE et al. (1974)
	0.75	13	
	1.0	23	
"Spontaneous recovery" (second dose required)	0.65	30.4 36.7	ROWLANDS (1975)
Second dose required	1.0–1.5	45–60	MEHTA and LEWIN et al. (1977)
"Spontaneous recovery"	1.0–1.5	40	HARTLEY and FIDLER (1977)
Second dose required	0.75	56	UNGERER and ERASMUS (1974)

doses of fazadinium or tubocurarine were given to achieve 90% inhibition of adductor pollicis muscle twitches; a further 25% of the total dose already given was injected to achieve maximal inhibition of the twitch. After 5 min, neostigmine 0.05 mg/kg, with atropine 0.02 mg/kg, reversed fazadinium more quickly than tubocurarine. The mean recovery times were 4.4 min for fazadinium and 5.3 min for tubocurarine. UNGERER and ERASMUS (1974) also showed complete reversal with neostigmine 2.5 mg with no subsequent indication of recurarisation. KEAN (1975) and SCHUH (1975) showed that the rate of reversal of fazadinium-induced muscle relaxation with neostigmine is dependent upon the extent of spontaneous recovery at the time of dosing, i.e. the greater the spontaneous recovery the faster the reversal of the residual paralysis. The effects of higher doses of fazadinium (1.0–1.5 mg/kg) with increments to produce paralysis lasting 1–3 h were also easily reversed with neostigmine 3.5 mg (COLEMAN et al. 1973). Similarly, ARORA et al. (1973), ROWLANDS and FIDDLER (1978), HARTLEY and FIDDLER (1977), and CANE and SINCLAIR (1976) all found fazadinium to be readily reversible with neostigmine. MEHTA et al. (1977) attempted an early reversal of fazadinium, 1–1.5 mg/kg, only 30 min after induction, but were unsuccessful; these doses of fazadinium normally lasted for 45–60 min in the same study. BARR and THORNLEY (1980) reported some difficulty in reversing 14 of 100 patients who received fazadinium and three of these showed signs of recurarisation in the recovery room. However, the remaining patients including some who had received up to 152 mg fazadinium reversed easily.

II. Selectivity of Action

During anaesthesia and surgery many factors affect cardiovascular stability and make it difficult to assess the effects induced by a neuromuscular blocking drug. The most likely effects to be produced are blockade of autonomic ganglionic transmission, histamine release and block of cardiac vagal muscarinic receptors.

1. Cardiovascular Effects

The cardiovascular effects of fazadinium were determined in a carefully controlled study in unpremedicated patients by SIMPSON et al. (1972) and SAVEGE et al. (1973) in which the effects of general anaesthesia, intubation and intermittent positive pressure ventilation were allowed to stabilise before administration of the relaxant. Under these conditions fazadinium 0.5 mg/kg slightly lowered systemic vascular resistance (28%) and central venous pressure (11%), but increased cardiac output (41%) which was accounted for primarily by an increase in heart rate (52%). However, there were no significant changes in mean systolic or diastolic blood pressure. Essentially similar results have been reported by others (COLEMAN et al. 1973; DELIGNE et al. 1973; PATSCHKE et al. 1974; LYONS et al. 1975; HUGHES et al. 1976). The severity of the tachycardia appears to be dependent upon the resting heart rate.

The resting bradycardia seen in anaesthetised humans, unlike in animals, may be due to high vagal tone. Thus, the effect of fazadinium on bradycardia induced by electrical stimulation of the vagus nerve in animals is probably mediated through the same antimuscarinic mechanism as that causing tachycardia in humans. High concentrations of fazadinium have been shown to enhance the effects of noradrenaline on isolated spontaneously beating guinea-pig atria MARSHALL and OJEWOLE (1979) probably through inhibition of noradrenaline uptake. However, these concentrations are probably too high to be significant clinically. In humans, PATSCHKE et al. (1974) reported tachycardia and increased dP/dt_{max} with fazadinium. They attributed the changes in dP/dt_{max} to a rate-driven inotropy rather than an effect on β_1-adrenoceptors since both preload and afterload were essentially unchanged.

ROWLANDS and FIDDLER (1978) in a large group of patients found that for fazadinium 0.67 mg/kg i.v. the most frequently observed side effect was tachycardia which was severe (i.e. resting >140 beats/min) in 22 patients and mild (<100 beats/min) in 304 patients. Arterial blood pressure changes were not common: hypotension was recorded in only 31 patients and hypertension in 11. However, not all of these changes in blood pressure could be attributed solely to the effects of fazadinium.

In summary, the most frequent side effect observed with fazadinium is mild tachycardia. However, blood pressure usually remains stable. In none of the studies referred to was there any report of abnormal electrocardiographic changes induced by fazadinium.

2. Histamine Release

Studies in the guinea-pig, dog and cat show that fazadinium does not cause a detectable release of histamine. In two studies in anaesthetised patients fazadinium 0.5 mg/kg i.v. did not increase plasma histamine levels (BLOGG et al. 1973; L. E. MARTIN and J. A. BELL 1973, unpublished work). In addition there is little clinical evidence of histamine release in adults. Mild urticaria (HARTLEY and FIDDLER 1977) and bronchospasm (ROWLANDS and FIDDLER 1978) have been reported in three patients receiving fazadinium, but none of these could be attributed directly to the relaxant drug.

III. Placental Transfer

Crash induction with fazadinium and a general anaesthetic in obstetric anaesthesia is also an important application of the drug. Prior to clinical evaluation in patients it was necessary to demonstrate that fazadinium did not cross the placental barrier in clinically significant amounts and affect the foetus. In anaesthetised pregnant rats nearing term, high intravenous doses of ^{3}H-fazadinium (5 mg/kg) caused complete cessation of spontaneous respiration in the dam and artificial ventilation was applied. At this time the motility of the foetuses in utero was unaffected and they continued to move when they were removed from the uterus with placentas intact. Determination of ^{3}H-fazadinium in the foetuses showed that only 0.08% of the dose given crossed the placental barrier to reach the foetus. Similar results were obtained with ^{3}H-fazadinium in pregnant bitches (TYERS 1975).

In early human pregnancy only minimal levels of fazadinium could be found in foetal blood (BLOGG et al. 1975) and in a larger study in patients undergoing elective caesarean section fazadinium had no adverse effects on foetal movements. In two patients undergoing termination hysterectomy only 0.005% and 0.008% of a dose of ^{3}H-fazadinium 0.75 mg/kg was detected in the foetus (BLOGG et al. 1975). These studies demonstrate that fazadinium does not cross the placental barrier in clinically significant amounts and it has subsequently been used without adverse effect in obstetric anaesthesia (CECCONELLO 1980).

C. Conclusions

Fazadinium has now been in clinical use for several years and has been given to many thousands of patients. It is clearly more rapid in onset than all other competitive neuromuscular blocking drugs and has a duration of action similar to that of pancuronium. Fazadinium is used for intubation where subsequent relaxation is required or where succinylcholine is contraindicated, such as in accident and emergency surgery. Unlike succinylcholine, no postoperative muscle pains are produced and the risk of vomiting is very low. Heart rate, however, may be raised, but usually this is short lasting. Fazadinium-induced muscle relaxation is easily reversed with anticholinesterase agents and the incidence of recurarisation is lower than for other drugs of this type. Muscle relaxation is not accompanied by increases in plasma histamine or potassium, and intraocular and intragastric pressure remain unchanged. Because fazadinium does not cross the placental barrier to affect the foetus, it has found a place particularly in obstetric surgery and caesarean section.

Acknowledgment. I am very grateful to Dr. J. A. Bell and Miss S. T. O'Driscoll for their assistance in the preparation of this manuscript.

References

Arora MV, Clark RSJ, Dundee JW, Moore J (1973) Initial clinical experience with AH 8165D, a new rapidly acting non-depolarizing muscle relaxant. Anaesthesia 28:188–191

Barr AM, Thornley BA (1980) Thiopentone and fazadinium crash induction. Anaesthesia 35:164–168

Bennetts FE, Khalil KI (1981) Reduction of post-suxamethonium pain by pretreatment with four non-depolarizing agents. Br J Anaesth 53:531–536

Blackburn CL, Morgan M (1978) Comparison of speed of onset of fazadinium, pancuronium, tubocurarine and suxamethonium. Br J Anaesth 50:361–364

Blogg CE, Savege TM, Simpson JC, Ross LA, Simpson BR (1973) A new muscle relaxant – AH 8165. Proc R Soc Med 66:1023–1027

Blogg CE, Simpson BR, Tyers MB, Martin LE, Bell (1975) Human placental transfer of AH 8165. Anaesthesia 30:23–29

Bracali AM (1980) Use of fazadinium in heart surgery. Minerva Anestesiol 46:1243–1248

Brittain RT, Tyers MB (1973) The pharmacology of AH 8165: a rapid-acting, short lasting, competitive neuromuscular blocking drug. Brit J Anaesth 45:837–843

Cane RD, Sinclair DM (1976) The use of AH 8165 for Caesarian section. Anaesthesia 31:212–214

Cecconello D (1980) Clinical experience with Fazadon in general surgery and caesarian section. Minerva Anestesiol 46:1249–1253

Churchill-Davidson HC (1973) A philosophy of relaxation. Anesth Analg Curr Res 52:495–501

Coleman AJ, O'Brien A, Downing JW, Jeal DE, Moyes DG, Leary WP (1973) AH 8165: a new non-depolarizing muscle relaxant. Anaesthesia 28:262–267

Deligne P, Bunodiere M, Prochiantz E, Loyagne F, Sicard JF, Goussard D (1973) Etude clinique prelimaire d'un nouveau curarisant de sythese, derive de l'azo-bis-arylimidazo (1,2-a) pyridinium (AH 8165). Ann Anesthiol Fr 14:407–417

Famewo CE (1981 a) Effect of fazadinium (Fazadon) on muscle fasciculations induced by succinylcholine. Can Anaesth Soc J 28:459–562

Famewo CE (1981 b) Clinical trial of fazadinium bromide (Fazadon). Can Anaesth Soc J 28:149–152

Feldman SA, Tyrrell MF (1970) A new theory of the termination of action of the muscle relaxants. Proc R Soc Med 63:692–695
Foldes FF (1960) The pharmacology of neuromuscular blocking agents in man. Clin Pharmacol Ther 1:345–395
Hartley JMF, Fiddler K (1977) Rapid intubation with fazadinium: a comparison of fazadinium with suxamethonium and alcuronium. Anaesthesia 32:14–20
Hashimoto Y, Shima T, Matsukawa S, Satou M (1979) Neuromuscular blocking potency of fazadinium in man. Anaesthesia 34:10–13
Hughes R, Payne JP, Sugai N (1976) Studies on fazadinium bromide (AH 8165): a new non-depolarizing neuromuscular blocking agent. Can Anaesth Soc J 23:36–47
Inoue K, Erdmann W, Stegbauer HP, Frey R (1974) Klinische Erfahrungen mit einem neuen Muskelrelaxans Fazadon – ein azobis-arylimidazo-pyridium Derivat. MMW 116:1839d–1844
Kean HMC (1975) The neuromuscular blocking properties of AH 8165 during halothane anaesthesia. Anaesthesia 30:333–337
Kienlen J, Ducailar J (1981) Pharmacology of fazadinium (AH 8165). Ann Anesth Francaise 6:631–639
Lyons SM, Clarke RSJ, Young HSA (1975) A clinical comparison of Fazadon and pancuronium as muscle relaxants. Br J Anaesth 47:725–729
Marshall RJ, Ojewole JAO (1979) Comparison of the autonomic effects of some currently used neuromuscular blocking agents. Br J Pharmacol 66:77P
Mehta S, Lewin K, Fiddler K (1977) Rapid intubation with fazadinium and suxamethonium. Can Anaesth Soc J 24:270–274
Minsaas B, Strovner J (1980) Artery-to-muscle onset time for neuromuscular blocking drugs. Br J Anaesth 52:403–407
Monks (1972) The reversal of non-depolarizing relaxants. Anaesthesia 27:313–318
Patschke D, Bruckner JB, Tarnow J, Weymar A (1974) Einfluß von Fazadon – eines neuen, nicht depolarisierenden Muskelrelaxans – auf die Hämodynamik des Menschen. Anaesthesist 23:430–433
Rowlands DE, Fiddler K (1978) Fazadinium in anaesthesia. Br J Anaesth 50:289–293
Savege TM, Blogg CE, Ross L, Lang M, Simpson BR (1973) The cardiovascular effects of AH 8165, a new non-depolarising muscle relaxant. Anaesthesia 28:253–261
Schuh FT (1975) Clinical neuromuscular pharmacology of AH 8165D, an azobis-arylimidazo-pyridinium compound. Anaesthesist 24:151–156
Simpson BR, Strunin L, Savege TM, Walton B, Blogg CE, Foley EL, Maxwell MP, Ross LA, Harris DM (1972) An azobis-arylomidazo-pyridinium derivative: a rapidly acting non-depolarizing muscle relaxant. Lancet I:516–519
Tyers MB (1975) Pharmacological studies on new, short-lasting, competitive neuromuscular blocking drugs. Ph.D thesis (Council for National Academic Awards)
Umeh BU (1980) Initial clinical experience with fazadinium bromide during anaesthesia in Nigerians. Med J Zambia 15:16–18
Ungerer MJ, Erasmus FR (1974) Clinical evaluation of a new non-depolarizing muscle relaxant. S Afr Med J 48:2561–2564
Williams NE, Webb SN, Calvey TN (1980) Differential effects of myoneural blocking drugs on neuromuscular transmission. Br J Anaesth 52:1111–1115
Vendramin (1980) Clinical applications of fazadinium bromide, a new non-depolarizing muscle relaxant, in paediatric anaesthesia. Act Anaesth Italica 31:16–19
Young HSA, Clarke RSJ, Dundee JW (1974) Tracheal intubation with AH 8165, a comparison with suxamethonium. Br J Anaesth 46:317–324
Young HSA, Clarke RSJ, Dundee JW (1975) Intubating conditions with AH 8165 and suxamethonium. Anaesthesia 30:30–33

CHAPTER 30

Atracurium

R. HUGHES and J. P. PAYNE

A. Introduction

Atracurium was first used in humans in January 1979; a preliminary report on its clinical properties appeared 1 year later (HUNT et al. 1980) and a more definitive account followed in 1981 (PAYNE and HUGHES 1981). More recently the present position of the drug has been assessed in the *Lancet* (ANNOTATION 1983) which concluded that atracurium seemed to fulfil all the requirements necessary for a neuromuscular blocking drug with an intermediate duration of action and as such offered a degree of flexibility in the management of anaesthesia that was previously lacking.

B. Quantitative Assessment

I. Neuromuscular Blocking Activity

The first quantitative evaluation of the neuromuscular blocking properties of atracurium was carried out by PAYNE and HUGHES (1981) based on techniques designed to compare the tetanic. and twitch responses of the adductor pollicis muscles (SUGAI et al. 1973; SUGAI 1974). An intravenous injection of atracurium 0.2 mg/kg completely blocks the tetanic response, but only partially blocks that of the single twitch (Fig. 1). However, full block of both responses is produced with doses in the range of 0.3–0.6 mg/kg. Endotracheal intubation is possible between 1 and 2 min after the injection of such doses and the block is readily reversed by neostigmine 2.5–5.0 mg preceded by atropine 1.2 mg (Fig. 2). It is also antagonised by edrophonium 0.75 mg/kg without subsequent muscle weakness (BAIRD and KERR 1983). Similar results were obtained by other workers (CALVEY et al. 1983; HUNTER et al. 1982a) who used train-of-four stimulation to assess the degree of neuromuscular block. In this connection it has been shown that there are no significant differences in the time course of the response to tetanic and train-of-four stimulation of the adductor pollicis muscle during neuromuscular block by atracurium (HUGHES and PAYNE 1983a).

The course of action of atracurium is appreciably shorter than that of the more established competitive blocking agents (PAYNE and HUGHES 1981) and although the total duration of action is dose related (Table 1), once recovery of muscle activity has begun the rate of recovery appears to be independent of dose (MADDEN et al. 1983). Moreover, with incremental doses of the order of 0.05–0.2 mg/kg the degree of block and its duration are remarkably consistent with the

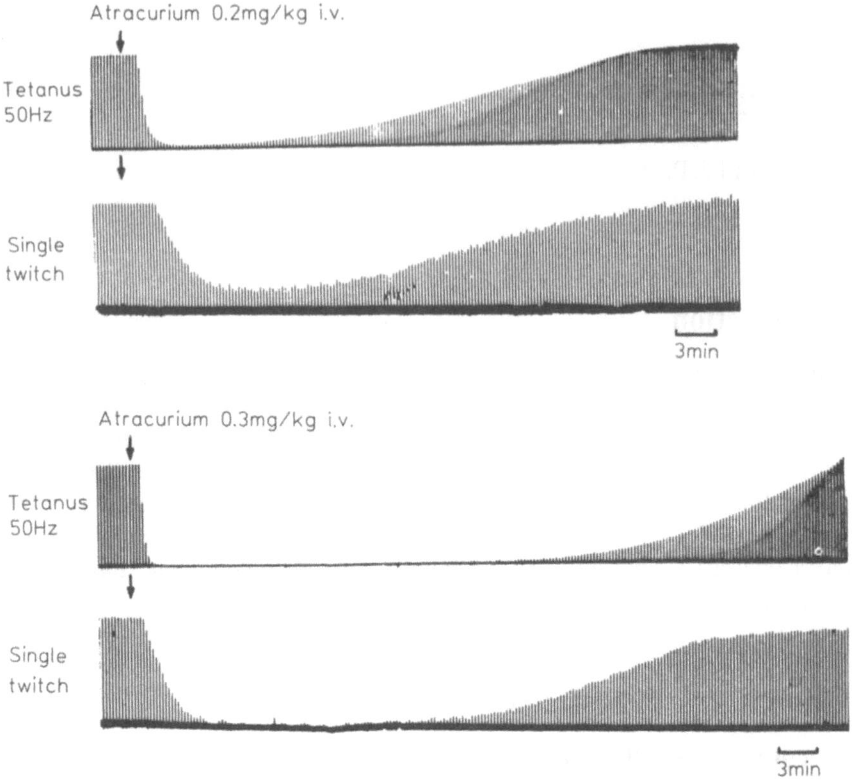

Fig. 1. Tracings from two anaesthetised patients showing neuromuscular paralysis of the tetanic and single twitch responses of the adductor pollicis muscle by atracurium 0.2 and 0.3 mg/kg i.v. (PAYNE and HUGHES 1981)

Table 1. Time course of neuromuscular blockade of the tetanic responses of the adductor pollicis muscle by atracurium. Mean results are quoted with standard error. Asterisks denote values significantly different from the 0.2 mg/kg dose at the 5% (*) and 0.1% (***) levels, respectively

Dose (mg/kg)	Onset of maximum block (min)	Duration of maximum block (min)	95% Recovery of peak contraction[a] (min)
0.2 (N=9)	2.8±0.3	5.6±0.6	28.9±2.3
0.3 (N=7)	* 1.9±0.1	*** 15.5±1.8	34.9±1.9
0.6 (N=6)	*** 1.2±0.1	*** 33.7±3.0	34.5±4.4
0.9 (N=5)	*** 1.0±0.1	*** 47.0±3.1	* 42.2±3.6

[a] Measured from time of onset of recovery

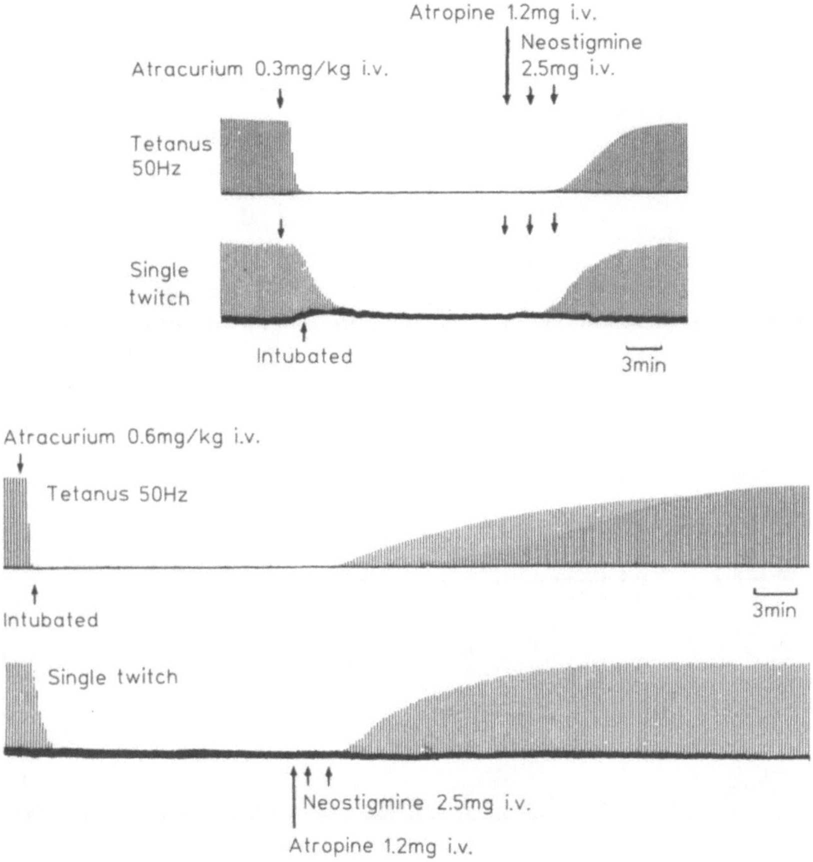

Fig. 2. Recordings of the tetanic and single twitch responses of the adductor pollicis muscle from two anaesthetised patients. *Upper tracings* show that intubation was accomplished within 2 min of administration of atracurium 0.3 mg/kg i.v. and, 20 min later, complete block was reversed by neostigmine 2 × 2.5 mg preceded by atropine 1.2 mg i.v. *Lower tracings* show that intubation was achieved within 1 min of administration of atracurium 0.6 mg/kg and neostigmine 2 × 2.5 mg preceded by atropine 1.2 mg i.v., given 20 min later, antagonised complete block

implication that within the clinical dose range cumulative effects are unlikely (Fig. 3). This interpretation is supported by the pattern of recovery after continuous infusion (Fig. 4). Given in this fashion, doses in the range 8–10 μg kg^{-1} min^{-1}, are sufficient to maintain complete block during 1 h infusion at the end of which recovery follows a similar pattern to that seen after bolus injections (HUGHES and PAYNE 1983 b).

II. Cardiovascular Effects

Cardiovascular studies by HUGHES and PAYNE (1983 b) indicated that mean arterial blood pressure and mean heart rate were not significantly changed by the doses of atracurium needed to produce complete block of the single twitch and

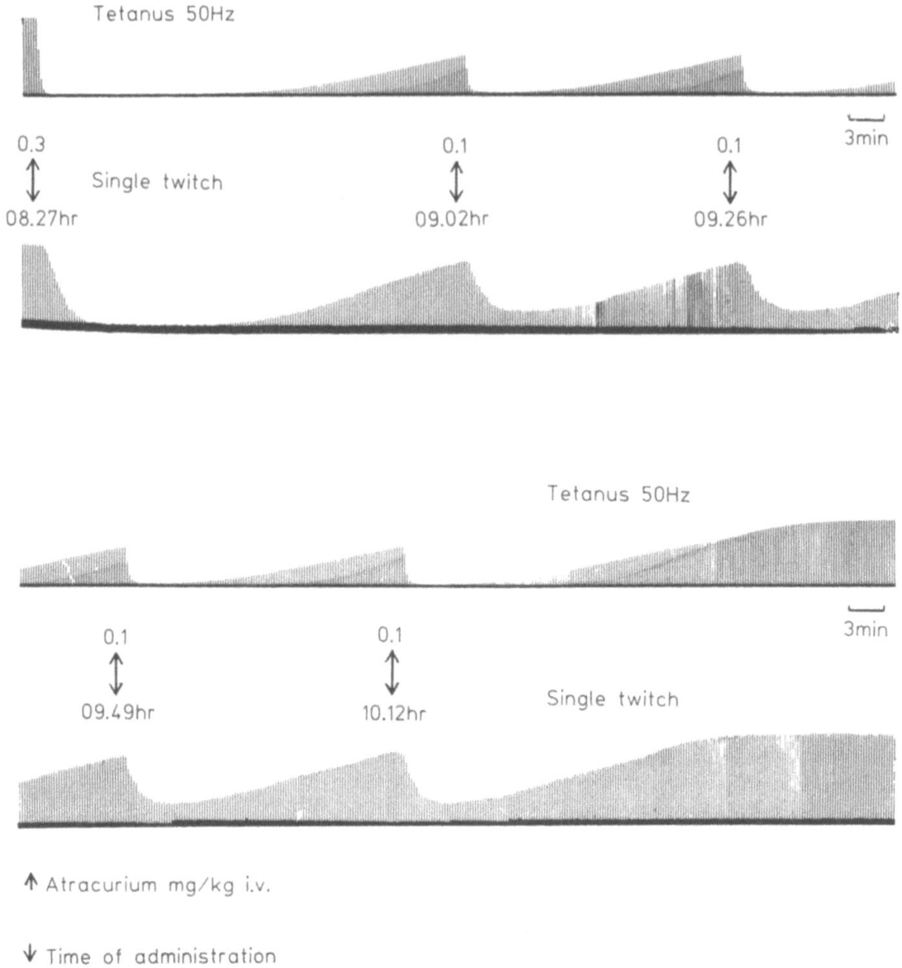

Fig. 3. Continuous recording of the tetanic and single twitch responses of the adductor pollicis muscle from a patient who received an initial dose of atracurium 0.3 mg/kg i.v. Incremental doses of 0.1 mg/kg were given when recovery of the peak tetanic contraction had reached 50%. *Arrows* indicate dose (mg/kg i.v.) and time of administration. (PAYNE and HUGHES 1981)

tetanic responses of the adductor pollicis muscle (Fig. 5). A similar absence of changes in arterial blood pressure and heart rate have been reported by other investigators (CALVEY et al. 1983; FLYNN et al. 1983; FOLDES et al. 1983; FRAGEN et al. 1982; HUNTER et al. 1982a, b, 1983; POKAR and BRANT 1983; ROWLANDS 1983a, b; RUPP et al. 1983). However, there is some disagreement about the cardiovascular effects of atracurium. BASTA et al. (1982a) found significant decreases in mean arterial pressure and increases in heart rate after doses of 0.5 and 0.6 mg/kg i.v. which they attributed to histamine release and BARNES et al. (1983) reported that atracurium 0.6 mg/kg i.v. produced a transient fall in mean arterial pressure ranging from 7% to 24% of control without significant changes in heart

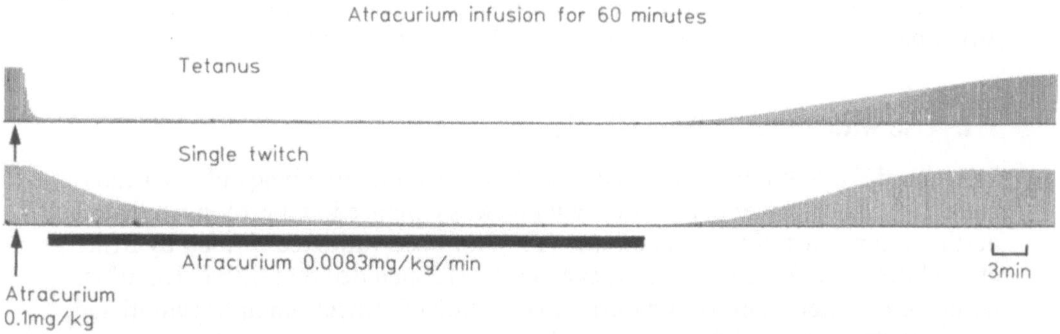

Fig. 4. Recording of the tetanic and single twitch responses of the adductor pollicis muscles from a patient who received an initial bolus dose of atracurium 0.1 mg/kg followed by an infusion at a rate of 0.0083 mg kg^{-1} min^{-1} for 60 min. (HUGHES and PAYNE 1983a)

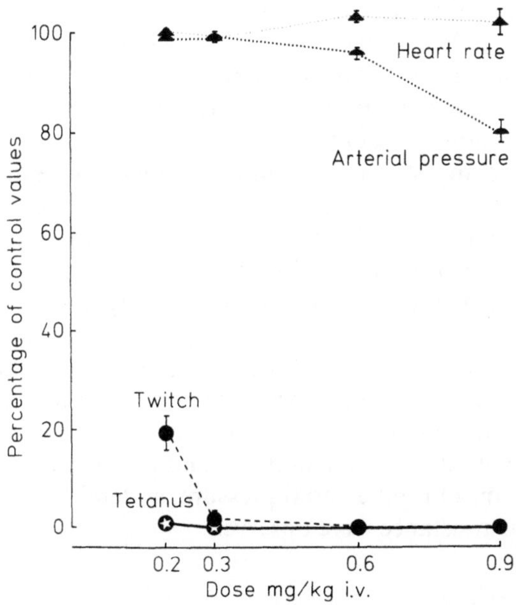

Fig. 5. Neuromuscular and cardiovascular effects of atracurium given at different doses to groups of 5–10 anaesthetised patients. Mean results are quoted and are expressed as percentages of the control values. *Vertical lines* indicate standard error. (HUGHES and PAYNE 1983a)

rate in 5 of 18 patients. HUGHES and PAYNE (1983b) found a similar fall in blood pressure, but only when the intravenous dose was raised to 0.9 mg/kg. The effects on the heart rate were not significant. In one patient in a group of 16 given atracurium 0.3 or 0.4 mg/kg i.v. PHILBIN et al. (1983) observed a typical histamine response with flushing, transient hypotension, a decrease in systemic vascular resistance and an increase in cardiac output. It remains to be established whether histamine release is more likely to occur when atracurium is administered with a particular combination of anaesthetic agents or after a particular mode of administration and whether or not it is of clinical significance. Nevertheless, it may be advisable to inject atracurium slowly or in divided doses to poor risk patients.

III. Use with Volatile Anaesthetics

Payne and Hughes (1981) demonstrated that neuromuscular blockade by atracurium, like that of other competitive agents, was enhanced and prolonged by anaesthetic concentrations of halothane (1%), an observation confirmed by Stirt et al. (1983). In contrast, Savarese et al. (1982a) concluded that the neuromuscular blocking potency, onset and duration of action of atracurium under halothane (0.7%–0.9%) did not differ significantly from data obtained under balanced anaesthesia.

There is evidence that isoflurane (0.75%) also increases the potency of atracurium (Rupp et al. 1982, 1983). In their study the dose–response curve was moved to the left when compared with nitrous oxide–fentanyl anaesthesia, but there was no difference in onset time and duration of neurmuscular block at comparable doses. Sokoll et al. (1983) studied the haemodynamic effects of atracurium 0.2 and 0.4 mg/kg in patients under nitrous oxide–isoflurane anaesthesia and found there were no significant changes in mean arterial pressure, systemic vascular resistance or central venous pressure.

During enflurane anaesthesia (minimum alveolar concentration 1.16%) the dose of atracurium can be reduced by approximately 25%–30% to obtain the same degree of neuromuscular block achieved during balanced anaesthesia (Ramsey et al. 1982a, b). Recovery from blockade by atracurium 0.18 mg/kg i.v. was significantly prolonged compared with that following a similar degree of block during balanced anaesthesia. However, after a higher dose of atracurium 0.36 mg/kg i.v. recovery was similar during enflurane and balanced anaesthesia. Other studies by Hilgenberg and Stoelting (1982) in patients during nitrous oxide–enflurane anaesthesia (1.0%–1.25%) demonstrated that atracurium 0.4 mg/kg provided adequate intubation conditions and no significant changes in heart rate, cardiac index, mean right arterial pressure, systemic mean arterial pressure or systemic vascular resistance were observed.

C. Pharmacokinetics

I. In Vitro Degradation

The work of Merrett et al. (1983) has demonstrated that at physiological pH and temperature the major route of inactivation of atracurium is by Hofmann elimination. When samples were incubated in Tris buffers at 37 °C a rise in pH from 6.9 to 7.6 resulted in a fourfold increase in the rate of degradation. In human plasma the rate of inactivation was about 1.8 times faster than in Tris buffer alone, but no difference was observed in the rate of degradation when atracurium was incubated in either normal plasma or in atypical plasma deficient in pseudocholinesterase. The effect of temperature was also demonstrated; the half-life in normal plasma was 18 min at 37 °C and 49 min when the temperature was reduced to 23 °C. In a clinical study the rate of recovery was slightly faster when the pH was raised by 0.16–0.28 units (HUGHES and PAYNE 1981).

II. Pharmacokinetic Profile

On the basis of plasma levels obtained from anaesthetised patients a complete pharmacokinetic and pharmacodynamic model for atracurium has been proposed by WEATHERLEY et al. (1983). Plasma concentrations of atracurium, together with simultaneous measurement of muscle twitch heights, indicated that the kinetics of atracurium could be described by a double exponential process. The plasma concentration data after intravenous bolus administration to humans were very reproducible and directly proportional to the dose which was varied over a threefold range. The half-life elimination for plasma was independent of dose. This model was used to predict the known plasma level after various rates of intravenous infusions and when tested in practice good agreement was obtained. Hence a very consistent model of the manner of distribution, elimination and physiological response to atracurium has been established which can be used to predict any desired neuromuscular block with minimal drug loading. Thus, extremes of plasma concentration or of neuromuscular block are avoided which may have particular application to lengthy surgical procedures with high risk patients. In anaesthetised patients WARD and WRIGHT (1983) found a good correlation between the plasma concentration of atracurium and its effects at the neuromuscular junction and the results demonstrated good reproducibility between different patients.

D. Comparative Studies with Other Neuromuscular Blocking Agents

I. Suxamethonium

Although atracurium is fairly rapid in onset, the occurrence of full block is not as fast as with suxamethonium (PAYNE and HUGHES 1981). In a direct comparison, using train-of-four monitoring, SCOTT and GOAT (1982) concluded that full peripheral paralysis took longer to achieve with atracurium 0.6 mg/kg i.v. than with suxamathonium 1 mg/kg i.v. No postoperative problems were encountered in those patients who received atracurium whereas most patients in the suxamethonium group complained of moderate to severe muscle pains. GERGIS et al. (1983) compared intubation conditions after administration of atracurium 0.4 and 0.5 mg/kg i.v. and suxamethonium 1 mg/kg i.v. Ease of intubation was rated as excellent in 90%–100% of all patients about 2 min after the administration of atracurium and 1 min after suxamethonium administration.

II. Tubocurarine and Dimethyltubocurarine

The rate of recovery of the tetanic response after neuromuscular block by atracurium 0.2 mg/kg i.v. was at least twice as fast as that after administration of tubocurarine 0.2 mg/kg ($P<0.01$) and four times faster than the rate of recovery after administration of dimethyltubocurarine 0.075–0.2 mg/kg ($P<0.001$) (PAYNE and HUGHES 1981). In dose which produced 97.7%–99.8% block of the twitch response SAVARESE et al. (1982b) found that there was no significant difference between the onset of action of atracurium 0.5 mg/kg i.v., tubocurarine 0.5 mg/kg i.v., and dimethyltubocurarine 0.3 mg/kg i.v. However, the time to 25% recovery of the twitch height was significantly shorter after administration of atracurium ($P<0.01$) than after tubocurarine and dimethyltubocurarine administration.

III. Pancuronium and Vecuronium

The work of PAYNE and HUGHES (1981) has shown that the rate of recovery from atracurium 0.2 mg/kg i.v. is about three times faster than the rate of recovery from pancuronium 0.05–0.06 mg/kg i.v. ($P<0.01$). In a randomised comparative study GRAMSTAD et al. (1982, 1983) using equipotent doses of atracurium 0.33 mg/kg i.v. and pancuronium 0.075 mg/kg i.v. found that although there were no significant differences in the onset times to 95% block of the twitch response, the duration of action to 25% recovery and the time course from 10% to 25% recovery were significantly shorter after atracurium than after pancuronium administration ($P<0.01$). These results were in agreement with those of SAVARESE et al. (1982a). A comparison of intubation conditions using atracurium 0.6 and 0.8 mg/kg i.v. and pancuronium 0.08 and 0.1 mg/kg has been carried out by TWOHIG et al. (1983). Smooth intubation was not possible before 60 s with either drug. There were no significant differences between the intubation conditions with atracurium 0.6 mg/kg i.v. and pancuronium 0.08 mg/kg i.v., but atracurium 0.8 mg/kg i.v. provided slightly better intubation conditions than pancuronium 0.1 mg/kg i.v. at 30 and 60 s after injection.

A comparison between atracurium and vecuronium has been carried out by FRAGEN et al. (1982) who calculated that the i.v. doses causing 90% twitch depression for atracurium and vecuronium were 0.189 mg/kg and 0.045 mg/kg i.v., respectively. Vecuronium had a quicker onset of action, shorter duration of action and a faster recovery index than atracurium at this dose when measured by the twitch response, but at three times the 90% paralysing dose the effects of both drugs were similar. GRAMSTAD et al. (1982, 1983) also reported that the duration of action to 25% recovery of the twitch height was significantly shorter after vecuronium than after atracurium administration ($P<0.01$) whereas the onset time to 95% block and the time course from 10% to 25% recovery were similar for both drugs. FOLDES et al. (1983) compared atracurium 0.4 or 0.5 mg/kg i.v. with vecuronium 0.1 mg/kg i.v. and concluded that there were no clinically significant differences in the neuromuscular effects of the two drugs although spontaneous recovery to 90%–95% of control was faster ($P<0.01$) with vecuronium (36 min) than with atracurium (54 min). No significant cardiovascular effects occurred after atracurium or vecuronium administration.

IV. Histamine-Releasing Potential of Atracurium, Dimethyltubocurarine and Tubocurarine

The histamine-releasing potential of atracurium has been compared with that of dimethyltubocurarine and tubocurarine (BASTA et al. 1982b). A dose of these drugs which increases plasma histamine to about 200% of control will result in clinically and statistically significant changes in arterial pressure and heart rate. Atracurium's ability to release histamine, relative to its neuromuscular blocking potency, was approximately one-half that of dimethyltubocurarine and less than one-third that of tubocurarine. Signs of histamine release were observed in 4 of 50 patients who were given atracurium 0.5 mg/kg i.v. (FOLDES et al. 1983). In three of these patients reddening and wealing occurred along the injected vein and the fourth patient also developed an expiratory wheeze over both lungs. The administration of repeat doses of atracurium was not accompanied by signs of further histamine release in any of the four patients.

V. Incremental Dosage of Atracurium and Vecuronium

As part of a broader study ALI et al. (1982, 1983), using the train-of-four technique, have studied the effects after incremental doses of atracurium 0.08 mg/kg and vecuronium 0.02 mg/kg i.v. administered on five occasions when the first response of the train-of-four had recovered to 25% of control. After the final incremental dose of vecuronium the train-of-four ratio, at the point of recovery of the first response to 95% of control height, was significantly greater than that of atracurium ($P<0.001$). Thus, vecuronium appears to show more residual fade and greater cumulative effects after repeated dosages than atracurium. Spontaneous decomposition of atracurium by Hofmann elimination may explain the apparent difference in recovery pattern from vecuronium which depends significantly upon biliary excretion for its clearance.

E. Specialised Uses

I. Obstetric Anaesthesia

In patients undergoing caesarean section, suxamethonium 1.5 mg/kg i.v. was used for endotracheal intubation; thereafter atracurium 0.3 mg/kg i.v. provided good surgical relaxation for about 30 min without cardiovascular disturbance (FLYNN et al. 1982; FRANK et al. 1983). Incremental doses of 0.1 or 0.2 mg/kg i.v. were given as required. Full recovery from neuromuscular blockade occurred spontaneously in most patients, but when necessary intravenous neostigmine 2.5 mg, preceded by atropine 1.2 mg, induced complete reversal in 1–3 min. In the neonates no adverse effects on Apgar scores or on the time to sustained respiration could be attributed to atracurium. Ratios of the umbilical vein to maternal vein plasma concentrations suggest that 5%–20% of the dose of atracurium crossed the placenta.

II. Paediatric Anaesthesia

Infants and young children appear to be more resistant than adult patients to the effects of atracurium. COOK et al. (1982) and BRANDOM et al. (1983) administered atracurium to children aged between 2 and 16 years. The doses based on total body weight required to produce 50% and 95% block of the train-of-four response during halothane anaesthesia (0.8%) in young children (2–10 years) were higher than those needed by older children (11–16 years). In the younger children doses of atracurium required to produce 50% and 90% block during halothane anaesthesia were comparable with those used during nitrous oxide–thiopentone–fentanyl anaesthesia. When the dosage was calculated on the basis of surface area there was no difference between the younger and older children. Following administration of atracurium 0.4 mg/kg i.v. about 99% neuromuscular block developed within 2–5 min. No important cardiovascular changes were observed and intubation conditions were excellent in most patients. One patient developed flushing of the neck and trunk, but did not develop hypotension. The time to 25% and 95% recovery following doses of atracurium that produced 95% block were 15 and 29 min, respectively. At equal multiples of the ED_{95} the times to 95% recovery in children were significantly shorter than those reported in adults (BASTA et al. 1982a; GRAMSTAD et al. 1982; PAYNE and HUGHES 1981; SAVARESE et al. 1982b). In another study by NIGHTINGALE and BUSH (1983) children in the age ranges 1–5, 6–10, and 11–15 years were given atracurium 0.6 mg/kg i.v. In most patients intubation conditions were excellent in 60 s, but in the age group 6–10 years conditions for intubation tended to be less satisfactory. The initial dose of atracurium lasted for almost 36 min with no clear difference between groups, and incremental doses of 0.1 mg/kg i.v. produced further clinical relaxation for 13–17 min. Most patients recovered spontaneously from neuromuscular blockade, but atropine and neostigmine were given when induced reversal was deemed necessary. A local "histamine-like" response was seen in most patients as a reddening of the vein proximal to the site of injection of atracurium. In all groups systolic blood pressure and heart rate increased slightly after injection of the drug, but these changes were transient and probably associated with endotracheal intubation.

III. Routine Anaesthesia in Elderly and Severely Ill Patients

ROWLANDS (1983a, b) has reported the use of atracurium in routine anaesthesia in 269 patients between the ages of 18 and 86 years, nearly half of whom were aged 65 years or over, and some of whom were severely ill. Provided that a dose of 0.5–0.6 mg/kg i.v. was used, in most patients endotracheal intubation was carried out within 90 s of the injection of atracurium. Surgical relaxation was excellent and arbitrary increments of 10 mg were given when indicated on clinical grounds. In most patients spontaneous breathing had returned before the routine administration of neostigmine and atropine. No problems were found in antagonism even when atracurium 20 mg i.v. had been given for closure of the peritoneum or if the operation had been completed in less than 15 min. Any changes in heart rate and arterial blood pressure were related to intubation, surgery of analgesia even for

patients giving a history of allergy or asthma. Nine patients developed a transient chest rash, only two of whom had given a previous history of allergy. There were no complications due to atracurium.

IV. Patients in Renal Failure

The fact that the inactivation of atracurium is independent of renal function implies that the drug may be of particular value in renal failure. Its use in such patients has been evaluated by UTTING et al. (1982) and HUNTER et al. (1982 b, 1983). Doses of atracurium 0.5 mg/kg i.v. provided adequate surgical relaxation in patients with impaired or absent renal function. Reappearance of the first response of the train-of-four in about 30 min was as rapid as that in patients with normal renal function. Even when up to nine incremental doses of atracurium 0.2 mg/kg i.v. were given at intervals of about 30 min no evidence of cumulative effects was observed. Reversal of neuromuscular block with neostigmine and atropine was as prompt as that in patients with normal renal function and no signs of recurarisation were seen.

V. Patients with Coronary Artery Disease

The haemodynamic effects of atracurium in patients with coronary artery disease have been studied by PHILBIN et al. (1983). In a group given atracurium 0.3 mg/kg i.v., mean and diastolic blood pressure fell significantly at 2 min after injection, but the changes were not significant 5 and 10 min later. No significant changes in cardiac output nor in systemic vascular resistance occurred. One patient exhibited a typical histamine response with flushing, moderate hypotension and an increase in cardiac output. If this patient was excluded the haemodynamic changes were no longer significant. In a second group of patients given atracurium 0.4 mg/kg i.v. none of the haemodynamic variables measured showed significant changes.

VI. Infusion for Long Procedures Including Cardiopulmonary Bypass

The feasibility of administering atracurium by continuous infusion was first tested by MADDEN et al. (1982) and its use for long surgical procedures has since been investigated by FLYNN et al. (1983). A single bolus dose of atracurium 0.6 mg/kg was given intravenously after which endotracheal intubation was accomplished. When the first response of the train-of-four had recovered to 10%–20% of the preinjection height, atracurium was infused at an average rate of 0.0068 mg kg^{-1} min^{-1} for 60–178 min. Surgical relaxation was good and cardiovascular changes were minimal. When the infusion was stopped full spontaneous recovery occurred in about 45 min. Administration of neostigmine and atropine when the train-of-four ratio had reached 50% reduced the total time for full recovery to about 27 min. In patients undergoing cardiopulmonary bypass an infusion of atracurium at an average rate of 0.0067 mg kg^{-1} min^{-1} was sufficient for satisfactory surgical relaxation before cooling. During hypothermia, when the body temperature was lowered to 25° or 26 °C for 1–2 h, a mean infusion rate of 0.0040 mg kg^{-1} min^{-1} i.v. provided adequate surgical relaxation; this rate was significantly slower than that during normothermia ($P<0.01$). The fact that less drug was required during hypothermia can be attributed mainly to the reduced inactivation of atracurium by Hofmann elimination at low temperatures.

F. Conclusions

Atracurium is a potent highly specific drug which acts competitively to produce full neuromuscular block within the dose range 0.3–0.6 mg/kg with virtually no cardiovascular side effects. Endotracheal intubation can be accomplished within 2 min after injection of the drug and paralysis is enhanced by halothane and other volatile anaesthetics. Neuromuscular block is antagonised by neostigmine and edrophonium. Atracurium's course of action is intermediate between that of suxamethonium and the established competitive blocking drugs. Although the total duration of action is dose related, once recovery of muscle activity has begun the rate of recovery seems to be independent of dose. With suitable increments the degree of block and its duration are remarkably consistent which suggests that cumulative effects are unlikely. Atracurium is equally effective by continuous infusion and rates of the order of 0.008 mg kg^{-1} min^{-1} will maintain complete block until the infusion is discontinued when recovery follows a similar course to that after bolus injections. Atracurium's ability to relese histamine relative to its neuromuscular blocking potency is approximately one-half that of dimethyltubocurarine and less than one-third that of tubocurarine.

Atracurium has been used to maintain surgical relaxation in patients undergoing caesarean section since it does not cross the placenta in clinically significant amounts. Infants and young children are most resistant than adults to the effects of atracurium. In elderly and severely ill patients there have been no problems with reversal and residual curarisation even after short operations. Atracurium has been used to advantage in patients in renal failure. Even at multiples of the usual dose no evidence of accumulation was detected, presumably a reflection of the drug's lack of dependence on renal mechanisms for its elimination. The fact that the inactivation of atracurium is temperature dependent has led to its effective use in open-heart surgery when the dose requirement was halved during induced hypothermia (25°–26 °C).

References

Ali HH, Savarese JJ, Basta SJ, Sunder N, Gionfriddo M (1982) Comparative patterns of recovery of three new non-depolarizing relaxants: BW444U, BW33A (atracurium) and ORG NC45 (vecuronium). Anesthesiology 57:A263

Ali HH, Savarese JJ, Basta SJ, Sunder N, Gionfriddo M (1983) Evaluation of cumulative properties of three new non-depolarizing relaxants: BW444U, atracurium and vecurium. Br J Anaesth 55:107–111S

Annotation (1983) Atracurium. Lancet 1:394–395

Baird WLM, Kerr WJ (1983) Reversal of atracurium with edrophonium. Br J Anaesth 55:63–66S

Barnes PK, Thomas VJE, Boyd I, Hollway T (1983) Comparison of the effects of atracurium and tubocurarine on heart rate and arterial pressure in anaesthetised man. Br J Anaesth 55:91–94S

Basta SJ, Ali HH, Savarese JJ (1982a) Clinical pharmacology of atracurium (BW33A): a new neuromuscular blocking agent. Anaesth Analg (Cleve) 61:723–729

Basta SJ, Savarese JJ, Ali HH, Moss J, Gionfriddo M (1982b) Histamine-releasing potencies of atracurium besylate (BW33A), metocurine and d-tubocurarine. Anesthesiology 57:A261

Brandom BW, Rudd GD, Cook DR (1983) Clinical pharmacology of atracurium in pediatric patients. Br J Anaesth 55:117–121S

Calvey TN, Macmillan RR, West DM, Williams NE (1983) Electromyographic assessment of neuromuscular blockade induced by atracurium. Br J Anaesth 55:57–62S

Cook DR, Rudd GD, Brandom BW (1982) Clinical pharmacology of atracurium (BW33A) in pediatric patients. Anesthesiology 57:A415

Flynn PJ, Frank M, Hughes R (1982) Evaluation of atracurium in Caesarian section using train-of-four responses. Anesthesiology 57:A286

Flynn PJ, Hughes R, Walton B, Jothilingham S (1983) Use of atracurium infusions for general surgical procedures including cardiac surgery with induced hypothermia. Br J Anaesth 55:135–138S

Foldes FF, Nagashima H, Boros M, Tassonyi E, Fitzal S, Agoston S (1983) Muscular relaxation with atracurium, vecuronium and Duador under balanced anaesthesia. Br J Anaesth 55:97–103S

Fragen RJ, Robertson EN, Booij LHDG, Crul JF (1982) A comparison of vecuronium and atracurium in man. Anesthesiology 57:A253

Frank M, Flynn PJ, Hughes R (1983) Atracurium in obstetric anaesthesia. A preliminary report. Br J Anaesth 55:113–114S

Gergis SD, Sokoll MD, Mehta M, Kemmostsu O, Rudd GD (1983) Intubating conditions after atracurium and suxamethonium. Br J Anaesth 55:83–86S

Gramstad L, Lilleaasen P, Minsaas B (1982) Onset time and duration of action for atracurium, Org NC45 and Pancuronium. Br J Anaesth 54:827–830

Gramstad L, Lilleaasen P, Minsaas B (1983) A comparative study of atracurium, vecuronium (Org NC45) and pancuronium. Br J Anaesth 55:95–96S

Hilgenberg JC, Stoelting RK (1982) Systemic vascular responses to atracurium during enflurane-nitrous oxide anaesthesia in healthy patients. Anesthesiology 57:A260

Hughes R, Payne JP (1983a) Atracurium: clinical assessment by tetanic and train-of-four responses of the adductor pollicis muscle. Br J Anaesth 55:239P

Hughes R, Payne JP (1983b) Clinical assessment of atracurium using the single twitch and tetanic responses of the adductor pollicis muscles. Br J Anaesth 55:47–52S

Hunt TM, Hughes R, Payne JP (1980) Preliminary studies with atracurium in anaesthetised man. Br J Anaesth 52:238–239P

Hunter JM, Jones RS, Utting JE (1982a) Use of atracurium during general surgery monitored by the train-of-four stimuli. Br J Anaesth 54:1243–1250

Hunter JM, Jones RS, Utting JE (1982b) Use of atracurium in patients with no renal function. Br J Anaesth 54:1251–1258

Hunter JM, Jones RS, Utting JE (1983) Atracurium in renal failure. Br J Anaesth 55:129S

Madden AP, Hughes R, Payne JP (1982) Recovery from neuromuscular blockade after infusion of atracurium. Br J Anaesth 54:226–227P

Madden AP, Hughes R, Payne JP (1983) Assessment of tetanic fade following atracurium. Br J Anaesth 55:53–55S

Merrett RA, Thompson CW, Webb FW (1983) In vitro degradation of atracurium in human plasma. Br J Anaesth 55:61–66

Nightingale DA, Bush GH (1983) Atracurium in paediatric anaesthesia. Br J Anaesth 55:115S

Payne JP, Hughes R (1981) Evaluation of atracurium in anaesthetised man. Br J Anaesth 53:45–54

Philbin DM, Machaj VR, Tomichek RC, Schneider RC, Alban JC, Lowenstein E, Lineberry CC (1983) Haemodynamic effects of bolus injections of atracurium in patients with coronary artery disease. Br J Anaesth 55:131–134S
Pokar H, Brandt L (1983) Haemodynamic effects of atracurium in patients after cardiac surgery. Br J Anaesth 55:139S
Ramsey FM, White PA, Stullken EH, Allen LL, Roy RC (1982a) Neuromuscular and haemodynamic effects of atracurium during enflurane anaesthesia. Anesthesiology 57:A254
Ramsey FM, White PA, Stullken EH, Allen LL, Roy RC (1982b) Enflurane potentiation of neuromuscular blockade by atracurium. Anesthesiology 57:A255
Rowlands DE (1983a) Atracurium in clinical anaesthesia. Br J Anaesth 55:125–128S
Rowlands DE (1983b) Atracurium in the severely ill. Br J Anaesth 55:123–145S
Rupp SM, Fahey MR, Miller RD (1982) Neuromuscular blocking effects of atracurium during N_2O-fentanyl or isoflurane anaesthesia. Anesthesiology 57:A257
Rupp SM, Fahey MR, Miller RD (1983) Neuromuscular and cardiovascular effects of atracurium during nitrous oxide-fentanyl and nitrous oxide-isoflurane anaesthesia. Br J Anaesth 55:67–70S
Savarese JJ, Basta SJ, Ali HH, Sunder N, Moss J (1982a) Neuromuscular and cardiovascular effects of BW33A (atracurium) in patients under halothane anaesthesia. Anesthesiology 57:A262
Savarese JJ, Ali HH, Basta SJ, Gionfriddo M, Sunder N (1982b) Comparative neuromuscular pharmacology of atracurium, pancuronium, tubocurarine and dimethyl tubocurarine in patients under nitrous oxide-narcotic barbiturate anaesthesia. In: Hughes R (ed) Abstracts atracurium symposium. Wellcome Foundation Ltd, London, p 45
Scott RPF, Goat VA (1982) Atracurium: its speed of onset. A comparison with suxamethonium. Br J Anaesth 54:909–911
Sokoll MD, Gergis SD, Mehta M, Kemmotsu O, Rudd D (1983) Haemodynamic effects of atracurium in surgical patients under nitrous oxide, oxygen and isoflurane anaesthesia. Br J Anaesth 55:77–79S
Stirt AJ, Katz RL, Murray AL, Schehl DL, Lee C (1983) Modification of atracurium blockade by halothane and by suxamethonium. A review of clinical experience. Br J Anaesth 55:71–75S
Sugai N (1974) Assessment of neuromuscular blockade in man during anaesthesia. PhD Thesis, University of London
Sugai N, Hughes R, Payne JP (1973) The use of tetanic and single twitch stimuli to assess neuromuscular block in man. Br J Anaesth 45:642–643
Twohig MM, Ward S, Corall IM (1983) Conditions for tracheal intubation using atracurium compared with pancuronium. Br J Anaesth 55:87–89S
Utting JE, Hunter JM, Jones RS (1982) Atracurium in patients with no renal function. Anaesthesiology 57:A252
Ward S, Wright D (1983) Combined pharmacokinetic and pharmacodynamic study of a single bolus dose of atracurium. Br J Anaesth 55:35–38S
Weatherley BC, Williams SG, Neill EAM (1983) Pharmacokinetics, pharmacodynamics and dose-response relationship of atracurium administered i.v. Br J Anaesth 55:39–45S

Antagonists of Neuromuscular Blocking Agents (Pharmacology and Clinical Use)

CHAPTER 31

Galanthamine

D. S. PASKOV

A. Introduction

The alkaloid galanthamine is a valuable product from the group of natural anticholinesterase substances. It was the first isolated by PROSKURNINA and YAKOVLEVA (1952) from *Galanthus woronowi* Los., family amaryllidaceae. UYEO and KOBAYASHI (1953) isolated the same alkaloid from *Lycoris radiata* Herb., and called it lycoremine. Later, galanthamine was found in more than 50 species of the family Amaryllidaceae (BOIT 1961). YUNUSOV and ABDUAZIMOV (1957) isolated galanthamine from various *Ungernia* species, *Ungernia victoris* Vved. proved to be the richest. In Bulgaria, galanthamine, isolated from *Galanthus nivalis* L. var. gracilis Čelak, family Amaryllidaceae, was studied and put into practice under the name Nivalin (PASKOV 1957; IVANOVA-BUBEVA 1957). Later, a high galanthamine content was found in the plant *Leucojum aestivum* L., also family Amaryllidaceae, growing within the territory of Bulgaria (IVANOVA-BUBEVA and IVANOV 1961; STEFANOV et al. 1974). At present, galanthaminum is manufactured industrially in the USSR from *Ungernia victoris* and in Bulgaria from *Leucojum aestivum.*

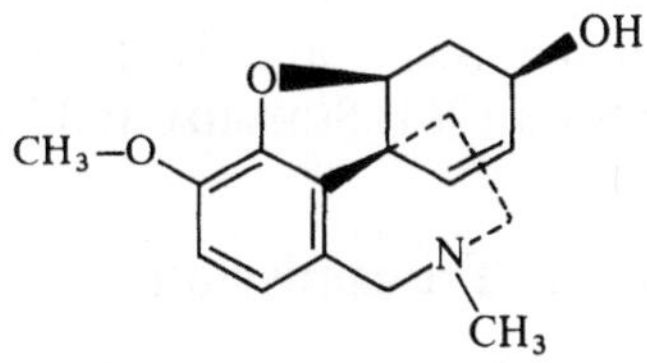

Fig. 1. Galanthamine

PROSKURNINA and YAKOVLEVA (1952, 1955) were the first to determine the physicochemical characteristics of galanthamine and to identify it as an alkaloid with a tertiary nitrogen atom in its molecule. It belongs to the alkaloids of the galanthamine group of the family Amaryllidaceae, including epigalanthamin, Narwedin, Narcissamin, etc. (DÖPKE 1976). The chemical structure of galanthamine (Fig. 1) was elucidated on the basis of total synthesis, biosynthetic studies via labeled precursors, and degradation (BARTON and KIRBY 1960; BARTON et al. 1962; KOIZUMI et al. 1964; MINAMI and UYEO 1964).

B. Pharmacology of Galanthamine

I. Anticholinesterase Activity

The first data on the anticholinesterase activity of galanthamine were reported by MASHKOVSKY and KRUGLIKOVA-LVOVA (1951) and PASKOV (1957, 1958, 1959). The inhibitory effect of galanthamine on acetylcholinesterase (AChE) of brain and skeletal muscles, as well as on serum cholinesterase, was found to be considerably lower than that of physostigmine and neostigmine. NESTERENKO (1965) observed in anesthetized cats in vivo after i.v. galanthamine application the inhibitory effect on brain AChE, about 10–12 times lower than that of physostigmine. These data of Nesterenko, together with the parallel electrophysiologic studies of ILIYUCHENOK (1965) revealed that galanthamine penetrated the blood–brain barrier, being a tertiary ammonium base. VASILENKO and TONKOPY (1974) characterized galanthamine as a reverse inhibitor of mixed type with predominantly competitive AChE inhibition.

IRWIN and SMITH (1960a) showed that galanthamine inhibited muscular cholinesterase more stronly than pyridostigmine, but was less potent than neostigmine. Those authors (IRWIN and SMITH 1960b) as well as MASHKOVSKY and ALTSHULER (1962) demonstrated that quaternization of the nitrogen atom in galanthamine by iodomethylation resulted in enhanced anticholinesterase activity of galanthamine methiodide, but, at the same time, its toxicity was increased. Galanthamine methiodide manifested a stronger effect on neuromuscular transmission, as compared with the hydrochloride and hydrobromide salts of galanthamine, but, owing to its poorer penetration through the blood–brain barrier, had a less pronounced effect on the central nervous system. Other modifications in the galanthamine molecule, with the exception of quaternization of the nitrogen atom, led to compounds with poorer anticholinesterase activity. Comparative studies on natural alkaloids from the galanthamine group revealed that galanthamine had the highest anticholinesterase activity and a potentiating effect on neuromuscular transmission (BOISSIER and LESBROS 1962; IRWIN and SMITH 1960b; LIEBMANN and MATTHIES 1961; SCHMIDT and MATTHIES 1961).

II. Effect on Neuromuscular Transmission

Galanthamine potentiates the effect of acetylcholine on frog rectus abdominis muscle (10^{-7} g/ml and higher concentrations) and in experiments in vivo on tibialis nerve–gastrocnemius muscle and fibularis nerve–tibialis anterior muscle preparations of the cat (1–3 mg/kg) (MASHKOVSKY and KRUGLIKOVA-LVOVA 1951; MASHKOVSKY 1955; PASKOV 1957, 1958). BOISSIER et al. (1960) have shown the potentiation of the toxic effect of acetylcholine in experiments on mice. At the same time, galanthamine potentiated the contractions of tibialis muscle and gastrocnemius muscle in response to stimulation of the fibularis nerve and tibialis nerve, respectively, when applied intravenously or close-arterially anesthetized cats (PASKOV 1959).

III. Antagonism Against Nondepolarizing Neuromuscular Blocking Agents

Galanthamine eliminated the neuromuscular block induced by nondepolarizing agents (MASHKOVSKY 1955; PASKOV 1957, 1958, 1959; CHEYMOL et al. 1964). This has been shown in experiments on neuromuscular preparations of cats in situ (Fig. 2), in experiments on rabbits by the head drop test, and in experiments in vitro on frog rectus abdominis muscle. COZANITIS (1977a) showed in rat phrenicodiaphragmatic preparations that neostigmine was about 18 times more active than galanthamine. BRETAGNE and VALETTA (1965) revealed that galanthamine, at a dose of 0.6 mg/kg i.v. and neostigmine 0.05 mg/kg i.v. antagonized tubocurarine-induced head drop in rabbits, i.e., a 12-fold stronger neostigmine activity was manifested.

Anticholinesterase activity of galanthamine plays the basic role in its antagonism against nondepolarizing neuromuscular blocking agents. Obviously, galanthamine induces a direct effect on acetylcholine receptors of the skeletal muscles after its close-arterial injection in high doses. This is proved by the results of ex-

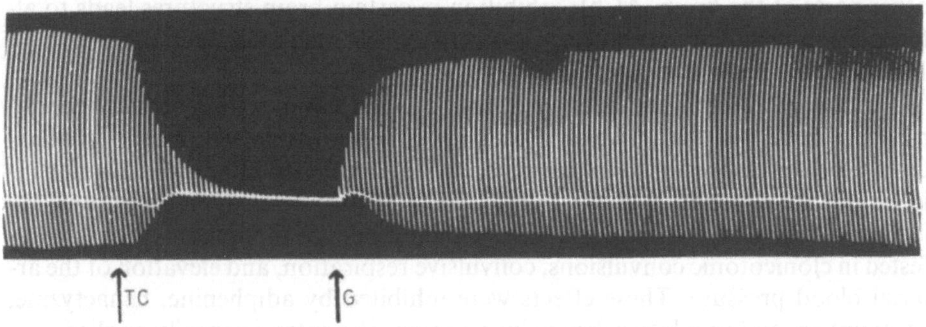

Fig. 2. Antagonistic effect of galanthamine vs tubocurarine on preparation in situ of cat sciatic nerve–tibialis anterior muscle under urethane anesthesia. Twitches were elicited every 10 s by stimulation of the motor nerves. At *TC* 0.4 mg/kg tubocurarine; at *G* 1 mg/kg galanthamine injected i.v.

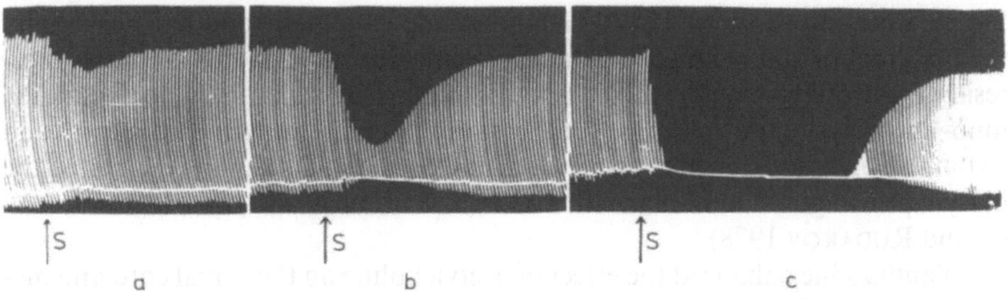

Fig. 3 a–c. Effect of galanthamine on preparation in situ of cat sciatic nerve–tibialis anterior muscle under urethane anesthesia. Twitches were elicited every 10 s by stimulation of the motor nerves. At *S* in **a** 10 μg/kg suxamethonium; at *S* in **b** 15 μg/kg suxamethonium; at *S* in **c** 15 μg/kg suxamethonium after 2 mg/kg galanthamine injected i.v.

periments on denervated gastrocnemius muscle. CHEYMOL et al. (1964) also admitted the existence of a double mechanism.

The duration of the anticurare effect of galanthamine was studied with a view to its practical application. The data from the experiments on nonanesthetized rabbits, by the head drop test, revealed that the anticurare effect of galanthamine 3–4 mg/kg lasted for about 2 h (PASKOV 1959). In experiments in situ on neuromuscular preparations in cats (Fig. 3) under urethane anesthesia, galanthamine lengthened succinylcholine-induced depolarization block (MASHKOVSKY 1955; PASKOV 1957, 1959; CHEYMOL et al. 1964). This effect was confirmed in patients by SAEV and TENEV (1963).

IV. Effect on the Central Nervous System

1. Effect on the Spinal Cord and Medulla Oblongata

Studies on the effect of galanthamine on the central nervous system revealed that it penetrates the blood–brain barrier, having an inhibitory effect on AChE of the brain cortex, thalamus, hypothalamus, mesencephalon, and medulla oblongata. The inhibitory effect of galanthamine on the cortex AChE is stronger than in other parts of the brain. AChE inhibition in certain brain structures leads to alterations in behavior and bioelectrical activity (NESTERENKO 1965; ILIYUCHENOK 1965).

Aiming at the elimination of a peripheral galanthamine effect, series of experiments were carried out in which it was intracisternally injected. Elevation of arterial blood pressure and respiratory excitation developed after galanthamine injection in the cisterna magna in doses of 0.4–0.5 mg. After the intracisternal injection of high doses, there was a sharp enhancement of the spinal reflexes, manifested in clonicotonic convulsions, convulsive respiration, and elevation of the arterial blood pressure. These effects were inhibited by adiphenine, benactyzine, caramiphen, and to a lesser degree by atropine, also intracisternally applied.

Series of electrophysiologic studies were conducted for a more thorough characteristization of the effect of galanthamine. With the record of spontaneous potentials of the efferent fibers of the sciatic nerve of thalamic frogs, the bioelectrical activity increased sharply after galanthamine injection. The frequency and amplitude of the skeletal muscle potentials were increased after intracisternal galanthamine injection (PASKOV 1959). Galanthamine increased spontaneous activity in the efferent preganglionic fibers of cat splanchnic nerve under urethane anesthesia. Electrophysiologic studies showed also that galanthamine facilitated mono- and polysynaptic reflexes, both by stimulation of the peripheral nerves and by stimulation of the dorsal roots of the spinal cord. Galanthamine facilitated the polysynaptic linguomandibular reflex as well (RUDAKOV and MITZOV 1977; MITZOV and RUDAKOV 1978).

Galanthamine enhanced the effect of acetylcholine on the spinal cord and medulla oblongata, both agents being intracisternally injected. Galanthamine also enhanced the spontaneous potentials of the afferent fibers of mesenteric nerves. All these data suggested both a direct and reflex effect of galanthamine on the centers in the spinal cord and medulla oblongata (PASKOV 1959).

2. Effect on the Brain

Experiments on rats with elaborated stereotype of conditioned reflexes revealed that galanthamine 0.5–2 mg/kg s.c. enhanced the excitation processes manifested by shortening of the latent period and the time of the rush. The conditioned reflex activity disturbed by antimuscarinic agents was normalized by galanthamine within certain limits (PASKOV 1959; ROSHCHINA 1969).

PASKOV (1959) has shown that galanthamine 0.1–3 mg/kg i.v. induced EEG desynchronization of nonanesthetized rabbits antagonized by adiphenine, caramiphen, and atropine. MASHKOVSKY and ILIYUCHENOK (1961) showed that the activation of cortical electrical activity resulted from the activating effect of galanthamine on the ascending activating system of the mesencephalon reticular formation. ILIYUCHENOK (1965) showed that benactyzine and some other central antimuscarinic substances antagonized this effect. BARAKA and HARIK (1977) observed in volunteers that galanthamine eliminated rapidly and effectively the central anticholinergic effect of 2 mg i.v. scopolamine. COZANITIS (1977 b) confirmed the therapeutic effectiveness of galanthamine in a patient intoxicated by scopolamine.

ILIYUCHENOK and PASTUKHOV (1968) found in cats that galanthamine increased, whereas benactyzine benzacine decreased the frequency of the discharges of the majority of the hippocampal neurons. These data, in the opinion of the authors, suggest the presence of muscarine-sensitive neurons in the hippocampus. KRAUS (1974) showed that improvement of short-term memory is conditioned by neurotropic substances with different mechanisms of central action (amphetamine, aethimizolum, strychnine, galanthamine) and is accompanied by elevation of excitation of the dorsal part of that structure.

The experimental data of COZANITIS et al. (1983) are of interest, concerning the analgesic activity of galanthamine, which clinicians have previously observed with the administration of that preparation in neuralgia of the peripheral nerves. These authors have established that the antinociceptive effect of galanthamine detected in the rat hot plate test had been partially reversed by low doses of naloxone. It has also been established that galanthamine inhibited the stretching response provoked by intraperitoneal administration of acetic acid, but this effect was not antagonized by naloxone. Galanthamine also potentiates the analgesic effect of morphine. The authors admit that the antinociceptive effcts of galanthamine are not mediated through opiate receptors. But having in mind the partial antagonism of galanthamine-induced analgesia by naloxone, they express the assumption that besides the anticholinesterase effect, galanthamine could as well interact, at least in the rat, with opiate receptors. There is also a difference between in vitro and in vivo data of galanthamine in rats. It has been presumed that galanthamine, in this species, is converted into an active metabolite. Finally, COZANITIS et al. (1983) do not exclude the possibility of liberation of endogenous opiate-like substance by galanthamine. It should be mentioned here that galanthamine is not an absolute synergist of morphine and its structural analogs. If a synergism is observed between galanthamine and morphine in their analgesic activity, then as regards the morphine-induced depression of respiration, galanthamine is a potent antagonist of that activity of morphine and its structural analogs (PASKOV et al. 1964). Evidently, further studies are necessary on the interaction between galanthamine on the one hand and morphine and its analogs on the other – first of all at the level of the central nervous system.

V. Effect on the Cardiovascular System

Galanthamine in doses of 0.1–3 mg/kg i.v. decreased the heart rate in rabbits. The PQ interval of the ECG was lengthened, and the T wave was enlarged. Galanthamine in concentrations of 10^{-7} g/ml and higher potentiated the effect of acetylcholine on isolated frog heart. In cats, in low doses, it induced a short-term hypotension, and in higher doses, after the initial hypotensive phase, a transitory hypertension developed. Galanthamine potentiated the hypotensive effect of acetylcholine and the effects of stimulation of the peripheral end of the vagus nerve. CHRUSCIEL and VARAGIČ (1966) observed only a hypertensive effect in rats, which they thought related to the central galanthamine stimulation of the sympathetic nervous system. GEORGIEV and JORDANOV (1961) reported a hypotensive effect of galanthamine in hypertensive patients.

VI. Effect on Respiration

Galanthamine has both direct and reflex stimulating effect on respiratory centers. A highly potentiated effect of acetylcholine and nicotine on nicotine receptors was observed in cat isolated glomus caroticum in situ, perfused by galanthamine solutions in concentrations of 2×10^{-7} g/ml and higher (PASKOV 1959). In experiments on cats and rabbits, galanthamine antagonized the depression of respiration induced by toxic doses of morphine (Fig. 4), pethidine, dextromoramide, and other narcotic analgesics (PASKOV et al. 1964; COZANITIS and ROSENBERG 1974). This antagonism was confirmed in patients by COZANITIS and TOIVAKKA (1974).

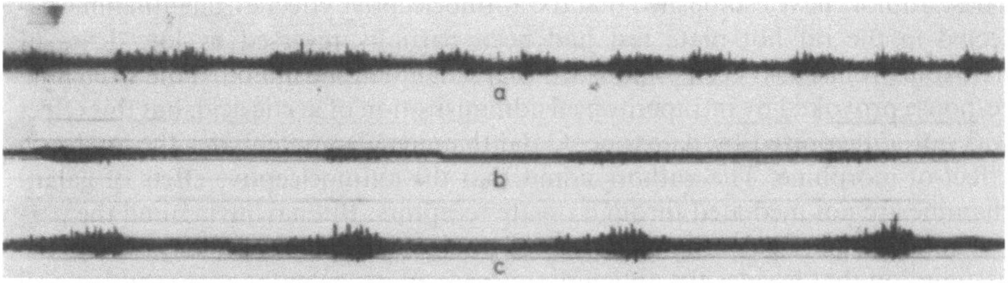

Fig. 4 a–c. Effect of morphine and galanthamine on the potentials of cat phrenic nerve under urethane anesthesia. Amplification 30 μV/cm; tape speed 30 mm/s. **a** control, **b** after application of morphine, **c** after application of galanthamine 3 mg/kg

VII. Effect on Smooth Muscles

Galanthamine, both in vitro and in vivo, activates the smooth muscles of the gastrointestinal tract, potentiates the effects of acetylcholine and the effect of the stimulation of the peripheral end of the vagus nerve (MASHKOVSKY and KRUGLI-

KOVA-LVOVA 1951; PASKOV 1959; BOISSIER et al. 1960). In experiments on the bladder in situ, galanthamine 1–3 mg/kg i.v. induced a contraction of the bladder, and potentiated the effect of acetylcholine, the effects of pelvic nerve stimulation, and the effect of nicotine (PASKOV 1959). These effects of galanthamine are, most likely, associated with its effect on parasympathetic ganglia and postganglionic neuromuscular junctions.

VIII. Effect on the Superior Cervical Ganglion and Adrenals

Galanthamine in doses of 0.1–3.0 mg/kg i.v. failed to evoke contractions of the nictitating membrane in urethane-anesthetized cats, but in the same doses, it enhanced the contractions of the nictitating membrane on i.v. acetylcholine application in doses of 1–10 μg in atropinized cats (1 mg/kg atropine), as well as on electrical stimulation of the preganglionic trunk of the superior cervical sympathetic ganglion. It also increased the amplitude of the evoked postganglionic potentials. Galanthamine antagonized the blocking effect of hexamethonium (PASKOV 1959). KOSTOWSKI and GUMULKA (1968) revealed that, after 100–250 μg galanthaminum was injected into the common carotid artery, an increase of the amplitude of the negative and positive potentials developed, as well as insignificant reduction of the negative afterpotential, antagonism of the blocking effect of hexamethonium, increase of the spontaneous postganglionic activity (in doses of 50–100 μg), and increase of amplitude and duration of the superficial ganglionic potential (in doses of 100–200 μg) evoked by 5–20 μg acetylcholine.

Galanthamine enhanced the liberation of catecholamines from the adrenal medulla, induced by acetylcholine or by the stimulation of the splanchnic nerve (PASKOV 1959). It was shown that galanthamine increased the 17-oxyketosteroid blood level (NAUMENKO et al. 1965) and the cortisol and ACTH content in human blood plasma (COZANITIS 1974; COZANITIS et al. 1973).

IX. General Effects and Toxicity

In experiments on conscious animals, galanthamine induced both muscarinic and nicotinic effects, characteristic of anticholinesterase substances. LD_{50} values for galanthamine, obtained by various authors are presented in Table 1 (MASHKOV-

Table 1. Comparative acute toxicity of galanthamine and neostigmine

Animal	Drug	LD_{50} (mg/kg)			
		Oral	Subcutaneous	Intraperitoneal	Intravenous
Mice	Galanthamine	25.0	11.7–13.3	15.2	4.9–5.2
Rats	Galanthamine	76.7	52.1	39.2	11.2
Rabbits	Galanthamine	70.0	27.0		LD_{100} = 10–12
Mice	Neostigmine	14.4	0.4–0.6	0.61	0.315–0.36
Rabbits	Neostigmine		0.5–0.75		0.2

SKY and KRUGLIKOVA-LVOVA 1951; PASKOV 1959; BOISSIER et al. 1960) as well as neostigmine LD_{50} values (LANSKAYA 1950; SPECTOR 1956). The toxicity of galanthamine is much lower that that of neostigmine. Hence, the LD_{50} ratio of these drugs in mice is 26.8 for s.c. administration and 16.6 for i.v. administration.

X. Distribution: Pharmacokinetics of Galanthamine in Animals and Healthy Volunteers

MIKHNO and KRAMARENKO (1970) studied the distribution of galanthamine in the organs of dogs after oral application of the lethal dose. Galanthamine was detected in liver, kidneys, heart, lungs, brain, urine, stomach, and intestines.

Lately, YAMBOLIEV (1984) studied the pharmacokinetic behavior of galanthamine in experiments with rats and cats, as well as in healthy volunteers. After intravenous injection of galanthamine to rats in doses of 1 and 3 mg/kg, plasma concentrations were dynamically determined. With a dose of 1 mg/kg, the maximum plasma concentration was reached after 5 min and was 1 μg/ml, by 15 min after the injection the concentration was 0.5 μg/ml. Later, up to 60 min, plasma galanthamine concentrations were more slowly reduced and by 60 min the concentration was 0.15–0.20 μg/ml.

After intravenous injections of galanthamine in a dose of 3 mg/kg, higher concentrations of galanthamine were observed in blood plasma over a longer time period. By 20 min, the concentration established was 1 μg/ml, by 90 min 0.5 μg/ml, and by 180 min 0.2 μg/ml. This reduction of plasma concentrations of galanthamine as a function of time corresponds to a two-compartment pharmacokinetic model. The plasma concentrations were also determined after oral administration of galanthamine in a dose of 5 mg/kg. A relatively quick absorption was established. The maximum plasma concentration, 2.21 μg/ml, was attained by 20 min after administration. By 40 min, the concentration was 1 μg/ml; by 120 min 0.15–0.20 μg/ml. From blood, galanthamine was distributed into the tissues of the organism and built up there. One of the organs where galanthamine builds up is the brain. This is suggested by the experimental studies on rats (YAMBOLIEV 1984), in which the galanthamine concentration was shown to be several times higher than in blood plasma, both after intravenous and oral administration. Thus, after intravenous injection of galanthamine in a dose of 3 mg/kg, its concentration in the brain was 4.49 μg/g by 2 min; 6.60 μg/g by 5 min; 3.25 μg/g by 60 min; and 1.75 μg/g by 180 min. A rapid increase of galanthamine in the brain tissue was also established with the oral administration in a dose of 5 mg/kg. Thus, for example, with a dose of 5 mg/kg its concentration in brain tissues was 1.97 μg/g; by 10 min 3.41 μg/g; by 45 min 6.10 μg/g; and 60 min 2.85 μg/g. These data correlate with the data from the studies on the effect of galanthamine on the central nervous system and determination of AChE activity in the various parts of the brain after intravenous injections of galanthamine to cats (PASKOV 1959; MASHKOVSKY and ILIYUCHENOK 1961; ILIYUCHENOK 1965; NESTERENKO 1965). The half-life of galathamine in rats is about 50 min.

The correlation between plasma concentrations of galanthamine and the contractions of m. tibialis in response to the stimulation of n. fibularis with square-wave electrical impulses was established in experiments with cats. A correlation was established between galanthamine plasma concentration and the increased amplitudes of the contractions of m. tibialis. It is worth mentioning that the half-life of galanthamine in cats in 1.5–2.0 h, i.e., about twice as long as that in rats.

It was established by studies on healthy volunteers with galanthamine administered subcutaneously in doses of 10 and 15 mg, that after a short lag time(10–15 min) a fast resorption followed with maximum plasma concentrations after 25–30 min. The half-life of galanthamine in healthy volunteers was 2.5 h. After s.c. injection of galanthamine in a dose of 15 mg, the plasma concentrations were as follows: 1.10 μg/ml by 15 min; 4.50 μg/ml by 30 min; 1.55 μg/ml by 60 min; 0.18 μg/ml by 120 min; 0.08 μg/ml by 240 min; and 0.04 μg/ml by 480 min. After the oral administration of galanthamine in a dose of 15 mg a fast resorption was established and galanthamine was present in plasma after 8 h. It could be concluded from the data obtained that galanthamine is rapidly resorbed both with s.c. and oral administration.

As compared with neostigmine and pyridostigmine, galanthamine is absorbed to a far higher degree with the oral administration of all three preparations. Galanthamine, at the same time, is eliminated more slowly from organisms. These facts explain both the rapid advancement of the effect of galanthamine and its longer duration. It was also established that galanthamine does not bind with plasma proteins of blood.

XI. Teratology and Embryotoxicity

The studies on the embryotoxic and teratogenic activity of galanthamine were carried out on Wistar rats and Belgian giant rabbits. Galanthamine was administered in various doses, orally and subcutaneously, in single doses and repeatedly, starting from the maximum 24-h dose for humans (0.5 mg/kg) up to 1/10 and 1/15 of LD_{50}. It was established that it induces no fetal malformations, but all doses exceeding 1/10 LD_{50}, had embryotoxic activity. These doses also have a general toxic effect on the pregnant mothers.

XII. Mutagenesis

The short-term testing of pharmaceuticals and other chemical compounds contribute to the protection of human health and the environment from chemical mutagens and carcinogens. Bearing in mind the high initial cost and the rather long testing of carcinogens, the information obtained from the short-term tests for mutagenic activity is of value in the determination of the potential risk of the compounds tested, including pharmaceuticals.

To have more reliable conclusions on the basis of mutagenicity testing, a battery of methods is used, as a rule, that could eventually detect the ability of the test compound to injure DNA and induce mutations, chromosomal aberration, reparative synthesis of DNA, or cellular transformation. The methods recommended by the World Health Organization are most frequently used in practice: determination of mutagenic activity on special HIS strains of *Salmonella typhi-*

murium (the so-called Ames test), determination of non-planned DNA synthesis, and detection of chromosomal aberrations, induced by the chemical agents tested, etc.

Galanthamine was tested for mutagenic activity by the following methods: (a) study on non-planned (reparative) synthesis of DNA in human peripheral lymphocytes; (b) testing of HIS strains of *S. typhimurium* (Ames test) (Maron and Ames 1983); (c) testing by the micronucleus test (Schmid 1980). In brief, the following results were obtained.

Galanthamine administered in nontoxic and slightly toxic concentrations (0.000015–0.15 mg/ml) for human lymphocytes induced no stimulation of non-planned synthesis of DNA in human peripheral lymphocytes after their incubation for 60 min at 37 °C with or without microsomal activating system S 9 mix. The data obtained from this test are evidence of the absence of any mutagenic effect of galanthamine.

The Ames test revealed that galanthamine in the concentrations used (0.003–3.0 mg) per Petri dish or 0.0015–1.5 mg/ml) had no mutagenic effect in the experiments with *S. typhimurium* TA 97, TA 100, and TA 102 both in the absence and in the presence of microsomal activating system S 9 mix.

The micronucleus tests were performed with experimental male and female BDF mice, F_1 hybrids between C_{57}Bl and DBA. Galanthamine was applied subcutaneously: (a) as a single dose; (b) repeatedly with an interval of 24 h between injections. It was injected in doses of 5.0 and 10.0 mg/kg, corresponding to 40% and 80% of LD_{50} for subcutaneous application. The animals were killed by dislocation of the cervical vertebrae at intervals. Then the bone marrow, taken from the bone marrow canal of the femur was studied. After treatment and staining the preparations in accordance with the instructions of Schmid (1980), the number of polychromatic erythrocytes containing micronuclei was determined. Separately, the number of normochromic erythrocytes was determined, corresponding to 200 polychromatic erythrocytes so as to determine the possible toxic effect of the test preparation.

The results obtained revealed that the dose of galanthamine (10 mg/kg, corresponding to 80% LD_{50}) induced a negligible reduction in the number of polychromatic erythrocytes, compared with normochromic erythrocytes – evidence of a slightly toxic effect. Therefore, the maximum galanthamine dose was selected accordingly, on the borderline of toxicity for the bone marrow. The data obtained reveal that galanthamine injected subcutaneously in a single dose or repeatedly in doses of 5.0 and 10 mg/kg, induces no increase of the number of polychromatic erythrocytes containing micronuclei within 24–72 h after the first injection of the preparation.

The results obtained with the mice treated with potassium bichromate in a dose of 50 mg/kg or with urethane in a dose of 12 mg/kg, i.e., in the positive controls, correspond to the data available in literature (Heddle et al. 1983). It is concluded that galanthamine has no mutagenic effct on the bone marrow of mice.

It is obvious from the experimental data presented that galanthamine, tested by the three basic tests for mutagenicity, has no mutagenic effect in these concentrations, doses and exposure times (P. M. Blagoeva; R. L. Balanski; Z. Y. Mircheva and D. S. Paskov 1985, unpublished work).

C. Clinical Application of Galanthamine

I. Clinical Application as a Decurarizing Agent

DEREDJAN and KRASTEVA (1960) first applied galanthamine to 14 patients for the elimination of residual muscular relaxation. Numerous papers followed, dealing with the use of galanthamine for decurarization of patients to whom tubocurarine, gallamine, diplacinum, Alloferin, and pancuronium were administered for muscle relaxation (PASKOV et al. 1962a, b, c; SALVINI et al. 1962; SAEV and TENEV 1963; STOJANOV and VULCHANOVA 1963; KRASTEV and PAMPULOV 1963; STOJANOV 1964a, b, 1965, 1971; BRETAGNE and VALETTA 1965; MITEV and ATANASSOV 1965; MAYRHOFER 1966, 1967; WISLICKI 1967; KOVANEV et al. 1967; SAEV 1968; COZANITIS 1971). In these papers, galanthamine was considered as an effective antidote of nondepolarizing neuromuscular blocking agents.

The observations of SAEV and TENEV (1963) revealed that after administration of galanthamine at 15–20 mg i.v., decurarization developed 2–3 min after the injection. Later experience with decurarization of more than 6,000 patients, as well as the observations of other authors, allowed SAEV (1968) to recommend galanthamine (Nivalin) for complete decurarization in doses of 15–25 mg i.v. Additional injection of 5–10 mg has only been necessary in rare cases. To children, doses of 2.5–15 mg were administered, depending on age and body weight.

STOJANOV and VULCHANOVA (1963) found that, usually after 1–1.5 min, there were the first signs of deep breathing, and if the patient was in apnea during the injection, after 2–3 min, the first attempts at spontaneous respiration appeared; after 4–5 min, respiration was restored. Recurarization phenomena were not observed. STOJANOV (1971), on the basis of his own experience and that of other investigators, suggested that the first galanthamine dose should range between 10 and 15 mg. In the majority of cases, three ampules of 5 mg were diluted in 5–10 ml 0.9% NaCl solution or 5% glucose and the mixture was slowly injected i.v. within 30–40 s. If after 4–5 min the respiration is not completely restored, 10–15 mg galanthamine was additionally injected. Usually these two doses were enough for the restoration of spontaneous respiration. The data presented suggest that the decurarizing doses of galanthamine recommended by Saev and Stojanov are similar. MITEV and ATANASSOV (1965) on the basis of volumetric studies on the decurarizing effect of galanthamine in two groups of patients – one with apnea and the other with residual neuromuscular block – revealed that the optimal decurarizing effect in the first group was observed after i.v. injection of 30–40 mg galanthamine. In the second group (patients with residual neuromuscular block), the optimal decurarizing effect was obtained with galanthamine doses of 20–30 mg. At 20 mg, respiration was often irregular, and the minute volume and expired air did not reach the initial values even after 10 min. The additional administration of 10–15 mg quickly normalized the respiratory indices. Therefore, it could be concluded that Mitev and Atanassov recommend the dose of 30 mg galanthamine for adults. KOVANEV et al. (1967) also consider galanthamine at a dose of 25–30 mg i.v. to be sufficiently effective for the elimination of the residual relaxation within 30–40 min. MAYRHOFER (1966, 1967) demonstrated in more than 1,000 cases that galanthamine in doses of 10–25 mg i.v. is sufficient for the elimination of residual neuromuscular block, induced by tubocurarine and Alloferin. SALVINI et al. (1962) and STOJANOV (1965) consider that the decurarizing activity of 1 mg neostigmine corresponds to an average of 15–20 mg galanthamine. BARAKA and COZANITIS (1973) revealed that pancuronium 0.03 mg/kg and tubocurarine 0.15 mg/kg were equivalent and their effect could be eliminated by galanthamine 0.3 mg/kg. All the authors mentioned reported the central stimulating effect of galanthamine and the rapid restoration of respiratory reflexes. Owing to the muscarinic effect of galanthamine, in the majority of cases decurarization can be performed without a preliminary atropine application, except for patients with bradycardia (STOJANOV and VULCHANOVA 1963; SAEV and TENEV 1963; MITEV and ATANASSOV 1965; MAYRHOFER 1966, 1967; KOVANEV et al. 1967; SAEV 1968; STOJANOV 1971; etc.). SALVINI et al. (1962) characterized galanthamine as an antidote of nondepolarizing neuromuscular blocking drugs with negligible adverse effects, with particular indications for patients with cardiopulmonary diseases. COZANITIS et al. (1973a) mentioned a slight effect of galanthamine on atrioventricular conduction and hence, they recommended atropine prior to decurarization. As compared with neostigmine and chinothylinum, galanthamine has a lower cardiotoxicity. Neostigmine evoked a pronounced and long-lasting bradycardia as well as gross disorders of cardiac rhythm and conduction, often necessitating urgent therapeutic intervention (KHMELEVSKY and GADALOV 1980). The patients decurarized by neostigmine and chinothylinum could be previously atropinized.

Galanthamine has the broadest therapeutic index as compared with neostigmine, Armine, phosphacol (Miotisal), and ambenonium (PROZOROVSKY et al. 1970) (Fig. 5). The value of galanthaminum is determined by the relatively easy elimination of its toxic effects by antimuscarinic substances.

KHMELEVSKY and GADALOV (1977) have shown that in the case of moderate residual muscle relaxation, induced by repeated injection of succinylcholine, galanthamine (0.39±0.02 mg/kg i.v.) manifested high decurarizing activity, according to EMG data. Its effect began within 5 min with the maximum attained

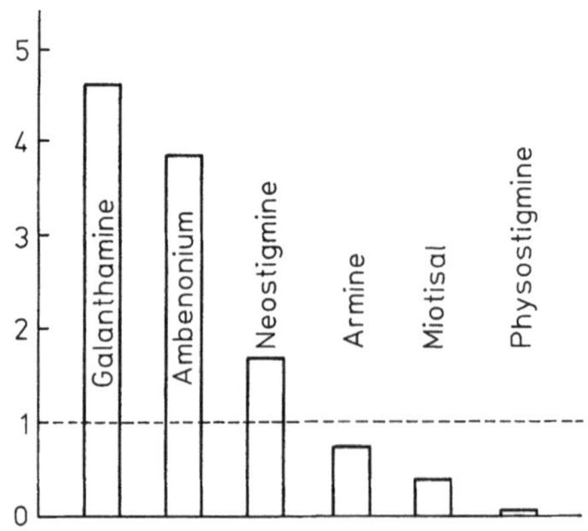

Fig. 5. Therapeutic index of choline-potentiating drugs with their application as anticurarizing agents. Ordinate: LD_{50}/ED_{50} of anticurarizing effect in experiments with frogs. (PROZOROVSKY et al. 1970)

after 35 min. By the end of the 60-min period of observation, the volume of pulmonary ventilation was equal to the initial indices in all the patients. KHMELEVSKY and GADALOV (1977) think that neostigmine application should be restricted owing to its cardiotoxic effect, and therefore, it could be successfully replaced by galanthamine and chinothylinum.

COZANITIS and TOIVAKKA (1971) found a histamine-releasing effect of galanthamine on intradermal injection, but eosinophils remained within the normal range. Only one case of anaphylactic reaction has been described so far, in a patient with actinomycosis, treated for a long time with penicillin, streptomycin, and potassium iodide (KILIMOV 1961 a), in tens of thousands of cases treated with galanthamine. Galanthamine induced no visible changes of the ventilation in asthmatic volunteers (COZANITIS et al. 1972). The authors reported that not a single case of bronchospasm was observed among 150 decurarization.

The experience of many authors (FERRARA et al. 1961; BOROMEI and DAIDONE 1962; KRASTEV and PAMPULOV 1963; SEDLOEV et al. 1964; KULOV 1967) indicates that galanthamine is effective for the elimination of intestinal and bladder atony in the postoperative period, its effect being lighter and longer than that of neostigmine.

II. Treatment of Diseases of the Central and Peripheral Nervous Systems

For two decades, galanthamine has been shown to be a valuable drug for numerous diseases of the central and peripheral nervous systems and for disorders of the neuromuscular apparatus. Its broad therapeutic index enabled individualization of the dosage and reduction of adverse effects. It could be administered orally as well as subcutaneously, intramuscularly, iontophoretically and, in certain cases, intravenously. In chronic diseases, the treatment started with a minimum dose which was increased every few days to the optimal dose administered up to the end of the therapeutic course. Usually, the treatment course lasts 30–60 days and, when necessary, the treatment course can be repeated two or three times, with intervals of 1–1.5 months. Iontophoretically, galanthamine is applied through the skin via the anode in case of diseases of the peripheral nervous system. The dose is also regulated according to age, with an average of 10–20 sessions, repeated every 2–3 months, if necessary.

1. Brain Diseases

EIDINOVA and PRAVDINA-VINARSKAYA (1957) applied galanthamine to children with residual motor disorders resulting from lesions of the brain in meningoencephalitis, intrauterine infections, and birth injury. The authors observed a reduction of the paretic and hypokinetic phenomena. Later PERNOV et al. (1963) treated 25 children with infantile cerebral paralysis and obtained improvement in 22 of the children, manifested in various degrees; in 5 of them improvement was very pronounced. NIKOLKOV (1980) analyzed the effect of treatment in 397 children with infantile cerebral paralysis, treated with galanthamine. The treatment lasted 25–45 days. The author found a distinct effect in more than 50% of the treated children with various forms of cerebral paralysis. Improvement of the motor disorders and psychic activity was observed in 75% of children with the hypotonic form of the disease. The author considers epilepsy to be a contraindication. DASKALOV and ATANASSOV (1980) observed a positive effect of galanthamine in aphasia, mainly owing to cerebrovascular processes. KILIMOV (1961 b) and CECCHINI (1966) observed a positive effect in patients with motor disorders due to brain vascular lesions.

2. Spinal Cord Diseases

SHENK et al. (1956), PASKOV (1960), DE RENZI (1961), LOMBARDO and ARENA (1962), and PORCELLI (1963) observed a positive effect not only in the initial, but also in the later and delayed phenomena of poliomyelitis. But the observations of REVELLI and GRASSO (1962) on the therapeutic potential of galanthamine in poliomyelitis sequelae were most significant. They treated 52 of 70 inpatients during the acute stage of poliomyelitis. The treatment started on the third or fourth day of the disease, i.e., after the febrile period, and lasted up to 50–100 days. Complete rehabilitation was attained in 27 patients, distinct improvement with a high probability of almost complete functional rehabilitation in 16 patients, with a possibility of further improvement in 16 patients; severe sequelae persisted in 3 patients. The authors reported high tolerance to the drug. REVELLI and GRASSO (1968) successfully treated with galanthamine a group of patients with peripheral motor neurons affected by other viruses. Galanthamine is also indicated in the poliomyelitis form of tick-borne encephalitis, in syringomyelia, myelitis, traumatic injuries of the spinal cord, amyotrophic sclerosis, etc.

3. Peripheral Nervous System Diseases

GEORGIEV et al. (1960), NASTEV et al. (1960), and KILIMOV (1961 c) applied galanthamine to patients treated for various diseases of the peripheral nervous system: neuritis of facial nerves, polyneuritis, radiculoneuritis, and neuralgia of the trigeminal nerve of various etiology. Often, galanthamine was combined with vitamins (B_1, B_2, B_{12}). Good or excellent therapeutic effects were obtained in the majority of cases.

4. Diseases with Probable Disorders of the Myoneural Junctions

After the recognition of galanthamine as an anticholinesterase drug, it has attracted the attention of neuropathologists interested in its application in myasthenia gravis. GEORGIEV et al. (1960), NASTEV et al. (1960), BERGAMINI and BAGGIORE (1960), and DE RENZI (1961) were the first to test the effect of galanthamine on this disease. The majority of these authors observed that an immediate effect developed after galanthamine administration in a greater part of the patients and lasted longer than the effect of neostigmine. In the rest of the patients, the effect developed slowly and was unsatisfactory. In some cases, galanthamine was reported to be more potent than pyridostigmine and neostigmine. Inconsistent effects obtained by the authors were, very likely, due to the varying severity of the disease and different dosages. Drugs with a faster effect could be combined with galanthamine which has a slower and longer effect.

Edelstein (1957) first applied galanthamine alone and in combination with the ganglion-blocking agent pachycarpinum for the treatment of progressive muscular dystrophy and obtained positive results. Georgiev et al. (1960), Begamini and Baggiore (1960), Nastev et al. (1960), Pernov et al. (1961), and De Renzi (1960) also reported positive effects after the application of galanthamine. Pernov et al. (1961) presented data that the best effects were obtained at high galathamine doses with a duration of treatment of 40 days. The ascending form of progressive muscular dystrophy is best affected in their opinion.

III. Application in Some Other Diseases

Galanthamine has been widely applied for the treatment of psychogenic sexual asthenia in males (Paskov and Trajkov 1975) and as a psychostimulant and for the treatment of enuresis in children (Tchakarov 1980).

D. Conclusions

Galanthamine belongs to the group of reverse inhibitors of AChE. It has both noncompetitive and competitive mechanisms of action. The competitive mechanism of inhibition of AChE is predominant. Among the galanthamine group of natural alkaloids of the family Amaryllidaceae, galanthamine has the highest anticholinesterase activity. On average, its toxicity is about 20 times lower than that of neostigmine.

The pharmacologic effect of galanthamine is explained by its anticholinesterase activity. It inhibits AChE in the neuromuscular junctions, at the level of sympathetic and parasympathetic ganglia, in adrenal medulla, at cholinergic junctions of the postganglionic parasympathetic endings, and effective cells.

Galanthamine enhances the effects of acetylcholine at the neuromuscular junctions and potentiates the contractions of the skeletal muscle in response to stimulation of the innervating nerves. It reverses the neuromuscular block induced by the nondepolarizing myorelaxants, and prolongs the neuromuscular block of succinylcholine.

Galanthamine enhances the stimulating processes in the central nervous system. The effect on the brain results, first of all, from the effect of galanthamine on the afferent activating system of the reticular formation of the mesencephalon and on some other structures containing cholinergic neurons. It also excites the reflex centers in the spinal cord and medulla oblongate. The effect of galanthamine on the central nervous system is a result of both the direct effect upon those structures of the central nervous system containing cholinergic synapses, since galanthamine crosses the blood–brain barrier, and from the reflex effect, owing to its effect on some peripheral reflexogenic zones.

Galanthamine antagonizes the depressive effects of morphine and its congeners on the respiration. It also antagonizes the effect of scopolamine, atropine, and other anticholinergic agents on the central nervous system. This antagonism is bilateral.

During more than two decades of clinical administration of galanthamine, it has been established that it is an effective preparation in the treatment of numerous diseases of the central and peripheral nervous systems, of diseases associated with disorders in the transmission of nervous impulses at the myoneural junctions, etc. Clinically, galanthamine has valuable qualities as an antagonist to the nondepolarizing neuromuscular blockers.

It is applied in anesthesiologic practice for elimination of residual neuromuscular block induced by nondepolarizing myorelaxants. Galanthamine, in decurarizing doses, has a rather weak cardiotoxic and bronchosecretory effect, whereas neostigmine, even under atropine protection, in decurarizing doses manifests a strong cardiotoxicity and some other muscarinic effects.

References

Baraka A, Cozanitis D (1973) Galanthamine versus neostigmine for reversal of non-depolarizing neuromuscular block in man. Anesth Analg Curr Res 52:832–836

Baraka A, Harik S (1977) Reversal of central anticholinergic syndrome by galanthamine. JAMA 238:2293–2294

Barton DHR, Kirby GW (1960) Synthesis of galanthamine. Proc Chem Soc 392–393

Barton DHR, Kirby GW, Taylor JB, Thomas GM (1962) Multiple labelling experiments in biosynthesis of Amaryllidaceae alkaloids. Proc Chem Soc 179

Bergamini V, Baggiore P (1960) La nivalinterapia in alcune malatile neurologiche. Minerva Med 51:3553–3558

Boissier JR, Lesbros J (1962) La galanthamine puissant cholinergique naturel. II – Activité anticholinesterasique de la gelanthamine et quelques derivés. Ann Pharm Fr 20:150–155

Boissier JR, Combes G, Pagny J (1960) La galanthamine puissant cholinergique naturel. I – Sources. Structure chimique. Characterization. Extraction. Toxicité. Actions sur les fibres lisses. Ann Pharm Fr 18:888–900

Boit HG (1961) Ergebnisse der Alkaloid-Chemie bis 1960. Akademie, Berlin, pp 410–475

Boromei A, Daidone R (1962) On the use of nivalin in postoperative urinary retention in neurosurgery. Minerva Chir 17, 19:947–956

Bretagne M, Valletta J (1965) Essais clinique en anesthesiologie d'un nouvel anticholinesterasique la galanthamine. Anesth Analg Reanimat 22:285–292

Cecchini M (1966) Il trattamento con galanthamina bromidrato dei postumi motori consecutivi a vasculopathie cerebrale. Minerva Med 57:3385–3386

Cheymol J, Boissier JR, Bourillet F, Fichelle-Pagny J, Roch-Arveiller M (1964) Activité neuromusculaire de la galanthamine, anticholinesterasique naturel. Ann Pharm Fr 22:1:41–48

Chrusciel M, Varagič V (1966) The effect of galanthamine on the blood pressure of the rats. Br J Pharmacol 26:295–301

Cozanitis DA (1971) Experiences with galanthamine hydrobromide as curare antagonist. Anaesthesist 20:226–229

Cozanitis DA (1974) Galanthamine hydrobromide versus neostigmine; a plasma cortisol study in man. Anaesthesia 29:163–168

Cozanitis DA (1977a) The potency ratio of galanthamine and neostigmine on the reversal of the tubocurarine block in the isolated rat diaphragm. Acta Anaesthesiol Belg 28:53–60

Cozanitis DA (1977b) Galanthamine hydrobromide, a longer acting anticholinesterase drug in the treatment of the central effects of scopolamine (Hyoscine). Anaesthesist 26:649–650

Cozanitis DA, Rosenberg P (1974) Preliminary experiments with galanthamine hydrobromide on depressed respiration. Anaesthesist 23:302–305
Cozanitis DA, Toivakka E (1971) A comparative study of nivaline and atropine/Neostigmine on conscious volunteers. Anaesthesist 20:416–421
Cozanitis DA, Toivakka E (1974) Treatment of respiratory depression with the anticholinesterase drug galanthamine hydrobromide. Anaesthesia 29:581–584
Cozanitis DA, Halttunen P, Edgren J (1972) A cinematographic study demonstrating the effect of galanthamine hydrobromide on conscious asthmatic volunteers. Anaesthesist 21:63–66
Cozanitis DA, Nuutila K, Karhunen P, Baraka A (1973 a) Changes in cardiac rhythm with galanthamine hydrobromide. Anaesthesist 22:457–459
Cozanitis DA, Toivakka E, Dessypris A (1973 b) Electroencephalographic and blood chemistry responses to galanthamine hydrobromide in epileptic volunteers. Anaesthesist 22:31–33
Cozanitis DA, Freidmann T, Fürst S (1983) Study of the analgesic effects of galanthamine, a cholinesterase inhibitor. Arch Intern Pharmacodyn Thér 266:229–238
Daskalov D, Atanassov A (1980) Nivalin application in rehabitation treatment of cerebral diseases with aphasic syndrome (in English and Russian). Med Biol Inform 3:10–12
Deredjan A, Krasteva E (1960) Our abservations on decurarization effect of nivalin (in Bulgarian). Khirurgia (Sofiia) 13:272–275
De Renzi E (1961) Il nivalin in Therapia Neurologica. Minerva Med 58:253–257
Döpke W (1976) Ergebnisse der Alkaloid-Chemie 1960–1968, vol 1. Akademie-Verlag, Berlin, pp 713–722
Edelstein EA (1957) On the application of new agents in the progressive muscular dystrophia (in Russian). Zh Nevropatol Psikhiatr 57:856–859
Eidinova MB, Pravdina-Vinarskaya EN (1957) Characteristic and treatment of residual motor disorders in children (in Russian). Zh Nevropatol Psikhiatr 57:814–819
Ferrara L, Vidili F, Aletti L (1961) Risultati clinici e considerationi sul un anticholinesterasico (Nivalina) nell trattamento dell'ileo postoperatoria. Minerva Anesthesiol 27, 2:74–78
Georgiev IV, Jordanov B (1961) Effetto ipotensive della nivaline. Cultura Medica 23:305–312
Georgiev IV, Pernov K, Dashin V, Jordanov B (1960) Our experiences in the treatment of myopathies and lesions of the peripheral nerves with the new Bulgarian preparation nivalin (in Bulgarian). Savr Med II,4:16–26
Heddle JA, Hite M, Kirkhart B, Mavournin K, MacGregor J, Newell G, Salamone M (1983) The induction of micronuclei as a measure of genotoxicity. A report of US Environmental Protection Agency GeneTox Program. Mutation Res 123:61–118
Iliyuchenok RJ (1965) Neurohumoral mechanism of reticular formation of brain stem (in Russian). Nauka, Moscow
Iliyuchenok RJ, Pastukhov JF (1968) Effect of some cholinolytic substances and galanthamine upon the activity of single neurons of hippocampus (in Russian). Farmakol Toksikol 31:137–141
Irwin RL, Smith HJ III (1960 a) Cholinesterase inhibition by galanthamine and lycoramine. Biochem Pharmacol 3:147–148
Irwin RL, Smith HJ III (1960 b) The activity of galanthamine and related compounds on muscle. Arch Int Pharmacodyn Ther 122:314–330
Ivanova-Bubeva L (1957) Phytochemical examination of *Galanthus nivalis* var *gracilis* (in Bulgarian). Farmacia 7, 2:23–26
Ivanova-Bubeva L, Ivanov V (1961) On the alkaloid content of *Leucojum aestivum* L (in Bulgarian). Trudove Nauchno-Issledov Institut Farmacia 3:89–91
Khmelevsky JM, Gadalov VP (1977) Comparative characteristics of chinothyline, galanthamine and prozerine as antagonists of the myorelaxants (in Russian). Anesteziol Reanimatol 4:32–37
Khmelevsky JM, Gadalov VP (1980) Influence of the antagonists of the myorelaxants on the cardiac activity (in Russian). Anesteziol Reanimatol 1:14–17

Kilimov N (1961a) Case of anaphylactic shock after use of Nivalin (in Bulgarian). Savr Med 11:11–112
Kilimov N (1961b) La cura dei postume paralitici da emorragia cerebrale mediante Nivalina. Cultura Medica 23:313–319
Kilimov N (1961c) Sul trattamento delle nevritti faciali con Nivalina. Cultura Medica 23:323–330
Kilimov N (1961d) Trattamento della nevralgia del nervo trigemino. Cultura Medica 23:331–341
Koizumi J, Kobayashi S, Uyeo S (1964) Galanthamine Chemistry. V. Formation of hydroxyapogalanthamine from Galanthamine and the synthesis of its trimethyl ether. Chem Pharm Bull 12:696–705
Kostowski W, Gumulka W (1968) Ganglion and central action of galanthamine. Int J Neuropharmacol 7:7–14
Kovanev VA, Khmelevsky JM, Ganina SS (1967) Galanthamine (nivalin) as antidote of non-depolarizing myorelaxants (in Russian). Exp Khirurg Anesthez 4:65–69
Krastev B, Pampulov Z (1963) Application of nivalin in the surgical practice (in Bulgarian). Khirurgiia (Sofiia) 16:257–265
Kraus VA (1974) Interrelations between ventral and dorsal hippocampus in improvement and deterioration of shortterm memory. Zh Vissh Nervn Deyateln 24:33–41
Kulov N (1967) Study upon the effect of nivalin stimulation in bladder atonias (in Bulgarian). Nauchni Trudove VMI Varna 6, 1:23–26
Lanskaya SS (1950) Comparative assessment of the physiological action of the anticholinesterasic substances on the nervous system (in Russian). Bull Biol Med 29:415–417
Liebmann H, Matthies H (1961) Die Wirkung der Amaryllidaceen-Alkaloide Galanthamin, Narwedin, Haemanthamin und Tazettin auf die neuromuskuläre Übertragung. Acta Biol Med Germ 7:411–419
Lombardo G, Arena G (1962) L'impiego del bromodrato de galanthamine (nivalina) nei postumi paralitica da póliomielite, da nevrassit e nella distrophia musculare. Minerva Med 14:724–728
Maron DM, Ames BN (1983) Revised methods for the Salmonella mutagenicity test. Mutation Res 113:173–215
Mashkovsky MD (1955) Influence of the galanthamine upon the sensibility of the skeletal musculature to the acetylcholine (in Russian). Farmakol Toksikol 18, 4:21–27
Mashkovsky MD, Al'tshuler RA (1962) Pharmacological properties of galanthamine iodmethylate (in Russian). Farmakol Toksikol 25:168–175
Mashkovsky MD, Iliyuchenok RJ (1961) A propos of the problem of galanthamine effect upon the central nervous system (in Russian). Zh Nevropatol Psikhiatr 61, 2:166–175
Mashkovsky MD, Kruglikova-L'vova RP (1951) On the pharmacology of the new alkaloid galanthamine (in Russian). Farmakol Toksikol 14, 6:26–30
Mayrhover O (1966) Clinical experience with diallylnortoxiferine and curare antidote galanthamine. South Med J 52:1364–1368
Mayrhover O (1967) Erfahrungen mit Galanthamine (Nivalin) als Antagonist der Relaxantien vom Curaretyp. Bull Schweiz Akad Med Wiss 23, 1–2:48–52
Mikhno VV, Kramarenko VF (1970) Distribution of galanthamine in organs of poisoned animals (in Ukrain). Farm Zh 25, 1:68–71
Minami S, Uyeo S (1964) Galanthamine chemistry. VI. The synthesis of deoxymethyllycoramine. Chem Pharm Bull 12:1012–1020
Mitev L, Atanassov D (1965) Volumetric studies in decurarization with nivalin. In: First National Conference on Anesthesiology and Reanimation, Sofia, pp 49–51
Mitzov V, Rudakov A (1978) Effect of home anticurare substances upon spinal cord reflexes (in Bulgarian). Farmacia 28, 2:42–46
Nastev G, Koynov R, Ovcharova P, Petrov A, Rashev R, Abadjiev M (1960) Nivalin treatment of patients with diseases of the nervous system. Cultura Med 15:87–97
Naumenko EV, Iliyuchenok RJ, Nesterenko LN (1965) A propos of the mechanism of nivalin effect upon the hypophyseal-adrenal system (in Russian). Farmakol Toksikol 28:659–662

Nesterenko LN (1965) Influence of the galanthamine on the activity of the acetylcholinesterase in different regions of the brain (in Russian). Farmakol Toksikol 28:413–414
Nikolkov V (1980) Nivalin – its application in the treatment of infantile cerebral paralysis (in English and Russian). Med Biol Inform 3:3–6
Paskov DS (1957) Nivalin effect on striated musculature (in Bulgarian). Izvest Otdel Biol Mediz Nauki 1:29–34
Paskov DS (1958) Pharmacological study on alkaloid nivalin from the plant *Galanthus nivalis* var. *gracilis* Čelax (in Russian). In: Anitchkov SV (ed) Gangliolytics and blockers of neuromuscular synapses. Inst Eksp Med AMN, Leningrad
Paskov DS (1959) Nivalin pharmacology and clinical application. Medizina i Fizkultura, Sofia
Paskov DS (1960) Una nuova sostanza nella terapia delle paralisi poliomielitiche la nivalina. Cultura Medica 16:1–6
Paskov DS, Traykov D (1975) Nivalin treatment of psychic sexual impotence. Med Biol Inform 3:5–8
Paskov DS, Stoyanov KA, Saev SK, Tenev KA, Mincheva ML (1962a) Clinical experience with nivalin as anticholinesterase drug on anaesthesiological practice. Proceedings of First European Congress of Anaesthesiology, 3–9 September, Wien
Paskov DS, Stoyanov KA, Saev SK, Tenev KA, Mincheva ML, Kristov HD (1962b) Clinical experience with nivalin as an anticholinesterase drug in anaesthesiological practice. Cultura Medica 30:222–224
Paskov DS, Stoyanov K, Saev S, Deredjan A, Krasteva E, Tenev K, Mincheva M (1962c) Azione anticurarizante della Nivalina. Minerva Anestesiol 28, 6:213–219
Paskov DS, Dobrev H, Nikiforov N (1964) Antagonistic action of Nivalin and morphine upon the respiratory center. In: Aviado DM, Palaček F (eds) Drugs and respiration. Proc 2nd int pharmacol meeting, 20–23 August 1963. Pergamon, New York
Pernov K, Samardjiev A, Nikolkov V (1961) Treatment of muscular dystrophies with high Nivalin doses. I comm (in Bulgarian). Trudove Nautchen Inst Nevrol Psikhiatr 7:1–11
Pernov K, Samardjiev A, Nikolkov W (1963) Über die zerebralen Kinderlähmungen mit Nivalin. Psychiatr Neurol Med Psychol Beih 11:425–428
Porcelli G (1963) Contributo clinico sull uso della nivalina neglii esivi di poliomielite anteriore acuta. Minerva Pediatr 15:1034–1036
Proskurnina NF, Yakovleva AP (1952) On the alkaloids of Galanthus woronovi. II. Isolation of a new alkaloid (in Russian). Zh Obshch Khim 22:1899–1902
Proskurnina NF, Yakovleva AP (1955) On the alkaloids of Galanthus Woronovi. III. On the structure of galanthamine (in Russian). Zh Obshch Khim 25:1035–1039
Prozorovsky VB, Khromova ON, Dubovitzkaya SI (1970) Comparative characteristics and analysis of the anticurare action of the cholinpotentiating agents (in Russian). Exp Khir Anesth 6:78–81
Revelli U, Grasso E (1962) Le traitment de la poliomyelite a la nivalin. Minerva Med 53, 24:881–882
Revelli U, Grasso E (1968) Therapeutic action of nivalin. Further contribution (in English). Med Biol Inform 1:17–20
Roshchina LF (1969) Galanthamine effect on the disorders of conditioned reflexes and behaviour of rats induced by central cholinolytics (in Russian). Farmakol Toksikol 32, 2:143–146
Rudakov A, Mitzov V (1977) Effect of pymadin, nivalin and their combination on polysynaptic lingual-mandibular reflex (in Bulgarian). Farmacia 27, 2:34–37
Saev S (1968) Application of nivalin in anesthesiology. Med Biol Inform 1:4–16
Saev S, Tenev KA (1963) Decurarization in the operations of patients with cardiovascular diseases by Nivalin (in Russian). Exp Khir Anesth 2:76–77
Salvini L, Frosali L, Pacet AM (1962) Valutazione clinica dell' antagonismo verso la d-tubocurarina di un nuovo anticholinesterasico (Nivalina). Minerva Anestesiol 28:201–204
Schmid W (1980) The micronucleus test: an in vivo bone marrow method. In: Hsu TC (ed) Cytogenetic Assays of Environmental Mutagens. Osmum, Allanheld, pp 221–229

Schmidt J, Matthias H (1961) Zur Pharmakologie der Amaryllidaceen-Alkalode Galanthamin, Narwedin, Haemanthamin und Tazettin. Acta Biol Med Germ 7:402–410
Sedloev S, Georgiev IV, Milev M (1964) On the effect of nivalin upon intestinal peristalsis. Acta Inst Sup Med (Sofia) 43, 1:69–73
Shenk NA, Eidinova MB, Mitbreit IM (1956) Therapeutic and diagnostic value of the galanthamine in the cases of patients with different stages of poliomyelitis (in Russian). Farmakol Toksikol 19, 4:36–41
Spector W (1956) Handbook of toxicology, vol 1. Saunder, Philadelphia, pp 210–211
Stefanov ZH, Slavchev P, Mitkov IV (1974) Qualitative and quantitative studies on alkaloid composition of wild-growing and introduced populations of *Leucojum aestivum* L. Farmacia Sofia 24/6:16–19
Stojanov E (1964a) Spirographic evidence of anticurare effect of nivalin. Acta Med Inst Sup (Sofia) 43, 4:39–46
Stojanov E (1964b) Galanthamine hydrobromide (Nivalin) ein neues Antidot der nichtdepolarisierenden Muskelrelaxantien. Anaesthesist 13, 7:217–220
Stojanov E (1965) Clinical experience with the new curare antidote galanthamine (Nivalin). In: International anaesthesiology clinics (European trends in anesthesiology 3) 4:675–685
Stojanov E (1971) Clinical use of nivalin in decurarization (A six year experience). Anesthesia Reanimacija 3, 1:17–23
Stojanov E, Vulchanova S (1963) The clinical application of Nivalin as an antidote of curare. Acta Med Inst Super (Sofia) 42, 4:1–4
Tchakarov IV (1980) Nivalin – comparative studies of enuresis treatment in children via electrophoresis and interference current (in English and Russian). Med Biol Inform 3:6–10
Uyeo S, Kobayashi (1953) Pharm Bull 1:139–142 (Ca, 1959, 13960)
Vasilenko ET, Tonkopy VD (1974) Characteristic of the galanthamine as reversal inhibitor of the cholinesterase (in Russian). Biokhimiia 39:701–703
Wislicki L (1967) Nivaline (Galanthamine hydrobromide) an additional decurarizing agent. Some introductory observations. Br J Anaesth 39:963–968
Yamboliev IA (1984) Experimental studies on some pharmacokinetic and pharmacobiochemical behaviours of nivalin. Dissertation, Medical Academy, Pharmaceutical Faculty, Sofia
Yunusov SY, Abduazimov KhA (1957) Study of the alkaloids of four species of Ungernia (in Russian). Zh Obshch Khim 27:3357–3361

CHAPTER 32

Chinothylinum

A. ĶIMENIS

A. Introduction

The dibromide of bis(β-quinolinoethyl)succinate which exhibits a strong anticholinesterase activity and relatively low acute toxicity (ĶIMENIS 1960a, 1961) has been synthesized in the course of a systemic study of heterocyclic analogs of suxamethonium (WASSERMANN 1961). Its diiodide was also prepared later, the pharmacologic properties of which were found identical to those of the dibromide. The new compound has been given the name chinothylinum (structure I) and recommended for clinical trial. The drug has been authorized for clinical use in the USSR.

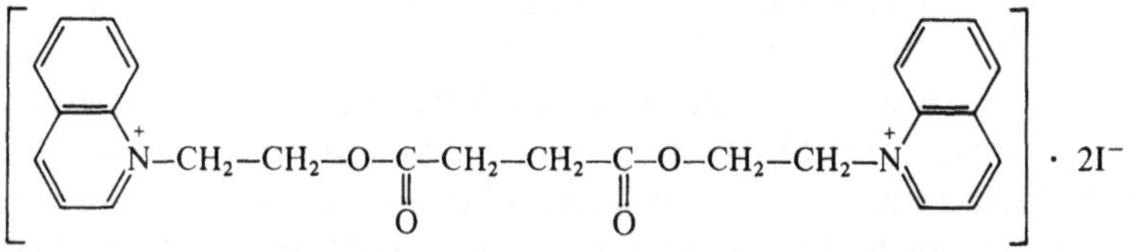
(I)

Chinothylinum is a greenish-yellow fine crystalline powder with a bitter taste and without odor, poorly soluble in water. It is manufactured as 0.05% and 0.2% solution in 2-ml ampules.

B. Experimental Findings

I. Anticholinesterase Activity

Anticholinesterase activity was assayed after the method of HESTRIN (1949) modified by AUGUSTINSSON (1957) using human blood serum as cholinesterase source. Acetylcholinesterase was obtained from human red blood cells. The activity of chinothylinum was compared with that of neostigmine. The pI_{50} values were determined.

It has been found that the pI_{50} of chinothylinum with respect to red blood cell acetylcholinesterase and serum cholinesterase is 7.27 (7.09–7.60) and 7.23 (6.98–7.80), respectively ($P = 0.05$). Neostigmine under these conditions was somewhat less active, the corresponding pI_{50} values being 6.57 (6.44–6.75) and 6.59 (6.50–7.70) (ĶIMENIS et al. 1982a).

Anticholinesterase activity of chinothylinum and neostigmine (0.2 mg/kg intraperitoneally) was measured in vivo in rats. The two drugs were found equally inhibitory with respect to brain acetylcholinesterase. However, their activity was rather weak, probably owing to the low permeability of the blood–brain barrier for quaternary ammonium bases. The peak of chinothylinum activity was observed 2 h after its administration, some decrease in brain acetylcholinesterase activity could still be noticed after 18 h.

The effect of chinothylinum on rat blood serum cholinesterase was somewhat less pronounced, as compared with neostigmine, but a significant decline in enzyme activity was already observable 20–30 min after chinothylinum administration. The maximum action was noted 1 h after application followed by almost complete restoration of initial cholinesterase activity after 2 h. The effect of neostigmine in these experiments was more protracted – activity of blood serum cholinesterase was still lowered after 18 h.

II. Acetylcholine-Potentiating Activity

When assayed on frog isolated rectus abdominis muscle, chinothylinum fails to elicit muscle contractions, but exerts a strong acetylcholine-potentiating effect. Used at 10^{-10} g/ml concentration, it increases manyfold the height of muscle contractions induced by acetylcholine (10^{-7} g/ml). At higher concentrations, the potentiating effect of chinothylinum on acetylcholine activity is still more pronounced. Maximal acetylcholine-potentiating effect of chinothylinum was noted at 10^{-6} g/ml. At higher chinothylinum concentrations, the neuromuscular blocking activity of the drug was manifested, resulting in the inhibition of acetylcholine-induced contractions (Ķimenis 1960b, 1961). Applied to neostigmine-pretreated (10^{-5} g/ml) rectus abdominis, i.e., when muscle acetylcholinesterase activity has been inhibited, chinothylinum failed to potentiate the effect of acetylcholine further. This indicates that the potentiation of acetylcholine effects is due to the anticholinesterase activity of chinothylinum. Chinothylinum (10^{-8} g/ml) exhibits a marked acetylcholine-potentiating effect on isolated small intestine of the guinea pig. On intravenous administration to anesthetized cats, chinothylinum (0.02–0.04 mg/kg) stimulates the depressor action of acetylcholine for 45–90 min (Ķimenis 1961).

III. Decurarizing Activity

The effect of chinothylinum on the action of tubocurarine was studied in urethane-anesthetized cats by measuring gastrocnemius contractions in response to peripheral sciatic nerve stimulation by square wave impulses (0.5 ms, 0.1 Hz). Partial blockade of neuromuscular transmission was achieved by intravenous injections of tubocurarine (0.15–0.21 mg/kg). Intravenous administration of chinothylinum (0.05 mg/kg) had a decurarizing effect. The decurarizing effect of the same dose of neostigmine applied after tubocurarine administration was not so marked and was characterized by delayed onset as compared with the effect of chinothylinum. Chinothylinum used at decurarizing doses either did not affect arterial pressure in cats or elicited a short-term decrease by 10–30 mmHg. At higher doses, chinothylinum (0.1–0.3 mg/kg) exerted a marked depressor effect (25–70 mmHg) (Ķimenis et al. 1982b).

IV. Toxicity

1. Acute Toxicity

Acute toxicity of chinothylinum was evaluated in randomly bred white mice weighing 18–24 g, after intraperitoneal, intravenous, and oral administration. Acute toxicity of neostigmine was determined for comparison. After intraperitoneal injection of toxic doses, the mice developed dyspnea followed by twitching of the extremities, seizures, and death. Animals usually died 5–15 min after the injection. On intravenous administration, all toxicity symptoms developed during and immediately after the injection. When given orally, symptoms of poisoning and death were observable after 10–30 min. The LD_{50} values are summarized in Table 1. These data demonstrate that chinothylinum is considerably less toxic than neostigmine. The acute toxicity of chinothylinum is 16.8 (intravenous), 8.75 (intraperitoneal), and 32 (oral) times lower than that of neostigmine (ĶIMENIS et al. 1982b).

During tolerance experiments performed in rabbits, chinothylinum was slowly infused intravenously and its toxic effects were observed. Increased salivation, as well as defecation and urination were noted in response to 0.3 mg/kg of the drug. At 1–2 mg/kg it caused skeletal muscle relaxation lasting 10–20 min. The average dose of chinothylinum eliciting head drop in rabbits is 1.37 (1.09–1.73) mg/kg (ĶIMENIS 1961). It should be noted that LD_{50} of neostigmine for intravenous injections in rabbits is 0.16 mg/kg (FROMHERZ and PELLMONT 1953).

2. Repeated Administration

Toxicologic experiments involving repeated administration of chinothylinum were conducted in rats and dogs. Rats were given subcutaneous injections (0.05 and 0.2 mg/kg) for 25 days. Chinothylinum has been shown not to have any appreciable effect on rat weight, as compared with the control group, or on blood cell composition. Alanine and aspartate aminotransferase activity in blood serum and rat liver homogenates was essentially the same in chinothylinum-treated and control rats. Histologic investigation also failed to reveal any toxicity of the drug with respect to the viscera.

Male dogs were given subcutaneous injections of chinothylinum (0.2 mg/kg) every other day for 60 days. Increased saliva secretion and repeated defecation

Table 1. Acute toxicity of chinothylinum and neostigmine in white mice

Drug	Route of administration	LD_{50} (mg/kg) ($P=0.05$)
Chinothylinum	Intravenous	2.2 (1.6–3.1)
	Intraperitoneal	5.6 (4.6–6.9)
	Oral	320 (263–381)
Neostigmine	Intravenous	0.131 (0.116–0.148)
	Intraperitoneal	0.64 (0.61–0.67)
	Oral	10.0 (8.5–11.8)

was observed 20–30 min after chinothylinum injection in dogs. The dogs receiving the drug showed some excitation 30–60 min after administration. Red blood cell counts were found to be somewhat lower toward the end of the second month. White cell counts and composition showed no appreciable variation in response to chinothylinum. Biochemical investigation of blood and urine revealed no hepatic or renal toxicity (ĶIMENIS et al. 1982b).

C. Clinical Findings

As an anticurare agent, chinothylinum has been applied during surgery in patients belonging to various age groups. Its effects on neuromuscular blockade induced by various neuromuscular blocking agents: tubocurarine, dioxonium, Lysthenon, etc., have been studied, revealing a distinct anticurare activity. The appropriate chinothylinum dose upon intravenous administration was 0.02–0.04 mg/kg, depending on the extent of neuromuscular blockade elicited by dioxonium or tubocurarine, i.e., decurarizing doses of chinothylinum and neostigmine are similar. However, muscle tone following chinothylinum administration was restored somewhat faster than after neostigmine injection. For example, in the case of partial restoration of skeletal muscle tone, chinothylinum caused a complete restoration of muscle tone after 2.2 ± 0.2 min (dioxonium) and after 2.4 ± 0.4 min (tubocurarine).

The corresponding values obtained for neostigmine were 3.6 ± 0.2 and 7.1 ± 0.3 min (mean $\pm$ standard error) (GINTERS et al. 1982). The maximal decurarizing effect of chinothylinum according to electromyographic findings was observed after 30–40 min (KHMELEVSKY and GADALOV 1976). The anticurare action of chinothylinum was most pronounced in patients with residual relaxation caused by repeated administration of suxamethonium (GADALOV and KHMELEVSKY 1976). Similar to neostigmine, administration of chinothylinum is not indicated in the case of apnea and strong antidepolarization block because of possible recurarization. To increase the duration of action of chinothylinum, it is recommended to be used in combination with small doses of neostigmine (0.5 mg) incapable of eliciting decurarization (GINTERS et al. 1982). The clinical findings indicate that, as compared with other anticholinesterase drugs, chinothylinum exerts a more selective neuromuscular action and causes less muscarinomimetic side effects. Therefore, administration of chinothylinum at doses under 2 mg requires no prior injection of atropine. At doses of 3–4 mg, hypersalivation and bradycardia can be observed (KHMELEVSKY and GADALOV 1980, 1981). A prior injection of atropine removes or considerably diminishes these side effects.

D. Conclusion

Unlike suxamethonium, its bisquinoline derivative, chinothylinum, bis(β-quinolinoethyl)succinate shows a weak neuromuscular blocking and strong anticholinesterase activity. As compared with neostigmine, it is characterized by considerably lower toxicity and shorter duration of action. Chinothylinum acts more selectively on the neuromuscular junctions of skeletal muscles than on smooth muscle organs and heart. This is revealed both in vitro using isolated organs and in vivo. For instance, the acetylcholine-potentiating effect is observed at much lower concentrations and is more pronounced in the rectus abdominis muscle of the frog than in isolated small intestine of the guinea pig. In intact animals, chinothylinum used at decurarizing doses has only a marginal effect of the cardiovascular system. These data have been supported by clinical observations that chinothylinum, while exhibiting a strong decurarizing action, fails to elicit any appreciable muscarinomimetic effects. It is possible that its low acute toxicity, as compared with neostigmine is partially due to its lower cardiac toxicity. It is interesting to note that the decurarizing effect of chinothylinum sets in more rapidly than that of neostigmine, especially during neuromuscular blockade caused by dioxonium. Of practical importance is its efficacy during neuromuscular blockade observed upon repeated administration of suxamethonium.

References

Augustinsson KB (1957) Assay methods for cholinesterases. In: Glick D (ed) Methods of biochemical analysis, vol 5. Interscience, New York, pp 1–63

Fromherz K, Pellmont B (1953) Pharmakologische Wirkungen des Mestinon „Roche“ (Dimethylcarbaminsäureester des 1-Methyl-3-oxypyridiniumbromid, Pyridostigmin-bromid. Schweiz Med Wochenschr 83:1187–1190

Gadalov VP, Khmelevsky YM (1976) Clinical characterization of a new myorelaxant antidote – chinothylinum (in Russian). Eksp Klin Farmakoter 6:90–99

Ginters JJ, Reinberga AJ, Zeberga ME (1982) Clinical study of chinothylinum (in Russian). Eksp Klin Farmakoter 11:125–128

Hestrin S (1949) The reaction of acetylcholine and other carboxylic acid derivatives with hydroxylamine, and its analytical application. J Biol Chem 180:249–261

Khmelevsky YaM, Gadalov VP (1976) Clinical characterization of chinothylinum, a new myorelaxant antagonist (in Russian). Vestn Akad Med Nauk SSSR 11:59–66

Khmelevsky YaM, Gadalov VP (1980) Effect of myorelaxants on cardiac activity (in Russian). Anest Reanim 1:14

Khmelevsky YaM, Gadalov VP (1981) Possible application of chinothylinum as antagonist of muscle relaxants (in Russian). Eksp Klin Farmakoter 10:113–120

Ķimenis AA (1960a) Anticholinesterase properties of dithyline and some of its heterocyclic derivatives (in Russian). LatvPSR Zin Akad Vestis 10:135–141

Ķimenis AA (1960b) Pharmacological characterization of quinolinium and pyridinium analogues of dithylinum (in Russian). LatvPSR Zin Akad Vestis 12:129–136

Ķimenis AA (1961) Kurarepodobnaya i antiholinesteraznaya aktivnostj nekotorih geterociklìčesskih analogov ditilina (diholinovogo efira yantarnoy kisloty) [Curare-like and anticholinesterase activity of some heterocyclic analogues of dithylinum (dicholine ester of succinic acid)]. Thesis, Riga

Ķimenis AA, Blüger NA, Klusa VE (1982a) Chinothylinum – an anticholinesterase drug (in Russian). Khim Farm Zh 2:247–249

Ķimenis AA,Klusa VE, Blüger NA (1982b) Pharmacological and toxicological characterization of chinothylinum (in Russian). Eksp Klin Farmakoter 11:113–124

Wassermann HM (1961) Preparative synthesis of cholinium – a new anticholinesterase agent (in Russian). Naučnaya sessiya. (Scientific session dedicated to the 10th anniversary of the Riga Medical Institute) (Abstracts). Riga Medical Institute, Riga, pp 29–30

[illegible] on the neuromuscular function of skeletal muscles [illegible] organs and tissue. This is revealed both in vitro using isolated organs [illegible] the action [illegible] is more pronounced in the rectus abdominis [illegible] in isolated [illegible] testing of the [illegible] has only a [illegible] effect on the [illegible] These data have been supported by [illegible]

CHAPTER 33

4-Aminopyridine Hydrochloride (Pymadin)

D. S. PASKOV, S. AGOSTON, and W. C. BOWMAN

A. Introduction

Some of the pharmacological actions of aminopyridines have been known for many years (DOHRN 1924; DINGEMANSE and WIBAUT 1928; VON HAXTHANSEN 1955; FASTIER and MACDOWALL 1958 a), but it is only since Bulgarian pharmacologists and anaesthetists, on the basis of their experimental and clinical studies, advocated the use of 4-aminopyridine hydrochloride (Pymadin) to facilitate neuromuscular transmission, and thereby antagonise neuromuscular blocking drugs (MITZOV 1967; PASKOV et al. 1969, 1973) that its effects have been studied in detail. Such is the current widespread interest in the actions of aminopyridines by membrane biologists, pharmacologists and clinicians, especially anaesthetists, that an international symposium on the subject was held in Paris in July 1981. The proceedings of this symposium have been published (LECHAT et al. 1982).

The pyridine nitrogen of 4-aminopyridine has a pK_a value of 9.1 so that at body pH values about 98% of the molecules are protonated to form the monocation. The charge on the cationic form is delocalised over both nitrogens, in the manner of an amidinium cation, and this effectively prevents the addition of a second proton (JOULE and SMITH 1972). The extremes of the delocalisation of charge are shown within the brackets below.

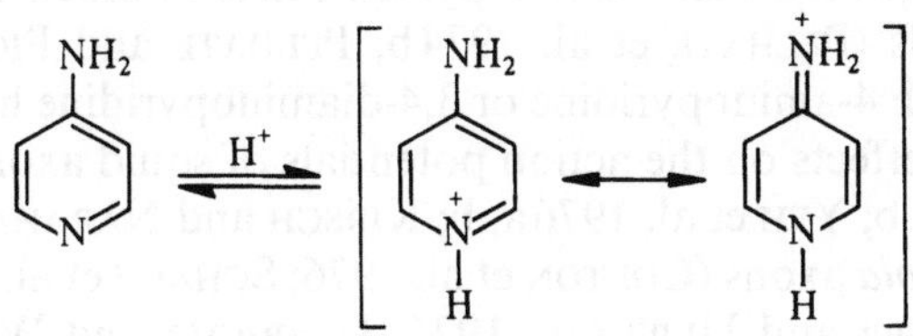

MOLGO et al. (1980) showed that when the pH of the solution bathing an isolated nerve–muscle preparation was raised, the potency of an aminopyridine in facilitating transmission was increased. An increase in pH suppresses the protonation of the compound, so that more molecules exist in the nonionised lipid-soluble form. In this form, the molecules would penetrate the lipids of cell membranes more effectively, so that the effect of pH may constitute evidence that the site of action is intracellular. Modification of extracellular pH has little effect on intracellular pH, so that once in the cytosol most of the molecules would again become protonated, suggesting that the monocationic forms are the active species once they have reached their site of action.

In addition to its ability to facilitate neuromuscular transmission in skeletal muscle, 4-aminopyridine produces a wide range of other effects. Thus, in vertebrates and invertebrates, it has been shown to facilitate transmission at synapses (both excitatory and inhibitory) in the brain and spinal cord, and at autonomic ganglia and autonomic neuroeffector junctions. Additionally, the release of certain hormones from endocrine glands may be increased, and the compound affects conduction in excitable membranes and increases muscle contractility. For recent reviews of the actions of aminopyridines, articles by THESLEFF (1980), BOWMAN and SAVAGE (1981), BOWMAN (1982), SONI and KAM (1982) and GLOVER (1982) may be consulted, and for the proceedings of a symposium, see LECHAT et al. (1982).

Most of the work on 4-aminopyridine has been carried out in relation to neuromuscular transmission. Nevertheless, since its important actions at all types of synapse probably depend upon a common mechanism, its actions on excitable membranes are considered first. Its actions at various synapses and junctions, followed by its effects on muscle in experimental animals are briefly dealt with, and finally its clinical actions, actual and potential, including its pharmacokinetics, are described.

B. Actions on Excitable Membranes

LEMEIGNAN et al. (1969) showed that 4-aminopyridine increased excitability and the amplitude and duration of extracellularly recorded action potentials of isolated frog and cat nerves. Similar effects were noted in invertebrate connectives (PELHATE et al. 1972). PELHATE et al. (1974a, b) analysed the effects on single giant axons of the cockroach, and found that 4-aminopyridine decreased delayed rectification thereby prolonging spike duration, and suggesting an action to decrease potassium conductance g_K. Experiments in which the technique of voltage clamp was used soon confirmed that 4-aminopyridine acts to block voltage-dependent potassium channels (PELHATE et al. 1974b; PELHATE and PICHON 1974). Subsequent studies with 4-aminopyridine or 3,4-diaminopyridine have demonstrated essentially similar effects on the action potentials of squid axons (MEVES and PICHON 1975, 1977a, b; YEH et al. 1976a, b; KIRSCH and NARAHASHI 1978; PICHON et al. 1982), *Myxicola* axons (COLTON et al. 1976; SCHAUF et al. 1976), frog nodes of Ranvier (WAGNER and ULBRICHT 1976; ULBRICHT and WAGNER 1976; ULBRICHT et al. 1982; DUBOIS 1982), lobster axons (WU et al. 1980), frog skeletal muscle (GILLESPIE and HUTTER 1975; GILLESPIE 1977; HORN et al. 1979) and mammalian nonmyelinated or demyelinated (with diphtheria toxin) nerve fibres (SHERRATT et al. 1980; BOSTOCK et al. 1981). Figure 1 illustrates the effect of 4-aminopyridine on the action potential of a space-clamped squid giant axon.

Mammalian myelinated nerve fibres, unlike those of amphibians, have few voltage-dependent potassium channels and therefore very little outward potassium current I_K (HORACKOVA et al. 1968; CHIU et al. 1979; BRISMAR 1980). Consequently, 4-aminopyridine has little effect on the action potentials of mammalian myelinated axons (SHERRATT et al. 1980). However, I_K appears when such axons are subjected to mechanical stress, osmotic shock or demyelinating toxins, presum-

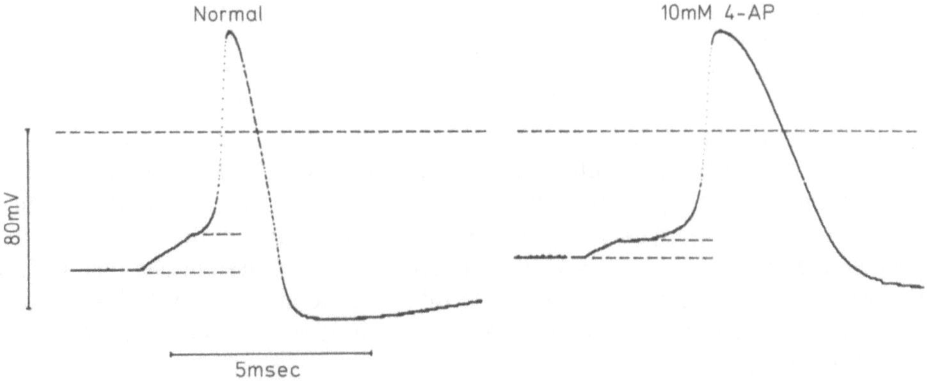

Fig. 1. Effects of 4-aminopyridine on resting potential and action potential in a space-clamped squid giant axon (temperature 8 °C). Note that the threshold is lower in the presence of the drug. (PICHON et al. 1982)

ably because potassium channels, present under the myelin, become exposed. After demyelination, 4-aminopyridine prolongs the action potentials and thereby facilitates conduction. SHERRATT et al. (1980) have suggested that a drug with this kind of mechanism of action might be of value in the symptomatic treatment of demyelinating diseases. Aminopyridine-sensitive potassium channels are also present in immature myelinated mammalian axons, and in axons that have regenerated after damage (KOCSIS and WAXMAN 1983).

PICHON et al. (1982) examined the effect of 4-aminopyridine on membrane current noise, generated by the potassium current, from small patches of nerve membrane of the cockroach (*Periplaneta americana*). As expected, the compound decreased the noise. In its presence, no high frequency component could be detected, suggesting that 4-aminopyridine does not bind to open potassium channels. Single-channel conductance was not modified, suggesting that 4-aminopyridine blocks potassium channels in an all-or-none manner.

The idea that 4-aminopyridine binds to the closed form of the potassium channel of invertebrate nerves is supported by the observation that membrane depolarisation, which opens the potassium gates, leads to unblocking of the channels, and the blocking effect is less pronounced for large depolarising pulses than for small (LLINAS et al. 1975, 1976; PELHATE et al. 1975; YEH et al. 1976a, b; MEVES and PICHON 1977a). Unblocking of open potassium channels means that this occurs to some extent during the overshoot of each action potential. At the nodes of frog axons, unblocking of the potassium channels during long depolarising pulses occurs to a lesser extent than in invertebrate axons. Unblocking in frog axons has a short time constant (0.2 s), whereas restoration of the block at the resting potential is much slower, with a time constant of more than 1 min. This explains why the block may be removed in a cumulative fashion by low frequency repetitive pulses. ULBRICHT et al. (1982) showed that, although strong depolarisation tends to reduce the block, weak depolarisation greatly enhanced the rate of reestablishment of the block, suggesting that the potassium gates of frog nodes of Ranvier must be open in order to expose the sites of block.

In squid axons, 4-aminopyridine acts when applied either inside or outside the membrane (MEVES and PICHON 1977a; YEH et al., 1976a, b). 3,4-Diaminopyridine is also effective from either side, but is much more potent when applied internally (KIRSCH and NARAHASHI 1978). This, and the effect of pH already described, have led to the conclusion that the receptor sites are nearer the intracellular end of the potassium channels. The results of experiments involving noise analysis on cockroach axon membrane, together with other results, led PICHON et al. (1982) to suggest that the 4-aminopyridine molecule binds with its lipid-soluble pyridine nucleus in the walls of the enlarged inner portion of the potassium channel and its polar head in the lumen of the channel. Following membrane depolarisation, the positively charged molecule would be driven out of the opened channel by the transmembrane field.

Although low concentrations of 4-aminopyridine (~10 μ*M*) prolong the spike, they have little or no effect on resting membrane potential (PELHATE et al. 1974a). However, high concentrations (~10 m*M*) both prolong the spike and produce membrane depolarisation, especially in squid axon (YEH et al. 1976a, b; PICHON et al. 1982). The membrane depolarising action on squid axon is evident in Fig. 1. 4-Aminopyridine also induces repetitive firing in nonstimulated squid axons (Fig. 2) and some other excitable membranes. The burst of action potentials is preceded and followed by oscillations of the membrane potential (Fig. 2). In relatively small concentrations of 4-aminopyridine (1–100 μ*M*), the frequency and duration of the bursts of spikes increase with time. With higher concentrations (1–10 m*M*) the bursts are transient, as in Fig. 2, and disappear as the membrane potential falls to around 50 mV (PICHON et al. 1982). The mechanisms underlying the depolarising action and the production of automaticity are not fully understood. According to GOLENHOFEN and MANDREK (1978), inhibition of g_K alone may not be sufficient to induce these effects, since other inhibitors of g_K do not produce them. However, more recent evidence shows that nerve cells possess a number of hitherto unsuspected and pharmacologically distinguishable

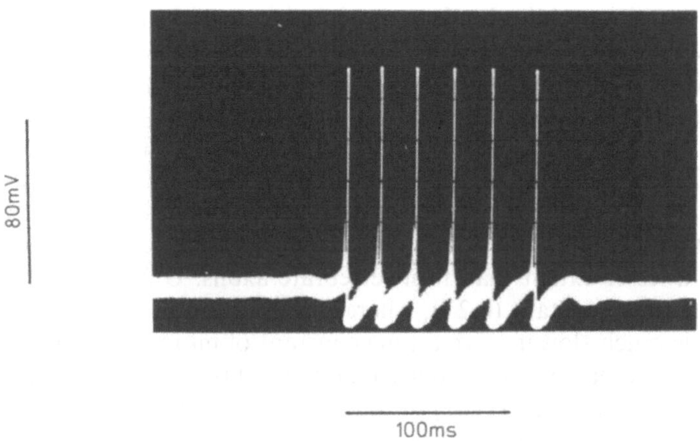

Fig. 2. Burst of activity recorded in an intact space-clamped squid giant axon 5 min after application of 10 m*M* 4-aminopyridine (temperature 8 °C). (PICHON et al. 1982)

conductances, including (1.) a Ca^{2+}-activated K^+ current that underlies the long-lasting spike after hyperpolarisation (KRNJEVIĆ 1978; HOTSON and PRINCE 1980), (2.) the so-called M current (ADAMS et al. 1981) which is a noninactivating K^+ current evoked by membrane depolarisation and blocked by muscarinic cholinoceptor agonists, and (3.) another transient outward K^+ current, resembling the I_A current of molluscan neurones, and exhibiting both voltage-dependent and time-dependent inactivation. This transient current comprises a pronounced outward rectification at the onset of electrotonic potentials, increasing the latency to directly evoked action potentials (GUSTAFSSON et al. 1982; GALVAN 1982). It may serve a nonsynaptic inhibitory role that delays or prevents action potential discharge, and thereby serves to suppress repetitive firing. This transient K^+ current in certain central neurones and in rat sympathetic neurones is blocked by 4-aminopyridine (GUSTAFSSON et al. 1982; GALVAN 1982), and this effect of the drug may explain its action to produce repetitive firing in some nerve fibres. Tetraethylammonium ions, although sharing some of the properties of 4-aminopyridine on g_K, do not block the transient, rapidly inactivating, rectifying K^+ current.

C. Actions on Neuromuscular Transmission

I. Evoked Acetylcholine Release

4-Aminopyridine, and related compounds, have been shown to antagonise the block produced by tubocurarine and other nondepolarising neuromuscular blocking drugs in mammalian and avian muscles in vivo and in vitro (VOHRA and PRADHAM 1964; LEMEIGNAN and LECHAT 1967; MITZOV 1967; PASKOV et al. 1969; SOBEK et al. 1968; PASKOV et al. 1973; BOWMAN et al. 1976, 1977a; FOLDES et al. 1976a, b; HARVEY and MARSHALL 1977a). Its anticurare action in the anaesthetised cat is illustrated in Fig. 3. 4-Aminopyridine does not antagonise neuromuscular block produced by the depolarising agents, decamethonium or suxamethonium (LEMEIGNAN and LECHAT 1967; FOLDES et al. 1976b). The compound is devoid of anticholinesterase activity in any dose that might be administered in vivo (SHAW and BENTLEY 1953; LEMEIGNAN and LECHAT 1967), and in fact a wide range of studies involving mechanical recording, electrophysiological analysis, collection and assay of acetylcholine, and electron microscope studies have demonstrated that the facilitatory action of 4-aminopyridine, and derivatives of it, on neuromuscular transmission is the result of a prejunctional effect on the nerve endings through which the evoked release of acetylcholine is increased (LEMEIGNAN and LECHAT 1967; SOBEK et al. 1968; MOLGO et al. 1975, 1977, 1979; BOWMAN et al. 1976; HARVEY and MARSHALL 1977a, b, c; LUNDH and THESLEFF 1977; HEUSER 1977; ILLES and THESLEFF 1978; LUNDH 1978a; JACOBS and BURLEY 1978; DURANT and MARSHALL 1978, 1980; KATZ and MILEDI 1979; HEUSER et al. 1979; THESLEFF 1980; KIM et al. 1980a; MAENO 1980; MOLGO 1982).

4-Aminopyridine increases the quantal content of the end-plate potential (and end-plate current) in amphibian and mammalian muscles after partial impairment of transmission with a curarising agent or with a bathing solution containing a high $[Mg^{2+}]:[Ca^{2+}]$ ratio (MOLGO et al. 1975, 1977; for reviews see THES-

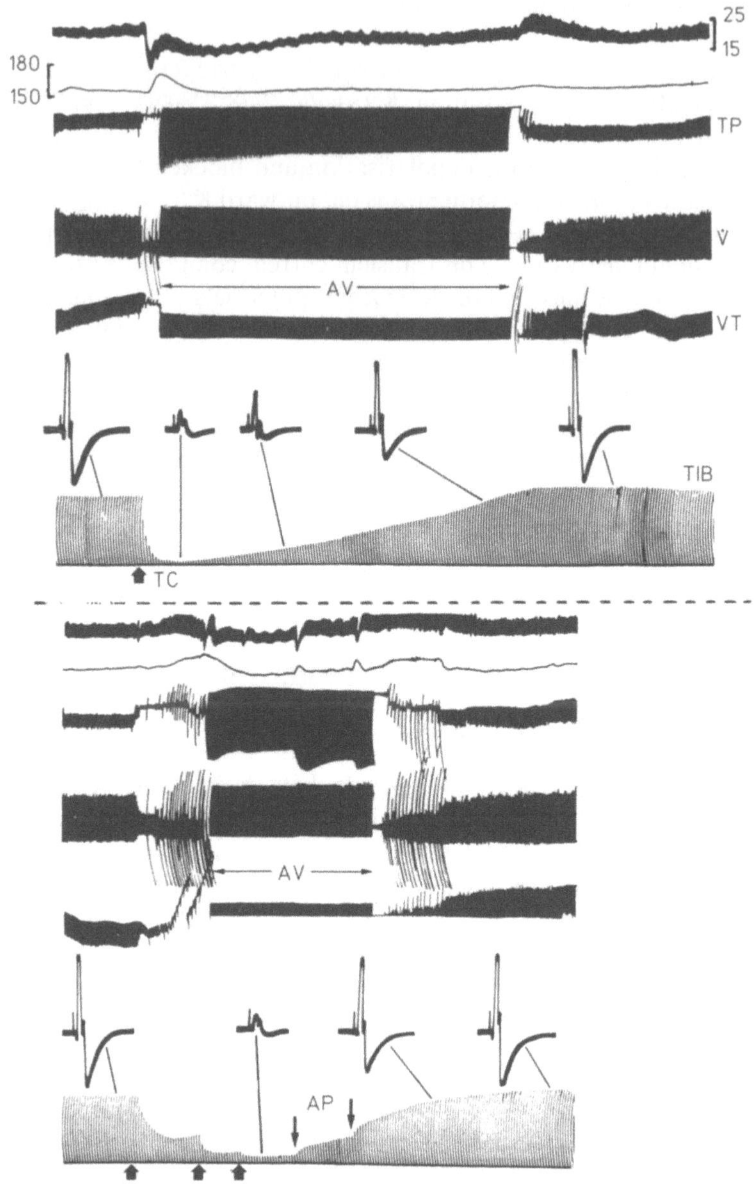

Fig. 3. Cat under chloralose anaesthesia. Records from above downwards: arterial blood pressure (kPas), heart rate (beats/min), transpulmonary pressure (*TP*), rate of airflow (*V̇*), tidal volume (*VT*), gross muscle action potentials associated with the indicated maximal twitches of a tibialis anterior muscle (*TIB*), which was stimulated through its motor nerve at a frequency of 0.1 Hz. Artificial ventilation was applied during periods marked *AV*. Above the *dashed line* shows a control response to tubocurarine (*TC* 0.4 mg/kg). Below the *dashed line* shows a later production of a similar degree of block (0.2, 0.05, 0.02 mg/kg tubocurarine at the bottom three *arrows*). At the *arrows* labelled *AP*, 4-aminopyridine (two doses each of 0.3 mg/kg) was injected. The block was antagonised. Each dose of 4-aminopyridine produced a transient fall in blood pressure and reflex rise in heart rate, and an increase in transpulmonary pressure representing a bronchoconstriction. Part of this figure has been published elsewhere (Agoston et al. 1982)

LEFF 1980; BOWMAN and SAVAGE 1981; MOLGO 1982; GLOVER 1982). Analysis of the junctional events indicates that 4-aminopyridine increases the average number of quanta m released in response to a nerve impulse by increasing the binomial parameter n, but without affecting the mean probability p of release (LUNDH 1979; MOLGO et al. 1979). It has been suggested that the physical counterpart of n may be the number of functional transmitter release sites at the nerve terminal, and morphological evidence supports this possibility. Thus, freeze-fracture replicas of the active zones in frog neuromuscular junctions have shown that 4-aminopyridine increases the number of structures thought to represent sites of fusion between the synaptic vesicles and the terminal axon membrane after a nerve impulse (HEUSER et al. 1979).

The facilitatory action of 4-aminopyridine on transmission is dependent on the presence of calcium ions in the bathing solution (MOLGO et al. 1977; KIM et al. 1980 a), and its anticurare action is blocked by the calcium slow channel blocking drugs verapamil and nifedipine (SAVAGE 1979), and by manganese ions (LUNDH and THESLEFF 1977; ILLES and THESLEFF 1978). 4-Aminopyridine cannot therefore substitute for calcium ions in the release process, but appears to act by increasing the influx of calcium ions from the extracellular fluid through the voltage-sensitive calcium channels in the nerve terminals in response to an action potential (MOLGO et al. 1980). Aminopyridines increase the width of the action potential in nerve endings at frog neuromuscular junctions (MOLGO 1982). Recent experiments on mouse neuromuscular junctions (MALLART and BRIGANT 1982; BRIGANT and MALLART 1982) have shown that sodium channels and action potential generation cease at the preterminal part of the axon which then depolarises the terminal arborisation by electrotonic spread. An inward calcium current I_{Ca} at the terminals is normally rapidly opposed by an outward potassium current I_K, which restores the resting membrane potential. Blockade of I_K with aminopyridine (or tetraethylammonium), as described in Sect. B), allows the full expression of I_{Ca} which is thereby greatly enhanced. LUNDH and THESLEFF (1977) and MOLGO et al. (1980) had also obtained evidence that 4-aminopyridine greatly enhances the inward calcium current at motor nerve terminals. The inward calcium current at nerve endings couples excitation to acetylcholine release (KATZ and MILEDI 1969), and the enhanced calcium ion influx in the presence of 4-aminopyridine therefore accounts for the enhanced release of acetylcholine. Whether, in addition to a secondary effect on calcium influx arising from potassium channel block, 4-aminopyridine also acts directly on voltage-sensitive calcium channels to facilitate calcium ion entry, as first suggested by LUNDH and THESLEFF (1977), remains a possibility that has not been definitely excluded, although the results of studies of the effect of 4-aminopyridine on miniature end-plate potential frequency in the presence of high potassium ion concentration provide indirect evidence against the suggestion (VAN DER KLOOT and MADDEN 1981; MOLGO 1982). If the effect of 4-aminopyridine on transmitter release is entirely the result of block of potassium channels, as seems likely, then it must be assumed that nerve terminal potassium channels differ from those of many other excitable membranes (Sect. B), since unlike the situation elsewhere, the effect of 4-aminopyridines on transmitter release is neither time dependent nor voltage dependent (MOLGO 1982).

In high concentrations, aminopyridines not only increase the amount of transmitter released by a nerve impulse, but also induce repetitive (at least double) nerve ending action potentials in response to single nerve stimuli. The repetitive nerve action potentials give rise to a train of rapidly diminishing end-plate potentials or currents (LUNDH 1978 a; MARSHALL et al. 1979; HEUSER et al. 1979; MOLGO 1982). Presumably the initial enhanced release depletes the terminals of readily releasable transmitter, so that a rapid fall-off in output occurs, giving rise to end-plate potential rundown.

High concentrations of aminopyridines may also induce repetitive firing in some nonstimulated motor nerves. The acetylcholine released then gives rise to pronounced muscle activity (VOHRA and PRADHAN 1964; BOWMAN et al. 1976, 1977 a; HARVEY and MARSHALL 1977 a; MARSHALL et al. 1979). The mechanism underlying the production of repetitive firing in some nerves is discussed in Sect. B. Similar mechanisms may be involved in motor nerves.

An unusual action of 4-aminopyridine was recorded by VITAL BRAZIL et al. (1983). They showed that the receptor desensitisation that normally follows prolonged application of carbachol to the end-plates of the isolated diaphragm of the rat was completely prevented by 4-aminopyridine. They also obtained evidence that once the desensitised state had developed in the presence of carbachol, the receptors could be restored to the normal state by 4-aminopyridine. The ability of 4-aminopyridine to normalise desensitised cholinoceptors could not be explained in terms of its action to block voltage-dependent K^+ channels.

II. Spontaneous Acetylcholine Release

A gradual increase in the frequency of spontaneous miniature end-plate potentials at a frog neuromuscular junction was detected by BOWMAN et al. (1976) and MARSHALL et al. (1979) in the presence of high concentrations of 4-aminopyridines, possibly as a consequence of terminal membrane depolarisation like that described in Sect. B. However, the effect was inconsistent and minor in nature (see also MOLGO 1982). Similar high concentrations of aminopyridines may induce the appearance of occasional giant miniature end-plate potentials at frog end-plates treated with tetrodotoxin (DURANT and MARSHALL 1978, 1980; KATZ and MILEDI 1979; MOLGO 1982). The underlying mechanism is unknown. Possibly they arise from exocytosis of oversized multiple vesicles formed during the vesicle recycling process, although, if this is the cause, the reason why aminopyridines should induce such premature membrane fusion between vesicle and axon terminal is obscure.

No effect on miniature end-plate potential frequency has been detected at normal or botulinus toxin-poisoned mammalian neuromuscular junctions (LUNDH 1978 a; KIM et al. 1980 a; MOLGO et al. 1980). However, in the presence of ouabain, which was itself without effect on frequency, 4-aminopyridine then produced bursts of high frequency miniature end-plate potentials (LUNDH et al. 1977 a). The explanation of this interaction is not known. Possibly terminal depolarisation, consequent upon sodium pump inhibition produced by ouabain, summates with an otherwise ineffective terminal depolarisation produced by 4-aminopyridine to reach a level that enhances miniature end-plate potential frequency.

However, 4-aminopyridine did not enhance the increased frequency evoked by terminal depolarisation in the presence of high $[K^+]_o$ (MOLGO 1982), which suggests that its interaction with ouabain is more complex than suggested above.

GUNDERSEN and JENDEN (1981) measured acetylcholine release from the rat phrenic nerve–diaphragm preparation by means of a gas chromatographic–mass spectrometric technique. They found that 4-aminopyridine was without detectable effect on spontaneous acetylcholine release. Since most of the spontaneous release of acetylcholine is nonquantal, and therefore undetectable by recordings of miniature end-plate potentials (KATZ and MILEDI 1977), the results of Gundersen and Jenden indicate that in mammalian muscles 4-aminopyridine is without effect on both the nonquantal leakage of acetylcholine and the quantal release.

A brief period of tetanic nerve stimulation, or a single electrotonic depolarisation of the nerve terminals, produces an increase in the frequency of miniature end-plate potentials. In the case of tetanic stimulation, this occurs both during and for a short period after the tetanus. Under these conditions, aminopyridines produce a further increase in the frequency of the miniature end-plate potentials (LEV-TOV and RAHAMIMOFF 1980; MOLGO 1982), probably as a consequence of facilitated Ca^{2+} influx occurring during the tetanus or during electrotonic depolarisation.

III. Repetitive Nerve Stimulation

When a motor nerve is stimulated with paired stimuli with an appropriate interval between each stimulus of the pair (up to about 60 ms), the amplitude of the second end-plate potential is augmented; that is, facilitation occurs. Facilitation is measured or defined as $(V\text{-}V_0)/V$ (MALLART and MARTIN 1967) where V_0 and V are the amplitudes of the first and second end-plate potentials of a pair. JACOBS and BURLEY (1978) found that 4-aminopyridine (5 μM) augmented both end-plate potentials equally, so that there was no change in the degree of facilitation. On the other hand, MOLGO (1982) found that facilitation at a curarised frog neuromuscular junction was enhanced to a small extent by 4-aminopyridine (1–5 μM). Facilitation of the second end-plate potential diminishes as the interval between the two stimuli is increased. 4-Aminopyridine did not alter the rate of decay of facilitation in MOLGO's (1982) experiments.

The reason for the difference between the results obtained by JACOBS and BURLEY (1978) and MOLGO (1982) is not known, but the effect is clearly concentration dependent and, although both groups of authors used similar concentrations, it may be that the conditions of Jacobs and Burley's experiments were such that the single concentration that they used acted like a higher concentration in Molgo's experiments. MOLGO (1982) found that higher concentrations of 4-aminopyridine (10–50 μM) caused depression, rather than facilitation, of the second end-plate potential of each pair in a curarised frog muscle. It seems likely that when release by the first stimulus is greatly enhanced (i.e. by a large dose of 4-aminopyridine or by the interaction of 4-aminopyridine with excess $[Ca^{2+}]_o$), mobilisation of more transmitter from the reserve to the readily releasable situation within the nerve terminal cannot occur rapidly enough to match the demand of the second stimulus, and so depression occurs. Depression of this sort, in any case in the presence of tubocurarine, is probably the physiologically or clinically important effect, since muscle movements are not evoked by single nerve action potentials, but by repetitive trains. When trains of stimuli at frequencies of 4–50 Hz were applied to a partially curarised frog nerve–muscle preparation only the first one or two end-plate potentials in each train were enhanced by 4-aminopyridine. The subsequent impulses in each train evoked rapidly diminishing responses (LUNDH 1978 a; ILLES and THESLEFF 1978; MOLGO et al. 1979; HEUSER et al. 1979), presumably because the demand by the facilitated release process exceeds the supply that can be met by mobilisation.

BOWMAN et al. (Chap. 5) discuss the likelihood that transmitter acetylcholine, in addition to depolarising the postjunctional membrane, functions in a positive feedback mechanism that enhances transmitter mobilisation to keep pace with the demands of high frequency stimulation. Tubocurarine blocks this feedback mechanism and so gives rise to tetanic rundown of end-plate responses and fade of tetanic tension. Such an action of tubocurarine would therefore be expected to enhance any rundown or mechanical tetanic fade arising from excessive initial release, and this has been found to be the case (GIBB et al. 1982). Different agents that block postjunctional receptors have different relative affinities for the prejunctional site, and so enhance fade to different degrees. For example, snake α-toxins have little or no prejunctional action and so do not enhance 4-aminopyridine-induced fade, whereas tubocurarine has a pronounced action in this respect (GIBB et al. 1982). Because of this prejunctional action of tubocurarine, the antagonistic ability of 4-aminopyridine is limited. When tetanic contractions, as distinct from twitches, are recorded, the limited increase in tetanic amplitude produced by 4-aminopyridine is not accompanied by any relief of tetanic fade (BOWMAN and SAVAGE 1981). Indeed, tetanic fade is increased. However, when a small dose of an anticholinesterase drug (e.g. neostigmine) is mixed with the 4-aminopyridine, the blockade of the prejunctional mechanism may be overcome by the preserved transmitter acetylcholine, so that mobilisation is then able to keep pace with the enhanced release mechanism, and tetanic fade, as well as depression of peak amplitude, is overcome. The Bulgarian anaesthetists frequently add galanthamine to 4-aminopyridine because of the improved anticurare effect that this achieves. The beneficial effects of combining 4-aminopyridine with anticholinesterase drugs have been studied in detail, both in animals (MILLER et al. 1978) and humans. The latter study is referred to again in Sect. J.II.

When transmitter release is depressed, for example by a high Mg^{2+}, low Ca^{2+} content of the bathing solution, 4-aminopyridine produces a sustained increase in evoked output throughout tetanic nerve stimulation (LUNDH 1978 a; MOLGO et al. 1979). Presumably, under these conditions, the overall depressed release mechanism prevents the enhancement of release produced by 4-aminopyridine reaching a level at which it depletes the readily available store below that which can be maintained by the mobilisation mechanism. This result indicates, as mentioned already, that at the motor nerve endings the potassium channel block is not removed by repetitive stimulation. It also suggests that, on theoretical grounds, 4-aminopyridine should be a better facilitator of depressed transmission when the depression is the result of a prejunctional defect (e.g. botulism, Eaton–

Lambert syndrome, excess Mg^{2+}, certain antibiotics) than when it arises from a postjunctional mechanism (e.g. myasthenia gravis, tubocurarine paralysis). Certainly, 4-aminopyridine seems to be a better antagonist of antibiotics that block transmission predominantly by a prejunctional mechanism (aminoglycosides, polymyxins) than of those that act by a postjunctional mechanism (tetracyclines, certain lincosamides) (SOBEK et al. 1968; SINGH et al. 1978a, b; LEE et al. 1978; MAENO and ENOMOTO 1980; ENOMOTO and MAENO 1981; UCHIYAMA et al. 1981; DURANT and LAMBERT 1981). Aminopyridines are also very effective against the depression of transmitter release in experimental botulism (LUNDH et al. 1977a; LUNDH 1978b; TAZIEFF-DEPIERRE et al. 1978; SIMPSON 1978; MOLGO et al. 1980), or that produced by crotoxin (VITAL-BRAZIL et al. 1979). They are less effective against β-bungarotoxin and taipoxin (HARVEY and MARSHALL 1977b; KAMENSKAYA and THESLEFF 1979), possibly because enhanced Ca^{2+} influx produced by aminopyridines not only increases acetylcholine release, but also enhances the opposing actions of the toxins. Aminopyridines improve transmission in the Eaton–Lambert syndrome, which is the result of a prejunctional lesion (LUNDH et al. 1977b; AGOSTON et al. 1978; SANDERS et al. 1980; MURRAY and NEWSON-DAVIS 1981). Further details of its action in this human disease are given in Sect. J.III. The aminopyridines also exert a beneficial effect on transmission in experimental (KIM et al. 1980b) and human myasthenia gravis (LUNDH et al. 1979; KIM et al. 1980b; MURRAY and NEWSON-DAVIS 1981). This disease is generally regarded as exclusively postjunctional in origin, but it is neither possible nor relevant to say whether 4-aminopyridine is less effective in this condition than in the Eaton–Lambert syndrome. Myasthenia gravis is referred to again in Sect. J.III.

D. Actions on Other Peripheral Synapses and Neuroeffector Junctions

4-Aminopyridine increases the release of acetylcholine from parasympathetic nerves innervating smooth muscles and the heart (VIZI et al. 1977; SHIWAKO et al. 1977; MORITOKI et al. 1978; AL-HABOUBI et al. 1978; KANNAN and DANIEL 1978; BIGALKE and HABERMAN 1980; WEIDE and LOFFELHOLZ 1980; BOWMAN et al. 1981; FOLDES et al. 1982), and of noradrenaline from sympathetic nerves (JOHNS et al. 1976; KIRPEKAR et al. 1976, 1977; LEANDER et al. 1977; GLOVER 1978; ITO et al. 1980; HARA et al. 1980; STONE 1981; NEDERGAARD 1981). It completely antagonised the depressant effect of guanethidine on evoked noradrenaline output from the perfused cat spleen (KIRPEKAR et al. 1978). The compound also facilitates transmission through automatic ganglia (DURANT et al. 1980; BOWMAN et al. 1981).

In high concentrations, 4-aminopyridine gives rise to hexamethonium-sensitive excitatory postsynaptic potentials in nonstimulated, rat isolated sympathetic ganglia (GALVAN et al. 1980). The authors proposed that the drug induces action potentials in the presynaptic terminals which lead to the release of acetylcholine. In the presence of 4-aminopyridine, a clear presynaptic effect of GABA is demonstrable. GABA greatly increases the frequency of the discharge evoked by 4-aminopyridine (GALVAN et al. 1980; GRAFE et al. 1981).

In experiments on anaesthetised cats in which various respiratory parameters were recorded concurrently with muscle contractions, 4-aminopyridine, in doses that antagonised tubocurarine, produced increases in transpulmonary pressure (see Fig. 3) that were attributed to constriction of respiratory bronchioles (W. C. BOWMAN, I. W. RODGER and A. O. SAVAGE 1978, unpublished work). The effect on transpulmonary pressure was blocked by atropine and so was probably the result of enhanced vagal activity. However, it is not known whether the enhanced

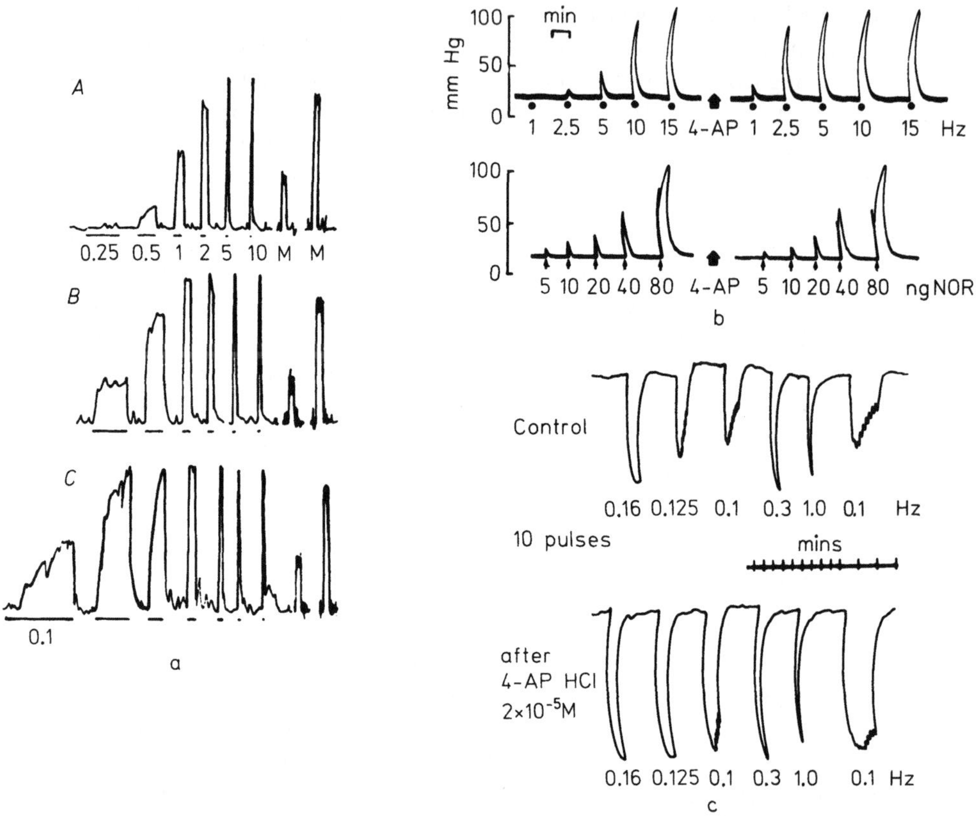

Fig. 4a–c. Effects of 4-aminopyridine on isolated autonomically innervated structures **a** the parasympathetically innervated chick oesophagus, **b** the sympathetically innervated perfused rabbit ear artery stimulated transmurally, **c** the nonadrenergically, noncholinergically innervated bovine retractor penis muscle (transmitter unknown) stimulated by field stimulation. **a** the nerve was stimulated at the frequencies (Hz) indicated with 50 stimuli in each case. Responses labelled *M* are to metacholine (5×10^{-7} and 10^{-6} *M*). *A* is before and *B* and *C* are in the presence of 5×10^{-5} and 10^{-4} *M* 4-aminopyridine respectively. (AL-HABOUBI et al. 1978) **b** the upper row shows responses to transmural (nerve) stimulation at the frequencies shown; the lower row shows responses to noradrenaline injected into the perfusate (rate of flow 4 ml/min). Responses are before (on the left) and in the presence (on the right) of 4-aminopyridine (10^{-5} *M*). Only the responses to nerve stimulation were enhanced. (SAVAGE 1979) **c** the inhibitory nerves were stimulated at the frequencies shown before (above) and in the presence of 4-aminopyridine (below). The bathing solution contained guanethidine and atropine to abolish any adrenergic and cholinergic responses. (A. BOWMAN and J. S. GILLESPIE, previously published in BOWMAN 1982a)

vagal activity was a consequence of central stimulation or of increased acetylcholine release at the periphery, or of both.

There is abundant evidence that, in addition to acetylcholine and noradrenaline, the peripheral autonomic nervous system makes use of a number of other transmitters. Axons that release such transmitters are collectively known as nonadrenergic, noncholinergic axons. The bovine retractor penis muscle is an example of a tissue that receives a nonadrenergic, noncholinergic innervation (KLINGE and SJÖSTRAND 1974). Stimulation of these nerves produces relaxation of the smooth muscle. The identity of the transmitter is unknown, but an extract of the smooth muscle contains an inhibitory factor that in many respects mimics the effects of nonadrenergic, noncholinergic nerve stimulation, and which may in fact be the transmitter (AMBACHE et al. 1975; GILLESPIE and MARTIN 1980). It seems that, at the present time, the inhibitory innervation of the bovine retractor penis is the only example of mammalian peripheral nonadrenergic, noncholinergic transmission to be studied in relation to the action of 4-aminopyridine. A. BOWMAN and J. S. GILLESPIE (1981, personal communication) showed that the inhibitory response to nerve stimulation in this tissue is enhanced by 4-aminopyridine (Fig. 4c), whereas the relaxation produced by the inhibitory extract is unchanged. The inhibitory response of the bovine retractor penis muscle is one of the few inhibitory mechanisms that have been shown to be enhanced by 4-aminopyridine. The effect of 4-aminopyridine on this tissue constitutes part of the evidence that the inhibitory response to field stimulation is a consequence of stimulation of intramural inhibitory nerve fibres.

Figure 4 illustrates the effects of 4-aminopyridine on parasympathetic transmission to the chick oesophagus (Fig. 4a), on sympathetic transmission to the isolated ear artery of the rabbit (Fig. 4b), and on nonadrenergic, noncholinergic transmission to the bovine retractor penis muscle (Fig. 4c).

4-Aminopyridine has also been shown to facilitate transmission at a number of invertebrate junctions, including the electric organ of *Torpedo marmorata* (DUNANT 1979; DUNANT et al. 1980, 1982; ISRAEL et al. 1982), the neuromuscular junctions of the lobster, the crayfish, the locust and *Drosophila* larva (SCHAUF et al. 1976; ANWYL 1977; JAN et al. 1977; ZUCKER and LARA-ESTRELLA 1979), the giant synapse of the squid (LLINAS et al. 1982), and the abdominal ganglion of the cockroach (HUE et al. 1975, 1976a, b, 1978).

E. Actions on the Spinal Cord and Brain

The convulsant actions of aminopyridines in experimental animals have been known for a long time (DOHRN 1924; DINGEMANSE and WIBAUT 1928; SHAW and BENTLEY 1949, 1952, 1955; CHARONNAT et al. 1953a, b; VON HAXTHAUSEN 1955; FASTIER and MCDOWALL 1958a, b; LEMEIGNAN 1971; MARTINEZ DE MUNOZ 1980; GOGOLAK et al. 1982), as has their ability to antagonise central depressants such as morphine and barbiturates (SHAW and BENTLEY 1949, 1952, 1955; HOTOVY and ROESCH 1958; VOHRA et al. 1965; RAO et al. 1977; TUNG and BRANDON 1982; SIA et al. 1982). Recently, they have been shown to be highly effective in antagonising ketamine–diazepam anaesthesia in monkeys (MARTINEZ-AGUIRRE and CRUL

1979) and, as described later, in humans (AGOSTON et al. 1980), although they have little effect against phencyclidine (S. AGOSTON 1980, unpublished work). 4-Aminopyridine also increases the respiratory drive by a central, probably cholinergic, mechanism (SEE et al. 1978; FOLGERING et al. 1979), and accordingly it stimulates breathing (SHAW and BENTLEY 1949, 1952, 1955; VON HAXTHAUSEN 1955; FASTIER and MCDOWALL 1958a, b). Its stimulant effect on breathing is evident from the records of Fig. 3.

When applied to the surface of the cat cortex, aminopyridines produce characteristic epileptiform discharges (BARANYI and FEHER 1979; SZENTE and PONGRÁCZ 1979); actually 3-aminopyridine was used in these experiments, but it seems likely that other aminopyridines, including 4-aminopyridine, will have similar effects. It has been suggested that this action of aminopyridines may provide a useful model of cortical epilepsy (BARANYI and FEHER 1979). 4-Aminopyridine also produces seizure-like activity when applied to slices of guinea-pig olfactory cortex (GALVAN et al. 1982). It increased the frequency and amplitude of spontaneous postsynaptic potentials, prolonged orthodromically evoked postsynaptic potentials and caused them to oscillate, induced seizure-like discharges accompanied by large rises in extracellular K^+ and falls in Ca^{2+}, and it prolonged the action potential of the lateral olfactory tract. The seizurelike discharges were blocked by the antiepileptic drug phenytoin, again suggesting that aminopyridine action may provide a useful model of epilepsy.

In the spinal cord, 4-aminopyridine exerts a marked convulsive action, and facilitates transmission at both excitatory and inhibitory synapses (CHANELET and LEMEIGNAN 1969; LEMEIGNAN 1972, 1973; SAADÉ et al. 1971a, b; JANKOWSKA et al. 1977; JANKOWSKA 1982; GALINDO and RUDOMIN 1978). It also increases the receptive field of cuneate and gracile neurones (SAADÉ et al. 1982). The facilitatory action on transmission appeared to be the result of increased transmitter release, since there were no changes in the passive electrical properties of the postsynaptic membrane (JANKOWSKA et al. 1977). Support for this conclusion comes from electron microscope studies of presynaptic membranes from the ventral horn of the spinal cord of rats. These studies have demonstrated that 4-aminopyridine enhances vesicular fusion with the presynaptic membrane as indicated by an increase in the number of open, crater-like, pits in freeze-etched replicas, and in omega profiles seen in thin sections (TOKUNAGA 1979a, b, c). 4-Aminopyridine has also been found to enhance electrotonic transmission in the perfused spinal cord of the frog after chemically mediated transmission has been blocked by manganese ions (SHAPOVALOV and SHIRIAEV 1978).

It is perhaps impossible to explain all of the actions of 4-aminopyridine on transmission and excitability of central synapses on the basis of a block of delayed rectifier K^+ channels, although clearly this action plays an important part. On the other hand, when more is known about the complex responses and interactions of central neurones, it may turn out that blockade of voltage-dependent K^+ channels is in fact the single basic mechanism of action. However, at the present time, insufficient knowledge is available to make speculation worthwhile. For a useful discussion of the problem, a paper by GALVAN et al. (1982) may be consulted. One additional possible mechanism that is worthy of consideration arises from the work of PERKINS and STONE (1980). They showed that, in rat cerebral cortex, 4-aminopyridine blocks the depression of firing rate produced by iontophoretic application of adenine nucleotides. They suggested that endogenously released adenine nucleotides may modulate synaptic transmitter release, and that 4-aminopyridine may facilitate transmission by blocking the action of the nucleotide. 4-Aminopyridine and adenosine do bear a chemical relationship to each other, and this may be significant in any interaction between them.

Another facet of the actions of aminopyridine on cholinergic neurones from brain is that they restore towards normal the depressed synthesis and release of acetylcholine that is consequent upon hypoxia. In conscious mice, they improve tight rope test performance that has been impaired by chemical hypoxia ($NaNO_2$-induced methaemoglobinaemia). Facilitated Ca^{2+} influx, consequent upon K^+ channel block, is probably the trigger for improved acetylcholine release and synthesis. These experiments were carried out with 3,4-diaminopyridine (PETERSON and GIBSON 1982), but it is likely that 4-aminopyridine has a similar effect, and the effect may not be confined to cholinergic terminals.

F. Actions on Endocrine Glands

Exocytosis of secretory granules in several types of endocrine cells is triggered by Ca^{2+}-dependent action potentials that have a marked delayed rectification caused by K^+ efflux. Examples of such cells include prolactin-secreting cells, calcitonin-secreting cells, and parathyroid hormone-secreting cells (SAND et al. 1982). 4-Aminopyridine increased the firing frequency and prolonged the action potentials in cultured rat pituitary tumour cells, leading to about a 100% increase in the secretion of prolactin; prolactin synthesis was not affected (SAND et al. 1980, 1982). 4-Aminopyridine also increased calcitonin and parathyroid hormone secretion from goat thyroid gland in vivo, and from cultured goat parathyroid cells respectively (SAND et al. 1982).

AHREN et al. (1981) showed that 4-aminopyridine increased plasma glucose concentration, without affecting insulin concentrations, in normal mice, but was without effect on glucose concentration in adrenalectomised chemically sympathectomised mice. 4-Aminopyridine inhibited glucose-induced insulin secretion in normal mice, but potentiated it in adrenalectomised chemically sympathectomised mice. The results suggested that 4-aminopyridine elevates plasma glucose and inhibits glucose-induced insulin secretion indirectly by stimulating the sympathoadrenal system. However, the observation that glucose-induced insulin secretion was enhanced by 4-aminopyridine after adrenalectomy and chemical sympathectomy suggested that the drug may also exert a direct stimulant action on the insulin-secreting cells. Evidence in support of such a direct action was provided in unpublished experiments by B. L. FURMAN (1981, personal communication) who showed that 4-aminopyridine increases the release of immunoreactive insulin from isolated islets of Langerhans.

High concentrations of 4-aminopyridine diminish, rather than increase, the secretion of corticosteroids from rat adrenal cortex slices, and impair the stimulatory effect of corticotrophin and cyclic AMP (LYMANGROVER and MARTIN 1981). The mechanism underlying these effects is not understood, although the authors assume it involves an altered distribution of K^+ across the cell membrane.

G. Actions on Muscle

4-Aminopyridine increases evoked twitches of certain skeletal muscles (rat diaphragm, rat toe muscle, cat tibialis anterior, frog semitendinosus), but not of others (mouse diaphragm, rabbit tibialis anterior, cat soleus, chick biventer cervicis). Even in muscles on which 4-aminopyridine produces a powerful direct effect on the contractility of twitches, it is without effect on maximal tetantic tension, showing that the drug does not increase the maximal tension that a muscle is capable of exerting (FASTIER and MCDOWALL 1958a, b; LEMEIGNAN and LECHAT 1967; MITZOV and PASKOW 1971; BOWMAN et al. 1977a, b; HARVEY and MARSHALL 1977a, b; MAETANI et al. 1979; KHAN and EDMAN 1979; AGOSTON et al. 1982; EDMAN and KAHN 1982). The dose requirement for the direct effect on contractility, and the time course of the effect, differ from those of the facilitatory effect on transmission.

In the rat diaphragm (BOWMAN et al. 1977b; HARVEY and MARSHALL 1977b), the cat tibialis anterior (AGOSTON et al. 1982) and even the chick biventer cervicis (HARVEY and MARSHALL 1977c), 4-aminopyridine is capable of antagonising the depressant effects of dantrolene sodium on contractility. The antagonism is not specific and may be classified as "physiological antagonism". It arises merely because the two drugs have opposing effects on contractility.

In view of the direct effect of 4-aminopyridine on contractility, an action which probably involves increased availability of calcium ions, it is of interest, and of importance in anaesthetic practice, that 4-aminopyridine failed to give rise to malignant hyperpyrexia in susceptible swine (HALL et al. 1980).

It remains to be determined whether 4-aminopyridine increases the contractility of human muscles. In preliminary unpublished experiments on human volunteers by one of us (S. A.), the drug, in a dose of 0.3 mg/kg, was found not to antagonise dantrolene nor to affect the gross action potential of the adductor pollicis muscle. However, it is not yet known whether the lack of effect was the result of nonreactivity of this muscle or of too small a dose being used. The dose used is an effective anticurare dose, but lies just below or somehwere near the threshold of effectiveness for the direct action on contractility in the cat.

An interesting observation made by HARVEY and MARSHALL (1977c) is that pretreatment of the chick biventer cervicis muscle with tetraethylammonium (TEA) ions uncovered a powerful direct effect of the aminopyridines on contractility, an effect which was otherwise only weakly evident in this muscle. HARVEY and MARSHALL (1977c) suggested that TEA and aminopyridine may each bind at separate sites in the K^+ channel of chick muscle to produce a partial conformational change from the open to the closed form, and that when TEA is already present, the binding of the aminopyridine is facilitated so that a more complete closure of the channel occurs.

KHAN and EDMAN (1979) and EDMAN and KHAN (1982) studied the effects of 4-aminopyridine on the time course of directly evoked contractions. The time to peak twitch tension was increased, but the time constant of tension decay, measured during a tetanus, was not affected. The increase in twitch tension occurred in the presence or absence of extracellular calcium ions. The mechanical threshold, that is, the level of depolarisation at which contraction is initiated, was not changed by 4-aminopyridine, nor was there any effect on K^+- or caffeine-induced contractures. Taken together these results suggest that 4-aminopyridine does not exert any specific effects on the mechanism of release of activator calcium ions, nor on the rate of calcium ion resequestration. The results are consistent with an increased amount of Ca^{2+} release from internal stores, and this is compatible with the observations that aminopyridines prolong the duration of muscle action potentials by blocking K^+ channels (GILLESPIE and HUTTER 1975; AUDIEBERT-BENOIT 1975; MOLGO et al. 1975; GILLESPIE 1977; FINK and WETTWER 1978; MOLGO 1978; DURANT and MARSHALL 1978; HORN et al. 1979; MAETANI et al. 1979; KHAN and EDMAN 1979; EDMAN and KHAN 1982). Prolongation of the action potential increases the release of Ca^{2+} from the sarcoplasmic reticulum (CARLSON and WILKIE 1974).

Membrane depolarisation facilitates the transport of 4-aminopyridine to its site of action, presumably the inner end of the membrane K^+ channel. EDMAN and KHAN (1982) found that no effect of 4-aminopyridine on either action potential or twitch was produced after 30 min exposure to the drug if the muscle fibre was left unstimulated during this period. However, on commencing stimulation after this time the effects of 4-aminopyridine began to build up to reach a maximum after about ten stimuli, just as occurs when stimulation is continually applied throughout the application of the drug. It is perhaps worth bearing in mind a related point made by BOWMAN et al. (1977b). Rather than the action potential being necessary to drive the aminopyridine to its site of action, they were concerned that the strong stimuli that are necessary to excite the muscle fibres directly might exert this effect, in which case the apparent direct action would be an artifact of the stimulation conditions. They made this suggestion for two reasons. First, 4-aminopyridine has been found not to affect the amplitude of twitches during partial neuromuscular block produced by depolarising drugs (LEMEIGNAN and LECHAT 1967; FOLDES et al. 1976a). The action of 4-aminopyridine on transmission would not be expected to antagonise this type of block, but any direct action exerted on the muscle fibres might have been expected to increase the contractility of those fibres that were still responding. Second, 4-aminopyridine is generally less effective in increasing the directly evoked contractions of chronically denervated muscles than in increasing those of an innervated muscle in which transmission has been blocked by a snake toxin or tubocurarine. After chronic denervation, electrical excitability is increased, and weaker stimuli (possibly too weak to drive in the 4-aminopyridine ions) are then adequate to excite the muscle.

AGOSTON et al. (1982) suggested that whether or not 4-aminopyridine affected contractility in a particular muscle might be determined by the degree of activation of the contractile elements of that muscle by a single action potential. If activation during a twitch is already maximal, as may be the case, for example, in the cat soleus muscle, then an increased availability of Ca^{2+} would not increase twitch tension, even though the action potential may be prolonged. An explanation of this sort may account for the observation of MAETANI et al. (1979) that 4-aminopyridine prolongs action potentials of the mouse diaphragm without modifying contractility.

It is of interest that although the contractions of mammalian slow fibres (soleus) are amongst those that are unaffected by 4-aminopyridine, the contractile apparatus of these fibres is increased in its sensitivity to Ca^{2+} by 4-aminopyridine, whereas that of fast fibres is unaffected (FINK and STEPHENSON 1982). The mechanism underlying this effect is unknown.

BARNARD et al. (1982) found that repeated injections of 4-aminopyridine produced some improvement in muscular performance in avian muscular dystrophy. They attributed the effect to chronic facilitation of transmission. However, ROUSSEV et al. (1982) found that 4-aminopyridine tended to normalise certain reactions undergone by myofibril proteins isolated from the muscles of dystrophic rats, and they considered that any beneficial effect on the disease process might be more complex than the familiar effect on neuromuscular transmission.

There is a considerable amount of confusion and controversy in the literature with regard to the actions of 4-aminopyridine on isolated cardiac muscle. Clearly, the drug exerts a positive inotropic effect in large concentrations (SOBEK 1970; FRANK et al. 1978; GLOVER 1979, 1981; WOLLMER et al. 1979; YANAGISAWA and TAIRA 1979, 1982); but part of this action is a consequence of catecholamine release (YANAGISAWA et al. 1978; WOLLMER et al. 1979) and, if the base is used instead of the hydrochloride in experiments in vitro, a substantial effect may be produced by the increase in pH of the bathing medium (RODGER and SHAHID 1981; SHAHID and RODGER 1982). The actions of 4-aminopyridine on cardiac action potentials have been studied. LEMEIGNAN et al. (1975) found an increase in the maximum velocity of depolarisation of guinea-pig ventricular muscle (i.e. an increase in the slope of phase 0). A prolongation of the falling phase (phase 3) of the action potential of guinea-pig atria, dog ventricle, and sheep Purkinje fibres has been described by FRANK et al. (1978), YANAGISAWA and TAIRA (1979, 1982), KENYON and GIBBONS (1979) and FREEMAN (1979). This effect on the repolarisation phase is attributed to block of K^+ channels.

Changes in pH are not responsible for the effects of 4-aminopyridine on nerve impulse-evoked transmitter release from mammalian nerve endings, since the concentrations required are 50–100 times smaller than those necessary to affect cardiac muscle, and the buffering capacity of physiological salt solutions is adequate to prevent any detectable pH change. In any case, the same effects are produced by 4-aminopyridine hydrochloride which does not cause changes in pH.

A more clear-cut effect of 4-aminopyridine has been described as occurring in the isolated sinus venosus of the frog heart. GUERRERO and NOVAKOVIC (1980) showed that in low concentrations (1–20 μM), the compound prolonged the sinus action potential by slowing the falling phase, and slowed the rate of diastolic (phase 4) depolarisation thereby depressing spontaneous automaticity. These effects were not modified by atropine, and the authors proposed that they were a consequence of blockade of potassium conductance. Potent direct cardiac actions of 4-aminopyridine of this sort have not been described for the mammalian heart.

At the present time, it has not been clarified to what extent and how 4-aminopyridine may influence mammalian cardiac muscle by actions that are independent of neurotransmitter release and of pH changes. In any case, it seems that such direct action requires larger concentrations than any that could be achieved by administration of tolerable doses to the intact organism.

GLOVER (1978) showed that, in concentrations higher than those necessary to facilitate noradrenergic transmission, 4-aminopyridine increases responses of the isolated ear artery of the rabbit to noradrenaline and histamine. 4-Aminopyridine was without effect on responses to noradrenaline, histamine or calcium ions when the cell membranes were depolarised by excess K^+. Glover concluded that the effect of 4-aminopyridine was exerted on the plasma membrane rather than on the contractile machinery. Essentially similar results on the pulmonary artery of the rabbit were obtained by PATON (1979), although there was little evidence of a direct effect on the smooth muscle of the rabbit portal vien. HARA et al. (1980) found that in the smooth muscle cells of the guinea-pig pulmonary artery, 4-aminopyridine in large concentrations depolarised the membrane, increased the membrane resistance, and suppressed the rectifying property of the membrane. A similar but less pronounced effect was produced on the smooth muscle of the guinea-pig portal vein.

Evidence of a direct stimulant effect of high concentrations of 4-aminopyridine on certain nonvascular smooth muscles has also been described. In chick upper oesophagus, the compound produced oscillating contractions that were not blocked by atropine or tetrodotoxin (AL-HABOUBI et al. 1978). In dog trachealis smooth muscle, similar mechanical effects were produced and, although they were suppressed by atropine, they were not affected by tetrodotoxin, and appeared not to be the result of acetylcholine release (KANNAN and DANIEL 1978). An interesting observation on the dog trachealis smooth muscle was that 4-aminopyridine increased the number of gap junctions between the smooth muscle cells, although this effect did not appear to be responsible for the mechanical activity (KANNAN and DANIEL 1978). In the smooth muscle of the vas deferens of the guinea-pig, high concentrations of 4-aminopyridine slightly depolarised the membrane, increased the membrane resistance and enhanced the spike amplitude (ITO et al. 1980). It seems likely that the main basic effect of 4-aminopyridine underlying its ability to cause or to enhance smooth muscle activity, as in other tissues, is the ability to block voltage-dependent K^+ channels and so to inhibit delayed rectification. The consequence of this effect in most smooth muscles would be to allow a more prolonged Ca^{2+} influx during the earlier phase of the action potential. As with skeletal muscle, there appears to be a differing sensitivity of different smooth muscles to 4-aminopyridine. For example, 4-aminopyridine is said not to produce contractions of rabbit vas deferens (JOHNS et al. 1976) or of guinea-pig ileum (MORITOKI et al. 1978).

H. Cardiovascular System

As described in Sect. G, 4-aminopyridine directly affects cardiac muscle and smooth muscle of blood vessels, but only in large concentrations. In doses tolerated by intact animals, or humans, the integrated effects on the cardiovascular system are almost entirely a consequence of facilitated junctional transmission.

Several workers have recorded moderate rises or biphasic changes in blood pressure and changes in heart rate following administration of aminopyridines to laboratory animals (DOHRN 1924; DINGEMANSE and WIBAUT 1928; VON HAXTHAUSEN 1955; FASTIER and MCDOWALL 1958a; FOLDES et al. 1976b). BOWMAN et al. (1981) studied the effects of 4-aminopyridine at two dose levels (0.5 and 2.0 mg/kg i.v.) in anaesthetised greyhounds. The drug was found to produce an initial transient atropine-sensitive fall in blood pressure (also evident in the cat) followed by a sustained rise. It also produced increases in left ventricular systolic pressure and dP/dt_{max}, in right atrial pressure, stroke volume, myocardial blood flow, myocardial oxygen consumption, external cardiac work, arterial oxygen content and blood haemoglobin. The use of appropriate blocking drugs indicated that these effects were attributable to facilitation of sympathetic transmission to the blood vessels, heart and spleen. Heart rate was not much affected because facilitation of vagal transmission to the sinoatrial node counteracted the increased sympathetic effect. There was no evidence of any direct inotropic action on the heart. Essentially similar effects on the canine (mongrel) heart were produced by 4-aminopyridine in the experiments of MARTINEZ-AGUIRRE et al. (1981), who studied its effects using a right heart bypass with extracorporeal circulation in order to control preload. The authors did not test the effects of antagonists of acetylcholine and noradrenaline, and no conclusions were reached with regard to mechanism of action.

In anaesthetised greyhounds, 4-aminopyridine also produced temporary cardiac dysrhythmias which were partly attributable to facilitation of sympathetic transmission (BOWMAN et al. 1981). The dysrhythmias were reminiscent of those associated with hyperkalaemia and the possibility was considered that 4-aminopyridine causes the release of liver K^+, possibly by facilitating adrenergic transmission to the liver cells.

The initial transient atropine-sensitive fall in blood pressure produced by intravenous 4-aminopyridine in both cat and dog was not accompanied by bradycardia and was therefore a consequence of vasodilatation. It seems unlikely that 4-aminopyridine should directly stimulate muscarinic receptors, and this pointed to the involvement of acetylcholine. Yet the transient nature of the effect was much briefer than any other facilitatory effect of 4-aminopyridine at a site of synaptic or neuroeffector transmission. BOWMAN et al. (1981) suggested that a bolus injection of 4-aminopyridine might briefly stimulate some afferent nerve endings that reflexly activate cholinergic vasodilator fibres.

AGOSTON et al. (1984) have shown that 4-aminopyridine reverses the cardiovascular collapse in cats produced by otherwise lethal doses of verapamil. All of the animals so treated survived. Verapamil inhibits calcium ion transport through the plasma membranes of smooth and cardiac muscle and at nerve endings. The effect at nerve endings impairs transmitter release, and it is likely that this effect contributes to the toxic effects of large doses. As mentioned in previous sections of this chapter, 4-aminopyridine facilitates transmission most powerfully at junctions at which transmitter release is already impaired. Consequently, it might be expected to restore towards normal the transmission defects produced by verapamil, and this action probably mainly accounts for its remarkable ability to prevent verapamil toxicity. It is possible that the direct effects of 4-aminopyridine on cardiac and smooth muscle cell membranes (Sect. G) also contribute, although, at least under normal conditions, these effects require doses of 4-aminopyridine larger than those that are normally tolerated. Whatever the precise mechanism, the action of 4-aminopyridine in protecting against verapamil toxicity (and probably that of other calcium slow channel blocking drugs) is worth bearing in mind for emergency use in humans.

J. Clinically Useful Effects

4-Aminopyridine has been clinically employed as a curare antagonist for more than 10 years, especially in Bulgaria (PASKOV et al. 1969, 1973). Most of the research on this compound has, therefore, been focused on its effects at the neuromuscular junction and on skeletal muscle. Its prejunctional effects, resulting in an increase in acetylcholine release, have been successfully utilised not only to reverse partial neuromuscular blockade produced by neuromuscular blocking agents, but also in patients with impaired neuromuscular transmission arising from various causes, including disorders of the peripheral nervous system. In addition to its actions at the neuromuscular junction, 4-aminopyridine produces a variety of other effects, as described in the previous sections, and some of these effects are potentially useful in clinical practice. There is a growing amount of experimental evidence concerning its central nervous system stimulating effects in humans. These are thought to be mediated by a presynaptic mechanism similar to that occurring in the peripheral nervous system. These central effects seem not to be restricted to any particular type of synapse or transmitter, as is demonstrated by the ability of the compound to reverse opiate-induced respiratory depression, and also coma resulting from toxic doses of benzodiazepines in humans. Although the anticurare effect of 4-aminopyridine is the only one which has been clinically employed to date, other beneficial, but as yet little investigated, effects in several other conditions (in which long-lasting administration of the oral forms of the compound might be required) may considerably extend the indications for its clinical use. For these reasons it seems appropriate to commence this section with a description of its bioavailability and its pharmacokinetics in humans after the administration of oral and parenteral forms before discussing its use in anaesthesia and in some other clinical conditions.

I. Human Pharmacokinetics

Pharmacokinetic studies with 4-aminopyridine were carried out in nine healthy volunteers each of whom received a 20-mg intravenous bolus of the compound (UGES et al. 1982a). Six of the subjects later received the same dose in the form of enteric-coated tablets, and four were given uncoated tablets. Each administration was separated by a period of at least 2 weeks. Blood, saliva and urine samples were assayed by high pressure liquid chromatography. The methods for the preparation and quality control of the various pharmaceutical forms of 4-amino-

Table 1. Pharmacokinetic variables after bolus i.v. injection of 20 mg 4-aminopyridine

Elimination half-life (h)	Distribution volume (l/kg)	Total serum clearance ($l\ kg^{-1}\ h^{-1}$)
3.6 ± 0.9	2.6 ± 0.9	0.61 ± 0.4

pyridine (solution for intravenous injection and tablets) have been described by UGES and HUIZINGA (1981) and UGES et al. (1982b).

After intravenous administration, the concentrations of 4-aminopyridine in saliva reached higher levels than those in serum and they remained higher for at least 5 h. There was, however, a good correlation between the serum and saliva levels with a mean correlation coefficient of 0.989. Consequently, no significant difference ($P > 0.05$) could be demonstrated between the pharmacokinetic variables (Table 1) calculated from serum and saliva concentrations.

The elimination half-lives after oral administration of enteric-coated and uncoated tablets were 3.3 and 4.6 h respectively. The ingestion of uncoated tablets caused gastric cramps in two of the volunteers and, therefore, this pharmaceutical form was not investigated further.

Bioavailability from the enteric-coated tablets, calculated from the area under the curve (corrected for terminal half-life), approximated $95\% \pm 29\%$, which is similar to the amount recovered ($98\% \pm 8\%$) from the urine in the same six subjects over 24 h.

After both parenteral and oral administration, the urinary excretion of 4-aminopyridine was similar. Approximately 65% of the administered dose was excreted in unchanged form in the urine in 6 h, 80% in 12 h, and the 24 h total recovery from the urine amounted to 99%. Elimination appears, therefore, to occur almost exclusively via the kidney, with an average renal clearance of 670 ml/min. As the normal glomerular filtration rate is approximately 130 ml/min it is deduced that active tubular secretion plays an important role in the elimination of 4-aminopyridine in humans.

A search for biotransformation products failed to show any evidence of glucuronidation or sulphonation of 4-aminopyridine. In addition, thin layer chromatography failed to detect any measurable amount of *N*-acetyl-4-aminopyridine which suggests that *N*-acetylation does not occur. Protein binding in human serum appeared to be negligible in a concentration range of 50–600 µg/l 4-aminopyridine using a method of ultrafiltration (UGES et al. 1982a).

II. Use of 4-Aminopyridine in Clinical Anaesthesia

4-Aminopyridine was introduced as a reversal agent for nondepolarising neuromuscular blocking drugs by PASKOV et al. (1973) and STOYANOV and VULCHEV (1975). In their early clinical studies 4-aminopyridine was shown to reverse partial neuromuscular blockade induced by tubocurarine. 4-Aminopyridine (20 mg i.v.), was injected without the preceding administration of atropine, at a time when the spontaneously recovering twitch response approximated 30% of its control value. After the administration of 4-aminopyridine complete recovery of the twitch response required 35–40 min, although clinical signs of recovery were already present after 10–20 min. At the same time a progressive increase in tidal volume was observed, reaching higher values than the controls at the end of the 35 min observation period. Apart from a slight elevation of the systolic blood pressure, no cardiovascular changes were observed. Further, the muscarinic effects which invariably occur after the use of the standard reversal agents (neostigmine and pyridostigmine) when atropine is not given concurrently were not seen with 4-aminopyridine, even in the absence of atropine.

After extensive clinical investigations 4-aminopyridine received official approval in Bulgaria and has become a standard reversal agent in clinical anaesthesia in that country.

The first clinical studies with 4-aminopyridine in Western Europe were started 10 years later, in 1980. These initial studies by one of us (S. A.) confirmed the findings reported by the Bulgarian investigators (PASKOV et al. 1969, 1973; STOYANOV et al. 1976). Using the same dose (0.3 mg/kg), complete reversal of pancuronium-induced neuromuscular blockade was achieved in 20–25 min. However, it should be noted that, under the same conditions, neostigmine, the most commonly used antagonist, usually produces complete reversal in less than 10 min, and edrophonium acts even faster than neostigmine (CRONNELLY and MORRIS 1982).

Because of the rather slow onset time of the initially suggested dose of 0.3 mg/kg, the effect of higher doses of 4-aminopyridine was investigated. However, larger doses (up to 0.5 mg/kg) were frequently associated with central nervous system stimulation, resulting in postoperative restlessness and confusion. In order to achieve fast and long-lasting reversal without using excessive doses of 4-aminopyridine, the compound may be combined with one of the routinely used reversal agents, such as neostigmine or pyridostigmine. The rationale for this combination is that 4-aminopyridine increases acetylcholine release, whereas cholinesterase inhibitors delay acetylcholine breakdown. Therefore, when administered together it would be expected that they will mutually potentiate each other (PASKOV et al. 1973). This possibility has been extensively studied by MILLER et al. (1979) who showed that in the presence of a small – by itself ineffective – dose of 4-aminopyridine the antagonistic doses of neostigmine or pyridostigmine could be reduced by 70% and 80% respectively because of the apparent synergism between these drugs. Therefore, the dose of one or both drugs in such a combination can be reduced, yet still provide increased antagonistic activity with, at the same time, a decreased risk of side effects. Drugs used in fixed dose combinations should act on the same target tissue by different mechanisms and should have similar pharmacokinetic profiles. Among the anticholinesterase drugs available, the pharmacokinetics of pyridostigmine (CRONNELLY et al. 1980) are most akin to those of 4-aminopyridine; pyridostigmine, therefore, is the drug of first choice when 4-aminopyridine is to be used together with an anticholinesterase agent for the reversal of the effects of neuromuscular blocking agents.

There are other, as yet less extensively explored, possibilities for the use of 4-aminopyridine in clinical anaesthesia. The central analeptic effects of the compound, described by STOYANOV and VULCHEV (1975), were more recently studied

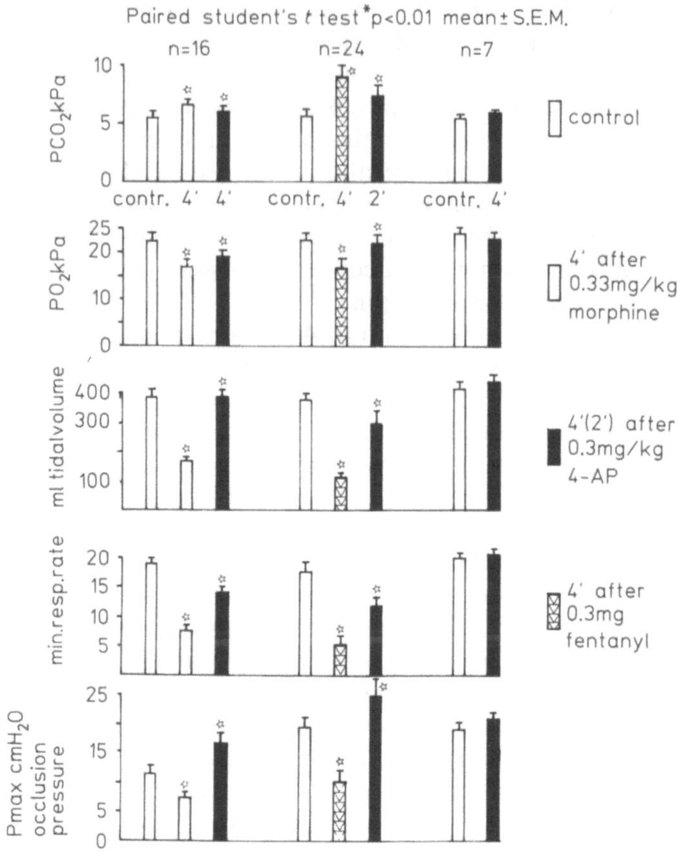

Fig. 5. Effects of 4-aminopyridine upon morphine- or fentanyl-induced respiratory depression. (SIA and ZANDSTRA 1981)

by SIA and ZANDSTRA (1981) in patients with opiate-induced respiratory depression. The findings of this study are summarised in Fig. 5. PCO_2, PO_2, tidal volume, respiratory rate and the occlusion pressure were measured before, 4 min after the administration of either morphine or fentanyl, and 2 and 4 min following the administration of 4-aminopyridine. All patients who received one of these opiates showed apnoea followed by a slow and shallow respiratory pattern. After the administration of 4-aminopyridine there was a statistically significant improvement in all measurements. It is important to note that a significant improvement was seen in the occlusion pressure, which has been reported by other investigators (DERENNE et al. 1976) to be a reliable index of respiratory drive in anaesthetised patients. Interestingly, in the absence of opiate-induced respiratory depression, 4-aminopyridine did not cause any change in these variables in a control group of seven patients (right-hand set of data in Fig. 5).

In another study (AGOSTON et al. 1980), 4-aminopyridine was found to increase significantly the recovery rate from ketamine–diazepam anaesthesia in healthy volunteers. The same interaction with ketamine–diazepam has been confirmed in children (MARTINEZ-AGUIRRE et al. 1980) as well as in adults (LANGREHR et al. 1984). Since diazepam was included in the anaesthetic regimens in these studies it is difficult to reach final conclusions as to whether the observed antagonism by 4-aminopyridine was predominantly of ketamine, or of diazepam, or of both equally. Both types of drug (ketamine, NGAI et al. 1978; benzodiazepines, HILL et al. 1977), have been shown to interfere with central cholinergic pathways, although this is probably not their main mechanism of action. 4-Aminopyridine is known to facilitate cholinergic transmission in both central and peripheral nervous systems as described already, and so it might reverse the inhibitory effects of these anaesthetics at cholinergic synapses in the brain. However, other synapses more specifically inhibited by ketamine might also be involved; this is supported by the lack of antagonistic effects of 4-aminopyridine on the depressant effects of phencyclidine (unpublished observation by S. A.), a compound which is a chemical precursor of ketamine. The possibility that a relatively specific interaction also occurs between diazepam (and other benzodiazepines) and 4-aminopyridine and that this contributed to the abrupt termination of the ketamine–diazepam anaesthesia in the study referred to, is reinforced by the case history of a nurse who, through an overdose of flunitrazepam, was not rousable postoperatively. The patient was scheduled for a minor surgical intervention and in her anxiety had ingested an excessive dose of flunitrazepam preoperatively (the total dose could not be verified). This remained unknown to the anaesthesiologist. After surgery and anaesthesia she remained in deep coma for 5 h, unresponsive to repeated administrations of naloxone and physostigmine. This state was promptly terminated by the administration of 0.15 mg/kg 4-aminopyridine. Extremely high concentrations of flunitrazepam (118 μg/ml) were detected by the hospital pharmacist in serum samples taken just before the administration of 4-aminopyridine. According to the standard values in our laboratory, the therapeutic concentrations of flunitrazepam are 5–20 μg/ml, and 50 μg/ml or more are considered to be toxic. The postoperative course after this episode was uneventful (SCHMUTZLER et al. 1984).

III. Experimental Clinical Use

Besides its application as a curare antagonist and its other uses in clinical anaesthesia, 4-aminopyridine has been shown to exert beneficial effects in a number of clinical conditions, particularly in certain neuromuscular diseases. PASKOV et al. (1976) reported marked therapeutic effects of electrophoretically applied 4-aminopyridine (0.5% solution) in 18 patients with traumatic nervus recurrens injuries after partial thyroidectomy, and also in 43 patients with nervus facialis and 7 patients with nervus fibularis injuries. The therapeutic efficacy appeared to be dependent on the duration of the disease and on the functional state of the muscles involved at the time when the therapy was started. Permanent improvement (sometimes complete cure) was achieved in most of the patients in whom this therapy was started within 2 years after the original injury.

The Eaton–Lambert Syndrome, a rather rare disorder, might be another indication for the therapeutic use of 4-aminopyridine. The disease is characterised by muscular weakness which is a consequence of a prejunctional lesion that results in a reduced release of acetylcholine from motor nerve endings. It is often associated with oat cell carcinoma of the lung (EATON and LAMBERT 1957) and may foreshadow neoplastic disease (CHERINGTON 1976). Recently, 4-aminopyridine has been shown to be effective in patients with Eaton–Lambert syndrome (LUNDH et al. 1977b; AGOSTON et al. 1978; KIM et al. 1980b). The drug produces an increase in the indirectly evoked muscle action potentials of as much as 300% after intravenous administration of 0.3–0.6 mg/kg (AGOSTON et al. 1978). There have been sporadic reports (unpublished observation by S. A. in two cases; and LUNDH et al. 1979) of the ability of intravenously injected 4-aminopyridine to restore neuromuscular transmission in patients with myasthenia gravis, although the lesion in this disease is most probably postjunctional (DRACHMAN 1978). Further studies of the Eaton–Lambert syndrome and myasthenia gravis are necessary in order to establish the therapeutic value of 4-aminopyridine. This research will be greatly facilitated by the availability of enteric-coated tablets (UGES et al. 1982b) which are well tolerated by the patients.

The first reports on the efficacy of 4-aminopyridine in experimental botulism (LUNDH et al. 1977a) and later in human botulism (BALL et al. 1979) evoked considerable interest in the therapeutic potential of this compound. Four patients (victims of the outbreak of botulism in Birmingham, England, in 1978) with severe paralytic botulism due to *Clostridium botulinum* type E were treated with intravenous injections of 4-aminopyridine (BALL et al. 1979). Single bolus doses ranging between 0.35 and 0.5 mg/kg were administered either alone or with neostigmine and atropine. In two patients 4-aminopyridine was given by intravenous infusion up to a total dose of 90 mg over 4 h. A marked progressive improvement in peripheral muscle power was noted within 5–10 min of the bolus injection, and this improvement persisted for 2–4 h. Rapid clinical improvement also occurred in those patients treated by intravenous infusion. The duration of the clinical improvement after 4-aminopyridine administration outlasted that of the EMG response in all patients, and in patients treated by i.v. infusion there was some permanent improvement in peripheral muscle power following treatment. The maximum effect was observed in the extrinsic ocular and limb muscles, and there was only a minor effect on the respiratory muscles, as judged by clinical observation and measurements of spontaneous tidal and minute volumes. In the two patients treated by high doses given by i.v. infusion, generalised convulsions developed requiring large doses (60–80 mg i.v.) of diazepam for control. Consequently, it would seem to be inadvisable to attempt to overcome the lack of response of the respiratory muscles by further increasing the 4-aminopyridine dosage. Apart from a slight increase in systolic blood pressure and heart rate, no other side effects were seen in these patients. The authors concluded that at present the benefits of 4-aminopyridine therapy are outweighed by its potential toxicity in the treatment of human botulism. It is hoped that analogues of 4-aminopyridine with reduced central neurotoxicity will produce the desired results in the future.

Another possible indication for a therapeutic application of 4-aminopyridine (alone or in conjunction with other drugs) might be Alzheimer's disease (BOWMAN 1982a, b). There is increasing evidence to show that in this condition the lesions are primarily in central cholinergic neurones and there is a corresponding choline acetyltransferase deficiency (BOWEN et al. 1976; DAVIES and MALONEY 1976;

DAVIES 1979). Since choline acetyltransferase deficiency is reflected in reduced acetylcholine turnover, a variety of therapeutic attempts have been made to increase brain acetylcholine in patients with Alzheimer's disease (THAL et al. 1981; SUMMERS et al. 1981; DAVIS and MOKS 1982). Not all cholinergic neurones are impaired to the same extent in the disease and so it is conceivable that a compound that promotes the release of acetylcholine from those neuronal terminals that are still functional may have beneficial effects. Increased release of acetylcholine produces a compensatory increase in synthesis. Although cholinergic dysfunction appears to be the main underlying disorder in the disease, it is probably not the only one. 4-Aminopyridine facilitates transmission at all chemically transmitting junctions and is more effective when transmitter release is subnormal than when it is normal, so that the drug may have a further beneficial effect even if it cannot be precisely defined at the present time. Any excessive peripheral autonomic cholinergic activity that might be produced could be selectively antagonised by a quaternary atropine-like drug that does not penetrate the blood–brain barrier. These considerations led WESSELING et al. (1984) to test 4-aminopyridine in patients with Alzheimer's disease. The study was carried out in 14 demented elderly residents of a nursing home under the most stringent experimental conditions, including randomisation and double-blind controlled crossover treatment periods of 6 weeks duration, either with placebo or 4-aminopyridine tablets (20 mg/day). Analysis of the results revealed that recent memory was significantly improved by 4-aminopyridine; other related functions – activity and recognition – showed the same tendency. Unwanted effects, such as irritability and nervousness, occurred more frequently during treatment with 4-aminopyridine than with the placebo. It was concluded that aminopyridines, either alone or in combination with centrally acting cholinesterase inhibitors, might open new perspectives in the treatment of Alzheimer's disease.

K. Conclusions

Our present knowledge of 4-aminopyridine indicates that its clinical use (either alone or in combination with cholinesterase inhibitors) is best restricted to the treatment of some relatively rare clinical conditions in which there is either a known failure of prejunctional transmission or there are theoretical grounds to believe that the compound might be effective where other measures have failed. Such conditions possibly include the Eaton–Lambert Syndrome, myasthenia gravis and neuromuscular blockade complicated by the concomitant use of certain antibiotics. In Huntington's chorea (WESSELING and LAKKE 1980) 4-aminopyridine may be effective in combination with neuroleptics in improving the state of mind of the patient, and in Alzheimer's disease, mental performance seems to be beneficially influenced. Another area of potential clinical application is in cases of intoxication with benzodiazepines, botulinus toxin or verapamil (see Sect. H). Most of the undesirable effects of 4-aminopyridine are of central origin – anxiety, restlessness, insomnia and convulsions. The risk of convulsions at doses above three times the safe therapeutic dose (0.3 mg/kg) suggests that the safety margin for 4-aminopyridine is not as great as is desirable. Nevertheless, the studies in patients with Alzheimer's disease or Huntington's chorea show that 4-aminopyridine can be administered safely in divided doses up to a total of 30 mg/day in enteric-coated tablets for several weeks without any major side effects. Convulsions arising from an absolute or relative overdose can be treated by high doses of diazepam, either alone, or in combination with nondepolarising neuromuscular blocking agents and artificial ventilation.

At present 4-aminopyridine cannot be considered to be a drug for routine clinical use. The synthesis and pharmacological testing of new derivatives are continuing in order to develop new drugs with an improved therapeutic index and more selective actions. The accumulated knowledge of the various pharmacological actions and therapeutic effects of 4-aminopyridine will substantially contribute to the achievement of this goal in the future.

References

Adams PR, Brown DA, Halliwell JV (1981) Cholinergic regulation of M-current in hippocampal pyramidal cells. J Physiol (Lond) 317:29P

Agoston S, van Weerden T, Westra P, Broekart A (1978) Effects of 4-aminopyridine in Eaton-Lambert syndrome. Br J Anaesth 50:383–385

Agoston S, Salt PJ, Erdmann W, Hilkemeijer T, Bencini A, Langrehr D (1980) Antagonism of ketamine-diazepam anaesthesia by 4-aminopyridine in human volunteers. Br J Anaesth 52:367–370

Agoston S, Bowman WC, Houwertjes MC, Rodger IW, Savage AO (1982) Direct action of 4-aminopyridine on the contractility of a fast-contracting muscle in the cat. Clin Exp Pharmacol Physiol 9:21–34

Agoston S, Maestrone E, van Hezik EJ, Ket JM, Houwertjes MC, Uges DRA (1984) Effective treatment of verapamil intoxication with 4-aminopyridine in the cat. J Clin Sci 73:1291–1296

Ahren BO, Leander S, Lundquist I (1981) Effects of 4-aminopyridine on insulin secretion and plasma glucose levels in intact and adrenalectomized-chemically sympathectomized mice. Eur J Pharmacol 74:221–226

Al-Haboubi HA, Bowman WC, Houston J, Savage AO (1978) Effects of 4-aminopyridine on the isolated parasympathetically-innervated oesophagus of the domestic fowl chick. J Pharm Pharmacol 30:517–518

Ambache N, Killick SW, Zar MA (1975) Extraction from ox retractor penis of an inhibitory substance which mimics its atropine-resistant neurogenic relaxation. Br J Pharmacol 54:409–410

Anwyl R (1977) The effect of foreign cations, pH and pharmacological agents on the ionic permeability of an excitatory glutamate synapse. J Physiol (Lond) 273:389–404

Audibert-Benoit M-L (1975) Effets de la 4-aminopyridine sur les proprietes membranaires de la fibre musculaire squelettique de Grenouille. J Physiol (Paris) 72:Al

Ball AP, Hopkinson RB, Farrell ID, Hutchison JGP, Paul R, Watson RDS, Page AJF, Parker RGF, Edwards CW, Snow M, Scott DK, Leone-Ganado A, Hastings A, Ghosh AC, Gilbert RJ (1979) Human botulism caused by clostridium botulinum Type E: the Birmingham outbreak. Q J Med 191:473–491

Baranyi A, Feher O (1979) Convulsive effects of 3-aminopyridine on cortical neurones. Electroencephalogr Clin Neurophysiol 47:745–751

Barnard PJ, Barnard EA, Allan ES (1982) Effects of aminopyridines on avian muscular dystrophy. In: Lechat P, Thesleff S, Bowman WC (eds) Aminopyridines and similarly acting drugs: effects on nerves, muscles and synapses. Advances in the Biosciences, vol 35. Pergamon, Oxford, p 323

Bigalke H, Habermann E (1980) Blockadé by tetanus and botulinum A toxin of postganglionic cholinergic nerve endings in the myenteric plexus. Naunyn Schmiedebergs Arch Pharmacol 312:255–263

Bostock H, Sears TA, Sherratt RM (1981) The effects of 4-aminopyridine and tetraethylammonium ions on normal and demyelinated mammalian nerve fibres. J Physiol (Lond) 313:301–315

Bowen DM, Smith CB, White P, Davison AN (1976) Neurotransmitter related enzymes and indices of hypoxia in senile dementia and other abiotrophics. Brain 99:459–496

Bowman WC (1982a) Conclusions. In: Lechat P, Thesleff S, Bowman WC (eds) Aminopyridines and similarly acting drugs: effects on nerves, muscles and synapses. Advances in the Biosciences, vol 35. Pergamon, Oxford, pp 335–341

Bowman WC (1982b) Aminopyridines: their pharmacological actions and potential clinical uses. Trends Pharmacol Sci 3:183–185

Bowman WC, Savage AO (1981) Pharmacological actions of aminopyridines and related compounds. Rev Pure Appl Pharmacol Sci 2:317–371

Bowman WC, Harvey AL, Marshall IG (1976) Facilitatory actions of aminopyridines on neuromuscular transmission. J Pharm Pharmacol 28:Suppl 79P

Bowman WC, Harvey AL, Marshall IG (1977a) The actions of aminopyridines on avian muscle. Naunyn-Schmiedebergs Arch Pharmacol 297:99–103

Bowman WC, Khan HH, Savage AO (1977b) Some antagonists of dantrolene sodium on the isolated diaphragm muscle of the rat. J Pharm Pharmacol 29:616–625

Bowman WC, Marshall RJ, Rodger IW, Savage AO (1981) Actions of 4-aminopyridine on the cardiovascular system of anaesthetized cats and dogs. Br J Anaesth 53:555–565

Bowman WC, Gibb AJ, Harvey AL, Marshall IG (1986) Prejunctional actions of cholinoceptor agonists and antagonists, and of anticholinesterase drugs. In: Kharkevich DA (ed) New curare-like agents. Handb Exp Pharmacol, vol 79. Springer, Berlin Heidelberg New York

Brigant JL, Mallart A (1982) Presynaptic currents in mammalian motor endings. J Physiol (Lond) 333:619–636

Brismar T (1980) Potential clamp analysis of membrane currents in rat myelinated nerve fibres. J Physiol (Lond) 298:171–184

Carlson FD, Wilkie DR (1974) Muscle physiology. Prentice-Hall, Englewood Cliffs

Chanelet J, Lemeignan M (1969) Effet d'une application microregionale de 4-aminopyridine au niveau de la moelle lombaire du chat. C R Soc Biol (Paris) 163:365–372

Charonnat R, Lechat P, Chareton J (1953a) Etude chimique de l'origine du choc thiaminique. Ann Pharm Fr 11:26–30

Charonnat R, Lechat P, Chareton J (1953b) Recherches biologiques sur la nature du choc thiaminique. Ann Pharm Fr 11:737–739

Cherington M (1976) Guanidine and germine in Eaton-Lambert syndrome. Neurology 26:944–946

Chiu SY, Ritchie JM, Rogart RB, Stagg D (1979) A quantitative description of membrane currents in rabbit myelinated nerve. J Physiol (Lond) 292:149–166

Colton JS, Schauf CL, Colton CA (1976) Effects of the aminopyridines and sparteine on Myxicola axons and the lobster neuromuscular junction. Biophys J 16:188a

Cronnelly R, Morris RB (1982) Antagonism of neuromuscular blockade. Br J Anaesth 54:183–191

Cronnelly R, Morris RB, Miller RD, Sheiner LB (1980) Pyridostigmine kinetics with and without renal function. Clin Pharmacol Ther 28:78–81

Davies P (1979) Neurotransmitter related enzymes in senile dementia of the Alzheimer type. Brain Res 171:319–327

Davies P, Maloney AJ (1976) Selective loss of central cholinergic neurones in Alzheimer's disease. Lancet ii:1403

Davis KL, Moks RC (1982) Enhancement of memory processes in Alzheimer's disease with multiple-dose intravenous physostigmine. Am J Psychiatry 139:1421–1424

Derenne JP, Contoure J, Iscoe S, Whitelaw WA, Milic-Emili J (1976) Occlusion pressures in man rebreathing CO_2 under methoxyflurane anaesthesia. J Appl Physiol 40:805–814

Dingemanse E, Wibaut JP (1928) Zur Pharmakologie von einigen Pyridylpyrrolen und einigen Abkömmlingen des α-Aminopyridins. Arch Exp Pathol Pharmakol 132:365–381

Dohrn M (1924) Pharmakologie einiger Pyridinderivate. Arch Exp Pathol Pharmakol 105:10–11
Drachman DB (1978) Myasthenia gravis. N Engl J Med 298:136–142, 186–193
Dubois JM (1982) Properties and physiological roles of K^+ currents in frog myelinated nerve fibres as revealed by 4-aminopyridine. In: Lechat P, Thesleff S, Bowman WC (eds) Aminopyridines and similarly acting drugs: effects on nerves, muscles and synapses. Advances in the Biosciences, vol 35. Pergamon, Oxford, pp 43–52
Dunant Y (1979) Acetylcholine changes analysed at short time intervals in the electric organ of Torpedo. J Physiol (Lond) 300:39 P
Dunant Y, Eder L, Servetiadis-Hirt L (1980) Acetylcholine release evoked by single or a few nerve impulses in the electric organ of *Torpedo*. J Physiol (Lond) 298:185–203
Dunant Y, Clostre F, Jones GJ, Loctin F, Muller D (1982) Aminopyridines as a tool to investigate the mechanism of acetylcholine release at the nerve electroplaque junction. In: Lechat P, Thesleff S, Bowman WC (eds) Aminopyridines and similarly acting drugs: effects on nerves, muscles and synapses. Advances in the Biosciences, vol 35. Pergamon, Oxford, pp 163–172
Durant NN, Lambert JJ (1981) The action of polymyxin B at the frog neuromuscular junction. Br J Pharmacol 72:41–48
Durant NN, Marshall IG (1978) The effects of 3,4-diaminopyridine on spontaneous and evoked transmitter release at the frog neuromuscular junction. J Physiol (Lond) 280:21 P
Durant NN, Marshall IG (1980) The effects of 3,4-diaminopyridine on acetylcholine release at the frog neuromuscular junction. Europ J Pharmacol 67:201–208
Durant NN, Lee C, Katz RL (1980) 4-Aminopyridine reversal of sympathetic ganglion blockade in the anesthetized cat. Anesthesiology 52:381–384
Eaton LM, Lambert EH (1957) Electromyography and electric stimulation of nerves in diseases of motor unit. J Am Med Assoc 163:1117–1124
Edman KAP, Khan AR (1982) Effects of 4-aminopyridine on contractile properties of skeletal muscle. In: Lechat P, Thesleff S, Bowman WC (eds) Aminopyridines and similarly acting drugs: effects on nerves, muscles and synapses. Advances in the Biosciences, vol 35. Pergamon, Oxford, pp 249–260
Enomoto K, Maeno T (1981) Presynaptic effects of 4-aminopyridine and streptomycin on the neuromuscular junction. Eur J Pharmacol 76:1–8
Fastier FN, McDowall MA (1958a) A comparison of the pharmacological properties of the three isomeric aminopyridines. Aust J Exp Biol 36:365–372
Fastier FN, McDowall MA (1958b) Analgesic activity of 4-methyl-2-aminopyridine and some related compounds. Aust J Exp Biol 36:491–498
Fink R, Stephenson DG (1982) Effects of 4-aminopyridine on isometric force – $[Ca^{++}]$ – relations in mammalian slow and fast skinned muscle fibres. In: Lechat P, Thesleff S, Bowman WC (eds) Aminopyridines and similarly acting drugs: effects on nerves, muscles and synapses. Advances in the Biosciences, vol 35. Pergamon, Oxford, p 275
Fink R, Wettwer E (1978) Modified K-channel gating by exhaustion and the block by internally applied TEA and 4-aminopyridine in muscle. Pflügers Arch 374:289–292
Foldes FF, Agoston S, Van Der Pol F, Amaki Y, Nagashima H, Crul J (1976a) The in vitro neuromuscular effects of 4-aminopyridine and its interaction with neuromuscular blocking agents. Proc Am Soc Anesth 179–180
Foldes FF, Braak G, Van Wezel H, Huynen R, Agoston S, Crul J (1976b) The interaction of 4-aminopyridine with neuromuscular blocking agents in the rat. Proc Am Soc Anesth 561–562
Foldes FF, Ohta Y, Shikawa Y, Vizi ES, Van Dijk JJ, Morita K (1982) The influences of 4-aminopyridine on parasympathetic transmission. In: Lechat P, Thesleff S, Bowman WC (eds) Aminopyridines and similarly acting drugs: effects on nerves, muscles and synapses. Advances in the Biosciences, vol 35. Pergamon, Oxford, pp 127–140
Folgering H, Rutten J, Agoston S (1979) Stimulation of phrenic nerve activity by an acetylcholine releasing drug: 4-aminopyridine. Pflügers Arch 379:181–185
Frank M, Flom L, Ffrench-Mullen JMH (1978) Effects of aminopyridine on electro mechanical properties of guinea-pig atrium. Fed Proc 37:863

Freeman E (1979) Cholinergic mechanisms in heart: Interactions with 4-aminopyridine. J Pharmacol Exp Ther 210:7–14

Galindo J, Rudomin P (1978) Facilitation of synaptic activity in the frog spinal cord produced by 4-aminopyridine. Neurosci Lett 10:299–304

Galvan M (1982) A transient outward current in rat sympathetic neurones. Neurosci Lett 31:295–300

Galvan M, Grafe P, Ten Bruggengate G (1980) Presynaptic actions 4-aminopyridine and γ-aminobutyric acid on rat sympathetic ganglion in vitro. Naunyn-Schmiedebergs Arch Pharmacol 314:141–147

Galvan M, Grafe P, Ten Bruggengate G (1982) Convulsant actions of 4-aminopyridine on the guinea-pig olfactory cortex slice. Brain Res 241:75–86

Gibb AJ, Marshall IG, Bowman WC (1982) Increased tetanic fade produced by 3,4-diaminopyridine in the presence of neuromuscular blocking agents. In: Lechat P, Thesleff S, Bowman WC (eds) Aminopyridines and similarly acting drugs: effects on nerves, muscles and synapses. Advances in the Biosciences, vol 35. Pergamon, Oxford, p 216

Gillespie JI (1977) Voltage-dependent blockage of the delayed potassium current in skeletal muscle by 4-aminopyridine. J Physiol (Lond) 273:64–65 P

Gillespie JI, Hutter OF (1975) The actions of 4-aminopyridine on the delayed potassium current in skeletal muscle fibres. J Physiol (Lond) 252:70–71 P

Gillespie JS, Martin W (1980) A smooth muscle inhibitory material from the bovine retractor penis and rat anococcygeus muscles. J Physiol (Lond) 309:55–64

Glover WE (1978) Potentiation of vasoconstrictor responses by 3- and 4-aminopyridine. Br J Pharmacol 63:577–585

Glover WE (1979) Differences in the mechanism of the positive inotropic actions of 4-aminopyridine and 4-methyl-2-aminopyridine on isolated rabbit atria. Proc Aust Physiol Pharmacol Soc 10:243 P

Glover WE (1981) Cholinergic effect of 4-aminopyridine and adrenergic effect of 4-methyl-2-aminopyridine in cardiac muscle. Eur J Pharmacol 71:21–31

Glover WE (1982) The aminopyridines. Gen Pharmacol 13:259–285

Gogolak G, Czech K, Stumpf C (1982) Effect of 4-aminopyridine on the electrical activity of the hippocampus and cerebellum. In: Lechat P, Thesleff S, Bowman WC (eds) Aminopyridines and similarly acting drugs: effects on nerves, muscles and synapses. Advances in the Biosciences, vol 35. Pergamon, Oxford, p 217

Golenhofen K, Mandrek K (1978) Slow automatic activity in squid axons induced by 4-aminopyridine. J Physiol (Lond) 284:68–70 P

Grafe P, Galvan M, Ten Bruggengate G (1981) 4-Aminopyridine reveals presynaptic GABA actions in rat sympathetic ganglia. In: DeFeudis FV, Mandel P (eds) Amino-acid neurotransmitters. Raven, New York, pp 257–261

Guerrero S, Novakovic L (1980) Effects of 4-aminopyridine on pacemaker activity of frog sinus venosus. Eur J Pharmacol 62:335–340

Gunderson CB, Jenden DJ (1981) Studies on the effects of agents which alter calcium metabolism on acetylcholine turnover in the rat diaphragm preparation. Br J Pharmacol 72:461–470

Gustafsson B, Galvan M, Grafe P, Wigstrom H (1982) A transient outward current in a mammalian central neurone blocked by 4-aminopyridine. Nature 299:252–254

Hall GM, Cooper GM, Lucke JN, Lister D (1980) 4-Aminopyridine fails to induce porcine malignant hyperthermia. Br J Anaesth 52:707

Hara Y, Kitamura K, Kuriyama H (1980) Actions of 4-aminopyridine on vascular smooth muscle tissues of the guinea-pig. Br J Pharmacol 68:99–106

Harvey AL, Marshall IG (1977a) The actions of three diaminopyridine on the chick biventer cervicis muscle. Eur J Pharmacol 44:303–309

Harvey AL, Marshall IG (1977b) The facilitatory actions of aminopyridines and tetraethylammonium on neuromuscular transmission and muscle contractility in avian muscle. Nauny-Schmiedebergs Arch Pharmacol 299:53–60

Harvey Al, Marshall IG (1977c) A comparison of the effects of aminopyridines on isolated chicken and rat skeletal muscle preparation. Comp Biochem Physiol 58c:161–165

Heuser JE (1977) Synaptic vesicle exocytosis revealed in quick-frozen frog neuromuscular junctions treated with 4-aminopyridine and given a single electrical shock. In: Cowan WM, Ferrendelli JA (eds) Approaches to the cell biology of neurons. Society for Neuroscience Symposium, vol 2. Society for Neuroscience, Bethesda, Maryland, pp 215–239

Heuser JE, Reese TS, Dennis MJ, Jan Y, Jan L, Evans L (1979) Synaptic vesicle exocytosis captured by quick freezing and correlated with quantal transmitter release. J Cell Biol 81:275–300

Hill GE, Stanley TH, Sentker CR (1977) Physostigmine reversal of postoperative somnolence. Can Anaesth Soc J 24:707–711

Horackova M, Nonner W, Stampfli R (1968) Action potentials and voltage clamp currents of single rat Ranvier nodes. Proc Int Union Physiol Sci 7:198

Horn AS, Lambert JJ, Marshall IG (1979) A comparison of the facilitatory actions of 4-aminopyridine methiodide and 4-aminopyridine on neuromuscular transmission. Br J Pharmacol 65:53–62

Hotovy R, Roesch E (1958) Zur Pharmakologie der Kreislaufwirkung von 5-Hydroxy tryptamin. Arch Exp Pathol Pharmakol 232:369–386

Hotson JR, Prince DA (1980) A calcium activated hyperpolarization follows repetitive firing hippocampal neurones. J Neurophysiol 43:409–419

Hue B, Pelhate M, Callec JJ, Chanelet J (1975) Modifications par la 4-aminopyridine de l'activite synaptique au niveau du dernier ganglion abdominal de la blatte (*Periplaneta americana*). C R Soc Biol (Paris) 169:876–883

Hue B, Pelhate M, Callec JJ, Chanelet J (1976a) Synaptic transmission in the sixth ganglion of the cockroach: Action of 4-aminopyridine. J Exp Biol 65:517–527

Hue B, Pelhate M, Callec JJ, Chanelet J (1976b) Effets pre- et postsynaptique de la 4-aminopyridine. J Physiol (Paris) 72:96A

Hue B, Pelhate M, Callec JJ, Chanelet J (1978) Postsynaptic effects of 4-aminopyridine in the sixth abdominal ganglion of the cockroach. Eur J Pharmacol 49:327–329

Illes P, Thesleff S (1978) 4-Aminopyridine and evoked transmitter release from motor nerve endings. Br J Pharmacol 64:623–629

Israel M, Manaranche R, Lesbats B, Gulik-Krzywicki T (1982) Intramembrane particles changes: a constant feature of the release mechanism. In: Lechat P, Thesleff S, Bowman WC (eds) Aminopyridines and similarly acting drugs: effects on nerves, muscles and synapses. Advances in the Biosciences, vol 35. Pergamon, Oxford, pp 173–182

Ito Y, Korenaga S, Tajima K (1980) Comparative study of the effects of 4-amino pyridine and tetraethylammonium on neuro-effector transmission in the guinea-pig vas deferens. Br J Pharmacol 69:453–460

Jacobs RS, Burley ES (1978) Nerve terminal facilitatory action of 4-amino pyridine: An analysis of the rising phase of the endplate potential. Neuropharmacology 17:439–444

Jan YN, Jan LY, Dennis MJ (1977) Two mutations of synaptic transmission in Drosophila. Proc R Soc B 1998:87–108

Jankowska E (1982) Effects of 4-aminopyridine on transmission in excitatory and inhibitory synapses in the spinal cord. In: Lechat P, Thesleff S, Bowman WC (eds) Aminopyridines and similarly acting drugs: effects on nerves, muscles and synapses. Advances in the Biosciences, vol 35. Pergamon, Oxford, pp 117–125

Jankowska E, Lundberg A, Rudomin P, Sykova E (1977) Effects of 4-aminopyridine on transmission in excitatory and inhibitory synapses in the spinal cord. Brain Res 136:387–392

Johns A, Golko DS, Lauzon PA, Paton DM (1976) The potentiating effects of 4-aminopyridine on adrenergic transmission in the rabbit vas deferens. Eur J Pharmacol 38:71–78

Joule JA, Smith GF (1972) Heterocyclic chemistry. Van Nostrand Reinhold, London, p 64

Kamenskaya MA, Thesleff S (1979) Effects of taipoxin on spontaneous and evoked transmitter release from motor nerve terminals. Toxicon [Suppl] 17:80

Kannan MS, Daniel EE (1978) Formation of gap junctions by treatment in vitro with potassium conductance blockers. J Cell Biol 78:338–348
Katz B, Miledi R (1969) Spontaneous and evoked activity of motor nerve endings in calcium Ringer. J Physiol (Lond) 203:689–706
Katz B, Miledi R (1977) Transmitter leakage from motor nerve endings. Proc R Soc Lond B 196:59–72
Katz B, Miledi R (1979) Estimates of quantal content during 'chemical potentiation' of transmitter release. Proc R Soc B 205:369–378
Kenyon JL, Gibbons WR (1979) 4-Aminopyridine and the early outward current of sheep cardiac Purkinje fibers. J Gen Physiol 73:139–157
Khan AR, Edman KA (1979) Effects of 4-aminopyridine on the excitation-contraction coupling in frog and rat skeletal muscle. Acta Physiol Scand 105:443–452
Kim YI, Goldner MM, Sanders DB (1980a) Facilitatory effects of 4-aminopyridine on normal neuromuscular transmission. Muscle Nerve 3:105–111
Kim YI, Goldner MM, Sanders DB (1980b) Facilitatory effects of 4-aminopyridine on neuromuscular transmission in disease states. Muscle Nerve 3:112–119
Kirpekar M, Kirpekar SM, Prat JC (1976) Effects of 4-aminopyridine (4-AP) on release of norepinephrine (NE) from the perfused cat spleen by nerve stimulation (NS) and potassium (K). Pharmacologist 18:208
Kirpekar M, Kirpekar SM, Prat JC (1977) Effect of 4-aminopyridine on release of noradrenaline from the perfused cat spleen by nerve stimulation. J Physiol (Lond) 272:517–528
Kirpekar M, Kirpekar SM, Prat JC (1978) Reversal of guanethidine blockade of sympathetic nerve terminals by tetraethylammonium and 4-aminopyridine. Br J Pharmacol 62:75–78
Kirsch GE, Narahashi T (1978) 3,4-Diaminopyridine. A potent new potassium channel blocker. Biophys J 22:507–512
Klinge E, Sjöstrand NO (1974) Contraction and relaxation of the retractor penis muscle and the penile artery of the bull. Acta Physiol Scand [Suppl] 420:1–88
Kocsis JD, Waxman SG (1983) Long-term regenerated nerve fibres retain sensitifity to potassium channel blocking agents. Nature 304:640–642
Krnjevic K, Puil E, Werman R (1978) EGTA and motoneuronal after-potentials. J Physiol (Lond) 275:199–223
Langrehr D, Agoston S, Sia R (1984) Ataranalgesia, A review. Acta Anaesth Belg 35:165–187
Leander S, Arner A, Johansson B (1977) Effects of 4-aminopyridine on mechanical activity and noradrenaline release in the rat portal vein in vitro. Eur J Pharmacol 46:351–361
Lechat P, Thesleff S, Bowman WC (1982) Aminopyridines and similarly acting drugs: effects on nerves, muscles and synapses. Advances in the Biosciences, vol 35. Pergamon, Oxford
Lee C, De Silva AJC, Katz RL (1978) Antagonism of polymyxin B-induced neuromuscular and cardiovascular depression by 4-aminopyridine in the anesthetized cat. Anesthesiology 49:256–259
Lemeignan M (1971) Abord pharmacologique de l'etude du mecanisme de l'action convulsivante de l'amino-4 pyridine. Therapie 26:927–940
Lemeignan M (1982) Analysis of the action of 4-aminopyridine on the cat lumbar spinal cord. I. Modification of the afferent volley, the monosynaptic discharge amplitude and the polysynaptic evoked responses. Neuropharmacology 11:551–558
Lemeignan M (1973) Analysis of the effects of 4-aminopyridine on the lumbar spinal cord of the cat. II. Modifications of certain spinal inhibitory phenomena, post-tetanic potentiation and dorsal root potentials. Neuropharmacology 12:641–651
Lemeignan M, Lechat P (1967) Sur l'action anticurare des aminopyridines. C R Seances Acad Sci (Paris) Ser D 264:169–172
Lemeignan M, Chanelet J, Saade NE (1969) Etude de l'action d'un convulsivant special la 4-aminopyridine sur les nerfs de Vertebres. C R Soc Biol (Paris) 163:359–365

Lemeignan M, Auclair MC, Rodallec A, Lechat P (1975) Analyse electro physiologique des effets de l'amino-4 pyridine sur le lambeau ventriculaire isole de coeur de Cobaye. Archs Int Pharmacodyn Ther 216:165–176

Lev-Tov A, Rahamimoff R (1980) A study of tetanic and post-tetanic potentiation of miniature end-plate potentials at the frog neuromuscular junction. J Physiol (Lond) 309:247–273

Llinas R, Walton K, Bohr V (1975) The action of 3- and 4-aminopyridine on synaptic transmission in the squid giant synapse. Biol Bull 149:435

Llinas R, Walton K, Bohr V (1976) Synaptic transmission in squid giant synapse after potassium conductance blockage with external 3- and 4-aminopyridine. Biophys J 16:83–86

Llinas R, Walton K, Sugimori M, Simon S (1982) 3- and 4-Aminopyridine in synaptic transmission at the squid giant synapse. In: Lechat P, Thesleff S, Bowman WC (eds) Aminopyridines and similarly acting drugs: effects on nerves, muscles and synapses. Advances in the Biosciences, vol 35. Pergamon, Oxford, pp 69–79

Lundh H (1978a) Effects of 4-aminopyridine on neuromuscular transmission. Brain Res 153:307–318

Lundh H (1978b) Paralysis in botulinum toxin poisoning. (An experimental study on pathophysiology and drug treatment). Thesis: Departments of Pharmacology and Neurology, University of Lund, Sweden

Lundh H (1979) Effects of 4-aminopyridine on statistical parameters of transmitter release at the neuromuscular junction. Acta pharmac tox 44:343–346

Lundh H, Thesleff S (1977) The mode of action of 4-aminopyridine and guanidine on transmitter release from motor nerve terminals. Eur J Pharmacol 42:411–412

Lundh H, Leander S, Thesleff S (1977a) Antagonism of the paralysis produced by botulinum toxin in the rat. The effects of tetraethylammonium, guanidine and 4-aminopyridine. J Neurol Sci 32:29–43

Lundh H, Nilsson O, Rosen I (1977b) 4-aminopyridine – a new drug tested in the treatment of Eaton-Lambert syndrome. J Neurol Neurosurg Psychiat 40:1109–1112

Lundh H, Nilsson O, Rosen I (1979) Effects of 4-aminopyridine in myasthenia gravis. J Neurol Neurosurg Psychiat 42:171–175

Lymangrover JR, Martin RA (1981) Effects of extracellular potassium and 4-aminopyridine on corticosteroid secretion. Mol Cell Endocrinol 21:199–210

Maeno T (1980) Kinetic analysis of a large facilitatory action of 4-aminopyridine on the motor nerve terminal of the neuromuscular junction. Proc Jpn Acad Ser B 56:241–245

Maeno T, Enomoto K (1980) Reversal of streptomycin induced blockade of neuromuscular transmission by 4-aminopyridine. Proc Jpn Acad Ser B 56:486–491

Maetani T, Shiba Y, Muneoka Y, Kanno Y (1979) Effects of Zn^{2+}, UO_2^{2+} and 4-aminopyridine on twitch contraction and action potential in the mouse skeletal muscles. Hiroshima J Med Sci 28:43–48

Mallart A, Brigant JL (1982) Action of 4AP and TEA on presynaptic currents in motor endings. In: Lechat P, Thesleff S, Bowman WC (eds) Aminopyridines and similarly acting drugs: effects on nerves, muscles and synapses. Advances in the Biosciences, vol 35. Pergamon, Oxford, p 223

Mallart A, Martin AR (1967) An analysis of facilitation of transmitter release at the neuromuscular junction of the frog. J Physiol (Lond) 193:679–694

Marshall IG, Lambert JJ, Durant NN (1979) Inhibition of aminopyridine-induced contractile activity in skeletal muscle by tetrodotoxin and by magnesium. Eur J Pharmacol 54:9–14

Martinez-Aguirre E, Crul JF (1979) Effect of tetrahydroaminoacridine and 4-aminopyridine on recovery from ketamine-diazepam anesthesia in the maccacus rhesus monkey. Acta Anaesth Belg 30:231–238

Martinez-Aguirre E, Wikinski JA, Bravo J (1980) 4-Aminopyridine antagonism of ketamine-diazepam anaesthesia in children. I. Recovery time. Abstract of communications No. 910, 7th World Congress of Anaesthesiologists, Excerpta Medica, Hamburg

Martinez-Aguirre E, Wikinski JA, Bello A, Izquerdo J, Garcia A, Velarde H (1981) Effect of 4-aminopyridine on cardiovascular functions in the dog. Can Anaesth Soc J 28:114–120

Martinez De Munoz D (1980) Anticonvulsant action of γ-hydroxy, γ-ethyl, γ-phenyl butyramide in mice treated with 4-aminopyridine. In: Canger R, Angeleri F, Penry JK (eds) XIth Epilepsy International Symposium. Advances in Epileptology. Raven, New York, pp 463–466

Meves H, Pichon Y (1975) Effects of 4-aminopyridine on the potassium current in internally perfused giant axons of the squid. J Physiol (Lond) 251:60–62P

Meves H, Pichon Y (1977a) The effect of internal and external 4-aminopyridine on the potassium currents in intracellularly perfused squid giant axons. J Physiol (Lond) 268:511–532

Meves H, Pichon Y (1977b) Modele d'action de la 4-aminopyridine au niveau des „canaux" potassium de l'axone geant de capalmar (*Loligo forbesi* L.). C R Seances Acad Sci (Paris) 284:1325–1328

Miller RD, Dennissen PAF, Van Der Pol F, Agoston S, Booij LHDJ, Crul JF (1978) Potentiation of neostigmine and pyridostigmine by 4-aminopyridine in the rat. J Pharm Pharmacol 30:699–702

Miller RD, Booij LHDJ, Agoston S, Crul JF (1979) 4-Aminopyridine potentiates neostigmine and pyridostigmine in man. Anesthesiology 50:416–420

Mitzov VZ (1967) Pharmacological characteristic and correlation between structure and action of a group of formamide and aminopyridine derivatives (in Bulgarian). Dissertation, VMI, Sofia

Mitzov V, Paskov D (1971) A propos of pymadin pharmacology (in Bulgarian). Mediz Arhiv 8:49–53

Molgo J (1978) Voltage-clamp analysis of the sodium and potassium currents in skeletal muscle fibres treated with 4-aminopyridine. Experentia 34:1275–1277

Molgo J (1982) Effects of aminopyridines on neuromuscular transmission. In: Lecahat P, Thesleff S, Bowman WC (eds) Aminopyridines and similarly acting drugs: effects on nerves, muscles and synapses. Advances in the Biosciences, vol 35. Pergamon, Oxford, pp 95–116

Molgo J, Lemeignan M, Lechat P (1975) Modifications de la liberation du transmetteur a la jonction neuromusculaire grenouille sous l'action de l'amino-4-pyridine. C R Seances Acad Sci (Paris) 281:1637–1639

Molgo J, Lemeignan M, Lechat P (1977) Effects of 4-aminopyridine at the frog neuromuscular junction. J Pharmacol Exp Ther 203:653–663

Molgo J, Lemeignan M, Lechat P (1979) Analysis of the action of 4-aminopyridine during repetitive stimulation at the neuromuscular junction. Eur J Pharmacol 53:307–311

Molgo J, Lundh H, Thesleff S (1980) Potency of 3,4-diaminopyridine and 4-aminopyridine on mammalian neuromuscular transmission and the effect of pH changes. Eur J Pharmacol transmission and the effect of pH changes. Eur J Pharmacol 61:25–34

Moritoki H, Takei M, Nakamoto N, Ishida Y (1978) Actions of aminopyridines on guinea-pig ileum. Arch Int Pharmacodyn Ther 232:28–41

Murray NMF, Newson-Davis J (1981) Treatment with oral 4-aminopyridine in disorders of neuromuscular transmission. Neurology 31:265–271

Nedergaard OA (1981) Facilitatory action of 4-aminopyridine on transmitter release from sympathetic nerve terminals in rabbit blood vessels. Br J Pharmacol 74:177–178P

Ngai SH, Cheney DL, Finck AD (1978) Acetylcholine concentrations and turnover in rat brain structures during anaesthesia with halothane, enflurane and ketamine. Anesthesiology 48:4–10

Paskov DS, Yovtchev AL, Djarov DN, Mitzov VZ, Stoyanov EA (1969) Agent for decurarization (in Bulgarian). Author's certificate (NPA 11092)

Paskov DS, Stoyanov EA, Mitzov VZ (1973) New anti-curare and analeptic drug Pymadin (4-aminopyridine hydrochloride) and its use in anaesthesia. Eksp Khir Anestesiol 18:48–52 (in Russian)

Paskov DS, Yonkov ST, Paunova M (1976) Pymadin electrophoretic application to certain diseases of the peripheral nervous system. Med Biol Inform 4:40–44

Paton DM (1979) Effect of 4-aminopyridine on adrenergic transmission in rabbit portal vein and pulmonary artery. Proc Aust Physiol Pharmacol Soc 10:174 P
Pelhate M, Pichon Y (1974) Selective inhibition of potassium current in the giant axon of the cockroach. J Physiol (Lond) 242:90–91 P
Pelhate M, Hue B, Chanelet J (1972) Effets de la 4-aminopyridine sur le systeme nerveux d'un Insecte: La Blatte (*Periplaneta americana* L.). C R Soc Biol (Paris) 166:1598–1605
Pelhate M, Hue B, Chanelet J (1974a) Modifications, par la 4-amino-pyridine, des caracteristiques electriques de l'axone geant isole d'un insecte, la Blatte (*Periplaneta americana* L.). C R Soc Biol (Paris) 168:27–34
Pelhate M, Hue B, Pichon Y, Chanelet J (1974b) Action de la 4-amino-pyridine sur la membrane de l'axone isole d'Insecte. C R Seances Acad Sci (Paris) 278:2807–2809
Pelhate M, Mony L, Hue B, Chanelet J (1975) Modifications des courants potassium par la 4-aminopyridine cas de l'axone geant isole de la Blatte (*Periplaneta americana*). C R Soc Biol (Paris) 169:1436–1441
Perkins MN, Stone TW (1980) 4-Aminopyridine blockade of neuronal depressant responses to adenosine triphosphate. Br J Pharmacol 70:425–428
Peterson C, Gibson GE (1982) 3,4-Diaminopyridine alters acetylcholine metabolism and behaviour during hypoxia. J Pharmacol Exp Ther 222:567–582
Pichon Y, Meves H, Pelhate M (1982) Effects of aminopyridines on ionic currents and ionic channel noise in unmyelinated axons. In: Lechat P, Thesleff S, Bowman WC (eds) Aminopyridines and similarly acting drugs: effects on nerves, muscles and synapses. Advances in the Biosciences, vol 35. Pergamon, Oxford, pp 53–68
Rao DBS, Nagashima H, Deery AM, Foldes FF (1977) Antagonism of barbiturate anesthesia by 4-aminopyridine in the rat. Proc Am Soc Anesth 749–750
Rodger IW, Shahid M (1981) Effects of increased pH on contractile force of rabbit isolated papillary muscles. Br J Pharmacol 73:295 P
Roussev GK, Paskov D, Jirkolova T (1982) Aminopyridine effects on molecular models in myopathic states. In: Lechat P, Thesleff S, Bowman WC (eds) Aminopyridines and similarly acting drugs: effects on nerves, muscles and synapses. Advances in the Biosciences, vol 35. Pergamon, Oxford, p 228
Saade NE, Chanelet J, Longchampt P (1971a) Action facilitatrice de microinjections de 4-aminopyridine sur les activites medullaires reflexes de la grenouille. C R Soc Biol (Paris) 165:2069–2077
Saade NE, Chanelet J, Longchampt P (1971b) Differents modes d'activation reflexe de la meme preparation de grenouille spinale: leur facilitation par la 4-aminopyridine. C R Soc Biol (Paris) 165:1242–1249
Saade NE, Banna NR, Khoury AJ, Jabbur SJ, Wall PD (1982) 4-aminopyridine-induced alteration in the receptive field of cuneate and gracile neurones. In: Lechat P, Thesleff S, Bowman WC (eds) Aminopyridines and similarly acting drugs: effects on nerves, muscles and synapses. Advances in the Biosciences, vol 35. Pergamon, Oxford, p 239
Sand O, Haug E, Gautvik KM (1980) Effects of thyroliberin and 4-aminopyridine on action potentials and prolactin release and synthesis in rat pituitary cells in culture. Acta Physiol Scand 108:247–252
Sand O, Hove K, Ozawa S, Gautvik K, Haug E (1982) Effects of 4-aminopyridine on endocrine cells. In: Lechat P, Thesleff S, Bowman WC (eds) Aminopyridines and similarly acting drugs: effects on nerves, muscles and synapses. Advances in the Biosciences, vol 35. Pergamon, Oxford, p 326
Sanders DB, Kim YI, Howard JF, Goetsch CA (1980) Eaton-Lambert syndrome – a clinical and electrophysiological study of a patient treated with 4-aminopyridine. J Neurol Neurosurg Psychiatr 43:978–985
Savage AO (1979) Pharmacological studies on aminopyridines and some analogues. Ph D Thesis, University of Strathclyde
Schauf CL, Colton CA, Colton JS, Davis FA (1976) Aminopyridines and sparteine as inhibitors of membrane potassium conductance: effects on Myxicola giant axons and the lobster neuromuscular junction. J Pharmacol Exp Ther 197:414–425

Schmutzler S, Uges D, Agoston S, Langrehr D (1984) Rohypnol-Intoxikation als Ursache eines postanaesthetischen Komas und Behandlungserfolg mit 4-Aminopyridin. Anaesthesist 33:294–295
See WR, Folgering M, Schlafke ME (1978) Central respiratory and cardiovascular effects of the ACh releaser 4-aminopyridine (4-AP). Pflügers Arch 377:R 20
Shahid M, Rodger IW (1982) Is the cardiotonic effect of aminopyrimidines in vitro a consequence of increased intracellular pH? In: Lechat P, Thesleff S, Bowman WC (eds) Aminopyridines and similarly acting drugs: effects on nerves, muscles and synapses. Advances in the Biosciences, vol 35. Pergamon, Oxford, p 278
Shapovalov AI, Shiriaev BI (1978) Modulation of transmission in different electronic junctions by aminopyridine. Experentia 34:67–68
Shaw FH, Bentley G (1949) Some aspects of the pharmacology of morphine with special reference to its antagonism by 5-aminoacridine and other chemically related compounds. Med J Aust 868–874
Shaw FH, Bentley G (1952) Morphine antagonism. Nature 169:712–713
Shaw FH, Bentley G (1953) The pharmacology of some new anticholinesterases. Aust J Exp Biol 31:573–576
Shaw FH, Bentley G (1955) Morphine antagonism. Aust J Exp Biol 33:143–151
Sherratt RM, Bostock H, Sears TA (1980) Effects of 4-aminopyridine on normal and demyelinated mammalian nerve fibres. Nature 283:570–572
Shiwako Y, Duncalf D, Nagashima H, Foldes FF (1977) Interactions of enkephalins, naloxone and 4-aminopyridine on the longitudinal muscle of the guinea-pig ileum. Proc Am Soc Anesth 467–468
Sia RL, Zandstra DF (1981) Use of 4-aminopyridine to reverse morphine induced respiratory depression in man. Br J Anaesth 53:865–868
Sia RL, Westra P, Wesseling H (1982) Naloxone, 4-aminopyridine and physostigmine as antagonists of morphine in rabbits. In: Lechat P, Thesleff S, Bowman WC (eds) Aminopyridines and similarly acting drugs: effects on nerves, muscles and synapses. Advances in the Biosciences, vol 35. Pergamon, Oxford, p 231
Simpson LL (1978) Pharmacological studies on the subcellular site of action of botulinum toxin type A[1]. J Pharmacol Exp Ther 206:661–669
Singh YN, Marshall IG, Harvey AL (1978 a) Some effects of the aminoglycoside antibiotics amikacin on neuromuscular and autonomic transmission. Br J Anaesth 50:109–117
Singh YN, Marshall IG, Harvey AL (1978 b) Reversal of antibiotic-induced muscle paralysis by 3,4-diaminopyridine. J Pharm Pharmacol 30:249–250
Sobek V (1970) On the pharmacology of 4-aminopyridine compared with adrenaline. Physiol Bohemoslovaca 19:417–419
Sobek V, Lemeignan M, Streichenberger G, Benoist J-M, Doguel A, Lechat P (1968) Etude sur le diaphragme isole de rat de l'antagonisme entre substances curarisantes et aminopyridines. Arch Int Pharmacodyn Ther 171:356–368
Soni N, Kam P (1982) 4-Aminopyridine – a review. Anaesth Intens Care 10:120–126
Stone TW (1981) The effects of 4-aminopyridine on the isolated vas deferens and its effects on the inhibitory properties of adenosine, morphine, noradrenaline and γ-aminobutyric acid. Br J Pharmacol 73:791–796
Stoyanov E, Vulchev P (1975) Pymadin. Electromyographic and electromechanographic studies. Med Biol Inform 5:21–23
Stoyanov E, Paskov D, Vulchev P, Marinova M (1976) Pymadin decurarization. Med Biol Inform 5:26–32
Summers WK, Viesselman JO, Marsh GM, Candelova K (1981) Use of THA in treatment of Alzheimer-like dementia: pilot study in twelve patients. Biol Psychiat 16:145–153
Szente M, Pongrácz F (1979) Aminopyridine-induced seizure activity. Electroencephalogr Clin Neurophysiol 46:605–608
Tazieff-Depierre F, Métézean P, Wunderer G (1978) Neuropharmacologie action de la toxine 2 isolée des tentacules d'anemone de mer sur la transmission neuro-musculaire Grenouille normale a inhibée par la toxine botulique. C R Seances Acad Sci (Paris) Ser D 286:655–658

Thal LJ, Rosen W, Sharpless NS, Crystal H (1981) Choline chloride fails to improve cognition in Alzheimer's disease. Neurobiol Aging 2:205–208
Thesleff S (1980) Aminopyridines and synaptic transmission. Neuroscience 5:1413–1419
Tokunaga A, Akert K, Sandri C (1979a) Three types of membrane modulations during transmitter release in rat spinal cord synapses. Neurosci Lett 12:147–152
Tokunaga A, Sandri C, Akert K (1979b) Ultrastructural effects of 4-aminopyridine on the presynaptic membrane in the rat spinal cord. Brain Res 163:1–8
Tokunaga A, Sandri C, Akert K (1979c) Increase of large intramembranous particles in the presynaptic active zone after administration of 4-aminopyridine. Brain Res 174:207–219
Tung AS, Brandon BW (1982) 4-Aminopyridines reversal of morphine analgesia. In Lechat P, Thesleff S, Bowman WC (eds) Aminopyridines and similarly acting drugs: effects on nerves, muscles and synapses. Advances in the Biosciences, vol 35. Pergamon, Oxford, p 233
Uchiyama T, Molgo J, Lemeignan M (1981) Presynaptic effects of bekanamycin at the frog neuromuscular junction. Reversibility by calcium and aminopyridines. Eur J Pharmacol 72:271–280
Uges DRA, Huizinga T (1981) 4-Aminopyridine; analysis and method for the preparation of a solution for injection in man. Pharm Acta Helv 56:158–162
Uges DRA, Sohn YJ, Greijdanus B, Scaf AHJ, Agoston S (1982a) 4-Aminopyridine kinetics. Clin Pharmacol Ther 31:587–593
Uges DRA, Versluis A, Ket JM, Wesseling H (1982b) 4-Aminopyridine tablets; a method for the preparation, in vitro and in vivo studies. Pharm Acta Helv 57:122–128
Ulbricht W, Wagner H-H (1976) Block of potassium channels of the nodal membrane by 4-aminopyridine and its partial removal on depolarization. Pflügers Arch 367:77–87
Ulbricht W, Wagner H-H, Schmidtmayer J (1982) Effects of aminopyridines on potassium currents of the nodal membrane. In: Lechat P, Thesleff S, Bowman WC (eds) Aminopyridines and similarly acting drugs: effects on nerves, muscles and synapses. Advances in the Biosciences, vol 35. Pergamon, Oxford, pp 29–42
Van der Kloot W, Madden KS (1981) 4-Aminopyridine does not increase m.e.p.p. frequencies at junctions depolarized by potassium. Brain Res 210:467–470
Vital Brazil O, Fontana MD, Heluany NF (1979) Effect of 4-aminopyridine on the neuromuscular blockade produced by crotoxin. Toxicon [Suppl] 17:16
Vital Brazil O, Fontana MD, Pavani NJP (1983) Effect of 4-aminopyridine on end-plate receptor desensitization caused by carbachol. Eur J Pharmacol 86:199–205
Vizi ES, Van Dijk J, Voldes FF (1977) The effect of 3,4-aminopyridine on acetylcholine release. J Neural Transm 41:265–274
Vohra MM, Pradhans SN (1964) Pharmacology of 3,4-diaminopyridine. Arch Int Pharmacodyn Ther 150:413–424
Vohra MM, Pradhan SM, Jain PC, Chatterjee SK, Anand N (1965) Synthesis and structure-activity relationships of some aminopyridines, imidazo-pyridines, and triazolopyridines. J Med Chem 8:296–304
Von Haxthausen EF (1955) Über Amino-Pyridin und seine Derivate. Arch Exp Pathol Pharmakol 226:163–171
Wagner H-H, Ulbricht W (1976) Voltage-dependent block of K channels by 4-aminopyridine. Pflügers Arch 362:R 31
Weide W, Loffelholz K (1980) 4-aminopyridine antagonises the inhibitory effect of pentobarbital on acetylcholine release in the heart. Naunyn-Schmiedebergs Arch Pharmacol 312:7–13
Wesseling H, Lakke JPWF (1980) Observations with 4-aminopyridine in Huntington's chorea. JRCS Med Sci 8:332–333
Wesseling H, Agoston S, Van Dam GBP, Pasma J, de Wit DJ, Havinga H (1984) Effects of 4-aminopyridine in elderly patients with Alzheimer's disease. N Engl J Med 310:988–989
Wollmer P, Wohlfart B, Khan R (1979) Effects of 4-aminopyridine on isolated papillary muscles of the rabbit. Proceedings of the XVI Scandinavian Congress of Physiology and Pharmacology. Acta Physiol Scand [Suppl] 473:30

Wu CH, Oxford GS, Narahashi T, Holan G (1980) Interaction of a DDT analog with the sodium channel of lobster axon. J Pharmacol Exp Ther 212:287–293
Yanagisawa T, Taira N (1979) Positive inotropic effect of 4-aminopyridine on dog ventricular muscle. Naunyn-Schmiedebergs Arch Pharmacol 307:207–212
Yanagisawa T, Taira N (1982) Effects of 4-aminopyridine on cardiac muscle. In: Lechat P, Thesleff S, Bowman WC (eds) Aminopyridines and similarly acting drugs: effects on nerves, muscles and synapses. Advances in the Biosciences, vol 35. Pergamon, Oxford, pp 261–274
Yanagisawa T, Satoh K, Taira N (1978) Excitation of autonomic nerves by 4-aminopyridine in the isolated, blood-perfused sino-atrial node preparation of the dog. Eur J Pharmacol 49:189–192
Yeh JZ, Oxford GS, Wu CH, Narahashi T (1976a) Interactions of aminopyridines with potassium channels of squid axon membranes. Biophys J 16:77–81
Yeh JZ, Oxford GS, Wu CH, Narahashi T (1976b) Dynamics of aminopyridine block of potassium channels in squid axon membrane. J Gen Physiol 68:519–535
Zucker RS, Lara-Estrella LO (1979) Is synaptic facilitation caused by presynaptic spike broadening? Nature 278:57–59

Nau H, [illegible] Holtel [illegible]
[illegible]
[illegible]
[illegible]
[illegible]
[illegible]
[illegible]
[illegible]
[illegible]
[illegible]
[illegible]
[illegible]
[illegible]
[illegible]
[illegible]

Subject Index

The following abbreviations are used, excepting as primary headings, throughout:

ACh: acetylcholine
AChE: acetylcholinesterase
AChR: acetylcholine receptors
anti-ChE: anticholinesterase
BuChE: butyrylcholinesterase
CCh: carbachol
ChE: cholinesterase
C.N.S.: central nervous system
MEPP: miniature end-plate potentials
NA: noradrenaline
NM: neuromuscular
NMB: neuromuscular blocking
NMBA: neuromuscular blocking agents
SCh: succinylcholine
TC: tubocurarine

Handbook of Experimental Pharmacology

Continuation of "Handbuch der experimentellen Pharmakologie"

Springer-Verlag
Berlin
Heidelberg
New York
Tokyo

Volume 38: Part 1
Antineoplastic and Immunosuppressive Agents I

Part 2
Antineoplastic and Immunosuppressive Agents II

Volume 39
Antihypertensive Agents

Volume 40
Organic Nitrates

Volume 41
Hypolipidemic Agents

Volume 42
Neuromuscular Junction

Volume 43
Anabolic-Androgenic Steroids

Volume 44
Heme and Hemoproteins

Volume 45: Part 1
Drug Addiction I

Part 2
Drug Addiction II

Volume 46
Fibrinolytics and Antifibronolytics

Volume 47
Kinetics of Drug Action

Volume 48
Arthropod Venoms

Volume 49
Ergot Alkaloids and Related Compounds

Volume 50: Part 1
Inflammation

Part 2
Anti-Inflammatory Drugs

Volume 51
Uric Acid

Volume 52
Snake Venoms

Volume 53
Pharmacology of Ganglionic Transmission

Volume 54: Part 1
Adrenergic Activators and Inhibitors I

Part 2
Adrenergic Activators and Inhibitors II

Volume 55
Psychotropic Agents

Part 1
Antipsychotics and Antidepressants

Part 2
Anxiolytics, Gerontopsychopharmacological Agents and Psychomotor Stimulants

Part 3
Alcohol and Psychotomimetics, Psychotropic Effects of Central Acting Drugs

Volume 56, Part 1+2
Cardiac Glycosides

Handbook of Experimental Pharmacology

Continuation of "Handbuch der experimentellen Pharmakologie"

Springer-Verlag
Berlin
Heidelberg
New York
Tokyo

Volume 57
Tissue Growth Factors

Volume 58
Cyclic Nucleotides

Part 1: **Biochemistry**

Part 2: **Physiology and Pharmacology**

Volume 59
Mediators and Drugs in Gastrointestinal Motility

Part 1: **Morphological Basis and Neurophysiological Control**

Part 2: **Endogenous and Exogenous Agents**

Volume 60
Pyretics and Antipyretics

Volume 61
Chemotherapy of Viral Infections

Volume 62
Aminoglycoside Antibiotics

Volume 63
Allergic Reactions to Drugs

Volume 64
Inhibition of Folate Metabolism in Chemotherapy

Volume 65
Teratogenesis and Reproductive Toxicology

Volume 66
Part 1: **Glucagon I**

Part 2: **Glucagon II**

Volume 67
Part 1
Antibiotics Containing the Beta-Lactam Structure I

Part 2
Antibiotics Containing the Beta-Lactam Structure II

Volume 68, Part 1+2
Antimalarial Drugs

Volume 69
Pharmacology of the Eye

Volume 70
Part 1
Pharmacology of Intestinal Permeation I

Part 2
Pharmacology of Intestinal Permeation II

Volume 71
Interferons and Their Applications

Volume 72
Antitumor Drug Resistance

Volume 73
Radiocontrast Agents

Volume 74
Antieliptic Drugs

Volume 75
Toxicology of Inhaled Materials

Volume 76
Clinical Pharmocology of Antiangial Drugs

Volume 77
Chemotherapy of Gastrointestinal Helminths

Volume 78
The Tetracyclines

MIX
Papier aus verantwortungsvollen Quellen
Paper from responsible sources
FSC® C105338

If you have any concerns about our products,
you can contact us on
ProductSafety@springernature.com

In case Publisher is established outside the EU,
the EU authorized representative is:
Springer Nature Customer Service Center GmbH
Europaplatz 3, 69115 Heidelberg, Germany

Printed by Libri Plureos GmbH
in Hamburg, Germany